D0857255

THE SMITHSONIAN
NATIONAL AIR AND SPACE MUSEUM

DIRECTORY OF AIRPLANES

THEIR DESIGNERS AND MANUFACTURERS

THE SMITHSONIAN
NATIONAL AIR AND SPACE MUSEUM

DIRECTORY OF AIRPLANES

THEIR DESIGNERS AND MANUFACTURERS

Edited by
DANA BELL

GREENHILL BOOKS, LONDON
STACKPOLE BOOKS, PENNSYLVANIA
in association with
THE SMITHSONIAN NATIONAL AIR AND SPACE MUSEUM,
WASHINGTON, D.C.

THE SMITHSONIAN NATIONAL AIR AND SPACE MUSEUM
DIRECTORY OF AIRPLANES, THEIR DESIGNERS AND MANUFACTURERS
First published 2002 by Greenhill Books, Lionel Leventhal Limited,
Park House, 1 Russell Gardens, London NW11 9NN
and
Stackpole Books, 5067 Ritter Road, Mechanicsburg, PA 17055, USA

Greenhill Books/Lionel Leventhal and Stackpole Books
in association with
the Smithsonian National Air and Space Museum Washington, D.C.

British Library Cataloguing in Publication Data
The Smithsonian National Air and Space Museum Directory of Airplanes, Their Designers and Manufacturers
1. Airplanes – History 2. Airplanes – Design and construction - History
3. Aircraft industry - History
I. Bell, Dana II. National Air and Space Museum III. Smithsonian Institution
629.1' 3334

ISBN 1-85367-490-7

Library of Congress Cataloging-in-Publication Data available

Printed and bound in Singapore by Kyodo Printing Company

CONTENTS

ILLUSTRATIONS

Between pages 192 and 193

1 Wright Brothers 1903 Flyer. *National Air and Space Museum, Smithsonian Institution (SI Neg. No. A-26767-B)*
2 Santos-Dumont 14*bis*. *National Air and Space Museum, Smithsonian Institution (SI Neg. No. 2001-11566)*
3 Blériot's Type XI. *National Air and Space Museum (NASM Videodisc No. 1A-10361), Smithsonian Institution*
4 Caproni Ca.1. *National Air and Space Museum (NASM Videodisc No. 1A-18377), Smithsonian Institution*
5 De Havilland (Airco) D.H.1. *National Air and Space Museum (NASM Videodisc No. 1A-34010), Smithsonian Institution*
6 Junkers J 1 (E.I). *National Air and Space Museum, Smithsonian Institution (SI Neg. No. A-2493)*
7 Junkers J 4 (J.I). *National Air and Space Museum, Smithsonian Institution (SI Neg. No. 76-2348)*
8 Farman F.50 Bn.2 (Far.13 Bn.2). *National Air and Space Museum (NASM Videodisc No. 1A-43768), Smithsonian Institution*
9 Breguet 19 B.2. *National Air and Space Museum (NASM Videodisc No. 1A-15727), Smithsonian Institution*
10 Curtiss NC-4. *National Air and Space Museum, Smithsonian Institution (SI Neg. No. 91-7085)*
11 Douglas O-2H. *National Air and Space Museum (NASM Videodisc No. 1A-39707), Smithsonian Institution*
12 Ryan Airlines NYP "Spirit of St Louis." *National Air and Space Museum, Smithsonian Institution (SI Neg. No. A-1598)*
13 Martin 130 Clipper. *Courtesy Lockheed Martin via National Air and Space Museum, Smithsonian Institution (SI Neg. No. 92-933)*
14 Piper J-3 Cub. *Courtesy Westinghouse via National Air and Space Museum (NASM Videodisc No. 1B-30727), Smithsonian Institution*
15 Waco YKS-7. *National Air and Space Museum, Smithsonian Institution (SI Neg. No. A-935-D)*
16 Short Mayo Composite Plane. *National Air and Space Museum, Smithsonian Institution (SI Neg. No. 90-8818)*
17 *Koolhoven F.K.56. National Air and Space Museum, Smithsonian Institution (SI Neg. No. 73-240)*
18 Douglas A-20A Havoc. *National Air and Space Museum (NASM Videodisc No. 1A-37478), Smithsonian Institution*
19 A6M5 Reisen Zeke (Navy Type 0 Carrier Fighter 52). *National Air and Space Museum, Smithsonian Institution (SI Neg. No. A-38634-C)*
20 Vought F4U-1 Corsair. *National Air and Space Museum, Smithsonian Institution (SI Neg. No. A-42334-H)*
21 Brewster F3A-1 Corsair. *National Air and Space Museum, Smithsonian Institution (SI Neg. No. 77-11435)*
22 Stearman N2S-5 Kaydet. *National Air and Space Museum (NASM Videodisc No. 9A-00272), Smithsonian Institution*
23 Supermarine Spitfire HF. Mark VII. *National Air and Space Museum, Smithsonian Institution (SI Neg. No. A-48644)*
24 Potez 63/11. *National Air and Space Museum, Smithsonian Institution (SI Neg. No. 2001-2586)*
25 Messerschmitt Bf 109 F. *National Air and Space Museum, Smithsonian Institution (SI Neg. No. 73-8699)*
26 North American XP-82 Twin Mustang. *National Air and Space Museum (NASM Videodisc No. 1B-27495), Smithsonian Institution*
27 Boeing WB-50A. *US Air Force photograph (Neg. No. 154089AC), courtesy US National Archives*
28 North American F-86D Sabre. *National Air and Space Museum, Smithsonian Institution (SI Neg. No. 74-3342)*

29 Fairchild C-123B Provider. *Courtesy of Fairchild via National Air and Space Museum, Smithsonian Institution (SI Neg. No. 73-5089)*

30 Sikorsky (USA) H-34A (CH-34A) Choctaw. *Courtesy United Technologies Corp. Archive via National Air and Space Museum (NASM Videodisc No. 1B-38807), Smithsonian Institution*

31 Bell YUH-1 (YHU-1, YH-40) Iroquois (Huey). *National Air and Space Museum, Smithsonian Institution (SI Neg. No. 83-15863)*

32 Antonov An-124 Condor. *Photograph by Paul Friend via National Air and Space Museum, Smithsonian Institution (SI Neg. No. 92-15349)*

33 Boeing 707-321B. *Courtesy Pan American World Airways*

34 Boeing 720-030B. *Courtesy Pan American World Airways*

35 Boeing C-135B. *US Air Force photograph (Neg No. KE 13121), courtesy US National Archives*

36 Douglas C-9A Nightingale. *National Air and Space Museum (NASM Videodisc No. 7A-32787), Smithsonian Institution*

37 British Aerospace 125-800XP (Raytheon Hawker 800XP). *Courtesy of Raytheon Company via National Air and Space Museum (NASM Videodisc No. 9A-00273), Smithsonian Institution*

38 General Dynamics YF-16 Fighting Falcon. *US Air Force photograph (Control No. CN 74-429 Neg C-661-74)*

INTRODUCTION

This volume is an attempt to create a single authoritative listing of aircraft names organized by designer or manufacturer. The listing grew from a National Air and Space Museum (NASM) effort to process and describe the Aircraft Reference Files in the Museum's Archives Division. Although we recognize that the 25,000 aircraft and 5,000 companies and designers featured here do not constitute a complete listing of aircraft types, we still hope that our efforts will prove useful to other museums and archives, historians, authors, publishers, and enthusiasts.

The project began in the early 1990s, when the archives staff began reorganizing a century's worth of aviation documents in their historical files. The files had become disorganized and difficult to access, with overstuffed folders causing damage to photos and papers. But before we could reorganize the folders, we needed to correctly identify the contents. Some aircraft appeared with multiple identifications; others were misidentified or not identified at all. Not finding a single reliable authority for aircraft names and designations, we began building an "authority list" for our own use. Over the years, that list has been refined and enhanced to produce this *Directory*. It remains a work in progress, and future editions will include additional designers, manufacturers, and aircraft as we learn of them.

The research for this *Directory* has focused on human-carrying, heavier-than-air vehicles that are supported primarily by dynamic lift. This includes *airplanes* (powered fixed-wing aircraft), *gliders* and *hang gliders* (unpowered, fixed-wing aircraft), *helicopters* (powered-rotor rotorcraft), *autogiros* (unpowered-rotor rotorcraft), and *ornithopters* (flapping wing designs). Following an internal NASM convention, we left certain other vehicle classes to separate lists. Thus, the *Directory* ignores lighter-than-air aircraft (such as balloons and dirigibles), missiles, rockets, and spacecraft, and any aircraft designed to fly as a drone, remotely piloted vehicle (RPV), or unmanned aerial vehicle (UAV). (The *Directory* does include piloted aircraft that were later converted or reconfigured to fly as unmanned vehicles.)

We tried to include every design that entered construction, even if the project never saw completion. Successful flight is not a prerequisite for inclusion in the *Directory*: these pages include many notable failures, as well as a few lesser-known disappointments. We also featured a number of significant designs that saw a considerable outlay of time and energy, only to be canceled before construction began. The *Directory* generally avoids simple studies or proposals: modern aircraft are often preceded by dozens of such plans, most of which are discarded as part of the design process. Still, the historical significance of a few of these unrealized proposals obliged us to include them too.

We have listed each aircraft under one firm or individual, even though many aircraft are associated with multiple designers or manufacturers. Subcontractors - those who build portions of another company's aircraft - are included only if they have also been responsible for design or construction of a complete airframe. In our research we found hundreds of confusing attributions, usually resulting from licensing agreements, company name changes, mergers, and buyouts, homebuilt and kit planes, or major modifications of existing aircraft. When given choices linking aircraft with makers, we have tried to follow the most commonly accepted historical practice, even though there are often several equally valid alternatives.

NAMES AND DESIGNATIONS

The proper name or designation of each aircraft type usually originates with the designer, manufacturer, and/or end user. Of course, there are exceptions such as the World War II Allies' code names for Japanese aircraft and the Cold War NATO code names for Soviet bloc aircraft. Each naming authority usually identifies aircraft within its own unique system of nomenclature. The rules and conventions of these systems may be firmly established at any given time, but are often revised or ignored as missions, organizational structures, or marketing strategies change.

In biological studies, each plant and animal has a universally accepted scientific name. Aviation history is a different matter, and finding the most accurate (or, in some cases, most commonly accepted) names and designations is subject to the accuracy and completeness of the historical record. For example, in 1946 a US air attaché reported at length on the new design from the Italian Velivolo company. Unaware that "velivolo" was the Italian word for "airplane," the attaché never noted that the aircraft was actually the Bestetti C.3. Syntax may also prove variable in the historical record, as in case of a French report that identified a Farman aircraft as the "F. 50 Bn 2," the "F. 50 Bn.2," the "F 50 Bn 2," and the "Far. 13 Bn.2." Our reliance on many secondary sources has been further complicated by the spurious creation and revision of designation systems by some writers and historians. While we have striven for accuracy, searching for authoritative names and designations, we recognize that we may have inadvertently repeated some earlier errors.

In our listing, we tried to include all of the "officially" assigned names and designations for each major aircraft type. In those cases where there are no identifiable names (as we often see when researching aircraft from the early 1900s), we have provided a simple description (such as "1905 Pusher Biplane"). Due primarily to space considerations (and to prevent confusion), we have also avoided most of the popular type names created and used by aircrew. Thus, for example, the British "Mossie," "Spit," and "Hurri" appear as Mosquito, Spitfire, and Hurricane, while the USAF "Bone," "Viper," and "Warthog" appear as B-1 Lancer, F-16 Fighting Falcon, and A-10 Thunderbolt II. We have limited ourselves to the level of detail that appears most useful to the majority of users. For example, we have listed the Boeing B-17F Flying Fortress but not such sub-variants as the B-17F-55-BO or B-17F-60-BO.

ORGANIZATION AND FORMAT

The heading for each manufacturer or designer featured in this *Directory* is organized by the most commonly recognized name. For instance, "Sociedade Aerotec Ldta" is found under ***Aerotec***, and "M. B. Arpin & Co" is seen under ***Arpin***, while "Ted Smith Aircraft Co, Inc" appears under ***Ted Smith***. To help prevent confusion, we posted cross reference notes throughout the *Directory*. We included all the foreign diacritical marks that were supported by our software, though several Czech, Polish, and Scandinavian letters were lost when we converted from WordPerfect to PageMaker shortly before publication. (English lettering was substituted in each of these instances.) When known, we listed the country or US city and state most frequently associated with the individual or firm. We also interfiled the headings for a number of miscellaneous subjects, such as ***Gliders, General***. These subject headings, specific to NASM's technical files, are included for those using this *Directory* as a guide to the Museum's collections.

The company and individual headings are set bold and italicized, and presented in English alphabetical order. When several entities have used similar names, we presented company names first, followed by individual names. Cross references are italicized, but not set bold.

When our research led us into complex corporate or individual histories, we prepared short explanatory notes, placed just beneath the headings. These notes present our best understanding of changes in organizational structures and assignment of production and design rights for aircraft through mergers, takeovers, licenses, and collapses. The notes also explain which conclusions we reached when deciding whether to list an aircraft under its designer, under the organization that started production, or under the organization that produced or sold the most airframes.

The entries for individual aircraft begin with the common name of the designer or manufacturer. Clarifying notes are included in parentheses, if needed, to distinguish between similarly named makers. For instance, the Williams Mark IV and Williams 601 HD are listed as Williams (Val D.) Mark IV and Williams (Vernon J.) 601 HD to lessen any possible confusion between the two builders. When aircraft have used multiple names or designations, the most widely used appears first, with alternatives in parentheses. Thus, the Fairchild R4Q-2 Packet, which was redesignated C-119F in 1962, appears in the *Directory* as "Fairchild R4Q-2 (C-119F) Packet." Translations of foreign names appear in parentheses: Libanski Jaskolka (Swallow), for example.

The code names devised by the Allies to identify Japanese aircraft during World War II are listed in italics (Aichi D1A *Susie*, for instance). The same applies to NATO reporting names used for Communist Bloc aircraft, such as the Beriev Be-30 *Cuff*.

In normal text, the nicknames of individual aircraft are distinguished from the names of aircraft types by the use of italics. For example, every B-17D built by Boeing was called a Flying Fortress, but only one of those aircraft was nicknamed *Swoose* by its crew. While only a few nicknames appear in the directory, they are listed in quotes to avoid confusion with any similar code names. Our example, then, appears as Boeing B-17D Flying Fortress "Swoose." In those cases where only one airframe was built, we made every effort to distinguish names from nicknames, though the designer or builder may have had different intentions. (All aircraft names and nicknames used in this directory are listed in the Index.)

For all other notes, parentheses are used to distinguish between terms that differentiate between aircraft and terms included for clarification. For instance, the listing "Lockheed HC-130H Hercules, USCG" distinguishes an aircraft used by the US Coast Guard from all other HC-130H variants, while the entry "Grumman JF-2 Duck (USCG)" clarifies that *all* JF-2 Ducks were built for the Coast Guard.

The term "NASM" appearing at the end of an aircraft entry denotes an aircraft which at one time was associated with the Smithsonian Institution. This includes aircraft which were, or are, a part of the National Collection and those which we have borrowed for display. (Our use of the modern term "NASM" is intended to encompass all earlier Smithsonian bureaus or departments that collected aviation artifacts.)

When sorting simple entries, we place letters first, in English alphabetical order. This is followed by numbers, with Arabic and Roman numbers mixed in strict numerical order. The words "Model," "Mark," and "Type" are ignored for ordering purposes. Spaces, hyphens, and commas, in that order, are treated as characters, ranked before letters or numbers. With these conventions, the following four companies appear in order as, "Aero Spacelines," "Aero-K Aviation," "Aerocar Co of America," and "Aerocar, Inc." Periods are ignored in the sorting of company acronyms, since their use varies so widely in corporate histories. For example, the German firm Allgemeinen Electrizitäts Gesellschaft is normally known as "A.E.G.," though the abbreviation "AEG" appears in some company records. The company's listings would appear in the same position in the *Directory*, regardless of which form of the abbreviation we used. In designations, however, periods are normally read as hyphens - though the two marks must not be considered interchangeable! Thus, the Fokker S-3 is found between the Fokker S.II and the Fokker S.IV.

Aircraft model designations often carry prefixes and suffixes to distinguish subtypes, variants, and modifications or to clarify aircraft features. In sorting, we have tried to follow the original order delineated by each designating authority. For example, early Heinkel aircraft appear in strict numerical order without regard for the "HE" ("Heinkel Eindecker") and "HD" ("Heinkel Doppeldecker") prefixes. US military designations start with prototypes, followed by each variant, with mission modification codes sorted within the variants. Using the Lockheed Hercules as an example, the YC-130 is followed by the C-130A, AC-130A DC-130A, C-130B, etc.

We have chosen to place pre-1962 US Navy designations first, followed by any similar Army, Air Force, or joint service designations. Following this convention, the Navy's Consolidated patrol series PY, P2Y, and P4Y are listed first, followed by the Army's pursuit series Consolidated P-25 and P-30, and then by the Consolidated patrol bomber series PBY, PB2Y, and PB4Y.

AMENDMENTS

While every effort has been made to keep this listing accurate and complete, we recognize that errors and omissions are unavoidable. We welcome all comments which will help us improve future editions of this *Directory*. Please send any documented corrections, additions, or comments to us at:

NASM Archives Division
Room 3100, Mail Code 322
National Air and Space Museum
PO Box 37012
Washington, DC 20013-7012

or e-mail us at: reference.desk@nasm.si.edu.

ACKNOWLEDGMENTS

At one time or another over the last decade, this Directory has been received the attentions of every member of the NASM Archives Division. This includes Henry "Hank" Brown, Tim Cronen, Marilyn Graskowiak, Kate Igoe, Kristine Kaske, Brian Nicklas, Mark Taylor, Barbara Weitbrecht, Patti Williams, and Larry Wilson. Melissa Keiser and David Schwartz dedicated significant blocks of time to this project, each completing the organization of several sections. Dan Hagedorn shared his vast knowledge of aviation history, and passed on hundreds of aircraft and manufacturers discovered during his research into Federal Aviation Administration aircraft registration records. Allan Janus and Paul Silbermann each spent several years on the project, helping to develop the formats and conventions we have used to present our findings, training volunteer researcher, and tracking down some of the answers to our thorniest questions. The Archives Division head, Dr. Thomas F. Soapes, gave this project the priority and support needed to see it through to completion, acting as a sounding board for policy decisions, offering his oft-needed advice, and promoting our efforts within the Museum.

Much of the day-to-day research for this project has been performed by volunteer researchers. As with so many projects, completion would have been difficult - or impossible - without their efforts. The unpaid staff members who have contributed to this Directory include Charles "Ned" Barnett, Fred Bruwelheide, Carl Eck, Shirley Eustis, Jerry Greenberg, John Heil, Frank Richardson, Christiane Sander, Ig Sargeant, Wes Smith, and Magda Wolfe.

Other volunteers have donated two weeks of their time over each of the last ten summers. They've come to Washington from around the country as part of our special archival research program, and added greatly to our efforts. Program members have included Ray Bartlett, Mark Behnke, Robert Benjamin, Rodney Benson, Paula Haynes Bosselman, Rich Carlson, Richard and Margaret Chandler, Dr. Donald Clark, Pat Colquette, Ann Dillon, Richard Dinda, Arthur W. Farr, Joanne Galanis, John M. Gallagher, James Hanson, Caroline Hicks, Michele Hoban, Doris Kneuer, Phil Lathrap, Karen Madej, Don Muller, Ken Pilpel, Dr. Audrey Roth, Ed Sachtleben, Cynthia Schieffelin, John Scott, Jim Sessa, Delmer Sparrowe, and Norland and Jean Wilson.

Several NASM colleagues have helped this work progress. Curators from the Aeronautics Department reviewed drafts and offered the benefits of their experiences. Among those directly involved in this *Directory*, we must recognize Tom Crouch, Peter Jakab, Russ Lee, Bob van der Linden and Dom Pisano. Trish Graboske, NASM's Director of Publications, helped change this *Directory* from a project into a publication.

Researchers at several other museums have also reviewed earlier drafts of this work, adding much to the final product. They include John Bolthouse from the San Diego Aerospace Museum (San Diego, CA), Brian J. Howard of the National Warplane Museum (Horseheads, NY), and John Merrell of the Frontiers of Flight Museum (Dallas, TX).

At Greenhill Books, the ever-wonderful Lionel Leventhal saw the merit of this work and chose to publish it. Roger Chesneau helped see us through the trauma of new software, and provided his editing and layout support. Jonathan North brought all the pieces together, and moved the work from its electronic format onto the printed pages you now hold.

And on the homefront, my wife Susan has offered the encouragement and support needed to see me through this project, particularly when it started coming home with me for evenings and weekends over the last year.

To all those who have helped I offer my profound thanks!

Dana Bell
NASM Archives
Washington, DC

A

A-K Aircraft Syndicate *(See: Knoll Brayton)*

AAA (Advanced Amphibious Aircraft) (International)
AAA Twin-turboprop Amphibian

AAC *(See: American Aviation Corp)*

AAC (Atelier Aéronautique de Colombes) AAC.1 *(See: Junkers Ju 52/3m)*

Aachen (Flugwissenschaftlichen Vereinigung Aachen) (Germany)
Aachen FVA-1 "Schwarze Teufel" ("Black Devil")
Aachen FVA-10B Rheinland
Aachen MS-11
Aachen Souris Bleue (Blue Mouse)

AAE *(See: Lecomte, Alfred)*

AAI *(See: American Aviation Industries)*

AAMSA (Aeronáutica Agrícola Mexicana SA) (Mexico)
AAMSA Quail (See: Callair A-9)
AAMSA Sparrow Commander

AASI *(See: Advanced Aerodynamics & Structures, Inc)*

AAT (Ameur Aviation Technologie) (France)
AAT Balbuzard L235
AAT Baljims 1A

A.B. Flygindustri (Halmstad, Sweden)
A.B. Flygindustri (Andreasson) BA-4
A.B. Flygindustri (Andreasson) BA-4B

A.B. Flygindustri (Malmö-Linham, Sweden) *(See: Junkers)*

A.B. Malmö Flygindustri (Sweden)
A.B. Malmö MFI-9

Abbott-Baynes Aircraft Ltd *(See: Baynes, L. E.)*

Abbruzzo, Onofrio
Abbruzzo Ornithopter (1868)

ABC Glider Club (Detroit, MI)
ABC Glider Sailplane

ABC Motors Ltd (UK)
ABC Motors Robin

Abel, Charles (Palatine, IL)
Abel 1929 Sailplane

Abic et Calas (France)
Abic et Calas 1909 Pusher Biplane
Abic et Calas 1909 Biplane Glider

ABL(Aéroplanes Bonnet-Labranche) *(See: Bonnet-Labranche)*

Abramov, V. V. (Russia)
Abramov VA-3
Abramov 48 Dnepr

Abrams Aircraft Corp (Lansing, MI)
Abrams Explorer
Abrams Explorer, NASM

Abreu (Joachim S.) Aeronautical Engineering (Abreu Patents Co, Inc) (San Francisco, CA)
Abreu A.E.5

Abrial (France)
Abrial No.17 Bagoas Flying Wing Glider (1933)

ACAZ *(See: Zeebrugge Aeronautical Construction Co)*

ACE (New York, NY) *(See: Aircraft Engineering Corp; also Keane, Horace)*

Ace Aircraft Composite Engineering (Australia)
Ace (Australia) Stingray Microlight

Ace Aircraft Manufacturing Co (Asheville, NC)
The **Ace Aircraft Manufacturing Co** of Asheville, NC was established to market kit planes originally designed by the **Corben Sport Plane and Supply Co.** Following the acquisition of the Corben rights by Paul Poberezny of the Experimental Aircraft Association (EAA), the airframes were substantially redesigned and, from mid-1950s on, were available through Ace, to keep the EAA from becoming directly involved in a profit-making concern. Ace later acquired the rights to the American Flea Ship and Heath Parasol. For all original Corben designs, see **Corben**. For all post-1953 products, including kit planes built from Ace kits, see **Ace**.
Ace (NC) American Flea Ship
Ace (NC) Baby Ace Model C
Ace (NC) Baby Ace Model D
Ace (NC) Junior Ace Model E

Ace Airplane Co (Los Angeles, CA) *(See: Essig)*

Aces High Light Aircraft Ltd (Canada)
Aces High CUBY 1
Aces High CUBY II

Acme (Air Craft Marine Engineering Co) (Los Angeles, CA)
Acme (Los Angeles) Anser Twin-jet Amphibian

Acme Aircraft Co (Torrance, CA)
Acme (Torrance) Centaur 101 (Convair L-13 Civil Conversion)

Acme Aircraft Corp (Rockford, IL)
Acme (IL) Model 21
Acme (IL) 1928 Parasol

Acro (Lyons, Wisconsin)
Acro Sport
Acro II

ACT (Aviation Composite Technology) (Philippines)
ACT Apache 1

ACV *(See: Air Cushion Vehicles)*

ADA (Aeronautical Development Agency) (India)
ADA TD-1 LCA (Light Combat Aircraft)

Adam Aircraft Industries LLC (Rick Adam) (Englewood, CO)
Adam (CO) M-309

Adam (Établissements Aéronautiques R. Adam) (France)
Adam (France) R.A.14 Loisirs (Leisure)
Adam (France) R.A.15 Major
Adam (France) R.A.17 Ag-plane

Adams Industries Inc *(See: Thorp Aircraft Co)*

Adams, Alex (Australia)
Adams (Alex) 1897 Ornithopter

ADC (Aircraft Disposal Co, Ltd) (UK)
ADC Martinsyde A.D.C.1 Disposalsyde
ADC Nimbus Martinsyde

Ader, Clément (France)
Ader 1873 Tethered Glider
Ader Avion I l'Eole (Aeolus) (1890)
Ader Avion II (1893)
Ader Avion III (1897)
Ader Avion IV (1897)

Adler Flugzeugbau (Adlerwerke GmbH; Adlerwerke vorm. Heinrich Kleyer AG) (Germany)
Adler G I
Adler G II.R

Admiralty Air Department (UK)
Admiralty A.D. Flying-Boat (1915/17)

Advance Aircraft Co *(See: Waco)*

Advanced Aerodynamics & Structures, Inc (AASI) (North Hollywood, CA)
AdvancedAerodynamics&Structures:
Jetcruzer 450
Jetcruzer 500
Jetcruzer 650 (M6-5)
Stratocruzer 1250
Stratocruzer 1250-ER

Advanced Amphibious Aircraft *(See: AAA)*

Advanced Aviation Inc (Orlando, FL)
Advanced Aviation Cobra
Advanced Aviation Hi-Nuski
Advanced Aviation Hi-Nuski Coyote
Advanced Aviation Hi-Nuski Huski
Advanced Aviation Hi-Nuski Huski II
Advanced Aviation Sierra

Advanced Soaring Concepts (ASC) (Camarillo, CA)
ASC Falcon (American Falcon)
ASC Spirit (American Spirit)

Advanced Technology Aircraft Co Inc (USA)
Advanced Technology Aircraft:
480 Predator
480 Turboprop Ag Aircraft

Adventure Air (USA)
Adventure Air Adventure Twin 2-Seat Amphibian
Adventure Air Adventurer 4-Place
Adventure Air Super Adventurer
Adventure Air 2+2 Amphibian

AEA *(See: Aerial Experiment Assoc)*

AEA Research Pty Ltd (Aeronautical Engineers Australia) (Australia)
AEA (Australia) Explorer 350
AEA (Australia) Explorer 500

A.E.G. (Allgemeinen Electrizitäts Gesellschaft) (Germany)
The Berlin engineering firm of **Allgemeinen Electrizitäts Gesellschaft** (**A.E.G.**) established its first aviation plant in 1913. The company focused primarily on military aircraft until the end of World War I, when it began work on civil transports. Under the terms of the Versailles treaty, A.E.G. left the aviation industry in October 1920. Company products are listed as **A.E.G.**
A.E.G. B.I
A.E.G. B.II
A.E.G. B.III
A.E.G. C.I
A.E.G. C.II
A.E.G. C.III
A.E.G. C.IV
A.E.G. C.IV N
A.E.G. C.V
A.E.G. C.VII
A.E.G. C.VIII
A.E.G. C.VIII Dr
A.E.G. D.I
A.E.G. Dr.I
A.E.G. DJ.I
A.E.G. E.I
A.E.G. G Series
A.E.G. G.I (K I)
A.E.G. G.II
A.E.G. G.III
A.E.G. G.IV
A.E.G. G.IVb
A.E.G. G.IVk
A.E.G. G.V
A.E.G. J.I
A.E.G. J.Ia
A.E.G. J.II
A.E.G. J.II (Civil)
A.E.G. N.I
A.E.G. P.E.
A.E.G. R.I
A.E.G. Wagner
A.E.G. Z.1
A.E.G. Z.2
A.E.G. Z.3
A.E.G. 1913 Flying Boat
A.E.G. 1915 Flying Boat
A.E.G. 1914 Float Plane

Aer Pegaso (Italy)
Aer Pegaso M 100S

Aer-Fabric S.R.L. (Argentina)
Aer-Fabric AF-3 Chuschin (Thrush)

Aerauto SA Costruzioni Aeronautiche e Meccaniche (Italy)
Aerauto PL.5C Roadable Monoplane

Aereon Corp (Princeton, NJ)
Aereon DYNAIRSHIP
Aereon VectoRotor
Aereon WASP/AF (Wide Aperture Surveillance Platform)
Aereon WASP/N (Wide Aperture Surveillance Platform)
Aereon III
Aereon 26

Aerfer (Italy)

Industrie Meccaniche e Aeronautiche Meridionali-Aerfer (**Aerfer**; later **Aerfer-Industrie Aerospaziali Meridionali SpA**) was established by the 1955 merger of **Officine di Pomigliano per Costruzioni Aeronautiche e Ferroviarie** and **Industrie Meccaniche e Aeronautiche Meridionali**. Through most of the 1960s, the company survived on subcontract work, before cooperating with **Aermacchi** on the AM.3 observation aircraft. In 1969, Aerfer, Fiat's aircraft section, and the avionics firm Salmoiraghi merged to form **Aeritalia**. Fiat and Aerfer continued to operate under their own names until 1971. Aerfer designs predating 1955 are filed under **Meridionali**; those which entered production between 1955 and 1971 are under **Aerfer**; subsequent designs are under **Aeritalia**.

Aerfer Sagittario (Archer) 2
Aerfer/Aermacchi AM.3 (MB.335, Bosbok)

Aerial Arts (UK)
Aerial Arts Chaser
Aerial Arts Chaser S
Aerial Arts Chaser S447
Aerial Arts Chaser S508

Aerial Distributors Inc (Wichita KS)
Aerial Distributors Distributor Wing DWI-1

Aerial Engineering Corp (USA)
Aerial Engineering Corp:
BR-1 Bee Line Racer (Booth Racer)
BR-2 Bee Line Racer (Booth Racer)
Thurston Monoplane

Aerial Experiment Assoc (Hammondsport, NY)
Aerodrome No 1 Red Wing
Aerodrome No 2 White Wing
Aerodrome No 3 June Bug
Aerodrome No 3 Loon
Aerodrome No 4 Silver Dart
Cygnet I
Cygnet II
Cygnet III
Gliders
Kites
1908 Biplane Glider

Aerial Navigation Co (Girard, KS)
Aerial Navigation Call Monoplane

Aerial Service Corp (Hammondsport, NY)
Aerial Service Corp:
Mercury Arrow
Mercury BT-120 Aerobat
Mercury Chick (Chic) T-2
Mercury Jr
Mercury Kitten
Mercury Night Mail
Mercury Primary Trainer
Mercury Racer
Mercury Standard
Mercury Training CW-4
Mercury Variable Camber (See: Schroeder-Wentworth 1929 Safety Plane)

Aeritalia (Italy)

Aeritalia was created by a November 1969 merger of **Fiat** and **Aerfer**. The two earlier companies continued operating under their own names until 1971. In 1990 Aeritalia merged with **Selenia** to form **Alenia**. With the exception of the F-104 and G 222 (which are both filed under **Aeritalia**), designs which entered production prior to 1971 are filed under **Fiat** or **Aerfer**.

Aeritalia F-104S
Aeritalia F-104Y
Aeritalia G 222

Aermacchi *(See: Macchi)*

Aero (France) *(See: Indraéro)*

Aero (Prague) (Letadla Továrny Aero) (Czechoslovakia)

Letadla Továrny Aero was established in 1919. The company was taken over during the German seizure of Czechoslovakia in 1938 and integrated into the German aircraft industry. At the end of World War II, the entire Czechoslovakian aviation industry was reorganized and Aero ceased to exist. All Letadla Továrny Aero products are listed under **Aero (Prague)**.

Aero (Prague) Ae 01
Aero (Prague) Ae 02
Aero (Prague) Ae 03
Aero (Prague) Ae 04
Aero (Prague) "Al-Ma"
Aero (Prague) A 8
Aero (Prague) A 10
Aero (Prague) A 11
Aero (Prague) A 11 C
Aero (Prague) A 11 HS
Aero (Prague) A 11 J
Aero (Prague) A 11 LD
Aero (Prague) A 11 N
Aero (Prague) A 211
Aero (Prague) Ab 11
Aero (Prague) Ab 11 N
Aero (Prague) Ab 111
Aero (Prague) Aš 11
Aero (Prague) A 12
Aero (Prague) A 14
Aero (Prague) A 15
Aero (Prague) A 17
Aero (Prague) A 18
Aero (Prague) A 18B
Aero (Prague) A 18C
Aero (Prague) A 19
Aero (Prague) A 20
Aero (Prague) A 21
Aero (Prague) A 22
Aero (Prague) A 23
Aero (Prague) A 24
Aero (Prague) A 25
Aero (Prague) A 125
Aero (Prague) A 26
Aero (Prague) A 27
Aero (Prague) A 29
Aero (Prague) A 30
Aero (Prague) A 130
Aero (Prague) A 230
Aero (Prague) A 330
Aero (Prague) A 430
Aero (Prague) A 30 HS
Aero (Prague) A 230
Aero (Prague) Ab 30
Aero (Prague) A 32
Aero (Prague) A 32 GR
Aero (Prague) A 32 IF
Aero (Prague) Ap 32
Aero (Prague) Apb 32
Aero (Prague) A 34 Kos
Aero (Prague) A 34 J
Aero (Prague) A 134
Aero (Prague) A 35
Aero (Prague) A 38
Aero (Prague) A 42
Aero (Prague) A 42b
Aero (Prague) A 42c

Aero (Prague) A 46
Aero (Prague) A 100
Aero (Prague) A 101
Aero (Prague) Ab 101
Aero (Prague) A 102
Aero (Prague) A 102 D
Aero (Prague) A 104
Aero (Prague) A 200
Aero (Prague) A 201
Aero (Prague) A 204
Aero (Prague) A 206
Aero (Prague) A 300
Aero (Prague) A 302
Aero (Prague) A 304
Aero (Prague) A 404
Aero (Prague) D.H.50 (License-Built)
Aero (Prague) MB200 (License-Built)

Aero (Vodochody) (Aero Vodochody Akciová Spolecnost) (Czechoslovakia)

In 1953, the Czech government established **Aero Vodochody Akciová Spolecnost** outside Prague. Following the fall of the Communist government in 1989, the factory continued operations under the Aero name. All Aero Vodochody Akciová Spolecnost products are filed under **Aero (Vodochody)**.

Aero (Vodochody) HC-2 Heli-Baby
Aero (Vodochody) HC-3
Aero (Vodochody) L 29 Delfin *Maya* (Dolphin)
Aero (Vodochody) L 39 Albatros (Albatross)
Aero (Vodochody) L 139 Albatros (Albatross)
Aero (Vodochody) Ae-45
Aero (Vodochody) Ae-45 S Super
Aero (Vodochody) Ae-145
Aero (Vodochody) Ae-245
Aero (Vodochody) Ae-345
Aero (Vodochody) Ae-148
Aero (Vodochody) Ae-50
Aero (Vodochody) Ae-53
Aero (Vodochody) L 59
Aero (Vodochody) L 159A ALCA (Advanced Light Combat Aircraft)
Aero (Vodochody) L 159B ALCA
Aero (Vodochody) K 60
Aero (Vodochody) XL 60 Brigadýr
Aero (Vodochody) L 60 Orlicanu
Aero (Vodochody) LB 60
Aero (Vodochody) K 160
Aero (Vodochody) L 160
Aero (Vodochody) K 75
Aero (Vodochody) Ae 270 Ibis

Aero Boero (Aero Talleres Boero SRL) (Hector & Cesar Ernesto Boero) (Argentina)

Aero Boero 95
Aero Boero 95A
Aero Boero 95B
Aero Boero AB.115
Aero Boero AB.115BS
Aero Boero AB.180AG
Aero Boero AB.180PSA
Aero Boero AB.180RV
Aero Boero AB.180RVR
Aero Boero AB.180SP
Aero Boero AB.210
Aero Boero AB.260AG

Aero Club du Royaume (Yugoslavia)

Aero Club du Royaume Bloudek XV

Aero Commander (Culver City, CA & Bethany, OK)

Aero Design and Engineering Co was founded in Culver City, CA in December 1944. In 1952, the company received a type certificate for its first aircraft - the Aero Commander - and moved to Bethany, OK. In 1958, the company was acquired by the **Rockwell-Standard Corp** and, in 1960 became **Aero Commander, Inc**, a subsidiary of Rockwell-Standard.

In July 1965, Aero Commander acquired the **Meyers Aircraft Co** and **Volaircraft, Inc** as its Tecumseh and Aliquippa divisions, respectively. In November 1965 Rockwell acquired the **Snow Aeronautical Co** as the **Olney Division of Aero Commander**. In December 1966 Rockwell acquired the rights to the **CallAir** line from **Intermountain Manufacturing Co (IMCO)** as the **Afton Division of Aero Commander**. Meyers and Volaircraft products were folded into the main Aero Commander line while Snow and CallAir products, primarily agricultural aircraft, were marketed under the "Ag Commander" name.

Following the 1967 merger between Rockwell-Standard and **North American Aviation**, Aero Commander became the **Aero Commander Division** of **North American Rockwell**. At the same time, the Aero Commander Jet Commander line was sold to **Israeli Aircraft Industries (IAI)**, which produced it as the IAI Commodore. In 1969, North American Rockwell combined its Aero Commander and **Remmert Werner** divisions into a new **General Aviation Division**. With the formation of **Rockwell International Corp** in 1973, the Bethany, OK, plant of the General Aviation Division became the **Commander Division of Rockwell International.** Rockwell sold the Commander line to **Gulfstream American** in 1981. In 1988 the Commander Aircraft Co of Bethany, OK purchased the rights to the single-engined aircraft developed after the 1969 reorganization.

For all Aero Commander product lines initiated before the 1969 reorganization, see **Aero Commander.** For all Meyers, Volaircraft, Snow, and CallAir products, including those folded into the Aero Commander line, see **Meyers, Volair, Snow,** and **Callair,** respectively. For all post-1967 Jet Commander production, see **IAI.** For all product lines begun after the 1969 reorganization, see **Commander.**

Aero Commander:
YL-26 (YU-9A)
YL-26A
L-26B (U-4A)
L-26B (U-9B)
L-26C (U-4B)
L-26C (U-9C)
NL-26D (NU-9D)
RL-26D (RU-9D)
L.3805 (Commander Prototype)
100 (Darter Commander)
200 (See: Meyers)
500
500A
500B
500S Shrike Commander
500U
520
560
560A
560E
560F
680 Super Commander
680E
680F
680FL Grand Commander
680FL(P) Pressurized Grand Commander
680FP
680T Turbo Commander
680V Turbo Commander
680W Turbo II Commander
690
720 Alti Cruiser
1121 Jet Commander

Aero Composite Technology Inc (Somerset, PA)
Aero Composite Sea Hawk
Aero Composite Sea Hawker

Aero Design and Engineering Co *(See: Aero Commander)*

Aero Design Associates (Opa-Locka, FL)
Aero Design Associates DG-1

Aero Designs Inc (San Antonio, TX)
Aero Designs Pulsar
Aero Designs Pulsar XP

Aero Diffusion (Spain) *(See: Jodel)*

Aero Dynamics Ltd (Seattle, WA)
Aero Dynamics Sparrow Hawk II

Aero Engine Services Ltd (New Zealand) *(See: AESL)*

Aero Flight Inc (Cody, WY)
Aero Flight B-17 Turboprop Conversion

Aero Import Co *(See: Ansaldo)*

Aero Industry Development Center (AIDC) (Taiwan)
Aero Industry Development Center AT-3

Aero International (Regional) *(See: ATR; also: British Aerospace)*

Aéro Navale *(See: SCAN)*

Aero Research, Ltd (UK)
Aero Research De Bruyne Snark
Aero Research De Bruyne-Maas Ladybird

Aero Resources Inc (USA)
Aero Resources J-2 (See: Jovair)
Aero Resources Super J-2 Gyroplane

Aero Spacelines, Inc (Van Nuys, CA & Santa Barbara, CA)
Aero Spacelines G-201A
Aero Spacelines Mini Guppy
Aero Spacelines Mini (Super Super) Guppy Turboprop (MGT)
Aero Spacelines Pregnant Guppy
Aero Spacelines Super Guppy

Aero Sport, Inc *(See: Extra Flugzeugbau GmbH)*

Aero Tech Aviation Ltd (Canada)
Aero Tech Canadian Skyrider

Aero Tek *(See: Applebay)*

Aero Wood Specialties Inc (USA)
Aero Wood Specialties Avocet 1

Aero-Club Der Schweiz (Switzerland) *(See: Extra)*

Aero-Craft Manufacturing Co (Detroit, MI)
Aero-Craft Aerocoupe

Aero-Difusión Ltd (Spain) *(See: Jodel)*

Aero-Ever (Hungary)
Aero-Ever R.1
Aero-Ever R.2
Aero-Ever R.3
Aero-Ever R.4
Aero-Ever R.5
Aero-Ever R.6
Aero-Ever R.7
Aero-Ever R.8
Aero-Ever R.9
Aero-Ever R.10
Aero-Ever R.11
Aero-Ever R.12
Aero-Ever R.13
Aero-Ever R.14
Aero-Ever R.15
Aero-Ever R.16
Aero-Ever R.17
Aero-Ever R.18
Aero-Ever R.19
Aero-Ever R.20
Aero-Ever R.21
Aero-Ever R.22

Aero-Flight Aircraft Corp (Buffalo, NY, Long Beach, CA)
Aero-Flight NC-1 Streak-85 (AFA-1)
Aero-Flight NC-2 Streak-125 (AFA-2)
Aero-Flight NC-3 Streak-165 (AFA-3)
Aero-Flight Streak 225

Aero-Jean SA (Spain)
Aero-Jean AJ.1 Serrania (Fournier RF-5 Variant)

Aero-Jodel (Germany) *(See: Jodel)*

Aero-K Aviation
Aero-K JetHawk

Aero-Kuhlmann (France)
Aero-Kuhlmann SCUB

Aero-M JSC (Russia)
Aero-M A-209 Transport
Aero-M A-230

Aero-Reek (Russia)
Aero-Reek Dingo

Aerobat Aircraft (Brownsville, TX)
Aerobat Aircraft Aerobat

Aerocad Inc (USA)
Aerocad AeroCanard
Aerocar Co of America (USA)
Aerocar Co of America Cloudster

Aerocar, Inc (Longview, WA)
Aerocar, Inc Coot Model A (Sooper-Coot)
Aerocar, Inc Coot Model B
Aerocar, Inc Imp
Aerocar, Inc Micro-Imp
Aerocar, Inc Mini-Imp
Aerocar, Inc Taylor Bullet (Bullet 2100)
Aerocar, Inc Ultra-Imp (Perigee)
Aerocar, Inc Model I
Aerocar, Inc Model II Aero-Plane
Aerocar, Inc Model III

Aerocar International Corp (USA)
Aerocar International Aerocar

Aérocentre *(See: SNCAC)*

Aerocomp (Merritt Island, FL)
Aerocomp Comp Monster
Aerocomp Comp Air 6
Aerocomp Comp Air 7
Aerocomp Comp Air 10
Aerocomp Merlin GT
Aerocomp Merlin E-Z

Aerodesign Desenvolvimentos Aeronáuticos (Brazil)
Aerodesign Pegasus

Aerodis America Inc (USA, Indonesia, Thailand)
Aerodis America AA200 Orion
Aerodis America AA300 Rigel
Aerodis America AA330 Theta

Aerodyn' (France)
Aerodyn' Skyranger

Aerodynamics & Structures Inc (ASI) *(See: Advanced Aerodynamics & Structures Inc)*

Aerodyne Systems Engineering Ltd (USA) *(See: Texas Helicopter Corp)*

Aerofab Corp (Sanford, ME)
Aerofab TSC-2 Explorer (See: Patchen)

Aerofan Aircraft Manufacturing Co (USA)
Aerofan SingleTwin Pusher Business Aircraft

Aerogypt (See: Helmy)
AeroLites Inc (Welsh, LA)
AeroLites AeroMaster AG
AeroLites Bearcat
AeroLites Ag Bearcat

Aeromarine Plane & Motor Co (Keyport, NJ)
The **Aeromarine Plane & Motor Co** originally entered the aviation industry during World War I. Following the War, however, the company was unable to build aircraft profitably and, in 1924, closed its aircraft branch.
Aeromarine AM-1 Night Mail Plane
Aeromarine AM-2 Night Mail Plane
Aeromarine AM-3 Night Mail Plane
Aeromarine AMC
Aeromarine AS
Aeromarine AS-1
Aeromarine AS-2
Aeromarine DH-4 Whitbeck Special
Aeromarine DH-4B
The British de Havilland D.H.4 was modified and built under license in the USA under the designation DH-4 by several manufacturers. For British-built aircraft, see **de Havilland D.H.4.** For American-built aircraft, see the individual manufacturers, including **Aeromarine**, **Fokker**, **Boeing**, **Dayton Wright**, **Gallaudet**, **Keystone**, and **Standard.** When the actual manufacturer cannot be determined from the available evidence, the photos and documents have been filed under **Dayton Wright.**
Aeromarine EO
Aeromarine Ice Boat
Aeromarine M-L
Aeromarine (Martin) NBS-1
Aeromarine Passenger-Cargo Land Airplane
Aeromarine PG-1
Aeromarine Sportsman
Aeromarine 39-A
Aeromarine 39-B
Aeromarine 40 (40-F)
Aeromarine 40-B
Aeromarine 40-C
Aeromarine 40-L
Aeromarine 40-T
Aeromarine 40-U
Aeromarine 43-L
Aeromarine 44-L
Aeromarine 50
Aeromarine 50-B-2
Aeromarine 50-C
Aeromarine 50-L-8
Aeromarine 50-S
Aeromarine 50-S-2
Aeromarine 50-U-8-D
Aeromarine 55-L-8
Aeromarine 60
Aeromarine 60-S-2
Aeromarine 75 (F-5L)
Aeromarine 80 (HS-2L)
Aeromarine 85 (HS-2L)
Aeromarine 700
Aeromarine 1915 Military Tractor

Aeromarine-Klemm Corp (Keyport, NJ & New York, NY)
In 1929, the **Aeromarine Plane & Motor Co** reconstituted its aviation branch as the **Aeromarine-Klemm Corp** to build light aircraft designed by the German firm of **Klemm Leichtflugzeugbau GmbH.** In February 1932 the company was purchased by the **Upperçu Corp.** For all Aeromarine Plane & Motor Co designs predating 1929, see **Aeromarine.** For all Aeromarine-built Klemm designs, see **Aeromarine-Klemm.** For all German-built Klemm designs, see **Klemm.**
Aeromarine-Klemm XIV
Aeromarine-Klemm AKL-20
Aeromarine-Klemm AKL-25
Aeromarine-Klemm AKL-25-A
Aeromarine-Klemm AKL-25-A-1
Aeromarine-Klemm AKL-26
Aeromarine-Klemm AKL-26-A
Aeromarine-Klemm AKL-26-B (AKL-85)
Aeromarine-Klemm AKL-60

Aeromercantil SA (Colombia) *(See also: Piper)*
Aeromercantil Gavilán (See: Gavilán)

Aeromere SpA (Italy)
Aeromere F.8 Falco (See: Aviamilano)
Aeromere M.100S Series V

Aeromirage Inc (Gainesville, FL)
Aeromirage TC-2

Aeromot (Aeronaves e Motores) (Brazil)
Aeromot AMT 100 Ximango
Aeromot AMT 100P Ximango
Aeromot AMT 100R Ximango
Aeromot AMT 200 Super Ximango
Aeromot AMT 300 Turbo Ximango

Aeron (Great Neck, NY)
Aeron Aircraft

Aeronasa (Constructora Aeronaval de Levante SA) (Spain) *(See: Piel Emeraude)*

Aeronautic Supply Co (USA) *(See: Benoist)*

Aeronáutica Agrícola Mexicana SA (Mexico) *(See: AAMSA)*

Aeronáutica Ansaldo SA (Italy) *(See: Ansaldo)*

Aeronáutica d'Italia SA (Italy) *(See: Fiat; See also Ansaldo)*

Aeronáutica Industrial, SA *(See: AISA)*

Aeronáutica Macchi Group (Italy) *(See: Macchi)*

Aeronáutica Militar Espanola *(See: A.M.E.)*

Aeronautical & Automobile College, Redhill, England (UK)
Aeronautical & Automobile College Skimmer Hovercraft

Aeronautical Corp of America *(See: Aeronca)*

Aeronautical Corp of Great Britain, Ltd (UK) *(See: Aeronca)*

Aeronautical Development Agency (India)
Aeronautical Development LCA

Aeronautical Engineers Australia *(See: AEA)*

Aeronautical Industrial Engineering & Project Management Co, Ltd *(See: AIEP)*

Aeronautical Products Inc (Detroit, MI)
Aeronautical Products A-3
Aeronautical Products 1A (Model NX1270)

***Aeronautical Research & Development Corp** (See: Brantly)*

***Aeronautical Syndicate Ltd** (See: ASL)*

***Aeronautics India Ltd (India)** (See: Hindustan)*

Aeronca (Aeronautical Corp of America) (Cincinnati & Middletown, OH)
The **Aeronautical Corp of American (Aeronca)** was established in 1928. The company became **Aeronca Aircraft Corp** in 1941. Aeronca stopped airframe production in the early 1950s to concentrate on subcontract work. **Champion Aircraft Corp** of Osceala, WI purchased the tooling and rights to build the Aeronca Model 7 series in 1954. For pre-1950 Aeronca products, see **Aeronca**. For post-1950 development, see **Champion**.
Aeronca C-1 Cadet
Aeronca C-2
Aeronca C-2 Scout
Aeronca C-2 Scout, NASM
Aeronca C-2 Scout, Roscoe Turner
Aeronca C-2, Model 100 (Aeronca UK)
Aeronca C-3 (Master)
Aeronca C-3 Collegian
Aeronca CF
Aeronca UC-64 (See: Noorduyn)
Aeronca 50 Chief (Pre-War)
Aeronca 65 Chief (Pre-War)
Aeronca 65 Super Chief (Pre-War)
Aeronca K
Aeronca KC
Aeronca KCA
Aeronca KF
Aeronca KM
Aeronca LA
Aeronca LB
Aeronca LC
Aeronca (L-3) YO-58 (65T)
Aeronca L-3A (O-58A)
Aeronca L-3B (O-58B)
Aeronca L-3C
Aeronca L-16 Prototype (7BC)
Aeronca L-16A (7BCM)
Aeronca L-16B (7CCM)
Aeronca (Fairchild) PT-19A Cornell
Aeronca (Fairchild) PT-19B Cornell
Aeronca (Fairchild) PT-23 Cornell
Aeronca T Trainer Family
Aeronca TA (Defender) Family
Aeronca TG-5
Aeronca 7AC Champion
Aeronca 7B-X Champion
Aeronca 7BCM Champion (See: L-16A)
Aeronca 7CCM Champion (See: L-16B)
Aeronca 7DC Champion
Aeronca 7EC Champion
Aeronca 8
Aeronca 9 Arrow
Aeronca 10 Eagle
Aeronca 11AC Chief
Aeronca 11ACS Scout
Aeronca 11BC Chief
Aeronca 11CC Super Chief
Aeronca 12 Chum
Aeronca 15AC Sedan
Aeronca 19

***Aeroneer** (See: Phillips Aviation Co)*

***Aeronics Pty Ltd (South Africa)** (See: Sequoia)*

Aeronova Costruzioni Aeronautiche (Italy)
Aeronova A.E.R.1

Aéroplanes Borel (See: Borel)

Aéroplanes de Beer (See: Beer)

Aéroplanes Goupy (See: Goupy)

***Aéroplanes Marcel Besson** (See: Besson)*

***Aéroplanes Morane-Saulnier** (See: Morane-Saulnier)*

Aeropract (Russia)
Aeropract A-21
Aeropract A-23M
Aeropract A-25 Breeze
Aeropract A-27 Twin-float

Aeroprakt Ltd (Ukraine)
Aeroprakt A-19
Aeroprakt A-20
Aeroprakt A-21M
Aeroprakt A-22
Aeroprakt A-23
Aeroprakt A-24
Aeroprakt A-25 Breeze
Aeroprakt A-26
Aeroprakt A-28

Aeroprogress (ROS-Aeroprogress, Roks-Aero Inc) (Russia)
Aeroprogress T-101 Gratch (Rook)
Aeroprogress T-106
Aeroprogress T-130 Fregat
Aeroprogress (Khrunichev) T-201 Sterkh
Aeroprogress T-203 Pchela (Bee)
Aeroprogress T-204 Griffon
Aeroprogress T-205 Vostochnyi Karavan (Eastern Caravan)
Aeroprogress (Khrunichev) T-230 Fregat
Aeroprogress T-274 Titan
Aeroprogress T-311 Kolibri (Hummingbird)
Aeroprogress T-401 Sokol (Falcon)
Aeroprogress T-407 Skvorets (Starling)
Aeroprogress (Khrunichev) T-411AIST-2
Aeroprogress T-417 Pegas
Aeroprogress T-420 Strizh
Aeroprogress T-430 Sprinter
Aeroprogress T-433 Flamingo
Aeroprogress T-440 Mercury
Aeroprogress T-501 Strizh (Martin)
Aeroprogress T-610 Voyage
Aeroprogress T-620 Korshun (Kite)
Aeroprogress T-710 Anaconda
Aeroprogress T-910 Kuryer (Courier)

Aeroric Nauchno-Proizvodstvennoye Predpriyatie OOO (Aeroric Science and Production Enterprise) (Russia)
Aeroric Dingo

Aerospace General Co (USA)
Aerospace General Mini-Copter

***Aerospace Industrial Development Corp (Taiwan)** (See: AIDC)*

Aérospatiale (Société Nationale Industrielle Aérospatiale, Aérospatiale SNI) (France)
In 1957 **Ouest-Aviation** merged with **SNCASE** to form **Sud-Aviation.** In 1970 Sud merged with **SNCAN** and **Sereb** to form **Aérospatiale.** Pre-1957 designs are listed under their original firms. Sud designs are under **Sud** or **Aérospatiale**.
Aérospatiale:
AS 316/319 (See: SNCASE SE 3160)
AS 332 Super Puma
AS 350 Ecureuil (Squirrel), Astar
AS 355 Twinecureuil (Twin Squirrel), Twinstar

SA 321 Super Frelon (Hornet)
SA 330 Puma
SA 340, 341, 342 Gazelle
SA 360 Dauphin (Dolphin)
SA 365 Dauphin 2 (Dolphin)
SA 365G Dauphin 2 (Dolphin) (HH-65A Dolphin)
SA 365M Panther
SN 600 Corvette Series
TB-9 Tampico
TB-10 Tobago
TB-20 Trinidad
TB-21 Trinidad

Aerosport (Holly Springs, NC)
Aerosport Flying Rail
Aerosport Scamp
Aerosport Quail
Aerosport Woody Pusher

AeroSPORT Pty Ltd (Australia)
AeroSPORT (Australia) Supapup Mk.4

Aerostar Aircraft Corp (Hayden Lake, ID)

In 1966, Ted Smith established **Ted Smith Aircraft Co, Inc** in Van Nuys, CA. In 1968 the company was purchased by the **American Cement Co**, which operated it for a year before selling it to **Butler Aviation International, Inc** in late 1969. In July 1970 Butler merged Ted Smith with **Mooney Aircraft Corp** of Wichita, KS as the **Aerostar Aircraft Corp** in Wichita. In early 1972, Butler suspended Aerostar production and Smith reacquired the right to the Aerostar design, establishing **Ted R. Smith & Associates** to manufacture the aircraft. In 1976, the company was renamed **Ted Smith Aerostar Corp.** In March 1978 **Piper Aircraft Corp** acquired the company as the **Santa Maria Division of Piper**, incorporating Aerostar designs into the Piper product line as the PA-60. In 1981 Piper closed the Santa Maria facility and moved Aerostar production to Vero Beach, FL until 1984, when it ended production. In 1991, the company reemerged as **Aerostar Aircraft Corp.** All Ted Smith designs prior to 1991 are filed under **Ted Smith.** Aerostar Aircraft Corp designs originating with Mooney, are filed under **Mooney.** Aerostar designs after 1991 are filed under **Aerostar Aircraft Corp.** Piper products, excluding Aerostar designs, are found under **Piper.**

Aerostar (ID) FJ-100
Aerostar (ID) 700 Aerostar

Aerostar SA (Romania)
Aerostar (Romania) Yak-52 (See: Yak)
Aerostar (Romania) Lancer (See: MiG-21)

Aerosystems (USA)
Aerosystems Cadet STF (See: Culver Cadet)

Aerotec (Sociedade Aerotec Ldta) (Brazil)
Aerotec A-132 Tangará (Tanager)
Aerotec A-135 Tangará II (Tanager II)
Aerotec T-23 Uirapuru

Aerotechnik CZ spol sro (Czechoslovakia)
Aerotechnik (Czech) L-13 Super Vivat
Aerotechnik (Czech) L-13SE
Aerotechnik (Czech) L-13SL Vivat
Aerotechnik (Czech) (Pottier) P 220 S Koala

Aerotechnik (Germany)
Aerotechnik (Germany) WGM-21 Helicopter

Aerotécnica SA (Spain)
Aerotécnica AC-11 (Matra MC.101 Variant)
Aerotécnica AC-12
Aerotécnica AC-13A
Aerotécnica AC-14

Aerotek (South Africa)
Aerotek Hummingbird

Aerotrade Sro (Czech Republic)
Aerotrade Racak 2

Aerovant Aircraft Corp (USA)
Aerovant Acroduster 1(See: Stolp SA-700)

Aerovel Aircraft Corp *(See: Miller, Howell)*

AeroVironment Pathfinder *(See: NASA)*

AESL (Aero Engine Services Ltd) (New Zealand)
AESL Air Cruiser CT/2
AESL Airtourer T1
AESL Airtourer T2
AESL Airtourer T3
AESL Airtourer T4
AESL Airtourer T5
AESL Airtourer T6 (Military T5)
AESL Airtourer T6/12
AESL Airtourer T6/24
AESL Airtourer AT.100 (See: Victa)
AESL Airtourer AT.115 (See: Victa)
AESL Airtourer 150 (AT.150)
AESL Airtourer Super 150
AESL Airtrainer CT/4 (Aircruiser)

Aetna Aircraft Corp *(See: Wally Timm Aerocraft)*

AFIC Pty Ltd (South Africa)
AFIC RSA.200 (Partenavia P.64 Variant)

Ag-Cat Corp *(See: Grumman G-164)*

AGO Flugzeugwerke GmbH (Aerowerke Gustav Otto) (Germany)

Gustav Otto established **AGO Flugzeugwerke GmbH (AGO)** in 1912. The company was heavily funded by **A.E.G.**, which took it over in 1917/8. A.E.G. renamed the company **Flugzeugwerke GmbH**, but it produced no aircraft after World War I. In 1937, a new company was organized in Berlin with the old AGO name. The new **AGO Flugzeugwerke GmbH** concentrated primarily on license production before the end of World War II forced it out of business.

AGO Ao 192 Kurier
AGO C.I
AGO C.IW
AGO C.II (2-Bay)
AGO C.II (3-Bay)
AGO C.IIW
AGO C.III
AGO C.IV
AGO C.VII
AGO C.VIII
AGO DV.3
AGO E.1 Eindecker (Monoplane)
AGO Pusher Seaplane (1914)
AGO S.I
AGO Sturmflug
AGO 1913 Biplane

Agostini, Livio *(See: Alaparma)*

Agrocopteros Ltda (Colombia)
Agrocopteros:
Scamp B (Aerosport Scamp Variant)
MXP-640 (Zenair CH-601 Variant)
MXP-740 (Zenair CH-701 Variant)

Agrolot Foundation (Poland)

Agrolot PZL-126P Mrówka (See: PZL Warsawa-Okecie)

Agusta (Construzioni Aeronautiche Giovanni Agusta) (Italy)

Agusta A 101
Agusta AB 102
Agusta A 105
Agusta A 106
Agusta A 109 Hirundo
Agusta A 109A Mk.II Widebody
Agusta A 109C/CM/EOA
Agusta A 109HA
Agusta A 109HO
Agusta A 109K/KM/KN/KZ
Agusta A 119 Koala
Agusta A 129 Mangusta (Mongoose)
Agusta AB 204
Agusta AB 204B
Agusta AB 204AS
Agusta AB 205
Agusta AB 206 Jet Ranger
Agusta (Bell) AB 212
Agusta AB 212
Agusta AB 212AS
Agusta AB 212ASW
Agusta AB 412
Agusta AB 412 Griffon
Augusta C22J Ventura
Agusta CP 110
Agusta EH101 (See: European Helicopter Industries Ltd)
Agusta EMA 124
Agusta SH-3D
Agusta HH-3F
Agusta P 111
Agusta S 211
Agusta SF 260
Agusta 47
Agusta 47G
Agusta 47J

Ahrens Aircraft Corp (Oxnard, CA)

Ahrens AR 404

***AIA** (See: Aviation Industries of Australia) (Australia)*

Aichi Kokuki KK (Aichi Aircraft Co Ltd) (Japan)

In 1920 the **Aichi Tokei Denki KK** (Aichi Clock and Electric Co Ltd) entered the aviation field, concentrating primarily on engines, seaplanes, and flying boats. In 1943, the company's aviation branch was detached as **Aichi Kokuki KK** (Aichi Aircraft Co Ltd). The company moved out of aviation at the end of World War II. For all products of Aichi Tokei Denki and Aichi Kokuki, see **Aichi**.

Aichi AB-1 (Transport)
Aichi AB-2 (Experimental Catapult-Launched Recon Seaplane)
Aichi AB-3 (Experimental Single-Seat Recon Seaplane)
Aichi AB-4 Experimental 6-Shi Night Recon Flying Boat
Aichi AB-4 Transport Flying Boat
Aichi AB-5 (Experimental Three-Seat Recon Seaplane) (HD 62)
Aichi AB-6 (Experimental 7-Shi Recon Seaplane)
Aichi AB-8 (Experimental 7-Shi Carrier Attack Aircraft)
Aichi AB-9 (Experimental 8-Shi Special Bomber)
Aichi AI-104 *Ione* (Navy Type 98 Recon Seaplane) (Fictional Type)
Aichi B7A Ryusei (Shooting Star) *Grace* (Navy Experimental 16-Shi Carrier Attack Bomber)
Aichi B7A1 Ryusei *Grace*, NASM
Aichi D1A *Susie* (Navy Type 94 Carrier Bomber)
Aichi D3A *Val* (Navy Type 99 Carrier Bomber)
Aichi E3A (Navy Type 90-1 Recon Seaplane) (HD 56)
Aichi E8A (Experimental 8-Shi Recon Seaplane) (AB-7)
Aichi E10A (Navy Type 96 Night Recon Seaplane) (AB-12)
Aichi E11A (Navy Type 98 Night Reconnaisance Seaplane)
Aichi E12A (Experimental 12-Shi Two-Seat Recon Seaplane)
Aichi E13A *Jake* (Navy Type 0 Recon Seaplane)
Aichi E16A Zuiun (Auspicious Cloud) *Paul* (Navy Experimental 16-Shi Recon Seaplane)
Aichi Experimental Type-H Carrier Fighter (HD 23)
Aichi Experimental Three-Seat Recon Seaplane (HD 28)
Aichi F1A1 (Experimental 10-Shi Observation Aircraft) (AB-13)
Aichi H1H3 (See: Hiro)
Aichi H2H1 (See: Hiro)
Aichi H9A1 (Navy Type 2 Training Flying-Boat Model 11)
Aichi M6A Seiran (Clear Sky Storm) (Navy Experimental 17-Shi Special Attack Bomber)
Aichi M6A1 Seiran, NASM
Aichi M6A1-K Nanzan (Southern Mountain) (Seiran Kai)
Aichi Navy Type 2 Single-Seat Recon Seaplane (HD 26)
Aichi Navy Type 2 Two-Seat Recon Seaplane (HD 25)
Aichi S1A Denko (Bolt of Light) (Navy Experimental 18-Shi Night Fighter)
Aichi Type 2 Transport (HD 25)
Aichi 15-Ko Experimental Recon Seaplane (Mi-Go)
Aichi Brandenburg W 29

AIDC-CAF (Aero Industry Development Center-Chinese AF) (Taiwan)

AIDC AT-3 Tzu Chiang
AIDC Ching-Kuo
AIDC F-5 (See: Northrop)
AIDC PL-1B Chieh-Shou (See: Pazmany PL-1B)
AIDC T-CH-1

Aiello, Eugenio (Argentina)

Aiello No.1
Aiello No.2
Aiello No.3
Aiello No.4
Aiello No.5

AIEP (Aeronautical Industrial Engineering & Project Management Co, Ltd) (Nigeria)

AIEP Air Beetle

AII (Aviation Industries of Iran) (Iran)

AII AVA-101 (IR-G1) Nasim (Glider)
AII AVA-202 (IR-02)
AII AVA-212
AII AVA-404 (IR-12)
AII AVA-505 Thunder (IR-H5)

Ailes Enghiennoises (France)

Ailes Enghiennoises Voisin L.V. 104

Aimé (Emmanuel) et Salmson (France)

Aimé et Salmson Autoplane (1909)

***Air & Space America, Inc** (See: Air & Space Manufacturing Inc)*

***Air & Space Manufacturing Inc (Muncie, IN)** (See also: Umbaugh)*

Air & Space Model 18-A Gyrocopter
Air & Space Twinstar

***Air + Up** (See: Arup Inc)*

Air Associates, Inc (Garden City, NY)
Air Associates Model 1 Glider (1930)

Air Command International Inc (Wylie, TX)
Air Command Commander 147A Autogyro

Air Craft Marine Engineering Co *(See: Acme)*

Air Creation (France)
Air Creation Rogallo-Winged Microlights

Air Cushion Vehicles, Inc (Troy, NY)
Air Cushion Vehicles, Inc, Air Cycle

Air King *(See: National Airways System)*

Air Light GmbH *(See: Air Max)*

Air Magic Ultralights (USA)
Air Magic Spitfire Microlight
Air Magic Spitfire Super Sport
Air Magic Spitfire II
Air Magic Spitfire II Elite

Air Max GmbH (Germany)
Air Max Wild Thing

Air Mechanics Inc *(See: Alexander Aircraft Corp)*

Air Muskoka (Canada)
Air Muskoka Aztek Nomad (Modified Piper Aztek)

Air Navigation & Engineering Co, Ltd *(See: ANEC)*

Air Nova Pty Ltd (South Africa)
Air Nova Reed Falcon
Air Nova Rooivalk (Kestrel)

Air Parts (NZ) Ltd (New Zealand)
Air Parts FU-24 (See: Fletcher)

Air Products Co, Ltd *(See: AirCoupe Division)*

Air Tractor Inc (Olney, TX) *(See also: Snow Aeronautical Corp)*
Air Tractor AT-301
Air Tractor AT-301A
Air Tractor AT-302
Air Tractor AT-302A
Air Tractor AT-400
Air Tractor AT-401B
Air Tractor AT-402
Air Tractor AT-402A
Air Tractor AT-402B
Air Tractor AT-502
Air Tractor AT-502A
Air Tractor AT-502B
Air Tractor AT-503A
Air Tractor AT-602
Air Tractor AT-802

Air Transport Manufacturing Co Ltd (Glendale, CA)
Air Transport B-6
Air Transport B-6-S
Air Transport B-8
Air Transport P-2 Meteor
Air Transport P-2-S Meteor
Air Transport T-6

Air-King *(See: National Airways System)*

Air-Light GmbH-Süd (Germany)
Air-Light Wild Thing (See: Air Max)

Air-Metal Flugzeugbau und Entwicklungs GmbH & Co KG (Germany)
Air-Metal AM-C111
Air-Metal STOL Transport

Air-Mod Engineering Co (Oklahoma City, OK) *(See: Doyn Aircraft Inc)*

Airbirde
Airbirde Flyer

Airborn Utility Cars (Seattle, WA)
Airborn Utility Cars Air/Car

Airborne Innovations LLC (Webster, TX)
Airborne Innovations Star Streak
Airborne Innovations Wizard

Airbus Industrie (Airbus Industrie Asia, AIA) (International)
Airbus A300
Airbus MRTT300
Airbus MRTT300-600
Airbus A300B
Airbus A310
Airbus MRTT A310
Airbus MRTT A310 AEW&C
Airbus AE316
Airbus AE317
Airbus AE318
Airbus A319
Airbus A320
Airbus A321
Airbus A330
Airbus A340 (Project TA11)

Aircar System Ltd Co (Palm Beach, FL)
Aircar System Aircar Jet

Airco (Aircraft Manufacturing Co Ltd) (UK) *(See also: de Havilland)*
Airco 1912 Farman-Type Biplane

Airconcept Flugzeug und Gerätebau GmbH und Co KG (Germany)
Airconcept VoWi 10 Airbuggy

Aircorp Pty Ltd (Australia)
Aircorp Bushmaster

AirCoupe Division (Air Products Co, Ltd) (Carlsbad, NM)
AirCoupe F-1A (See: Fourney)

Aircraft Cooperative Mechta (Russia)
Aircraft Cooperative Mechta AC-4

Aircraft Corp of America *(See: Westbrook)*

Aircraft Designs Inc (USA)
Aircraft Designs (USA) Bumblebee
Aircraft Designs (USA) Sportster
Aircraft Designs (USA) Stallion

Aircraft Designs Ltd (UK)
Aircraft Designs (UK) Sheriff

Aircraft Disposal Co, Ltd *(See: ADC)*

Aircraft Engineering Corp (New York, NY) *(See also: Keane, Horace)*
The **Aircraft Engineering Corp** was established in 1919 to build aircraft designed by N.W. Dalton, the chief engineer. The company failed in late 1919, succeeded by **Horace Keane Aeroplanes, Inc.**
Aircraft Engineering Ace (1919)

Aircraft Hydro-Forming Corp *(See: Stout Bushmaster 2000 Trimotor)*

Aircraft Improvement Corp (New York, NY)
Aircraft Improvement Monobiplane

Aircraft Industries, Inc (San Leandro, CA)
Aircraft Industries (CA) Primary Training Glider (1929)

Aircraft Industries, Inc (Orlando, FL)
Aircraft Industries (FL) Sierra
Aircraft Industries (FL) Sierra BL-1
Aircraft Industries (FL) Sierra BLW-2
Aircraft Industries (FL) III, Model 1

Aircraft Investment Corp Ltd (UK)
Aircraft Investment Meteor I (See: Blackburn Segrave; See also: Saunders-Roe A.22 Meteor)

Aircraft Manufacturing & Development Co Inc (Eastman, GA) *(See: Zenair)*

Aircraft Manufacturing Co (USA) Texas Bullet *(See: Johnson Bullet)*

Aircraft Manufacturing Co Ltd (UK)
(See: de Havilland)

Aircraft Manufacturing Corp (Pasadena, CA)
Aircraft Manufacturing (CA) Californian

Aircraft Manufacturing Depot (India)
Aircraft Manufacturing (India) HS.748 (See: Hawker Siddeley)

Aircraft Mechanics, Inc *(See: Alexander)*

Aircraft Methvin Inc (Atlanta, GA)
Aircraft Methvin XP-101 (Methvin Safety Wing)

Aircraft Repair Works Kbely *(See: LOK)*

Aircraft Research Corp
Aircraft Research Corp XBT-11

Aircraft Sales and Parts *(See: ASAP)*

Aircraft Spruce & Specialty Inc (USA)
Aircraft Spruce & Specialty:
DR.107 One Design
DR.109 Rihn
Wittman W.10 Tailwind

Aircraft Technologies, Inc (Lilburn, GA)
Aircraft Technologies, Inc Acro 1

Aircraft Technology Industries
(See: Airtech)

Airdisco
Airdisco Glider (1922)

Aire Kraft, Inc (Washington, PA)
(See: Snyder Motor Gliders Inc)

Airgo Manufacturing Co (Guthrie, OK)
Airgo A-1

Airhoppers Glider Club (New York, NY)
Airhoppers Glider #1 (1929 Glider)

Airight Inc (USA) *(See: Swearingen)*

Airlifts Inc (Miami, FL) *(See: Burnelli CanCargo CBY-3)*

Airliner Engineering Corp
(See: Burnelli)

Airmark Ltd (UK)
Airmark Special IIIM (See: Cassutt)

Airmaster Helicopters, Ltd (UK)
Airmaster Helicopters H2-B1

Airmaster Inc (Renton, WA)
Airmaster A-1200 Guardian
Airmaster Avalon Twinstar 1000
Airmaster Avalon 680

AiRover Co (Burbank, CA)
AiRover Unitwin Model 2 Starliner

Airplane Construction Co (Boston, MA)
Airplane Construction Co CM-2

Airplane Development Corp *(See: Vultee V-1)*

Airplane Engineering Dept *(See: Engineering Division)*

Airport Transport and Travel Ltd (UK)
Airport Transport and Travel D.H.10 Conversion

Airspeed Ltd (UK)
Airspeed Ltd was formed in 1931. It joined the **de Havilland Group** in 1940 and merged as a de Havilland division in 1951. Airspeed lost its corporate identity in a 1955 de Havilland reorganization. Airspeed aircraft, including those designed while a de Havilland division, are listed as **Airspeed**.
Airspeed Ambassador (AS.57)
Airspeed Courier (AS.5)
Airspeed Courier, Consul
Airspeed Consul (AS.65)
Airspeed Elizabethan
Airspeed Envoy (AS.6)
Airspeed Ferry (AS.4)
Airspeed Fleet Shadower (AS.39)
Airspeed Horsa Mk.I (AS.51)
Airspeed Horsa Mk.II (AS.58)
Airspeed Oxford AS.10
Airspeed Oxford AS.40
Airspeed Oxford AS.41
Airspeed Oxford AS.42
Airspeed Oxford AS.43
Airspeed Oxford AS.46
Airspeed Queen Wasp (AS.30)
Airspeed Viceroy (AS.8)

Airtech (Aircraft Technology Industries) (International)
Airtech is a cooperative of Indonesia's **IPTN (Industri Pesawat Terbang Nusantara)** and Spain's **CASA (Construcciones Aeronáuticas, SA)**.
Airtech CN-235

AirUtility Cargo Co (USA)
AirUtility AU18-150

Airworthy Airplane Co (Chicago, IL)
Airworthy Terrier

AISA (Aeronautica Industrial, SA) (Spain)
AISA AC-12
AISA AC-14
AISA AVD-12
AISA GN Autogiro
AISA HM-1
AISA HM-1B
AISA HM-2
AISA HM-3
AISA HM-5
AISA HM-6
AISA HM-7
AISA HM-9
AISA I-11A
AISA I-11B
AISA I-115 (E.6)

AISA I-122
AISA I-123
AISA I-124
AISA I-18
AISA-SIAI S.205

Ajep (A. J. E. Perkins) (UK) *(See: Wittman Tailwind)*

AJI *(See: American Jet Industries)*

Akadflieg München (Germany)
Akadflieg München Mü 4
Akadflieg München Mü 10 Milan
Akadflieg München Mü 13
Akadflieg München Mü 13 D3
Akadflieg München Mü 30 Schlacro

Akasamitra Homebuilt Aircraft Assoc (Indonesia)
Akasamitra ST-220

Akerman, Professor John D. (Minneapolis, MN)
Akerman JDA-8
Akerman Tailless
Akerman Tailless, NASM

Akron
Akron Condor (Sailplane)

Akron Aircraft Co *(See: Funk)*

Akron Aircraft Inc *(See: Funk)*

Akrotech Aviation, Inc (Scappoose, OR)
Akrotech Giles G-200
Akrotech Giles G-202
Akrotech Giles G-300

Akrotech Europe (France)
Akrotech Europe CAP 10 B
Akrotech Europe CAP 10 R
Akrotech Europe CAP 222
Akrotech Europe CAP 231 EX
Akrotech Europe CAP 232

AKS-Invest (Russia) *(See: MiG TA4)*

Alamo Aero Service (San Antonio, TX) *(See: Howard Aero Manufacturing Division)*

Alan Muntz & Co *(See: Baynes)*
Alaparma SpA (Italy)
Alaparma AM-6
Alaparma AM-8
Alaparma AM-10 Tucano (Toucan)
Alaparma AM-75 Baldo (AP-75)
Alaparma AP-65

Alaska International Corp (Albuquerque, NM)
Alaska International Silvaire (See: Luscombe)

Albatros (Germany)
Albatros A.1
Albatros Antoinette (See: Antoinette)
Albatros B.I (L1)
Albatros B.II (L2)
Albatros B.II Racer (1914)
Albatros B.III (L5)
Albatros Biplane Amphibian (Typ Bodensee, 1913)
Albatros C.I (L6)
Albatros C.II (L8)
Albatros C.III (L10)
Albatros C.IV (L12)
Albatros C.V (L14) Series
Albatros C.V/16 (L14)
Albatros C.V/17 (L14)
Albatros C.V
Albatros C.V Experimental (L14)
Albatros C.VI (L16)
Albatros C.VII (L18)
Albatros C.VIII N (L19)
Albatros C.IX (L23)
Albatros C.X (L25)
Albatros C.XII (L27)
Albatros C.XIII
Albatros C.XIV
Albatros C.XV (L47)
Albatros D.I (L15)
Albatros D.II (L17)
Albatros D.II (Oef)
Albatros D.III (L20)
Albatros D.III (Oef)
Albatros D.III/D.V
Albatros D.IV (L22)
Albatros D.V (L24)
Albatros D.Va (L24)
Albatros D.Va (L24), NASM
Albatros "D.Vb" (Fictitious Type)
Albatros D.VII (L34)
Albatros D.IX (L37)
Albatros D.X (L38)
Albatros D.XI (L41)
Albatros D.XII (L43)
Albatros Dr.I (L36)
Albatros Dr.II
Albatros G
Albatros G.I
Albatros G.II (L11)
Albatros G.III
Albatros J.I (L40)
Albatros J.II
Albatros L.3
Albatros L.9
Albatros L30
Albatros L57
Albatros L57b
Albatros L58
Albatros L59
Albatros L60
Albatros L65
Albatros L66
Albatros L68c Alauda
Albatros L68d
Albatros L69
Albatros L72a
Albatros L72b
Albatros L72c Albis
Albatros L73
Albatros L73b
Albatros L74 Adlershof (Eagle's Nest)
Albatros L75a Ass
Albatros L76a
Albatros L79 Kobold (Goblin)
Albatros L82
Albatros L84
Albatros L100
Albatros L101
Albatros L102 (See also: Focke-Wulf Fw 55)
Albatros W102 (See also: Focke-Wulf Fw 55)
Albatros ME (L9)
Albatros Military Biplane (1912)
Albatros Taube (Dove) Amphibian (Typ Bodensee)
Albatros Taube (Dove) Biplane EE
Albatros Taube (Dove) Biplane FT
Albatros Taube (Dove) Biplane (1912)
Albatros Taube (Dove) Biplane (1913)
Albatros Taube (Dove) Biplane (1913) Pfeil (Arrow)
Albatros W 1
Albatros W 2
Albatros W 3
Albatros W 4
Albatros W 5
Albatros W 8
Albatros Wahl
Albatros XVII (Russian-Built)
Albatros Series 22 (OAW) Knoller-Albatros
Albatros 1910 Farman Type
Albatros 1911 Biplane (Farman Type)
Albatros 1911 Biplane (Gnome)
Albatros 1911 Biplane (Pusher)
Albatros 1912 Biplane (Farman Type, Floats)

Albatross Aircraft Corp (Long Beach, CA) *(See: American Albatross Co)*

Albert Aéronautique (France)
Albert A.10 Hirondelle (Swallow)
Albert A.20
Albert A.60
Albert A.61
Albert A.62
Albert A.100 (T.E.I)
Albert A.140 R.V.
Albert Avionnette "Baby Columbia"

Albert, Jean (France)
Albert (Jean) 1910 Monoplane

Alberta Aerospace Corp (Canada)
Alberta Phoenix Fanjet

Albessard (France)
Albessard Autostable (1912)
Albessard Triavion (Triplane) (1929)
Albessard Quadriavion (Quadruplane)

Albree, George (East Boston, MA)
Albree Pigeon (Scout)

Alco (Allison Airplane Co) (Lawrence, KS)
Alco Commercial LM2B
Alco Coupster
Alco Junior Coupe
Alco Sport Biplane
Alco Utility MT 3

Alco Hydro-Aeroplane Co (Allan Loughead Co) *(See: Loughead)*

Alcock, J. W. (UK)
Alcock A.1 Scout

Alcor Aircraft Corp (Oakland, CA)
Three years after the failure of the **Loughead Brothers Aircraft Corp** in 1934, Allan Loughead established the **Alcor Aircraft Corp** in Oakland, CA. The company failed following the crash of its only design in 1938 and Loughead left the aviation industry.
Alcor C.6.1 (C-6-1, Duo-6) Junior Transport

Alcor Aviation, Inc (San Antonio, TX)
Modified Helio Courier with Tricycle Gear

Aleksandrov-Kalinin (Russia)
Aleksandrov-Kalinin AK-1

Alenia (Italy)
Alenia was formed by the 1990 merger of **Aeritalia** and **Selenia**.
Alenia G.222 (See: Fiat)

Alexander Aircraft Corp (Aircraft Mechanics, Inc) (Colorado Springs, CO)
Alexander Eaglerock A-1
Alexander Eaglerock A-2
Alexander Eaglerock A-2 Combination Wing
Alexander Eaglerock A-2 Float Plane
Alexander Eaglerock A-2 Long Wing
Alexander Eaglerock A-3
Alexander Eaglerock A-4
Alexander Eaglerock A-5
Alexander Eaglerock A-7
Alexander Eaglerock A-8
Alexander Eaglerock A-12
Alexander Eaglerock A-13
Alexander Eaglerock A-14
Alexander Eaglerock A-15
Alexander Eaglerock Model C
Alexander Eaglerock C-1 Bullet
Alexander Eaglerock C-3 Bullet
Alexander Eaglerock D-1 Flyabout
Alexander Eaglerock D-2 Flyabout
Alexander Eaglerock Float Plane
Alexander Eaglerock Glider
Alexander Eaglerock Sedan (Transport)
Alexander Eaglerock Trainer

Alexander Film Co (Denver, CO)
Alexander Combowing Eaglerock
Alexander Long Wing Eaglerock

Alexander, Patrick (UK)
Alexander 1899 Helicopter

Alexandre (France)
Alexandre Garaix/ACR Monoplane

Alexandria Aircraft Corp (Alexandria, VA)
Alexandria Aircraft Corp Flying Boat

Alfa-M Nauchno-Proizvodstvennoye Predpriyatie AOOT (Alpha-M Scientific-Production Enterprise Co Ltd) (Russia)
Alfa-M A-211 Gzhelka (Little Gzhel)

Alfaro, Heraclio (Cleveland, OH)
Alfaro Cabin Monoplane
Alfaro 5A Cabin Monoplane
Alfaro 8

Alhambra Airport and Air Transport Co *(See: Alcor)*

Alkan, Oscar (France)
Alkan le Enfin (1910)

All American Aircraft Inc (Long Beach, CA)
All American Aircraft Ensign

All-American Aircraft Co Inc (Akron, OH)
All-American Aircraft L.C.11

Allard & Carbonnier (Belgium & France)
Allard & Carbonnier le Vautour (Vulture)

Allen, Edmund T. (Salt Lake City, UT)
Allen (Edmund T.) 1928 Biplane

Allen, Edward J. (East Paterson, NJ)
Allen (Edward J.) Sport (1929)

Allenbaugh (USA)
Allenbaugh "Californian" Midget Racer

Alliance Aeroplane Co Ltd (UK)
Alliance P.1
Alliance P.2 Seabird

Alliance Aircraft Corp (Alliance, OH)
Alliance Aircraft Argo (Model A)

Allied Aero Industries *(See: Waco Aircraft Co)*

Allied Aviation Corp (Baltimore, MD)
Allied Aviation Corp was established in 1941 to produce molded aircraft structures. The company became involved in the Navy glider program, developing the LRA-1. The Trimmer amphibian was designed in 1944, with production rights sold to **Commonwealth Aircraft, Inc** in 1945. For all Allied products, see **Allied Aviation Corp.**
Allied Aviation Corp LRA-1
Allied Aviation Corp Trimmer

Allgemeinen Electrizitäts Gesellschaft *(See: A.E.G.)*

Allison Airplane Co *(See: Alco)*

Allison Division of General Motors (Indianapolis, IN)
Allison C-131H (VC-131H)
Allison NC-131H TIFS
Allison Prop-Jet Super Convair (Convair 580)
Allison Super 580 (580A)
Allison Turbine (Beech) Bonanza
Allison Turbine (Beech) Mentor
Allison Turbo Flagship (ATF) 580S

Allman, William H. (Blythedale, MO)
Allman 1920 Flying-Machine Patent

Alon Inc (McPherson, KS)
Alon A2 AirCoupe (See: Fourney)
Alon A-4

Alpavia (France) *(See: Jodel)*

Alpha Aviation Co (Greenville, TX)
Alpha IID

Alpha Jet (International)
Alpha Jet Prototypes
Alpha Jet Advanced Trainer/Light Attack (Alpha Jet E) (MS1)
Alpha Jet Close Support (Alpha Jet A)
Alpha Jet DSFC (Direct Side-Force Control)
Alpha Jet NGEA (Nouvelle Génération Entraînement et Appui) (MS2)

Alpha-M (NPP Alpha-M) (Russia)
Alpha-M A-211
Alpha-M A-211K
Alpha-M SL-A

Altenrhein Group (Switzerland) *(See: F.W.A.)*

Alter (Germany)
Alter A.I

Alula *(See: Commercial Aeroplane Wing Syndicate; also: Martinsyde Semiquaver; also de Havilland D.H.6)*

Alvarez et de Condé (France)
Alvarez et de Condé:
Seaplane
High-wing Flyingboat
Armored Amphibian

Alvarez, Joe *(See: Polliwagen)*

AMA (Anczutin, Malinowski, and Aleksandrowicz) (Poland)
AMA 1934 Powered Glider

Aman, C. (France)
Aman 1910 Etrich Taube Copy
Aman & Kerchone 1910 Etrich Taube Copy

Amax (Australia)
Amax Double Eagle
Amax Eagle

Ambrosini (Società Aeronautica Italiana) (Italy) *(See also: SAI, Italy)*
Ambrosini Allievo Cantu Glider
Ambrosini A.R.4 Flying Bomb
Ambrosini Asiago 2 C.V.V.2
Ambrosini Canguro (Kangaroo) C.V.V.6
Ambrosini Grifo S.1001
Ambrosini P.512
Ambrosini Rondone G.F.4
Ambrosini Rondone F.7
Ambrosini S.7
Ambrosini S.7*bis* (Super S.7)
Ambrosini Sagittario (Sagittarius)
Ambrosini Sagittario II (See: Aerfer)
Ambrosini Trasimenus S.1002

AMD *(See: Zenair)*

A.M.E. (Aeronautica Militar Española) (Spain)
A.M.E. VI
A.M.E. VI-A

Ameco-Hawk International *(See: Hawk Industries)*

American Aero (Edinburg, IL)
American Aero Phoenix II

American Aerolights, Inc (Albuquerque, NM)
American Aerolights Double Eagle
American Aerolights Double Eagle, NASM
American Aerolights Eagle Glider Rider
American Aerolights Eagle XL
American Aerolights Eagle 2-Place
American Aerolights Falcon
American Aerolights Falcon XP

American Aeronautical Corp (Long Island City, NY)
American Aeronautical S-55
American Aeronautical S-56 Baby Amphibion
American Aeronautical S-62

American Aeroplane Co (Wilmington, NC)
American Aeroplane (Palmgren) 1911 Twin-Engined Monoplane

American Aeroplane Supply House (AASH) (Garden City, NY)
AASH Cross Country Type Blériot Monoplane
AASH Passenger Type Blériot Monoplane

American Air Racing, Inc (Portland, OR)
American Air Racing Parker JP001 "Wild Turkey"

American Aircraft Builders, Inc (Portland, OR)
American Aircraft Builders Student Prince

American Aircraft Co (Long Beach, CA) *(See: Security)*

American Aircraft Co of Indiana (USA)
American Aircraft Co of Indiana Model D

American Airmotive Corp (Miami, FL)
American Airmotive NA-75 (Stearman 75 Ag Mod)

American Airplane & Engine Co (Corp) (Farmingdale, NY) *(See: Fairchild)*

American Albatross Co (New York, NY)
American Albatross B (Model B-1)
American Albatross B "Pride of Hollywood"

American Amphibian Corp (Burlington, NC)
American Amphibian Twin-Engined Amphibian

American Armament Corp *(See: Miller Aircraft Corp)*

American Aviation Corp (AAC) (Cleveland, OH)
American Aviation AA-1 Trainer
American Aviation AA-1 Yankee
American Aviation AA-1 Yankee,

NASA Spin Test Aircraft
American Aviation AA-5 Traveler

American Aviation Industries (AAI) (USA)
American Aviation Industries FanStar (Re-engined Lockheed JetStar)

American Champion Aircraft Corp (ACAC) (Rochester, WI)
In 1954 **Flyers Service Inc** purchased the rights to the Aeronca Model 7 from **Aeronca Manufacturing Corp**, establishing the **Champion Aircraft Corp** to undertake production. Champion developed this basic model and also created new designs. In 1970 **Bellanca Sales Co** (originally known as **International Aircraft Manufacturing Inc** or **Inter-Air**) purchased Champion's assets and became **Bellanca Aircraft Corp**. Bellanca ceased Champion production in 1980. Rights, tooling, and partially completed aircraft were sold to **Champion Aircraft Co, Inc** in 1982. In 1989 the rights were again sold to what became **American Champion Aircraft**. All pre-1954 development appears under **Aeronca**. Subsequent development is filed under **Champion Aircraft Corp**, **Bellanca**, **Champion Aircraft Co**, and **American Champion Aircraft**.
American Champion:
Citabria Adventure (7GCAA)
Citabria Aurora (7ECA)
Citabria Explorer (7GCBC)
Decathlon (8KCAB)
Decathlon CS (8KCAB)
Super Decathlon (8KCAB, 8KCAB-80)
Scout (8GCBC)

American Eagle Aircraft Corp (Kansas City, MO) (See also: Wallace Aircraft Co)
American Eagle A-27
American Eagle A-101 (A-1)
American Eagle A-129
American Eagle A-130
American Eagle A-201
American Eagle A-229
American Eagle A-230 Eaglet
American Eagle A-251 Phaeton
American Eagle A-329
American Eagle A-330 (See also: Wallace C-2 Touroplane)
American Eagle A-429
American Eagle A-430
American Eagle A-430 D
American Eagle A-430 E
American Eagle F-430
American Eagle Twin Engine Cabin Monoplane

American Eagle-Lincoln Aircraft Corp (See: American Eagle Aircraft Corp)

American Eaglecraft Co (Ft Worth, TX)
American Eaglecraft Eaglet

American General Aircraft Corp (See: Grumman American)

American Glider Assoc (Dearborn, MI)
American Glider All-American Glider

American Gyro Co (See: Crusader)

American Helicopter Co, Inc (Manhattan Beach, CA & Mesa, AZ)
American Helicopter:
XA-5 Top Sergeant
XA-6 Buck Private
XH-26 Jet Jeep

American Homebuilts Corp (Hebron, IL)
American Homebuilts John Doe STOL

American Hoppi-Copters, Inc (See: Hoppi-Copters Inc)

American Jet Industries, Inc (Van Nuys, CA) (See also: Gulfstream Aerospace)
American Jet Model 400 Turbo-Star (Hustler 400)
American Jet Model 500 Hustler
American Jet A-610 Super Pinto
American Jet T-610 Super Pinto

American Microflight Inc (See: Sadler)

American Moth (See: Vulcan)

American Motorless Aircraft Co (Port Washington, NY)
American Motorless Aircraft Amphibian Glider

American Multiplane Co (See: John, Herbert F.)

American Regional Aircraft Industries Inc (See: AMRAI)

American Sportscopter International, Inc (ASI) (Newport News, VA)
American Sportscopter:
Ultrasport 254
Ultrasport 331
Ultrasport 496
Ultrasport 500 Vigilante
Ultrasport 600 Vigilante

American Sunbeam Co (Los Angeles, CA)
American Sunbeam LP-1 Pup

American Tool & Die Co
American Tool de Chappedelaine Helicopter

American Transit Co (Pittsburgh, PA) (See: Mattullath)

American Utilicraft Corp (See: AUC)

AmeriPlanes Inc (Truro, IA)
AmeriPlanes:
A-10B (See: Mitchell Aircraft Corp)
A-10D (Deluxe Mitchell Aircraft A-10)
T-10D (Deluxe Mitchell Aircraft T-10)

Ameur Aviation Technologie (See: AAT)

AMF Aviation Enterprises Ltd (UK)
AMF Chevron 2-32 C
AMF Super Chevron 2-45 CS

Amiot (Avions Amiot) (France) (See also: SECM)
In 1916, Félix Amiot founded the **Société d'Emboutissage et de Constructions Méchaniques (SECM)** at Colombes, a suburb of Paris, France. This metal pressing and stamping firm built Moraine, Sopwith, and Breguet aircraft under license. In 1921 the firm introduced the Type XX series, the first aircraft produced under the SECM name; aircraft of this series are filed under **SECM**. In 1929 SECM amalgamated with **Latham & Cie**, taking over Latham's flying-boat works at Caudebec-en-Caux, France. The combined company was known as **SECM Avions Amiot**; the company's Type 100 series aircraft are filed under **SECM Amiot**. The SECM name was dropped by the mid-1930s, the company then being known simply as **Avions Amiot**. The Type 300 series

aircraft of this period are filed under **Amiot**. With the German occupation of France during World War II, Avions Amiot built Ju 52 aircraft under the control of Junkers. After the war, Amiot was nationalized as Atelier Aéronautique de Colombes, and continued to produce Ju 52s under the designation AAC.1; these aircraft are filed under **Junkers**.

Amiot 340
Amiot 341
Amiot 350
Amiot 354
Amiot 370

Amouroux (France)
Amouroux Flying Machine

Amphibions Inc *(See: Ireland Aircraft Inc)*

Amphibious Aircraft Corp (Niagara Amphibious Aircraft Corp) (Buffalo, NY)
Amphibious Aircraft Pusher

AMRAI (American Regional Aircraft Industries Inc) (Cincinnati, OH)
AMRAI N-270 (See: IPTN)

AMX International Ltd (Aeritalia/Alenia, Aermacchi, EMBRAER) (Italy & Brazil)
AMX A-1 (Single-Seat)
AMX-E
AMX-T (AMX A-1B)
AMX Super AMX

Anahuac (Fabrica de Aviones Anahuac SA) (Mexico)
Anahuac Tauro (Bull) 300
Anahuac Tauro (Bull) 350

Anakle, D. S. (Russia)
Anakle Anasal

Anatra, A. A. (Russia)
Anatra C.I
Anatra Model D Anade (Anatra-Dekan)
Anatra Model D.I Anadis (Anispano, Bimonokok, Bi-kok)
Anatra Model DE
Anatra Model DM Anamon (Anatra-Monocoque)
Anatra Model DS Anasal (Anatra-Salmson)
Anatra Model DSS Anasal (Anatra-Salmson)
Anatra Model VI (Voisin Ivanov) (Revised Voisin 5)
Anatra Model VKh Anadva Twin Tractor (Dvukhvostka, Anatra-Khioni)

Anbo, Gustaitis (Lithuania)
Anbo I
Anbo II
Anbo III
Anbo IV
Anbo V
Anbo VI
Anbo VII
Anbo VIII
Anbo 41
Anbo 51

Anciens Établissements Aéronautiques Maurice Mallet *(See: Zodiac)*

Andermat Aeroplane Co (Sunnyvale, CA)
Andermat Biplane Bomber
Andermat Cabin Biplane

Anderson & Gewert (C. Lee Anderson & E. R. Gewert) (Reno, NE)
Anderson & Gewert Racing Model

Anderson Aircraft Corp (Raymond, ME)
Anderson EA-1 Kingfisher

Anderson Greenwood & Co (Houston, TX)
Anderson Greenwood AG-14

Anderson, William H. (Andiz Inc) (Los Angeles, CA)
Anderson (William H.) Monoplane

Andiz *(See: Anderson, William H.)*

Andover Kent Aviation Corp *(See: Langley Aviation Corp)*

d'Andre (France)
d'Andre Aérovoile (1908 Hang Glider)

Andreae (Central Valley, NY)
Andreae Multiplane (1909)

Andreasson, Björn (San Diego, CA) *(See also: A.B. Malmö)*
Andreasson BA-4B
Andreasson BA-6
Andreasson BA-7
Andreasson BA-11

ANEC (Air Navigation & Engineering Co, Ltd) (UK)
ANEC I (1922 Monoplane)
ANEC IA (1923 Monoplane)
ANEC II (See also: Handasyde H.2)
ANEC III (7-Seat Biplane)
ANEC IV Missel Thrush

ANF-Mureaux *(See: Mureaux)*

Angkatan (Indonesia)
Angkatan Belalang (Grasshopper) (Lipnur Model 90)
Angkatan LT-200 (See: Pazmany PL-2)

Angel Aircraft Corp (Orange City, IA)
Angel Aircraft Angel

Angus, A. L. (UK)
Angus Aquila (1931 Ultralight)

Anjou Aéronautique *(See: SEA)*

ANPK MiG *(See: MiG)*

Ansaldo, Ansaldo S.V.A. (Savoia, Verduzio, Ansaldo) (Italy)
Ansaldo AC2
Ansaldo AC3
Ansaldo AC4
Ansaldo Am
Ansaldo AP
Ansaldo A1 Balilla
Ansaldo A3
Ansaldo A5
Ansaldo A115
Ansaldo A115*bis*
Ansaldo A120
Ansaldo A120ady
Ansaldo A120*bis*
Ansaldo A200
Ansaldo A201
Ansaldo A202
Ansaldo A300
Ansaldo A300/2
Ansaldo A300/3
Ansaldo A300/4 (A304)
Ansaldo A300/5
Ansaldo A300/6
Ansaldo A300C
Ansaldo A300T
Ansaldo A400
Ansaldo O (1917 Biplane)
Ansaldo Sport Biplane
Ansaldo S.V.A. 5 (SV.5)
Ansaldo S.V.A. 9
Ansaldo S.V.A. 10

Ansell (Sweden)
Ansell Pusher Biplane

Antoinette (Société Anonyme Antoinette) (France) *(See also: Ferber; Levavasseur)*
Antoinette Monobloc (Blindé, 1911)
Antoinette I (1906)
Antoinette Gastambide-Mengin (Antoinette II)
Antoinette III (See: Ferber IX)
Antoinette IV
Antoinette IV, First Channel Attempt
Antoinette V
Antoinette VI
Antoinette VII
Antoinette 1910 Type

Antoni (Societa Italiana Brevetti Antoni) (Italy)
Antoni Type 25

Antonov, K. T. (Russia)
Antonov A-1
Antonov A-2
Antonov A-7 (Glider)
Antonov A-9
Antonov A-11
Antonov A-13
Antonov A-15 Glider
Antonov A-15 Glider, Modified
Antonov Amur
Antonov An-2 *Colt*
Antonov An-2 *Colt*, NASM
Antonov An-3
Antonov An-4
Antonov An-6
Antonov An-8 *Camp*
Antonov An-10 Ukraina *Cat*
Antonov An-12 *Cub*
Antonov An-13
Antonov An-14 Pchelka (Little Bee) *Clod*
Antonov An-22 Antheus, Antei *Cock*
Antonov An-24 *Coke*
Antonov An-26 *Curl*
Antonov An-28 *Cash*
Antonov An-30 *Clank*
Antonov An-32 *Cline*
Antonov An-38
Antonov An-70
Antonov An-72 *Coaler*
Antonov An-77
Antonov An-124 *Condor*
Antonov An-140
Antonov An-174
Antonov An-180
Antonov An-218
Antonov An-225 Mriya (Dream) *Cossack*
Antonov AT-1 (Glider)
Antonov Bich-22
Antonov Bk-6 Neringa
Antonov Bro-9
Antonov Bro-11
Antonov Bro-12
Antonov Kryl'yatanka (Flying Tank)
Antonov Mak-15
Antonov N-2 (Copy of Fieseler Fi 156)
Antonov OKA-33 (LEM-2)
Antonov Pai-6
Antonov Pai-7
Antonov Pk-4 Kaunus
Antonov Sh-16
Antonov Sh-18
Antonov SS (OKA-38) (Copy of Fieseler Fi 156)
Antonov Slavutich-Sport Hang Glider
Antonov Sukhanov Diskoplan
Antonov Vega 1
Antonov Vega 2

Anzani, Alexandre (France)
Anzani (de Mas) 1909 Monoplane

AOI (Arab Organization for Industrialization) (Egypt) *(See: Heliopolis; also: Helwan)*

Apollo Aircraft Corp *(See: Hunt, Ulys H.)*

Applebay (George) Sailplanes (Rio Rancho, NM)
Applebay GA II Chiricahua
Applebay Mescalero
Applebay Zia (Homebuilt)
Applebay Zuni (Aero Tek Model V) (Homebuilt)
Applebay Zuni II
Applebay Zuni II, NASM

Applegate and Weyant (Quincy, MA; Elkhart, IN; Tecumseh, MI)
Applegate and Weyant acquired the rights to the Culver Dart-G following the bankruptcy of the **Culver Aircraft Corp** in 1946. For all Culver-built Darts, see **Culver**. For all Applegate and Weyant-built Darts, see **Applegate and Weyant.**
Applegate and Weyant A-3
Applegate and Weyant Dart

Applegate (Ray) Piper (Goshen, IN)
Applegate Duck Amphibian
Applegate Piper AP-1 Duck Amphibian (See also: Piper P-1)

Aquaflight, Inc (Wilmington, DE)
Aquaflight Aqua I W-6
Aquaflight Aqua II W-6

A.R. *(See: Dorand)*

Arab Organization for Industrialization (Egypt) *(See: Heliopolis; also: Helwan)*

Arado (Germany)
Arado Ar S.I
Arado Ar SC.IArado Ar V.I
Arado Ar 64
Arado Ar 65
Arado Ar 66
Arado Ar 68
Arado Ar 69
Arado Ar 76
Arado Ar 77
Arado Ar 79
Arado Ar 80
Arado Ar 81
Arado Ar 95
Arado Ar 96
Arado Ar 98
Arado Ar 99
Arado Ar 196
Arado Ar 196, NASM
Arado Ar 197
Arado Ar 198
Arado Ar 199
Arado Ar 231
Arado Ar 232
Arado Ar 232 A
Arado Ar 232 B
Arado Ar 233
Arado Ar 234 A Blitz (Lightning)
Arado Ar 234 B Blitz (Lightning)
Arado Ar 234 B Blitz, NASM
Arado Ar 234 C Blitz (Lightning)
Arado Ar 240
Arado Ar 340
Arado Ar 396 (SIPA 10)
Arado Ar 440
Arado L I
Arado L 2a
Arado L 11
Arado L 11a
Arado L 101A
Arado L 111
Arado V.1
Arado W.II

Arc Atlantique Aviation (France)
Arc Atlantique RF 47 (Fournier Design)

Archdeacon, Ernest (France)
Archdeacon's First Glider (1904)
Archdeacon's Second Glider (1904/05)
Archdeacon's Third Glider (1905)

ARCO *(See: Johnson Bullet)*

Arctic Aircraft Co (Arctic Aviation) (Anchorage, AK)
Arctic Tern (Interstate S-1 Development)

Arctur (Germany)
Arctur Flying Wing (1946)

ARDC (Aeronautical Research & Development Corp)*(See: Brantly)*

Ardea (Letadla Prikryl-Blecha) (Czechoslovakia)
Ardea PB 1
Ardea PB 2
Ardea PB 3
Ardea PBL 3
Ardea PB 4 Racek
Ardea PB 5 Racek
Ardea PB 6 Racek
Ardea PB 7

Argonaut Aircraft Co (North Tonawanda, NY)
Argonaut Crusader
Argonaut Pirate

Ariel (Société Ariel)(France)
Ariel (Wright) Type A

Ariel Aircraft Inc (Coffeyville, KS)
Ariel (USA) 2-Seater Light Monoplane

Arizona Airways, Inc (Phoenix, AZ)
Arizona Airways Whitewing

Arkansas Aircraft Co *(See: Command-Aire, Inc)*

Arkhangel'sky (Russia)
Arkhangel'sky Ar-2 (SB-RK)
Arkhangel'sky MMM
Arkhangel'sky SBB-1

Arlington Aircraft Co (Arlington, TX)
Arlington Sisu 1 Sailplane
Arlington Sisu 1A Sailplane
Arlington Sisu 1A Sailplane, NASM

Armel, M. C. (Athens, GA)
Armel White Dove Commercial

Armella-Senemaud (Société Armella-Senemaud)(France)
Armella-Senemaud A.S.10 Mistral

Armitage, George H. (Cumberland, RI)
Armitage 88 Racer

Armstrong Whitworth (UK)
Sir W.G. Armstrong, Whitworth & Co Ltd was formed in 1897 and produced aircraft between 1913 and 1919. A separate **Armstrong Whitworth Development Co Ltd** was registered in 1919; assets would eventually include the **Sir W.G. Armstrong Whitworth Aircraft Co Ltd** which formed in 1920. The development company became the **Armstrong Siddeley Development Co** in 1927 and acquired **A.V. Roe & Co Ltd** in 1928. The **Hawker Siddeley Aircraft Co** was formed in 1935 with **Armstrong Siddeley Development Co** as a subsidiary. Armstrong Whitworth designed and produced aircraft until a 1961 merger with **Gloster Aircraft Co Ltd** resulted in the formation of **Whitworth Gloster Aircraft Ltd** within the Hawker Siddeley group. A 1963 reorganization formed the **Avro Whitworth Division of Hawker Siddeley Aviation**, but within the year the Whitworth name was permanently dropped. Products of the different Whitworth organizations are listed by company designation under **Armstrong Whitworth**.

Armstrong Whitworth:
- A.W.I Awana
- A.W.XIV Starling
- A.W.XV Atalanta
- A.W.16
- A.W.17 Aries
- A.W.19
- A.W.22
- A.W.23
- A.W.27 Ensign
- A.W.29
- A.W.35 Scimitar
- A.W.38 Whitley
- A.W.41 Albemarle
- A.W.52 Flying Wings
- A.W.55 Apollo
- A.W.650 Argosy
- A.W.660 Argosy
- Ajax
- Ape
- Ara
- Argosy (1920s)
- Atlas
- F.K.1 Sissit
- F.K.2
- F.K.3
- F.K.5
- F.K.6
- F.K.7
- F.K.8
- F.K.9
- F.K.10 Quadruplane
- F.M.4 Armadillo
- R.T.1
- Sinaia
- S.R.2 Siskin (See: Siddeley Deasy)
- Siskin II
- Siskin IIIA
- Siskin IIIB
- Wolf

Arnaud (France)
Arnaud 1911 Biplane

Arnet Pereyra (Italy))
Arnet Pereyra:
- Aventura (Adventure) UL 447
- Aventura (Adventure) HP 503
- Aventura (Adventure) HP 582
- Aventura (Adventure) II 582
- Aventura (Adventure) II 912
- Aventura (Adventure) 912
- Jet Fox JF 91
- Jet Fox JF 97
- Ultralight Trainer

Arnoldi (USA)
Arnoldi 1927 2-Place Monoplane

Arnoux, René (France)
Arnoux "Modèle No.1" (1910)
Arnoux "Modèle No.2" (1912)
Arnoux "Modèle No.3" (1912)
Arnoux "Modèle No.4" (1912)
Arnoux "Modèle No.5" (1914)
Arnoux "Modèle No.6"
Arnoux "Modèle No.7" (1921 Biplane Flying Wing)
Arnoux "Madon-Carmier" (1922 Monoplane Flying Wing)
Arnoux 1923 Moto Aviette

Arnoux-Simplex *(See: Arnoux)*

Area de Material Cordoba (Argentina) *(See: FMA)*

Arondel, Paul (France)
Arondel 1911 Monoplane

Arpin (M. B. Arpin & Co)(UK)
Arpin A-1 Pusher Safety Pin

Arrow Aircraft (UK)
Arrow (UK) Active I
Arrow (UK) Active II

Arrow Aircraft & Motors Corp (Havelock, NE)
Arrow Model L-P-2B
Arrow Sport A2-60
Arrow Sport Model F
Arrow Sport Model M
Arrow Sport Pursuit
Arrow Sport Pursuit, Modified
Arrow Sport 2A
Arrow Five (L-5)

Arrowing Co (Newark, NY)
Arrowing A-2 Chummy

Arsenal (France)
In 1937 France nationalized its military aircraft production factories and organized them geographically into six companies: **Société Nationale de Constructions Aéronautiques du Nord** (**SNCAN**); **Société Nationale de Constructions Aéronautiques du Centre** (**SNCAC**); **Société Nationale de Constructions Aéronautiques du Sud-Est** (**SNCASE**); **Société Nationale de Constructions Aéronautiques du Sud-Ouest** (**SNCASO**); **Société Nationale de Constructions Aéronautiques du Ouest** (**SNCAO**); and **Société Nationale de Constructions Aéronautiques du Midi** (**SNCAM**). An additional company, the **Aéronautiques Arsenal** (**Arsenal**) began designing and producing aircraft just prior to World War II. French designs continued to be produced under their original company names and designations for several years. In this listing, French aircraft are organized by designer (whether private or government facility) rather than production factory.
Arsenal V-100 (Air 100) Sailplane
Arsenal VB 10
Arsenal VG 10
Arsenal VG 33
Arsenal VG 36
Arsenal VG 70
Arsenal VG 90
Arsenal VG O-101

Artigala, Enrique (Argentina)
Artigala "Argentino I" (1911 Biplane)

Artigau, Pedro (Argentina)
Artigau "Coronel Pringles"

Artois (Chantiers de l'Artois; d'Artois)(France)
Artois Aéro-Torpille (Torpedo)(1912)
Artois Hydroaéroplane (Hydravion à Cocque)(1912)
Artois Hydroaéroplane (Hydravion à Cocque)(1913)

Arup Inc (Air + Up)(South Bend, IN)
Arup S-35
Arup 1
Arup 2
Arup 3
Arup 4

ARV Aviation Ltd (UK)
ARV ARV-1 Super2

Arzenkoff (Russia)
Arzenkoff "Moskva"

ASAP (Aircraft Sales and Parts) (Canada)
ASAP Beaver RX 28
ASAP Beaver RX 550
ASAP Beaver RX 550 Plus
ASAP Beaver SS
ASAP Chinook Plus 2
ASAP Chinook 2 XS (912)

Asboth, Oskar von (Hungary, UK)
Asboth 1917 Helicopter

***ASC** (See: Advanced Soaring Concepts)*

Ashley David D. (Freeport, NY)
Ashley Bob-O-Link

Ashmussen Manufacturing Co (Omaha, NE)
Ashmussen Biplane
Ashmussen Blue Bird
Ashmussen Mail Plane

***ASI** (See: Advanced Aerodynamics & Structures Inc)*

***ASI** (See: American Sportscopter International Inc)*

Asiago (Denmark)
Asiago G.P.2

ASL (Aeronautical Syndicate Ltd)(UK)
ASL Valkyrie
ASL Valkyrie Type B
ASL Valkyrie Type C
ASL Viking

***ASL Hagfors Aero AB** (See: Hagfors)*

Asquier-Régence (France)
Asquier-Régence 1910 Monoplane

Associate Air (Woodland, WA)
Associate Air Liberty 181

Associated Aircraft Co (Corp)(Wichita, KS)
Associated Aircraft 1929 Cabin

AST (Aviaspetstrans OAO)(Russia)
AST Yamal Amphibian

***Astanières** (See: Constantin-d'Astanières)*

Asteria (Italy)
Asteria M-B Monoplane
Asteria 1912-13 Monoplane

Astoux-Védrines (France)
Astoux-Védrines Experimental Triplane

Astra (Société de Constructions Aéronautiques Astra)(France)
Astra Bomber (3 x 220-HP Renault)
Astra Type C
Astra Type C Hydro
Astra Type CM
Astra Type CM Hydro-Avion (Marin)
Astra Triplane (See: Voisin)
Astra 1908 Tandem Monoplane
Astra-Paulhan Flying Boat
Astra-Wright Type A
Astra-Wright Type BB (Bébé)
Astra-Wright Type E
Astra-Wright Type L
Astra-Wright Type 1910

Astruc, Edmond (France)
Astruc 1908 Tailless Biplane
Astruc 1909 Monoplane
Astruc 1910 Aircraft

Atec Vos (Czech Republic)
Atec Vos Zephyr 2

***Ateliers Aéronautiques de Colombes AAC.1** (See: Junkers Ju 52/3m)*

***Ateliers Aéronautiques de l'Est (AAE)** (See: Lecomte, Alfred)*

***Ateliers de Construction Aéronautique Zeebrugge** (See: Zeebrugge)*

Ateliers de Construction de Nord de la France et des Mureaux *(See: Mureaux)*

Ateliers des Mureaux *(see: Mureaux)*

Ateliers et Chantiers de Dunkerque *(See: Dunkerque)*

Ateliers et Chantiers de la Loire (See: *Gourdou-Leseurre*)

Ateliers de Constructions Aéronautiques Belges (LACAB)(Belgium)
LACAB GR.8
LACAB T.7

Atlanta Aircraft Corp (Atlanta, GA)
Atlanta Prudden-Whitehead Low-Wing Tri-Motor

Atlantic Aircraft Co (New York, NY)
Atlantic 1914 Twin Tractor Biplane
Atlantic (Heinrich) 1916 Battleplane

Atlantic Aircraft Corp (Hasbrouck Heights, NJ)*(See: Fokker)*

Atlas Aircraft Co (Hemet, CA)
Atlas H-10

Atlas Aircraft Corp of South Africa Ltd (South Africa)
Atlas C4M Kudu

Atomic Powered Aircraft *(See: Nuclear Powered Aircraft)*

ATR (Avions de Transport Regional)(International)
ATR ATR 42
ATR ATR 42 Petrel
ATR ATR 42F
ATR ATM 42R
ATR SAR 42
ATR ATR 72
ATR ATR 72 Petrel

Aubaud (France)
Aubaud 1851 Flying Machine Proposal

Aubert (Avions Paul Aubert)(France)
Aubert PA-204S Super-Cigale

Aubry (France)
Aubry Météore (Meteor)

AUC (American Utilicraft Corp) (Sterling, VA)
AUC FF-1080-100 Freight Feeder
AUC FF-1080-200 Freight Feeder
AUC FF-1080-500 Freight Feeder

Audenis, Charles (France)
Audenis C2
Audenis E.P.2

Audineau (France)
Audineau 1910 Monoplane

Auffm-Ordt, Clement (France)
Auffm-Ordt 1908 Monoplane (Voisin)
Auffm-Ordt 1909 Monoplane

Auger, Alfred (France)
Auger Vautour (Vulture) 1928 Sailplane
Auger 1925 Monoplane Glider

AUI (Aviazione Ultraleggera Italiana)(Italy)
AUI Leone (Lion) Ultralight Family

Auster (UK)
In 1938, A.L. Wykes purchased the British rights to manufacture Taylorcraft designs, founding **Taylorcraft Aeroplanes Ltd** in Thurmaston, Leics. The name of the company was changed to **Auster Aircraft Ltd** in 1946. The company was purchased, along with Miles Aircraft by the Pressed Steel Company to form **British Executive and General Aircraft, Ltd (Beagle)**. In 1966, Beagle was sold to the British government. Plans to resume production were not realized, and operations were suspended in 1968.
Auster B.4 Ambulance/Freighter
Auster B.8 Agricola
Auster (Taylorcraft) Plus C
Auster (Taylorcraft) Plus C.2
Auster C.6 Atlantic
Auster (Taylorcraft) D.4 (D.4/108)
Auster (Taylorcraft) D.6 (D.6/160)
Auster (Taylorcraft) Plus D
Auster (Taylorcraft) Plus EY
Auster J.1 Autocrat
Auster J.1 Kingsland
Auster J.1B Aiglet
Auster J.1N Alpha
Auster J.1U Workmaster
Auster J.2 Arrow
Auster J.3 Atom
Auster J.4
Auster J.5 Autocrat (Adventurer)
Auster J.5A
Auster J.5B Autocar
Auster J.5E
Auster J.5F Aiglet Trainer
Auster J.5G Cirrus Autocar
Auster J.5H Autocar
Auster J.5K Aiglet Trainer
Auster J.5L Aiglet Trainer
Auster J.5P Autocar
Auster J.5Q Alpine
Auster J.5R Alpine
Auster J.5V Autocar
Auster J.8F Aiglet Trainer
Auster Model P Avis
Auster 'S' Prototype
Auster V.J.1
Auster V.J.1A
Auster A.O.P. Mk.I
Auster A.O.P. Mk.II
Auster A.O.P. Mk.III
Auster 3
Auster A.O.P. Mk.IV
Auster 4
Auster A.O.P. Mk.V
Auster 5
Auster 5C
Auster 5D
Auster 5M
Auster Alpha 5
Auster A.O.P. Mk.6
Auster 6
Auster 6A Tugmaster
Auster T.Mk.7
Auster 7
Auster A.O.P. Mk.9
Auster 9
Auster 9M

Austflight Ula Pty Ltd (Australia)
Austflight Drifter (Maxair Variant)

Austin (UK)
Austin AFT3 Osprey Triplane
Austin Ball
Austin Greyhound
Austin Kestrel
Austin Whippet

Australian Autogyro Co (Australia)
Australian Autogyro Skyhook

Australian Aviation Works (Australia)
Australian Aviation Aerolite
Australian Aviation Aeromax 1700 Sport
Australian Aviation Karatoo C
Australian Aviation Spacewalker

Australite Inc (Ventura, CA)
Australite Ultrabat (Ultralite)

AUT *(See: Umbra)*

Auto-Aero (Hungary)
Auto-Aero Góbé R-26S

Autogiros, General

Autogiro Company of America (Philadelphia, PA)*(See also: Pitcairn; Pitcairn-Cierva)*
Autogiro Co of America AC-35
Autogiro Co of America AC-36

Autogyro Design Bureau (Russia)
Autogyro Design Bureau Ariel 211
Autogyro Design Bureau Ariel 212
Autogyro Design Bureau Boomerang
Autogyro Design Bureau Pelegrim

Automedia sro (Czech Republic)
Automedia JK-1

Autoplan *(See: Pischoff)*

Avalon, Richard (California)*(See: Mitchell Aircraft Corp)*

AVCO Lycoming (Lycoming Division, AVCO Corp)(Stratford, CT)
AVCO Lycoming Flying Duck (Hydrofoil)

Avgur Aerostat Center (Russia)
Avgur Krechet VTOL

AVI SA Industrial, Comercial y Financiera (Argentina)
AVI 205 (E-185/9, HF2-185)

Avia Akciova Spolecnost pro Prumysl Letecky (Czechoslovakia)
Avia (Czech) BH-Exp
Avia (Czech) BH-1
Avia (Czech) BH-1 Exp
Avia (Czech) BH-1*bis*
Avia (Czech) BH-2
Avia (Czech) BH-3
Avia (Czech) BH-4
Avia (Czech) BH-5 Boska
Avia (Czech) BH-6
Avia (Czech) BH-7 A
Avia (Czech) BH-7 B
Avia (Czech) BH-8
Avia (Czech) BH-9
Avia (Czech) BH-10
Avia (Czech) BH-11
Avia (Czech) BH-11 B
Avia (Czech) BH-11 C
Avia (Czech) BH-11 E
Avia (Czech) BH-12
Avia (Czech) 14 (See: Ilyushin Il-14)
Avia (Czech) BH-16
Avia (Czech) BH-17
Avia (Czech) BH-19
Avia (Czech) BH-20
Avia (Czech) BH-21
Avia (Czech) BH-21 J
Avia (Czech) BH-21 R
Avia (Czech) BH-22
Avia (Czech) B 122
Avia (Czech) Ba 122
Avia (Czech) BH-23
Avia (Czech) BH-25
Avia (Czech) BH-25 J
Avia (Czech) BH-25 L
Avia (Czech) BH-26
Avia (Czech) BH-28
Avia (Czech) BH-29
Avia (Czech) Ba 33
Avia (Czech) BH-33 (BH-33 E)
Avia (Czech) BH-33 L
Avia (Czech) BH-34
Avia (Czech) B 534
Avia (Czech) Bk 534
Avia (Czech) 35
Avia (Czech) 135
Avia (Czech) F.39 (See: Fokker F.IX)
Avia (Czech) B 46
Avia (Czech) 51
Avia (Czech) 52
Avia (Czech) B. 354
Avia (Czech) 56
Avia (Czech) 156
Avia (Czech) 57
Avia (Czech) 158
Avia (Czech) 57
Avia (Czech) 71 (B 71)
Avia (Czech) 36
Avia (Czech) E.44
Avia (Czech) S 92
Avia (Czech) CS 92
Avia (Czech) S 99 (C 10)
Avia (Czech) S 99 (C 10), Israel
Avia (Czech) CS 99 (C 110)
Avia (Czech) Bs 122
Avia (Czech) Ba 222
Avia (Czech) Ba 322
Avia (Czech) Ba 422
Avia (Czech) S 199 (C 210) Mezek

Avia (Ateliers Vosgieus)(France, Pre-WWI)
Avia (France, Pre-WWI):
Biplanes, 1909 Scott Biplane
Biplanes, 1st Design
Biplanes, 2nd Design
Biplanes, 3rd Design (See: Bonnet-Lebranche No.5)
Biplanes, 4th Design (Tractor)
Gliders
Monoplanes, 1st Design
Monoplanes, 2nd Design
Monoplanes, 3rd Design
Monoplanes, 4th Design

Avia (France, Inter-War)
Avia (France, Inter-War):
Zogling (Pupil) O 1 (1930 Sailplane)
Zogling (Pupil) 10A (1930 Sailplane)
IIA Vautour (Vulture)
X A
XI A
XV A
20 A
30A Rapace
30E
32E Glider (1931)
41P
50 Glider (1933)
51 Glider (1933)
60 Glider (1933)

Avia (Azionari Vercellese Industrie Aeronautiche)(Italy)
Avia (**Azionari Vercellese Industrie Aeronautiche**) produced aircraft designed by Italian aviator Francis Lombardi. In 1948 Avia was absorbed by **Francis Lombardi & C.** Several designs were also built under license by **Meteor SpA Costruzioni Aeronautiche** in Trieste. All Avia and Lombardi designs are filed under **Avia (Italy)**.
Avia (Italy) FL.3
Avia (Italy) LM.5 Aviastar
Avia (Italy) LM.7
Avia (Italy) Meteor FL.53
Avia (Italy) Meteor FL.55

Avia Baltika Aviation Ltd (Lithuania)
Avia Baltika LAK-X

Avia Nauchno-Proizvodstvennoye Obedinenie (Avia Scientific-Production Assoc)(Russia)
Avia (Russia) Accord Jet
Avia (Russia) Accord, Lightweight
Avia (Russia) Accord-201

Aviabellanca Aircraft Corp *(See: Bellanca)*

Aviacomplex *(See: Aviakompleks AO)*

Aviacor Mezhdunarodnaya Aviatsionnaya Korporatsiya OAO (Aviacor International Aircraft Corp JSC)(Russia)
Aviacor M-12 Kasatnik (Darling)

Aviafiber AG (Switzerland)
Aviafiber Canard-2FL (Sailplane)

Aviakit (France)
Aviakit Hermes

Aviakompleks AO (Russia)
Aviakompleks AS-2

Aviamilano Costruzioni Aeronautiche (Italy)*(See also: Aeromere)*
Aviamilano F.8 Falco
Aviamilano F.8L Falco III
Aviamilano F.8L Falco IV
Aviamilano F.14 Nibbio
Aviamilano F.250 (See also: SIAI-Marchetti SF.260)
Aviamilano P.19 Scricciolo

Avian Aircraft Ltd (Canada)
Avian Aircraft Ltd 2/180 Gyroplane

Aviaspetstrans OAO *(See: AST)*

Aviastar Ulyanovsky Aviatsionnyi Promshlennyi Kompleks (Ulyanovsk Aviation Industrial Complex "Aviastar")(Russia)
Aviastar Module

Aviasud Industries (France)
Aviasud AE 206 Mistral
Aviasud AE 207 Twin Mistral
Aviasud AE 209 Albatros
Aviasud AE 210 Alizé (Tradewind)
Aviasud Sirocco

Aviat Aircraft, Inc (Afton, WY)
Aviat Eagle II (See: Christen)
Aviat Husky A-1
Aviat Millennium Swift
Aviat Pitts (See: Pitts)
Aviat 110 Special

Aviatehnologia (Moldova)
Aviatehnologia Favorit (Favorite)

Aviatik (Germany)
Aviatik A.6
Aviatik A.K.1
Aviatik A.K.2
Aviatik A.K.15
Aviatik B.I
Aviatik B.II
Aviatik B.III
Aviatik B.231 (13)
Aviatik C.I
Aviatik C.II
Aviatik C.III
Aviatik C.V
Aviatik C.VI (D.F.W. C.V)
Aviatik C.VIII
Aviatik C.IX
Aviatik D.I
Aviatik D.II
Aviatik D.III
Aviatik D.IV
Aviatik D.VI
Aviatik D.VII
Aviatik Dr.1
Aviatik G.III
Aviatik Gr.II (30.17)
Aviatik Gr.III (30.18)
Aviatik Type H Monoplane
Aviatik P.14
Aviatik P.G.20
Aviatik Pfeil (Arrow) Biplane
Aviatik Pfeil (Arrow) Monoplane
Aviatik R.III
Aviatik Taube (Dove)
Aviatik W.P.18
Aviatik W.P.18 Hydro
Aviatik Type III Biplane
Aviatik Type IV
Aviatik 30.01
Aviatik 30.02
Aviatik 30.03
Aviatik 30.04
Aviatik 30.06
Aviatik 30.07
Aviatik 30.08 (35.02)
Aviatik 30.09
Aviatik 30.10
Aviatik 30.11
Aviatik 30.12
Aviatik 30.13
Aviatik 30.14 Fighter Prototype
Aviatik 30.14 (New)
Aviatik 30.19
Aviatik 30.20
Aviatik 30.21
Aviatik 30.22
Aviatik 30.23
Aviatik 30.24
Aviatik 30.25
Aviatik 30.26 & 30.28
Aviatik 30.27 & 30.29
Aviatik 30.30
Aviatik 30.31 & 30.32
Aviatik 30.34
Aviatik 30.38
Aviatik 30.39
Aviatik 30.40
Aviatik 30.42
Aviatik 1910 Biplane
Aviatik 1911 Monoplane
Aviatik 1912 Hydro
Aviatik 1912 Monoplane
Aviatik 1913 Biplane

Aviatika Layt (Russia)*(See also: KB MAI)*
Aviatika 890
Aviatika 900
Aviatika 960

Aviation Assoc of Technical University Students in Lwów *(See: ZASPL)*

Aviation Boosters (Kansas City, MO)
Aviation Boosters Skyhopper

Aviation Composite Technology *(See: ACT)*

Aviation Composites (UK)*(See: INAV)*

Aviation Construction Engineers (Chicago, IL)
Aviation Construction Crossland Ace

Aviation Engineering Corp *(See: Aviation Engineering School)*

Aviation Engineering School (New York, NY)
Aviation Engineering School 3-place Open Land Biplane,1928

Aviation Farm Ltd (Poland)
Aviation Farm J-5 Marco (See: Marko-Elektronik)

Aviation Franchising International *(See: Prescott)*

Aviation Industries Inc (Omaha, NE)
Aviation Industries 75C

Aviation Industries of Australia (AIA) (Australia)
AIA MA-2 Mamba

Aviation Industries of Iran *(See: AII)*

Aviation Scotland Ltd (UK)*(See: ARV)*

Aviation Specialties Inc (Mesa, AZ)
Aviation Specialties S-55-T (Modified Sikorsky)

Aviation Traders, Ltd (UK)
Aviation Traders ATL.90 Accountant 1
Aviation Traders AT(E)L.98 Carvair Car Ferry (DC-4 Conversion)

Aviation Traders AT(E)L.98 Carvair Cargo (DC-4 Conversion)

Aviaton Nauchno-Proizvodstvennaya Aviatsionnaya Firma (Russia)
Aviaton Merkury (Mercury)

Aviator (Belgium)
Aviator 1910 Monoplane

Aviator Scientific-Production Enterprise (Russia)
Aviator Scientific M-9 Marathon

Avibras (Sociedade Avibras Ltda) (Brazil)
Avibras A-80 Falcao

Avid Aircraft Inc (Caldwell, ID)
Avid Aircraft Bandit
Avid Aircraft Catalina
Avid Aircraft Champion
Avid Aircraft Flyer (See: Light Aero Avid Flyer)
Avid Aircraft Magnum
Avid Aircraft Speedwing

Aviette *(See: Human-Powered Flight)*

Avileo *(See: Lioré et Olivier)*

Aviméta (France)

Between 1922 and 1926, the French industrial concern **Établissements Schneider** produced a small number of all-metal aircraft using their aluminum alloy "alférium." Georges Lepère (who had designed the LUSAC 11 in the US) was chief of the aviation department. In mid-1926 Schneider formed a subsidiary, the **Société pour la Construction d'Avions Métalliques Aviméta** (commonly known as **Société Aviméta**). Lepère departed early in 1928, and Établissements Schneider discontinued the Aviméta subsidiary. Aircraft produced between 1922 and 1926 are filed under **Schneider**; later aircraft are filed under **Aviméta.**

Aviméta A.V.M. 88
Aviméta A.V.M. 92
Aviméta A.V.M. 130
Aviméta A.V.M. 132

Aviolight Srl (Italy)
Aviolight P86 Mosquito (See: Partenavia)
Aviolight P66D Delta (See: Partenavia)

Avion Corp *(See: Northrop)*

Avionnerie de Levallois *(See: de Lesseps, Robert)*

Avionnerie Lac St-Jean Inc *(See: Lac St-Jean)*

Avions Automobiles Philippe Moniot *(See: Moniot)*

Avions C. T. Weymann *(See: Weymann)*

Avions Claude Piel *(See: Piel)*

Avions Croses *(See: Croses)*

Avions de Colombia SA (Colombia)
Avions de Colombia AC-05 Pijad

Avions de Transport Regional *(See: ATR)*

Avions Fairey *(See: Fairey)*

Avions H. Nicollier *(See: Nicollier)*

Avions Jacques Coupé *(See: Coupé)*

Avions Marcel Dassault-Breguet Aviation *(See: Dassault)*

Avions Mauboussin*(See:Mauboussin)*

Avions Max Holste *(See: Max Holste)*

Avions Mudry et Cie *(See: Mudry)*

Avions Robin *(See: Robin)*

Avions Simplex *(See: Arnoux)*

Avions Weymann *(See: Weymann)*

Aviotechnica *(See: Interavia)*

AVPK Sukhoi *(See: Sukhoi)*

Avi SA Industrial Comercial y Financiera (Argentina)
Avi 205

Avis Flugzeugwerke und Autowerke GmbH (Germany)
Avis B.G.VI

Avro (A. V. Roe & Co Ltd)(UK)

Alliott Verdon Roe built his first airplane in 1907. In 1910 he formed **A.V. Roe and Co Ltd** (or **Avro**). Avro was acquired by the **Armstrong Siddeley Development Co** in 1928, and then by the **Hawker Siddeley Aircraft Co** in 1935. Aircraft with Avro designations were produced until 1963 when the company became the **Avro-Whitworth Division of Hawker Siddeley Aviation.** A.V. Roe aircraft designed prior to 1963 are listed by design number under **Avro.**

Avro Burga Monoplane (1912)
Avro Curtiss-Type Biplane (1911) (Lakes Water Bird)
Avro Type D Biplane
Avro Type D Seaplane
Avro Duigan Biplane (1911)(See also: Lakes Sea Bird)
Avro Type E (See: Avro 500)
Avro Type Es (See: Avro 502)
Avro Type F Monoplane
Avro Farman-Type Biplane (1910)
Avro Type G Biplane
Avro Type H (See: 503 Seaplane)
Avro Roe I Biplane
Avro Roe I Triplane "Bull's-eye"
Avro Roe II Triplane "Mercury"
Avro Roe III Triplane
Avro Roe IV Triplane
Avro Roe 5 Triplane
Avro Roe X Triplane
Avro 500 (Type E)
Avro 501
Avro 502 (Type Es)
Avro 503 Seaplane (Type H)
Avro 504
Avro 504A
Avro 504B
Avro 504C
Avro 504D
Avro 504E
Avro 504F
Avro 504G
Avro 504H
Avro 504I
Avro 504J
Avro 504K
Avro 504K Mk.II
Avro 504L
Avro 504M
Avro 504N
Avro 504O
Avro 504Q
Avro 504R Gosport
Avro 508
Avro 510
Avro 511 Arrowscout
Avro 514
Avro 519
Avro 519A
Avro 521
Avro 521A

Avro 522
Avro 523 Pike
Avro 527
Avro 527A
Avro 528 Silver King
Avro 529
Avro 529A
Avro 530
Avro 531 Spider
Avro 533 Manchester
Avro 534 Baby
Avro 534A Water Baby
Avro 534B Baby
Avro 534C Baby
Avro 534D Baby
Avro 534E Baby
Avro 534F Baby
Avro 534G Baby
Avro 536
Avro 538
Avro 539
Avro 539A
Avro 539B
Avro 540
Avro 543 Baby
Avro 544 Baby
Avro 545
Avro 546
Avro 547 Triplane
Avro 548 Tourist
Avro 548A Tourist
Avro 549 Aldershot
Avro 551
Avro 552
Avro 552A
Avro 552B
Avro 554 Antarctic Baby
Avro 555 Bison
Avro 555A Bison
Avro 555B Bison
Avro 557 Ava
Avro 558
Avro 560
Avro 561 Andover
Avro 562 Avis
Avro 563 Andover
Avro 566 Avenger
Avro 567 Avenger II
Avro 571 Buffalo I
Avro 572 Buffalo II
Avro 574 (See: Cierva C.6C)
Avro 575 (See: Cierva C.8L Mk.I)
Avro 576 (See: Cierva C.9)
Avro 581 Avian
Avro 584 Avocet
Avro 594 Avian
Avro (USA) 594 Avian
Avro 604 Antelope
Avro 605 Avian (Float-equipped 594 Avian)
Avro 605A Avian
Avro 605B Avian
Avro 611 (Cierva C.8L Mk.II)
Avro 612 (Cierva C.17 Mk.I)
Avro 616 Avian
Avrlo 617 (Cierva C.8L Mk.III)
Avro 618 Ten
Avro 619 Five
Avro 620 (Cierva C.17 Mk.II)
Avro 621 Tutor
Avro 624 Six
Avro 625 Avian Monoplane Racer
Avro 626
Avro 627 Mailplane
Avro 631
Avro 636
Avro 636A
Avro 637
Avro 638 Club Cadet
Avro 639 Club Cadet
Avro 640 Club Cadet
Avro 641 Commodore
Avro 642
Avro 643 Cadet
Avro 646 Sea Tutor
Avro 652 "Avalon" & "Avatar"/"Ava"
Avro 652A Anson Mk.I
Avro 652A Anson Mk.II
Avro 652A Anson Mk.III
Avro 652A Anson Mk.IV
Avro 652A Anson Mk.V
Avro 652A Anson Mk.VI
Avro 652A Anson Mk.X
Avro 652A Anson Mk.11
Avro 652A Anson Mk.12
Avro 652A Anson Mk.18
Avro 652A Anson Mk.18C
Avro 652A Anson Mk.19
Avro 652A Anson Mk.20
Avro 652A Anson Mk.21
Avro 652A Anson Mk.22
Avro 654 Mailplane
Avro 661 (See: Parnall Parasol)
Avro 667
Avro 667A
Avro 671 Rota (Cierva C.30)
Avro 674 (See: Hawker Audax)
Avro 679 Manchester Mk.I
Avro 679 Manchester Mk.IA
Avro 679 Manchester Mk.II
Avro 679 Manchester Mk.III
Avro 683 Lancaster B.Mk.I
Avro 683 Lancaster B.Mk.I (Special)
Avro 683 Lancaster PR.Mk.I
Avro 683 Lancaster B.Mk.II
Avro 683 Lancaster B.Mk.III
Avro 683 Lancaster ASR.Mk.3
Avro 683 Lancaster GR.Mk.3
Avro 683 Lancaster MR.Mk.3
Avro 683 Lancaster B.Mk.VI
Avro 683 Lancaster B.Mk.VII
Avro 683 Lancaster B.Mk.X
Avro 685 York C.Mk.1
Avro 685 York C.Mk.2
Avro 688 Tudor Mk.1
Avro 688 Tudor Mk.3
Avro 688 Tudor Mk.4
Avro 688 Tudor Mk.4B
Avro 688 Tudor Mk.8 (Turbojet)
Avro 689 Tudor Mk.2
Avro 689 Tudor Mk.5
Avro 689 Tudor Mk.7
Avro 691 Lancastrian C.Mk.1
Avro 691 Lancastrian C.Mk.2
Avro 691 Lancastrian C.Mk.3
Avro 691 Lancastrian C.Mk.4
Avro 694 Lincoln B.Mk.1
Avro 694 Lincoln B.Mk.2 (695 Lincolnian)
Avro 694 Lincoln B.Mk.4
Avro 694 Lincoln B.Mk.15
Avro 694 Lincoln Mk.30 (B)
Avro 694 Lincoln Mk.31 (NR)
Avro 696 Shackleton G.R.Mk.1
Avro 696 Shackleton M.R.Mk.1A
Avro 696 Shackleton M.R.Mk.2
Avro 696 Shackleton M.R.Mk.3
Avro 698 Vulcan B.Mk.1
Avro 698 Vulcan B.Mk.1A
Avro 698 Vulcan B.Mk.2
Avro 698 Vulcan B.Mk.2MRR
Avro 698 Vulcan K.Mk.2
Avro 701 Athena T.Mk.1
Avro 701 Athena T.Mk.1A
Avro 701 Athena T.Mk.2
Avro 706 Ashton Mk.1
Avro 706 Ashton Mk.2
Avro 706 Ashton Mk.3
Avro 706 Ashton Mk.4
Avro 707
Avro 707A
Avro 707B
Avro 707C
Avro 720 Rocket Interceptor
Avro 729 All-Weather Fighter
Avro 733 (See: Armstrong Whitworth A.W.650 Argosy)
Avro 748 (See: Hawker Siddeley HS.748; also: Hawker Siddeley Andover)

Avro (Canada)

Avro (Canada) Avrocar (VZ-9V)
Avro (Canada) Avrocar (VZ-9V), NASM
Avro (Canada) C-102 Jetliner
Avro (Canada) CF-100 Canuck

Avro (Canada) CF-105 Arrow

Avtek Corp (Camarillo, CA)
Avtek 400
Avtek 400A
Avtek 419 Express

Awazu Hiko Kenkyusho (Minoru Awazu; Awazu Flight Research Studio)(Japan)
Awazu Number 2 Seicho-Go

Ayres Corp (Albany, GA)
Ayres Corp LM200 Loadmaster
Ayres Corp LM200 Searchmaster
Ayres Corp LM250 Loadmaster
Ayres Corp Turbo-Thrush S2R (See also: Rockwell)
Ayres Corp V-1A Vigilante
Ayres Corp 660 Turbo-Thrush (See also: Rockwell)

Ayres, Dr. W. O. (New Haven, CT)
Ayres (Dr. W. O.) 1885 Aerial Machine

Azcarate (Mexico)
Azcarate O-E (Observacion y Entrenamiento)
Azcarate O-E (Observacion y Entrenamiento) Hydro

***Azionari Vercellese Industrie Aeronautiche** (See: Avia, Italy)*

B

***B & C** (See: Bristol M.R.1)*

B & F Technik Vertriebs GmbH (Germany)
B & F Technik FK.6
B & F Technik FK.9
B & F Technik FK.9 Mk.3
B & F Technik FK.12 Comet

B & G Aircraft Co Inc (Hartville, OH)
B & G Sparrow (Ultralight)

B & L Hinz (Germany)
B & L Hinz BL.1 Kéa

***B. Taylorcraft Aerospace, Inc** (See: Taylorcraft)*

B W Rotor Co, Inc (Tonowanda, KS)
B W Rotor Sky Cycle

B-B (Kenneth Bowser and Gailard Barker)(Phoenix, AZ)

***B.A.** (See: British Aircraft Manufacturing Co)*

***Baade** (See: VEB Dresden)*

Baatz (Germany)
Baatz Glider

Babcock, Verne Clifton "Bab" (USA)
Verne Babcock's first design was a 1905 copy of the Wright Flyer built at Benton Harbor, Michigan. In 1907 he moved to Seattle, Washington, where, in 1909, he built a second Wright copy. That same year, he and a partner formed **Babcock-Breininger Aeroplane Supply Co** and built several more derivative aircraft. Babcock moved on to several other aviation training and engineering activities before forming **Babcock Aircraft Co** in Akron, Ohio, in 1924. The company experienced several reorganizations during the early 1930s, apparently known as **Babcock-Vlcek Co** and **Taubman Aircraft Co** at various times. By the late 1930s, Babcock formed a new **Babcock Airplane Corp**, which soon moved to DeLand, Florida. The company built assault gliders during WWII. Babcock left the aviation industry soon after the War (with rights to produce his LC-13 passing to **Bartlett Aircraft Corp**). Aircraft designed and built by Babcock are all listed as **Babcock**.
Babcock Airmaster
Babcock Airster
Babcock Gull Wing Taube Type
Babcock LC-7 (See: United States Airplane & Motor Engineering Co)
Babcock LC-11 Cadet
Babcock LC-13 Taube
Babcock LC-13A Taube
Babcock Light Deperdussin Type
Babcock Light Parasol
Babcock "Potlatch Bug"
Babcock Ranger
Babcock Swastika Model A (Series I)
Babcock Teal
Babcock 1905 Wright Biplane Copy
Babcock 1909 Wright Biplane Copy
Babcock 1913 Curtiss Pusher Type

Babinski, Zbigniew (Poland)
Babinski 1912 First Glider
Babinski 1912 Second Glider
Babinski 1913 Glider

***B.A.C. (1928-36)**(See: British Aircraft Co)*

BAC (British Aircraft Corp)(UK)
British Aircraft Corp (BAC) was established in 1960 with the nationalization of **Bristol Aircraft Ltd**, **English Electric Aviation Ltd**, **Vickers-Armstrong (Aircraft) Ltd**, and **Hunting Aircraft Ltd.** In 1977 BAC was absorbed into **British Aerospace**. Aircraft designed after 1960, or whose main production occurred after 1960, are listed under **BAC** by their BAC designations. Earlier or later products are filed by their original companies.
BAC Lightning (See: English Electric Lightning)
BAC TSR-2
BAC VC10
BAC VC10 Series 1100
BAC VC10 C.Mk.1
BAC Super VC10 (Series 1150)
BAC One-Eleven Family
BAC One-Eleven Series 200
BAC One-Eleven Series 300
BAC One-Eleven Series 400
BAC One-Eleven Series 475
BAC One-Eleven Series 500
BAC 167 Strikemaster
BAC Two-Eleven Series

***BAC (Buchanan Aircraft Corp)** (See: Buchanan)*

***BACC (Business Aircraft Corp)** (See: Howard Aero Manufacturing Division)*

Bach Aircraft Co Inc (Los Angeles, CA)
Bach Air Yacht 3-CT-2
Bach Air Yacht 3-CT-4
Bach Air Yacht 3-CT-6 Air Transport
Bach Air Yacht 3-CT-8
Bach Air Yacht 3-CT-9
Bach Air Yacht 3-CT-9S

Bachelard (France)
Bachelard 1910 Monoplane

Bachelier-Dupont-Baudrin (France)
Bachelier-Dupont-Baudrin 1913 Flying Boat

Bachem (Germany)
Bachem Ba 349 Natter (Viper)(BP-20)
Bachem Ba 349 B-1 Natter, NASM

Back Forty Developments Ltd (Canada)
Back Forty Tundra

Backstrom, Al (Ft Worth, TX; Lubbock TX)
Backstrom EPB-1 (Easley, Powell, Backstrom) Plank
Backstrom EPB-1A (Easley, Powell, Backstrom) Plank
Backstrom WPB-1 (White, Powell, Backstrom) Flying Plank II

Bacon (Erle L. Bacon Corp)(Santa Monica, CA)
Bacon Super AT-6

de Bacqueville, Marquis (France)
de Bacqueville 1742 Ornithopter

Badaire (France)
Badaire 1913 Monoplane

Baddeck Biplane *(See: Canadian Aerodrome Co)*

Baden-Powell, Fletcher Smythe (UK)
Baden-Powell Water-Chute Glider

Badgley, H. (USA)
Badgley (H.) Aerial Machine (1879)

Bahnbedarf AG (Germany)
Bahnbedarf D.1
Bahnbedarf 1924 Moto Aviette

Baikov-Kanenev (A. A. Baikov & S. I. Kanenev (Russia)
Baikov-Kanenev OSO-1

Bailey, Dick (USA)
Bailey Bitty Bipe

Bailey, P. P.
Bailey Flying Machine (1880)

Ballard Aircraft Corp *(See: Burnelli CanCargo CBY-3)*

Baillod (France)
Baillod 1912 Seaplane

Bailly (France)
Bailly 1902 Helicopter

B.A.J. (Boncourt-Audenis-Jacob) (France)
B.A.J. C2

Bakeng (Edmond, WA)
Bakeng Duce

Baker, Art (Kansas City, MO)
Baker (Art) B-2

Baker. Gil (USA)
Baker (Gil) BCA-1-3 Amphibian

Baker, Marion "Jim" (Baker Air Research)(Huron, OH)
Baker (Marion) Aquarius Midget Racer
Baker (Marion) Boo Ray
Baker (Marion) MB-1 Delta Kitten

Baker, Ray W. and Al (Independence, MO)
Baker (Ray & Al) Special "Miss Kansas City" (Rebuilt Howard "Pete")

Baker & Collins (B. B. Baker & M. W. Collins)(Van Nuys, CA)
Baker & Collins Primary Glider

Baker-McMillen Co (Akron, OH)
Baker-McMillen Akron Condor
Baker-McMillen Cadet II

Bakshaev (Russia)
Bakshaev LIG-4 (LK)
Bakshaev LIG-7 (RK)
Bakshaev LIG-8 (MP)

Balaban-Bloudek (Austria-Hungary)
Balaban-Bloudek 1917 Helicopter

Balassian de Manawas (France)
Balassian de Manawas 1914 Monoplane

Balaye (France)
Balaye Robur (1912)

Baldwin Aircraft Corp *(See: Orenco)*

Baldwin Aircraft International (Gary Baldwin)(USA)
Baldwin Aircraft ASP-XJ

Baldwin, Thomas (Buffalo, NY)
Baldwin 1909 Tractor Biplane (Curtiss-built, Red Devil)
Baldwin 1910 Pusher Biplane (Curtiss-built, Red Devil II)
Baldwin 1911 Red Devil III (Wittemann-built, Steel)
Baldwin Red Devil, Later Models
Baldwin Red Devil, Later Model, NASM

Balmer, Reginald (Los Angeles, CA)
Balmer "Constance"

Balsan, Jacques (France)
Balsan 1911 Monoplane (Built by Lioré et Olivier)

Ball, Clifford (McKeesport, PA)
Ball S-T Biplane (1929)

Ball-Bartoe (Boulder, CO)
Ball-Bartoe JW-1 Jetwing

Ballouth & Beckley (E. E. Ballouth & E. Beckley)(USA)
Ballouth & Beckley Bristol Fighter (Rebuilt as Monoplane, Clark Y Airfoil)

Baltzell & Miller (Ft Wayne, IN)
Baltzell & Miller1928 Experimental Sport

Bancroft (UK)
Bancroft (1917)

Baoshan Iron & Steel Complex (China)*(See: Venga)*

Barcala, Cierva, y Diaz *(See: B.C.D.)*

Bardin (France)
Bardin 1923 Monoplane Glider
Bardin 1926 Monoplane Glider

Barillon, Pierre (France)
Barillon 1908/09 Monoplane, 12hp
Barillon 1909 Monoplane, 25hp
Barillon 1910 Monoplane, Juvisy
Barillon 1910 Monoplane, 50hp
Barillon 1911 Monoplane Racer, 65hp

Barker (Barkers et Lefevre)(France)
Barker 1909 Twin-Pusher Biplane

Barker, Gailard *(See: B-B)*

Barkley-Grow Aircraft (Detroit, MI)
Barkley-Grow Aircraft T8P-1

Barlatier et Blanc (Emile Barlatier & Henri Blanc) (France)
Barlatier & Blanc Kites

Barlatier & Blanc Powered Kite Models
Barlatier & Blanc 1906 Monoplane
Barlatier & Blanc 1908 Monoplane

Barling Aircraft Corp (St Joseph, MO)

Aircraft bearing Walter H. Barling's name were produced by three organizations. The Barling Bomber was designed by Barling for the Army Air Service's **Engineering Division** and manufactured by **Witteman** as the NBL-1. In 1927 Barling began work with **Nicholas-Beazley Aircraft Co, Inc** where that company's NB-3 was marketed as a Barling product. In 1929 Barling formed his own **Barling Aircraft Corp**. Barling aircraft designs are listed under **Barling Aircraft Corp**, **Engineering Division**, and **Nicholas-Beazley** sections.

Barling B-6 (Model A, Model A-1)

Barnett Rotorcraft (K. J. Barnett Rotorcraft Co)(Olivehurst, CA)
Barnett Rotorcraft J-3M
Barnett Rotorcraft J-4B

Barnett, Jerrie (Olivehurst, CA)
Barnett J-3M Gyroplane

Barnett, F. (Keokuk, IA)
Barnett (F.) Flying-Machine (1877)

Barney (Barney Snyder)(East San Diego, CA)
Barney S-1

Barney Oldfield *(See: Oldfield)*

Barnhart (Pasadena, CA)
Barnhart BT-15 Wampus-kat

Barnwell Brothers (R. Harold and Frank S. Barnwell)(UK)
Barnwell Brothers 1909 Biplane

Barnwell, Frank Sowter (UK)
Barnwell (Frank S.) B.S.W. (Barnwell, Scott, Whitchurch) Mk.1 (1938)

Barnwell, R. Harold (UK)
Barnwell (R. Harold) 1911 Monoplane

Baron, A. (France)
Baron Aéro Ramo-Planeur

Barr Aircraft (Montoursville, PA)
Barr Aircraft BarrSix

Barrón (Spain)
Barrón Delta "Conejo"

Barros, A. A. *(See: Schiller and Barros)*

Bartel, Ryszard (Poland)
Bartel 1909 Glider
Bartel 1911 Glider
Bartel 1918 Monoplane

Bartelt, F. T. (UK)
Bartelt 1911 Ornithopter

Bartini (Russia)
Bartini DAR
Bartini Stal-6
Bartini Stal-7
Bartini Stal-8

Bartlett Aircraft Corp (Rosemead, CA)
Bartlett Blue Zephyr (Developed from Babcock Taube)
Bartlett LC-13A Zephyr 150 (Developed from Babcock Taube)

Barton, Ernest (Mahaska, KS)
Barton Model 100 Biplane

Baserga, Horacio (Argentina)
Baserga HB-1
Baserga Tacuara

Basler Turbo Conversions Inc (Oshkosh, WI)
Basler Turbo Conversions Turbo 67 (DC-3)

Bassan-Gué (France)
Bassan-Gué BN4 Bomber

Bassou, Antoine (France)
Bassou "Rubis" (1933)

Bastianelli (La Compagnia Bastianelli)(Italy)

La Compagnia Bastianelli, with engineers Pegna, Rossi, and Bastianelli, collapsed in 1922. The PRB design rights were acquired by **Pegna-Bonmartini Construzioni Navali-Aeronautiche**. Pegna-Bonmartini was acquired by **Società Anonima Piaggio & Co** in 1923.

Bastianelli PRB1
Bastianelli PRB2

Bastier (France)
Bastier 1911/12 Biplane

B.A.T. (British Aerial Transport)(UK)
B.A.T. F.K.22
B.A.T. F.K.22 Bantam Mk.II
B.A.T. F.K.23 Bantam
B.A.T. F.K.24 Baboon
B.A.T. F.K.25 Basilisk
B.A.T. F.K.26 5-Seater Biplane
B.A.T. F.K.27 Sporting Biplane
B.A.T. F.K.28 Crow

Bates, Carl (Chicago, IL)
Bates 1909 Glider (Homebuilt)

Bathiat-Sanchez (France)
Bathiat-Sanchez 1st 1913 Floatplane
Bathiat-Sanchez 2nd 1913 Floatplane
Bathiat-Sanchez 1-Seat 1913 Monoplane Type E (Sommer Type F Variant)
Bathiat-Sanchez 2-Seat 1913 Monoplane
Bathiat-Sanchez 1913 Land Monoplane

Battey, Dr S. B. (New York, NY)
Battey 1892 Air Ship

Batwing Aircraft Corp (Alameda, CA)
Batwing 2000 (Tailless 2-Seater)

Baudot, François (France)
Baudot "Flying Bicycle" (1936)

Baumann Aircraft Corp (Burbank, CA)
Baumann Brigadier 250
Baumann B-290 Brigadier
Baumann B-360 Delux Brigadier
Baumann Single Engine

Baümer Aero (Germany)
Baümer B.II Monoplane
Baümer B.III 2-Seat Sport
Baümer B.IV Sausewind (Southwind)
Baümer Udet Sportsplane
Baümer IX (1928)
Baümer B.XV (1930)

Bay Aviation Services Co (Oakland, CA)
Bay Aviation Services Co Super-V

Bayerische Aero Klub (Germany)
Bayerische Aero Klub 1922 Sailplane

Bayerische Flugzeugwerke (See: BFW)

Baylac, Jean (France)
Baylac 1909 Helicopter

Baynes, L. E. (UK)
Baynes Auxiliary
Baynes B-3

Baynes Bee
Baynes Carrier Wing (Bat)
Baynes Heliplane
Baynes Scud III Glider

Bazin, Alfred (France)
Bazin Gliders
Bazin 1907 Ornithopter

B.C.D. (Barcala, Cierva, y Diaz) (Spain)
B.C.D. 2

BD Micro-Technologies Inc *(See: Bede)*

Beach, Irl (Wichita, KS)
Beach B-5 Biplane

Beach, William J. (Australia)
Beach (William J.) 1920 Helicopter

Beachey, Lincoln (USA)
Beachey Little Looper
Beachey Monoplane (Eaton-Beachey Tractor)(1915)

Beadle, F. P. H. *(See: Perry, Beadle & Co)*

Beagle (UK)
British Executive and General Aviation Ltd (**Beagle**) was formed in October 1960 as a branch of **Pressed Steel Co Ltd.** Beagle took **Auster Aircraft Ltd** as a subsidiary which was immediately reorganized as **Beagle-Auster Aircraft Ltd.** A technical and manufacturing relationship with **F.G. Miles Ltd** was formalized through the creation of **Beagle-Miles Aircraft Ltd** that November. Beagle-Miles and Beagle-Auster were both absorbed into **Beagle Aircraft Ltd**, which formed in 1962 and went bankrupt in 1970. Production of the Pup and Bulldog passed to **Scottish Aviation Ltd.** Most Auster designs are found under **Auster**; the Auster A.O.P. 11 was renamed the Beagle 11 in production and is filed under **Beagle**. F. G. Miles designs are filed under **Miles (F. G.)**. All Beagle designs are filed under **Beagle**.
Beagle Mark Eleven (A.O.P. Mk.11)
Beagle A.61 Terrier
Beagle A.61 Tugmaster
Beagle A.109 Airedale
Beagle A.111 Airedale
Beagle M.117
Beagle B.121 Pup
Beagle B.125 Bulldog (See also: Scottish Aviation)
Beagle B.206 (Basset CC.Mk.1)
Beagle M.218
Beagle B.242

Beagle-Auster Aircraft Ltd *(See: Beagle)*

Beagle-Miles Aircraft Ltd *(See: Beagle)*

Beagle-Wallis *(See: Wallis Autogyros)*

Beal Airplane (Manufacturing) Co (Ralph A. Beal)(Kansas City, MO)
Beal B-P-2 Biplane
Beal B-M-3 Monoplane
Beal C-M-4 Monoplane

Beard Brothers Garage (Lois Franklin Beard & John Otis Beard (St Petersburg, FL)
Beard Brothers Model B Biplane

Beardmore (William Beardmore & Co) (UK)
Beardmore Inflexible
Beardmore Typhoon
Beardmore W.B.I
Beardmore W.B.II
Beardmore W.B.III
Beardmore W.B.IV
Beardmore W.B.V
Beardmore W.B.VIb
Beardmore W.B.VIc
Beardmore W.B.VId
Beardmore W.B.VIII
Beardmore W.B.IX
Beardmore W.B.X
Beardmore W.B.XXIV Wee Bee I

Béarn (Constructions Aéronautiques du Béarn, C.A.B.) (France)
Béarn Minicab GY-201
Béarn Supercab GY-30

Beatty, George W. (Long Island, NY)
Beatty 1910 Demoiselle Type
Beatty 1912 Wright Type

Beaufeist (France)
Beaufeist 1909 Aircraft

Beaurin (France)
Beaurin 1909 Biplane

Beauyard-Viradelle (France)
Beauyard-Viradelle 1924 Mono-Aviettes

Bebin (France)
Bebin 1911 Aircraft

Becher and Wolf
Becher and Wolf Glider Triplane

Bêchéreau *(See: S.A.B.)*

Bede Aircraft Corp (Bede Jet Corp) *(See also: American Aviation)*
Bede BD-1
Bede BD-2
Bede BD-3
Bede BD-4
Bede BD-5
Bede BD-5A
Bede BD-5B
Bede BD-5B, NASM
Bede BD-5D
Bede BD-5J
Bede BD-6
Bede BD-7
Bede BD-8
Bede BD-9
Bede BD-10
Bede BD-12
Bede BD-14
Bede XBE-2

Bedek Aircraft Corp *(See: Israel Aircraft Industries)*

Bédelia (Bourbeau et Devaux) (France)
Bédelia 1912 Flyingboat

Bedunkovich (Russia)
Bedunkovich P-3 (LIG-5)
Bedunkovich SKh-1 (LIG-10)

Bee Aviation Associates, Inc (San Diego, CA)
Bee Aviation Honey Bee
Bee Aviation Queen Bee
Bee Aviation Wee Bee

Bee Line Aircraft Co *(See: Aerial Engineering Corp)*

Beebe, Emmett W. (Muskegon, MI)
Beebe Parasol Cabin (1928)

Beech Aircraft Corp (Wichita, KS)
In 1932, Walter H. and Olive Ann Beech formed the **Beech Aircraft Corp** to produce Walter's Model 17 Staggerwing aircraft. The company became a

Raytheon Co subsidiary in 1980. In 1994, Beech Aircraft was merged with **Raytheon Corporate Jets Inc** to form **Raytheon Aircraft Co.** All aircraft designed or produced by Beech Aircraft Corp are listed under **Beech**. Aircraft designed after the 1993 creation of Raytheon Aircraft Co are listed under **Raytheon**.

Beech XA-38 Grizzly (Model 28)
Beech AT-7 Navigator
Beech AT-7A Navigator
Beech AT-7C Navigator
Beech AT-10 Wichita
Beech AT-10 Wichita, Butterfly Tail
Beech AT-11 Kansan
Beech Badger (See: Beech Travel Air)
Beech Baron 55
Beech Baron 95-55 (C95A)
Beech Baron A55
Beech Baron B55
Beech Baron C55
Beech Baron D55
Beech Baron E55
Beech Baron 56TC
Beech Baron 58
Beech Baron 58P
Beech Baron 58TC
Beech Baron 59P
Beech Turbo Baron
Beech Beechjet 400A
Beech Bonanza E33
Beech Bonanza E33A
Beech Bonanza E33B
Beech Bonanza E33C
Beech Bonanza F33
Beech Bonanza F33A
Beech Bonanza F33C
Beech Bonanza G33
Beech Bonanza 35
Beech Bonanza 35 "Waikiki Beach," NASM
Beech Bonanza A35
Beech Bonanza A35TC
Beech Bonanza B35
Beech Bonanza C35
Beech Bonanza D35
Beech Bonanza E35
Beech Bonanza F35
Beech Bonanza G35
Beech Bonanza H35
Beech Bonanza J35
Beech Bonanza K35
Beech Bonanza M35
Beech Bonanza N35
Beech Bonanza O35
Beech Bonanza P35
Beech Bonanza S35
Beech Bonanza V35
Beech Bonanza V35A
Beech Bonanza V35ATC
Beech Bonanza V35B
Beech Bonanza V35BTC
Beech Bonanza 36
Beech Bonanza A36
Beech Bonanza A36C
Beech Bonanza A36TC
Beech Bonanza B36TC
Beech Bonanza T36TC
Beech VC-6A
Beech C-12A Huron
Beech C-12B Huron
Beech C-12C Huron
Beech C-12D Huron
Beech C-12F Huron
Beech RC-12K Huron
Beech RC-12N Huron
Beech RC-12P Huron
Beech C-12R Huron
Beech C-29 (See: British Aerospace 125-800FI)
Beech YC-43 Traveller
Beech UC-43B Traveller
Beech C-45 Expeditor
Beech C-45A Expeditor
Beech C-45B Expeditor
Beech C-45F Expeditor
Beech C-45G Expeditor
Beech C-45H Expeditor
Beech C-45J Expeditor
Beech RC-45J Expeditor
Beech TC-45J Expeditor
Beech CQ-3
Beech Debonair 33
Beech Debonair A33
Beech Debonair B33
Beech Debonair C33
Beech Debonair C33A
Beech Debonair D33
Beech Duchess (Model 76, PD 289)
Beech Duke 60
Beech Duke A60
Beech Duke B60
Beech Expeditor Mk.II
Beech Expeditor Mk.3
Beech Expeditor Mk.3N
Beech Expeditor Mk.3T
Beech Expeditor Mk.3TM
Beech F-2
Beech F-2A
Beech F-2B
Beech GB (17 Staggerwing) Traveller
Beech GB-1 Traveller
Beech GB-2 Traveller
Beech Hawker 125 Family

Beechcraft Hawker Corp, a **Beech Aircraft Corp** subsidiary, was formed in 1970 to help market, assemble, and modify the Hawker Siddeley HS.125 business jet as the BH 125. Beechcraft Hawker continued in this role after Hawker Siddeley was absorbed by British Aerospace in 1977. In 1994, Beechcraft Hawker lost its corporate identity when Beech merged into **Raytheon Aircraft Co.** HS.125/BH 125 versions through the 700-series are listed under **Hawker Siddeley**. Versions from the 800- through 1000-series are listed under **British Aerospace.** The Horizon and all subsequent 125 developments appear under **Raytheon.**

Beech Horizon (See: Raytheon)
Beech JB-1 (C17R Staggerwing) Traveller
Beech Jet Mentor (Model 73)
Beech JRB-1 Expeditor
Beech JRB-2 Expeditor
Beech JRB-3 Expeditor
Beech JRB-4 Expeditor
Beech JRB-6 Expeditor
Beech King Air 90
Beech King Air 65-90
Beech King Air 65-90, NASM
Beech King Air A90
Beech King Air B90
Beech King Air C90
Beech King Air C90A
Beech King Air C90B
Beech King Air C90SE
Beech King Air E90
Beech King Air F90
Beech King Air 100
Beech King Air A100
Beech King Air B100
Beech Super King Air 200
Beech Super King Air B200
Beech Super King Air B200C
Beech Super King Air B200T
Beech Super King Air 300
Beech Super King Air 350
Beech Super King Air 350C
Beech YL-23 Seminole
Beech L-23A Seminole
Beech L-23B Seminole
Beech XL-23C Seminole
Beech L-23D (U-8D) Seminole
Beech RL-23D (RU-8D) Seminole
Beech L-23E (U-8E) Seminole
Beech L-23F (U-8F) Seminole
Beech Lightning (38P)
Beech (Morane-Saulnier) M-S 760
Beech Musketeer (23)
Beech Musketeer II (A23)
Beech Musketeer Custom (B23)
Beech Musketeer Custom III (A23A)
Beech Musketeer Sport (19A)

Beech Musketeer Sport III (A23-19)
Beech Musketeer Super (A24)
Beech Musketeer Super III (A23-24)
Beech Musketeer Super R (A24R) (Sierra)
Beech Premier (See: Raytheon)
Beech Queen Air 65
Beech Queen Air A65
Beech Queen Air 70
Beech Queen Air 79
Beech Queen Air 80
Beech Queen Air A80
Beech Queen Air B80
Beech Queen Air 88
Beech Sierra 200 (B24R)
Beech Sierra (C24R)
Beech Skipper (Model 77, PD285)
Beech SNB-1 Kansan
Beech SNB-2 Navigator
Beech SNB-2C Navigator
Beech SNB-3 Navigator
Beech SNB-3Q Navigator
Beech SNB-4 Navigator
Beech SNB-5 Navigator
Beech SNB-5P Navigator
Beech Sport 150 (B19)
Beech Sport 160 (Experimental)
Beech Starship (Turbojet Proposal)
Beech Starship 1 Prototype
Beech Starship 1 Model 2000
Beech Starship 1 Model 2000A
Beech Sundowner 180 (C23)
Beech T-1A Jayhawk (Beech 400A)
Beech T-6A Texan II
Beech (T-6A) CT-156 Harvard II
Beech T-6A Texan II, Export Variants
Beech (T-34) Model 45 Mentor
Beech YT-34 Mentor
Beech T-34A Mentor
Beech T-34B Mentor
Beech YT-34C Turbine Mentor
Beech T-34C Turbine Mentor
Beech AT-34 Turbine Mentor 420
Beech T-36A
Beech T-42A Cochise
Beech T-44A
Beech Travel Air (Badger) Model 95
Beech Travel Air Model B95
Beech Travel Air Model B95A
Beech Travel Air Model D95A
Beech Travel Air Model E95
Beech Traveller Mk.I (RAF Model 17 Staggerwing)
Beech Twin Quad (Model 34)
Beech Twin Bonanza 50
Beech Twin Bonanza B50
Beech Twin Bonanza C50
Beech Twin Bonanza D50
Beech Twin Bonanza E50
Beech Twin Bonanza F50
Beech Twin Bonanza G50
Beech Twin Bonanza H50
Beech Twin Bonanza J50
Beech NU-8F Seminole
Beech U-8G Seminole
Beech U-21A Ute
Beech RU-21D Ute
Beech RU-21E Ute
Beech U-21F Ute
Beech QU-22B Pave Eagle
Beech 17R Staggerwing
Beech A17F Staggerwing
Beech B17E Staggerwing
Beech B17L Staggerwing
Beech SB17L Staggerwing
Beech B17R Staggerwing
Beech C17 Staggerwing
Beech C17B Staggerwing
Beech SC17B Staggerwing
Beech C17E Staggerwing
Beech C17L Staggerwing
Beech C17L Staggerwing, NASM
Beech C17R Staggerwing
Beech D17 Staggerwing
Beech D17A Staggerwing
Beech D17R Staggerwing
Beech D17S Staggerwing
Beech SD17S Staggerwing
Beech D17W Staggerwing
Beech E17B Staggerwing
Beech SE17B Staggerwing
Beech E17L Staggerwing
Beech F17D Staggerwing
Beech G17S Staggerwing
Beech A18A
Beech D18C
Beech D18CT
Beech D18R
Beech D18S
Beech D18S, NASM
Beech E18S
Beech G18S
Beech H18
Beech H18 Turboprop
Beech M18
Beech M18R
Beech S18A
Beech 18A
Beech 18B
Beech 18D
Beech 18R
Beech 18S
Beech 40
Beech 81
Beech 99
Beech 99A
Beech B99
Beech C99
Beech 1900C
Beech 1900D

Beechcraft Hawker Corp (Wichita, KS)

Beechcraft Hawker Corp, a **Beech Aircraft Corp** subsidiary, was formed in 1970 to help market, assemble, and modify the Hawker Siddeley HS.125 business jet as the BH 125. Beechcraft Hawker continued in this role after Hawker Siddeley was absorbed by **British Aerospace** in 1977. In 1994, Beechcraft Hawker lost its corporate identity when Beech merged into **Raytheon Aircraft Co.** HS.125/BH 125 versions through the 700-series are listed under **Hawker Siddeley**. Versions from the 800- through 1000-series are listed under **British Aerospace**. The Horizon and all subsequent 125 developments appear under **Raytheon**.

Beech-Willard (Stanley Y. Beech & Charles F. Willard)(USA)

Beech-Willard Monoplane (1908)

Beecraft Associates Inc (See: Bee Aviation Associates, Inc)

de Beer, B. (Aéroplanes de Beer) (Belgium, France)

de Beer Type 1
de Beer Type 2
de Beer Type 3 Variable Incidence Monoplane
de Beer Type 4 Variable Incidence Monoplane

Beets, Glenn (Riverside, CA)

Beets G/B Special

Beijing Keyuan Light Aircraft Industrial Co Ltd (China)

Beijing Keyuan AD-200 Blue Eagle)

Bekas (See: Kasper, Withold)

Belcher Aerial Manufacturing Co Inc (D. S. Belcher)(Los Angeles, CA)

Belcher "California"

Bell (USA)

In 1935, Lawrence D. Bell and several other former Consolidated employees formed **Bell Aircraft Corporation** in Buffalo, NY. The corporate structure saw few changes until 1957, when **Bell Aerosystems** was formed; divisions included Bell Aircraft Corp (though the company was already moving out of the airplane-building business) and a new **Bell Helicopter Corp.** In 1960, Bell Aerosystems was reorganized as **Bell Aerospace**, a wholly owned subsidiary of **Textron, Inc**, and in 1982, the Bell Helicopter was renamed **Bell Helicopter Textron, Inc.** All aircraft of the various Bell companies are listed under **Bell**.

Bell ACLS (Air Cushion Landing System)
Bell Air Scooter (ACV)
Bell B-23 [ACV; AALC JEFF (B) Test]
Bell Bat (Bell Advanced TiltRotor; LHX)
Bell D-188A
Bell D-292 ACAP
Bell D-326
Bell XFL-1 Airabonita
Bell XFM-1 Airacuda
Bell YFM-1 Airacuda
Bell YFM-1A Airacuda
Bell YFM-1B Airacuda
Bell XUH-1 (XH-40) Iroquois (Huey)
Bell YUH-1 (YH-40) Iroquois (Huey)
Bell UH-1A (HU-1A) Iroquois (Huey)
Bell UH-1B (HU-1B) Iroquois (Huey)
Bell UH-1C (HU-1C) Iroquois (Huey)
Bell UH-1D (HU-1D) Iroquois (Huey)
Bell UH-1E (HU-1E) Iroquois (Huey)
Bell T/UH-1F (H-48) Iroquois (Huey)
Bell AH-1G Cobra
Bell UH-1H Iroquois (Huey)
Bell UH-1H Iroquois (Huey), NASM
Bell AH-1J Seacobra
Bell HH-1K Iroquois (Huey)
Bell UH-1L Iroquois (Huey)
Bell UH-1M Iroquois (Huey)
Bell UH-1N (CUH-1N/CH-135) Iroquois (Huey)
Bell UH-1P Iroquois (Huey)
Bell AH-1Q TOW Cobra
Bell AH-1S Cobra
Bell AH-1S (Upgraded) Cobra (AH-1E/F/P)
Bell AH-1T Seacobra
Bell AH-1W (AH-1T+) SuperCobra
Bell UH-1(X)(UH-1H Upgrade Proposal) Iroquois (Huey)
Bell YOH-4
Bell XH-12 (XR-12)
Bell YH-12B (YR-12B)
Bell YH-13 (YR-13) Sioux
Bell YH-13A (YR-13A) Sioux
Bell H-13B Sioux
Bell H-13C Sioux
Bell H-13D Sioux
Bell H-13E Sioux
Bell XH-13F Sioux
Bell H-13G Sioux
Bell H-13H Sioux
Bell H-13J (VH-13J) Sioux
Bell H-13J (VH-13J) Sioux, NASM
Bell TH-13M Sioux
Bell TH-13N Sioux
Bell OH-13S Sioux
Bell TH-13T Sioux
Bell XH-15
Bell TH-57A Sea Ranger
Bell OH-58A Kiowa
Bell OH-58C Kiowa
Bell OH-58D Aeroscout
Bell YAH-63 (Model 409)
Bell XHSL-1
Bell HSL-1
Bell HTL-1 Sioux
Bell HTL-2 Sioux
Bell HTL-3 Sioux
Bell HTL-4 Sioux
Bell HTL-5 Sioux
Bell HTL-6 Sioux
Bell HTL-7 Sioux
Bell HUL-1 Sioux
Bell Hypersonic Transport (Proposal)
Bell L-39
Bell LACV-30 (ACV)
Bell TP-39 Airacobra
Bell XP-39 Airacobra
Bell YP-39 Airacobra
Bell XP-39B Airacobra
Bell P-39C Airacobra
Bell P-39D Airacobra
Bell XP-39E Airacobra
Bell P-39F Airacobra
Bell P-39K Airacobra
Bell P-39L Airacobra
Bell P-39M Airacobra
Bell P-39N Airacobra
Bell P-39Q Airacobra
Bell P-39Q Airacobra, NASM
Bell P-39Q Airacobra Racer
Bell (P-39) Airacobra I (Caribou)
Bell P-400 Airacobra
Bell XP-59 (Model 16)
Bell XP-59A Airacomet
Bell XP-59A Airacomet, NASM
Bell YP-59A Airacomet
Bell P-59A Airacomet
Bell P-59B Airacomet
Bell P-63 Kingcobra, Butterfly Tail
Bell XP-63 Kingcobra
Bell XP-63A Kingcobra
Bell P-63A Kingcobra
Bell P-63A Kingcobra, NASM
Bell RP-63A Kingcobra
Bell P-63C Kingcobra
Bell P-63C Kingcobra Racer
Bell RP-63C Kingcobra
Bell P-63D Kingcobra
Bell P-63E Kingcobra
Bell P-63F Kingcobra
Bell P-63F Kingcobra Racer
Bell RP-63G Kingcobra
Bell XP-77
Bell XP-83
Bell Quad TiltRotor
Bell SES-100B (ACV)
Bell SK-1 (ACV)
Bell SK-3 Carabao (ACV)
Bell SK-5 (ACV)
Bell SK-6C (ACV)
Bell SK-9 (ACV)
Bell SK-10 (ACV)
Bell SKMR-1 (ACV)
Bell Trailing-Rotor Transport Proposal
Bell XV-3
Bell XV-15
Bell X-1
Bell X-1 "Glamorous Glennis"
Bell X-1 "Glamorous Glennis," NASM
Bell X-1A
Bell X-1B
Bell X-1C
Bell X-1D
Bell X-1E
Bell X-2
Bell X-5
Bell X-14
Bell X-16
Bell X-22A
Bell 9 (Attack Bomber Proposal)
Bell 30 #1
Bell 30 #2
Bell 30 #3
Bell 42
Bell 47
Bell 47A
Bell 47B
Bell 47B-3
Bell 47D
Bell 47D-1
Bell 47G
Bell 47G "Wing Ding"
Bell 47G-1
Bell 47H
Bell 47J
Bell 65 ATV (Air Test Vehicle)(Bell VTOL)
Bell 200 (Mock-UP)
Bell 204B
Bell 205A
Bell 206A JetRanger
Bell 206B JetRanger
Bell 206L LongRanger
Bell 206L LongRanger "Spirit of Texas," NASM
Bell 207 Sioux Scout
Bell 208 Twin Delta (Modified UH-1D; UH-1H Prototype)

Bell 209 (AH-1G Prototype)
Bell 212 Twin Two-Twelve
Bell 214B
Bell 214ST
Bell 222
Bell 222B
Bell 222UT
Bell 308 Kingcobra
Bell 400
Bell 406CS Combat Scout
Bell 406L-400 LongRanger IV
Bell 406LT TwinRanger IV
Bell 407
Bell 412
Bell 427
Bell 430
Bell 533 (Modified YUH-1/YH-40)
Bell 608 LASH (Light Antisubmarine Helicopter)
Bell (Canada) 7380 "Voyageur" (ACV)
Bell (Canada) 7501 "Viking" (ACV)
Bell Boeing Civil TiltRotor Proposal
Bell Boeing V-22 Osprey

Bell Aeromarine (UK)
Bell Aeromarine Flitzer

Bell, Alexander Graham (USA/ Canada)*(See also: Aerial Experiment Assoc)*
Bell (Alexander Graham) Hydrofoil

Bellamy (France)
Bellamy (France) 1906 Biplane Hydro
Bellamy (France) 1908/09 Monoplane

Bellamy, James Exum (Enfield, NC)
Bellamy (NC) 2-Seat Biplane

Bellanca (USA)

After helping to design the first Italian aircraft to fly, Giuseppe M. Bellanca emigrated to the US in 1911, soon building his first American aircraft in Brooklyn, NY, and teaching himself to fly. He worked with several US companies before forming **Bellanca Aircraft Corporation of America** in the 1920s, and **Bellanca Aircraft Corporation** in 1927. Bellanca produced numerous innovative and record-setting designs through the 1920s, 1930s, and 1940s, but by 1955 the company assets were acquired by **Northern Aircraft**, with the Bellanca name dropped altogether in 1959. Northern would become **Downer Aircraft Industries**, then **International Aircraft Manufacturing Inc** (or **Inter-Air**). Inter-Air formed **Bellanca Sales Co**, which acquired **Champion Aircraft Corp** and reorganized as **Bellanca Aircraft Corp** in 1970. Bellanca quickly became a subsidiary of **Anderson, Greenwood, and Co**, taking control of **Eagle Aircraft Co** in 1979. Bellanca fell upon hard times in the 1980s, in 1982 selling rights, tooling, and partially completed Champion series aircraft to **Champion Aircraft Co, Inc**, and Viking series aircraft to **Viking Aviation**. The final Bellanca company, **Bellanca, Inc**, formed out of Viking in 1984, closing its doors in 1988.

A smaller company, **Bellanca Aircraft Engineering Inc**, which was formed to explore the use of fiberglass in aircraft design, operated concurrently from 1956 through 1986.

Anderson, Greenwood, and Co, and Eagle Aircraft Co designs are filed under their respective companies. Pre-1954 development of the highly successful Champion series of aircraft appears under **Aeronca**, with subsequent development filed under **Champion Aircraft Corp**, **Bellanca, Champion Aircraft Co**, and **American Champion Aircraft**. All designs of the various Bellanca companies appear under **Bellanca**.

Bellanca Airbus P-100
Bellanca Airbus P-200
Bellanca Airbus P-300
Bellanca Aircruiser 66-67
Bellanca Aircruiser 66-70
Bellanca Aircruiser 66-75
Bellanca Aircruiser 66-76
Bellanca Aries T-250
Bellanca Y1C-27
Bellanca C-27A
Bellanca C-27B
Bellanca C-27C
Bellanca C.D.
Bellanca C.E.
Bellanca C.F. Air Sedan
Bellanca C.F. Air Sedan, NASM
Bellanca Champ 7ACA
Bellanca Citabria
Bellanca Cruisair 14-12-F3
Bellanca Senior Cruisair 14-13
Bellanca Senior Cruisair 14-13, NASM
Bellanca Cruisemaster 14-19, 260
Bellanca de Havilland DH-4 Mailplane, Modified
Bellanca Decathlon
Bellanca Experimental Fighter Design
Bellanca J
Bellanca J "Green Flash"
Bellanca J "North Star"
Bellanca J "Pathfinder"
Bellanca J "Santa Rosa Maria"
Bellanca J-2, 1931 Endurance Flight
Bellanca J-2 "Abyssinia"
Bellanca J-3
Bellanca J-3 "Hochi-Hinomaru"
Bellanca J-3-500 "Santa Lucia" (Special Long Distance)
Bellanca J-300
Bellanca J-300 "Cape Cod"
Bellanca J-300 "Leonardo da Vinci"
Bellanca J-300 "Liberty"
Bellanca J-300 "Olympia"
Bellanca J-300 "Warsaw"
Bellanca JE-1
Bellanca Junior 14-7
Bellanca Junior 14-9 (Cruisair Jr)
Bellanca Junior 14-9L (Cruisair Jr)
Bellanca Junior 14-10 (Cruisair Jr)
Bellanca Junior 14-12 (Cruisair Jr)
Bellanca K
Bellanca K "Enna Jettick"
Bellanca K "Roma"
Bellanca YO-50
Bellanca Pacemaker
Bellanca Pacemaker "Tradewind"
Bellanca Pacemaker (Diesel)
Bellanca Pacemaker CH (CH-200)
Bellanca Pacemaker CH (CH-200) "Reliance"
Bellanca Pacemaker CH-300
Bellanca Pacemaker CH-300, Kinney Blind Flight, 20 Mar 1933
Bellanca Pacemaker CH-300 "Lituanica"
Bellanca Pacemaker E
Bellanca Pacemaker Freighter PM-300
Bellanca Pacemaker 300-W
Bellanca Pacemaker 300-W, Bjorkvall
Bellanca Senior Pacemaker 31-42 (Pacemaker Series Eight)
Bellanca Parasol
Bellanca XRE-1
Bellanca XRE-2
Bellanca XRE-3
Bellanca RE-3
Bellanca Scout
Bellanca XSE-1
Bellanca XSE-2
Bellanca Skyrocket CH-400 (Pacemaker 400)
Bellanca Skyrocket CH-400 "American Nurse"
Bellanca Skyrocket CH-400 "Miss Veedol"
Bellanca Skyrocket D
Bellanca Skyrocket F
Bellanca Skyrocket F-2
Bellanca Senior Skyrocket 31-50
Bellanca Senior Skyrocket 31-55

Bellanca 19-25 Skyrocket II
Bellanca (AviaBellanca) 19-25 Skyrocket III
Bellanca XSOE-1
Bellanca Tandem "Blue Streak"
Bellanca T.14-14 Trainer
Bellanca Viking
Bellanca Super Viking
Bellanca Turbo Viking
Bellanca Viking 300
Bellanca WB-1 (CG)
Bellanca WB-2 "Miss Columbia"
Bellanca 28-70 "Irish Swoop"
Bellanca 28-90 Flash "Dorothy"
Bellanca 28-90 (28-110) Flash
Bellanca 28-92 "Alba-Iulia 1918"
Bellanca 77-140 (77-143, 77-147, 77-320)

Bellanger Frères (France)
Bellanger Bille Variable Wing Biplane
Bellanger (Denhaut) Twin-Engined Flying Boat
Bellanger 1922 Biplane Glider

Belyaev (Russia)
Belyaev DB-LK

Ben Jansson *(See: BJ)*

Bendix Aviation Corp (Detroit, MI)
Bendix Tractor Monoplane
Bendix Model 51 (Pusher Monoplane)

Bendix Helicopter, Inc (Stratford, CT)
Helicopters, Inc was established by Vincent Bendix in 1943, renamed **Bendix Helicopter, Inc** (not affiliated with **Bendix Aviation**) in 1944, and renamed **Helicopters, Inc** in 1948. The company was purchased by **Gyrodyne Co of America** in 1949. All Helicopters, Inc and Bendix Helicopter products are listed under **Bendix Helicopter**. The post-1949 development of Bendix Helicopter Model J-2 is listed under **Gyrodyne GCA-2**.
Bendix Helicopter Model G
Bendix Helicopter Model J
Bendix Helicopter Model J-1
Bendix Helicopter Model J-2 (See also: Gyrodyne GCA-2)
Bendix Helicopter Model J-3
Bendix Helicopter Model K

Bénégent (France)
Bénégent Sirius #1 (1910 Monoplane)
Bénégent Sirius #2 (1910 Monoplane)

Beneš-Mráz Továrna Letadel (Beneš-Mráz Aircraft Factory) (Czechoslovakia)
Beneš-Mráz Be 50 Beta-Minor
Beneš-Mráz Be 51 Beta-Minor
Beneš-Mráz Be 52 Beta-Major
Beneš-Mráz Be 55
Beneš-Mráz Be 56 Beta-Major
Beneš-Mráz Be 60 Bestiola
Beneš-Mráz Be 62
Beneš-Mráz Be 64
Beneš-Mráz Be 74
Beneš-Mráz Be 75
Beneš-Mráz Be 87
Beneš-Mráz Be 150 Beta-Junior
Beneš-Mráz Be 156 Beta-Major
Beneš-Mráz Be 250 Beta-Major
Beneš-Mráz Be 251 Beta-Major
Beneš-Mráz Be 252 Beta-Scolar
Beneš-Mráz Be 252C Beta-Scolar
Beneš-Mráz Be 350 Beta-Minor
Beneš-Mráz Be 500 Bibi
Beneš-Mráz Be 501 Bibi
Beneš-Mráz Be 502 Bibi
Beneš-Mráz Be 520 Sokolu
Beneš-Mráz Be 550 Bibi
Beneš-Mráz Be 555 Super-Bibi
Beneš-Mráz Be 580

Bennett (Grover Bennett & Son) (Keosauqua, IA)
Bennett (IA) Seraph

Bennett Aircraft (Dallas, TX)
Bennett (TX) Executive

Bennett Aircraft Corp *(See: Globe Aircraft Corp)*

Bennett Aviation Ltd (New Zealand)
Bennett (New Zealand) PL.11 Airtruck

Bennett-Carter (George Bennett & Richard Carter)(MS)
Bennett-Carter Dottie S

Benoist, Thomas W. (St Louis, MO)
Benoist 1912 Headless
Benoist 1912 Tractor
Benoist 1915 Flying Boat
Benoist 1916 Flying Boat
Benoist 1917 Flying Boat (#106)
Benoist Type XII
Benoist Type XII #32
Benoist Type XIII
Benoist Type XIV
Benoist Type XIV #43 "Lark of Duluth"
Benoist Type E-17
Benoist Type F-17 (#103)
Benoist Type G-17
Benoist Type H-17

Bensen Aircraft Corp (Raleigh, NC)
Bensen B-2
Bensen B-4 Sky-Scooter
Bensen B-5
Bensen B-6
Bensen B-6, NASM
Bensen B-7M
Bensen B-7MC
Bensen B-7W
Bensen B-8 Gyro-Glider
Bensen B-8M Gyro-Copter
Bensen B-8M Gyro-Copter "Spirit of Kitty Hawk," NASM
Bensen B-8MW Hydro-Copter
Bensen B-8V Gyro-Copter
Bensen B-8W Hydro-Glider
Bensen B-9 Little Zipster
Bensen B-10 Prop-Copter
Bensen B-11M Kopter-Kart
Bensen B-12 Sky-Mat
Bensen B-13
Bensen X-25B Gyro-Glider

Benton, John Frederick "Johnny" (UK)
Benton B1 Air-Car
Benton B2 Air-Car
Benton B3 Tractor
Benton B4 Tractor
Benton B5 Tractor
Benton B6 Tractor
Benton B7 Tractor

Benz (Hermann)(Czechoslovakia)
Benz (Hermann) Sailplane

Berckmans Aeroplane Co (New York, NY)
Berckmans B-2
Berckmans B-3
Berckmans BL-12
Berckmans Speed Scout

Bereznyak & Isaev (Russia)
Bereznyak & Isaev RP-318 (BI-1)

Berg *(See: Aviatik)*

Berg & Storm (Denmark)
Berg & Storm B&S I Monoplane
Berg & Storm B&S II Monoplane
Berg & Storm B&S III Monoplane

Bergamaschi *(See: CAB)*

Berger (France)
Berger-Gardey Glider

Berger 1913 Monoplane

Bergholt (Minneapolis, MN)
Bergholt Airsport
Bergholt Glider

Bergousi (Italy)
Bergousi Aidea

Beriev (Russia)
Beriev A-50 *Mainstay*
Beriev Be-R-1
Beriev Be-2 (KOR-1)
Beriev Be-4 (KOR-2)
Beriev Be-6 *Madge*
Beriev Be-8
Beriev Be-10 *Mallow* (M-10)
Beriev Be-12 Tchaika (Seagull) *Mail*
Beriev Be-30 *Cuff*
Beriev Be-32 *Cuff*
Beriev Be-40 Albatross
Beriev Be-42 Mermaid
Beriev Be-200
Beriev GST (MP-7; License-Built Consolidated Model 28)
Beriev MBR-2
Beriev MBR-2*bis*
Beriev MBR-7 (MS-8)
Beriev MDR-5 (MS-5)
Beriev MP-1 (Civil MBR-2)

Berkel *(See: van Berkel)*

Berkshire Manufacturing Co (Oakwood, NJ)
Berkshire Concept 70 (Sailplane)

Berliaux et Salètes (France)
Berliaux et Salètes 1909 Glider #1
Berliaux et Salètes 1909 Glider Pourquoi Pas? II (Why Not? II)

Berlin Doman Helicopters, Inc
(See: Doman)

Berliner, Emile (Alexandria, VA)
Berliner (Emile) Demoiselle Type
Berliner (Emile) Flying Boat
Berliner (Emile) Pusher Biplane
Berliner (Emile) Helicopter, 1910
Berliner (Emile) Helicopter, 1919/20
Berliner (Emile) Helicopter, 1921
Berliner (Emile) Helicopter, 1922
Berliner (Emile) Helicopter, 1923 Triplane
Berliner (Emile) Helicopter, 1925 Sesquiplane
Berliner (Emile) Rocket Plane (1903)

Berliner, Henry (Alexandria, VA)
Berliner (Henry) Metal High-Wing Cabin Monoplane
Berliner (Henry) Metal Low-Wing Cabin Monoplane

Berliner & Warner Helicopter
(See: Berliner Aircraft Co, Inc)

Berliner & Williams Helicopter
(See: Berliner, Emile)

Berliner Aircraft Co, Inc (Alexandria, VA)
Berliner Cabin Transport
Berliner CM-4
Berliner CM-4B
Berliner CM-5W
Berliner SC-M2

Berliner Segelflug Verein (Germany)
Berliner (BSV) "Luftikus"

Berliner-Joyce Aircraft Corp (B/J Aircraft Corp)(Dundalk, MD)
Berliner-Joyce XFJ-1
Berliner-Joyce XFJ-2
Berliner-Joyce XF2J-1
Berliner-Joyce XF2J-2
Berliner-Joyce XF3J-1
Berliner-Joyce XOJ-1
Berliner-Joyce OJ-2
Berliner-Joyce XOJ-3
Berliner-Joyce XP-16
Berliner-Joyce Y1P-16 (PB-1)
Berliner-Joyce 29-1

Bernard (Établissements Adolphe Bernard, Société Avions Bernard) (France)
Bernard A.B.1 Bn2
Bernard A.B.2 Bn2
Bernard A.B.3 Mail Plane
Bernard A.B.4 Airliner
Bernard CPA 1M (See: CPA)
Bernard S.A.B. C1 Fighter
Bernard H 52 C1 Hydravion
Bernard H.V. Hydravion Series
Bernard H.V.47
Bernard H.V.120
Bernard H.V.220
Bernard S.I.M.B. V-1
Bernard S.I.M.B. V-2
Bernard 12 C1
Bernard 14 C1
Bernard 15 C1
Bernard 18 T
Bernard 20 C1
Bernard 60 T
Bernard 80 G.R.
Bernard 80 G.R. "Jean-Hubert"
Bernard 80 G.R. "Oiseau-Tango" (Bird)
Bernard 81 G.R.
Bernard 82 B3
Bernard 190 T
Bernard 191 T No.01 "France"
Bernard 191 Gr
Bernard 192 T
Bernard 193 T
Bernard 197 Gr
Bernard 200 Series
Bernard 201 T

Bertelsen (Neponset, IL)
Bertelsen Aeromobile

Berthaud (France)
Berthaud Monoplane W
Berthaud Moreau-Berthaud Monoplane
Berthaud Prini-Berthaud 1908 Biplane

Bertin, Léonce (France)
Bertin Helicoplan
Bertin 1907 Helicopter
Bertin 1910 Monoplane
Bertin 1912 Monoplane
Bertin-Lieber Helicopter

Bertrand, René (France)
Bertrand:
1909 Monoplane (Unic-Bertrand)
No.2 Monoplane (Unic-Bertrand)
No.3 Monoplane (Unic-Bertrand)

Besler (Seattle, WA)
Besler (Modified Travel Air 2000)

Besnier (France)
Besnier Ornithopter (1678)

Besson (Aéroplanes Marcel Besson) (Société de Constructions Aéronautiques et Navales Marcel Besson)(France)
Besson H 3 Triplane Hydravion
Besson H 5 Triplane Hydravion
Besson H 6 Triplane Hydravion
Besson MB 26
Besson MB 35 Hydravion
Besson MB 41/411

Besson, Marcel (France)
Besson 1911 Canard Monoplane
Besson 1912 Canard Floatplane
Besson 1912 Flyingboat
Besson et Pajo 1912 Biplane

Bestetti (Italy)
Bestetti C.3

***Beta Air Ltd** (See: Beriev Be-200)*

Bethlehem Aircraft Corp (BACO) (Bethlehem, PA)
Bethlehem 2-Seater (1921)
Bethlehem Skylark

Bezobrazov, A. A. (Russia)
Bezobrazov 1914 Experimental Triplane

BFMW (Bayeriche Flug und Motorenwerke)(Germany)
BFMW Monoplane (1912)

BFW (1916)(Bayerische Flugzeug-Werke)(Germany)
In 1916, **Albatros** established the **Bayerische Flugzeug Werke** (**BFW**) to subcontract production. With the end of World War I and the Treaty of Versailles restrictions on German aviation, BFW was forced to close its doors. This BFW should not be confused with the unrelated **Bayerische Flugzeugwerke AG** (**BFW**) which was established in 1926. For products of the original BFW, see **BFW (1916)**. For products of the later company, see **BFW (1926)**.
BFW (1916) CL.I
BFW (1916) CL.II
BFW (1916) CL.III
BFW (1916) Monoplane (1918)
BFW (1916) N.I

BFW (1926) (Bayerische Flugzeugwerke AG) (Germany)
In July 1926 the **Bayerische Flugzeugwerke AG** (**BFW**) was established to develop the designs of the moribund **Udet Flugzeugbau GmbH**. Under pressure from the Bavarian government in 1927, BFW and **Messerschmitt Flugzeugbau GmbH** (est. 1923) pooled their resources, with Messerschmitt providing the design and development office and BFW providing manufacturing facilities; Messerschmitt retained all design and patent rights. In 1931, BFW went into receivership. In 1933, Messerschmitt amassed sufficient capital to revive BFW under his direction, effectively merging the two companies. In 1938, the companies reorganized as **Messerschmitt AG.** This BFW should not be confused with the unrelated **Bayerische Flugzeug Werke** established in 1916. For products of the original BFW, see **BFW (1916)**. For products of the later company predating 1933, see **BFW (1926)**. For designs of the Udet company preceding the U 12a, see **Udet**. For designs originating at Messerschmitt, see **Messerschmitt**.
BFW (1926):
BFW-1 Sperber (Sparrowhawk)
(Udet) U 12A Flamingo
(Udet) U 13A Bayern (Bavaria)

BG Aviation Associates (Springfield, MA)
BG BeeGee Baby Sportster (1948)

Bharat Heavy Electricals Ltd (India)
Bharat LT-IIM Swati

Biedenmeister, Karl A. (Indianapolis, IA)
Biedenmeister 1926 3-Seat

Bielany School Students (Poland)
Bielany School Students 1924 Glider

Biggs (Moore, OK)
Biggs A Special Racer

Bikle, Paul (USA)
Bikle T-6 (Modified Schreder HP-14)

Billard (France)
Billard 1911 Military Aircraft

***Bille (S.A.C.A.N.A.)(France)**(See also: Bellanger)*
Bille S.A.C.A.N.A. Triplane Bombers

Billie Aero Marine (France)
Billie Petrel

Billing, Eardley (UK)
Billing 1911 Farman-Type Biplane

Bilski (Poland)
Bilski Mewa (Gull)

Binder Aviatik (Germany)
Binder CP 301 Smaragd (Emerald) (License-built Piel Emeraud)

Biot, Gaston (France)
Biot 1861 Conical Tailless Kite
Biot 1880 Stabilized Kite
Biot-Massia 1879 Glider
Biot-Massia 1882 Glider

Birchan (UK)
Birchan "Beetle"

Bird Aircraft...
During the 1920s and 1930s, two separate and unrelated companies operated under the name Bird Aircraft. The first of these companies operated briefly from San Diego, California as the **Bird Aircraft Co** in 1926. This company's 2-Seater touring biplane is listed under **Bird Aircraft Co (San Diego)**. In 1928 the **Brunner-Winkle Aircraft Co** was established in New York. Brunner-Winkle was renamed **Bird Aircraft Corp** in 1930. In 1933 all rights were sold to **Perth Amboy Title Co**; no further aircraft were produced. Listings for this second company appear as **Bird Aircraft Corp (NY)**. In the 1960s, a third Bird company modified Consolidated PBY Catalinas in their California factory.

Bird Aircraft Co (San Diego, CA)
Bird (Aircraft Co) Two-Seater Touring Biplane

Bird Aircraft Corp (New York, NY)
Bird (Aircraft Corp) Model A
Bird (Aircraft Corp) Model B
Bird (Aircraft Corp) Model BK
Bird (Aircraft Corp) Model C
Bird (Aircraft Corp) Model C-J
Bird (Aircraft Corp) Model E
Bird (Aircraft Corp) Model F
Bird (Aircraft Corp) Model 59

Bird Corp (Palm Springs, CA)
Bird (Corp) Innovator (PBY-6A Conversion)

Bird Wing Commercial Aircraft Corp (St Joseph, MO)
Bird Wing Imperial 6
Bird Wing Imperial 10

Birdman Aircraft Inc (Daytona Beach, FL)
Birdman Aircraft TL-1 (Powered Sailplane Kit)

Birdman Enterprises Inc (Canada)
Birdman Enterprises WT-11 Chinook

Biro de Ditro, Desiderio (Argentina)
Biro de Ditro "Patoruzu"

Biro de Ditro "Regina" (Glider)
Biro de Ditro 1928 Monoplane (Rebuilt "Regina")

Bishop-Barker Aeroplanes (Canada)
Bishop-Barker JN-4Can Rebuilds

Bisnovat (Russia)
Bisnovat SK
Bisnovat SK-2

B/J (See: Berliner-Joyce)

BJ (Ben Jansson)(Sweden)
BJ-1 Dynamite
BJ-1b Duster

BKLAIC (Beijing Keyuan Light Aircraft Industrial Co, Ltd)(China)
BKLAIC AD-2000

Blackburn (UK)
Blackburn Aeroplane Co built its first aircraft in 1910. The company registered as **Blackburn Aeroplane and Motor Co Ltd** in 1914 and reregistered as **Blackburn Aircraft Ltd** in 1936. The absorption of **General Aircraft Ltd** in 1949 resulted in the formation of **Blackburn and General Aircraft Ltd.** In 1959 The **Blackburn Group Ltd** was organized with its aircraft subsidiary known as **Blackburn Aircraft Ltd.** The Blackburn Group became part of the **Hawker Siddeley Group** in 1960, and in 1963 Blackburn Aircraft Ltd became part of a new **Hawker Blackburn Division.** The Blackburn name was dropped after a 1965 Hawker Siddeley reorganization. Most Blackburn products are listed under **Blackburn** with some later aircraft (such as the Buccaneer) also appearing in the **Hawker Siddeley** and **British Aerospace** sections.
Blackburn Airedale (R.2)
Blackburn B-2
Blackburn B-3 (M.1/30)
Blackburn B-5 Baffin
Blackburn B-6 Shark
Blackburn B-7
Blackburn B-9 (H.S.T.10, C.A.21A)
Blackburn B-20 Seaplane
Blackburn B-24 Skua
Blackburn B-25 Roc
Blackburn B-26 Botha
Blackburn B-37 Firebrand
Blackburn B-48 (Y.A.1) Firecrest
Blackburn B-54 (Y.A.5)
Blackburn B-88 (Y.B.1)
Blackburn B-101 Beverley
Blackburn B-103 Buccaneer (N.A.39)
Blackburn Beagle (B.T.1)
Blackburn Blackburd
Blackburn Blackburn (R.1)
Blackburn Bluebird (L.I)
Blackburn Bluebird II (L.IA)
Blackburn Bluebird III (L.IB)
Blackburn Bluebird IV (L.IC)
Blackburn C.A.15
Blackburn C.A.15C
Blackburn Cubaroo (T.3)
Blackburn Dart (T.2)
Blackburn Type E Monoplane (1912)
Blackburn F.3 (F.7/30)
Blackburn G.P. (General Purpose)
Blackburn Iris (R.B.1)
Blackburn Kangaroo (R.T.I, Reconnaissance Torpedo I)
Blackburn Type L (2-Seat Seaplane)
Blackburn Lincock (F.2)
Blackburn Mercury
Blackburn Monoplane, 1909
Blackburn Monoplane, 1911
Blackburn Monoplane, 1913, Single-Seat
Blackburn N.1B
Blackburn N.2/42
Blackburn Nautilus (2F.1)
Blackburn Nile (C.B.2)
Blackburn "Pelican" (See: de Havilland D.H.50J)
Blackburn Pellet
Blackburn Perth Flying Boat (R.B.3A)
Blackburn Ripon (T.5)
Blackburn Segrave (B.1)
Blackburn Sidecar
Blackburn Sprat (T.R.1)
Blackburn Swift (T.1)
Blackburn Sydney (R.B.2)
Blackburn T.B. (twin-engine, twin-hull)
Blackburn T.7B (3MR4)
Blackburn Triplane (Pusher, 1916)
Blackburn Turcock (F.1)
Blackburn Universal (G.A.L.60)
Blackburn Velos (T.3)
Blackburn "White Falcon" (1915 Monoplane)
Blackburn Type I Monoplane, two-Seat

Blaicher, Capt Michal (Poland)
Blaicher B 1 Glider
Blaicher B-38 Glider

Blanc, Henri (France)*(See also: Barlatier & Blanc)*
Blanc (Henri) Monoplane

Blanc, Maurice (France)
Blanc (Maurice) 1914 Monoplane

Blanchard (Constructions Aéronautiques Blanchard) (France)
Blanchard B3 Hydravion (Blanchard BrdlB3)(1923)
Blanchard C1 Hydravion (1923)

Blanchard, James F. (USA)
Blanchard (J. F.) Monoplane, Military A

Blanchard, Jean-Pierre (France)
Blanchard (Jean-Pierre) Vaisseau Volant Ornithopter (1781)

Blanchard et Viriot (Ferdinand Blanchard et Jack Viriot)(France)
Blanchard et Viriot:
1909 Monoplane No.1
1909 Monoplane No.2

Bland, Edward Floyd (Seattle, WA)
Bland (Edward) 1947 Amphibian

Bland, Miss Lilian (UK)
Bland (Lilian) Mayfly (1911 Biplane)

Blard, Lt (France)
Blard 1912 Canard Monoplane

Blazynski, Alojzy (Poland)
Blazynski Polon Glider

Bleecker *(See: Curtiss-Bleecker Helicopter)*

Blériot (France)
Louis Blériot began his aviation experiments in 1906. By World War I his **Blériot-Aéronautique** was one of France's leading aeronautical firms. Company factories were nationalized in 1937, with the design organization continuing independent operations until German occupation in World War II. Designs of Blériot and his company are listed as **Blériot**. (See also **SPAD.**)
Blériot (1) Ornithopter (1903)
Blériot II (Voisin-Built)
Blériot II
Blériot-Voisin Blériot III
Blériot-Voisin Blériot IV
Blériot V
Blériot VI ("La Libellule")("Dragonfly")
Blériot VII
Blériot VII*bis*

Blériot VIII
Blériot VIII*bis*
Blériot VIII*ter*
Blériot IX
Blériot X
Blériot XI, Original Configuration (1909)
Blériot XI, Cable Hook-on System (Adolphe Pégoud)
Blériot XI, Blériot Cross-Channel
Blériot XI, 1910, Léon Morane
Blériot XI, 1910, 2nd Channel Crossing, de Lesseps
Blériot XI, 1910 Quimby Cross Channel
Blériot XI Militaire
Blériot XI Pingouin (Penguin)
Blériot XI Racer
Blériot XI Racer, Domenjoz No.340, NASM
Blériot XI-2
Blériot XI-2 Artillerie
Blériot XI-2 Artillerie, Civil Version
Blériot XI-2 Hydroaeroplane
Blériot XI-2 Military
Blériot XI-2 Military Trainer
Blériot XI-2*bis*
Blériot XI-2*bis* Militaire
Blériot XI-2 BG (Blériot-Gouin) Vision Totale
Blériot XI-3
Blériot XI-3, 1911, Reims, No. 26
Blériot XII, Original Configuration
Blériot XII Racer
Blériot XII, Triple-Stabilizer Tail Mod
Blériot XIII (l'Aérobus)
Blériot XXI
Blériot XXIII (XI*bis*)
Blériot XXIII (XI*bis*), 1911 Eastchurch
Blériot XXIV (Aéronef, Berline)(Coach)
Blériot XXVII
Blériot XXXIII Canard
Blériot XXXVI Blindé (Armored)
Blériot XXXIX
Blériot XL
Blériot XLII Canard (Type 42)
Blériot XLIII
Blériot XLV
Blériot LIII (Type 53)
Blériot Type 56
Blériot LXVII (Type 67)
Blériot LXXI (Type 71)
Blériot LXXIII Bn3 (Type 73)
Blériot 74 "Mammouth" (Mammoth)
Blériot 75 ("Aérobus")
Blériot 76 (Never Built)
Blériot 105
Blériot 110 No.01 "Joseph le Brix"
Blériot 111
Blériot 111 "Le Sagittaire" (Sagitarius)
Blériot 111*bis*
Blériot 115
Blériot 115B
Blériot 117
Blériot 125
Blériot 127
Blériot 135
Blériot 137
Blériot 152
Blériot 155
Blériot 165
Blériot 165 "Léonard de Vinci"
Blériot 165 "Octave Chanute"
Blériot 165
Blériot 175
Blériot 195
Blériot 290
Blériot 5190 Hydravion Santos-Dumont
Blériot Guillemin J.G. 10
Blériot Guillemin J.G. 10, Open Cockpit
Blériot Guillemin J.G. 10, Enclosed Cockpit
Blériot Guillemin J.G. 40
Blériot SPAD (See: SPAD)
Blériot Hydravions Blanchard (See: Blanchard)

Blériot-SPAD (See: SPAD)

Blindermann
Blindermann Monoplane (c.1911)

Blindermann-Mayeroff
Blindermann-Mayeroff Monoplane

Bloch (Avions Marcel Bloch)(France)
Bloch 7
Bloch 8
Bloch 9
Bloch 10
Bloch 11
Bloch XV
Bloch 60
Bloch 61
Bloch 70
Bloch 71
Bloch 80
Bloch 81
Bloch 90
Bloch 120
Bloch 130
Bloch 131
Bloch 133
Bloch 141
Bloch 150 (MB 150)
Bloch 151 (MB 151)
Bloch 152 (MB 152)
Bloch 155 (MB 155)
Bloch 157 (MB 157)
Bloch 160
Bloch 161 (See: SNCASE 161)
Bloch 162 (See: SNCASE 162)
Bloch 174
Bloch 175
Bloch 200
Bloch 210
Bloch 211
Bloch 220
Bloch 300
Bloch 700 (MB 700)(See: SNCASO)

Blohm und Voss (Germany)
Hamburger Flugzeugbau GmbH was established in July 1933 as a subsidiary of Blohm und Voss shipyards; aircraft were designated with "Ha" prefixes. In September 1937 the company was renamed **Blohm und Voss**, with new production carrying "Bv" designations. Blohm und Voss ceased aircraft production at the end of World War II. In 1955 **Hamburger Flugzeugbau GmbH (HFB)** was reestablished as a subsidiary of Blohm und Voss, building aircraft with "HFB" designations. The company merged with **Messerschmitt-Bölkow** in May 1969 to form **Messerschmitt-Bölkow-Blohm (MBB)**. Aircraft built before 1946 (using "Ha" or "Bv" designations) appear under **Blohm und Voss**, with post-WWII production under **HFB**.

Blohm und Voss Bv 40
Blohm und Voss Ha 135
Blohm und Voss Ha 136
Blohm und Voss Ha 137
Blohm und Voss Bv 138
Blohm und Voss Ha 139
Blohm und Voss Ha 139 A
Blohm und Voss Ha 139 B
Blohm und Voss Ha 140
Blohm und Voss Bv 141
Blohm und Voss Bv 142
Blohm und Voss Bv 144
Blohm und Voss Bv 155
Blohm und Voss Bv 155, NASM
Blohm und Voss Bv 170
Blohm und Voss Bv 222 Wiking
Blohm und Voss Bv 237
Blohm und Voss Bv 238
Blohm und Voss P.188-01
Blohm und Voss P.194-01
Blohm und Voss P.215-02

Blom, Conrad C. (Schenetady, NY)
Blom Sport Coupe
Blom 1928 Three-Place Monoplane

Blomqvist and Nyberg (Atle Blomqvist and Emil Nyberg)
Blomqvist and Nyberg 1926 Monoplane

Blondin (J. A.)(Los Angeles, CA)
Blondin (J. A.), Patents

Blue Bird (Morgantown, NC)
Blue Bird Parasol (1920s)

Blue Yonder Aviation (Canada)
Blue Yonder E-Z Flyer Ultralight

Blume (Germany)
Blume 1924 Glider

BMW (Bayerische Motorenwerke) (Germany)
BMW Biplane

Bobenreith (France)
Bobenreith Biplane

Boccaccio (Avions Paul Boccaccio) (France)
Boccaccio B.S. 2 (c.1924)

Bodiansky (France)
Bodiansky 16
Bodiansky 20

Boeing (Seattle, WA)

William E. Boeing founded **Pacific Aero Products** in 1915, reforming the company as **Boeing Airplane Co** in 1917. By 1929, all Boeing stock was owned by **United Aircraft and Transport Corp (UATC)**. With the dissolution of UATC in 1934, Boeing regained its independence as **Boeing Aircraft Co**; this changed back to **Boeing Airplane Co** in 1947. In 1961 the parent company was renamed Boeing Co, reflecting Boeing's more diversified corporate interests. In 1996, Boeing merged with McDonnell Douglas, the two companies reforming under the Boeing name.

The long history of Boeing has included numerous subsidiary companies; their products are found under the original company names, such as **de Havilland (Canada)**, **Stearman**, and **McDonnell Douglas**. The helicopters of Boeing's Vertol Division are filed under **Boeing-Vertol**. All other Boeing products appear under **Boeing**.

Boeing A-5 (Radio-Control P-12)
Boeing AT-3 (Model 68)
Boeing B & W (Boeing & Westervelt) (Model 1)
Boeing B-1 (Model 6)
Boeing B-1C
Boeing B-1D (Model 6D)
Boeing B-1E (Model 6E)
Boeing YB-9 (Model 215, XB-901)
Boeing Y1B-9 (Model 214)
Boeing Y1B-9A (Model 246)
Boeing XB-15 (Model 294, XBLR-1)
Boeing B-17 Model 299 Flying Fortress ("XB-17")
Boeing Y1B-17 Flying Fortress (299B, YB-17, B-17)
Boeing Y1B-17A Flying Fortress (299F)
Boeing B-17B Flying Fortress
Boeing B-17C Flying Fortress
Boeing B-17D Flying Fortress
Boeing B-17D Flying Fortress "Swoose," NASM
Boeing B-17E Flying Fortress
Boeing B-17E Mapping Fortress
Boeing B-17E, Reed Project
Boeing B-17F Flying Fortress
Boeing B-17G Flying Fortress
Boeing B-17G Flying Fortress, NASM
Boeing SB-17G (B-17H) Flying Fortress
Boeing TB-17H Flying Fortress
Boeing QB-17L Flying Fortress
Boeing QB-17N Flying Fortress
Boeing DB-17P Flying Fortress
Boeing (B-17) Fortress Mk.I (299U)
Boeing (B-17) Fortress Mk.II
Boeing (B-17) Fortress Mk.IIA
Boeing (B-17) Fortress Mk.III
Boeing Y1B-20 (Model 294, Service Test B-15)
Boeing XB-29 Superfortress
Boeing YB-29 Superfortress
Boeing B-29 Superfortress
Boeing B-29 Superfortress, Silverplate
Boeing B-29 Superfortress, Silverplate "Enola Gay," NASM
Boeing QB-29 Superfortress
Boeing SB-29 Superfortress
Boeing WB-29 Superfortress
Boeing B-29A Superfortress
Boeing EB-29A Superfortress
Boeing B-29B Superfortress
Boeing B-29B Superfortress Carrier Aircraft
Boeing EB-29B "Monstro"
Boeing B-29C Superfortress
Boeing B-29D Superfortress
Boeing XB-29E Superfortress
Boeing B-29F Superfortress
Boeing XB-29G Superfortress
Boeing XB-29H Superfortress
Boeing YB-29J Superfortress
Boeing YB-29J Superfortress "Pacusan Dreamboat"
Boeing YKB-29J Superfortress
Boeing FB-29J (RB-29J) Superfortress
Boeing CB-29K Superfortress
Boeing KB-29M Superfortress
Boeing B-29MR (B-29L) Superfortress
Boeing KB-29P Superfortress
Boeing YKB-29T Superfortress
Boeing (B-29) Washington Mk.I
Boeing XB-38
Boeing XB-39
Boeing XB-40
Boeing YB-40
Boeing TB-40
Boeing XB-44
Boeing XB-47 Stratojet
Boeing B-47A Stratojet
Boeing B-47B Stratojet
Boeing B-47B-II Stratojet
Boeing B-47B (CL-52) Stratojet
Boeing YDB-47B Stratojet
Boeing DB-47B Stratojet
Boeing RB-47B Stratojet
Boeing TB-47B Stratojet
Boeing WB-47B Stratojet
Boeing YB-47C (YB-56) Stratojet
Boeing XB-47D Stratojet
Boeing B-47E Stratojet
Boeing YDB-47E Stratojet
Boeing DB-47E Stratojet
Boeing EB-47E Stratojet
Boeing QB-47E Stratojet
Boeing RB-47E Stratojet
Boeing YB-47F Stratojet
Boeing KB-47G Stratojet
Boeing RB-47H Stratojet
Boeing ERB-47H Stratojet
Boeing YB-47J Stratojet
Boeing RB-47K Stratojet
Boeing EB-47L Stratojet
Boeing B-50A
Boeing B-50A "Lucky Lady II"
Boeing TB-50A
Boeing B-50B
Boeing EB-50B
Boeing RB-50B
Boeing B-50D
Boeing DB-50D
Boeing JB-50D
Boeing KB-50D
Boeing TB-50D
Boeing WB-50D
Boeing RB-50E
Boeing RB-50F

Boeing RB-50G
Boeing TB-50H
Boeing KB-50J
Boeing KB-50K
Boeing XB-52 Stratofortress
Boeing YB-52 Stratofortress
Boeing B-52A Stratofortress
Boeing NB-52A Stratofortress
Boeing B-52B Stratofortress
Boeing RB-52B Stratofortress
Boeing B-52C Stratofortress
Boeing B-52D Stratofortress
Boeing B-52E Stratofortress
Boeing NB-52E Stratofortress
Boeing B-52F Stratofortress
Boeing B-52G Stratofortress
Boeing B-52H Stratofortress
Boeing XB-55 (Model 474)
Boeing XB-56 (YB-57C, Model 450-19-10)
Boeing XB-59
Boeing BB-1 (Model 7)
Boeing BB-L6 (Model 8)
Boeing BQ-7 (Castor, Aphrodite)
Boeing XBT-17, Model X-90 (See: Stearman Model X90)
Boeing XBT-17, Model X-91 (See: Stearman XBT-17)
Boeing YC-14
Boeing C-18A
Boeing EC-18B
Boeing EC-18D
Boeing C-19A
Boeing C-22A
Boeing C-22B
Boeing VC-25A
Boeing VC-32A
Boeing C-40A Clipper (737-700C)
Boeing C-73 (Model 247D)
Boeing C-75 Stratoliner (Model 307)
Boeing XC-97 Stratofreighter
Boeing YC-97 Stratofreighter
Boeing YC-97A Stratofreighter
Boeing C-97A Stratofreighter
Boeing JC-97A Stratofreighter
Boeing KC-97A Stratofreighter
Boeing YC-97B Stratofreighter
Boeing C-97C Stratofreighter
Boeing MC-97C Stratofreighter
Boeing C-97D Stratofreighter
Boeing VC-97D (C-97D) Stratofreighter
Boeing KC-97E Stratofreighter
Boeing KC-97F Stratofreighter
Boeing C-97G Stratofreighter
Boeing EC-97G Stratofreighter
Boeing HC-97G Stratofreighter
Boeing KC-97G Stratofreighter
Boeing KC-97H Stratofreighter
Boeing YC-97J (YC-137) Stratofreighter
Boeing C-97K Stratofreighter
Boeing KC-97L Stratofreighter
Boeing KC-97L Stratofreighter, NASM
Boeing C-98 (Model 314A)
Boeing XC-105 (Model 294)
Boeing XC-108
Boeing YC-108
Boeing XC-108A
Boeing XC-108B
Boeing C-135A
Boeing EC-135A
Boeing KC-135A Stratotanker
Boeing JKC-135A
Boeing NKC-135A
Boeing GNKC-135A
Boeing NC-135A
Boeing RC-135A
Boeing C-135B
Boeing C-135B (TRIA)
Boeing KC-135B
Boeing OC-135B Open Skies
Boeing RC-135B
Boeing VC-135B
Boeing WC-135B
Boeing C-135C
Boeing EC-135C
Boeing RC-135C
Boeing KC-135D
Boeing RC-135D
Boeing C-135E
Boeing EC-135E
Boeing KC-135E
Boeing NKC-135E
Boeing RC-135E
Boeing C-135F
Boeing C-135FR
Boeing EC-135G
Boeing EC-135H
Boeing GEC-135H
Boeing EC-135J
Boeing EC-135K
Boeing EC-135L
Boeing RC-135M
Boeing C-135N
Boeing EC-135N
Boeing EC-135P
Boeing KC-135Q
Boeing KC-135R (KC-135RE)
Boeing RC-135R
Boeing RC-135S
Boeing TC-135S
Boeing KC-135T
Boeing RC-135U
Boeing RC-135V
Boeing RC-135W
Boeing TC-135W
Boeing RC-135X
Boeing EC-135Y
Boeing C-137, Germany
Boeing KC-137
Boeing VC-137A
Boeing VC-137B
Boeing C-137C
Boeing VC-137C
Boeing CC-137C
Boeing EC-137D
Boeing CALF (Common Affordable Light-Weight Fighter)
Boeing XCO-7 (Model 42)
Boeing XCO-7A (Model 42)
Boeing XCO-7B (Model 42)

The British de Havilland D.H.4 was modified and built under license in the United States under the designation DH-4 by several manufacturers. For British-built aircraft, see **de Havilland D.H.4**. For American-built aircraft, see the individual manufacturers, including **Aeromarine**, **Boeing**, **Dayton Wright**, **Fokker**, **Gallaudet**, **Keystone**, and **Standard**. When the actual manufacturer cannot be determined from the available evidence, the photos and documents have been filed under **Dayton Wright**.

Boeing DH-4 Mail
Boeing DH-4B
Boeing Cuban DH-4B
Boeing XDH-4M
Boeing DH-4M-1
Boeing E-3A Sentry (AWACS)
Boeing KE-3A Sentry (AWACS)
Boeing E-3B Sentry (AWACS)
Boeing E-3C Sentry (AWACS)
Boeing E-4A (NEACP)
Boeing E-4B (NEACP)
Boeing E-6A Hermes (TACAMO)
Boeing E-8A J-Stars
Boeing E-8B J-Stars
Boeing E-8C J-Stars
Boeing EX (Electronics Experimental)
Boeing FB-1 (Model 15)
Boeing FB-2 (Model 53)
Boeing FB-3 (Model 55)
Boeing FB-4 (Model 54)
Boeing FB-5 (Model 67)
Boeing FB-5 (Model 67), NASM
Boeing FB-6 (Model 54)
Boeing XF2B-1 (Model 69)
Boeing F2B-1 (Model 69)
Boeing XF3B-1 (Model 74)
Boeing F3B-1 (Model 77)
Boeing XF4B-1 (Model 83)
Boeing XF4B-1 (Model 89)
Boeing F4B-1 (Model 99)
Boeing F4B-1A (Model 99)
Boeing F4B-2 (Model 223)
Boeing F4B-3 (Model 235)
Boeing F4B-4 (Model 235)

Boeing F4B-4 (Model 235), NASM
Boeing F4B-4A (Model 234)
Boeing XF5B-1 (Model 205)
Boeing XF6B-1 (XBFB-1, Model 236)
Boeing XF7B-1 (Model 273)
Boeing XF8B-1 (Model 400)
Boeing F-9
Boeing F-9A
Boeing F-9B
Boeing F-9C (RB-17G)
Boeing F-13 (R-13, RB-29)
Boeing F-13A (R-13A, RB-29A)
Boeing Fresh I (ACV)
Boeing G.A.X. (See: Engineering Division)
Boeing GA-1
Boeing GA-2
Boeing (Curtiss) HS-2L
Boeing HSCT (Hypersonic Commercial Transport)
Boeing Hydrofoil "High Point" (PCH-1)
Boeing Hydrofoil "Pegasus" (PHM-1)
Boeing Hydrofoil "Plainview (AGEH-1)
Boeing Hydrofoil "Tucumcari" (PCH-2)
Boeing Joint Strike Fighter (JSF)
Boeing XL-15 Scout
Boeing YL-15 Scout
Boeing L-15A Scout
Boeing LIT (Light Intratheater Transport)
Boeing (Thomas-Morse) MB-3A
Boeing (Thomas-Morse) MB-3M
Boeing VNB-1 (Model 21)
Boeing NB-1 (Model 21)
Boeing NB-2 (Model 21)
Boeing NB-3 (Model 21)
Boeing NB-4 (Model 21)
Boeing XN2B-1 (Model 81)
Boeing NOTAIL ATT (Advanced Theater Transport)
Boeing O2B-1
Boeing O2B-2
Boeing PB-1 (Model 50) Flying Boat
Boeing XPB-2 (Model 50) Flying Boat
Boeing PB-1 (B-17)
Boeing PB-1G (B-17)
Boeing PB-1W (B-17)
Boeing P2B-1S (B-29)
Boeing XP3B-1
Boeing XP-4 (Model 58)
Boeing XP-7 (Model 93)
Boeing XP-8 (Model 66)
Boeing XP-9 (Model 96)
Boeing P-12 (Model 102)
Boeing XP-12A (Model 101)
Boeing P-12B (Model 102B)
Boeing P-12C (Model 222)
Boeing P-12D (Model 227)
Boeing P-12E (Model 234)
Boeing P-12F (Model 251)
Boeing XP-12G (Model 102B)
Boeing XP-12H (Model 227)
Boeing P-12J (Model 234)
Boeing YP-12K (Model 234)
Boeing XP-12L (Model 234)
Boeing XP-15 (Model 202)
Boeing XP-26 (Model 248, XP-936, Y1P-26, P-26)
Boeing P-26A (Model 266)
Boeing P-26A (Model 266), NASM
Boeing P-26B (Model 266A)
Boeing P-26C (Model 266)
Boeing P-29, XP-940
Boeing YP-29 (Model 264)
Boeing YP-29A (Model 264, P-29A)
Boeing YP-29B (Model 264)
Boeing XPBB-1 Sea Ranger (Model 344)
Boeing XPW-9 (Model 15)
Boeing PW-9 (Model 15)
Boeing PW-9A (Model 15A)
Boeing PW-9B (Model 15B)
Boeing PW-9C (Model 15C)
Boeing PW-9D (Model 15D)
Boeing T-5 (2PCLM, Boeing School of Aeronautics)
Boeing T-43A (CT-43A)
Boeing TB-1 (Model 63)
Boeing Model 2 (C-4)
Boeing Model 3
Boeing Model 3 (C-5)
Boeing Model 3 (C-6)
Boeing Model 3 (C-11)
Boeing Model 4 (EA)
Boeing Model 5 (C-650-700)
Boeing Model 5 (C-1F)
Boeing Model 5 (C-L4-S)
Boeing Model 40
Boeing Model 40A
Boeing Model 40B
Boeing Model 40B-2
Boeing Model 40B-4
Boeing Model 40B-4A
Boeing Model 40C
Boeing Model 40H-4
Boeing Model 40X
Boeing Model 40Y
Boeing Model 64
Boeing Model 69-B (Export F2B)
Boeing Model 72
Boeing Model 80
Boeing Model 80 Special
Boeing Model 80A
Boeing Model 80A-1
Boeing Model 80B
Boeing Model 81A
Boeing Model 81B
Boeing Model 81C
Boeing Model 95
Boeing Model 95A
Boeing Model 100
Boeing Model 100A
Boeing Model 100D
Boeing Model 100E
Boeing Model 100F
Boeing Model 200 Monomail
Boeing Model 203
Boeing Model 203A
Boeing Model 203B
Boeing Model 204
Boeing Model 204A
Boeing Model C-204 Thunderbird
Boeing Model 218
Boeing Model 218 (XP-925)
Boeing Model 218 (XP-925A)
Boeing Model 221 Monomail
Boeing Model 221A Monomail
Boeing Model 226
Boeing 247
Boeing 247A
Boeing 247D
Boeing 247D, NASM
Boeing 247D, RAF
Boeing 247D, RCAF
Boeing 247E
Boeing 247Y
Boeing 256 (Export F4B-4)
Boeing 267 (Export F4B)
Boeing 281 (Export P-26)
Boeing 299-Z
Boeing 299AB
Boeing 306A Flying Wing Bomber
Boeing S-307 Stratoliner, NASM
Boeing S-307 (PAA-307) Stratoliner
Boeing SA-307B Stratoliner
Boeing SA-307B-1 Stratoliner
Boeing SB-307B Stratoliner
Boeing 314 Clipper
Boeing 314A Clipper
Boeing B-314 Clipper (USN)
Boeing 367-80 (707 Prototype)
Boeing 367-80, NASM
Boeing 377-10-19 Stratocruiser Prototype
Boeing 377 Stratocruiser
Boeing 377 Stratocruiser Cargo Conversion
Boeing 417
Boeing 450
Boeing 474
Boeing 707-100
Boeing 707-138 (707 Short Body)
Boeing 707-100B
Boeing 707-200
Boeing 707-300
Boeing 707-300B
Boeing Advanced Model 707-300B
Boeing 707-300C

Boeing 707-300F
Boeing 707-400
Boeing 707-700
Boeing 707 Peace Station (Iran)
Boeing 717-200 (See: McDonnell Douglas MD-95)
Boeing 720 (707-020, 717-020)
Boeing 720B
Boeing 727QC
Boeing 727UDF
Boeing 727-100
Boeing 727-100, NASM
Boeing 727-100C
Boeing 727-100F
Boeing 727-200
Boeing Advanced Model 727-200
Boeing 727-200F
Boeing 727-300
Boeing 733 SST
Boeing 735 (Swing-Tail 707 Project)
Boeing 737-100
Boeing 737-130 (NASA Transport Systems Research Vehicle)
Boeing 737-200
Boeing Advanced Model 737-200
Boeing Executive Model 737-200
Boeing 737-200C
Boeing 737-200F
Boeing 737-200QC
Boeing 737-2X9 (Indonesia)
Boeing 737-300
Boeing 737-400
Boeing 737-500
Boeing 737-700 Wedgetail (AWACS)
Boeing 737-700C
Boeing 737-700ER Special Air Mission
Boeing 737-800
Boeing 737 Commander in Chief (CINC) Support
Boeing 747SP (747-100SP)
Boeing 747SR
Boeing 747-100
Boeing 747-100, Iran
Boeing 747-100 NASA Shuttle Carrier
Boeing 747-100B
Boeing 747-100F
Boeing 747-100M
Boeing 747-100SL
Boeing 747-200B
Boeing 747-200B Combi
Boeing 747-200C
Boeing 747-200F
Boeing 747-200M
Boeing 747-300
Boeing 747-300BC
Boeing 747-300ER
Boeing 747-300LR
Boeing 747-300M
Boeing 747-300SR
Boeing 747-400
Boeing 747-400M
Boeing 757-200
Boeing 757-200C
Boeing 757-200PF
Boeing KC-767 Tanker Transport
Boeing 767/AOA (Airborne Optical Adjunct)
Boeing 767-200
Boeing 767-200ER
Boeing 767-300
Boeing 767-300ER
Boeing 767-400ER
Boeing E-767 AWACS
Boeing 777 Prototype
Boeing 929 Jetfoil
Boeing 2707 SST
Boeing Sikorsky LHX
Boeing Sikorsky RAH-66 Comanche (LHX)
Boeing (Canada) Canso A (PBY-5A)
Boeing (Canada) Canso A (PBY-5A) "Explorer I"
Boeing (Canada) PB2B-1 Catalina
Boeing (Canada) PB2B-2 Catalina
Boeing (Canada)(PB2B) Catalina IVB
Boeing (Canada)(PB2B) Catalina VI
Boeing (Canada) Steel Truss Glider
Boeing (Canada) Totem

Boeing-Vertol (USA)

P-V Engineering Forum was organized in 1941 and incorporated in 1943. In 1946 the company became **Piasecki Helicopter Corp.** In 1955 a second company, **Piasecki Aircraft Corp (PiAC)**, was formed. (Piasecki Aircraft remains in business at this writing.) Piasecki Helicopter Corp went on to become **Vertol Aircraft Corp** in March 1955, the **Vertol Division of Boeing (Boeing-Vertol)** in 1960, **Boeing Vertol Co** (a Boeing subsidiary) in 1972, and finally, **Boeing Helicopters** in 1985. Aircraft designed by the Piasecki companies are listed under **Piasecki**. Aircraft designed by Vertol or Boeing-Vertol are under **Boeing-Vertol**.

Boeing-Vertol (MBB) BO 105C
Boeing-Vertol CH-113 Voyageur (RCAF Model 107)
Boeing-Vertol CH-113A Voyageur (Canadian Army Model 107)
Boeing-Vertol CH-46A (HRB-1) Sea Knight
Boeing-Vertol HH-46A (HRB-1) Sea Knight
Boeing-Vertol UH-46A (HRB-1) Sea Knight
Boeing-Vertol (YCH-46C) YHC-1
Boeing-Vertol CH-46D Sea Knight
Boeing-Vertol UH-46D Sea Knight
Boeing-Vertol CH-46E Sea Knight
Boeing-Vertol (H-46) HKP-4 (Sweden)
Boeing-Vertol YCH-47A (YHC-1B) Chinook
Boeing-Vertol CH-47A Chinook
Boeing-Vertol ACH-47A Chinook
Boeing-Vertol CH-47B Chinook
Boeing-Vertol CH-47C Chinook
Boeing-Vertol CH-47D Chinook
Boeing-Vertol CH-47SD Super D Chinook
Boeing-Vertol (H-47) Chinook HC.1
Boeing-Vertol (H-47) Model 414 Chinook
Boeing-Vertol YUH-61A UTTAS
Boeing-Vertol YUH-61A UTTAS LAMPS III
Boeing-Vertol XCH-62A Heavy Lifter
Boeing-Vertol Tilt-Wing
Boeing-Vertol Tilt-Wing Intratheater Transport
Boeing-Vertol VZ-2A
Boeing-Vertol VZ-2A, NASM
Boeing-Vertol X-Wing
Boeing-Vertol Model 107
Boeing-Vertol Model 179(Civil UTTAS)
Boeing-Vertol Model 234
Boeing-Vertol Model 347
Boeing-Vertol Model 360

Boerner (Germany)

Boerner High-Altitude Aircraft (Höhenflugzeug)
Boerner Propellerless Aircraft (Propellerloses Flugzeug)

Boero, Hector & Cesar Ernesto

(See: Aero Talleres Boero)

Bohatyrew, Michal (Poland)

Bohatyrew Kaczka-Nadzieja
Bohatyrew Mis (Teddy Bear) Glider
Bohatyrew Motyl (Butterfly) Glider

Bohemia Air sro (Czechoslovakia)

Bohemia Air Bohem 1
Bohemia Air Bohem 3

Boisavia (France)

Boisavia B-50 Muscadet
Boisavia B-60 Mercurey
Boisavia B-260 Anjou

Boishardy (France)

Boishardy 1923 Moto Aviette

BOK (Byuro Opytnikh Konstruktsii; Bureau of Special Design)(Russia)

BOK-1 (SS)

BOK-2
BOK-3 (TsKB-9)
BOK-5
BOK-7
BOK-8
BOK-11
BOK-15

Bokor, Morris (New York)
Bokor 1909 Pusher Triplane

Boland Aeroplane Motor Co (Rahway, NJ)
Boland 1908 Tailless
Boland 1909
Boland 1911 Tailless
Boland 1912
Boland 1912 Tailless
Boland 1913 Tailless
Boland 1914 Tailless Flying Boat

Bolkhovitinov, V. F. (Russia)
Bolkhovitinov DB-A
Bolkhovitinov Spartak (BBS-1)

Bölkow (Germany)

Bölkow-Entwicklungen KG, established in 1956, was renamed **Bölkow GmbH** following a minority interest acquisition by **Boeing** in 1965. The company merged with **Messerschmitt AG** in 1968 to form **Messerschmitt-Bölkow GmbH** (MB). (Bölkow operations remained under control of **Bölkow-Apparatebau GmbH**, a wholly-owned subsidiary of **Bölkow GmbH**.) A merger with **Hamburger Flugzeugbau GmbH** formed **Messerschmitt-Bölkow-Blohm GmbH (MBB)** in 1969. (Bölkow operations fell under a Helicopter and Transport Systems Division.) In 1989 MBB merged with **Dornier**, **MTU Motoren- und Turbinen-Union München**, and **Telefunken SystemTechnik (TST)** to form **Deutsche Aerospace AG (DASA)**. In 1991/92 the helicopter divisions of **Aérospatiale** and DASA formed **Eurocopter SA**; DASA's component was **Eurocopter Deutschland**. Designs originating with Bölkow carried "Bo" (no umlaut) designations and are filed under **Bölkow** and **Boeing-Vertol**. Aircraft produced with Eurocopter's "EC" designations are filed under **Eurocopter**. Products produced with MBB/Kawasaki "BK" designations are found under **MBB/Kawasaki**.

Bölkow Bo 46
Bölkow Bo 103
Bölkow F 207 (4-seat Kl 107)
Bölkow FS 24 Phönix (Sailplane)
Bölkow/Klemm:
- Kl 107 B (Post-war Rebuild)
- Kl 107 C (Post-war Rebuild)

Bölkow/MBB:
- Bo 104 (2-Seat Helicopter)
- Bo 105 Series
- Bo 105 Prototype
- Bo 105 C
- Bo 105 CBS
- Bo 105 M (Rotor Mast Sight Testbed)
- Bo 105 P (PAH-1)
- Bo 108 (See: Eurocopter EC-155)
- Bo 207
- Bo 208
- Bo 208 A Junior
- Bo 208 C Junior 3
- Bo 209 Monsun (Monsoon)
- L-252 Phoebus B1 Sailplane
- L-252 Phoebus C Sailplane
- Fliegender Jeep (Flying Jeep)

de Bolotoff, Serge (France)
de Bolotoff 1909 Triplane (Voisin-Built)

Bolte Aircraft Corp (Henry Bolte) (Des Moines, IA)
Bolte LW-1 Fliver
Bolte LW-2

Bombardier Regional Aircraft (Canada)(See Also: Canadair; de Havilland Canada)
Bombardier Continental
Bombardier Global Express (BD701)

Bonamy (France)
Bonamy 1914 Tandem-Wing

***Bondy, Milos, & Co** (See: Avia Akciová Spolecnost pro Prumysl Letecky)*

***Bone (R. O. Bone Co)**(See: Golden Eagle Aircraft Corp)*

***Bone & Campbell (R. O. Bone & Mark M. Campbell)**(See: Golden Eagle Aircraft Corp)*

Bonnet, Charles (France)
Bonnet 1922 Glider
Bonnet 1933 Tandem Monoplane Glider

Bonnet-Labranche (France)
Bonnet-Labranche 1869 Helicoplane

Bonnet-Labranche (Aéroplanes Bonnet-Labranche, ABL)(France)
Bonnet-Labranche No.1 (1908)
Bonnet-Labranche No.2 (1908)
Bonnet-Labranche No.3 (1909)
Bonnet-Labranche No.4 (1909)
Bonnet-Labranche No.5 (1910)
Bonnet-Labranche No.6 Monoplane
Bonnet-Labranche No.7 2-Seat Racer (Course Deux Places)
Bonnet-Labranche No.8 3-Seat Military (Militaire Trois Places)

Bonnet-Miguet (France)
Bonnet-Miguet 1923 Monoplane Glider

Bonnetti (Keene, NH)
Bonnetti 1908 Biplane (Curtiss Type)

Bonney, Leonard W. (Long Island, NY)
Bonney 1909 Monoplane
Bonney Gull (1927)

Bonomi (Italy)
Bonomi 1930 Monoplane
Bonomi 1932 Glider

Boomerangs, General

***Booth Racer** (See: Aerial Engineering Corp)*

Borcuki, Stefan (Poland)
Borcuki 1911 Biplane
Borcuki 1911 Monoplane

***Bordelaise (Société Aérienne Bordelaise)(France)**(See also: Dyle et Bacalan)*
Bordelaise A.B. 15
Bordelaise A.B. 20
Bordelaise A.B. 21
Bordelaise SEMA 12
Bordelaise D.B. 70 (See: Dyle et Bacalan)
Bordelaise D.B. 80 (See: Dyle et Bacalan)

Bordier, A. M. (France)
Bordier 1909 Aircraft

Borel (Société Anonyme des Établissements Borel)(France)
Borel Aeroyacht Type Denhaut I
Borel Aeroyacht Type Denhaut II
Borel Aeroyacht Type Denhaut III
Borel (Borel-Odier) B.O.2 Hydravion
Borel B.O.3 Hydravion
Borel B.O.92 Hydravion
Borel C.A.n.2

Borel C.A.p.2
Borel l'Obus (Artillery Shell)(1912)
Borel 1911 Monoplane Series
Borel 1912 Monoplane Series
Borel 1912 2-Seat Hydravions
Borel 1912 2-Seat Hydravions, 80hp Gnome
Borel 1912 Monoplane, 1-Seat Type École (Trainer), Anzani
Borel 1912 Monoplane, 1-Seat Type École (Trainer), Grégoire-Gyp
Borel 1912 Monoplane, 1-Seat Type Circuit Européen (Gnome 70hp)
Borel 1912 Monoplane, 1-Seat Type Militaire (Gnome 50hp)
Borel 1912 Monoplane, 1-Seat Type "Paris-Madrid" (Gnome 50hp)
Borel 1912 Monoplane, 1-Seat Type "No. 37" (Gnome 70hp)
Borel 1912 Monoplane, 2-Seat Type Militaire Renforcé (Gnome 50hp)
Borel 1912 Monoplane, 2-Seat Type Militaire (Gnome 70hp)
Borel 1912 Monoplane, 2-Seat Type Course (Racer), Gnome 50 hp
Borel 1912 Monoplane, 2-Seat Type Course (Racer), Gnome 70hp
Borel 1912 Monoplane, Multi-Seat, Gnome 100hp
Borel 1912 Monoplane, Multi-Seat, Gnome 140hp
Borel 1913 Monoplane Series
Borel 1913 Hydravion Monoplanes, 2-Seaters
Borel 1913 Hydravion Monoplane, Chemet 2-Seater
Borel 1913 Hydravion Monoplane, Navy 2-Seater
Borel 1913 Pusher Monoplane Type Militaire
Borel 1918 Tri-Motored Biplane Flying Boat
Borel 1918 Twin-Engined Hydravion Triplane (Odier)
Borel 1922-23 Trimotor Monoplane Transport (Modèle 11)
Borel 1923 All-Metal Day Fighter
Borel 1924 2-Seat Flyingboat
Borel 1924 3-Engined Passenger Aircraft
Borel-Boccaccio C1 Flandre
Borel-Boccaccio C2, Type 3000
Borel-Boccaccio C2, 1922
Borel-Boccaccio Type Gordon Bennett
Borel-Ruby Torpille (Torpedo)

Borello, Antonio Guido (Argentina)
Borello I "El Argentino" ("The Argentine")
Borello II "El Colorado" ("The Red One")

Borgfeldt, Nicholas H. (Brooklyn, NY)
Borgfeldt 1894 Ornithopter Patent

Borgnis et Desbordes de Savignon (France)
Borgnis et Desbordes de Savignon 1909 Pusher Triplane

Borgward (Germany)
Borgward Kolibri I (Hummingbird)

Borovkov-Florov (Russia)
Borovkov-Florov I-207

Bossi (Italy)
Bossi Flying Boat (c.1913)
Bossi Man-Powered Aircraft (1936)

Bothy, Léopold (France)
Bothy 1910 Monoplane

Bothy & Soufrogel *(See: Bothy)*

Bothy de Namur *(See: Bothy)*

Botts, R. H. (Baltimore, MD)
Botts (R. H.) 1904 Flying Machine

Boucheron (France)
Boucheron 1910 Pusher Biplane

Boughton, John W. (Philadelphia, PA)
Boughton (John W.), General

Bougie, Yuan C. (Canada)
Bougie Hanscat

Boullay (France)
Boullay 1913 Float Monplane

Boulton & Paul *(See: Boulton Paul)*

Boulton Paul (UK)
Boulton & Paul, Ltd was an established firm which began producing aircraft in 1917. In June 1934 all aircraft activities were transferred to a new company, **Boulton Paul Aircraft Ltd (BPA)** and, in 1936, all links between Boulton & Paul and Boulton Paul Aircraft were severed. BPA withdrew from aircraft production in 1954, but continued to produce aviation-related components. In 1969, the company was absorbed into **Dowty Boulton Paul**, which continues component production. All Boulton & Paul designs and Boulton Paul Aircraft designs are listed under **Boulton Paul**.
Boulton Paul Atlantic (P.8)
Boulton Paul Balliol P.108 Prototype
Boulton Paul Balliol T.Mk.1 (P.108)
Boulton Paul Balliol T.Mk.2 (P.108)
Boulton Paul Sea Balliol T.Mk.21 (P.108)
Boulton Paul Bittern (P.31)
Boulton Paul Boblink (P.3)
Boulton Paul Bodmin (P.12)
Boulton Paul Bolton (P.15)
Boulton Paul Bourges Mk.I (P.7)
Boulton Paul Bourges Mk.IA (P.7)
Boulton Paul Bourges Mk.II (P.7)
Boulton Paul Bourges Mk.IIA (P.7)
Boulton Paul Bugle Mk.I (P.25)
Boulton Paul Bugle Mk.II (P.25A)
Boulton Paul Defiant Mk.I (P.82)
Boulton Paul Defiant Mk.II (P.82)
Boulton Paul Defiant TT.Mk.I (P.82)
Boulton Paul Defiant TT.Mk.III (P.82)
Boulton Paul Mail Carrier (P.64)
Boulton Paul Overstrand (P.75)
Boulton Paul P.6
Boulton Paul P.9
Boulton Paul P.10
Boulton Paul P.32
Boulton Paul P.71A
Boulton Paul P.92/2
Boulton Paul P.106
Boulton Paul P.109
Boulton Paul P.111
Boulton Paul P.111A
Boulton Paul P.112
Boulton Paul P.116
Boulton Paul P.119
Boulton Paul P.120
Boulton Paul P.124
Boulton Paul P.125
Boulton Paul P.131
Boulton Paul Partridge (P.33)
Boulton Paul Phoenix (P.41)
Boulton Paul Sidestrand Mk.I (P.29)
Boulton Paul Sidestrand Mk.II (P.29)
Boulton Paul Sidestrand Mk.III (P.29)
Boulton Paul Sidestrand Mk.IIIS (P.29)

Bounsall, Curtis (Mesquite, NV)
Bounsall (Curtis) Prospector
Bounsall (Curtis) Super Prospector

Bounsall, Edward W. "Eddie" (Logandale, NV)
Bounsall ("Eddie") Treasure Hawk
Bounsall ("Eddie") Treasure Hawk SP.1

Bourcart (Italy)
Bourcart 1866 Ornithopter

Bourcart, J. J. (Germany)
Bourcart (J. J.) 1908 Ornithopter

Bourdariat, Edouard (France)
Bourdariat 1908/1910 Pusher

Bourdon Aircraft Corp (Hillsgrove, RI)
Bourdon Aircraft Corp built the Kittyhawk light aircraft from 1928. **Viking Flying Boat Co**, of New Haven, CT, was established in 1928 for the licensed production of **FBA-Schrek** flying boats. In 1930, Viking bought Bourdon and all manufacturing rights to the Kittyhawk. All Viking production is listed under **Viking**, except for Kittyhawks, which are filed under **Bourdon**.
Bourdon Kittyhawk B-2
Bourdon Kittyhawk B-4
Bourdon (Viking) Kittyhawk B-8

Bourgoin et Kessels (France)
Bourgoin et Kessels:
Aérobus (1911/12)
Parachute Monoplane (1913)

Bourhis (France)
Bourhis 1914 Aircraft

Bouriau-Chapautau (France)
Bouriau-Chapautau 1925 Monoplane Glider

Bousson, Firmin (France)
Bousson 1909 Triplane Glider

Bousson-Borgnis (France)
Bousson-Borgnis Auto-Aviateur (1908)

Boutard *(See: M-B)*

Boutaric (France)
Boutaric 1909 Tandem Biplane

Bowdler Aviation Inc (Beavercreek, OH)
Bowdler Supercat

Bowers, Peter M. (Seattle, WA)
Bowers Fly Baby
Bowers Fly Baby 1A
Bowers Fly Baby 1B
Bowers Wright EX "Vin Fiz" Replica

Bowlby, Dick (Wichita, KS)
Bowlby Sunbeam

Bowlus, Hawley (Bowlus Sailplane Co, ltd; Bowlus-duPont Sailplane Co)(San Diego, CA)
Bowlus Albatross
Bowlus Albatross II
Bowlus BA-100 Baby Albatross
Bowlus BA-100 Baby Albatross, NASM
Bowlus BS-100 Super Albatross
Bowlus BTS-100 (BA-102) Two-Place Baby Albatross
Bowlus Bumblebee (See: Nelson Aircraft Corp)
Bowlus XCG-7
Bowlus XCG-8
Bowlus XCG-16
Bowlus G-100
Bowlus(Bowlus/Nelson) Dragonfly
Bowlus MC-1 Glider
Bowlus MIPU3 Sailplane
Bowlus Senior Albatross II
Bowlus SPD Glider
Bowlus SP-1 Paper Wing Glider
Bowlus S1000
Bowlus XTG-12
Bowlus-duPont 1-S-2100 Senior Albatross
Bowlus-duPont 1-S-2100 Senior Albatross "Falcon," NASM
Bowlus #1 (1911) Monoplane Hang Glider
Bowlus #2 (1912) Biplane Glider
Bowlus #3 (1913) Biplane Hang Glider
Bowlus #4 (1913) Biplane Hang Glider
Bowlus #5 (1913) Biplane Hang Glider
Bowlus #6 (1913) Biplane Hang Glider
Bowlus #7 (1913) Biplane Hang Glider
Bowlus #13 (1922) Glider
Bowlus #14 (1922) Glider
Bowlus #15 (1922) Glider
Bowlus #16 (1929) Soaring Plane
Bowlus #17
Bowlus #18
Bowlus #19

Bowlus, Michael (USA)
Bowlus (Michael) BZ-1

Bowser, Kenneth *(See: B-B)*

Boyd (W. Hunter A. Boyd & C. M. Boyd)(Baltimore, MD)
Boyd Model 1 Monoplane (Variable Camber Monoplane)
Boyd Model 2 Monoplane
Boyd 1924 Monoplane
Boyd 1925 Monoplane
Boyd 1928 Monoplane
Boyd 1932 Monoplane (Model C)

Boyd, Millard *(See: Martin-Boyd)*

Boyenval et Jouhan (France)
Boyenval et Jouhan 1904 Sesquiplane Glider

de Boyesson (France/Japan)
de Boyesson A3 (1922 Recon Biplane)

Brabazon *(See: Moore-Brabazon)*

Bradley Aerospace (Chico, CA)
Bradley Aerospace Aerobat BA100

de Brageas (France)
de Brageas:
1911 Aircraft
1912 Monoplane, Darracq-Powered
1912 Monoplane, Multiplace
1912 Monoplane, Russian Army
1912 Monoplane, Tourer
1912 Monoplane, Trainer
1913 Two-Seat Pusher Monoplane

Braley Aircraft Co (Corp)(Thomas E. Braley)(Wichita, KS)
Braley (Aircraft) B1 Beezle Bug
Braley (Aircraft) B2-C6 Breezle Bug
Braley (Aircraft) B2-C6

Braley Glider Corp (Wichita, KS)
Braley (Glider) BG-1 Skysport

Brandeis (Omaha, NE)
Brandeis 1929 Glider (822H)

Brandenburg (Hansa und Brandenburgische Flugzeugwerke AG)(Germany)
Brandenburg B.I
Brandenburg C.I
Brandenburg C.I (U)
Brandenburg C.I (Ph)
Brandenburg C.II
Brandenburg CC
Brandenburg D
Brandenburg D.I
Brandenburg D.I (Ph)
Brandenburg DD
Brandenburg FB 1915
Brandenburg FD (B.1)
Brandenburg G.I
Brandenburg GDW
Brandenburg GNW
Brandenburg GW
Brandenburg KDW
Brandenburg KW
Brandenburg L.14
Brandenburg L.15
Brandenburg L.16

Brandenburg LW
Brandenburg NW
Brandenburg W
Brandenburg W 11
Brandenburg W 12
Brandenburg W 13
Brandenburg W 16
Brandenburg W 17
Brandenburg W 18
Brandenburg W 19
Brandenburg W 20
Brandenburg W 25
Brandenburg W 26
Brandenburg W 27
Brandenburg W 29
Brandenburg W 32
Brandenburg W 33
Brandenburg 05.01 thru 05.04
Brandenburg 05.05
Brandenburg 05.06
Brandenburg 05.07
Brandenburg 05.08
Brandenburg 60.54
Brandenburg 60.55
Brandenburg 60.56 & 60.57
Brandenburg 60.58

Brannen-Smith
Brannen-Smith Biplane

Brant (UK)
Brant 1931 Glider

Brantly (Brantly Helicopter, Brantly-Hynes Helicopter) (Frederick, OK & Vernon, TX)
Brantly B-1
Brantly B-2
Brantly B-2 Turbine
Brantly B-2A
Brantly B-2B
Brantly B-2E
Brantly YHO-3 (B-2)
Brantly 305
Brantly 305 Turbine

Bratukhin, I. P. (Russia)
Bratukhin B-5
Bratukhin B-9
Bratukhin B-10
Bratukhin B-11
Bratukhin 2MG Omega Helicopter
Bratukhin Omega II Helicopter
Bratukhin G-3
Bratukhin G-4
Bratukhin 5-EA
Bratukhin 11-EA
Bratukhin 11-EA-PV

Bratukhin-Kuznetsov (I. P. Bratukhin & V. A. Kuznetsov (Russia)
Bratukhin-Kuznetsov 2-EA

Brazier (France)
Brazier Tandem-Wing Seaplane Glider

Brazil Aircraft Corp (Brazil, IN)
Brazil Lion

Brea (Air Club of Brea, CA)
Brea Humming Bird

Bréant (France)
Bréant 1854 Ornithopter

Breda (Società Italiana Ernesto Breda)(Italy)
Breda A.2
Breda A.3
Breda A.4
Breda A.7
Breda A.8
Breda A.9
Breda A.10
Breda A.14
Breda Ba.15
Breda A.16
Breda Ba.19
Breda Ba.25
Breda Ba.27
Breda Ba.28
Breda Ba.33
Breda Ba.39
Breda Ba.42
Breda Ba.44
Breda Ba.64
Breda Ba.65
Breda Ba.75
Breda Ba.82
Breda Ba.88 Lince Attack
Breda Ba.201 Picchiatelli (License-built Ju 87 B)
Breda BI 5
Breda CC 20
Breda CC 3000
Breda SC.4
Breda (Pensuti License) 1919 Triplane
Breda 2 (Wind Tunnel Model)
Breda 3
Breda 4 HS
Breda 7
Breda 9
Breda 9*bis*
Breda 15
Breda 15S
Breda 16
Breda 16 Seaplane
Breda 19
Breda 25
Breda 26
Breda 27
Breda 28
Breda 32
Breda 32d
Breda 33
Breda 39
Breda 42
Breda 44
Breda 46
Breda 64
Breda 65
Breda 65*bis*
Breda 79S
Breda 82
Breda 88
Breda 2-Engine, Sesquiplane Day Bomber
Breda-Pittoni BP.471
Breda-Zappata BZ 308
Breda-Zappata BZ 309

BredaNardi (BredaNardi Costruzioni Aeronautiche)(Italy)
BredaNardi NH-300C
BredaNardi NH-500MD

Breese Aircraft Co (Vance Breese) (Portland, OR)
Breese Aircraft Junior
Breese Aircraft LT-1
Breese Aircraft Land Monoplane
Breese Aircraft Model 5 (See: Breese-Wilde)
Breese Aircraft Model R-6
Breese Aircraft Model R-6-C

Breese, Sydney S. (Southhampton, NY)
Breese (Sydney S.) Penguin (1918)

Breese-Dallas (Breese & Dallas, Inc) (Vance Breese & Marion M. Dallas) (Detroit, MI)*(See: Lambert 1344)*

Breese-Wilde Aircraft Co (San Francisco, CA)
Breese-Wilde "Aloha"
Breese-Wilde "Pabco Pacific Flyer"
Breese-Wilde Model 5

Breezy Aircraft Co (Palos Park, IL)
Breezy RLU1 (Roloff-Liposky-Unger)

Breguet (Bréguet)(France)
Société Anonyme des Ateliers d'Aviation, Louis Bréguet was a

major producer of French military aircraft during World War I. In 1971 **Breguet Aviation** (which the company had been called since 1966) merged with **Dassault** to form **Avions Marcel Dassault-Breguet Aviation**. All Breguet designs appear under **Breguet**.

Breguet Pre-1914 Designations

Breguet No.1 Biplane (Breguet-Richet No.3)(1909)
Breguet No.2 Biplane (Breguet-Richet No.4)(1909)
Breguet No.3 Biplane (1910)
Breguet C.1 (Chenu 40hp)
Breguet C.2 (Chenu 80hp)
Breguet D.2 (Dansette 100hp)
Breguet G (G.1)(Gnôme 50hp)
Breguet G.2 (Gnôme 70hp)
Breguet G.2*bis*, Gnôme 80 hp
Breguet G.2*bis*, 9-Cyl, Gnôme 100hp
Breguet G.3 (14-Cyl, Gnôme 100hp)
Breguet G.4 (Gnôme 140 or 160hp)
Breguet A-G.4 (Gnôme 140 or 160hp)
Breguet L.1 (Renault 50/60 hp)
Breguet L.2 (Renault 70 hp)
Breguet L.2*bis* (Renault 90 hp)
Breguet L.3 (Renault 100 hp)
Breguet O.1 (Le Rhône 80hp)
Breguet R.1 (REP 50/60hp)
Breguet R.2 (REP 70hp)
Breguet U.1 (Canton-Unné 7 Cyl)
Breguet U.2 (Canton-Unné 9 Cyl)
Breguet H-U.2 Seaplane
Breguet A-U.2 Hydroplane "La Marseillaise"
Breguet B-U.3 (Canton-Unné 200hp)
Breguet H-U.3 (Canton-Unné 200hp)

Breguet Post-1913 Designations

Breguet AV 1 (Bre.14 Prototype, Bre.13)
Breguet AV 2 (Bre.14 Prototype)
Breguet BU 3
Breguet Gyroplane (1936)
Breguet Gyroplane G.IIE
Breguet Gyroplane G.III
Breguet SN 3
Breguet BR.1 (BLC)
Breguet-Michelin BM.2 (BLM)
Breguet-Michelin BM.3 (BAM)
Breguet-Michelin BM.4
Breguet BR.5
Breguet BR.6
Breguet BR.7 (BUC)
Breguet BR.8 (BC)
Breguet BR.9 (BAC)
Breguet-Michelin BM.10 (BUM)
Breguet BR.11 Corsaire (Corsair)
Breguet BR.12
Breguet Bre.13 (See: AV 1)
Breguet Bre.14 AE
Breguet Bre.14 A2
Breguet Bre.14 Ap2
Breguet Bre.14 B1
Breguet Bre.14 B2
Breguet Bre.14 C
Breguet Bre.14 E2
Breguet Bre.14 H
Breguet Bre.14 S
Breguet Bre.14 T
Breguet Bre.14 T*bis*
Breguet Bre.15
Breguet Bre.16 Bn2
Breguet Bre.17 C2
Breguet Bre.18 T
Breguet Bre.19 A2
Breguet Bre.19 A2 "Nungesser-Coli"
Breguet Bre.19 B2
Breguet Bre.19 GR "Point d'Interrogation" ("Question Mark")
Breguet Bre.20 Leviathan
Breguet Bre.21 Leviathan
Breguet Bre.22 Leviathan
Breguet Bre.23
Breguet Bre.23-0
Breguet Bre.25 C2
Breguet Bre.26 T
Breguet Bre.27 Prototype
Breguet Bre.27-0 A2
Breguet Bre.27-1 A2
Breguet Bre.27-3 (A2, R2, & B2)
Breguet Bre.27-4
Breguet Bre.280 T
Breguet Bre.281 T
Breguet Bre.284 T
Breguet Bre.33 Joe III
Breguet Bre.33-0 R2
Breguet Bre.390 T
Breguet Bre.391 T
Breguet Bre.393 T
Breguet Bre.41-0 M3
Breguet Bre.41-1 M3
Breguet Bre.41-1 M3
Breguet Bre.41-3 M3
Breguet Bre.41-4 M3
Breguet Bre.46-0 M5 Vultur (Vulture)
Breguet Bre.46-2 B4 Vultur (Vulture)
Breguet Bre.47-0 T Fulgur
Breguet Bre.500 T Colmar
Breguet Bre.52-1 Bizerte (See also: Short S.8/2)
Breguet Bre.53-0 Saïgon (See also: Short S.8/2)
Breguet Bre.670
Breguet Bre.69-0
Breguet Bre.693
Breguet Bre.730
Breguet Bre.731 Bellatrix
Breguet Bre.76-1 Deux-Ponts (Two Bridges)
Breguet Bre.76-1 S Deux-Ponts (Two Bridges)
Breguet Bre.76-3 Deux-Ponts/ Provence (Two Bridges)
Breguet Bre.76-5 Deux-Ponts/ Sahara (Two Bridges)
Breguet Bre.890 H Mercure (Mercury)
Breguet Bre.891 R Mars
Breguet Bre.892 S Mercure (Mercury)
Breguet Bre.900
Breguet Bre.901 S Mouette (Seagull) (Sailplane)
Breguet Bre.940 Integral
Breguet Bre.941 Prototype (McDonnell 188 STOL Demo)
Breguet Bre.941 S
Breguet Bre.941 S, McDonnell Douglas 188 STOL Demo
Breguet Bre.942 Integral
Breguet Bre.960 Vultur (Vulture)
Breguet Bre.1001 Taon (Gadfly)
Breguet Bre.1011
Breguet Bre.1050 Alizé (Tradewind)
Breguet Bre.1150 Atlantic
Breguet Bre.1150 Atlantic 2
Breguet 1916 Twin-Engined Bomber
Breguet 1922 Limousine
Breguet 1923 Moto Aviettes

Breguet, Ltd (Breguet Aeroplanes, Ltd)(UK)

Breguet, Ltd (UK) 1912 Biplane
Breguet, Ltd (UK) 1913 Biplane

Breguet-Richet (France)

Breguet-Richet:
Gyroplane No.1 (1907)
Gyroplane No.2 (1908)
Gyroplane No.2*bis* (1908/09)
No.3 (See: Breguet No.1)
No.4 (See: Breguet No.2)

Breguet-Wibault (France)

Breguet-Wibault 670T21

Brennan, Louis (UK)

Brennan 1919 Helicopter

Breslau (Germany)

Breslau Charlotte (Tailless Glider)

Breslauer (Akaflieg Breslauer) (Germany)

Breslauer "Schlesien in Not"

du Breuil, Marquis Picat (France)
du Breuil Parasol Monoplane (1911)
du Breuil Modwing Monoplane

Brewster Aeronautical Corp (Long Island City, NY; Johnsville, PA)
Brewster XA-32
Brewster XA-32A
Brewster XF2A-1 Buffalo
Brewster F2A-1 Buffalo
Brewster XF2A-2 Buffalo
Brewster F2A-2 Buffalo
Brewster F2A-3 Buffalo
Brewster Buffalo 239 (Finland)
Brewster Buffalo 339B (Belgium)
Brewster Buffalo 339D (Netherlands)
Brewster Buffalo I (339E, UK)
Brewster Buffalo 439 (Netherlands)
Brewster F3A-1 Corsair
Brewster F3A-1D Corsair
Brewster XSBA-1
Brewster XSB2A-1 Buccaneer
Brewster SB2A-2 Buccaneer
Brewster SB2A-4 Bermuda (340D)
Brewster Bermuda I (340E, SB2A-1B)
Brewster-Fleet 10, B-1 (See: Fleet (Canada) Model 16F)

Bridoulot (France)
Bridoulot Monoplane Glider (1909)
Bridoulot Powered Aircraft

Briegleb Aircraft Co Inc (Van Nuys, CA)
Briegleb BG-6
Briegleb BG-7
Briegleb BG-8
Briegleb BG-12
Briegleb BG-12A
Briegleb BG-12B
Briegleb BG-12BD
Briegleb BG-12C
Briegleb BG-12/16
Briegleb XTG-13

Briggiler, Luis F. (Argentina)
Briggiler 1911 Monoplane

***Briggs Aeroplane Co** (See: Alexandria Aircraft Corp)*

Bright (France)
Bright 1859 Helicopter

Brimm, Daniel J. (Garden City, NY)
Brimm Special Biplane (1928)

Brissard, L. A. (France)
Brissard 1914 Monoplane

Brissaud (France)
Brissaud 1909 Biplane

Bristol (UK)
British and Colonial Aeroplane Co was formed in 1910 by George and Samuel White. The company name changed to **Bristol Aeroplane Co** in 1920, building aircraft and aircraft engines. In 1956, the company split into **Bristol Aircraft** and **Bristol Aero Engines**. In 1960, Bristol Aircraft merged with **English Electric** and **Vickers Armstrong Ltd** to form **British Aircraft Corp (BAC)**. All Bristol aircraft designs are listed under **Bristol**.
Bristol Advanced Trainer (Jupiter Engine)
Bristol Airliner
Bristol Babe
Bristol Badger
Bristol Badminton (Type 99)
Bristol Bagshot (Type 95)
Bristol Beaufighter Prototype (Type 156)(F.17/39)
Bristol Beaufighter Mk.IC
Bristol Beaufighter Mk.IF
Bristol Beaufighter Mk.IIF
Bristol Beaufighter Mk.III (Type 158)
Bristol Beaufighter Mk.IV (Type 158)
Bristol Beaufighter Mk.V
Bristol Beaufighter Mk.VI (I.T.F)
Bristol Beaufighter Mk.VIC
Bristol Beaufighter Mk.VIF
Bristol Beaufighter Mk.VII
Bristol Beaufighter Mk.VIII
Bristol Beaufighter Mk.IX
Bristol Beaufighter T.F.Mk.X
Bristol Beaufighter T.T.Mk.10
Bristol Beaufighter Mk.XIC
Bristol Beaufighter Mk.XII
Bristol Beaufighter Mk.21
Bristol Beaufort (Type 152)
Bristol Beaver
Bristol Berkeley
Bristol Bisley
Bristol Blenheim
Bristol Bloodhound (Type 84)
Bristol Boarhound
Bristol Bolingbroke
Bristol Bombay (Type 130)
Bristol Boxkite
Bristol Brabazon (Type 167)
Bristol Braemar
Bristol Brandon (Type 62)
Bristol Brigand (Type 164)
Bristol Britannia (Type 175)
Bristol Britannia 100 Series
Bristol Britannia 300 Series
Bristol Brownie I (Type 91)
Bristol Brownie II
Bristol Brownie III (Type 98)
Bristol Buckingham (Type 163)
Bristol Buckmaster (Type 166)
Bristol Bulgarian Tourer
Bristol Bulldog
Bristol Bullet
Bristol Bullfinch
Bristol Bullpup
Bristol Coupé (Converted D.H.4)
Bristol Fighter F.2A, F.2B (Brisfit)
Bristol Fighter F.2B (Brisfit)
Bristol Fighter F.2C (Brisfit)
Bristol Freighter
Bristol Greek Tourer
Bristol Gordon England G.E.1
Bristol Gordon England G.E.2
Bristol Gordon England G.E.3
Bristol High Altitude Monoplane (138A)
Bristol M.1A Monoplane Scout
Bristol M.1B Monoplane Scout
Bristol M.1C Monoplane Scout
Bristol M.1D Monoplane Scout
Bristol M.R.1 Metal Biplane (Type 13)
Bristol Pullman (Type 26)
Bristol Racer, 1911 (Biplane No.33, Grand Seigne)
Bristol Racer Type 72 (1922)
Bristol S.2A
Bristol Scout A
Bristol Scout B
Bristol Scout C
Bristol Scout D
Bristol Scout E
Bristol Scout F
Bristol Seely
Bristol S.S.A. (Single-Seat, Armoured)
Bristol Sycamore (Type 171)
Bristol Type T Biplane (1911)
Bristol T.T.A.
Bristol Taxiplane (Type 73)
Bristol Taxiplane Trainer
Bristol Ten-Seater (Type 75)
Bristol Ten-Seater (Jupiter Engine)
Bristol Tourer
Bristol Tourer Seaplane (Type 28)
Bristol Tramp
Bristol Wayfarer (Type 170)
Bristol Zodiac
Bristol Type 92
Bristol Type 93
Bristol Type 93A
Bristol Type 93B
Bristol Type 101
Bristol Type 109
Bristol Type 110
Bristol Type 118
Bristol Type 120

Bristol Type 123 (F.7/30)
Bristol Type 133 (F.7/30)
Bristol Type 138A High Altitude Monoplane
Bristol Type 142 "Britain First"
Bristol Type 143
Bristol Type 146 (F.5/34)
Bristol Type 147
Bristol Type 148 (A.39/34)
Bristol Type 159 Heavy Bomber
Bristol Type 173 Helicopter
Bristol Type 188
Bristol Type 191 Helicopter
Bristol Type 192 Belvedere Helicopter
Bristol Type 200
Bristol 1910 Glider
Bristol-Burney Flying Boat
Bristol-Burney X.1
Bristol-Burney X.2
Bristol-Burney X.3
Bristol-Coanda B.R.7
Bristol-Coanda B.R.8
Bristol-Coanda G.B.75
Bristol-Coanda P.B.8
Bristol-Coanda T.B.8
Bristol-Coanda T.B.8H Seaplane
Bristol-Coanda Monoplane
Bristol-Coanda 1913 Canard
Bristol-Prier Monoplanes
Bristol-Prier Single-Seat School Type

Bristol Aeronautical Corp (Bristol, CT)
Bristol Aeronautical XLRQ-1 Glider

Bristow, W. A. *(See: Short-Bristow Crusader)*

British & Colonial Aeroplane Co *(See: Bristol)*

British & Colonial Aviation Co, Ltd *(See: Bristol)*

British Aerial Transport *(See: B.A.T.)*

British Aerospace (UK)
British Aerospace (BAe) was established 29 April 1977 by the merger of **British Aircraft Corp (BAC)**, the **Hawker Siddeley Group** (including **Hawker Siddeley Aviation**), and **Scottish Aviation**. Only products developed after 1977, or whose main production occurred after this date, are listed under **British Aerospace** by their BAe designation. For earlier products, see the previous companies.

British Aerospace:
Buccaneer (See: Blackburn B-103)
Bulldog (See: Scottish Aviation Bulldog)
EAP (L)
Harrier (See: Hawker Siddeley)
Hawk T.Mk.1/1A
Hawk 50
Hawk 60
Hawk 100
Hawk 200
(Hawk) T-45A Goshawk (See: McDonnell Douglas)
Jetstream ATP
Jetstream 31
One-Eleven (See: BAC One-Eleven)
P.132
Sea Harrier (See: Hawker Siddeley Sea Harrier)
Trident (See: Hawker Siddeley HS.121 Trident)
VC-10 (See: BAC VC-10)
111 (See: BAC One-Eleven)
125 Family
Based on de Havilland's initial design for the D.H.125, the HS.125 was developed and produced by Hawker Siddeley. **Beechcraft Hawker Corp**, a US-based **Beech Aircraft Corp** subsidiary, was formed in 1970 to help market, assemble, and develop the aircraft as the BH 125. Beechcraft Hawker continued in this role after Hawker Siddeley was absorbed by **British Aerospace** in 1977. In June 1993 **Raytheon Co** purchased **British Aerospace's Corporate Jet Division**, merging this with **Beech Aircraft Corp** into a new **Raytheon Aircraft Co** in 1994. Early HS.125/BH 125 versions, through the 700-series, are listed under **Hawker Siddeley**. Versions from the 800- through 1000-series are listed under **British Aerospace**. The Horizon and all subsequent developments of the 125 are listed under **Raytheon**.
125-800 (BAe 800, Raytheon Hawker 800)
(125-800FI) C-29
(125-800FI) U-125
125-800RA
125-800SIG
(125-800SM) U-125A
125-800XP (Raytheon Hawker 800XP)
125-900
125-1000 (Raytheon Hawker 1000)
146-100
146-200
146-300
748 (See: Hawker Siddeley HS.748)
780 Andover (See: Hawker Siddeley Andover)

British Aircraft Co (B.A.C.)(UK)
British Aircraft Co Super Drone
British Aircraft Co II (Glider)

British Aircraft Corp *(See: BAC)*

British Aircraft Mfg Co, Ltd (B.A.) (UK)
British Klemm Aeroplane Co Ltd imported Klemm L 25s from Germany starting in 1929. The British-built version, known as the BK (British Klemm) Swallow, first flew in 1933. The firm changed its name to **The British Aircraft Manufacturing Co Ltd** in 1935.
British Aircraft Mfg BK.1 (2L)
British Aircraft Mfg Cupid (B.A.3)
British Aircraft Mfg Double Eagle
British Aircraft Mfg Eagle
British Aircraft Mfg Swallow

British Burnelli *(See: Cunliffe-Owen)*

British Caudron Co Ltd *(See: Caudron)*

British Executive and General Aviation Ltd *(See: Beagle)*

British Hovercraft Corp (UK)
British Hovercraft SR.N4

British Klemm Aeroplane Co *(See: British Aircraft Mfg Co, Ltd)*

British Taylorcraft (UK)*(See: Auster Aircraft Ltd)*

Britten Norman (UK)
Established 1953 as **Britten Norman Ltd**, the company went into receivership in 1971 and reorganized as **Britten Norman (Bembridge) Ltd.** It was sold to the **Fairey Group** in Aug 1972 and came under the control of **Fairey Britten Norman Ltd. Pilatus Flugzeugwerke** assumed ownership in 1979 and the company continued operations as **Pilatus Britten Norman.** All Britten Norman, Fairey Britten Norman, and Pilatus Britten Norman aircraft are listed as **Britten Norman.**
Britten Norman BN-1
Britten Norman BN-2A Islander
Britten Norman BN-2A Dowty Ducted Propulsor Testbed

Britten Norman BN-2A-III Trislander
Britten Norman BN-2B Defender
Britten Norman BN-2T Turbine Islander
Britten Norman BN-3 Nymph

BRNO (Czechoslovakia)
BRNO Aeron XA-66

Broadsmith (Australia)
Broadsmith B-1
Broadsmith B-4

Brochet (Avions Maurice Brochet) (France)
Brochet "Le Petit Brochet" ("The Little Brochet")(c.1936)
Brochet M.B.50 Pupistrelle
Brochet M.B.60
Brochet M.B.70
Brochet M.B.71
Brochet M.B.80
Brochet M.B.84
Brochet M.B.100
Brochet M.B.101
Brochet M.B.110
Brochet M.B.120

Brochet-Poinsard (Avions Maurice Brochet)(France)
Brochet-Poinsard Monoplane (1934)

Brochocki, S. K. *(See: Kasper, Witold)*

Brock (Ken Brock Mfg)(Stanton, CA)
Brock Rogallo Wing Hang Glider

Brokaw (Brokaw-Jones)(Leesburg, FL)
Brokaw Bullet (Brokaw-Jones BJ-520)

Bromon Aircraft Corp (Ramey, PR)
Bromon BR2000

Bronislawski, Boleslav (France)
Bronislawski Control System Biplane (Modified H. Farman)

Brookland (UK)
Brookland Mosquito Mk.1 Hornet
Brookland Mosquito Mk.2
Brookland Mosquito Mk.3

Brooklands Aircraft Co Ltd (UK)
Brooklands Optica Scout (See: Edgley Aircraft)

Brooks Aeroplane Co (Saginaw, MI)
Brooks Outfit #1 (1911 Kit Monoplane)
Brooks Outfit #2 (1911 Kit Monoplane)
Brooks Outfit #3 (1911 Kit Monoplane)
Brooks Outfit #4 (1911 Kit Monoplane)
Brooks Outfit #5 (1911 Kit Monoplane)

Broughton Blayney Aircraft Co, Ltd (UK)
Broughton Blayney Brawny

Brown Metalplane Co (Spokane, WA)
Brown Metalplane Metalark
Brown Metalplane Silver Streak

Brown, Alden (Los Angeles, CA)
Brown (Alden) Special Monoplane Racer

Brown, Ben E. (Lawrence, KS)
Brown (Ben E.) Monoplane

Brown, Lawrence (Lawrence W. Brown Aircraft Co)(Los Angeles, CA)
Brown (L. W.) B-1 "Brown Special"
Brown (L. W.) B-2 "Miss Los Angeles"
Brown (L. W.) B-3
Brown (L. W.) L-5
Brown (L. W.) Miles-Atwood Special
Brown (L. W.) Parasol (c.1925)

Brown, Louis H. (Toledo, OH)
Brown (Louis) 1926 Biplane

Brown, Walter H. (Topeka, KS)
Brown (W. H.) 1928 Training Biplane

Brown-Mercury Aircraft Co (Los Angeles, CA)
Brown-Mercury C-2 Monoplane
Brown-Mercury C-3

Brown-Young Aircraft Co *(See: Columbia Aircraft, Inc)*

Browning, J. B. (Oceanside, CA)
Browning C-1

BRT (France)
BRT Biplane

Bruckner, Clayton (Terre Haute, IN)
Bruckner (Clayton)

Brügger, Max (Switzerland)
Brügger MB 2 Kolibri (Hummingbird)

Brulé-Girardot (France)
Brulé-Girardot Biplane (1911/12)

Brun-Cottan (France)
Brun-Cottan H.B.2 Flying Boat
Brun-Cottan 1918 Patrol Flying Boat

Brunet (France)
Brunet 1910 Tandem-Wing Biplane

Brunet-Descamps (France)
Brunet-Descamps A.2 (1924 2-seat Scout)

Brüning (Fahrzeugbau Brüning GmbH)(Forssman)(Germany)
Brüning Giant Triplane Transport (1917)(Poll Giant)

Brunner-Winkle *(See: Bird)*

Brunswick Technical High School (Germany)
Brunswick Zaunkönig (Wren) LF-1
Brunswick Zaunkönig (Wren) II

Brustmann (Germany)
Brustmann Glider (1933)

Brutsche Aircraft Corp (Neal H. Brutsche)(Salt Lake City UT)
Brutsche Freedom Sport Utility
Brutsche Freedom 40
Brutsche Freedom 180

de Bruyère (France)
de Bruyère 1917 Canard

Bruyère et Sarazin (France)
Bruyère et Sarazin 1909 Monoplane

Bryan Aircraft Co (Bryan, OH)
Bryan Phoenix SL (Homebuilt)
Bryan II Roadable (Homebuilt)
Bryan III

Bryant Aircraft Syndicate (Los Angeles, CA)
Bryant (Aircraft Syndicate) Monoplane (1927)

Bryant, Alys McKey (USA)
Bryant (Alys) Curtiss-Type Biplane

Bryant, Frank (San Francisco, CA)
Bryant (Frank) 1912 Twin Tractor Biplane

Bryant, John (San Francisco, CA)
Bryant (John) Pusher Biplane

Bryant (John) Pusher Hydroaeroplane

BTA Top-Air Sro (Czech Republic)
BTA Top-Air Tango (See: S-Wing Swing)

BUAA (Beijing Universitry of Aeronautics and Astronautics)(China)
BUAA M-16

Buchanan Aircraft Corp Ltd (BAC) (Australia)
Buchanan (Aircraft) BAC-204

Buchanan, William O. (USA)
Buchanan (W. O.) Zipper (Rebuilt Gotch & Brundage Special)

Bucher (USA)
Bucher Special "Flying Dutchman"

Buchet (France)
Buchet 1906/08 Helicopter

Bücker (Bücker Flugzeugbau GmbH) (Germany)
Bücker Bü 131 A Jungmann (See also: CASA 1.131)
Bücker Bü 131 B Jungmann
Bücker Bü 133 A Jungmeister
Bücker Bü 133 C Jungmeister
Bücker Bü 133 C Jungmeister, NASM
Bücker Bü 180 Student
Bücker Bü 181 Bestmann
Bücker Bü 181 Bestmann, NASM
Bücker Bü 182 Kornett

Buckeye Industries Inc (Argos, IN)
Buckeye Dream Machine 503
Buckeye Dream Machine 582
Buckeye Eagle 447
Buckeye Eagle 503
Buckeye Falcon 582
Buckeye Millennium
Buckeye SkyHawk

Buckley Aircraft Co (R. B. Buckley) (Wichita, KS)
Buckley F1 (1929 4-place Monoplane)
Buckley Wichcraft

Budd (Edward G. Budd Mfg Co) (Philadelphia, PA)
Budd BB-1 Pioneer
Budd RB-1 Conestoga

Budig, F. (Germany)
Budig Motor Glider

Budil'nik (Russia)
Budil'nik (Alarm Clock) 1927 Homebuilt

Bueno et Demaurex (France)
Bueno et Demaurex Biplane (1910)
Bueno et Demaurex Monoplane

***Buente. Benjamin E. , Jr** (See: Shannon & Buente)*

Buers, E. J. (Cicero, IL)
Buers (E. J.) Tractor Biplane (1915)

Buethe (W. B. Buethe Enterprises, Inc)(Cathedral City, CA)
Buethe Barracuda (Homebuilt)

Bugatti (France)
Bugatti Rocket Plane (100 Derivative)
Bugatti 100

Buhl (Buhl-Verville Aircraft Co, Buhl Aircraft Co)(Marysville, MI)
Buhl Airsedan CA-3B (Junior Airsedan)
Buhl Airsedan CA-3C (Special Airsedan)
Buhl Airsedan CA-3D (Sport Airsedan)
Buhl Airsedan CA-3E (Diesel Airsedan)
Buhl Airsedan CA-5
Buhl Airsedan CA-5 "Miss Doran"
Buhl Airsedan CA-5A
Buhl Airsedan CA-6 (Standard Airsedan)
Buhl Airsedan CA-8A (Senior Airsedan)
Buhl Airsedan CA-8B (Senior Airsedan)
Buhl Airster CA-3 (J4 Airster)
Buhl Airster CA-3A (J5 Airster)
Buhl Airster CW-3 (OX-5 Airster)
Buhl Autogiro
Buhl Bull Pup (LA-1)
Buhl Bull Pup (LA-1A)
Buhl Mailplane CA-1
Buhl Monoplane
Buhl Primary Trainer

Bulaero Ultra Leger Aviation (France)
Bulaero Zùlù (Ultralight)

Bulot, Walther (Belgium/France)
Bulot 1909 Triplane
Bulot 1909 Biplane
Bulot 1911 Biplane la Mouette
Bulot 1911 Monoplane
Bulot 1911 Monoplane la Mouette

Bunting, John F. (Seattle, WA)
Bunting Model A

Bunyard (Flushing, NY)
Bunyard BAX-3 Amphibian
Bunyard Sportsman Amphibian (Prototype)

Bünzli (France)
Bünzli Glider (1916)

***Burdette** (See: Whites Burdette S-30)*

***Burga, Lt Romulo (Peru)**(See: Avro Burga Monoplane)*

Burgess, Walter K. (Chanute Field, IL)
Burgess (Walter K.) Amphibian

Burgess (Marblehead, MA)
W. Starling Burgess Co Ltd began aeronautical work in 1909. In 1910, the company's aeronautical work was transferred to **Burgess Co and Curtis**, which dissolved in January 1914 to reform as **Burgess Co.** Burgess Co was sold to **Curtiss Aeroplane & Motor Co** in February 1916, but operations at Burgess' Marblehead plant remained under the direction of Burgess and products were sold under the Burgess or Burgess-Dunne names. All W. S. Burgess, Burgess Co and Curtis, and Burgess Co aircraft (including license-built Wright and Grahame-White aircraft) are listed under **Burgess.** Aircraft built under the Dunne license are listed as **Burgess-Dunne.** The Burgess Model F modified by Burgess and Howard Gill to compete for the 1912 Gould Prize is listed as **Burgess-Gill.**

Burgess A (Flying Fish)
Burgess B
Burgess Type BP (Primary Trainer)
Burgess C
Burgess D
Burgess E (Grahame-White Baby)
Burgess F (Wright B) Moth
Burgess F (Wright B) Hydro
Burgess Gordon Bennett Racer
Burgess H (Military Tractor)
Burgess H (Military Tractor) Modified
Burgess Type HT-A (Speed Scout A)
Burgess Type HT-B (HT-2)(Speed Scout B)
Burgess I (Coastal Defence Hydro)
Burgess J
Burgess K (Navy Flying Boat D-1)
Burgess M (Collier Flying Boat)
Burgess Navy Flying Boat D-2
Burgess O (Gunbus)
Burgess S
Burgess Twin-Engined Hydro (1917)
Burgess U
Burgess-Dunne Astor Seaplane
Burgess-Dunne BD
Burgess-Dunne BDF

Burgess-Dunne BDH
Burgess-Dunne No.1
Burgess-Dunne No.2 (Navy AH-7)
Burgess-Dunne No.3 (Army)
Burgess-Dunne No.4
Burgess-Dunne No.5
Burgess-Dunne No.6 (Military)
Burgess-Dunne No.7
Burgess-Dunne No.8
Burgess-Dunne Navy A-55
Burgess-Dunne Navy AH-10
Burgess-Dunne 1914 (Canadian)
Burgess-Dunne 1915 (Russian)
Burgess-Gill (Gould Prize Aircraft)

Burgfalke Flugzeugbau (Germany)
Burgfalke Lo 100 Zwergreiher (Heron)
Burgfalke Lo 150 Sailplane
Burgfalke Lo 150 B Sailplane
Burgfalke N.150 Schulmeister (Schoolmaster)

Burke, A. L. (USA)
Burke (A. L.) Monoplane

Burke, Carroll F. (Los Angeles, CA)
Burke (Carroll) Condor Wing No 1
Burke (Carroll) Condor Wing No 2

Burkhart Grob Flugzeugbau *(See: Grob)*

Burlington Airplane Co, Inc (Burlington, IA)
Burlington:
H-L (A. T. Hartman/A. P. Logan)
H.L.1 (A. T. Hartman/A. P. Logan)

Burnelli (Keyport, NJ)
Vincent Burnelli's first aircraft design was the **Burnelli-Carisi** 1915 Biplane. In 1916, Burnelli established the **Continental Aircraft Corp** in Keyport, NJ. This was followed by **Airliner Engineering Corp** in 1920 and **Remington-Burnelli Co** in 1921. This company became **Upperçu-Burnelli Co** in 1929 and **Burnelli Aircraft Inc** in 1936. All Burnelli designs, including aircraft built under Burnelli's direction by **Cancargo** (Canada), are listed under **Burnelli**. British-license Burnelli aircraft manufactured by **Cunliffe-Owen Aircraft, Ltd** are listed under **Cunliffe-Owen (Burnelli)**.
Burnelli A-1
Burnelli B-1000
Burnelli CB-8
Burnelli CB-16
Burnelli CB-35
Burnelli (CanCargo) CBY-3 Loadmaster
Burnelli CU-16
Burnelli GB-171
Burnelli (Continental) GR-1 (1916)
Burnelli (Continental) KB-1 (1916)
Burnelli (Continental) KB-3T (1916)
Burnelli (Remington-Burnelli) RB-1
Burnelli (Garvan-Burnelli) RB-2
Burnelli (Upperçu-Burnelli) UB-14
Burnelli (Upperçu-Burnelli) UB-14B
Burnelli (Upperçu-Burnelli) UB-16
Burnelli (Upperçu-Burnelli) UB-20 (UB-18)
Burnelli (Upperçu-Burnelli) UBSS
Burnelli (Upperçu-Burnelli) X-3 (GX-3) (Guggenheim Prize)
Burnelli-Carisi Biplane (1915)

Burney, Lt Charles Dennistoun *(See: Bristol-Burney)*

Burns Aircraft Corp (Starkville, MS)
Burns Aircraft Corp BA-42

Buscaylet et Cie. (France)
Buscaylet Amphibian (1922)
Buscaylet-Bêchéreau Sesquiplane

Buscaylet-de Monge (France)*(See also: Monge)*
Buscaylet-de Monge:
7.4
7.5 (2-Engined Flying Wing Racer)
52 C1 (5./2)
T.75
1925 2-Seat, 2-engined Monoplane
1926 Reconnaissance Aircraft

Bush Welding Works (Piqua, OH)
Bush (OH) B-1 Monoplane

Bush, Eldon and Gilbert (UK)
Bush, Eldon and Gilbert Motorplane

Bushby Aircraft Inc (Glenwood, IL & Minooka, IL)
Bushby MM-1 Midget Mustang
Bushby M-II Mustang II

Bushey-McGrew (Los Angeles, CA)
Bushey-McGrew Racer (Modified Rider "Bumble Bee")

Bushmaster Aircraft Corp 2000 Trimotor *(See: Stout Bushmaster 2000 Trimotor)*

Business Aircraft Corp (BACC) *(See: Howard Aero Manufacturing Division)*

Butler Aircraft Corp (Kansas City, MO)
Butler Black Hawk
Butler Coach

Butterfly Aero (Oakville, WA)
Butterfly Banty

Butusov, William (Chicago, IL)
Butusov 1896 Glider (Chanute Experiments)

Buxton (Texas)
Buxton Transporter (Glider)

Buxton, Jay (Hawthorne CA)
Buxton (Jay) B-4-A Powered Glider

BX Aviation (Max Brändli) (Switzerland)
BX-2 Cherry

C

C. & A. Witteman *(See: Witteman, C. & A.)*

C. W. Aircraft (C. R. Chronander & J. I. Waddington)(UK)*(See: G.A.L. Cygnet)*

CAARP (Cooperative des Ateliers Aéronautiques de la Région Parisienne)(France)
CAARP CAP 10
CAARP CAP 20

CAB (Cantieri Aeronautica Bergamaschi, Caproni Aeronautica Bergamascha) (Italy)
CAB AP.1 Prototype
CAB AP.1
CAB AP.1*bis*
CAB C.1
CAB C.2
CAB C.4
CAB PL.3
CAB PS.1

C.A.B. (Constructions Aéronautiques du Béarn)*(See: Béarn)*

Cabrinha, Richard (USA)
Cabrinha RC-412 Free Spirit (Mk I)

C.A.C. *(See: Civilian Aircraft Co)*

CAC (Commonwealth Aircraft Corp, Ltd)(Australia)
"Encouraged" by the Australian government, **Commonwealth Aircraft Corp, Ltd** (**CAC**) was established in 1936 to manufacture military aircraft and engines in Australia. CAC built a variety of native and licensed designs until 1985, when it became a wholly-owned subsidiary of **Hawker de Havilland Ltd** (**HDH**) of Bankstown, NSW. In 1986, the company was renamed **Hawker de Havilland Victoria Ltd** (**HDHV**). In 1991/92, airframe work was transferred to HDH while HDHV concentrated on engine manufacture as Australia's only native aircraft engine company. For all CAC aircraft and license work having Commonwealth ("CA-") designations, see **CAC**. For all license work not having CA- designations, see the original manufacturer.
CAC Wirraway (CA-3)
CAC Boomerang (CA-12)
CAC Boomerang (CA-13)
CAC Boomerang (CA-14)
CAC Boomerang (CA-19)
CAC CA-15
CAC CA-23
CAC CA-24 (P1081)
CAC CA-31
CAC Ceres (CA-28)
CAC JetRanger (Bell A206B-1)(CA-32)
CAC LADS (Fokker F-27 Laser Airborne Depth Sounder)(CA-35)
CAC Macchi MB.326H (CA-30)
CAC Mirage III (CA-29)
CAC Mustang (CA-17)

CAC Mustang (CA-18)
CAC Mustang (CA-21)
CAC Orion P-3C (CA-33)
CAC Pilatus PC-9 (CA-36)
CAC Sabre (CA-26)
CAC Sabre (CA-27)
CAC Wackett (CA-2)
CAC Wackett (CA-6)
CAC Wamira (CA-34)
CAC Winjeel (CA-23)
CAC Winjeel (CA-25)
CAC Wirraway (CA-1)
CAC Wirraway (CA-5)
CAC Wirraway (CA-7)
CAC Wirraway (CA-8)
CAC Wirraway (CA-9)
CAC Wirraway Dive Bomber (CA-10)
CAC Wirraway Dive Bomber (CA-16)
CAC Wirraway (CA-20)
CAC Woomera (CA-4)
CAC Woomera (CA-11)

Cadillac Aircraft Corp (Detroit, MI)
Cadillac Voyageur (Amphibian)

Cage, John M. (Denver, CO)
Cage (John M.) Tiltrotor (c.1909)

Cagny, Raymond (France)
Cagny Performance 2000

Caille (France)
Caille 1911 Type Militaire

Cain Aircraft Corp (Detroit, MI)
Cain Sport

Cairns Aircraft (Naugatuck, CT)
The **Cairns Aircraft Syndicate** was established in October 1928 to develop low-wing all-metal monoplanes. In August 1929, the Syndicate was taken over by the **Cairns Development Co**, which continued to work on the same basic airframe until the company faded from the aviation scene in the immediate pre-WWII years. All **Cairns Aircraft Syndicate** and **Cairns Development Co** products are listed under **Cairns**.
Cairns Model A

Cairns Model AC6
Cairns Model C2
Cairns Pusher Cabin Monoplane .

C.A.L. (Columbia Air Liners, Inc) (New York, NY)
C.A.L. Mailplane "Uncle Sam" (1929)
C.A.L. CAL-1 Triad
C.A.L. CAL-2 (5-Place Amphibian)

Calderara, Mario (Italy)
Calderara 1912 Hydroaeroplane

California Aero Co (Tracy, Ca)*(See: Helton)*

California Aero Glider Co (USA)
California Aero Glider "Skyway Express"

California Aircraft Corp (Los Angeles, CA)
California Cub D-1
California Cub D-2

California Glider Club *(See: Walters Brothers)*

California Institute of Technology (Pasadena, CA)
California Institute of Technology Merrill Type Stagger-Decalage

Call Monoplane (c.1911)*(See: Aerial Navigation Co)*

CallAir (Call Aircraft Co)(Afton, WY)

The **Call Aircraft Co** (**CallAir**) was established in Afton, WY by the Call family in the late 1930s. In 1962, the company was purchased at auction by **Intermountain Manufacturing Co** (**IMCO**), which continued to manufacture aircraft under the CallAir name. IMCO sold its CallAir assets to the **Aero Commander** division of **Rockwell-Standard Corp**, which operated CallAir as its **Aero Commander-Afton Division** and marketed CallAir products under the "Ag Commander" name. After Rockwell's merger with **North American Aviation** to form **North American Rockwell Corp**, CallAir's products were marketed as the "Quail Commander" series. For all CallAir-originated designs, see **CallAir** (with later market names added in parentheses).

CallAir A
CallAir A-2
CallAir A-3
CallAir A-9 (AG Commander A-9, Quail Commander)

Calvignac (France)
Calvignac 1913 Monoplane

Camair Aircraft Corp (Remsenburg, NY)
Camair 480 Twin Navion A (CTN-A)
Camair 480 Twin Navion B (CTN-B)
Camair 480 Twin Navion C (CTN-C)
Camair 480 Twin Navion D (CTN-D)

Camal, Victor (France)
Camal 1911 Flying Machine (Patent)

CAMCO *(See: Chicago Aircraft Manufacturing Co)*

Cameron
Cameron Bottle Tuborg - Version 52
Cameron N-56
Cameron O-56
Cameron N-77
Cameron O-77
Cameron V-77
Cameron N-90

Cameron Iron Works, Inc *(Galveston, TX)(See: Camair)*

Cammacorp (El Segundo, CA)
Cammacorp DC-8-71 (DC-8 Super 70)
Cammacorp DC-8-71CF (DC-8 Super 70)
Cammacorp DC-8-72 (DC-8 Super 70)
Cammacorp DC-8-72AF (DC-8 Super 70)
Cammacorp DC-8-72CF (DC-8 Super 70)
Cammacorp DC-8-73 (DC-8 Super 70)
Cammacorp DC-8-73AF (DC-8 Super 70)
Cammacorp DC-8-73CF (DC-8 Super 70)
Cammacorp DC-8-73PF (DC-8 Super 70)

Campbell Aircraft Co (St Joseph, MO)
Campbell Model F

Campbell Aircraft Ltd (UK)
Campbell Cricket (License-Built Bensen Autogyro)
Campbell Curlew

CAMS (Chantiers Aéro-Maritimes de la Seine)(France)

Chantiers Aéro-Maritimes de la Seine (**CAMS**), established in 1921, specialized in the design and construction of flying boats. The company was acquired by **Aéroplanes Henry Potez** in 1933, but continued work under its own name until 1937. With the nationalization of the French aviation industry, the CAMS works were split between the **Société Nationale de Constructions Aéronautiques du Nord** (**SNCAN**) and the **Société Nationale de Constructions Aéronautiques de Sud-Est** (**SNCASE**). Some prototype development continued under the Potez name until the fall of France in 1940. For all CAMS products, including those produced under the Potez name, see **CAMS**.

CAMS C.9 (License-Built Savoia S.9)
CAMS C.13 (License-Built Savoia S.13)
CAMS 30
CAMS 30e
CAMS 30E
CAMS 30^E
CAMS 30^T
CAMS 31
CAMS 31-M
CAMS 31-P
CAMS 32R
CAMS 33
CAMS 33B
CAMS 33T
CAMS 35
CAMS 36 (1922 Schneider Race)
CAMS 36*bis* (1923 Schneider Race)
CAMS 37
CAMS 37-2
CAMS 37-10
CAMS 37A
CAMS 37GR
CAMS 38 (1923 Schneider Cup Racer)
CAMS 41
CAMS 41*bis*
CAMS 43
CAMS 44
CAMS 45
CAMS 46 Et2
CAMS 51
CAMS 51 C
CAMS 51 R3
CAMS 53
CAMS 53-1
CAMS 53-2
CAMS 53-3
CAMS 53-4
CAMS 54
CAMS 54 GR
CAMS 55-1
CAMS 55-2
CAMS 55-6
CAMS 56

CAMS 57 (53-R)
CAMS 58
CAMS 58-0
CAMS 58-1A
CAMS 58-1M
CAMS 58-2
CAMS 58-3
CAMS 60
CAMS 80
CAMS (Potez-CAMS) 110
CAMS (Potez-CAMS) 120
CAMS (Potez-CAMS) 141
CAMS (Potez-CAMS) 160
CAMS (Potez-CAMS) 161
CAMS (Potez-CAMS) 170

Can-Car *(See: CCF)*

Canada Air RV Inc (Canada)
Canada ARV Griffin

Canadair (Canada)
Canadair CL-1 Canso (Licensed PBY)
Canadair CL-2 North Star
Canadair CL-4 Argonaut (C-4)
Canadair CL-5 (C-5)
Canadair CL-13 Sabre (Licensed F-86)
Canadair CL-21
Canadair CL-28 Argus (CP-107)
Canadair CL-30 Silverstar (T-33)
Canadair CT-133 (T-33AN) Silver Star 3
Canadair CL-41 Tutor (CT-114)
Canadair CL-44 (CC-106 Yukon)
Canadair CL-66 (CC-109 Cosmopolitan) (Licensed Convair 540)
Canadair CL-84 (V/STOL)
Canadair CL-90 (CF-104) Starfighter
Canadair CL-201 (CF-104) Starfighter
Canadair CL-215
Canadair CL-219 (Northrop CF-5D)
Canadair CL-226 (Northrop CF-5)
Canadair CL-415
Canadair CL-540
Canadair CL-600 Challenger (CC-144, CE-144A)
Canadair CL-600 Challenger E
Canadair CL-601 Challenger
Canadair CL-604 Challenger
Canadair Regional Jet (RJ)
Canadair Special Edition (Regional Jet)

Canadian Aerodrome Co (Canada)
Canadian Aerodrome:
Baddeck Biplane #1
Baddeck Biplane #2
Hubbard Monoplane
Silver Dart (See Aerial Experiment Assoc)

Canadian Aeroplanes Ltd (Canada)
Canadian Aeroplanes F-5L

Canadian Car and Foundry Co Ltd *(See: CCF)*

Canadian Vickers Ltd *(See: Vickers Canada)*

Canadian Wooden Aircraft Co (Canada)
Canadian Wooden Aircraft Robin

Canaero Dynamics Aircraft Inc (Canada)
Canaero Dynamics Toucan

Canamerican Helicopter Manufacturing Co (Canada)
Canamerican S.G. VI

Cancargo Aircraft Manufacturing Co *(See: Burnelli)*

Cannon, Burrell *(See: Ezekiel Air Ship Manufacturing Co)*

Cannon, Walter (Los Angeles, CA)
Cannon (Walter) Biplane (1911)
Cannon (Walter) Monoplane (1911)

C.A.N.S.A. (Construzioni Aeronautiche Novaresi, SA)(Italy)
Gabardini, a former artist from Turin, began building aircraft in 1909, establishing **Aeronautica Gabardini SA** in 1913. In May 1936 the company became **Construzioni Aeronautiche Novaresi SA (C.A.N.S.A.)**, continuing production of the Gabardini Lictor series of light aircraft. C.A.N.S.A. was absorbed by **Fiat** in 1939, but continued to design and produce aircraft under its own name until forced out of business at the end of World War II. For all pre-1936 designs, including the Lictor series, see **Gabardini**. For all C.A.N.S.A.-originated designs, including those after 1939, see **C.A.N.S.A.**
C.A.N.S.A. C.5
C.A.N.S.A. FC.12

Cant (CNT)(Cantieri Navali Triestino) (Italy)*(See also: Cant, CRDA)*
In 1923, the shipbuilding firm of **Cantieri Navali di Monfalcone** in Trieste established **Cantieri Navale Triestino (CNT**, normally called **Cant)** as an aircraft manufacturing branch to concentrate on flying boats and seaplanes. In 1931 Cant reorganized as **Cantieri Riuniti dell'Adriatico (CRDA**, also commonly called Cant) and acquired Filippo Zappata, a well-known Italian aeronautical engineer, from Blériot. The company closed after the Armistice in 1943. For all CNT products, see **Cant (CNT)**. For all CRDA products having Zappata ("Z.") designations, see **Cant (CRDA)**.
Cant (CNT) 6
Cant (CNT) 6*ter*
Cant (CNT) 7
Cant (CNT) 7*bis*
Cant (CNT) 7*ter*
Cant (CNT) 10
Cant (CNT) 10*ter*
Cant (CNT) 10M-R1
Cant (CNT) 13
Cant (CNT) 21
Cant (CNT) 21*bis*
Cant (CNT) 22
Cant (CNT) 22R.1
Cant (CNT) 23
Cant (CNT) 25
Cant (CNT) 26

Cant (CRDA)(Cantieri Riuniti dell'Adriatico)(Italy)*(See also: Cant, CNT)*
In 1923, the shipbuilding firm of **Cantieri Navali di Monfalcone** in Trieste established **Cantieri Navale Triestino (CNT**, normally called **Cant)** as an aircraft manufacturing branch to concentrate on flying boats and seaplanes. In 1931 Cant reorganized as **Cantieri Riuniti dell'Adriatico (CRDA**, also commonly called **Cant)** and acquired Filippo Zappata, a well-known Italian aeronautical engineer, from Blériot. The company closed down after the Armistice in 1943. For all CNT products, see **Cant (CNT)**. For all CRDA products having Zappata ("Z.") designations, see **Cant (CRDA)**.
Cant (CRDA) Z.501
Cant (CRDA) Z.505
Cant (CRDA) Z.506
Cant (CRDA) Z.506A
Cant (CRDA) Z.506B
Cant (CRDA) Z.506C
Cant (CRDA) Z.506S
Cant (CRDA) Z.508
Cant (CRDA) Z.509
Cant (CRDA) Z.511

Cant (CRDA) Z.1007*bis*
Cant (CRDA) Z.1010
Cant (CRDA) Z.1011
Cant (CRDA) Z.1012
Cant (CRDA) Z.1018

***Cantieri Aeronautica Bergamaschi** (See: CAB)*

***Cantieri Navali Triestino** (See: Cant, CNT)*

***Cantieri Riuniti dell'Adriatico** (See: Cant, CRDA)*

***Cantilever Aero Co** (See: Christmas)*

Canton (France)
Canton S2 (Armored Attack Biplane)

Canton et Unné (France)
Canton et Unné:
1909 Tandem Triplane
1910 Monoplane
1910 Monoplane, Third Aircraft

CAP (Companhia Aeronáutica Paulista)(Brazil)
CAP 1 Planalto
CAP 4 Paulistinha Tourer
CAP 4B Ambulance
CAP 4C Military Observation
CAP 5 Carioca

Capelis Safety Airplane Corp Ltd (Hesperia, CA)
Capelis C-12

Capella Aircraft Corp (Austin, TX)
Capella Fastback
Capella Javelin
Capella Javelin II
Capella T-Raptor
Capella TD
Capella TR

Capen Aircraft Corp (Ernest J. Capen)(Lincoln, IL)
Capen Skyway 100
Capen 1929 Monoplane

Capital Aircraft Co Inc (Lansing, MI)
Capital Aircraft Air Trainer

Capital Copter Corp (Capital Helicopter Corp)(St Paul, MN)
Capital Copter:
C-1 Firefly (Hoppi-Copter, Strap-on Helicopter)
C-1L (Hoppi-Copter, Strap-on Helicopter)
3C (Autogyro)

Capon (France)
Capon 1911 Monoplane

Caproni (Italy)
Gianni Caproni began designing aircraft in 1908 and established **Ingg. De Agostini & Caproni Aviazione** in 1911. De Agostini soon withdrew and, in November 1911, Caproni formed **Caproni & C.** with Carlo Comitti. In September 1912 engineer Luigi Faccanoni joined, and the company became **Società Ingegneri Caproni e Faccanoni**. This company was soon bought by the Italian government, which retained Caproni as technical director. In March 1915 Caproni formed **Società per lo Sviluppo dell'Aviazione** in Italia, a cooperative, to build aircraft at the old Caproni plant at Vizzola (which Caproni leased back from the government). The company restructured in 1929, becoming **Aeroplani Caproni SA**; based at Taliedo, it was popularly known as **Caproni Taliedo**. The Vizzola factory became **Scuola Aviazione Caproni** then, in March 1937, **Caproni Vizzola SA**. Caproni also acquired **Cantieri Aeronautica Bergamschi (CAB)**, which became **Caproni Aeronautica Bergamasca (CAB)** in July 1938. Design and construction were shared by the three branches (Caproni Taliedo, Caproni Vizzola, and CAB), as well as other members of the Caproni group as such **Aeronautica Predappio Nuova (Caproni-Predappio)** and **Aeroplani Caproni-Cantiere Aeronautico di Trento (Caproni-Trento)** until April 1939 when design and prototype testing were centralized at CAB and production and development centered at Taliedo and Vizzola. During the 1930s, Caproni also acquired **Avio Industrie Stabiensi Catello Coppola fu Antonio (AVIS)**, **Compagnia Nazionale Aeronautica (CNA)**, and **Officine Meccanische Italiane Reggiane**, each of which produced aircraft under their own names. Wartime destruction and dislocation undermined the company. Although Caproni Vizzola remained solvent, it withdrew from airframe manufacture.

In 1968 Caproni Vizzola began producing sailplanes, eventually moving to jet-powered motor gliders and light trainers. In 1983 Caproni Vizzola was acquired by Agusta, which continued development under its own name.

For products having Caproni ("Ca") designations, see **Caproni**. For products developed by subsidiaries, but not having Caproni designations, see **AVIS**, **CAB**, **CNA**, or **Reggiane**. For aircraft built at Caproni plants, but having designations based on the designing engineer (such as Stipa, Campini, etc.), see **Caproni-Campini**, **Caproni-Chiodi**, **Caproni-Fabrizi**, **Caproni-Pensuti**, **Caproni-Stipa**, and **Caproni-Trigona**. For aircraft designed after 1968, see **Caproni Vizzola**. For aircraft designed after 1983 see **Agusta**.

Caproni Biplane Glider (1908)
Caproni Unbuilt Glider Studies
Caproni Ca.1 (First Flight 1910)
Caproni Ca.2 (Rebuilt Ca.1)
Caproni Ca.3
Caproni Ca.4
Caproni Ca.5
Caproni Ca.6 Tractor
Caproni Ca.7
Caproni Ca.8 (Cm 1)
Caproni Ca.9
Caproni Ca.10
Caproni Ca.11 (Cm 5)
Caproni Ca.12 (Cm 6)
Caproni Ca.13 (Cm 7)
Caproni Ca.14 (Cm 9)
Caproni Ca.15
Caproni Ca.16 (Cm 12)
Caproni Ca.17
Caproni Ca.18
Caproni Ca.19
Caproni Ca.20
Caproni Ca.21
Caproni Ca.22
Caproni Ca.23
Caproni Ca.24
Caproni Ca.25
Caproni Ca.31 (Military Ca.1)
Caproni Ca.32 (Military Ca.1, Ca.300)
Caproni Ca.33
Caproni Ca.34
Caproni Ca.35
Caproni Ca.36
Caproni Ca.36 M
Caproni Ca.36 S (Military Ca.3, Ca.450)
Caproni Ca.37

Caproni Ca.38
Caproni Ca.40 (Military Ca.4)
Caproni Ca.41 (Military Ca.4)
Caproni Ca.42 (Military Ca.4)
Caproni Ca.43 (Military Ca.4 Floatplane)
Caproni Ca.44 (Military Ca.5, Ca.450)
Caproni Ca.45 (Military Ca.5)
Caproni Ca.46 (Military Ca.5, US-Built)
Caproni Ca.47 Idrovolante Caproni (Caproni Hydroplane, I.Ca.)
Caproni Ca.48
Caproni Ca.50
Caproni Ca.51 (Military Ca.4)
Caproni Ca.52 (Military Ca.4)
Caproni Ca.53
Caproni Ca.56
Caproni Ca.56a
Caproni Ca.57
Caproni Ca.58
Caproni Ca.59
Caproni Ca.60 Transaereo
Caproni Ca.61a
Caproni Ca.64 (MC1)
Caproni Ca.66 (LB.4)
Caproni Ca.67
Caproni Ca.70
Caproni Ca.71 (Ca.70)
Caproni Ca.72
Caproni Ca.73
Caproni Ca.73*bis*
Caproni Ca.73*ter*
Caproni Ca.74
Caproni Ca.79
Caproni Ca.80
Caproni Ca.80 S
Caproni Ca.82
Caproni Ca.82 Co-C
Caproni Ca.87
Caproni Ca.88
Caproni Ca.89 (Ca.74)
Caproni Ca.90
Caproni Ca.95
Caproni Ca.97 (Prototype)
Caproni Ca.97 C.Mo
Caproni Ca.97 C.Tr
Caproni Ca.97 Co
Caproni Ca.97 Idro
Caproni Ca.97 M
Caproni Ca.97 Ri
Caproni Ca.100 Capronicino
Caproni Ca.101
Caproni Ca.101 C
Caproni Ca.101*bis*
Caproni Ca.101 E
Caproni Ca.102
Caproni Ca.102*quat*
Caproni Ca.103
Caproni Ca.104
Caproni Ca.105
Caproni Ca.109
Caproni Ca.111 Prototype
Caproni Ca.111
Caproni Ca.111 C
Caproni Ca.113
Caproni Ca.113 AQ
Caproni Ca.114
Caproni Ca.120
Caproni Ca.122
Caproni Ca.123
Caproni Ca.124
Caproni Ca.125
Caproni Ca.127
Caproni Ca.131
Caproni Ca.132
Caproni Ca.133
Caproni Ca.133 S
Caproni Ca.134
Caproni Ca.135 Prototype
Caproni Ca.135
Caproni Ca.135*bis*
Caproni Ca.135/P.XI
Caproni Ca.137
Caproni Ca.140
Caproni Ca.142
Caproni Ca.146
Caproni Ca.148
Caproni Ca.161
Caproni Ca.161*bis*
Caproni Ca.163
Caproni (Predappio) Ca.164
Caproni Ca.165
Caproni Ca.166
Caproni Ca.167
Caproni Ca.169
Caproni Ca.193
Caproni (Bergamaschi) Ca.308 (Ca.306) Borea
Caproni Ca.309 Ghibli
Caproni Ca.310 Libeccio
Caproni Ca.310*bis* Libeccio
Caproni Ca.311
Caproni Ca.312
Caproni Ca.312*bis*
Caproni Ca.313
Caproni Ca.313 RPB.1
Caproni Ca.313 RPB.2
Caproni Ca.314
Caproni Ca.316
Caproni Ca.331
Caproni Ca.335
Caproni Ca.355
Caproni Ca.405
Caproni Ca.602
Caproni Ca.603

Caproni Bulgaria (Bulgare)(Kaproni Bulgarski)(Bulgaria)
Caproni Bulgaria:
KB-1 Papillon (Butterfly) (CaB100UO)
KB-2A Tchutchuliga (Lark)
KB-2UT (Ca 113 Derivative)
KB-3 Tchutchuliga I (Lark I)
KB-4 Tchutchuliga II (Lark II)
KB-5 Tchutchuliga III (Lark III)
KB-11 Fazan (Pheasant)
KB-309 Papagal
KB-311 Kvazimodo (Quasimodo)

Caproni Vizzola (Italy)
Caproni Vizzola A-12
Caproni Vizzola A-21 Calif
Caproni Vizzola A-21J Calif
Caproni Vizzola A-21S Calif
Caproni Vizzola A-21SJ Calif
Caproni Vizzola C-22J

Caproni-Campini (Italy)
Caproni-Campini CC.2

Caproni-Chiodi (Italy)
Caproni-Chiodi CH.1

Caproni-Fabrizi (Italy)
Caproni-Fabrizi (Vizzola) F.4
Caproni-Fabrizi (Vizzola) F.5
Caproni-Fabrizi (Vizzola) F.6M
Caproni-Fabrizi (Vizzola) F.6Z

Caproni-Pensuti (Italy)
Caproni-Pensuti Triplane (c.1918)

Caproni-Stipa (Italy)
Caproni-Stipa Proof-of-Concept
Caproni-Stipa 2-Engined Bomber
Caproni-Stipa 3-Engined Bomber
Caproni-Stipa 403(4-Engine Bomber)

Caproni-Trigona (Italy)
Caproni-Trigona TR.
Caproni-Trigona Tricap Sauro 1

Carden-Baynes Aircraft Ltd (Sir John Carden)*(See: Baynes, L. E.)*

Cardoen (Industrias Cardoen Ltda)(Chile)
Cardoen C 206L-III (Modified Bell JetRanger)

Cardoza-Parso (Herbert J. Cardoza & Harry Parso)(San Jose, CA)
Cardoza-Parso PC-1

Caribe Doman Helicopters, Inc
(See: Doman)

Carley Aircraft Manufactory (J. D. "Joop" Carley)(Netherlands)
Carley C.II
Carley C.III Tandem Trainer
Carley C.12
Carley S.I
Carley Single-Seat Biplane (c1922)

Carlier, René (France)
Carlier 1911 Biplane

Carlson Aircraft Inc (Ernie Carlson) (East Palestine, OH)
Carlson Criquet
Carlson Skycycle
Carlson Sparrow Sport Special
Carlson Sparrow Ultralight
Carlson Sparrow II
Carlson Sparrow II-XTC

Carlton AeronauticaL Research Laboratory (Farmington, NY)
Carlton Convertiplane

Carma Manufacturing Co (Tucson, AZ)
Carma VT-1 Weejet

Carman (USA)
Carman 3000 Trimotor (See also: Stinson SM-6000)

Carmier, Pierre (France)
Carmier 1924 Monoplane

Carolina Aeronautical Corp (Burlington, NC)
Carolina Thiadchues No.1

Carolina Aviation (Greenville, SC)
(See: Johnson, Luther C.)

Carpenter, Merrell L. (Joplin, MO)
Carpenter MLC-3 (1933)

Carr, Walter (Lansing, IL)*(See also: Paramount Aircraft Corp)*
Carr Special "Saginaw Junior"

Carroll (France)
Carroll A2

Carson Helicopters Inc (Perkasie, PA)
Carson Super C-4 (Modified Bell 47G)

Carstedt Inc (USA)
Carstedt Jet Liner 600 (Turboprop D.H.104 Dove)

Carton-Lachambre (France)
Carton-Lachambre Biplane
Carton-Lachambre Monoplane

CASA (Construcciones Aeronáuticas, SA)(Spain)
CASA C-101
CASA C-212
CASA 1.131E
CASA 1.131L
CASA III ("Avioneta")
CASA 201 Alcotan
CASA 202 Halcón
CASA 207 Azor
CASA CN-235 (See: Airtech)
CASA 352L
CASA 352L, NASM
CASA-Breguet 19GR "Jesus del Gran Poder"

Cascade Ultralights (Issaquah, WA)
Cascade Kasperwing
Cascade Kasperwing, NASM

Casey Jones School of Aeronautics (Newark, NJ)
Casey Jones Amphibian

Casmuniz *(See: Muniz, Col Antonio)*

Caspar Werke, GmbH (Germany)
Carl Caspar originally established **Zentrale für Aviatik** in 1911. In 1913 the company was renamed **Hansa Flugzeugwerke**, which merged with **Brandenburgische Flugzeugwerke** in 1914 to form **Hansa und Brandenburgische Flugzeugwerke GmbH**. Caspar separated his operation from Hansa-Brandenburg in 1916 to form **Hanseatische Flugzeugwerke Carl Caspar**, which survived on subcontract work for the rest of WWI, before being liquidated in 1922. Caspar then formed **Caspar Werke, GmbH**, which produced a variety of aircraft until the company was liquidated in 1927. For aircraft produced by Zentrale für Aviatik and Hansa, see **Hansa.** For aircraft produced by Hansa-Brandenburg, see **Hansa-Brandenburg.** For aircraft produced by Caspar Werke, see **Caspar**.
Caspar C.16
Caspar C.23
Caspar C.24
Caspar C.26
Caspar C.35 Priwall
Caspar C.36 Bayern (Bavaria)
Caspar Cl.E.II
Caspar C.L.E.12
Caspar C.L.E.17
Caspar (Caspar-Theis) C.T.1
Caspar (Caspar-Theis) C.T.2
Caspar (Caspar-Theis) C.T.3
Caspar LE.11
Caspar U.1 (V-1)

Cassel *(See: Kassel)*

Cassio Muniz SA *(See: Muniz, Antonio)*

Cassutt, Tom (Roslyn Heights, NY)
Cassutt Special I
Cassutt Special II
Cassutt Special IIIM

Castagne et Cambageon (France)
Castagne et Cambageon 1911 Biplane

Castaibert, Pablo (Argentina)
Castaibert 910-I
Castaibert 911-II
Castaibert 912-III
Castaibert 913-IV
Castaibert 914-V
Castaibert 915-VI
Castaibert 915-VII

CAT (Costruzioni Aeronautiche Taliedo)(Italy)
CAT QR.14 Leviero
CAT Tm.2 Glider

CATA (Construction Aéronautique de Technologie Avancee)(France)
CATA LMK.1 Oryx

Caters *(See: Voisin de Caters)*

Cathelin (France)
Cathelin 1909 Biplane

CATIC (China National Aero-Technology Import & Export Corp)(China)
CATIC A-5C (See: Nanchang)
CATIC B-6D (See: Xian)
CATIC F-7M (See: Shenyang)
CATIC FT-7 (See: Shenyang)
CATIC F-8 II (See: Shenyang)
CATIC Petrel (See: Nanchang)
CATIC SH-5 (See: Harbin)

CATIC Y-7-100 (See: Xian)
CATIC Y-8 (See: Shaanxi)
CATIC Y-12 (See: Harbin)
CATIC Z-8 (See: Changhe)
CATIC Z-9 (See: Harbin)

Cato (Stockton, CA)
Cato 1912 Biplane

Cato Aircraft and Engine Co (Joseph L. Cato)(Downey, CA)
Cato

Catron & Fisk Airplane & Engine Co (J. W. Catron and Edward M. Fisk)(Ocean Park, CA)*(See also: International Aircraft Corp)*
Catron and Fisk:
CF-10 Triplane
CF-10 "Pride of Los Angeles"
CF-11 Biplane
Twin-Engine Triplane No.1
Twin-Engine Triplane No.2
1921 Single-Seat Triplane

Caudron (Société Anonyme des Avions Caudron)(France)
After several years of building aircraft as **S.A.F.A.**, Gaston and René Caudron established **Caudron Frères** around 1912. With Gaston's death in 1915, the company continued under René Caudron until about 1926, when it became **Société Anonyme des Avions Caudron**. Avions Caudron avoided nationalization in 1936 but established a close relationship with Renault, merging by 1939 to form **Société Anonyme des Avions Caudron-Renault**. During World War II, Caudron-Renault built aircraft on Vichy and German contracts. After the war, Caudron-Renault was nationalized as part of **Société Nationale de Constructions Aéronautiques du Nord (Nord)**, eventually losing its identity. For all S.A.F.A., Caudron, and Caudron-Renault designs and all Nord designs bearing Caudron ("C.") designations, see **Caudron**.
Caudron Type A
Caudron Type B
Caudron Type C
Caudron C.17 A2
Caudron C.20
Caudron C.21
Caudron C.22
Caudron C.23
Caudron C.25
Caudron C.27
Caudron C.27*ter*
Caudron C.33
Caudron C.37
Caudron C.39
Caudron C.43
Caudron C.51
Caudron C.59
Caudron C.60
Caudron C.61
Caudron C.65 Tourisme
Caudron C.67
Caudron C.68
Caudron C.74
Caudron C.77
Caudron C.81
Caudron C.91
Caudron C.92
Caudron C.97
Caudron C.98
Caudron C.101
Caudron C.103
Caudron C.104
Caudron C.107
Caudron C.109
Caudron C.110
Caudron C.113
Caudron C.125
Caudron C.127
Caudron C.128
Caudron C.140
Caudron C.161
Caudron C.168
Caudron C.180
Caudron C.183
Caudron C.190
Caudron C.191
Caudron C.192
Caudron C.193
Caudron C.230
Caudron C.232
Caudron C.270 Luciole (Firefly)
Caudron C.272 Luciole (Firefly)
Caudron C.274 Luciole (Firefly)
Caudron C.276 Luciole (Firefly)
Caudron C.277 Luciole (Firefly)
Caudron C.278 Luciole (Firefly)
Caudron C.280 Phalène (Moth)
Caudron C.282 Phalène (Moth)
Caudron C.286 Super Phalène (Moth)
Caudron C.344
Caudron C.362
Caudron C.366
Caudron C.400 Phalène (Moth)
Caudron C.430
Caudron C.440 Goéland (Seagull)
Caudron C.441 Goéland (Seagull)
Caudron C.445 Goéland (Seagull)
Caudron C.449 Goéland (Seagull)
Caudron C.450
Caudron C.461
Caudron C.480 Frégate (Frigatebird)
Caudron C.510 Pélican (Pelican)
Caudron C.520 Simoun (Simoom) 4
Caudron C.520 Simoun (Simoom) 6
Caudron C.530 Rafale (Squall)
Caudron C.570 Kangourou (Kangaroo)
Caudron C.600 Aiglon (Eaglet)
Caudron C.601 Aiglon (Eaglet)
Caudron C.620 Simoun (Simoom)
Caudron C.630 Simoun (Simoom)
Caudron C.631 Simoun (Simoom)
Caudron C.635 Simoun (Simoom)
Caudron C.640 Typhon (Typhoon)
Caudron C.641 Typhon (Typhoon)
Caudron C.660
Caudron C.670
Caudron C.684
Caudron C.690
Caudron C.713
Caudron C.714 Cyclone
Caudron C.800 Sailplane
Caudron C.860
Caudron CRB
Caudron Type D
Caudron Type E
Caudron Type F
Caudron Type G
Caudron Type G.2
Caudron Type G.3
Caudron Type G.4
Caudron Type G.4, NASM
Caudron Type G.5
Caudron Type G.6
Caudron Type H (Monaco)
Caudron Type J
Caudron Type K
Caudron Type L
Caudron Type M
Caudron Type Monaco
Caudron Type N
Caudron O2
Caudron Type Populaire (Popular)
Caudron P.V.200 (de Vizcaya License)
Caudron Type R
Caudron Type R.3
Caudron Type R.4 ("R.8," "R.19")
Caudron Type R.5
Caudron Type R.9
Caudron Type R.11
Caudron Type R.12
Caudron Type R.14
Caudron Type R.15
Caudron 1909 Glider

Caudron 1917 Heavy Bomber Project

Caux et Camboullive (France)
Caux et Camboullive 1910 Aeroplane

Cavalier Aircraft Corp (Sarasota, Fl)
Cavalier (NAA) F-51D Mustang

Cavanah-Geppert (Cecil Cavanah) (St Louis, MO)
Cavanah-Geppert C-G-1 Sailplane

Caye, Maurice (France)
Caye 1912 Monoplane

Cayley, Sir George (UK)
Cayley 1796 Toy Helicopter
Cayley 1843 Helicopter
Cayley 1852 Glider

Cayol (France)
Cayol 1912 Monoplane

Cayre, Edmond et Ernest (France)
Cayre 1910 Monoplane

Cayrol-Castagnat (France)
Cayrol-Castagnat 1878 Flying Machine

de Caze, Viscount (France)
de Caze Helicoplane

CCF (Canadian Car and Foundry Co Ltd)(Canada)
The **Canadian Car and Foundry Co, Ltd** (**CCF**, later **Can-Car**), Canada's largest railroad car manufacturer, entered the aviation field in 1937 by acquiring a license to build Grumman G-23s. In 1945 the company set up **Cancargo Aircraft Manufacturing Co** to produce the Burnelli Load-master and further develop that aircraft. In 1946, CCF acquired the assets of **Noorduyn Aviation Ltd** and continued production of the Noorduyn Norseman until 1953, when the Can-Car disposed of its Noorduyn interests to Mr. R. B. C. Noorduyn. CCF became a subsidiary of **A. V. Roe, Canada, Ltd** in 1956. Most CCF production was devoted to license and subcontract work. For most CCF complete air-frame construction, see **CCF**. For all Noorduyn designs, see **Noorduyn**. For Cancargo work, see **Burnelli**. For all subcontract work, see the prime contractor.

CCF F.A.T.1 Maple Leaf I
CCF F.A.T.2 Maple Leaf II
CCF F.D.B.1
CCF (Grumman) G-23
CCF Harvard IV (License-Built AT-6)
CCF SBW-1 Helldiver
CCF SBW-3 Helldiver
CCF SBW-4E Helldiver
CCF (Beech) T-34A Meteor

CEA (Centre Est Aeronautique) *(See: Robin, Pierre)*

CECA (Compañia Española de Construcciones Aeronáuticas)(Spain)
CECA BB

Cedric Lee Co *(See: Lee, Cedric)*

CEI (Auburn, CA)
CEI Free Spirit Mk II

Ceita (France)
Ceita 1911 Monoplane

Celair (Peter Cellier)(South Africa)
Celair GA-1 Celstar

de la Celle (France)
de la Celle 1914 (Modified Deperdussin)

Centrair (France)
Centrair 101 Pegase (Pegasus) A
Centrair 101 Pegase (Pegasus) B
Centrair 101 Pegase (Pegasus) D
Centrair 101 Pegase (Pegasus) Club

Central Aircraft (Mahaska, KS)
Central (Mahaska,KS) 1927 Monoplane

Central Aircraft Co (UK)
The **Central Aircraft Co** was established in London in 1916 as a subsidiary of the **R. Cattle Ltd** woodworking firm. The company closed in May 1926, and the assets, primarily completed aircraft, were sold to various operators. For all aircraft produced by Central Aircraft, see **Central (UK)**.

Central (UK) Centaur 1a
Central (UK) Centaur 2a
Central (UK) Centaur 2b
Central (UK) Centaur 2e
Central (UK) Centaur 4
Central (UK) Centaur 4a
Central (UK) Centaur 4b
Central (UK) Centaur 5
Central (UK) Centaur 8 (VIII)

Central Aircraft Co (Yakima, WA) *(See: Roberts Sport Aircraft)*

Central Aviation Engineering Co (Wichita, KS)
Central (Wichita, KS) PR-2 Centaur

Central States Aero Co, Inc *(See Monocoupe)*

Central Workshops of the Polish Army Air Service, Warsaw *(See: Zalewski, Wladyslaw)*

Central-Lamson *(See Lamson)*

Centralne Studium Samolotów *(See: CSS)*

Centre Est Aeronautique (CEA) *(See: Robin, Pierre)*

Centro di Volo a Vela de Torino *(See: CVT)*

Centro Técnico de Aeronáutica (CTA) (Brazil)
Centro Técnico de Aeronáutica:
BF-1 Beija-Flor (Hummingbird)
FG-8 Guanabara
IPD/PAR-6505 Urupema

Centro Técnico Aerospacial (CTA) (Brazil)
Centro Técnico Aerospacial Paulistinha 65 (Re-engined CAP 4)

Centro Volo a Vela, Politecnico di Milano *(See: CVV)*

Centrul de Projiectare Si Consulting Pentru Aviatie SA *(See: CPCA SA)*

Century Aerospace Corp (Albuquerque, NM)
CenturyAerospaceCA-100 CenturyJet

Century Aircraft Corp (Amarillo, TX)
Century (TX) Jetstream 3 (Reengined Handley Page)

Century Aircraft Corp (Chicago, IL)
Century (IL) Sea Devil

Century Aircraft Corp (Kansas City, MO)
Century (MO) CBM-4
Century (MO) Centurion

Certonciny et James (France)
Certonciny et James 1912 Monoplane

CERVA (Consortium Européen de Réalisation de Ventes d'Avions) (France)
CERVA CE.43 Guépard (Cheetah)(See: Wassmer)
CERVA CE.44 Cougar (See: Wassmer)
CERVA CE.45 Léopard (See: Wassmer)

César (France)
César 1910 Tandem Biplane

Ceskomoravská Kolben-Danek SA
(See: Praga)

Cessna (Wichita, KS)
Cessna A-37 (OA-37, Cessna 318)
Cessna YAT-37
Cessna AA
Cessna AC
Cessna AF
Cessna AS
Cessna AT-8
Cessna AT-17 Bobcat (T-17, T-50)
Cessna AW
Cessna BW
Cessna C-34
Cessna C-37 Airmaster (C-77, UC-77)
Cessna C-37 Airmaster, Prototype
Cessna C-38 Airmaster
Cessna UC-78 Bobcat (C-78, T-50)
Cessna C-106 Loadmaster (P-260)
Cessna XC-106 Loadmaster (P-260)
Cessna LC-126 (C-126, Cessna 195)
Cessna LC-126A (C-126A, Cessna 195)
Cessna LC-126B (C-126B, U-20B, 195)
Cessna LC-126C (C-126C, U-20C, 195)
Cessna C-145
Cessna C-165 Airmaster (C-94, UC-94)
Cessna CG-1 Glider
Cessna CG-2 Glider
Cessna CM-1 Racer
Cessna Comet (1917)
Cessna CR-1
Cessna CR-2
Cessna CR-3
Cessna CPW-6 "Goebel Special"
Cessna Crane (Cessna T-50)
Cessna CS-1 Glider
Cessna CW-5
Cessna CW-6
Cessna DC-6
Cessna DC-6A Chief
Cessna DC-6B Scout
Cessna EC-1
Cessna EC-2
Cessna FC-1
Cessna GC-1
Cessna GC-2
Cessna CH-1 Seneca (Skyhook)
Cessna CH-1A Seneca (Skyhook)
Cessna CH-1B (YH-41) Seneca (Skyhook)
Cessna CH-1C Seneca (Skyhook)
Cessna CH-1D Seneca (Skyhook)
Cessna NH-41A Seneca (Skyhook)
Cessna JRC-1 (Cessna T-50)
Cessna L-19A (O-1A)(Cessna 305)
Cessna (L-19A) O-1A, NASM
Cessna L-19A-IT (Cessna 305)
Cessna TL-19A (Cessna 305)
Cessna XL-19B (Turboprop)
Cessna XL-19C (Turboprop)
Cessna TL-19D (TO-1D)
Cessna L-19E (O-1E)
Cessna TL-19E (TO-1E)
Cessna (L-19) O-1F
Cessna (L-19) O-1G
Cessna MW-1
Cessna OE-1 (305, O-1B)
Cessna OE-2 (321, O-1C)
Cessna O-2A (M337, MC337)
Cessna O-2A, NASM
Cessna O-2B
Cessna P-7
Cessna P-10
Cessna Silverwing
Cessna T-37 (Cessna 318)
Cessna XT-37 (Cessna 318)
Cessna T-41 Mescelaro
Cessna T-41A Mescelaro
Cessna T-41B Mescelaro
Cessna T-41C Mescelaro
Cessna T-41D Mescelaro
Cessna T-47A (Cessna 552)
Cessna T-47B (Cessna 552)
Cessna T-50
Cessna Travel Aire 5000 Prototype
Cessna U-3 Blue Canoe (Cessna 310)
Cessna U-17 (Cessna 185)
Cessna U-27A (Cessna 208)
Cessna Design #1 (Phantom)
Cessna Design #2
Cessna Design #2, Modified (1927)
Cessna 120
Cessna 140
Cessna 140A
Cessna 150
Cessna 150L Commuter, NASM
Cessna 150, Reims F150
Cessna 152
Cessna 160
Cessna 170
Cessna 170, Experimental (Belt-Drive)
Cessna 172 Skyhawk
Cessna R172K Hawk XP
Cessna 172L Skyhawk
Cessna 172N Skyhawk 100
Cessna 172RG Cutlass RG
Cessna 172Q Cutlass II
Cessna 175 Skylark
Cessna 177 Cardinal
Cessna 177 Redhawk (Robertson-Modified Cardinal)
Cessna 177RG Cardinal RG
Cessna 180 Skywagon
Cessna 180 Skywagon "Spirit of Columbus," NASM
Cessna 182 Skylane
Cessna 182 Skylane RG
Cessna 182 Skylane II
Cessna 182M Skylane
Cessna 185 AgCarryall
Cessna 185 Skywagon
Cessna 187
Cessna 188 AgWagon
Cessna 188 AgWagon A
Cessna 188 AgWagon Prototype
Cessna 188 230 AgWagon
Cessna 188 300 AgWagon
Cessna 188A AgWagon B
Cessna 188B AgPickup
Cessna A188B AgTruck
Cessna A188B AgWagon
Cessna T188C AgHusky
Cessna 190, P-780
Cessna 195
Cessna 195B
Cessna 205
Cessna 206 Stationair
Cessna 206 Stationair 6
Cessna 206 Super Skywagon
Cessna 206 Turbo Skywagon
Cessna 206 Turbo Super Skywagon
Cessna 206 Turbo Stationair
Cessna 206 Turbo Stationair 6
Cessna 207 Skywagon
Cessna 207 Stationair 7
Cessna 207 Stationair 8
Cessna 207 Skywagon
Cessna 207 Stationair 7
Cessna 207 Stationair 8
Cessna 208 Caravan I
Cessna X-210 (1949)
Cessna 210 Centurion
Cessna P210 Pressurized Centurion
Cessna T210 Turbo Centurion
Cessna 303, T303 Clipper
Cessna T303 Clipper
Cessna T303 Crusader
Cessna 305
Cessna 308, Experimental
Cessna 310
Cessna 310B "Song Bird"
Cessna 310C

Cessna 310P
Cessna 310Q
Cessna 319
Cessna 320 Executive Skyknight
Cessna 320 Skyknight
Cessna 320 Riley Conversion
Cessna 335
Cessna 336 Skymaster
Cessna 337 Super Skymaster
Cessna P337 Pressurized Skymaster
Cessna T337 Super Skymaster
Cessna T337 Pressurized Super Skymaster
Cessna T337G Pressurized Super Skymaster
Cessna 337 Pressurized Super Skymaster
Cessna 340
Cessna 340A
Cessna 401
Cessna 402A
Cessna 402B Businessliner
Cessna 402B Utililiner
Cessna 404 Titan Series
Cessna 405 Jet
Cessna 406 Caravan II
Cessna 407 Jet
Cessna 411
Cessna 414
Cessna 414A Chancellor
Cessna 421
Cessna 421A
Cessna 421B Golden Eagle
Cessna 421C Golden Eagle
Cessna 425 Conquest I
Cessna 425 Corsair
Cessna 441 Conquest
Cessna 441 Conquest II
Cessna 500 Citation I
Cessna 500 Citation Fanjet
Cessna 501 Citation I SP
Cessna 525 Citationjet
Cessna 526 J-PATS
Cessna 550 Citation II
Cessna S550 Citation S/II
Cessna 551 Citation II/SP
Cessna S551 Citation S/II SP
Cessna 560 Citation V
Cessna 620
Cessna 650 Citation III/VI/VII
Cessna 700 Citation III Trijet
Cessna 750 Citation X
Cessna 1014 XMC
Cessna 1034 XMC
Cessna 1914 Model
Cessna 1916 Model

CFA (Compagnie Française d'Avaition) (France)

CFA D.7 Cri-Cri (Cricket) Major
CFA D.21T-4 Super Phryganet
CFA D.57 Phryganet

CFM Aircraft Ltd (UK)

CFM Shadow
CFM Shadow D
CFM Shadow E
CFM Star Streak
CFM Streak
CFM Streak Shadow
CFM Streak SLA

C.G.C.A. (Compagnie Generale de Constructions Aéronautiques) (France)

C.G.C.A. (Compagnie Generale de Constructions Aéronautiques) was founded in 1920, designing aircraft for **Chantiers Aero-Maritimes de la Seine (CAMS)** In 1925 C.G.C.A. registered the trademark "Météore" and sold its production rights to **Société Provençale de Constructions Aéronautiques (SPCA)**. For all C.G.C.A. and Météore designs, see **CAMS** and **SPCA.**

CGNA (Compagnie Général de Navigation Aérienne)(France)

CGNA (Wright) Type A Transitional

CGS Aviation Inc (Cleveland, OH)

CGS Aviation Hawk (Ultralight)

CH-7 Helicopters Heli-Sport Srl (Italy)

CH-7 Angel
CH-7 Kompress

Chabeau et Biffu (France)

Chabeau et Biffu 1908 Aeroplane

Chadwick Helicopters, Inc (Sherwood, OR)

Chadwick C-122S Rainbow

Chalfant, Brantly (Philadelphia, PA)

Chalfant 1909 Monoplane (Patent)

Challenger, George (UK)

Challenger Boxkite (Zodiac-Farman Hybrid, 1910)

Chamberlin Aeronautical Corp (Jersey City, NJ)

The **Chamberlin Aeronautical Corp** was established in 1929 by Clarence Chamberlin. Shortly after introducing two high-wing monoplanes, the company became the **Crescent Aircraft Corp.** For the original two designs by Chamberlin, see **Chamberlin.** For later work, see **Crescent.**

Chamberlin C-2
Chamberlin C-81
Chamberlin C-82 "Green River Whiskey"

Chambers, Albert A. (Hornell, NY)

Chambers (Albert) 1928 High-Wing Monoplane

Chambers, Russell (Armonk, NY)

Chambers (Russell) R-1 Special

Champel, Florentin (France)

Champel early-1912 Biplane
Champel mid-1912 Biplane
Champel late-1912 Biplane (Champel No.5)

Champion Aircraft Co, Inc (Tomball, TX)

In 1954 **Flyers Service Inc** purchased the rights to the Aeronca Model 7 from **Aeronca Manufacturing Corp,** establishing the **Champion Aircraft Corp** to undertake production. Champion developed this basic model and also created new designs. In 1970 **Bellanca Sales Co** (originally known as **International Aircraft Manufacturing Inc** or **Inter-Air**) purchased Champion's assets and became **Bellanca Aircraft Corp.** Bellanca ceased Champion production in 1980. Rights, tooling, and partially completed aircraft were sold to **Champion Aircraft Co, Inc** in 1982. In 1989 the rights were again sold to what became **American Champion Aircraft.** All pre-1954 development appears under **Aeronca.** Subsequent development is filed under **Champion Aircraft Corp, Bellanca, Champion Aircraft Co,** and **American Champion Aircraft.**

Champion (Texas):
- Citabria Standard (7ECA)
- Citabria 150 (7GCAA)
- Citabria 150S (7GCBC)
- Decathlon (8KCAB)
- Decathlon CS
- Super Decathlon (8KCAB-180)
- Scout (8GCBC)

Champion Aircraft Corp (Osceola, WI)

In 1954 **Flyers Service Inc** purchased the rights to the Aeronca Model 7 from **Aeronca Manufacturing Corp**, establishing the **Champion Aircraft Corp** to undertake production. Champion developed this basic model and also created new designs. In 1970 **Bellanca Sales Co** (originally known as **International Aircraft Manufacturing Inc** or **Inter-Air**) purchased Champion's assets and became **Bellanca Aircraft Corp**. Bellanca ceased Champion production in 1980. Rights, tooling, and partially completed aircraft were sold to **Champion Aircraft Co, Inc** in 1982. In 1989 the rights were again sold to what became **American Champion Aircraft**. All pre-1954 development appears under **Aeronca**. Subsequent development is filed under **Champion Aircraft Corp**, **Bellanca**, **Champion Aircraft Co**, and **American Champion Aircraft**.

Champion (Wisconsin):
- 7EC Traveler
- 7ECA Citabria
- 7FC Tri-Traveler
- 7GC Sky Trac
- 7GCA Sky Trac (Agricultural)
- 7GCAA Citabria
- 7GCB Challenger
- 7GCBA Challenger
- 7GCBC Citabria
- 7HC DX'er
- 7JC Tri-Con
- 7KC Olympia
- 7KCA Citabria
- 7KCAB Citabria
- 8KCAB Citabria
- 8KCAB Citabria Pro
- 402 Lancer

Champion, Kenneth R. (Gobles, MI)

Champion (Ken) Freedom Falcon
Champion (Ken) J-1 Jupiter
Champion (Ken) Jupiter K-2

Change, Oliver K. (Minneapolis, MN)

Chance (Oliver K.) Flying Machine

***Chance Vought** (See: Vought)*

Changhe Aircraft Manufacturing Corp (China)

Changhe Z-8 (Zhi-8)

***Chantiers Aéro-Maritimes de la Seine** (See: CAMS)*

***Chantiers Aéronautiques de Normandie** (See: Amiot)*

***Chantiers Aéronavals E. Romano** (See: Romano)*

***Chantiers de Provence-Aviation-Hydravions A. Bernard** (See: CPA)*

***Chantiers de l'Artois** (See: Artois)*

Chanute, Octave (Chicago, IL)

Chanute Kites
Chanute Multiplane (5 Wings) Glider (Model)
Chanute Triplane Glider (Model)
Chanute 1896 Biplane Glider
Chanute 1896 Lilienthal-Type Glider
Chanute 1896 Multiplane (4-Wing) Glider
Chanute 1896 Pivot-Wing Glider
Chanute 1896 Pivot-Wing Glider, No.1 Multiplane Modification
Chanute 1896 Pivot-Wing Glider, No.2 Multiplane Modification
Chanute 1897 Airplane (Moy-Chanute-Herring Design)
Chanute 1897 Glider
Chanute 1902 Triplane Glider

Chapeaux, Emile (France)

Chapeaux Type Ch 23 Glider (1933)

Chapiro (France)

Chapiro Biplane No.1 (1909)
Chapiro Biplane No.2 (1910)

Chardon (Switzerland)

Chardon 1922 Biplane Hang Glider

***Charles Ward Hall Inc** (See: Hall-Aluminum Aircraft Corp)*

Charles, Paul D. (Gettysburg, PA)

Charles (Paul) R 1

Charles, Ralph (Zanesville, OH)

Charles (Ralph) Model A Monoplane
Charles (Ralph) E-1 Racer
Charles (Ralph) 5-Place Biplane

Charlett Flugzeugwerke (Germany)

Charlett Parasol Monoplane

Charmiaux (France)

Charmiaux 1910 Monoplane

Charpentier (France)

Charpentier 1909 Seaplane Design

Chase Aircraft Co, Inc (Trenton, NJ)

Chase Aircraft Co, Inc was established in 1943 to produce assault cargo aircraft for military use. In 1953, the company was purchased by **Willys Motors, Inc**, a wholly-owned subsidiary of **Kaiser-Frazer Corp**. In June 1953, the USAF canceled production contracts for the Chase C-123 and transferred production to **Fairchild Engine and Aircraft Corp**. Kaiser-Frazer phased out its Chase operations, while the Chase design staff established **Stroukoff Aircraft Corp**. For all Chase aircraft including the XC-123 and XC-123A, see **Chase**. For production C-123s, see **Fairchild**. For post-1953 development work on the basic C-123 airframe, see **Fairchild**.

Chase YC-122 Avitruc
Chase YC-122A Avitruc
Chase YC-122B Avitruc
Chase YC-122C Avitruc
Chase C-122 Avitruc Civil Conversions
Chase XC-123
Chase XC-123A
Chase XCG-14A
Chase XCG-18A
Chase YCG-18A
Chase XCG-20

Chase-Sisley (Robert Chase)(USA)

Chase-Sisley C100-S
Chase-Sisley C101
Chase-Sisley C102

Chasle (France)

Chasle YC-12 Tourbillon

Chassagny et Constantin (France)

Chassagny Demoiselle Copy
Chassagny et Constantin Nieuport-Type Monoplane
Chassagny et Constantin 1910 Small Monoplane

Chaussée (France)

Chaussée 1909 Biplane

Chauvière, Lucien (France)

Chauvière Gyrocopter
Chauvière 1909 Monoplane
Chauvière 1909 Monoplane Glider
Chauvière 1909 Penteado Biplane

Chazal, Henry (France)

Chazal-Gourgas l'Aiglon (Eaglet)(1911)

Chèdeville, Georges (France)
Chèdeville 1909 Monoplane
Chèdeville 1911 Monoplane

Cheetah Light Aircraft Co Ltd (Canada)
Cheetah Light Aircraft Cheetah
Cheetah Light Aircraft Super Cheetah

Chelm Secondary School (Poland)
Chelm Secondary School 1927 Glider

Chelmiky, Prince (Russia)
Chelmiky 1907 Helicopter (Voisin-Built)

Chengdu (China)
Chengdu F-7 (See: Xian)
Chengdu FC-1
Chengdu J-5 (See: Shenyang)
Chengdu JJ-5 (See: Shenyang)
Chengdu J-7 (See: Xian)
Chengdu J-10
Chengdu J-12

Chepaux, Pascal (France)
Chepaux Biplane

Chepaux et de Bussac (Pascal Chepaux)(France)
Chepaux et de Bussac 1910 Monoplane

Cheranovsky, B. I. (Russia)
Cheranovsky BICh.1 Tailless Glider
Cheranovsky BICh.2 Tailless Glider
Cheranovsky BICh.3
Cheranovsky BICh.4 Tailless Glider
Cheranovsky BICh-7A Parabole
Cheranovsky BICh-14 Parabole
Cheranovsky BICh.20
Cheranovsky BICh.21

Chernikhovsky, S. D. (Russia)
Chernikhovsky S.Ch.I

Chesley-Caruso (Ray Chesley & James Caruso)(Long Beach, CA)
Chesley-Caruso C&C Glider (1929)

Chesnay (France)
Chesnay 1910 Monoplane

Chester, Art (Los Angeles, CA)
Chester (Art) Goon
Chester (Art) Jeep
Chester (Art) Swee'Pea
Chester (Art) Swee'Pea II

Chester, Milton A. (Bristol, PA)
Chester (Milton) 1929 Monoplane

Chetverikov, Igor Vvyacheslavovich (Russia)
Chetverikov ARK-3
Chetverikov MDR-3
Chetverikov MDR-6 (Che-2)
Chetverikov OSGA-101 (SPL)
Chetverikov TA-1

Chevallier, Yves (France)
Chevallier 1914 Pusher Monoplane

Chevallier et de Clèves (France)
Chevallier et de Clèves Monoplane

de Chevigny, Hubert (France)
de Chevigny Explorer

Chicago Aircraft Manufacturing Co (Chicago, IL)
Chicago CAMCO Model 2 Biplane

Chicago Aviation Co (Chicago, IL)
Chicago (Aviation) Biplane (1927)

Chicago Helicopters Ltd (See: Perry Helicopter)

Chichester-Miles Consultants Ltd (UK)
Chichester-Miles Leopard

Chilton Aircraft Ltd (UK)
Chilton D.W.1
Chilton D.W.1A

China National Aero-Technology Import & Export Corp (See: CATIC)

China Naval Air Establishment (China)
China Naval Air Establishment Chiang Hung Seaplane

Chotia, John (See: Weedhopper)

Chrislea Aircraft Co Ltd (UK)
Chrislea LC.1 Airguard
Chrislea CH.3 Ace
Chrislea CH.3 Super Ace
Chrislea CH.4 Skyjeep

Christen Industries Inc (Bealeton, VA)
Christen Eagle I
Christen Eagle II

Christensen, Harvey A. "Chris" (Minneapolis, MN)
Christensen Zipper

Christiansen, Julius (New York, NY)
Christiansen (Julius) 1911 Patent

Christmas, William Whitney (Washington, DC)
Dr. William Whitney Christmas established the **Christmas Aeroplane Co Inc** in 1911 to market his own aircraft designs. The company became the **Durham Christmas Aeroplane Sales and Exhibition Corp, Inc** in 1913 without any change in product. The company folded during World War I and Christmas began to work with the **Continental Aircraft Corp** of Amityville, NY on the construction of the Christmas Bullet. By 1919 Christmas formed the **Cantilever Aero Co** to build and market the Bullet, but there was no interest in the aircraft. In the late 1920s, Christmas designed a large Freight Monoplane through the **General Development Co, Inc**, which also failed to attract interest. For all Christmas designs, including the work done on the Bullet at Continental and the Freight Monoplane at General Development, see **Christmas**. For all Continental work not involving Christmas, see **Burnelli**.
Christmas Biplane (1912)
Christmas Bullet (1918)
Christmas Freight Monoplane (1930)

Christofferson (San Francisco, CA)
Silas and Harry Christofferson established the **Christofferson Aircraft Manufacturing Co** of San Francisco. In mid-1915, the company was renamed the **Christofferson Aeroplane and Engine Corp** and moved to Redmond, CA. In 1916, the Christofferson engine interests were transferred to the **Christofferson Motor Corp of New York**, while the aircraft manufacturing interests remained with the **Christofferson Aircraft Co.** For all aircraft built by Silas Christofferson or by the various Christofferson companies, see **Christofferson**.
Christofferson:
1911 Monoplane
1912 Biplane
1912 Biplane, Covered
1912 Biplane, Dual Control
1912 Biplane, Headless
1912 Biplane, Twin-Tail
1912 Seaplane

1914 Tractor Biplane
1915 Flying Boat (Model D)
1915 Tractor Biplane
1916 Tractor Biplane

Chrysler Corp (Detroit, MI)
Chrysler VZ-6 Flying Jeep

Chrysostomides, Stavros (Colunbia, SC)
Chrysostomides Jaguar

Chrzanowski Secondary School (Poland)
Chrzanowski 1912 Glider

Chu (Taiwan)
Chu CJC-3

Chupp (USA)
Chupp Rotor-Flyer Model H (Captive Autogiro Toy)

Church Airplane & Manufacturing Co (Chicago, IL)
Church Mid-Wing Monoplane

Cia Aeronautica Constructora de Baja California SA *(See: Cia Aeronautica Constructora y de Transport SA de Tijuana)*

Cia Aeronautica Constructora y de Transport SA de Tijuana)(Mexico)
Cia Aeronautica Constructora y de Transport SA de Tijuana:
BC-1
BC-2
BC-3

Cicaré Aeronautica (Argentina)
Cicaré I
Cicaré II
Cicaré CH-III (C.K.1)

de la Cierva, Juan (Spain)
Cierva C.1 Autogiro
Cierva C.2 Autogiro
Cierva C.3 Autogiro
Cierva C.4 Autogiro
Cierva C.5 Autogiro
Cierva C.6 Autogiro
Cierva C.6C
Cierva C.7 Autogiro
Cierva C.8 Autogiro
Cierva C.8 Mk.IV (C.8W) Autogiro
Cierva C.8 Mk.IV (C.8W), NASM
Cierva C.8L
Cierva C.8L Mk.I
Cierva C.9 Autogiro
Cierva C.10
Cierva C.11
Cierva C.12 Autogiro
Cierva C.17 Autogiro
Cierva C.18 Autogiro
Cierva C.19 Autogiro
Cierva CL.20
Cierva C.24 Autogiro
Cierva C.24
Cierva C.25
Cierva C.26
Cierva C.30 Autogiro
Cierva C.30A Rota
Cierva C.30P Autogiro
Cierva C.40 Autogiro
Cierva CL.20
Cierva W.9
Cierva W.11 Airhorse
Cierva W.14 Skeeter

Cijan, Boris (Yugoslavia)
Cijan C-3 Trojka

CINA (France)*(See: Roux; See also: Filippi)*

Circa Reproductions (Canada)
Circa Nieuport 11 Replica
Circa Nieuport 12 Replica

Cirigliano, Serafin (New Castle, DE) *(See also: Smith-Cirigliano)*
Cirigliano 1927 Biplane ("Baby Hawk")

Cirrus Design Corp (Baraboo, WI; Duluth, MN)
Cirrus SR-20
Cirrus ST-50
Cirrus VK-30 (Homebuilt)

Civilian Aircraft Co (C.A.C.)(UK)
Civilian Aircraft Co Coupé Mk.I
Civilian Aircraft Co Coupé Mk.II

CKD *(See: Praga)*

Clancy Brothers (Australia)
Clancy Sky Baby

Clark & Fitzwilliams
Clark & Fitzwilliams Cycleplane

Clark Aircraft Corp (Hagerstown, MD)
To better account for time spent developing the Duromold wood/resin bonding process and the Model 46 aircraft, in 1936 **Fairchild Engine and Airplane Co** founded the subsidiary **Duromold Aircraft Corp.** In 1938, the majority interest in Duromold was bought by a group of investors (including process inventor Col. Virginius E. Clark), who formed the **Clark Aircraft Corp.** Fairchild kept a minority interest in Clark, retaining Duromold as a holding company. In September 1938, Fairchild renamed its Duromold division **Fairchild Airplane Investment Corp**, and Clark created a subsidiary called **Duramold Aircraft Corp** (note the spelling change). In 1938 Duramold was renamed **Molded Aircraft Corp.** In 1939, Fairchild Engine and Airplane Corp bought back a controlling interest in Clark and renamed Molded Aircraft **Duramold Aircraft Manufacturing Corp.** The Duramold and Clark companies disappeared during one of Fairchild's World War II reorganizations. Virginius Clark's early work at General Aviation appears under **General Aviation.** The F-46 (or Fairchild 46), which was the sole production aircraft for Clark Aircraft, Duramold, and Duramold, appears under **Fairchild.**
Clark (MD) F-46 (See: Fairchild 46)

Clark Aircraft Inc (Marshall, TX)
Clark (TX) 12
Clark (TX) 1000C

Clark, Earl H. (Buffalo, NY)
Clark (Earl H.) M-1 Monoplane (1926)

Clarke (T. W. K. Clarke and Co)(UK)
Clarke Laking No I (1910 Biplane)
Clarke 1909 Glider (Wright Copy)
Clarke 1910 Biplane Glider
Clarke 1910 Monoplane

Classic Aero Enterprises (USA)
Classic (Aero) H-2 Honey Bee
Classic (Aero) H-3 Pegasus
Classic (Aero) HP-40 Warhawk

Classic Aircraft Corp (Lansing, MI)
Classic (Aircraft) Waco F-5

Claude, Léon (France)
Claude (Léon) Canard

Cleary Aircraft Corp (USA)
Cleary CL-1 Zipper

Clem, John Wesley (Wichita, KS)
Clem (John) Goldbug

Clément, Louis (France)
Clément (Louis) 1911 Biplane
Clément (Louis) 1911 All-Metal Taube Copy
Clément (Louis) "Moineau" (1920)
Clément (Louis) 1922 Triplane Glider

Clément, Maurice (France)
Clément (Maurice) Pusher Biplane
Clément (Maurice) Tractor Monoplane

Clément-Bayard, Gustave-Adolphe (Clément-Bayard Constructors) (Paris, France)
Clément-Bayard:
1908 Twin-Boom Monoplane (Comte de la Vaulx)
1909 Demoiselle (See: Santos-Dumont)
1910 Pusher Biplane
1911 Tractor Biplane
1911 Tractor Monoplane
1912 Tractor Biplane, Exposed Aft Fuselage
1912 Tractor Monoplane, Exposed Aft Fuselage
1912 Tractor Monoplane, Skids on Main Landing Gear
1912 Tractor Monoplane, V-Leg Main Landing Gear
1913 Armored Military Monoplane
1913 Pusher Biplane Flyingboat
1913 Tractor Monoplane
1913 Tractor Monoplane, Military
1914 Single-Seat 50hp Military Monoplane
1914 Single-Seat 70hp Military Monoplane
1914 Two-Seat Military Monoplane
1914 Three-Seat Military Biplane
1915 Twin-Engined Bomber

Clemson Aero Club (Clemson College, SC)
Clemson Special (1929)

Cleone Motors Co (St Louis, MO)
Cleone 5M Paroquet
Cleone 7M

le Clère (France)
le Clère Big Bird

Clerget & Cie. (Pierre Clerget) (France)
Clerget Marquézy (CAM, Clerget-Archdeacon-Marquézy)
Clerget 1910 Monoplane, 200 hp

Clifford Aircraft Corp (Seattle, WA) (See also: Northwest)
Clifford MG Glider (1929)

Cloud Coupe Aircraft Corp (Milan, IN)
Cloud Coupe EXP-1
Cloud Coupe M-7
Cloud Coupe MP (M-P)
Cloud Coupe SQ-2

Cloud Dancer Aeroplane Works Inc (Columbus, OH)
Cloud Dancer Jenny (Ultralight)

Cloudbuster Ultralights Inc (Sarasota, FL)
Cloudbuster Ultralights Cloudbuster

Clutton, Eric (UK)
Clutton FRED (Flying Run-about Experimental Design)Mk.I
Clutton FRED Mk.II
Clutton FRED Mk.III

Cluzan (France)
Cluzan 1909 Biplane

CMASA (Construzioni Meccaniche Aeronàutiche SA)(Italy)
CMASA BGA (See: Fiat)
CMASA CS
CMASA G.8 (See: Fiat)
CMASA JS.54
CMASA MF.4 (See: Fiat)
CMASA MF.5 (See: Fiat)
CMASA MF.6 (See: Fiat)
CMASA MF.10 (See: Fiat)
CMASA RS.14
CMASA Wal (Whale)(Dornier License)

CMR Sail-Plane Co (Paul Chamberlain, Joseph McKenzie, & Curt Ratzer) (Santa Monica, CA)
CMR Motorless Sailplane (1929)

CNA (Compagnia Nazionale Aeronàutica)(Italy)
The **Compagnia Nazionale Aeronàutica (CNA)** was established in 1920. The company concentrated mainly on light aircraft for touring and training operations but also built aircraft under license from other manufacturers. In the 1930s, CNA was acquired by the **Caproni** interests, but continued to operate under its own name. The company disappeared during the dislocations of Italian industry in 1943. For all CNA designed aircraft, see **CNA**. For all license-built aircraft, see the original designer firms.
CNA ETA
CNA PM.1
CNA 15
CNA 25

CNNA (Companhia Nacional de Navegaçao Aérea)(Brazil)
CNNA HL-1 (See: Muniz M-11)
CNNA HL-1 Series B
CNNA HL-2
CNNA HL-4
CNNA HL-6

CNNC (Companhia Nacional de Navegaçao Costiera)(Brazil)
CNNC M 7 (See: Muniz)
CNNC M 9 (See: Muniz)

CNT (Cantieri Navali Triestino) *(See: Cant, CNT)*

Co-Z Development Corp (Mesa, AZ)
Co-Z Cosy Classic
Co-Z Mark III
Co-Z Mark IV

Coanda, Henri
Rumanian Henri Coanda began designing aircraft in Austria in 1907. By 1911, he had moved to France, where he developed several ideas under the name **Aeroplanes Henri Coanda**. In 1912, he joined the **British and Colonial Aeroplane Co (Bristol)**, where he remained until September 1914, when he returned to France. For all Coanda designs not associated with Bristol, see **Coanda**. For all Coanda work at Bristol, see **Bristol** and **Bristol-Coanda**.
Coanda 1907 Mixed-Propulsion Aircraft
Coanda 1910 Turbo-Propulsion Sesquiplane
Coanda 1911 Turbo-Propulsion Sled
Coanda 1911 Twin-Motored Sesquiplane

Coanda (Coanda-Ernoult) 1911 Racing Biplane
Coanda 1914 Biplane
Coanda 1915 Observation Biplane
Coanda 1916 Bomber Biplane
Coanda 1916/17 Delaunay-Belleville Biplane (See: Delaunay-Belleville)
Coanda 1917 BN2 Bomber (See: SIA, France)
Coanda 1935 "Coanda-Effect" Jet

Cobelavia (Compagnie Belge d'Aviation)(Belgium)*(See: Fairey Tipsy Nipper)*

Cocke & Scott (Lt William A. Cocke Jr & Lt William J. Scott) (Schofield Barracks, HA)
Cocke & Scott "Nighthawk" Glider

Codock (Australia)
Codock 1933 Monoplane

Cody, Samuel F. (Cody and Sons Aerial Navigation Co) (UK)
Cody, Kites
Cody 1903 Biplane Gliders
Cody 1908 Army Airplane No.1
Cody 1909 Biplane ("Cathedral")
Cody 1910 Biplane, Michelin
Cody 1911 Biplane
Cody 1911 Biplane, Circuit of Britain (Cody III)
Cody 1912 Biplane
Cody 1912 Large Biplane
Cody 1912 Monoplane (Cody IV)
Cody 1912 Pusher Biplane (Cody V)
Cody 1912 Sixty-Foot Hydro Biplane
Cody 1913 Transatlantic Monoplane

Coffman (Oklahoma City, OK)
Samuel Coffman began designing aircraft in the mid-1920s, establishing **Coffman-Strong Aircraft Co, Inc** in 1928. In 1929 the company was renamed **Coffman Monoplanes, Inc**; it was acquired by the **Ranger Aircraft Co** in 1930. For all Coffman designs, see **Coffman**.
Coffman Air Coupe
Coffman Ranger A (Model A)
Coffman Model 3 Monoplane
Coffman Three-Place Monoplane (1928)

Coggin-Pliska (Gray Coggin & John V. Pliska) (Midland, TX)
Coggin-Pliska 1912 Biplane

Colditz Prison (Germany)
Colditz Cock Glider

Cole
Cole Parasol Monoplane (c.1910)

Cole (William) & Sons (UK)
Cole & Sons Tandem Monoplane (1911)

Cole Aircraft Corp (Cleveland, OH)
Cole Sport Commercial (c.1930)

Cole School of Aviation (Cleveland, OH)
Cole (School of Aviation) Model 1 Monoplane (1928)

Cole, J. Raymond (Oklahoma City, OK)
Cole (J. Raymond) 1932 Glider
Cole (J. Raymond) 1933 Biplane

Cole, Ross A. (Dallas, TX)
Cole (Ross) Circular-Wing Aircraft

Coler (Leutnant) (Germany)
Coler F.E.G. (Flugzeug-und Explosionmotoren Gesellschaft)

Colgate-Larsen Aircraft Co (Amityville, NY)
Established in 1937 as **Spencer-Larsen Aircraft Corp** to develop the designs of **P. H. Spencer**. The company became the **Colgate-Larsen Aircraft Corp**, Inc in 1940 and the **Colgate Aircraft Corp** in 1941. Work stopped on in-house development in 1942 when the company switched to war-related work. The company did not survive the industry draw-downs in 1945 and may have been acquired by the **Republic Aviation Corp** at that time. For development predating 1940, see **Spencer-Larsen**. For all work in 1940-42, see **Colgate-Larsen**.
Colgate-Larsen CL-15 Amphibian

Collier & Son Aeronautical Aircraft Experimenters (Wilmington, DE)
Collier & Son C-4 Powered Glider

Collier Aircraft Inc (Wichita, KS & Tulsa, OK)
Collier CA-1 Ambassador
Collier Cabin Monoplane (1928)
Collier Trainer

Collier-Combs Aircraft Co (W. S. Collier & L. A. Combs) (Ponca City, OK)
Collier-Combs Commercial Cabin Plane

Collin de Laminières (France)
Collin de Laminières 1911 Aeroplane

Collins Aero (USA)
Collins (Aero) Dipper Amphibian

Colliver (Charles Olliver) (France)
Colliver CO.01 (Rebuilt Robin DR.250)
Colliver CO.02
Colliver CO.3 (Robin DR.300 Variant)

Collomb (France)
Collomb 1904 Ornithopter
Collomb 1909 Ornithopter

Colomban, Michel (France)
Colomban MC-10 Cri-Cri ("Cricket")
Colomban MC-12 Cri-Cri ("Cricket")
Colomban MC-15 Cri-Cri ("Cricket")
Colomban MC-100 Banbi

Colombes (Ateliers Aéronautiques de Colombes *(See: Junkers Ju 52/3m)*

Colombo *(See: Bestetti)*

Colonial Aircraft Corp (Sanford, ME)
Colonial Aircraft Corp was established in 1946 by David B. Thurston and Herbert P. Lindblad. The company went bankrupt in 1959, with the assets acquired by the **Lake Aircraft Corp** of Sanford, ME, which was established by Lindblad in 1960. In 1962, the company was taken over by **Consolidated Aeronautics, Inc.** Consolidated Aeronautics established **Aerofab Inc** at Sanford to produce Lake designs which were then marketed through the **Lake Aircraft Division** of Consolidated Aeronautics. For all Colonial designs predating the 1959 bankruptcy, see **Colonial.** For post 1959 development, see **Lake**.
Colonial C-1 Skimmer
Colonial C-2 Skimmer Tach IV

Columbia *(See: Washington Aeroplane Co)*

Columbia Air Liners, Inc (See: C.A.L.)

Columbia Aircraft Corp (New York, NY) *(See: Bellanca)*

Columbia Aircraft Corp (Valley Stream, NY)
The **Columbia Aircraft Corp**

produced Grumman J2F-6 Ducks during World War II. Following the War, Columbia worked on a J2F replacement designated XJL-1. **Commonwealth Aircraft Corp** acquired the company in 1946 and switched the facilities to Commonwealth Skyranger production. Commonwealth apparently failed in 1947 or 1948. For all Columbia designs, see **Columbia**. For Commonwealth projects, see **Commonwealth**. For the J2F, see **Grumman**.

Columbia (Valley Stream, NY) XJL-1

Columbia Aircraft, Inc (Tulsa, OK)

Columbia (OK) Brown-Young BY-1

Combes (France)

Combes Aeroplane

Command-Aire Inc (Little Rock, AR)

The **Arkansas Aircraft Co** organized in March 1928 to build Heinkel designs under license. In November 1928 the company was reorganized and renamed **Command-Aire, Inc**. The company folded in 1931. For all Arkansas Aircraft and Command-Aire products, see **Command-Aire**.

Command-Aire BS-14
Command-Aire BS-16
Command-Aire CX-3 3-Place Commercial Biplane
Command-Aire MR-1 "Little Rocket"
Command-Aire Model 3-C-2
Command-Aire Model 3-C-3
Command-Aire Model 3-C-3-T
Command-Aire Model 3-C-3A
Command-Aire Model 3-C-3A-T
Command-Aire Model 3-C-3B
Command-Aire Model 4-C-3
Command-Aire Model 5-C-3
Command-Aire Model 5-C-3 Crop Duster
Command-Aire Model 5-C-3A
Command-Aire Model 5-C-3B
Command-Aire Model 5-C-3C

Commander Aircraft Co (Bethany, OK)

In 1988, the **Commander Aircraft Co** of Bethany, OK purchased the rights to the single-engined aircraft developed by the **Commander Division of Rockwell International Corp**. For all Aero Commander product lines initiated before the 1969 reorganization, see **Aero Commander**. For all Meyers, Volaircraft, Snow, and CallAir products, including those folded into the Aero Commander line, see **Meyers**, **Volair**, **Snow**, and **Callair**, respectively. For all post-1967 Jet Commander production, see **Israel Aircraft Industries**. For all product lines begun after the 1969 reorganization, see **Commander**.

Commander 112
Commander 114
Commander 114AT
Commander 114B
Commander 114TC

Commercial Aeroplane Wing Syndicate (UK)

Commercial Aeroplane Wing Syndicate:
Alula Fighter
Pelican
Semi-Quaver Alula Racer (See: Martinsyde)

Commercial Aircraft Co (Aircraft Corp, Ltd) (Van Nuys, CA)

Commercial C-1 Sunbeam
Commercial 102

Commonwealth Aircraft Corp Ltd (Australia) *(See: CAC)*

Commonwealth Aircraft, Inc (Kansas City, KS)

In May 1929, R. A. Rearwin and his sons established **Rearwin Airplanes, Inc** in Kansas City, KS. In December 1937 the company acquired the rights and properties of the **Le Blond Aircraft Corp** of Cincinnati, OH, and was renamed **Rearwin Aircraft and Engines, Inc**. In October 1942, the company was acquired by Charles Dolan who, in January 1943, changed the name to **Commonwealth Aircraft, Inc**. During World War II, the company was primarily involved in glider production. In 1945, Commonwealth acquired production rights to **Allied Aviation Corp's** Trimmer light amphibian. In 1946 Commonwealth acquired the **Columbia Aircraft Corp** of Valley Stream, NY and **Cairn Manufacturing Co**, intending to convert both plants to production of Commonwealth aircraft. The company appears to have failed in late 1946 or early 1947. For all aircraft originally designed by Rearwin, see **Rearwin**. For World War II glider production, see **Commonwealth**. For the Trimmer amphibian, see **Allied Aviation Corp**.

Commonwealth (USA) (Waco) CG-3A

Commuter Aircraft Corp (USA)

Commuter CAC-100

Comp *(See: Aerocomp)*

Compagnie Belge d'Aviation (Cobelavia) (Belgium) *(See: Fairey Tipsy Nipper)*

Compagnie Française d'Avaition *(See: CFA)*

Compagnie Général de Navigation Aérienne *(See: CGNA)*

Compagnie Internationale de Navigation Aérienne (France) *(See: Roux; See also: Filippi)*

Companhia Aeronautica Paulista *(See: CAP)*

Companhia Nacional de Avioes Ltda *(See: Conal)*

Companhia Nacional de Navegaçao Aérea *(See: CNNA)*

Companhia Nacional de Navegaçao Costiera *(See: CNNC)*

Compañia Española de Construcciones Aeronáuticas *(See: CECA)*

Comper Aircraft Co Ltd (UK)

After leaving the RAF (where he had designed the C.L.A. series of light aircraft for the **Cranwell Light Aeroplane Club**), Flt.Lt. Nicholas Comper established the **Comper Aircraft Co, Ltd** in March 1929. In August 1934, the company was reorganized due to financial problems, and the assets passed to the **Heston Aircraft Co, Ltd.** For all Nicholas Comper designs at Cranwell, see **Cranwell**. For all Comper Aircraft Co designs, see **Comper**.

Comper Autogiro

Comper C.L.A.7 Swift
Comper Kite
Comper Mouse
Comper Streak

***Composite Aircraft Corp** (See: Windecker)*

Composite Aircraft Design Inc (USA)
Composite Aircraft Design CADI 2001

Composite Aircraft Industries (South Africa)
Composite Aircraft Industries SE-86

Compton, Arthur Hally (Chicago, IL)
Compton Triplane Glider

Comte (Flugzeugbau A. Comte, Schweizerische Flugzeugfabrik) (Switzerland)
By 1927, Alfred Comte had expanded his Zürich flying school and repair operations to build original designs as **Alfred Comte Aviation et Constructions Aéronautiques**. The company was later renamed **Flugzeugbau A. Comte** and later still **Schweizerische Flugzeugfabrik**. The company apparently left the airframe construction industry in 1935/36. For all Comte products, see **Comte**.
Comte A.C.1
Comte A.C.2
Comte A.C.3
Comte A.C.4 Gentleman
Comte A.C.8
Comte A.C.11
Comte A.C.12 Moskito
Comte Wild X (See: Wild)

Comte, Gournay, et Michaux (Captain Comte, Henry Gournay, & César Michaux) (France)
Comte, Gournay, et Michaux 1909 Biplane Glider

Conal (Companhia Nacional de Avioes Ltda) (Brazil)
Conal W-151 Sopocaba

Concorde (International)
Concorde 001 (French Prototype)
Concorde 002 (British Prototype)
Concorde 01 (British Pre-Production)
Concorde 02 (French Pre-Production)
Concorde, Production

Condit, Clifford L. (Parkridge, IL)
Condit Experimental 1928 Homebuilt

Condor Aero Inc (Vero Beach, FL)
Condor Aero Shoestring (Homebuilt)

Condor Aircraft (Miami, FL)
Buddy Head established **Seahawk Industries Inc** in 1982 to produce ultralights. In November 1982 the company became **Condor Aircraft**. For all Seahawk and Condor products, see **Condor Aircraft**.
Condor Aircraft Condor II

Connor, A. L., Jr (Long View, TX)
Connor 1928 3-Seat Monoplane

Conord, Luis Aristide (Argentina)
Conord 1920 Monoplane

Conroy Aircraft Corp (Jack Conroy) (Santa Barbara, CA)
Conroy Stolifter (Cessna 337 Mod)
Conroy Turbo Three

Consolidated Aircraft Corp (USA)
In 1923, Reuben Fleet founded **Consolidated Aircraft Corp** to build Dayton Wright TW-3s in space leased from the **Gallaudet Aircraft Corp** in East Greenwich, RI. In February 1929, Fleet formed **Fleet Aircraft, Inc** to build the Consolidated Model 14 Husky Junior in Consolidated's Buffalo (NY) plant. Consolidated purchased Fleet Aircraft in August 1929, but continued production under the Fleet name. At the same time, Consolidated acquired **Thomas-Morse Aircraft Corp** as a subsidiary. In 1930, Reuben Fleet established **Fleet Aircraft of Canada, Ltd** to build Fleet aircraft in Canada, passing control to Canadian-owned **Fleet Aircraft, Ltd** in 1936. In 1940, Consolidated purchased **Hall-Aluminum**.

In 1941 Fleet sold his Consolidated shares to **Vultee Aircraft Inc**, an **AVCO** subsidiary, and in 1943 the companies merged as **Consolidated Vultee Corp** (known internally as **Convair**). In 1947 **AVCO** transferred Consolidated Vultee to **Atlas Corp**; Atlas passed control to **General Dynamics Corp** in 1953, with the **Convair Division of General Dynamics** formed in 1954. A 1961 reorganization split the Convair facilities between the Convair and Fort Worth divisions of General Dynamics, with the latter responsible for airframe fabrication.

Aircraft which began development before the merger with Vultee, including those with Consolidated project numbers and those produced in the US as Fleets, are listed as **Consolidated**. Aircraft designed after the 1943 merger are under **Convair**. Aircraft designed after the 1961 reorganization are under **General Dynamics**. For aircraft subsidiaries having independent identities are listed as **Thomas-Morse** and **Hall-Aluminum**. Aircraft produced by Fleet in Canada are under **Fleet (Canada)**.
Consolidated XA-11
Consolidated XBY-1 (Fleetster 18)
Consolidated XB2Y-1
Consolidated XB-24 Liberator
Consolidated YB-24 Liberator
Consolidated B-24A Liberator
Consolidated XB-24B Liberator
Consolidated B-24C Liberator
Consolidated B-24D Liberator
Consolidated B-24E Liberator
Consolidated B-24E Liberator, B-29 Turret Trainer
Consolidated XB-24F Liberator
Consolidated B-24G Liberator
Consolidated B-24H Liberator
Consolidated B-24H Liberator, Bell Nose Turret Testbed
Consolidated B-24J Liberator
Consolidated XB-24J, B-17G Nose Installation
Consolidated B-24J Liberator, J35 Engine Testbed
Consolidated XB-24K Liberator (B-24ST Single-Tail Prototype)
Consolidated B-24L Liberator
Consolidated B-24M Liberator
Consolidated B-24M Liberator, Eagle Radar Testbed
Consolidated EZB-24M (Icing Research)
Consolidated XB-24N Liberator
Consolidated XB-24P Liberator
Consolidated XB-24Q Liberator
Consolidated (B-24) LB-30 Liberator
Consolidated (B-24) LB-30A Liberator
Consolidated (B-24) Liberator I

Consolidated (B-24) Liberator II
Consolidated (B-24) Liberator III
Consolidated (B-24) Liberator V
Consolidated (B-24) Liberator VI
Consolidated XB-32 Terminator, Dominator
Consolidated B-32 Terminator, Dominator
Consolidated TB-32 Terminator, Dominator
Consolidated XB-36
Consolidated YB-36
Consolidated B-36A
Consolidated B-36B
Consolidated B-36B, Mod to B-36D
Consolidated B-36C
Consolidated B-36D
Consolidated RB-36D
Consolidated GRB-36D (FICON)
Consolidated RB-36E
Consolidated B-36F
Consolidated RB-36F
Consolidated GRB-36F (FICON)
Consolidated B-36H
Consolidated DB-36H
Consolidated NB-36H (Airborne Nuclear Reactor Testbed)
Consolidated RB-36H
Consolidated B-36J
Consolidated XB-41 Liberator
Consolidated BQ-8 (Liberator Drone Conversion)
Consolidated (XBT-6) XBT-937
Consolidated Y1BT-6 (BT-6)
Consolidated Y1C-11
Consolidated Y1C-22
Consolidated C-87 Liberator Express Prototype
Consolidated C-87 Liberator Express
Consolidated C-87A Liberator Express
Consolidated (C-87) Liberator Mk.I
Consolidated (C-87) Liberator Mk.II
Consolidated (C-87) Liberator Mk.II, BOAC
Consolidated (C-87) Liberator Mk.II, Mackenzie King
Consolidated (C-87) Liberator Mk.II, "Commando" (Churchill)
Consolidated (C-87) Liberator Mk.VII
Consolidated XC-99
Consolidated XC-109 Liberator
Consolidated C-109 Liberator
Consolidated Commodore (Model 16)
Consolidated XF-7 Liberator
Consolidated F-7A Liberator
Consolidated F-7B Liberator
Consolidated Fleet Family (See also: Fleet, Canada)
Consolidated Fleet Model 1 (Husky Junior, Model 14)
Consolidated Fleet Model 1/3 Racer
Consolidated Fleet Model 2
Consolidated Fleet Model 2X Special
Consolidated Fleet Model 3
Consolidated Fleet Model 7
Consolidated Fleet Model 7 DeLuxe
Consolidated Fleet Model 7C
Consolidated Fleet Model 8 (Sport)
Consolidated Fleet Model 9 (Sport)
Consolidated Fleet Model 10
Consolidated Fleet Model 11
Consolidated Fleet Model 11, Mexico
Consolidated Fleet Model 12
Consolidated Fleetster Model 17
Consolidated Fleetster Model 20
Consolidated Liberator-Liner (Model 39, Convair 104, R2Y)
Consolidated NY-1 (Husky)
Consolidated NY-2 (Husky)
Consolidated NY-2 (Husky), Doolittle Blind Flight Aircraft
Consolidated NY-2A (Husky)
Consolidated NY-3 (Husky)
Consolidated (Fleet) XN2Y-1
Consolidated (Fleet) N2Y-1
Consolidated XN3Y-1
Consolidated N4Y-1
Consolidated XO-17 Courier
Consolidated O-17 Courier
Consolidated XO-17A Courier
Consolidated (O-17) Courier Model 15
Consolidated OA-10 Catalina
Consolidated (Vickers Canada) OA-10A Catalina
Consolidated OA-10B Catalina
Consolidated XPY-1 Admiral
Consolidated XP2Y-1 Ranger
Consolidated P2Y-1 Ranger
Consolidated XP2Y-2 Ranger
Consolidated P2Y-2 Ranger
Consolidated P2Y-3 Ranger
Consolidated P2Y-3A Ranger
Consolidated XP4Y-1 Corregidor (Model 31)
Consolidated Y1P-25
Consolidated P-30 (PB-2)
Consolidated P-30A (PB-2A)
Consolidated P-30A (PB-2A) Single-Seat Conversion
Consolidated XPBY-1 (XP3Y-1) Catalina
Consolidated PBY-1 Catalina
Consolidated PBY-2 Catalina
Consolidated PBY-3 Catalina
Consolidated PBY-4 Catalina
Consolidated PBY-5 Catalina
Consolidated PBY-5 Catalina, NASM
Consolidated PBY-5 Catalina Civil
Consolidated XPBY-5A Catalina
Consolidated PBY-5A Catalina
Consolidated PBY-5A Catalina Civil
Consolidated PBY-6A Catalina
Consolidated PBY-6A Catalina Civil
Consolidated (PBY) 28-1 Catalina
Consolidated (PBY) 28-1 Catalina "Guba"
Consolidated (PBY) 28-2 Catalina
Consolidated (PBY) 28-2 Catalina "Guba"
Consolidated (PBY) 28-4 Catalina
Consolidated (PBY) Catalina Mk.I
Consolidated (PBY) Catalina Mk.IIA
Consolidated (PBY) Catalina Mk.IIIA
Consolidated XPB2Y-1 Coronado
Consolidated PB2Y-2 Coronado
Consolidated XPB2Y-3 Coronado
Consolidated PB2Y-3 Coronado
Consolidated PB2Y-3 Coronado, Civil
Consolidated XPB2Y-4 Coronado
Consolidated XPB2Y-5 Coronado
Consolidated PB2Y-5 Coronado
Consolidated PB2Y-5 Coronado, Civil
Consolidated XPB2Y-6 Coronado
Consolidated (PB2Y) Coronado GR.Mk.I
Consolidated PB4Y-1 Liberator
Consolidated PB4Y-1 Liberator, NAMU Drone Carrier
Consolidated XPB4Y-2 Privateer
Consolidated PB4Y-2 Privateer
Consolidated PB4Y-2B (Bat Bomb Mother Ship)
Consolidated PB4Y-2M (Meteorology)
Consolidated PB4Y-2P (Photo Recon)
Consolidated PB4Y-2S (Anti-Sub)
Consolidated (PB4Y-2) Privateer Civil
Consolidated XPT-1 (Trusty)
Consolidated PT-1 (Trusty)
Consolidated XPT-2 (XP-469)
Consolidated XPT-3 (XP-470)
Consolidated PT-3
Consolidated PT-3A
Consolidated XPT-5
Consolidated (Fleet) XPT-6
Consolidated (Fleet) YPT-6
Consolidated (Fleet) YPT-6A
Consolidated (Fleet) PT-6A
Consolidated XPT-8
Consolidated XPT-8A
Consolidated (XPT-11) XPT-933 (Model 21-A)

Consolidated Y1PT-11
Consolidated Y1PT-11A (PT-11A)
Consolidated Y1PT-11B (PT-11B)
Consolidated Y1PT-11C (PT-11C)
Consolidated Y1PT-11D (PT-11D)
Consolidated Y1PT-12 (PT-12, Y1BT-7, BT-7)
Consolidated RY-1 Liberator Express
Consolidated RY-2 Liberator Express
Consolidated RY-3 Liberator Express
Consolidated (RY-3) Liberator IX
Consolidated Model 10
Consolidated Model 11 (Monoplane Bomber Study)
Consolidated Model 21-A
Consolidated Model 21-C

Consortium Européen de Réalisation de Ventes d'Avions *(See: CERVA)*

Constantin (France)
Constantin Aéromobile

Constantin-d'Astanières (Louis Constantin and François d'Astanières) (France)
Constantin-d'Astanières 1912 Monoplane

Construcciones Aeronáuticas, SA *(See: CASA)*

Construction Aéronautique de Technologie Avancee *(See: CATA)*

Construction Aéronautique Edmond de Marçay *(See: Marçay)*

Constructions Aéronautiques du Béarn *(See: Béarn)*

Constructions Aéronautiques Blanchard *(See: Blanchard)*

Constructions Aéronautiques Latécoère *(See: Latécoère)*

Constructions Aéronautiques Stampe et Renard *(See: Stampe et Renard)*

Constructions de Planeurs à Moteur Auxiliare *(See: PAMA)*

Constructora Aeronaval de Levante SA (Spain) *(See: Piel Emeraude)*

Construzioni Aeronautiche Giovanni Agusta *(See: Augusta)*

Construzioni Meccaniche Aeronàutiche SA *(See: CMASA)*

Contal (France)
Contal Monoplane, Water-Cooled Inline Engine
Contal 1910 Monoplane
Contal 1910 Protin-Contal Monoplane

Contender Aircraft Inc (Sunnyvale, CA)
Contender 202
Contender 303
Contender 606

Contenet (France)
Contenet 1910 Monoplane

Continental Aircraft Corp (Amityville, NY) *(See: Burnelli)*

Continental Copters Inc (Fort Worth, TX)
Continental Copters El Tomcat (Bell 47 Ag Mod)
Continental Copters JC-1 Jet-Cat (Bell JetRanger Ag Mod)

Continental, Inc (Danbury, CT)

Continental, Inc was established in 1945 to develop and produce the roadable "Airphibian" designed by Robert E. Fulton. The company collapsed in 1954. For all Continental, Inc and Robert Fulton designs, see **Continental Inc**.

Continental FA-1 Airphibian
Continental FA-2 Prototype
Continental FA-2 Airphibian
Continental FA-3 Airphibian

Convair (Consolidated Vultee Aircraft Corp) (USA)

The first Vultee aircraft was produced in 1932 by **Airplane Development Corp**, a subsidiary of **Cord Corp.** In 1934, following Cord's successful bid for control of **AVCO**, Airplane Development Corp was reorganized as a division of AVCO's **Aviation Manufacturing Corp** subsidiary; the division became **Vultee Aircraft Division** in November 1937. Vultee acquired control of AVCO's **Stinson Aircraft Corp** subsidiary in October 1939, creating the **Stinson Division of Vultee. Vultee Aircraft Inc** was established in November 1939, acquiring the assets of Aviation Manufacturing Corp but remaining an AVCO subsidiary.

Vultee purchased a controlling interest in **Consolidated Aircraft Corp** in November 1941, forming **Consolidated Vultee Aircraft Corp (Convair)** in 1943. In 1947 AVCO passed control of Consolidated Vultee to **Atlas Corp.** Atlas passed control to **General Dynamics Corp** in 1953, with the **Convair Division** of General Dynamics formed in 1954. A 1961 reorganization split the Convair facilities between the General Dynamics Convair and Fort Worth divisions, with the latter responsible for airframe fabrication.

Aircraft which began development before the 1943 Consolidated Vultee merger are listed under **Consolidated** or **Vultee.** Aircraft designed after the 1943 merger are under **Convair.** Aircraft designed by Stinson and Vultee's Stinson Division are under **Stinson.** Aircraft designed after the 1961 General Dynamics reorganization are listed under **General Dynamics.**

Convair XB-46
Convair XB-58 Hustler
Convair YB-58A (RB-58A) Hustler
Convair B-58A Hustler
Convair TB-58A Hustler
Convair YB-60
Convair C-131A Samaritan
Convair C-131B Samaritan
Convair YC-131C Samaritan
Convair C-131D Samaritan
Convair C-131E Samaritan
Convair EC-131E Samaritan
Convair RC-131F Samaritan
Convair EC-131G Samaritan
Convair C-131H (VC-131H) (See: Allison Division of General Motors)
Convair NC-131H TIFS (See: Allison Division of General Motors)
Convair Charger (Model 48)
Convair XFY-1 Pogo
Convair XFY-1 Pogo, NASM
Convair XF2Y-1 Seadart
Convair XF2Y-1 Seadart, NASM
Convair (XF-92) XP-92 (Ramjet)
Convair XF-92A
Convair YF-102 Delta Dagger
Convair YF-102A Delta Dagger
Convair F-102A Delta Dagger
Convair TF-102A Delta Dagger
Convair (F-102) PQM-102B (SPAD) (Drone Modification)

Convair F-106A Delta Dart
Convair F-106B Delta Dart
Convair XL-13
Convair L-13A
Convair L-13B
Convair L-13 Civil Conversions (See also: Acme Centaur 101)
Convair NX-2 (Nuclear-Powered Bomber Program)
Convair XP5Y-1
Convair XP-81
Convair R3Y-1 Tradewind
Convair R3Y-2 Tradewind
Convair R4Y-1 (C-131F) Samaritan
Convair R4Y-2 (C-131G) Samaritan
Convair XT-29 (XAT-29)
Convair T-29A
Convair T-29B
Convair VT-29B
Convair T-29C
Convair T-29D
Convair YT-32
Convair TBY Seawolf
Convair (Stout) Model 6 Helicopter
Convair Model 37
Convair Model 103 (Spratt-Stout Model 8 Skycar)
Convair Model 110
Convair Model 111 Air-Car
Convair (Theodore Hall) 116 Flying Car
Convair (Theodore Hall) 118 Convaircar
Convair 240 Convair-Liner
Convair 240 Convair-Liner "Caroline," NASM
Convair 240 Turbo-Liner
Convair 340 Convair-Liner
Convair 340 Freighter
Convair 440 Metropolitan
Convair 540 Cosmopolitan
Convair 580 (See: Allison Division of General Motors)
Convair 600
Convair 640 (340/440 RR Dart Turboprop Conversion)
Convair 880, Golden Arrow Proposal
Convair 880 Convair-Jet (Skylark 600)
Convair 990 Coronado (Convair 600, Model 30)
Convair 990 "Galileo" (NASA 711)
Convair 990 "Galileo II" (NASA 712)

Convertawings, Inc (Amityville, NY)
Convertawings Model A Quadrotor
Convertawings Model E
Convertawings Model F

Convertible Aircraft (General)

Convertoplane Corp *(See: Herrick, Gerardus Post)*

Cook Aircraft Corp (John Cook) (Rancho Palos Verdes, CA)
Cook JC-1 Challenger

Cook, Clarence N. (Kansas City, KS)
Cook (Clarence N.) 1929 Glider

Cooke, G. Carlyle (Winston-Salem, NC)
Cooke (Carlyle) 1928 Monoplane

Cooke, Weldon B. (Weldon B. Cooke Aeroplane Co) (Sandusky, OH)
Cooke (Weldon) 1913 Tractor Airboat
Cooke (Weldon) 1913 Tractor Biplane

Cooley & Stroben (Cooley & A. W. Stroben) (Woodlake, CA)
Cooley & Stroben Model A Monoplane

Coonley, Harold A. "Hal" (Miami, FL)
Coonley Special "Little Toot"

Cooper, J. B. (Bridgeton, MO)
Cooper S-A-1

Cooperative des Ateliers Aéronautiques de la Région Parisienne *(See: CAARP)*

Copin, Georges (France)
Copin 1909 Biplane (Popp-Copin)
Copin 1911 Monoplane

Copland, Harry Depew (Detroit, MI)
Copland Biplane (1911)

Corbadec (France)
Corbadec Ornithopter

Corben Sport Plane and Supply Co (Peru, IN)
The **Corben Sport Plane and Supply Co** was established in 1923, specializing in kit planes. Later, the company became the **Corben Aircraft Co.** In 1953 the Corben assets were acquired by Paul Poberezny of the **Experimental Aircraft Association**, who redesigned the aircraft. To prevent the EAA from becoming involved in a profit-making venture, the rights were sold to the **Ace Aircraft Manufacturing Co** of Asheville, NC. For all original Corben products, see **Corben.** For all post-1953 products, including kit-built aircraft after the EAA redesign, see **Ace.**
Corben Baby Ace
Corben Cabin Ace
Corben Junior Ace
Corben Super Ace

Cord Corp *(See: Vultee)*

Cordas, Al C. (Spring Valley, CA)
Cordas SCS-1

Corignan, Pierre (France)
Corignan 1910 Seaplane

Corman Aircraft Inc (Detroit, MI)
Corman Aircraft Inc, established in 1928, was the first aeronautical venture of the **Cord Corp.** In August 1929, Cord acquired a controlling interest in the **Stinson Aircraft Corp** and, in early 1930, merged Corman into Stinson. For all pre-1930 Corman designs, see **Corman.** For all Stinson designs and post-1930 work, see **Stinson.** For other Cord-related products, see also **Vultee.**
Corman 3000 Airliner

Cornelius (USA)
George Cornelius established **Cornelius Aircraft Corp Ltd** in the late 1920s in the San Francisco, CA area, later moving to the Glendale, CA area in the 1930s and the Dayton, OH area in the 1940s.
Cornelius XBG-1
Cornelius Fre-Wing
Cornelius Mallard
Cornelius Pilotless Target Glider

Cornell Aero Club (Alliance, OH)
Cornell 1911 Glider

Cornet Autoplane *(See: Pischoff)*

Cornu, Paul (France)
Cornu (Paul) 1904 Helicopter Model
Cornu (Paul) 1907 Helicopter

Correa, Federico (Argentina)
Correa 1911 Parasol

Cosmic Aircraft Corp (Tulsa, OK)
The **Cosmic Aircraft Corp** was established in 1970 to produce the Funk F-23 agricultural aircraft, designed by **D. D. Funk Aviation Co** in 1962. For all post-1970 aircraft, see **Cosmic.** For all pre-1962 production, see **Funk.**
Cosmic F-23

Cosmic Wind *(See: LeVier)*

Cosmos (France)
Cosmos Atlas 21
Cosmos Bidulm
Cosmos Bidulm 43
Cosmos Bidulm 46
Cosmos Bidulm 50
Cosmos Bidulm 53
Cosmos Dragster 43

de Coster, Charles (France)
de Coster Flugi (1910 Monoplane)

Costruzioni Aeronautiche Taliedo *(See: CAT)*

Costruzioni Aeronautiche Tecnam Srl (Italy)
Costruzioni Aeronautiche Tecnam:
P92 Echo
P96 Golf

Cosy Europe (Germany) *(See: Co-Z)*

Cothran, Edward Everett (Santa Clara, CA)
Cothran V/STOL Monoplane Patent

Cotten, Virgil T. (Highland Park, MI)
Cotten 1929 Sport Monoplane

Couade, Captaine (France)
Couade 1913 Monoplane

Council for Scientific and Industrial Research *(See: CSIR)*

Coupé (Avions Jacques Coupé, Coupé-Aviation) (France)
Coupé JC-01
Coupé JC-2
Coupé JC-200

Coupet (France)
Coupet 1922 Monplane Glider

Courier Monoplane Co Inc (Long Beach, CA)
Courier BP-1 (1929 Cabin Monoplane)
Courier TK-100 (1933 Parasol)

Courneuve (Société des Avions Bernard) *(See: Bernard)*

Courrejou (France)
Courrejou 1910 Monoplane

Coursier (France)
Coursier 1907 Aeroplane

Court (Germany)
Court Monoplane A
Court 1913 Monoplane

Courtet (France)
Courtet 1910 Biplane

Courtney, Frank (New York, NY)
Courtney (Curtiss-Caproni) CA-1 Commuter

Courtois-Suffit Lescop (Roger Courtois-Suffit & Captaine Lescop) (France)
Courtois-Suffit Lescop (C.S.L.) C1

Couse, K. W. (Newark, NJ)
Couse Transport Aircraft System

Cousin, Dr Georges (France)
Cousin 1909 Glider
Cousin 1914 Monoplane

Coutant (France)
Coutant 1917 Flying Boat

Couture and Fletcher (Oscar Couture & Bert Fletcher (Oakland, CA)
Couture and Fletcher C 2 M

Couzinet (Société des Avions Réné Couzinet) (France)
Réné Couzinet established the **Société des Avions Réné Couzinet** in 1928 to produce his design for a trans-Atlantic **mailplane**. In 1934 the **Société des Ateliers d'Aviation Louis Breguet** acquired controlling interest, but Couzinet continued to operate under its own name. In 1935 the French government and Breguet withdrew support from Couzinet's new bomber design. Couzinet continued to work until 1937 when forced out of business by the nationalization of the French aviation industry. All Couzinet designs are listed as **Couzinet.**
Couzinet:
Single-Engine Monoplane
10 "l'Arc-en-Ciel" ("The Rainbow")
33
70 "Arc-en-Ciel ("Rainbow") No.2"
71 "Arc-en-Ciel ("Rainbow") No.3"
110
1928 Hydravion "La France"
1928 Transatlantic Trimotor

Coventry Ordnance Works Ltd (COW) (UK)
Coventry Ordnance Works:
F.29/27 (See: Vickers 161 and Westland COW)
1910 Biplane (See: Wright, Howard)
1910 Monoplane (See: Wright, Howard)
1912 Chenu-powered Biplane
1912 Gnome-powered Biplane
1913 Monoplane

Cowan Aviation Co *(See: Pruett/Cowan)*

Coward & Associates *(See: Bee Aviation Associates, Inc)*

Cox Air Resources Ltd (Canada)
Cox Turbo Otter (Converted DHC-3)

Cox-Klemin Aircraft Corp (College Point, NY; Baldwin, NY)
Cox-Klemin Aircraft Corp was established in 1921 by L. Charles Cox and Alexander Klemin. The company generally produced aircraft designed by others, including Ernst Heinkel, the US Navy and US Army Air Service. The company went into receivership in 1926.
Cox-Klemin XA-1
Cox-Klemin Air Service Type XII Bomber Proposal
Cox-Klemin Air Service Type XIII Bomber Proposal
Cox-Klemin CO-1
Cox-Klemin CO-2
Cox-Klemin XO-4
Cox-Klemin CK-1 Amphibian
Cox-Klemin CK-3 Night Observation Aircraft
Cox-Klemin CK-19 Amphibian
Cox-Klemin Curtiss MF Rebuild
Cox-Klemin PS (Heinkel HD 22)
Cox-Klemin Nighthawk Mailplane
Cox-Klemin Standard J-1 Rework
Cox-Klemin Stelling (Nungesser) Amphibian (CK-18)
Cox-Klemin TW-2 (CK-2)
Cox-Klemin XS-1

CPA (Chantiers de Provence-Aviation-Hydravions A. Bernard) (France)
CPA 1M

CPC (France)
CPC 1912 Monoplane

CPCA SA (Centrul de Projiectare Si Consulting Pentru Aviatie SA) (Romania)
CPCA ADC-H1
CPCA ADC-X0
CPCA DK-10 Dracula

Craft AeroTech (USA)
Craft AeroTech 200
Craft AeroTech 200 FW

Crane, James A. (Pequabuck, CT)
Crane 1925 Patent
Crane 1928 Eagle Ornithopter

Cranmer, William S. (Fresno, CA)
Cranmer 1917 Ornithopter Patent

Cranwell Light Aeroplane Club (UK)
Cranwell C.L.A.2
Cranwell C.L.A.3
Cranwell C.L.A.4
Cranwell C.L.A.7 Swift(See: Comper)
Cranwell Pixie II (1925 Glider)

Crawford All-Metal Airplane Co (Seal Beach, CA)
Crawford (Seal Beach) A-1
Crawford (Seal Beach) CLM
Crawford (Seal Beach) Runabout

Crawford Airplane Co (Venice, CA)
Crawford (Venice, CA) Commercial
Crawford (Venice, CA) Courier

Crawford, William F. (Crawford Motor and Airplane Manufactory, Inc) (Long Beach, CA)
Crawford (William F.):
Model A-1 Monoplane
Powered Glider
Special Trimotor Monoplane

CRDA (Cantieri Riuniti dell' Adriatico) (See: CANT (CRDA))

Creese-Dederich (A. G. Creese) (UK)
Creese-Dederich 1909 Monoplane

Crescent Aircraft Corp (Jersey City, NJ)
The **Chamberlin Aeronautical Corp** was established in 1929 by trans-Atlantic record-holder Clarence Chamberlin. Shortly after introducing two monoplanes, the company became **Crescent Aircraft Corp**. Original Chamberlin designs are under **Chamberlin**; later work is under **Crescent**.
Crescent Model A

CRO Holding spol sro (Czechoslovakia)
CRO Metallica LG2 Motorglider
CRO Metallica LG2 Ultralight

Crocco, Gaetano Arturo (Italy)
Crocco Telebomba (Pilotless Aircraft)

Crocker-Hewitt (UK)
Crocker-Hewitt 1920 Helicopter

Croplease PLC (UK)
Croplease Fieldmaster (See: NDN)
Croplease Firemaster

Cropmaster Aircraft Pty Ltd (Australia) (See: Yeoman)

Crosby Aviation Corp Ltd (UK) (See: Andreasson BA-4B)

Crosby, Harry (Los Angeles, CA)
Crosby C6R-2 Crosby Special (Racer)
Crosby C6R-3 Crosby Special (Racer)
Crosby CR-4 (Racer)

Croses (Avions Croses, Emilien Croses) (France)
Croses Airplume
Croses EC-1 Pouplume
Croses EC-3 Pouplume
Croses EC-6 Criquet (Locust)
Croses EC-7 Tout-Terrain
Croses EC-8 Tourisme
Croses EC-9 Paras-Cargo
Croses-Laubie 10 Criquet (Locust)

Crosley Aircraft Corp (Cincinatti, OH)
Crosley CF "The Crosley Flea"
Crosley Moonbeam 1 (C-1) Monoplane
Crosley Moonbeam 2 (C-2) Monoplane
Crosley Moonbeam 3 Biplane
Crosley Moonbeam Five Monoplane

Crossland, E. (See: Aviation Construction Engineers)

Crouch-Bolas Aircraft Corp (Providence, RI)
Crouch-Bolas B-37
Crouch-Bolas Dragonfly

Crowley, Walter A. (USA)
Crowley Hydro-Air (ACV)

Crown Motor Carriage Co, Aircraft Division (Los Angeles, CA)
Crown Custombilt B3

Crump, Thomas Henry (Chicago, IL)
Crump (Thomas Henry) Ornithopter

Crusader Aircraft Corp (Glendale, CA)
Originally **Gyro Airlines, Inc**, **Crusader Aircraft Corp** was established in September 1933. In 1935 the company was acquired by the **American Gyro Co** of Denver, CO, but continued to operate under the Crusader name. For all Crusader and American Gyro products, see **Crusader**.
Crusader AG-4
Crusader AG-7

CSIR (Council for Scientific and Industrial Research) (South Africa)
CSIR SARA II
CSIR SARA III

CSS (Centralne Studium Samolotów) (Poland)
CSS 10A
CSS 10C
CSS 11
CSS 12

CTA (See: Centro Técnico de Aeronáutica)

CTA (See: Centro Técnico Aerospacial)

Cub Aircraft Corp Ltd (Canada)
Cub (Piper) J3C-65
Cub (Piper) L-4B Prospector

Culp's Specialties (Steve Culp) (Shreveport, LA)
Culp Sopwith Pup

Culver Aircraft Corp (Wichita, KS)
In 1938, K. K. Culver and Al Mooney purchased rights to build the Monocoupe Monosport. They formed **Dart Manufacturing Corp** of Columbus,

OH, to build the aircraft as the Dart Model G. In 1939 Dart was reorganized as **Culver Aircraft Co**, moving to Wichita, KS. In 1946 Culver Aircraft went bankrupt, and **Applegate and Weyant** acquired the rights to the Dart. In mid-1956, **Superior Aircraft Co** (a division of **Priestley Hunt Aircraft Corp**) was established to acquire Culver's remaining assets. Superior developed the Culver Model V into the Superior Satellite, but folded in 1962/63. **The California Aero Co** of Tracey, CA acquired the rights and tooling for the Culver Cadet/PQ-8 series in December 1961 and began development of an improved version called the Lark 95. **Lark Aircraft Co** was established to build the aircraft, but folded before completing any airframes. The **Helton Manufacturing Co** of Mesa, AZ acquired the rights in 1965. Culver and Dart products are listed under **Culver**. Remaining production is listed under **Applegate and Weyant**, **Superior**, or **Helton**, as appropriate.

Culver Cadet (LCA, LFA)
Culver Dart-G
Culver XPQ-8 Cadet
Culver PQ-8 Cadet
Culver PQ-8A Cadet
Culver PQ-8A Cadet Civil Conversions
Culver XPQ-9
Culver XPQ-10
Culver PQ-14 Series
Culver XPQ-14
Culver PQ-14A
Culver PQ-14B
Culver XPQ-15
Culver "Rigid Midget" (Glider)
Culver "Screamin Wiener" (Glider)
Culver TDC-2 Cadet
Culver TD2C-1 Cadet
Culver TD2C-1 Cadet, NASM
Culver XTD4C-1
Culver Model V

Cunliffe-Owen (Cunliffe-Owen Aircraft, Ltd) (UK)

Cunliffe-Owen Concordia
Cunliffe-Owen (Burnelli) O.A.1
Cunliffe-Owen (Burnelli) O.A.2

Cunningham-Hall Aircraft Corp (Rochester, NY)

The **Cunningham-Hall Aircraft Corp** was established in 1928 as a subsidiary of **James Cunningham, Son & Co**, manufacturers of the Cunningham car, to develop aircraft designed by Randolf F. Hall. In 1941 Hall left to join **Bell Aircraft Corp** and Cunningham left the aviation field. For all Cunningham-Hall designs, see **Cunningham-Hall**.

Cunningham-Hall, Hall Wing (Fairchild 22 Wind Tunnel Test)
Cunningham-Hall GA-21M
Cunningham-Hall GA-36
Cunningham-Hall PT-6 (CHPT-6)
Cunningham-Hall PT-6F (CHPT-6F)
Cunningham-Hall Model X (1929 Guggenheim Entry)

Curran, Edward (Alhambra, CA)

Curran Monoplane (1928)

Currie, J. R. (UK)

Currie WOT Biplane

Curry, Maj. John F. (McCook Field, OH)

Curry (John F.) Corps Observation Airplane (1924)

Curtis, Richard D. (Wichita, KS)

Curtis 3/4-Scale Curtiss Pusher

Curtiss (USA)

Glenn Hammond Curtiss built lightweight engines that were used by many aviation experimenters at the beginning of the 20th century. In 1907, he helped found the **Aerial Experiment Association**, with whom he designed the successful "June Bug" biplane. In 1909 he launched the **Herring-Curtiss Co**; this was replaced by the **Curtiss Aeroplane Co** in 1910. In 1916 Curtiss Aeroplane merged with the **Curtiss Motor Co** to form the **Curtiss Aeroplane and Motor Co**, which included **Curtiss Aeroplanes and Motors Ltd** as its Canadian subsidiary. In 1929, Curtiss merged with the **Wright Aeronautical Corp** to form the **Curtiss-Wright Corp**. Based in Buffalo, New York, the Curtiss Aeroplane and Motor Co remained a principal Curtiss-Wright division; it was renamed the **Curtiss Aeroplane Division** in 1936.

In St Louis, Missouri, the **Curtiss-Robertson Aircraft Corp** had been formed in 1928; it became the **Curtiss-Robertson Division** upon the formation of Curtiss-Wright. The **Travel Air Manufacturing Co** became a Curtiss-Wright subsidiary in 1930, and was merged with Curtiss-Robertson to form the **Curtiss-Wright Airplane Co**. This, in turn, became the **St Louis Airplane Division** of the Curtiss-Wright Corp in 1936.

Following World War II, Curtiss-Wright moved all of its aircraft production to its Columbus, Ohio, plant. Unable to secure post-war contracts, Curtiss-Wright sold that factory to **North American Aviation**. A final attempt to resume aircraft production ended in 1959. All Curtiss, Curtiss-Robertson, and Curtiss-Wright, designs are listed under **Curtiss**. Products developed by Travel Air before the 1930 merger are listed under **Travel Air**; later aircraft marketed with the "Travel Air" name appear under **Curtiss**.

Among the other Curtiss and Curtiss-Wright subsidiaries: the **Burgess Co** was acquired in 1916; all Burgess aircraft are found under **Burgess**. The **Moth Manufacturing Co** was acquired in the later 1920s and merged into the Curtiss-Robertson Division; its products appear under **Moth**. A merger with the **Reid Aircraft Co** in Canada formed the **Curtiss-Reid Aircraft Co** in 1929; Reid and Curtiss-Reid aircraft appear under **Reid**. The **Keystone Aircraft Corp** was acquired in 1930, operating under its own name until closed in 1936; company designs are listed with **Keystone**.

Curtiss A-1 (AH-1) Type
Curtiss A-1 (AH-1)
Curtiss A-2 (Owl, E-1, AX-1)
Curtiss A-3 (AH-3)
Curtiss A-4 (AH-4)
Curtiss A-3 Falcon
Curtiss A-3A Falcon
Curtiss A-3B Falcon
Curtiss XA-4 Falcon
Curtiss XA-7 Proposal
Curtiss XA-8 Shrike
Curtiss YA-8 Shrike
Curtiss Y1A-8 Shrike
Curtiss Y1A-8A Shrike
Curtiss YA-10 Shrike

Curtiss A-12 Shrike
Curtiss A-12 Export Shrike
Curtiss XA-14 (Model 76) Shrike
Curtiss Y1A-18 (A-18, Model 76A) Shrike
Curtiss A-25A Shrike, Helldiver
Curtiss A-25A Helldiver, Australia
Curtiss AH-8
Curtiss AH-9
Curtiss AH-11
Curtiss AH-12
Curtiss AH-13
Curtiss AH-14
Curtiss AH-15
Curtiss AH-16
Curtiss AH-17
Curtiss AH-18
Curtiss Aircar Hovercraft (Model 2500)
Curtiss XAT-4
Curtiss AT-4
Curtiss AT-5
Curtiss AT-5A
Curtiss AT-9 Jeep (Fledgling) (CW-25)
Curtiss AT-9A Jeep (Fledgling) (CW-25)
Curtiss Autoplane (Model 11)
Curtiss XB-2 Condor
Curtiss B-2 Condor
Curtiss XB-908A
Curtiss Baldwin Red Devil
Curtiss BAP (Model 14)
Curtiss BAT (Model 13)
Curtiss Beachey Special
Curtiss Beachey Tractor
Curtiss XBFC-1 (See: XF11C-1)
Curtiss BFC-2 (See: F11C-2)
Curtiss XBF2C-1 (See: XF11C-3)
Curtiss BF2C-1
Curtiss Bleeker Helicopter
Curtiss Bristol Fighter (USAO-1)
Curtiss BT Flying Boat
Curtiss BT Flying Boat, Modified
Curtiss XBTC-1
Curtiss XBTC-2
Curtiss XBT2C-1
Curtiss XBT-4 Falcon
Curtiss C-1 (AB-1)
Curtiss C-2 (AB-2)
Curtiss C-3 (AB-3)
Curtiss C-4 (AB-4)
Curtiss C-5 (AB-5)
Curtiss C-6 (Commercial 6-Seat Development of MF Seagull)
Curtiss XC-10 Robin
Curtiss YC-30 Condor
Curtiss C-46 Commando
Curtiss C-46A Commando
Curtiss XC-46B Commando
Curtiss XC-46C Commando
Curtiss C-46D Commando
Curtiss C-46E Commando
Curtiss C-46F Commando
Curtiss C-46F Commando, NASM
Curtiss C-46G Commando
Curtiss C-46H Commando
Curtiss XC-46K Commando
Curtiss XC-46L Commando
Curtiss C-55 (See: CW-20A)
Curtiss YC-76 Caravan
Curtiss C-76 Caravan
Curtiss YC-76A Caravan
Curtiss XC-113 Commando
Curtiss Canada C-1
Curtiss Canoe Plane
Curtiss Canoe Plane, Headless
Curtiss Carrier Pigeon (Model 40)
Curtiss Carrier Pigeon II
Curtiss CB Battleplane (Liberty Battler)
Curtiss CO-X
Curtiss Condor CO Transport (Condor 18)
Curtiss Condor T-32 (Condor II)
Curtiss Condor BT-32 Export Bomber (Condor II)
Curtiss Courtney Amphibian (CA-1)
Curtiss CR-1 Racer (Model 23, L-17-1)
Curtiss CR-2 Racer (Model 23, L-17-2)
Curtiss CR-3 Racer
Curtiss CR-1 Skeeter
Curtiss CR-2 Coupe
Curtiss Crane Amphibian (Model 20)
Curtiss CS-1 (SC-1) Scout
Curtiss CS-2 (SC-2) Scout (Model 31)
Curtiss CS-3 Scout
Curtiss CS-4 Scout
Curtiss CS-5 Scout
Curtiss CT-1 Torpedoplane
Curtiss CW-1 (See: Junior)
Curtiss CW-2
Curtiss CW-3 (See: Duckling)
Curtiss CW-4 (See: Condor T-32)
Curtiss CW-5
Curtiss CW-6 (See: Travel Air 6000)
Curtiss CW-7 (See: Travel Air 7000)
Curtiss CW-8 (See: Travel Air 8000)
Curtiss CW-9 (See: Travel Air 9000)
Curtiss CW-10 (See: Travel Air 10)
Curtiss CW-11 (See: Travel Air 11)
Curtiss CW-12 Sport Trainer
Curtiss CW-14 Osprey
Curtiss CW-14 Speedwing
Curtiss CW-14 Sportsman
Curtiss CW-15 Sedan
Curtiss CW-16 Light Sport
Curtiss CW-17R Pursuit Osprey
Curtiss CW-18 Army Trainer
Curtiss CW-19L Coupe (Sparrow)
Curtiss CW-19R
Curtiss CW-A19R
Curtiss CW-B19R
Curtiss CW-19W Coupe
Curtiss CW-20 (CW-20A, C-55, St Louis)
Curtiss CW-20T (Condor III)
Curtiss CW-21 Demon
Curtiss CW-21A Demon
Curtiss CW-21B Demon
Curtiss CW-A22 Falcon
Curtiss CW-22 Falcon
Curtiss CW-22A Falcon
Curtiss CW-22B Falcon
Curtiss CW-22N Falcon (See: SNC)
Curtiss CW-23
Curtiss CW-24 (See: XP-55)
Curtiss CW-24B
Curtiss CW-25 (See: AT-9)
Curtiss CW-27 (See: C-76)
Curtiss CW-29 (See: P-87)
Curtiss CW-32 Transport
Curtiss Model D
Curtiss Model D, Ely
Curtiss Model D Headless
Curtiss Model D Headless, NASM
Curtiss Model D Headless, Ruth Law
Curtiss Model D Hydro
Curtiss Model D Hydro Headless
Curtiss Model D Hydro Triplane
Curtiss Duckling (CW-3, CW-3L, CW-3W)
Curtiss Model E
Curtiss Model E Triplane
Curtiss Model E Hydro
Curtiss Model E Hydro Headless
Curtiss Model E Hydro Headless "Bumble Bee"
Curtiss E-Boat (SC #15)
Curtiss Eagle I
Curtiss Eagle II
Curtiss Eagle III
Curtiss Eagle III Ambulance
Curtiss F Boat
Curtiss F Boat, 1913
Curtiss F Boat, 1914
Curtiss F Boat, 1917
Curtiss F Boat Triplane
Curtiss (Curtiss-Hall) F4C-1
Curtiss F6C-1 Hawk
Curtiss F6C-2 Hawk
Curtiss XF6C-3 Hawk
Curtiss F6C-3 Hawk
Curtiss XF6C-4
Curtiss F6C-4 Hawk
Curtiss XF6C-5 Hawk
Curtiss XF6C-6 Page Racer
Curtiss XF6C-7
Curtiss XF7C-1 Seahawk
Curtiss F7C-1 Seahawk
Curtiss XF8C-1 (OC-1) Falcon
Curtiss F8C-1 (OC-1) Falcon
Curtiss XF8C-2 Helldiver
Curtiss F8C-3 (OC-2)
Curtiss XF8C-4 Helldiver

Curtiss F8C-4 Helldiver
Curtiss F8C-5 (O2C-1) Helldiver
Curtiss XF8C-6 Helldiver
Curtiss XF8C-7 (XO2C-2, O2C-2, Helldiver Cyclone Command)
Curtiss F8C-7 Cyclone Helldiver
Curtiss XF8C-8 Cyclone Military Helldiver (Helldiver A-3, O2C-2)
Curtiss XF8C-8 Helldiver
Curtiss XF9C-1 Sparrowhawk
Curtiss XF9C-2 Sparrowhawk
Curtiss F9C-2 Sparrowhawk
Curtiss F9C-2 Sparrowhawk, NASM
Curtiss XF10C-1 (XS3C-1)
Curtiss XF11C-1 (XBFC-1) Hawk
Curtiss XF11C-2 Hawk
Curtiss F11C-2 (BFC-2) Goshawk
Curtiss XF11C-3 (XBF2C-1) Hawk
Curtiss XF12C-1
Curtiss XF13C-1
Curtiss XF13C-2
Curtiss XF13C-3
Curtiss XF14C-1
Curtiss XF14C-2
Curtiss XF15C-1
Curtiss F-5L
Curtiss F-6
Curtiss Falcon Conqueror Demonstrator
Curtiss Falcon, Export, Chile
Curtiss Falcon, Export, Cyclone
Curtiss Falcon, Export, Model 37F
Curtiss Falcon Lindbergh Special
Curtiss Falcon Conqueror Mailplane
Curtiss Falcon D-12 Mailplane
Curtiss Falcon Liberty Mailplane
Curtiss Falcon PAA Cyclone Mailplane
Curtiss Falcon Wright Mailplane
Curtiss Falcon II (Model 72)
Curtiss FL (Model 7)
Curtiss Fledgling (Model 51)
Curtiss Fledgling Guardsman (Model 51)
Curtiss Fledgling J-1 (Model 51)
Curtiss Fledgling J-2 (Model 51)
Curtiss Fledgling Jr (Model 51)
Curtiss Flying-Boat No.1
Curtiss Flying-Boat No.2 "Flying Fish"
Curtiss Freak Boat
Curtiss G (Army 1913 Tractor)
Curtiss G (Pusher R)
Curtiss Goupil "Dalton's Duck"
Curtiss GS-1
Curtiss GS-2
Curtiss Model H "America" (H-1)
Curtiss Model H "America" - 2-Engine (H-1)
Curtiss Model H "America" - 3-Engine (H-1)
Curtiss Model H-2
Curtiss Model H-3
Curtiss Model H-4 America
Curtiss Model H-5
Curtiss Model H-6
Curtiss Model H-7
Curtiss Model H-8 Small America
Curtiss Model H-9
Curtiss Model H-10 Small America
Curtiss Model H-11
Curtiss Model H-12 Large America
Curtiss Model H-12A Large America
Curtiss Model H-12B Large America
Curtiss Model H-12L Large America
Curtiss Model H-13
Curtiss Model H-14
Curtiss Model H-15
Curtiss Model H-16
Curtiss Model H-16-1
Curtiss Model H-16-2
Curtiss Model H-16A
Curtiss HA Dunkirk Fighter
Curtiss HA-1 Dunkirk Fighter
Curtiss HA-2 Dunkirk Fighter
Curtiss HA Mail
Curtiss Hawk, Export, Japan
Curtiss Hawk 1
Curtiss Hawk 1 "Doolittle Hawk"
Curtiss Hawk 1A
Curtiss Hawk 1A "Gulfhawk," NASM
Curtiss Hawk II Model 35, Goshawk
Curtiss Hawk II Model 35, Germany, Udet
Curtiss Hawk II Model 47, Goshawk
Curtiss Hawk III (Model 68)
Curtiss Hawk IV (Model 79)
Curtiss Hawk 75
Curtiss Hawk H75A
Curtiss Hawk H75A, France
Curtiss Hawk H75A Mohawk (UK)
Curtiss Hawk 75B
Curtiss Hawk 75D ("XP-36")
Curtiss Hawk 75E (See: Y1P-36)
Curtiss Hawk 75H
Curtiss Hawk 75I (See: P-37)
Curtiss Hawk 75J
Curtiss Hawk 75K
Curtiss Hawk 75L (See: P-36A thru P-36F)
Curtiss Hawk 75M
Curtiss Hawk 75N
Curtiss Hawk 75O
Curtiss Hawk 75P (See: XP-40)
Curtiss Hawk 75Q
Curtiss Hawk 75R
Curtiss Hawk 75S (See: XP-42)
Curtiss Helldiver A-4 (See: XF8C-7)
Curtiss HS-1
Curtiss HS-1L
Curtiss HS-2L
Curtiss HS-3
Curtiss Hudson Flyer
Curtiss Hummingbird
Curtiss Hydro, Original
Curtiss Hydro, 2d Modification
Curtiss Hydro, 3rd Modification
Curtiss Hydro Triplane
Curtiss Ice Boat/Sled (1907)
Curtiss J
Curtiss Janin Patent Boat
Curtiss JN-2 Jenny
Curtiss JN-3 Jenny
Curtiss JN-4 Jenny Ambulance Mods
Curtiss JN-4 Jenny Civil Mods
Curtiss JN-4 Jenny Hollywood Mods
Curtiss JN-4 Jenny
Curtiss JN-4A Jenny
Curtiss JN-4B Jenny
Curtiss JN-4C Jenny
Curtiss JN-4Can Canuck
Curtiss JN-4Can Oriole Special (Robertson Oriole/JN-4 Combination)
Curtiss JN-4D Jenny
Curtiss JN-4D Jenny, NASM
Curtiss JN-4D2 Jenny
Curtiss JN-4H Hisso Jenny
Curtiss JN-4HB Hisso Jenny
Curtiss JN-4HG Hisso Jenny
Curtiss JN-4HT Hisso Jenny
Curtiss JN-5 Twin Jenny
Curtiss JN-5H Jenny
Curtiss JN-6H Jenny
Curtiss JN-6HB Jenny
Curtiss JN-6HG-1 Jenny
Curtiss JN-6HG-2 Jenny
Curtiss JN-6HO Jenny
Curtiss JN-6HP Jenny
Curtiss JNS Jenny
Curtiss Judson Triplane
Curtiss June Bug (See: Aerial Experiment Assoc)
Curtiss Junior (CW-1)
Curtiss Junior (CW-1), NASM
Curtiss K
Curtiss Kingbird, 1st Prototype
Curtiss Kingbird, 2nd Prototype
Curtiss Kingbird, 3rd Prototype
Curtiss Kingbird D-2 (Model 55)
Curtiss Kingbird D-3 (Model 55)
Curtiss L
Curtiss L-1
Curtiss L-2
Curtiss Ladybird
Curtiss Lark (Model 41)
Curtiss Model M (Morris) Boat
Curtiss "McCormick Boat"
Curtiss MF Seagull (Model 18)
Curtiss (Curtiss-Robertson) Moth (See: Moth Aircraft Corp)
Curtiss MX-955 Advanced Trainer

Curtiss N
Curtiss N-6
Curtiss N-8
Curtiss N-9
Curtiss N-9C
Curtiss N-9H
Curtiss N-9H, NASM
Curtiss N-10
Curtiss XN2C-1 Fledgling
Curtiss N2C-1 Fledgling
Curtiss N2C-2 Fledgling
Curtiss NBS-1 (Martin MB-2)
Curtiss XNBS-4
Curtiss NC-1
Curtiss NC-2 (P2N-1)
Curtiss NC-3 (P2N-1)
Curtiss NC-4 (P2N-1)
Curtiss NC-4, NASM
Curtiss NC-5 to NC-10 (See: Naval Aircraft Factory)
Curtiss Night Mail
Curtiss O
Curtiss OC-1 (See: XF8C-1 & F8C-1)
Curtiss OC-2 Falcon (See: F8C-3)
Curtiss XOC-3 Falcon
Curtiss O2C-1 Helldiver (See: F8C-5)
Curtiss O2C-2 (See: XF8C-7 & XF8C-8)
Curtiss XO3C-1 Seagull
Curtiss XO-1 Falcon
Curtiss O-1 Falcon
Curtiss O-1A Falcon
Curtiss XO-1B Falcon
Curtiss O-1B Falcon
Curtiss O-1B Special Falcon
Curtiss O-1C Falcon
Curtiss O-1E Falcon
Curtiss O-1F Falcon
Curtiss XO-1G Falcon
Curtiss O-1G Falcon
Curtiss XO-11 Falcon
Curtiss O-11 Falcon
Curtiss O-11A Falcon
Curtiss XO-12 Falcon
Curtiss XO-13 Falcon
Curtiss XO-13A Falcon
Curtiss O-13B Falcon
Curtiss YO-13C (O-13C) Falcon
Curtiss YO-13D Falcon
Curtiss XO-16 Falcon
Curtiss XO-18 Falcon
Curtiss Y1O-26 (XO-26) Falcon
Curtiss O-39 Falcon
Curtiss YO-40 Raven
Curtiss YO-40A Raven
Curtiss Y1O-40A Raven
Curtiss Y1O-40B (O-40B) Raven
Curtiss O-52 Owl (Model 85)
Curtiss Orenco D (Model 26)
Curtiss Oriole (Model 17, Experiment 519, Design L-72)
Curtiss Oriole Clipped-Wing Racer (Clipped-Wing Oriole)
Curtiss Osprey (See: CW-14)
Curtiss P-1 Hawk
Curtiss P-1 Hawk, Allison Air-Cooled Engine Modification
Curtiss XP-1A Hawk
Curtiss P-1A Hawk
Curtiss P-1A Hawk, Export
Curtiss XP-1B Hawk
Curtiss P-1B Hawk
Curtiss P-1B Hawk, Export
Curtiss XP-1C Hawk
Curtiss P-1C Hawk
Curtiss P-1D Hawk
Curtiss P-1E Hawk
Curtiss P-1F Hawk
Curtiss XP-2 Hawk
Curtiss P-2 Hawk
Curtiss XP-3 Hawk
Curtiss XP-3A Hawk
Curtiss P-3A Hawk
Curtiss P-5 Hawk
Curtiss XP-6 Hawk
Curtiss P-6 (YP-6) Hawk
Curtiss P-6 Hawk, Export
Curtiss XP-6A Hawk
Curtiss P-6A Hawk
Curtiss XP-6B Hawk
Curtiss P-6C Hawk
Curtiss XP-6D Hawk
Curtiss P-6D Hawk
Curtiss XP-6E Hawk
Curtiss P-6E Hawk
Curtiss XP-6F Hawk
Curtiss XP-6G (P-6G) Hawk
Curtiss XP-6H Hawk
Curtiss P-6S Hawk (Export)
Curtiss XP-10
Curtiss P-11 Hawk
Curtiss XP-17 Hawk
Curtiss YP-20 Hawk
Curtiss XP-21 Hawk
Curtiss XP-21A Hawk
Curtiss XP-22 Hawk
Curtiss Y1P-22 Hawk
Curtiss XP-23 Hawk
Curtiss YP-23 Hawk
Curtiss XP-31 (XP-934) Swift
Curtiss Y1P-36 (P-36) Hawk
Curtiss P-36A Hawk
Curtiss P-36B Hawk
Curtiss P-36C Hawk
Curtiss XP-36D Hawk
Curtiss XP-36E Hawk
Curtiss XP-36F Hawk
Curtiss P-36G Hawk
Curtiss XP-37
Curtiss YP-37
Curtiss XP-40 Warhawk
Curtiss P-40 Warhawk
Curtiss P-40B Warhawk
Curtiss P-40C Warhawk
Curtiss P-40C Warhawk, 2-Engine Mockup
Curtiss P-40D Warhawk
Curtiss P-40E Warhawk
Curtiss P-40E Warhawk, NASM
Curtiss XP-40F Warhawk
Curtiss YP-40F-3 Warhawk
Curtiss P-40F Warhawk
Curtiss XP-40K Warhawk
Curtiss P-40K Warhawk
Curtiss TP-40K Warhawk
Curtiss P-40L Warhawk
Curtiss P-40M Warhawk
Curtiss XP-40N Warhawk
Curtiss P-40N Warhawk
Curtiss TP-40N Warhawk
Curtiss P-40P Warhawk
Curtiss XP-40Q Warhawk
Curtiss XP-40Q-1 Warhawk
Curtiss XP-40Q-2 Warhawk
Curtiss P-40R Warhawk
Curtiss (P-40) Kittyhawk Mk.I
Curtiss (P-40) Kittyhawk Mk.IA
Curtiss (P-40) Kittyhawk Mk.II
Curtiss (P-40) Kittyhawk Mk.III
Curtiss (P-40) Kittyhawk Mk.IV
Curtiss (P-40) Tomahawk Mk.I
Curtiss (P-40) Tomahawk Mk.II
Curtiss XP-42
Curtiss XP-46
Curtiss XP-46A
Curtiss P-47G (See: Republic)
Curtiss XP-53
Curtiss XP-55 Ascender
Curtiss XP-55 Ascender, NASM
Curtiss XP-60
Curtiss XP-60A
Curtiss YP-60A
Curtiss P-60A
Curtiss XP-60B
Curtiss XP-60C
Curtiss XP-60D
Curtiss XP-60E
Curtiss YP-60E
Curtiss XP-62
Curtiss XP-71
Curtiss XP-87 (XF-87) Blackhawk
Curtiss P-307 2-Engined Transport
Curtiss Pfitzner Monoplane
Curtiss PN-1
Curtiss PN-5 (See: F-5L)
Curtiss PN-7
Curtiss PN-8
Curtiss PN-9 thru PN-12 (See: Naval Aircraft Factory)

Curtiss P2N-1 (See: NC Boats)
Curtiss Prudden CP-2 1932 Transport
Curtiss XPW-8 Hawk
Curtiss PW-8 Hawk
Curtiss XPW-8A Hawk
Curtiss XPW-8B Hawk
Curtiss RC-1 Kingbird
Curtiss R2C-1 Racer
Curtiss R2C-2 Racer
Curtiss R3C-1 Racer
Curtiss R3C-2 Racer
Curtiss R3C-2 Racer, NASM
Curtiss R3C-3 Racer
Curtiss R3C-4 Racer
Curtiss R4C-1 Condor
Curtiss R5C-1 Commando
Curtiss R (Model 2)
Curtiss Pusher R
Curtiss Twin R
Curtiss R-2
Curtiss R-2A
Curtiss R-3
Curtiss R-3 (AH-62, A66)
Curtiss R-3 (AH-65, A67)
Curtiss R-4
Curtiss R-4L
Curtiss R-4LM
Curtiss R-6 (Model 2A)
Curtiss R-6L
Curtiss R-6 Army Racer
Curtiss R-7 "New York Times"
Curtiss R-8 Army Racer
Curtiss R-9
Curtiss Reid Rambler (See: Reid Aircraft Co)
Curtiss Robin (Model 50)
Curtiss Robin B
Curtiss Robin B-2
Curtiss Robin C
Curtiss Robin C-1
Curtiss Robin C-1 "St Louis Robin"
Curtiss Robin C-2
Curtiss Challenger Robin (Model 50A)
Curtiss Comet Robin
Curtiss Robin CR
Curtiss Robin J-1 (Model 50H)
Curtiss Robin J-1, Douglas "Wrong-Way" Corrigan
Curtiss Robin J-1 "Ole Miss," NASM
Curtiss Robin J-2 (Model 50I)
Curtiss Robin J-3
Curtiss Robin M
Curtiss Robin W
Curtiss Robin 4C (Model 50E)
Curtiss Robin 4C-1
Curtiss Robin 4C-1A (Model 50G)
Curtiss Robin 4C-2
Curtiss XSC-1 Seahawk
Curtiss SC-1 Seahawk
Curtiss XSC-2 (XSC-1A) Seahawk
Curtiss SC-2 Seahawk
Curtiss XSC-6 (SC-6) Scout (See: Martin)
Curtiss XSC-7 Scout
Curtiss XS2C-1 Shrike
Curtiss XS3C-1 (See: XF10C-1)
Curtiss XS4C-1 (See: XSBC-1)
Curtiss S-1 Baby Scout, Speed Scout
Curtiss S-2 Wireless
Curtiss S-3
Curtiss S-8
Curtiss XSBC-1 (XS4C-1)
Curtiss XSBC-2
Curtiss XSBC-3
Curtiss SBC-3 Helldiver
Curtiss XSBC-4 Helldiver
Curtiss SBC-4 Helldiver
Curtiss SBC-4 Helldiver, Export
Curtiss (SBC-4) Cleveland
Curtiss XSB2C-1 Helldiver
Curtiss SB2C-1 Helldiver
Curtiss SB2C-1A Helldiver
Curtiss SB2C-1C Helldiver
Curtiss XSB2C-2 Helldiver
Curtiss XSB2C-3 Helldiver
Curtiss SB2C-3 Helldiver
Curtiss SB2C-4 Helldiver
Curtiss SB2C-4E Helldiver
Curtiss XSB2C-5 Helldiver
Curtiss SB2C-5 Helldiver
Curtiss SB2C-5 Helldiver, NASM
Curtiss XSB2C-6 Helldiver
Curtiss XSB3C-1 Helldiver
Curtiss SE-5
Curtiss SNC-1 Falcon
Curtiss XSOC-1 (XO3C-1) Seagull
Curtiss SOC-1 Seagull
Curtiss SOC-2 Seagull
Curtiss SOC-3 Seagull
Curtiss SOC-4 Seagull
Curtiss XSO2C-1 Seagull
Curtiss XSO3C-1 Seagull
Curtiss SO3C-1 Seagull
Curtiss SO3C-2 Seagull
Curtiss SO3C-2C Seamew
Curtiss SO3C-3 Seagull
Curtiss Speedwing (See: CW-14)
Curtiss Sportsman (See: CW-14)
Curtiss SX4-1 Water Glider
Curtiss T Wanamaker Triplane, Model 3
Curtiss Tadpole
Curtiss Tanager (Model 54)
Curtiss TC-1
Curtiss Teal (Model 57)
Curtiss Thrush (Model 56)
Curtiss Tractor Hydro
Curtiss Travel Air (See: Travel Air)
Curtiss Triad
Curtiss TS-1
Curtiss TS-2
Curtiss VZ-7AP VTOL
Curtiss X-1
Curtiss X-19 VTOL
Curtiss X-100 VTOL
Curtiss No.1 Gold Bug, Golden Flier
Curtiss 18-B Hornet (Curtiss-Kirkham)
Curtiss 18-T Wasp (Curtiss-Kirkham)
Curtiss-Bleecker Helicopter
Curtiss-Cox Racer "Cactus Kitten"
Curtiss-Cox Racer "Texas Wildcat"
Curtiss-Herring No.1 "Reims Racer"

Curtiss-Robertson Airplane Manufacturing Co *(See: Curtiss)*

Curtiss-Wright *(See: Curtiss)*

Cussac (France)
Cussac Aeroplane

Custer Channel Wing Corp (Hagerstown, MD)

In 1947, Willard R. Custer established the **National Aircraft Corp** in Hagerstown, MD, to develop aircraft based on his "channel wing" theory of lift. By the early 1950s the company had been renamed **Custer Channel Wing Corp**. Following construction of working prototypes, Custer became involved in a long battle to obtain type certification for the aircraft and in a series of patent infringement suits against a number of other aircraft manufacturers. For all National and Custer designs, see **Custer**.

Custer Airbus
Custer CCW-1 (National H.S.1)
Custer CCW-1, NASM
Custer CCW-2
Custer CCW-5
Custer CCW-8
Custer CCW-12
Custer Light Aircraft Proposal
Custer Proof-of-Concept Model
Custer Prototype (1948)

Cvjetkovic, A. (Newbury Park, CA)
Cvjetkovic CA-61 Mini Ace
Cvjetkovic CA-65 Skyfly
Cvjetkovic CA-65A Skyfly

CVT (Centro di Volo a Vela de Torino) (Italy)
CVT M-200

CVV (Centro Volo a Vela, Politecnico di Milano) (Italy)
CVV P.110

CVV PM.280 Tartuca (Tortoise)

Cycloplane Ltd (Los Angeles, CA)
Cycloplane C-1 Flying Trainer
Cycloplane Ground Trainer

Cygnet *(See: Sisler Aircraft Corp)*

Cywinski, Stanislaw *(See: Lublin)*

Czech Military Aircraft Works *(See: Letov)*

Czechowski, Lieutenant (Poland)
Czechowski "Spiesz sie powoli" ("Hasten slowly") Glider

Czerwinski, Sergiusz (Poland)
Czerwinski (Sergiusz) 1917 Glider

Czerwinski, Waclaw (Poland)
Czerwinski (Waclaw):
CW I Glider
CW II Glider
CW III Glider
CW IV Glider
CW V Glider
CW 7 Glider
CW 8 Glider
W.W.S.1 Salamandra (Salamander) Glider (1936) (See also: de Havilland Canada)
W.W.S.1 Salamandra 48
W.W.S.1 Salamandra 49
W.W.S.1 Salamandra 53
W.W.S.1 Salamandra 53A
W.W.S.2 Zaba (Frog)
W.W.S.3 Delfin (Dolphin)

Czerwinski & Jaworski (Waclaw Czerwinski & Wladyslaw Jaworski) (Poland)
Czerwinski & Jaworski CWJ Glider
Czerwinski & Jaworski CWJ*bis* Glider
Czerwinski & Jaworski CWJ 2 (See: ITS)

D

***D. D. Funk Aviation Co, Inc** (See: Funk Aviation)*

***D 2 Inc (Bend, OR)** (See: Davis, Leeon)*

d'Apuzzo, Nicholas D. (Blue Bell, PA)
d'Apuzzo D-200 Freshman
d'Apuzzo D-201 Sport Wing
d'Apuzzo D-295 Senior Aero Sport

***d'Artois** (See: Artois)*

d'Ascanio, Corridino (Italy)
d'Ascanio Helicopter (1929)

***d'Astanières** (See: Constantin-d'Astanières)*

***d'Equevilly** (See: Equevilly)*

***d'Oplinter** (See: Oplinter, Jean)*

da Vinci (Leonardo da Vinci) (Italy)

Leonardo da Vinci is more properly called "Leonardo," as "da Vinci" is an epithet rather than a proper name. Material relating to Leonardo is filed under **da Vinci**, however, as this is the manner by which he is more commonly known. Identification of Leonardo's various designs is based on C.H. Gibbs-Smith, *Leonardo da Vinci's Aeronautics* (London: HM Stationary Office, 1975).

da Vinci Type A Ornithopter (c.1486)
da Vinci Type A Ornithopter with Head Harness (c.1486)
da Vinci Type A Ornithopter (c.1487)
da Vinci Type C Ornithopter (c.1486-90)
da Vinci Type I Helicopter (c.1486-90)
da Vinci Flap-Valve Wings (c.1486-90)
da Vinci Parachute

da Zara, Leonino (Italy)
da Zara Biplane (Farman Type)

Dabrowski, Jerzy (Poland)
Dabrowski Cykacz (Ticker)

Daedlus & Icarus (Greek Legend)
Daedlus & Icarus, Legend c.1100 BC

Daedalus Project (MIT, MA)
Daedalus 88 Human-Powered Aircraft
Daedalus 88, NASM

Daewoo Heavy Industries Ltd (South Korea)
Daewoo KTX-1 Woong-Bee
Daewoo KTX-2

Dagling (UK)
Dagling Sparrow Glider

***Dai-Ichi Kaigun Koku Gijitsusho** (See: Yoko- suka)*

***Dai-Ichi Rikugun Kokusho** (See: Rikugun)*

Dailey, H. H. (Berwyn, IL)
Dailey Center-Drop (Gull-Winged) Biplane

Daimler Motoren Gesellschaft Werke, Daimler-Werke AG (Germany)

In July 1915, **Daimler Motoren Gesellschaft Werke** established an aircraft division, concentrating on licensed production of other companies' designs through World War I. The company suspended aircraft production between 1919 and 1922 due to Versailles Treaty restrictions. After 1922, Daimler began development of a series of light aircraft designed by Hanns Klemm. In 1926, Daimler merged with **Benz and Cie** to form **Daimler-Benz AG**, which concentrated on the production of aircraft engines. Klemm continued to design aircraft, establishing **Leichtflugzeugbau Klemm** as a successor to Daimler in 1927. For all original pre-1927 Daimler designs, see **Daimler**. For post-1926 work, see **Klemm**. For World War I license work, see the original design firm.

Daimler L 6 (D.I)
Daimler L 8
Daimler L 9
Daimler L 11
Daimler L 14
Daimler L 15
Daimler R.I (G.I)
Daimler R.II (G.II)

Daimler-Benz AG (Daimler-Benz Aerospace) (Germany)

Daimler-Benz AG was formed in 1926 by the merger of **Daimler Motoren Gesellschaft** and **Benz and Cie**, both manufacturers of aircraft engines. The aircraft work that had been carried out by Daimler was continued by **Lichtflugzeugbau Klemm** and the new Daimler-Benz concern concentrated on aircraft engine production. During World War II, Daimler-Benz develop several aircraft proposals, none of which were actually built. For pre-1926 designs, see **Daimler**. For post-1926 light aircraft work, see **Klemm**. For World War II projects, see **Daimler-Benz**.

Daimler-Benz AT-2000 Mako

DaimlerChrysler Aerospace AG (International)

DaimlerChrysler AT-2000 Mako (See: Daimler-Benz)

Dale (Oklahoma City, OK)

Dale Monoplane

Dale, Eldon L. (Sylvia, KS)

Dale (Eldon) Model A Glider

Dallas Aviation School (Dallas, TX)

Dallas 1 Experimental Biplane Trainer

Dallas Motor Sales, Inc (Charles Dallas) (Detroit, MI) *(See: Lambert 1344)*

Damblanc, Louis (France)

Damblanc Alerion Helicopter

Dana Aviation Inc *(See: Bede)*

Danard et Nayot (France)

Danard et Nayot 1912 Biplane

Dandrieux (France)

Dandrieux Ornithopter (1879)

Danish Navy *(See: Orlogsvaerftet)*

Dansaire Corp (Dansville, NY)

Dansaire Coupe

Danton *(See: Denhaut)*

Danville Aircraft Co (USA)

Danville 1924 Sport Plane

Daphne Airplanes (Bedford, OH)

Daphne SDI-A (Homebuilt)

DAR (Darzhavna Aeroplanna Robotilnista) (Bulgaria)

DAR-UI
DAR-1
DAR-2
DAR-3
DAR-4
DAR-5
DAR-6
DAR-7
DAR-8
DAR-9
DAR-10
DAR-10A
DAR-10F

Dardelet (France)

Dardelet 1909 All-Metal Aeroplane

Dare Aircraft Corp (Detroit, MI)

Dare DVC-4
Dare 1928 Variable Camber Monoplane

Darjavna Aeroplanna Rabotilnitza *(See: DAR)*

Darmstadt (Akad. Fliegergruppe Darmstadt; Darmstadt University Flying Club) (Germany)

Darmstadt D.9 Konsul
Darmstadt D.18
Darmstadt D.22
Darmstadt D.36
Darmstadt "Darmstadt" (1928 Glider)
Darmstadt 1
Darmstadt Type 4 Glider
Darmstadt Type 6 Glider "Geheimrat"

Darracq (France)

Darracq F.B.24G (Developed Vickers F.B.24)

Dart Aircraft Ltd (UK)

Dart (UK) Flittermouse
Dart (UK) Kitten
Dart (UK) Pup (Dunstable Dart)

Dart Manufacturing Co (USA) *(See: Culver)*

Darzhavna Aeroplanna Robotilnista *(See: DAR)*

Dassault (France)

Following his release from a German concentration camp in 1945, Marcel Bloch returned to **Société des Aviation Marcel Bloch**, the majority of which had been nationalized by the French government. The Bloch family assumed the resistance code name of Marcel's brother (d'Assault) and, in 1949, the Bloch company became **Avions Marcel Dassault (AMD)**. During the late-1950s and early-1960s, the company was known as **Générale Aéronautique Marcel Dassault (GAMD)**, reverting to AMD by 1967, when Dassault acquired control of **Breguet Aviation**. The two merged in 1971 to form **Avions Marcel Dassault-Breguet Aviation**. In 1977, the company was partially nationalized with the French government controlling one half of the stock by 1981. In April 1990, the company was renamed **Dassault Aviation**. In September 1992, Dassault and **Aérospatiale SNI** pooled resources under a new state-owned holding company, **Société de Gestion de Participations Aéronautiques (SOGEPA)**, but each continued to operate under its own name. For Block designs pre-dating the end of World War II, see **Bloch**. For all post-World War II Bloch/Dassault designs, see **Dassault**. For pre-1971 Breguet designs, see **Breguet**. For post-1971 designs, see **Dassault**.

Dassault Balzac
Dassault Étendard (Banner) II
Dassault Étendard IV
Dassault Étendard IV M
Dassault Étendard IV P
Dassault Étendard VI
Dassault Super Étendard
Dassault Flamant (Flamingo) (MD 315)
Dassault Gardian (Cowboy)
Dassault Hirondelle (Swallow) (MD 320)
Dassault MD 316 T
Dassault Mercure (Mercury)
Dassault Milan (Kite/Bird)
Dassault Mirage F1
Dassault Mirage F1 B
Dassault Mirage F1 E
Dassault Mirage F2
Dassault Mirage G
Dassault Mirage G8
Dassault Mirage I
Dassault Mirage III Series
Dassault Mirage III B
Dassault Mirage III C
Dassault Mirage III D
Dassault Mirage III E
Dassault Mirage III O
Dassault Mirage III R
Dassault Mirage III S
Dassault Mirage III V
Dassault Mirage IV
Dassault Mirage IV A
Dassault Mirage 5

Dassault Mirage 50
Dassault Mirage 2000
Dassault Mirage 2000B
Dassault Mirage 4000
Dassault Mystère (Mystery) I
Dassault Mystère II
Dassault Mystère III
Dassault Mystère IV
Dassault Mystère IV A
Dassault Mystère IV B
Dassault Mystère IV N
Dassault Super Mystère B1
Dassault Super Mystère B2
Dassault Mystère/Falcon 10
Dassault Mystère/Falcon 20
Dassault Mystère/Falcon 20 Cargo Falcon
Dassault Mystère/Falcon 20 Cargo Falcon "Wendy," NASM
Dassault Mystère/Falcon 30
Dassault Mystère/Falcon 50
Dassault Mystère/Falcon 200
Dassault Mystère/Falcon 900B
Dassault Mystère/Falcon 900EX
Dassault Mystère/Falcon 2000
Dassault Ouragan (Hurricane) (MD 450)
Dassault Rafale (Squall)
Dassault HU-25 Guardian (USCG)

Dätwyler, Max (MDC Max Dätwyler AG) (Switzerland)
Dätwyler Lerche
Dätwyler MD-3 SwissTrainer
Dätwyler MDC-Trailer

Daucourt et Pourrain (France)
Daucourt et Pourrain 1911 Flying Boat

Dauheret (France)
Dauheret Helicopter

Daugherty, Earl S. (Chicago, Il)
Daugherty-Stupar Tractor Biplane

David, R. E. (USA)
David 1914 Chanute-Type Glider
David 1914-16 Glider

Davidson, G. L. O. (Scotland)
Davidson (G. L. O.) Glider
Davidson (G. L. O.) Gyropter (1906)

Davidson, Jesse (USA)
Davidson (Jesse) 1910/11 BlériotType

Davis Wing Ltd (Nampa, ID)
Davis Wing Starship Alpha

Davis, Emory (New York, NY)
Davis (Emory) Monoplane (1898)

Davis, John H. (Philadelphia, PA)
Davis (John) VTOL Aircraft Patent

Davis, Leeon (Davis Aircraft Corp) (Lake Village, IN)
Leeon Davis began designing light aircraft in the late 1950s at the **Davis Aircraft Corp** of Lake Village, IN. Through the 1960s and 1970s he produced a series of homebuilt designs marketed under the Davis name. In the 1980s, **D 2 Inc** of Bend, OR acquired the rights to several of Davis' designs, which it marketed under the D 2 name. For all Leeon Davis designs, see **Davis, Leeon.**
Davis (Leeon) DA-1A
Davis (Leeon) DA-2A
Davis (Leeon) DA-5A
Davis (Leeon) DA-6
Davis (Leeon) DA-7

Davis, Robert N. (USA)
Davis (Robert N.) Acro-Pro II

Davis, Walter C. (Davis Aircraft Corp) (Richmond, IN)
In January 1929 Walter Davis established the **Davis Aircraft Corp** in Richmond, IN to take over the **Vulcan Aircraft Co** of Portsmouth, OH. Davis reengineered the Vulcan American Moth into the Davis D-1. For all Walter Davis designs, see **Davis (Walter)**. For all Vulcan designs, see **Vulcan.**
Davis (Walter) D-1
Davis (Walter) D-1-K
Davis (Walter) D-1-W
Davis (Walter) D-1-85
Davis (Walter) V-3

Davis-Douglas Corp *(See: Douglas)*

Dawson, Clarence Richard "Cal" (with Clayton Henley and Glenn Johnson) (Coeur d'Alene, ID)
Dawson "Pistol Ball" (Racer)

Day, Charles Healy (Los Angeles, CA)
Charles Healy Day designed his first aircraft in 1909. In 1911 he worked as a designer for the new **Glenn L. Martin Co.** In 1912-13, Day managed his own shop, building tractor biplanes. He returned to Martin in mid-1913, and moved to **Sloane Aeroplane Co** in 1914. Sloane was taken over as part of the **Standard Aero Corp** in 1916; it was there that Day designed the Standard J-1; Standard closed in 1919. In 1927 Day and Ivan Gates formed **Gates-Day Aircraft Corp**, which was incorporated as **New Standard Aircraft Corp** in 1928. Day resigned in 1931, independently designing one further aircraft - his Model A. Day's independent designs are listed under **Day**; his designs for Martin, Sloane, and Standard are found under those companies. All Gates-Day and New Standard designs are found under **New Standard.**
Day (Chas) 1931 Biplane "The Errant"
Day (Chas) 1909 Tractor Biplane
Day (Chas) 1912 Tractor Biplane
Day (Chas) 1913 Pusher Biplane
Day (Chas) 1914 Gnôme Tractor
Day (Chas) 1914 Gyro Tractor

Day, Curtis LaQ (Paxton, IL)
Day (Curtis LaQ) 1910 Biplane Glider

Dayton Air Racing Assoc (Dayton, OH)
Dayton Racer "Miss Dara"

Dayton Aircraft Corp (Dayton, OH)
Dayton Overmount X

Dayton Wright Airplane Co (Dayton, OH)
In 1916, the **Wright Co** merged with the **Glenn L. Martin Co**, the **Simplex Automobile Co**, **Wright Flying Field, Inc**, and the **General Aeronautic Co of America** to form the **Wright-Martin Aircraft Corp.** (This firm became the **Wright Aeronautical Corp** after Martin left in 1917.) Orville Wright also began an informal association with Charles F. Kettering of the **Dayton Engineering Laboratory Co** (Delco). In April 1917, Wright, Kettering, and Col. H. E. Talbott, president of the **Dayton Metal Products Co** established the **Dayton Wright Airplane Co** around Dayton Metal Products, Delco, and the original Wright experimental station

in Dayton. **General Motors Corp** acquired the firm in 1919, later renaming it the **Dayton Wright Division** of GM. In 1923 GM dissolved the company and sold the rights to Dayton Wright trainer designs to **Consolidated Aircraft Corp**. For all Dayton Wright designs, see **Dayton Wright**. For all Wright Brothers designs, see **Wright**. For all Wright-Martin designs, see **Wright-Martin**. For all Wright Aeronautical designs, see **Wright Aeronautical**.

Dayton Wright XB-1A (USXB-1A)
Dayton Wright "Bull Head" Biplane
Dayton Wright D-1
Dayton Wright DH-4 Series

The British de Havilland D.H.4 was modified and built under license in the United States under the designation DH-4 by several manufacturers. For British-built aircraft, see **de Havilland D.H.4**. For American-built aircraft, see the individual manufacturers, including **Aeromarine**, **Atlantic**, **Boeing**, **Dayton Wright**, **Gallaudet**, **Keystone**, and **Standard**. When the actual manufacturer cannot be determined from the available evidence, the photos and documents have been filed under **Dayton Wright**.

Dayton Wright DH-4
Dayton Wright DH-4, NASM
Dayton Wright DH-4 Civil Conversions
Dayton Wright DH-4 Mailplane
Dayton Wright DH-4 Mailplane, NASM
Dayton Wright DH-4 Mailplane, Bellanca Modification
Dayton Wright DH-4Amb-1
Dayton Wright DH-4Amb-2
Dayton Wright DH-4B
Dayton Wright DH-4B, Engine Testbed
Dayton Wright DH-4B, Floatation Gear Testbed
Dayton Wright DH-4B-5 (Transport Conversion)
Dayton Wright DH-4BD (USDA Crop Dusting Modification)
Dayton Wright DH-4BS (Supercharged Engine Testbed)
Dayton Wright DH-4C
Dayton Wright DH-4K Honeymoon Express
Dayton Wright DH-4R Nine-Hour Cruiser
Dayton Wright (Douglas) DT-2
Dayton Wright FP-2 (Forestry Patrol Seaplane)
Dayton Wright FS-1
Dayton Wright (Standard) J-1
Dayton Wright K-T Prototype
Dayton Wright K-T
Dayton Wright O-W Aerial Coupé
Dayton Wright PS-1
Dayton Wright RB (Racer)
Dayton Wright SDW-1
Dayton Wright Sedan
Dayton Wright T-4 Messenger
Dayton Wright (TA-3) Chummy
Dayton Wright TA-3 (Prototype)
Dayton Wright TA-3
Dayton Wright TA-5
Dayton Wright TA-5, Single-Wheel Landing Gear Test
Dayton Wright TW-3 Series
Dayton Wright TW-3 Prototype
Dayton Wright (Consolidated) TW-3
Dayton Wright USD-9A (D.H.9)
Dayton Wright Model W Series
Dayton Wright WA
Dayton Wright WS
Dayton Wright Model 2

***Daytona Aircraft Construction Inc** (See: Jamieson)*

***de Bacqueville, Marquis** (See: Bacqueville)*

***de Beer, B.** (See: Beer)*

***de Bolotoff, Serge** (See: Bolotoff)*

de Bothezat, George (US AAS Engineering Division, McCook Field, OH)
de Bothezat Helicopter

***de Boyesson** (See: Boyesson)*

***de Brageas** (See: Brageas)*

de Brik (Netherlands)
de Brik 1914 Biplane (Farman-Type)

***de Bruyère** (See: Bruyère)*

***De Bruyne** (See: Aero Research, Ltd)*

***de Caters** (See: Voisin)*

***de Caze** (See: Caze)*

***de Chappedelaine, Jean (France)** (See also: American Tool & Die Co)*
de Chappedelaine 1928 Giroptére
de Chappedelaine 1934 Aérogyre

De Chenne Aeroplane & Motor Co (Monett, MO)
De Chenne 1911 Aluminum Biplane

***de Chevigny, Hubert** (See: Chevigny)*

***de Coster** (See: Coster)*

***de Dion** (See: Dion)*

***de Fabrègue** (See: Fabrègue)*

***de Goué** (See: Resnier de Goué)*

De Groof (Belgium/UK)
De Groof 1864 Ornithopter

de Havilland Aircraft Co Ltd (UK)

Geoffrey de Havilland began designing aircraft in 1908, producing two biplanes, the second of which was purchased by the British War Office at Farnborough. In 1911 de Havilland joined **H.M. Balloon Factory** at Farnborough (later renamed the **Army Aircraft Factory** and, in 1912, the **Royal Aircraft Factory**). In June 1914 de Havilland joined the **Aircraft Manufacturing Co Ltd** (**Airco**) of Hendon as Chief Designer. In 1920, Airco was sold to **Birmingham Small Arms Co Ltd** (**BSA**), which closed down the operation. In September 1920, de Havilland formed de **Havilland Aircraft Co Ltd**, using key men from the Airco operation. De Havilland extended operations into the Commonwealth during the 1920s and 1930s, opening **de Havilland Aircraft Pty Ltd** (1927, Australia), **de Havilland Aircraft of Canada Ltd** (1928), and further operations in India (1929), South Africa (1930), and New Zealand (1939). In 1948, de Havilland acquired **Airspeed Ltd**, merging completely in June 1951 with Airspeed becoming the **Airspeed Division** of de Havilland. In 1960, de Havilland's parent company, **de Havilland Holdings** (established 1955) was purchased by **Hawker Siddeley Group**. De Havilland operated as the **de Havilland Division** of Hawker Siddeley until 1963, when the Hawker Siddeley aviation interests were reorganized as **Hawker Siddeley Aviation** and all de Havilland products were redesignated with Hawker Siddeley (H.S.) designations

For all Geoffrey de Havilland designs predating 1911, Airco designs having D.H. designations, or de Havilland Aircraft Co designs, see **de Havilland.** For all other Airco designs, see **Airco.** For aircraft designed by the Canadian or Australian branches of de Havilland and having D.H.C. or D.H.A. designations, see **de Havilland (Canada)** or **de Havilland (Australia)**, respectively. For aircraft designed by the de Havilland Technical School, see **de Havilland (School).** For all aircraft designed by de Havilland while employed at Farnborough (H.M. Balloon Factory, Army Aircraft Factory, or Royal Aircraft Factory), see **Royal Aircraft Factory.** For all Airspeed designs, see **Airspeed.** For de Havilland designs postdating 1961 with designations above D.H.120, see **Hawker Siddeley.**

de Havilland:
- Biplane No.1
- Biplane No.2 (F.E.1)
- (Cierva) C.24 Autogyro
- (Airco) D.H.1
- (Airco) D.H.1A
- (Airco) D.H.2 Prototype
- (Airco) D.H.2
- (Airco) D.H.3
- (Airco) D.H.3A

The British de Havilland D.H.4 was modified and built under license in the United States under the designation DH-4 by several manufacturers. For British-built aircraft, see **de Havilland D.H.4.** For American-built aircraft, see the individual manufacturers, including **Aeromarine**, **Atlantic**, **Boeing**, **Dayton Wright**, **Gallaudet**, **Keystone**, and **Standard.** When the actual manufacturer cannot be determined from the available evidence, the photos and documents have been filed under **Dayton Wright**.

- (Airco) D.H.4 Prototype
- (Airco) D.H.4
- (Airco) D.H.4 US License Pattern
- (Airco) D.H.4 Civil Conversions
- (Airco) D.H.4 Floatation Testbed
- (Airco) D.H.4A
- (Airco) D.H.4R (Racer)
- (Airco) D.H.5 Prototype
- (Airco) D.H.5
- (Airco) D.H.6 Prototype
- (Airco) D.H.6
- (Airco) D.H.9
- (Airco) D.H.9, HP Slotted Wing
- (Airco) D.H.9, Civil Conversion
- (Airco) D.H.9A Prototype
- (Airco) D.H.9A
- (Airco) D.H.9AJ Stag
- (Airco) D.H.9B
- (Airco) D.H.9C
- (Airco) D.H.9J
- (Airco) D.H.9R (Racer)
- (Airco) D.H.10 Amiens I
- (Airco) D.H.10 Amiens II
- (Airco) D.H.10 Amiens III
- (Airco) D.H.10A Amiens IIIA
- (Airco) D.H.10C Amiens IIIC
- (Airco) D.H.11 Oxford
- (Airco) D.H.12
- (Airco) D.H.14 Okapi
- (Airco) D.H.14A
- (Airco) D.H.15 Gazelle
- (Airco) D.H.16
- (Airco) D.H.18
- (Airco) D.H.18A
- (Airco) D.H.18B
- D.H.27 Derby
- D.H.29
- D.H.29 Doncaster
- D.H.34
- D.H.37
- D.H.37A
- D.H.38
- D.H.42 Dormouse
- D.H.42A Dingo I
- D.H.42B Dingo II
- D.H.50
- D.H.50A
- D.H.50J (Sir Alan Cobham)
- D.H.51
- D.H.51A
- D.H.52 "Sibylla" (Glider)
- D.H.52 "Margon" (Glider)
- D.H.53 Prototype
- D.H.53 Humming Bird
- D.H.53 Humming Bird, Airship Experiments
- D.H.54 Highclere
- D.H.56 Hyena
- D.H.60 Family, HP Slots
- D.H.60 Moth Prototype
- D.H.60 Cirrus I Moth
- D.H.60 Cirrus II Moth
- D.H.60 Genet Moth
- D.H.60G Gipsy Moth
- D.H.60GIII Moth Major
- D.H.60M Moth
- D.H.60T Moth Trainer
- D.H.60X Moth
- D.H.61 Giant Moth
- D.H.65 Hound
- D.H.65A Hound
- D.H.65J Hound
- D.H.66 Hercules
- D.H.71 Tiger Moth
- D.H.72
- D.H.75 Hawk Moth Prototype
- D.H.75A Hawk Moth
- D.H.75B Hawk Moth
- D.H.77
- D.H.80 Moth Three (Puss Moth Prototype)
- D.H.80A Puss Moth
- D.H.81 Swallow Moth
- D.H.81A Swallow Moth
- D.H.82 Tiger Moth
- D.H.82A Tiger Moth Mk.II
- D.H.82B Queen Bee (Drone)
- D.H.82C Tiger Moth
- D.H.83 Fox Moth
- D.H.84 Dragon 1
- D.H.84 Dragon 1 "Seafarer"
- D.H.84 Dragon 2
- D.H.84 Rapide
- D.H.84M Dragon
- D.H.85 Leopard Moth Prototype
- D.H.85 Leopard Moth
- D.H.86, Single Pilot Version
- D.H.86, Two Pilot Version
- D.H.86A
- D.H.86B
- D.H.87 Hornet Moth
- D.H.87A Hornet Moth
- D.H.87B Hornet Moth
- D.H.88 Comet (Racer)
- D.H.89 Dragon Rapide Prototype
- D.H.89 Dragon Rapide
- D.H.89A Dragon Rapide
- D.H.89B Dominie Mk.I
- D.H.89B Dominie C.Mk.II
- D.H.89B Dominie, Civil
- D.H.89M Dragon Rapide
- D.H.90 Dragonfly
- D.H.91 Albatross Prototype
- D.H.91 Albatross
- D.H.92 Dolphin
- D.H.93 Don
- D.H.93 Don, Communications
- D.H.94 Moth Minor
- D.H.95 Flamingo Prototype
- D.H.95 Flamingo
- D.H.95 Hertfordshire
- D.H.98 Mosquito Prototype
- D.H.98 Mosquito, Prototype Fighter
- D.H.98 Mosquito PR.Mk.I
- D.H.98 Mosquito NF.Mk.II
- D.H.98 Mosquito T.Mk.III

D.H.98 Mosquito B.Mk.IV
D.H.98 Mosquito B.Mk.V
D.H.98 Mosquito FB.Mk.VI
D.H.98 Mosquito B.Mk.VII
D.H.98 Mosquito B.Mk.VIII
D.H.98 Mosquito B.Mk.IX
D.H.98 Mosquito PR.Mk.IX
D.H.98 Mosquito NF.Mk.XII
D.H.98 Mosquito FB.Mk.XIII
D.H.98 Mosquito NF.Mk.XV
D.H.98 Mosquito B.Mk.XVI
D.H.98 Mosquito PR.Mk.XVI
D.H.98 Mosquito FB.Mk.XVII
D.H.98 Mosquito FB.Mk.XVIII
D.H.98 Mosquito B.Mk.XX
D.H.98 Mosquito FB.Mk.21
D.H.98 Mosquito T.Mk.22
D.H.98 Mosquito B.Mk.23
D.H.98 Mosquito FB.Mk.24
D.H.98 Mosquito B.Mk.25
D.H.98 Mosquito FB.Mk.26
D.H.98 Mosquito T.Mk.27
D.H.98 Mosquito T.Mk.29
D.H.98 Mosquito NF.Mk.30
D.H.98 Mosquito NF.Mk.31
D.H.98 Mosquito PR.Mk.32
D.H.98 Sea Mosquito TR.Mk.33
D.H.98 Mosquito PR.Mk.34
D.H.98 Mosquito B.Mk.35
D.H.98 Mosquito TT.Mk.35
D.H.98 Mosquito TT.Mk.35, NASM
D.H.98 Mosquito NF.Mk.36
D.H.98 Sea Mosquito TR.Mk.37
D.H.98 Mosquito NF.Mk.38
D.H.98 Sea Mosquito TT.Mk.39
D.H.98 Mosquito FB.Mk.40
D.H.98 Mosquito PR.Mk.40
D.H.98 Mosquito PR.Mk.41
D.H.98 Mosquito FB.Mk.42
D.H.98 Mosquito T.Mk.43
(D.H.98) F-8 Mosquito (US AAF)
D.H.98 Mosquito, Civil
D.H.100 Vampire Prototype
D.H.100 Vampire F.Mk.I
D.H.100 Vampire Mk.II
D.H.100 Vampire F.Mk.3
D.H.100 Vampire FB.Mk.5
D.H.100 Vampire FB.Mk.9
D.H.100 Sea Vampire Mk.10 Prototype
D.H.100 Vampire F.Mk.20
D.H.100 Vampire F.Mk.21
D.H.100 Vampire T.Mk.22
D.H.100 Vampire FB.Mk.25
D.H.100 Vampire FB.Mk.30
D.H.100 Vampire F.Mk.31
D.H.100 Vampire F.Mk.32
D.H.100 FB.Mk.52
D.H.103 Hornet Prototype
D.H.103 Hornet F.Mk.1
D.H.103 Hornet PR.Mk.2
D.H.103 Hornet F.Mk.3
D.H.103 Hornet F.Mk.4
D.H.103 Sea Hornet F.Mk.20 Prototype
D.H.103 Sea Hornet F.Mk.20
D.H.103 Sea Hornet NF.Mk.21 Prototype
D.H.103 Sea Hornet NF.Mk.21
D.H.104 Dove Prototype
D.H.104 Dove
D.H.104 Dove Mk.8 ("Custom 800")
D.H.104 Devon C.Mk.1
D.H.104 Devon C.Mk.2
D.H.104 Sea Devon C.Mk.20
D.H.106 Comet Prototypes
D.H.106 Comet Mk.1
D.H.106 Comet Mk.1A
D.H.106 Comet Mk.2
D.H.106 Comet Mk.2X
D.H.106 Comet C.Mk.2
D.H.106 Comet Mk.3
D.H.106 Comet Mk.4
D.H.106 Comet Mk.4A
D.H.106 Comet Mk.4B
D.H.106 Comet Mk.4C
D.H.108, 1st Prototype
D.H.108, 2nd Prototype
D.H.108, 3rd Prototype
D.H.110 Sea Vixen Prototypes
D.H.110 Sea Vixen F.A.W.Mk.1
D.H.110 Sea Vixen F.A.W.Mk.2
D.H.112 Venom Prototypes
D.H.112 Venom FB.Mk.1
D.H.112 Venom NF.Mk.2 Prototype
D.H.112 Venom NF.Mk.2
D.H.112 Venom NF.Mk.3
D.H.112 Venom FB.Mk.4
D.H.112 Sea Venom F.A.W.Mk.20 Prototype
D.H.112 Sea Venom F.A.W.Mk.20
D.H.112 Sea Venom F.A.W.Mk.21
D.H.112 Venom FB.Mk.50
D.H.112 Venom NF.Mk.51
D.H.113 Vampire NF.Mk.10 Prototype
D.H.113 Vampire NF.Mk.10
D.H.113 Vampire NF.Mk.10, Laminar Flow Tests
D.H.113 Vampire NF.Mk.54
D.H.114 Heron Prototype
D.H.114 Heron Mk.1
D.H.114 Heron Mk.2
D.H.114 Heron C.Mk.3
D.H.114 Heron C.Mk.(VVIP)4
D.H.114 Sea Heron C.Mk.20
D.H.115 Vampire Trainer Prototype
D.H.115 Vampire T.Mk.11
D.H.115 Sea Vampire T.Mk.22
D.H.115 Vampire T.Mk.33
D.H.115 Sea Vampire T.Mk.34
D.H.115 Vampire T.Mk.35
D.H.115 Vampire T.Mk.55
D.H.121 Trident (See: Hawker Siddeley HS.121 Trident)
D.H.125 (See: Hawker Siddeley HS.125; Also: British Aerospace 125)

de Havilland (Australia) Aircraft Pty Ltd (Australia)

The **de Havilland Aircraft Pty Ltd (DHA)** was established by **de Havilland Aircraft Co Ltd** in 1927 to assemble prefabricated parts of de Havilland Moth airframes, a role later expanded to include the assembly of other de Havilland designs. The company undertook limited design work on its own rights, using the DHA designation for home-grown designs. De Havilland (Australia) was acquired by the **Hawker Siddeley Group** when Hawker Siddeley purchased its parent company, **de Havilland Holdings** in 1960. In 1962 DHA acquired **Bristol Aeroplane Co (Australia) Pty Ltd** and **Bristol Aviation Services Pty Ltd.** When Hawker Siddeley reorganized its aviation interests in 1963, DHA became **Hawker de Havilland Australia Pty Ltd (HDH)**. In 1985, HDH acquired **Commonwealth Aircraft Corp, Ltd (CAC)**, which became **Hawker de Havilland Victoria Ltd (HDHV)** in 1986. DHA and HDH have concentrated primarily on subassembly and engine manufacture. For all native de Havilland (Australia) designs having DHA designations, see **de Havilland (Australia)**. For original de Havilland designs, see **de Havilland**. For Bristol designs, see **Bristol**. For all aircraft having CAC (CA-) designations, see **CAC**.

de Havilland (Australia):
DHA.G1 (Glider)
DHA.G2 (Glider)
DHA.3 Drover Prototype
DHA.3 Drover

de Havilland (Canada) Aircraft of Canada Ltd (Canada)

De Havilland Aircraft of Canada Ltd (DHC) was established by **de**

Havilland Aircraft Co Ltd in 1928 to assemble and manufacture aircraft based on de Havilland designs. Following World War II, DHC developed its own designs, using the DHC designation. DHC was acquired by the **Hawker Siddeley Group** when Hawker Siddeley purchased its parent company, **de Havilland Holdings** in 1960. In June 1974, ownership of the company was transferred to the Canadian government as a temporary measure until "responsible Canadian investors" could assume control. In January 1986, the DHC was acquired by **Boeing of Canada Ltd** as the **de Havilland Division** of Boeing Canada. In February 1992, the company was sold to **Bombardier Aerospace North America** and the Ontario government and renamed **de Havilland Inc**. For all native designs by DHC, the de Havilland Division of Boeing Canada, and de Havilland Inc having DHC designations, see **de Havilland (Canada)**. For all aircraft designed by de Havilland Aircraft Co, see **de Havilland.**

de Havilland (Canada):
DHC-1 Chipmunk Prototype
DHC-1A Chipmunk
DHC-1A Chipmunk, NASM
DHC-1A Chipmunk 1 (RCAF)
DHC-1B Chipmunk
DHC-1B Chipmunk Mk.2 (RCAF)
DHC-1B Cheekee Chipmunk
DHC-1B Chipmunk, Aerobatic Conversions
DHC-1 Chipmunk T.Mk.10
DHC-1 Chipmunk T.Mk.20
DHC-1 Chipmunk Mk.21
DHC-1 Chipmunk Mk.22
DHC-1 Chipmunk Mk.23
DHC-2 Beaver Prototype
DHC-2 Beaver Mk.I
DHC-2 Beaver Mk.II
DHC-2 Turbo-Beaver Mk.III
(DHC-2) YL-20 Beaver
(DHC-2) L-20A (U-6A) Beaver
(DHC-2, L-20) TU-6A Beaver
(DHC-2) L-20B (U-6B) Beaver
DHC-3 Otter Prototype
DHC-3 Otter
DHC-3 Otter (Water Bomber)
DHC-3 Otter (RCAF)
DHC-3-T Turbo-Otter
(DHC-3) UC-1 Otter
(DHC-3) YU-1 Otter
(DHC-3) U-1A Otter
DHC-4 Caribou Prototype
DHC-4 Caribou (Prototype T-64 Testbed)
DHC-4 Caribou
(DHC-4) YAC-1 Caribou
(DHC-4) AC-1A (CV-2A, C-7A) Caribou
(DHC-4) VC-7A Caribou
(DHC-4) AC-1B (CV-2B, C-7B) Caribou
(DHC-4) CC-108 Caribou Mk.1
(DHC-4) CC-108 Caribou Mk.1A
(DHC-4A) CC-108 Caribou Mk.1B
DHC-5 Buffalo
DHC-5 Augmentor Wing Buffalo
DHC-5A Buffalo
(DHC-5) YCV-7 (YAC-2, CV-7A) Buffalo
(DHC-5) C-8A Buffalo
(DHC-5) C-8B Buffalo
(DHC-5) CC-115 Buffalo
(DHC-5) CC-115 Buffalo ACLS
(DHC-5A) CC-115 Buffalo
DHC-6 Twin Otter 1 (Prototype)
DHC-6 Twin Otter Series 100
DHC-6 Twin Otter Series 200
DHC-6 Twin Otter Series 300
(DHC-6) CC-138 Twin Otter
(DHC-6) V-18 Twin Otter Series
(DHC-6) UV-18A Twin Otter
(DHC-6) UV-18A Twin Otter
DHC-7 Dash 7 Series 100
DHC-7 Dash 7 Series 150
DHC-7 Dash 7 IR (Ice Reconnaissance)
(DHC-7) "RC-7B" ARL-C (E-5)
(DHC-7) "RC-7B" ARL-I (E-5)
(DHC-7) RC-7B ARL-M
(DHC-7) CC-132 Dash 7
DHC-8 Dash 8 Series 100
DHC-8 Dash 8 Series 200
DHC-8 Dash 8 Series 300
DHC-8 Dash 8 Series 300M Triton (ASW)
(DHC-8M) CC-142
(DHC-8M) CT-142
(DHC-8M) E-9A
CWA Robin Glider (See also; Czerwinski (Waclaw) W.W.S.1)
Sparrow Glider (See also; Czerwinski (Waclaw) W.W.S.1)

de Havilland (School) (de Havilland Aeronautical Technical School) (UK)
The **de Havilland Aeronautical Technical School** was established in 1928 to expand the existing apprentice system of the **de Havilland Aircraft Co Ltd** due to the expansion of the company following the success of the D.H.60 Moth. In addition to building de Havilland designs, the students designed and built several original aircraft, which received the T.K. designation for **"Tekniese Kollege,"** a title applied to the first such design by the Dutch student responsible for the drawings of the aircraft. For all original Technical School designs, see **de Havilland (School)**. For all aircraft designed by de Havilland Aircraft Co, see **de Havilland.**
de Havilland (School) T.K.1
de Havilland (School) T.K.2
de Havilland (School) T.K.3
de Havilland (School) T.K.4
de Havilland (School) T.K.5

De Hesse, Phillips (Chicago, IL)
De Hesse 1930 Primary Glider

de Korvin *(See: Korvin)*

de la Celle *(See: Celle)*

de la Hault (Belgium)
de la Hault Flyer (1908 Ornithopter)

de la Landelle, Gabriel (France)
de la Landelle 1863 Steam Air Liner

de la Vaulx-Tatin *(See: Clément-Bayard 1908 Twin-Boom Monoplane)*

de Lailhacar, Jacques Albert *(See: J.A.L)*

de Laminières *(See: Collin de Laminières)*

de Lesseps, Paul (France)
de Lesseps (Paul) Taris Monoplane (1910)

de Lesseps, Robert (France)
de Lesseps (Robert) Frégate (Frigatebird) Monoplane:
First Design
Second Design
Third Design

de Lestage, Arthur-Austin (France)
de Lestage 1907 Gyclogiro

de Marçay *(See: Marçay)*

de Mas *(See: Anzani)*

de Monge *(See: Monge; See also: Buscaylet-de Monge)*

de Muth (Brazil)
de Muth Flying Boat Transport Proposal

de Pescara *(See: Pescara)*

de Pischoff *(See: Pischoff, Pischoff-Kœchlin, Kœchlin)*

de Poix et de Roig *(See: Poix et de Roig)*

de Rouge, Charles *(See: Rouge)*

de Rue *(See: Ferber)*

de Schelde (N.V. Koninklijke Maatschappij de Schelde) (Netherlands)
de Schelde:
de Schelde S.12
de Schelde S.21 (See: Focke-Wulf Fw 198)
de Schelde Scheldemeeuw (Schelde Gull)
de Schelde ScheldeMusch (Schelde Sparrow)

de Scórzewski, Bernard (USA)
de Scórzewski Lynx

de Vore Aviation (Albuquerque, NM)
de Vore "Affordable Airplane"

De Wolf, Henry (Ferndale, MI)
De Wolf 1927 Cabin Monoplane

Dean, Herbert F. (Flint, MI)
Dean Delt-Air 250

Debort, Serge and Jack (France)
Debort 1908 Monoplane Glider
Debort 1908 Biplane Glider
Debort No.3 Monoplane
Debort No.4 Monoplane

Deboughy & Puthey (France)
Deboughy & Puthey Biplane Glider

Debougnies (Belgium/France)
Debougnies 1911 Monoplane Glider

Decatur Aircraft Co (Decatur, IL)
Decatur T-W-1 Sopwith Cruiser

Decazes et Basançon (France)
Decazes et Basançon 1902 Helicopter

Deconde (France)
Deconde 1918 Flying Boat

Deekay Aircraft Corp Ltd (UK)
Deekay Knight

Deflers (France)
Deflers 1909 Biplane

Defries (France)
Defries Biplane

Degen (Austria)
Degen 1812 Ornithopter

Degn, Paul Frederik (Germany)
Degn 1908 Flying Apparatus Patent

Dekellis-Olson (Oroville, CA)
Dekellis-Olson AT-1

Del Mar Engineering Laboratories (Los Angeles, CA)
Del Mar DH-2C
Del Mar DHT-1 Helicopter Whirlymite
Del Mar DHT-2
Del Mar DHT-2A

Delabrosse et Christollet (France)
Delabrosse et Christollet Variable Area Monoplane (1910)

deLackner Helicopters, Inc (deLackner Helicopter Co) (Mount Vernon, NY)
deLackner Clouduster
deLackner DH-4 Heli-Vector (Aerocycle)
deLackner DH-5 Aerocycle
deLackner HZ-1 (YHO-2) Aerocycle
deLackner Model 450

Delagrange, Léon (France)
Delagrange No.1 (See: Voisin No.2)
Delagrange No.2 (See: Voisin No.3)
Delagrange No.3 (See: Voisin No.20)

Delaisis (France)
Delaisis 1910 Monoplane

Delalandre (France)
Delalandre 1914 Aeroplane

Delamotte (France)
Delamotte 1909 Glider

Delanne Aircraft Corp (New York, NY)
Maurice Henry Delanne developed a series of tandem-wing aircraft in pre-World War II France. Following the end of the war, he emigrated to the United States, where he continued his work, establishing the **Delanne Aircraft Corp.** For all aircraft designed by Delanne in pre-War France, see **Delanne.** For all aircraft designed by Delanne Aircraft Corp, see **Delanne Aircraft.**
Delanne Aircraft DL-240
Delanne Aircraft DL-250

Delanne, Maurice-Henri (France)
Maurice-Henri Delanne developed a series of tandem-wing aircraft in pre-World War II France. Following the end of the war, he emigrated to the United States, where he continued his work, establishing the **Delanne Aircraft Corp.** For all aircraft designed by Delanne in pre-War France, see **Delanne.** For all aircraft designed by Delanne Aircraft Corp, see **Delanne Aircraft.**
Delanne D.II Ibis Bleu (Blue Ibis) (See also: Moreau JM 10)
Delanne 10 C2
Delanne 60 E1

Delasalle, Louis (France)
Delasalle 1911 Gliders

Delattre (France)
Delattre BN3 Night Bomber

Delaunay-Belleville (France)
Delaunay-Belleville 1908 Aeroplane

Delaurier, James D. (Canada) & Harris, Jeremy M. (Columbus, OH)
Delaurier & Harris 1991 Ornithopters
Delaurier & Harris Model 1 (See: Project Ornithopter, Inc)

Delauries (France)
Delauries l'Aérien (Ornithopter)

Delaygue (Pablo L. & Augusto Jose Delaygue) (Argentina)
Delaygue 1910 Monoplane
Delaygue 1913 Monoplane

Delest, Juan Alberto (Argentina)
Delest "Porteno"

Delfosse, Arthur (Germany)
Delfosse Monoplane (1909)

Delgado (Isaac) Central Trades School (New Orleans, LA)
Delgado Flash (Racer)
Delgado Maid (Racer)

Delhamende (Belgium) *(See: Fairy Tipsy)*

Delta Dart Flugzeugbau (Germany)
Delta Dart II

Delta System-Air AS (Czech Republic)
Delta (Czech) Pegass

Delta Wing Kites (Van Nuys, CA)
Delta Wing Kites Delta Wing
Delta Wing Kites Rogallo
Delta Wing Kites Trike
Delta Wing Kites Viper

Delta-V SRL (Moldova)
Delta-V AK-21 Autogyro

Deltour (France)
Deltour Helix

Demazel, Lucien and Paul (France)
Demazel 1912 Biplane Trainers
Demazel 1913 Civil Biplane
Demazel 1913 Military Biplane, Type M2

Demblon (Belgium)
Demblon 1925 Monoplane Glider

Demonty (Belgium)
Demonty 1924 Moto Aviette

Démouveaux (France)
Démouveaux 1901 Tandem-Wing Glider

Den-Gro (A. V. Denehy & F. H. Gross) (San Francisco, CA)
Den-Gro 1912 Monoplane

Denehie, W. A. (Los Angeles, CA)
Denehie 1911 Pusher Biplane
Denehie 1916 Tractor Biplane

Denhaut, François-Victor (France)
François-Victor Denhaut designed his several gliders and powered aircraft between 1907 and 1911. In 1912, he became chief designer for **Société des Hydroaéroplanes Donnet-Lévêque**, leaving in 1913 after a split between Donnet and Lévêque. Denhaut worked with **Morane-Saulnier** (1913), **Aéroplanes Borel** (1913), and **Établissements Aéronautics Ambroise Goupy** (1914), before joining Donnet to form **Donnet-Denhaut**. Denhaut left in 1919 to join **Bellanger Frères**. By the mid-1920s, Denhaut had joined **France-Aviation**. Denhaut's independent work is listed under **Denhaut**, with subsequent work under each of the companies using his designs.
Denhaut 1907 Gliders
Denhaut 1908 Denhaut-Bouyer-Mercier Canard Biplane
Denhaut Danton 1910 Racing Biplane
Denhaut Danton 1910 Racing Biplane, Elliptical Wings
Denhaut 1912 Flying Boat

Denien, Ralph R. (Cape Coral, FL)
Denien B-2 Wren B
Denien D-6 Sparrow Hawk
Denien Hocker-Denien Sparrow Hawk

Denight, William B. "Bart" (Bristol, PA)
Denight Special DDT

Denissel et Godville (France)
Denissel et Godville 1909 Helicoplane

Denny (William) & Brothers Ltd (Scotland)
Denny D.1 Hovercraft
Denny Helicopter (c.1914)

Denny Aerocraft Co (Boise, ID)
Denney Aerocraft Kitfox (Homebuilt)

Departemen Angkatan Udara Republik Indonesia, Lembaga Industri Penerbangan Nurtanio (Indonesia) *(See: Angkatan)*

Departmento de Aeronaves (Brazil) *(See: Centro Tecnico de Aeronautica)*

Deperdussin (France)
In 1910, Armand Deperdussin formed **Établissements A. Deperdussin** to produce the designs of Louis Béchereau and André Herbemont. In August 1913, Deperdussin was arrested for embezzlement and the company placed in receivership. Louis Blériot soon took control of the company, which was renamed **"SPAD"** (erroneously thought to mean "speed" in the then-fashionable international language *Volapuk*). Contemporary company literature justified the new name as an acronym for **Société Anonyme pour l'Aviation et ses Derives**. Following the war, Blériot combined the SPAD interests with his own company, **Blériot Aéronautique**, although designs continued to receive SPAD designations for some years. For all designs pre-dating Blériot's acquisition of the company in 1914, see **Deperdussin**. For post-1914 designs, see **SPAD**. For aircraft designed by the **British Deperdussin Aeroplane Co Ltd**, see **Deperdussin (UK)**.
Deperdussin-de Feure Canard (1910)
Deperdussin-de Feure Canard, Second Example (1910)
Deperdussin:
- Type A Monoplane (40hp Clerget, 1 Seat)
- Type B Monoplane (50hp Gnome, 1 Seat)
- Type C Monoplane (50/70hp Gnome, 1-2 Seats)
- Type D Monoplane (70hp Daimler or Panhard, 70/100hp Gnome, 2 Seat)
- Type E Monoplane (70/100hp Gnome, 3 Seats)
- Type F Monoplane (70/100hp Gnome, 2-3-4 Seats)
- 1910 Monoplane, 25hp Anzani
- 1910 Monoplane, 40hp Clerget
- 1910 Monoplane, Six-Bladed Prop, Paris Salon
- 1911 European Circuit

1911 Military, 70hp Gnôme
1911 Military, Reims Trials
1911 Racer, 50hp Gnôme
1912 Military, Single Seat
1912 Military, Two Seat
1912 Monocoque Racer, Anjou
1912 Monocoque Racer, Gordon Bennet Aircraft
1912 Monoplane, Four Seat
1912 Racer, Aéro Club de France Grand Prix
1913 Monocoque Racer
1913 Monocoque Racer, Schneider Trophy
1914 Military, Single Seat
1914 Military, Two Seat

Deperdussin (British Deperdussin Co Ltd) (UK)

The **British Deperdussin Co Ltd** was established in early 1912 to market and manufacture aircraft designed by **Établissements A. Deperdussin.** Although the company created several unique aircraft based on the basic Deperdussin work, it largely ceased to exist by World War I. For all British-designed aircraft, see **Deperdussin (UK)**. For all original (French) Deperdussin aircraft, see **Deperdussin.**

Deperdussin (UK) 1912 Military, Salisbury Trials
Deperdussin (UK) 1912 Military, 2-Seat
Deperdussin (UK) 1913 Racer
Deperdussin (UK) 1913 Hydro-Monocoque

Deremer, Arnold, & (J. C.) McNary *(See: Deremer, Arnold, & Trowbridge)*

Deremer, Arnold, & Trowbridge (F. E. Deremer, C. F. Arnold, & O. H. Trowbridge) (Detroit, MI)

Deremer, Arnold, & Trowbridge 2-place Monoplane (1928)

Dergint (France)

Dergint 1914 Aeroplane

Dernaut (France)

Dernaut 1908 Monoplane

Descamps (Elisée Alfred Descamps) (France)

E. Descamps was established in the early 1920s by Elisée Alfred Descamps. The company, general known as **Descamps-Brunet**, was absorbed by **Avions Caudron** late in 1927. All Descamps and Descamps-Brunet designs are filed under **Descamps**. For later designs, see **Caudron.**

Descamps 16
Descamps (Descamps-Brunet) 17 A2
Descamps 27 C1

Deschamps et Blondeau (France)

Deschamps et Blondeau 1909 Aeroplanes

Desert Aviation (USA)

Desert Aviation Staggerlite

Desfons (France)

Desfons Twin-Prop Seaplane

Desfontaines (France)

Desfontaines Aeroplane

Desforges, Abbot (France)

Desforges 1772 Ornithopter

Deshayes (France)

Deshayes 1922 Sailplane

Desmonceaux, de Givray, et Gallisti (France)

Desmonceaux, de Givray, et Gallisti 1908 Glider

Desoutter Aircraft Co Ltd (Marcel Desoutter) (UK)

Desoutter D.A.C.1 Sports Coupe
Desoutter D.A.C.2 Sports Coupe
Desoutter D.A.C.3 Sports Coupe

Dessau (Flugtechnischen Vereins Dessau E.V.) (Germany)

Dessau "Der Dessauer" (Sailplane)

Dessieux, Cliff (Alhambra, CA)

Dessieux Sport Monoplane (1929)

Désusclade (France)

Désusclade 1910 Monoplane

Detable et Tabary (Pierre Detable, his Son, and Tabary) (France)

Detable et Tabary:
Gliders (1892)
1908 Tailless
1912 Automatically Stable Aircraft
1914 Tractor Biplane

Detroit Aeronautical Construction Co (Detroit, MI)

The Detroit Aeronautical Construction Co was established in April 1911.

Detroit High-Wing Monoplane

Detroit Aircraft Corp (Detroit, MI)

The **Detroit Aircraft Corp** was established in June 1929 by the merger of **Lockheed Aircraft Corp** (Los Angeles, CA), **Ryan Aircraft Corp** (St Louis, MO), **Aircraft Development Corp** (Detroit, MI), **Eastman Aircraft Corp** (Detroit, MI), **Gliders, Inc** (Detroit, MI), **Blackburn Aircraft Corp** (Detroit, MI), **Marine Aircraft Corp** (Detroit, MI), **Winston Aviation Engine Co** (Cleveland, OH), **Parks Air College, Aircraft Parts Co, Aviation Tool Co, Inc** (Detroit, MI), and **Grosse Ile Airport, Inc** (Detroit, MI). The various companies continued to operate as independent divisions of Detroit under their original names until 1931 when the company went bankrupt. For all aircraft designed by the various divisions of Detroit, see under the original company name.

Detroit Aircraft TE-1 (See: Great Lakes TG-2)

Detroit Boat Co (Detroit, MI)

Detroit Boat Flying Fish

Deutsche Bristol-Werke (Germany) *(See: Halberstadt)*

Deutsche Flugzeug Werke *(See: DFW)*

Deutsche Forschungsinstitut für Segelflug *(See: DFS)*

Deutsche Sommer Werke (Germany)

Deutsche Sommer Werke Sesquiplane

Deutschen Versuchsanstalt für Luftfahrt *(See: D.V.L.)*

Deutschland (Flugwerkes Deutschland) (Germany)

Deutschland 1912 Biplane

Devaux (France)
Devaux 1911 Monoplane

Dewailly et Vertadier (France)
Dewailly et Vertadier 1911 Aeroplane

Dewey Airplane Co (Dewey, OK)
(See also: Will D Parker)
Dewey 1918 Tractor Biplane

DeWitt, Albert H. (Gary, IN)
DeWitt T-1

Dewoitine (France)
Emile Dewoitine established **Constructions Aéronautiques E. Dewoitine (CAD)** in 1920. In 1927, CAD shut down and Dewoitine reorganized the company as **Société Aéronautique Française (Avions Dewoitine) (SAF)**. During the nationalization of the French aviation industry in 1936, SAF became **Société Nationale de Constructions Aéronautiques du Midi (SNCAM)** until December 1940, when it was absorbed by the **Société National de Constructions Aéronautiques du Sud-Est (SNCASE)**. For all aircraft designed by CAD and SAF and all SNCAM and SNCASE having Dewoitine ("D.") designations, see **Dewoitine.**
Dewoitine D.1 C1
Dewoitine D.3 (Glider)
Dewoitine D.4 Bn2
Dewoitine D.7 (Sport Aviette)
Dewoitine D.7 (Sport Aviette)
Dewoitine D.8 C1
Dewoitine D.9 C1
Dewoitine D.10
Dewoitine D.12 C1
Dewoitine D.13 C1
Dewoitine D.14 (L), D.1 C1 (R)
Dewoitine D.15 C1
Dewoitine D.19 C1
Dewoitine D.21 C1
Dewoitine D.25 C2
Dewoitine D.26 C1
Dewoitine D.27 C1
Dewoitine D.27 III-R Racer
Dewoitine D.30
Dewoitine D.33 "Le Trait d'Union" ("The Hyphen")
Dewoitine D.332
Dewoitine D.338
Dewoitine D.342
Dewoitine D.35
Dewoitine D.37 (D.370)
Dewoitine D.371 C1
Dewoitine D.373 C1
Dewoitine D.412 (HD.412) (1931 Schneider Cup Racer)
Dewoitine D.43 (D.430)
Dewoitine D.48
Dewoitine D.500
Dewoitine D.501
Dewoitine D.510
Dewoitine D.510R
Dewoitine D.511
Dewoitine D.520
Dewoitine D.53
Dewoitine D.550
Dewoitine D.560
Dewoitine D.570 (Rebuilt D.560)
Dewoitine D.620
Dewoitine 1922 Sailplane
Dewoitine 1923 Monoplane (Aviette)

DFE Ultralights (Vanderbilt, PA)
DFE Ultralights Ascender III-A
DFE Ultralights Ascender III-B
DFE Ultralights Ascender III-C

DFL Holdings Inc (Gainesville FL)
DFL Foxtrot-4
DFL Tango-2

DFS (Deutsche Forschungsinstitut für Segelflug) (Germany)
DFS Habicht (Goshawk) (Sailplane)
DFS Kranich (Crane) (Sailplane)
DFS Meise (Titmouse) "Olympia"
DFS Reiher (Heron) (Sailplane)
DFS Storch (Stork) (Tailless)
DFS Weihe (Kite) (Sailplane) (See also: Jacobs Schweyer)
DFS 39
DFS 228
DFS 230
DFS 332
DFS 346 (Supersonic Research)
DFS 582 (See: D.V.L.)
DFS 611 (See: D.V.L.)

DFW (Deutsche Flugzeug Werke) (Germany)
DFW B.I (MD.14)
DFW B.II
DFW C Type
DFW C.I
DFW C.II
DFW C.IV
DFW C.V
DFW C.VI
DFW D.I
DFW Dr.I
DFW F 26
DFW F 34 (D.II)
DFW F 37 (C.VII)
DFW Farman Biplane
DFW Floh (Flea)
DFW Gitterschwanz (Lattice Tail) Pusher Biplane
DFW High-Speed Scout (1914)
DFW Hydroaeroplane (1913)
DFW Mars Biplane
DFW Mars Monoplane
DFW P.I
DFW R.I
DFW R.II
DFW Taube (Dove)

DG Flugzeugbau (Germany)
DG Flugzeugbau DG-100
DG Flugzeugbau DG-101
DG Flugzeugbau DG-200
DG Flugzeugbau DG-202
DG Flugzeugbau DG-300
DG Flugzeugbau DG-400
DG Flugzeugbau DG-500
DG Flugzeugbau DG-600
DG Flugzeugbau DG-800

Di Lorenzo, John (Pittsburgh, PA)
Di Lorenzo Airship (1907 Patent)

Diablo Aircraft Co (Stockton, CA)
Diablo One Place Bi-Plane (1929)

Diamond Aircraft Corp (Canada)
(See: Diamond Aircraft Industries)

Diamond Aircraft Industries GmbH (Austria) *(See also: Hoffmann)*
Diamond DA20-A1 Katana 100 (DV 20)
Diamond DA20-A2 Katana
Diamond DA20-B1 Katana
Diamond DA20-C_1 Katana
Diamond DA20-C_1 Katana Eclipse
Diamond DV 22 Speed Katana
Diamond DA40 Katana
Diamond DA40-180 Diamond Star
Diamond DA42 Diamond Star
Diamond HK 36R Super Dimona
Diamond HK 36TC Katana Xtreme
Diamond HK 36TS Katana Xtreme
Diamond HK 36TTC Katana Xtreme
Diamond HK 36TTS Katana Xtreme
Diamond MPX Motorglider
Diamond Super Dimona TC
Diamond Super Dimona TS
Diamond Super Dimona TTC
Diamond Super Dimona TTS

Diamond Xtreme Motorglider

Diamond Airplane Co (Pittsburg, California) *(See: Maupin-Lanteri)*

Dienaide, E. (France)
Dienaide 1881 Aeroplane Proposals

Diepen (Frits) Vliegtuigen N.V. (Netherlands)
Diepen-Difoga 421

Dietrich-Gobiet (Germany)
Dietrich D.P.I
Dietrich D.P.II
Dietrich D.P.III
Dietrich D.C.VI
Dietrich D.P.VII

Dietz Manufacturing Co (John W. Dietz, Dietz Airplane Co) *(See: General Aeronautical Corp)*

Dietz, Howard J. (New York, NY)
Dietz (Howard J.) Paraplane

Dietz-Schriber (Garden City, NY)
Dietz-Schriber Skylark (1910 Biplane)

DINFIA *(See: FMA)*

Dinoird (France)
Dinoird No.2 Monoplane
Dinoird 1910 Biplane
Dinoird 1911 Monoplane

Dirección General de Reparaciones y Construciones Navales (Mexico)
Dirección General de Reparaciones y Construciones Navales: "Tonatiuh II"
"El Barcenas" ("The Dapple") B-01

Dirección Nacional de Fabricaciónes e Investigaciónes Aeronáuticas *(See: FMA)*

de Dion (Albert, Marquis de Dion) (France)
de Dion 1909 Multiplane
de Dion 1911 Biplane (de Dion-Bouton Biplane)

Ditro *(See: Biro de Ditro)*

Dits-Moineau (Henri Dits & Réné Moineau) (France)
Dits-Moineau 1923 Moto Aviette

Dittmar, Heini (Germany)
Dittmar Condor
Dittmar Condor 2
Dittmar Condor 2A
Dittmar H.D.53 Möwe (Gull)
Dittmar H.D.153 Motor-Möwe (Gull)
Dittmar H.D.156

Dixon, Harry S. (UK)
Dixon Nipper (1911 Canard)

Dixon, Tom C. (USA)
Dixon Special (1948) (Racer)

D.K.D. (Dzialowski, Kruger, Dzialowski) (Poland)
D.K.D.1
D.K.D.3
D.K.D.4
D.K.D.5
D.K.D.7
D.K.D.8
D.K.D.10 Aeromobil (Aerocar)

DM Aerospace Ltd (UK) *(See: Thorp T-211 AeroSport)*

D.N.F. (Duperron-Niepce-Fetterer) (France)
D.N.F. Aérobus
D.N.F. 1916 Three-Engined Bomber

Doak Aircraft Co, Inc (Torrance, CA)
Doak DRD-1 (Trainer)
Doak VZ-4 (Model 16)

Doak-Deeds (Edmond R. Doak & W. C. Deeds) (Los Angeles, CA)
Doak-Deeds Sportsman (1928)

Dobkewitsch, Lt. G. (Lithuania)
Dobkewitsch D.1 Monoplane
Dobkewitsch D.2 Monoplane
Dobkewitsch 1924 Sport Monoplane

Doblhoff *(See: WNF)*

Dodge, William de Leftwich (New York, NY)
Dodge Flying Machine (c.1900)

Doflug Altenrhein (Switzerland)
Doflug Altenrhein D-3802
Doflug Altenrhein D-3802A
Doflug Altenrhein D-3803

Doman Helicopters, Inc (Danbury, CT)
Glidden S. Doman established **Doman Helicopters, Inc** in 1945 to exploit his helicopter-related patents. In January 1966, the company's assets, including all plant fixtures and certificates, were purchased by **Caribe Doman Helicopters, Inc** and moved to Bayamon, Puerto Rico. In September 1967, Doman joined with Don R. Berlin to revive the original Doman company, now renamed **Berlin Doman Helicopters, Inc** and moved to Toughkenamon, PA. Berlin Doman purchased the original Doman assets back from Caribe Doman. Berlin Doman faded from the aviation industry in the early 1970s. For all Doman, Caribe Doman, and Berlin Doman designs, see **Doman**.
Doman BD-68
Doman D-10B (LZ-5A)
Doman YH-31
Doman LZ-1A (Modified Sikorsky R-6B)
Doman LZ-2A Pelican
Doman LZ-4A
Doman LZ-5

Domenjoz, John (Old Orchard, ME)
Domenjoz Glider

Domingo, J. C. (France)
Domingo 1914 Aéraptère

Dominion Aircraft Corp (Canada; Renton, WA) *(See Also: Skytrader)*
Dominion Skytrader 800

Domrachev-Vil-Dgrub (Russia)
Domrachev-Vil-Dgrub LIG-6 (LEM-3)

Don S. Mitchell Co *(See: Mitchell)*

Dong Xue Ping (Dong In Industries) (China)
Dong Wizard

Donnet (Hydravions J. Donnet) (France)
Hydravions J. Donnet was formed by Jérôme Donnet in 1919, after François Denhaut left Donnet-Denhaut. The new firm continued

operations until about 1923. Pre-1919 designs appear under **Donnet-Denhaut**, with later designs under **Donnet**.

Donnet HB 3

Donnet-Denhaut (Société des Établissements Donnet-Denhaut) (France)

In 1914 Jérôme Donnet and François Denhaut, who had worked together at **Société des Hydroaéroplanes Donnet-Lévêque**, formed **Société des Établissements Donnet-Denhaut**. In 1919, Denhaut left, and the company was renamed **Hydravions J. Donnet**. For pre-1919 designs, see **Donnet-Denhaut**. For later designs, see **Donnet**.

Donnet-Denhaut D.D.1
Donnet-Denhaut D.D.2
Donnet-Denhaut D.D.8
Donnet-Denhaut D.D.9
Donnet-Denhaut D.D.10
Donnet-Denhaut G.L.400
Donnet-Denhaut P.10
Donnet-Denhaut P.15

Donnet-Lebranche (France)

Donnet-Lebranche 1910 Biplane

Donnet-Lévêque (Société des Hydroaéroplanes Donnet-Lévêque) (France)

Jérôme Donnet and M. Lévêque formed **Société des Hydroaéroplanes Donnet-Lévêque** in July 1912 to build François Denhaut's flying boat designs. In 1913, the company became **Société des Hydroaéroplanes Lévêque**, which was soon absorbed by **Franco-British Aviation**. Early designs are under **Donnet-Lévêque**, with later work under **FBA**.

Donnet-Lévêque Type A Flyingboat
Donnet-Lévêque Type B Flyingboat
Donnet-Lévêque Type C Flyingboat
Donnet-Lévêque Type PD (Paris-to-Deauville)

Dorand, Captaine Jean (France)

Dorand Type A.R.1 A2
Dorand Type A.R.1 D2
Dorand Type A.R.2 A2
Dorand BU Bomber (1915/16)
Dorand Do 1 Armored Biplane (1913)
Dorand Flying Boat
Dorand Type 7
Dorand 1894-1908 Kites (Powered, Manned, and/or Steerable)
Dorand 1908 Quadruplane/Triplane/Biplane
Dorand 1910 Biplan-Labratoire
Dorand 1912 Biplan de Place Forte (Fortress Biplane)
Dorand 1913 Armored Interceptor

Dormay (France)

Dormay 1916 Hydroplane

Dormoy, Etienne (Dayton, OH)

Dormoy Bathtub Monoplane
Dormoy Bathtub Monoplane Replica
Dormoy Bathtub 1924 Monoplane

Dorna (H. F. Dorna Co) (Iran)

Dorna Blue Bird

Dornier, Claude (Germany)

Claude Dornier became an airship designer for Zeppelin in 1910, designing and building numerous aircraft at that company's Lindau works. In 1919 he formed **Dornier Metallbauten GmbH**, and in 1926 he created a second company in Switzerland (**AG für Dornier-Flugzeuge**) to circumvent post-WWI restrictions on German aviation. In 1933 the German company restructured as **Dornier-Werke GmbH**, with several subsidiary companies. The company again reformed under this name in 1954, and under **Dornier GmbH** in 1972. All designs of Claude Dornier and his companies are listed together under **Dornier**.

Dornier Aerodyne (VTOL)
Dornier (Zeppelin-Lindau) C.I (Cl I)
Dornier (Zeppelin-Lindau) C.II (Cl II) (Ja)
Dornier (Zeppelin-Lindau) C 3 Komet I (Comet I)
Dornier (Zeppelin-Lindau) C 3 Komet II (Comet II)
Dornier (Zeppelin-Lindau) C 3 Komet III (Comet III)
Dornier CD2 Seastar
Dornier Cs I
Dornier (Zeppelin-Lindau) D.I
Dornier (Zeppelin-Lindau) Do A Libelle I (Dragonfly I)
Dornier (Zeppelin-Lindau) Do A Libelle II (Dragonfly II)
Dornier Do B Merkur I (Mercury I)
Dornier Do B Merkur Is (Mercury I Seaplane)
Dornier Do C4 (Do 10)
Dornier Do D
Dornier Do E
Dornier Do F (Do 11)
Dornier Do G Greif (Griffin) Project
Dornier Do H Falke (Falcon)
Dornier Do H Seefalke (Sea Falcon)
Dornier Do J Wal (Whale) (Do 16)
Dornier Do J II Wal (Whale) (Do 16)
Dornier Do K 1
Dornier Do K 2
Dornier Do K 3
Dornier Do K 4
Dornier Do L1 Delphin I (Dolphin I) (Cs 2)
Dornier Do L2 Delphin II (Dolphin II)
Dornier Do L3 Delphin III (Dolphin III)
Dornier Do N
Dornier Do P
Dornier Do P 59-04
Dornier Do P 59-05
Dornier Do P 231/1-01
Dornier Do P 231/2-01
Dornier Do P 231/2-03
Dornier Do P 231/3-01
Dornier Do P 232/2-02
Dornier Do P 232/3-06
Dornier Do P 238/1-01
Dornier Do P 247/6-01
Dornier Do P 252/3-01
Dornier Do P 256/1-01
Dornier Do R2 Superwal 1 (Super Whale)
Dornier Do R4 Superwal 2 (Super Whale)
Dornier Do S
Dornier Do T
Dornier Do U
Dornier Do V Komet II
Dornier Do W
Dornier Do X
Dornier Do Y (Dornier Do 15)
Dornier Do 12 Libelle (Dragonfly) III
Dornier Do 13
Dornier Do 14
Dornier Do 16 Militar Wal (Military Whale)
Dornier Do 17
Dornier Do 17 E
Dornier Do 17 F
Dornier Do 17 L
Dornier Do 17 M
Dornier Do 17 P
Dornier Do 17 Z
Dornier Do 18 Family
Dornier Do 18 E
Dornier Do 18 F
Dornier Do 19
Dornier Do 20
Dornier Do 21

Dornier Do 22
Dornier Do 22 L
Dornier Do 23
Dornier Do 24
Dornier Do 24 A
Dornier Do 24/72
Dornier Do 25
Dornier Do 26 Seefalke (Sea Falcon)
Dornier Do 27
Dornier Do 28 Skyservant
Dornier Do 29
Dornier Do 30
Dornier Do 31
Dornier Do 32
Dornier Do 34 Kiebitz (Lapwing)
Dornier Do 36
Dornier Do 128
Dornier Do 128-2
Dornier Do 128-6
Dornier Do 212
Dornier Do 214
Dornier Do 215
Dornier Do 216
Dornier Do 217
Dornier Do 217 E
Dornier Do 217 K
Dornier Do 217 M
Dornier Do 217 J Night Fighter
Dornier Do 228
Dornier Do 231
Dornier Do 252
Dornier Do 254
Dornier Do 317
Dornier Do 318
Dornier Do 324
Dornier Do 335 Pfeil (Arrow)
Dornier Do 335 A-02 Pfeil, NASM
Dornier Do 335 Z Zwilling Pfeil (Twin Arrow)
Dornier Do 417
Dornier Do 435
Dornier (Zeppelin-Lindau) Gs I
Dornier (Zeppelin-Lindau) Gs II
Dornier (Zeppelin-Lindau) Rs I
Dornier (Zeppelin-Lindau) Rs IIa
Dornier (Zeppelin-Lindau) Rs IIb
Dornier (Zeppelin-Lindau) Rs III
Dornier (Zeppelin-Lindau) Rs IV
Dornier (Zeppelin-Lindau) RsN IV
Dornier (Zeppelin-Lindau) Rs V Project
Dornier (Zeppelin-Lindau) Rs VI
Dornier Spatz (Sparrow)
Dornier TNT

DOSAAF/Komsomol Group (Russia)

DOSAAF/Komsomol TR-2 Komsomolets

***Doswiadczalne Warsztaty Lotnicze** (See: RWD-8)*

Dottori (France)

Dottori 1912 Biplane

Douglas (Santa Monica, CA)

Donald Douglas and David R. Davis established the **Davis-Douglas Co** in July 1920. By July 1921, Davis lost interest in aircraft manufacturing and Douglas reorganized the company with new financial backing as the **Douglas Co.** In November 1928, he restructured the company again, forming **Douglas Aircraft Co Inc.** In January 1932, Douglas Aircraft financed 51% of John Northrop's second company, **Northrop Corp**; when Northrop experienced serious labor problems in 1937, Douglas Aircraft acquired the remainder of Northrop's stock and, in September 1937, dissolved the company, absorbing Northrop's operations as the **El Segundo Division** of Douglas Aircraft. In April 1967, the financially troubled **Douglas Aircraft Corp** merged with the **McDonnell Company** to form the **McDonnell Douglas Corporation.** For aircraft designed by Davis-Douglas, Douglas, and Douglas Aircraft prior to 1967, see **Douglas.** For all aircraft designed by Northrop Corp, see **Northrop.** For all products postdating the 1967 merger, see **McDonnell Douglas.**

Douglas AD-1 (BT2D-1) Skyraider
Douglas BT2D-1 (AD-1Q) Skyraider
Douglas XAD-1W Skyraider
Douglas XAD-2 (BT2D-2) Skyraider
Douglas AD-2 Skyraider
Douglas AD-2D Skyraider
Douglas AD-2Q Skyraider
Douglas AD-2QU Skyraider
Douglas AD-2W Skyraider
Douglas AD-3 Skyraider
Douglas AD-3E Skyraider
Douglas AD-3N Skyraider
Douglas AD-3Q Skyraider
Douglas AD-3QU Skyraider
Douglas AD-3S Skyraider
Douglas AD-3W Skyraider
Douglas AD-4 Skyraider
Douglas AD-4B Skyraider
Douglas AD-4L Skyraider
Douglas AD-4N Skyraider
Douglas AD-4NA (A-1D) Skyraider
Douglas AD-4NL Skyraider
Douglas AD-4Q Skyraider
Douglas AD-4W Skyraider
Douglas AD-5 (A-1E) Skyraider
Douglas AD-5N (A-1G) Skyraider
Douglas AD-5Q (EA-1F) Skyraider
Douglas AD-5S Skyraider
Douglas AD-5W (EA-1E) Skyraider
Douglas AD-6 (A-1H) Skyraider
Douglas AD-6 (A-1H) Skyraider, NASM
Douglas AD-7 (A-1J) Skyraider
Douglas XA2D-1 Skyshark
Douglas A2D-1 Skyshark
Douglas XA3D-1 Skywarrior
Douglas YA3D-1 Skywarrior
Douglas A3D-1 (A-3A) Skywarrior
Douglas YA3D-1Q (YEA-3A) Skywarrior
Douglas A3D-1Q (EA-3A) Skywarrior
Douglas YA3D-1P (YRA-3A) Skywarrior
Douglas A3D-1P (YRA-3A) Skywarrior
Douglas A3D-2 (A-3B) Skywarrior
Douglas YA3D-2P Skywarrior
Douglas (A3D-2) ERA-3B Skywarrior
Douglas (A3D-2) NA-3B Skywarrior
Douglas (A3D-2) NRA-3B Skywarrior
Douglas (A3D-2) UA-3B Skywarrior
Douglas A3D-2P (RA-3B) Skywarrior
Douglas A3D-2Q (EA-3B) Skywarrior
Douglas A3D-2T (TA-3B) Skywarrior
Douglas A3D-2Z (VA-3B) Skywarrior
Douglas EKA-3B Skywarrior
Douglas KA-3B Skywarrior
Douglas XA4D-1 Skyhawk
Douglas A4D-1 (A-4A) Skyhawk
Douglas A4D-2 (A-4B) Skyhawk
Douglas A4D-2N (A-4C) Skyhawk
Douglas A4D-2N (A-4C) Skyhawk, NASM
Douglas A4D-3 Skyhawk
Douglas A4D-4 Skyhawk
Douglas A4D-5 (A-4E) Skyhawk
Douglas A4D-6 Skyhawk
Douglas TA-4E Skyhawk
Douglas A-4F Skyhawk
Douglas TA-4F Skyhawk
Douglas A-4G Skyhawk
Douglas TA-4G Skyhawk
Douglas A-4H Skyhawk
Douglas TA-4H Skyhawk
Douglas TA-4J Skyhawk
Douglas A-4K Skyhawk
Douglas TA-4K Skyhawk
Douglas A-4KU Skyhawk
Douglas TA-4KU Skyhawk
Douglas A-4L Skyhawk
Douglas A-4M (A-4Y) Skyhawk

Douglas OA-4M Skyhawk
Douglas A-4N Skyhawk
Douglas A-4P Skyhawk
Douglas A-4PTM Skyhawk
Douglas TA-4PTM Skyhawk
Douglas A-4Q Skyhawk
Douglas A-4S Skyhawk
Douglas TA-4S Skyhawk
Douglas XA-2 (1925)
Douglas A-4 Drone (1940) (See: BT-2BG)
Douglas A-6 Drone (Modified O-38)
Douglas A-20, Civil Use
Douglas A-20 Havoc
Douglas A-20A Havoc
Douglas XA-20B Havoc
Douglas A-20B Havoc
Douglas A-20C Havoc
Douglas A-20C Boston IIIA
Douglas A-20D Havoc
Douglas A-20E Havoc
Douglas XA-20F Havoc
Douglas A-20G Havoc
Douglas A-20H Havoc
Douglas A-20J Havoc
Douglas A-20J Boston IV
Douglas A-20K Havoc
Douglas A-20K Boston V
Douglas A-24 Dauntless
Douglas A-24A Dauntless
Douglas RA-24A (QF-24A) Dauntless
Douglas A-24B (F-24B) Dauntless
Douglas DF-24B Dauntless
Douglas XA-26 Invader
Douglas XA-26A Invader
Douglas A-26A (B-26K) Counter Invader
Douglas XA-26B Invader
Douglas A-26B (B-26B) Invader
Douglas CB-26B Invader
Douglas TB-26B Invader
Douglas VB-26B Invader
Douglas VB-26B Invader, NASM
Douglas XA-26C Invader
Douglas A-26C (B-26C) Invader
Douglas RB-26C Invader
Douglas XA-26D Invader
Douglas XA-26E Invader
Douglas XA-26F (XB-26F) Invader
Douglas A-26G Invader
Douglas A-26H Invader
Douglas B-26J Invader (See: JD-1)
Douglas YB-26K Counter Invader
Douglas A-26Z Invader
Douglas A-33 (See: Northrop)
Douglas XA-42 (See: XB-42)
Douglas BD-1 (A-20)
Douglas BD-2 (A-20)
Douglas XB-7
Douglas Y1B-7 (B-7)
Douglas YB-11 (See: YOA-5)
Douglas B-18 Bolo
Douglas B-18A Bolo
Douglas B-18AM Bolo
Douglas B-18B Bolo
Douglas B-18E Bolo
Douglas B-18M Bolo
Douglas B-18 Digby (RCAF)
Douglas XB-19 (XBLR-2)
Douglas XB-19A
Douglas B-22
Douglas B-23 Dragon
Douglas XB-31
Douglas XB-42 (XA-42) Mixmaster
Douglas XB-42A Mixmaster
Douglas XB-42A Mixmaster, NASM
Douglas XB-43
Douglas XB-43, NASM
Douglas RB-66A Destroyer
Douglas B-66B Destroyer
Douglas EB-66B Destroyer
Douglas NB-66B Destroyer
Douglas RB-66B Destroyer
Douglas EB-66C Destroyer
Douglas RB-66C Destroyer
Douglas WB-66D Destroyer
Douglas EB-66E Destroyer
Douglas Bomber Proposal (1920)
Douglas BTD-1 Destroyer
Douglas XBT2D-1 Dauntless II
Douglas XBT2D-1N Dauntless II
Douglas XBT2D-1P Dauntless II
Douglas XBT2D-1Q Dauntless II
Douglas BT2D-1 (See: AD Skyraider)
Douglas BT-1 (Modified O-2K)
Douglas BT-2 (Modified O-32)
Douglas BT-2A (Modified O-32A)
Douglas BT-2B
Douglas BT-2BI
Douglas BT-2BG (A-4)
Douglas BT-2BR
Douglas BT-2C
Douglas BT-2CI
Douglas BT-2CR
Douglas C-1
Douglas C-1A
Douglas C-1B
Douglas C-1C
Douglas C-9A Nightingale
Douglas C-9B Skytrain II
Douglas VC-9C
Douglas Y1C-21 Dolphin (C-21) (See: OA-3)
Douglas EC-24A
Douglas Y1C-26 Dolphin (C-26) (See: OA-4)
Douglas Y1C-26A Dolphin (C-26A) (See: OA-4A)
Douglas C-26B Dolphin (See: OA-4B)
Douglas C-29 Dolphin (FP-2B)
Douglas XC-32
Douglas C-32A
Douglas C-33
Douglas YC-34 (C-34)
Douglas C-38
Douglas C-39
Douglas C-41
Douglas C-41A
Douglas C-42
Douglas C-47 Skytrain
Douglas C-47A Skytrain
Douglas RC-47A Skytrain
Douglas SC-47A Skytrain (HC-47A)
Douglas VC-47A Skytrain
Douglas C-47B Skytrain
Douglas TC-47B Skytrain
Douglas VC-47B Skytrain
Douglas XC-47C Skytrain
Douglas C-47C Skytrain
Douglas C-47D Skytrain
Douglas AC-47D Puff the Magic Dragon, Spooky
Douglas EC-47D Skytrain (AC-47D Airways Check Aircraft)
Douglas RC-47D Skytrain
Douglas SC-47D Skytrain (HC-47D)
Douglas TC-47D Skytrain
Douglas VC-47D Skytrain
Douglas C-47E Skytrain
Douglas YC-47F Skytrain (YC-129 Super DC-3)
Douglas C-47M Skytrain
Douglas EC-47N Skytrain
Douglas EC-47P Skytrain
Douglas EC-47Q Skytrain
Douglas (C-47) Dakota I
Douglas (C-47A) Dakota III
Douglas (C-47B) Dakota IV
Douglas (C-47B) Dakota IV, Dart Turboprop Testbed
Douglas (C-47B) Dakota IV, Mamba Turboprop Testbed
Douglas C-48
Douglas C-48A
Douglas C-48B
Douglas C-48C
Douglas C-49
Douglas C-49A
Douglas C-49B
Douglas C-49C
Douglas C-49D
Douglas C-49E
Douglas C-49F
Douglas C-49G
Douglas C-49H
Douglas C-49J
Douglas C-49K

Douglas C-50
Douglas C-50A
Douglas C-50B
Douglas C-50C
Douglas C-50D
Douglas C-51
Douglas C-52
Douglas C-52A
Douglas C-52B
Douglas C-52C
Douglas C-52D
Douglas C-53 Skytrooper
Douglas XC-53A Skytrooper
Douglas C-53B Skytrooper
Douglas C-53C Skytrooper
Douglas C-53D Skytrooper
Douglas (C-53) Dakota II
Douglas C-54 Skymaster
Douglas C-54A Skymaster
Douglas C-54B Skymaster
Douglas VC-54C Skymaster "Sacred Cow"
Douglas VC-54C Skymaster "Sacred Cow," NASM
Douglas C-54D Skymaster
Douglas C-54D (All-Weather Flight Center Autopilot Aircraft)
Douglas EC-54D Skymaster (AC-54D Airways Check Aircraft)
Douglas JC-54D Skymaster
Douglas SC-54D (HC-54D) Skymaster
Douglas TC-54D Skymaster
Douglas VC-54D Skymaster
Douglas C-54E Skymaster
Douglas XC-54F Skymaster
Douglas C-54G Skymaster
Douglas VC-54G Skymaster
Douglas XC-54K Skymaster
Douglas C-54L Skymaster
Douglas C-54M Skymaster
Douglas MC-54M Skymaster
Douglas EC-54U Skymaster
Douglas (C-54D) Skymaster I
Douglas C-58
Douglas UC-67
Douglas C-68
Douglas C-74 Globemaster I
Douglas C-74 Globemaster I Civil Conversions
Douglas C-84
Douglas C-110
Douglas XC-112A
Douglas XC-114
Douglas XC-115
Douglas XC-116
Douglas C-117A Skytrooper
Douglas C-117B Skytrooper
Douglas C-117C Skytrooper
Douglas C-117D Skytrooper (See: R4D-8)
Douglas VC-118 Liftmaster "Independence"
Douglas C-118A Liftmaster
Douglas MC-118A Liftmaster
Douglas VC-118A Liftmaster
Douglas DC-118B Liftmaster
Douglas YC-124 Globemaster II
Douglas YC-124A Globemaster II
Douglas C-124A Globemaster II
Douglas YC-124B Globemaster II
Douglas C-124C Globemaster II
Douglas JC-124C Globemaster II
Douglas C-132
Douglas C-133A Cargomaster
Douglas C-133B Cargomaster
Douglas XCG-17
Douglas Cloudster I
Douglas Cloudster II
Douglas Commuter
Douglas D-205 (Attack Bomber)
Douglas D-558-1 Skystreak
Douglas D-558-2 Skyrocket
Douglas D-558-2 Skyrocket, NASM
Douglas D-558-3
Douglas D-906
Douglas D-920
Douglas D-966
Douglas D-974
Douglas D-1107 (Transport Proposal)
Douglas D-1221 (Long-Range Bomber)
Douglas D-1226 (High-Altitude Research Aircraft Proposal)
Douglas DA-1 Ambassador
Douglas DAM-1 Mailplane (See: M-1)
Douglas DB-1
Douglas DB-2
Douglas DB-7
Douglas DB-7, Twin Tail Modification
Douglas DB-7 Boston I
Douglas DB-7 Havoc I (Boston II)
Douglas DB-7 Havoc I Pandora (Havoc III)
Douglas DB-7 Havoc I Turbinlite
Douglas DB-7A
Douglas DB-7A Boston II
Douglas DB-7A Havoc II
Douglas DB-7A Havoc II Turbinlite
Douglas DB-7B Boston III
Douglas DB-7B Boston III Turbinlite
Douglas DB-7C
Douglas DB-73
Douglas DC-1
Douglas DC-2
Douglas DC-2A
Douglas DC-2B
Douglas (DC-3) DST (Douglas Sleeper Transport)
Douglas (DC-3) DST-A (Douglas Sleeper Transport)
Douglas DC-3
Douglas DC-3, NASM
Douglas DC-3A
Douglas DC-3B
Douglas DC-3C
Douglas DC-3D
Douglas DC-3S (Super DC-3)
Douglas DC-4 Experimental (DC-4E)
Douglas DC-4
Douglas DC-4, Sabena Swing-Tail Cargo Modification
Douglas DC-5
Douglas DC-6
Douglas DC-6A
Douglas DC-6B
Douglas DC-6C
Douglas DC-7 (Civil C-74 Proposal)
Douglas DC-7
Douglas DC-7, NASM
Douglas DC-7B
Douglas DC-7C Seven Seas
Douglas DC-7CF Seven Seas
Douglas DC-7D (Turboprop Proposal)
Douglas DC-7 Amor (French Space Program Support Modification)
Douglas DC-8 Skybus (Pusher Transport Proposal)
Douglas DC-8 Series 10
Douglas DC-8 Series 20
Douglas DC-8 Series 30
Douglas DC-8 Series 40
Douglas DC-8 Series 50
Douglas DC-8 Series 50AF (DC-8F)
Douglas DC-8 Series 50CF
Douglas DC-8-55F (Armée de l'Air Electronic Recon Aircraft)
Douglas DC-8 Super Sixty Series
Douglas DC-8-61 (Super DC-8)
Douglas DC-8-61CF (Super DC-8)
Douglas DC-8-62 (Super DC-8)
Douglas DC-8-62AF (Super DC-8)
Douglas DC-8-62CF (Super DC-8)
Douglas DC-8-63 (Super DC-8)
Douglas DC-8-63AF (Super DC-8)
Douglas DC-8-63CF (Super DC-8)
Douglas DC-8-63PF (Super DC-8)
Douglas DC-8 ORBIS (Ophthamological Teaching Hospital)
Douglas DC-9 (Twin-Prop Transport Proposal)
Douglas DC-9 (Four Turbojet Transport Proposal)
Douglas DC-9 Series 10

Douglas DC-9 Series 20
Douglas DC-9 Series 30
Douglas DC-9 Series 40
Douglas DC-9 Series 50
Douglas DC-9 Super 80 (MD-80 Prototype)
Douglas DC-9 Super 80 Unducted Fan (UDF) Testbed
Douglas DC-9-81 (MD-81)
Douglas DC-9-82 (MD-82)
Douglas DC-9-83 (MD-83)
Douglas DC-9-87 (MD-87)
Douglas DC-9-88 & subsequent series (See: McDonnell Douglas MD-88, etc.)
Douglas DC-10 (Four Turboprop Transport Proposal)
Douglas DC-10 3-Engined Airliner (See: McDonnell Douglas)
Douglas DF
Douglas Dolphin
Douglas DT-1
Douglas DT-2
Douglas DT-2B
Douglas DT-3
Douglas DT-4
Douglas DT-5
Douglas DT-6
Douglas XFD-1
Douglas XF3D-1 Skyknight
Douglas F3D-1 (F-10A) Skyknight
Douglas F3D-1M (MF-10A) Skyknight
Douglas F3D-2 (F-10B) Skyknight
Douglas F3D-2B Skyknight
Douglas F3D-2M (MF-10B) Skyknight
Douglas F3D-2Q (EF-10B) Skyknight
Douglas F3D-2T Skyknight
Douglas F3D-2T2 (TF-10B) Skyknight
Douglas F3D-3 Skyknight
Douglas XF4D-1 Skyray
Douglas F4D-1 (F-6A) Skyray
Douglas F5D-1 (F4D-2N) Skylancer
Douglas F6D Missileer
Douglas XF-3
Douglas YF-3
Douglas F-3A
Douglas FP-1 Dolphin (See: OA-3)
Douglas FP-2 Dolphin (See: OA-4)
Douglas FP-2B (See: C-29 Dolphin)
Douglas JD-1 (UB-26J) Invader
Douglas JD-1D (DB-26J) Invader
Douglas M-1 (DAM-1) Mailplane
Douglas M-2 Mailplane
Douglas M-2 Mailplane, NASM
Douglas M-3 Mailplane
Douglas M-4 Mailplane
Douglas M-4A Mailplane
Douglas M-4S Mailplane
Douglas M-5 Mailplane
Douglas XNO-1
Douglas XNO-2
Douglas OD-1
Douglas XO2D-1
Douglas XO-2
Douglas O-2
Douglas O-2A
Douglas O-2B
Douglas O-2BS
Douglas XO-2C
Douglas O-2C
Douglas O-2D
Douglas O-2E
Douglas O-2H
Douglas O-2J
Douglas O-2K
Douglas O-2M (Mexico)
Douglas O-2MC (China)
Douglas O-5 (DOS)
Douglas O-7
Douglas O-8
Douglas O-9
Douglas XO-10 (Proposal)
Douglas XO-14
Douglas O-22 (YO-22)
Douglas O-25
Douglas XO-25A
Douglas O-25A
Douglas O-25B
Douglas O-25C
Douglas O-29
Douglas Y1O-29A (O-29A)
Douglas XO-31
Douglas YO-31
Douglas YO-31A (XYO-31A, O-31A)
Douglas Y1O-31A (See: Y1O-43)
Douglas YO-31B
Douglas Y1O-31C (See: Y1O-43)
Douglas YO-31C (XYO-31C, O-31C)
Douglas O-32
Douglas O-32A
Douglas XO-34
Douglas XO-35
Douglas Y1O-35 (O-35)
Douglas XO-36 (See: XB-7)
Douglas O-38
Douglas O-38A
Douglas O-38B
Douglas O-38C
Douglas O-38D
Douglas O-38E
Douglas O-38F
Douglas O-38P
Douglas O-38S
Douglas Y1O-43 (Y1O-31A, Y1O-31C, O-43)
Douglas O-43A
Douglas YO-44 (See: YOA-5)
Douglas XO-46
Douglas O-46A
Douglas YO-48
Douglas O-53
Douglas OA-3 Dolphin (Y1C-21, C-21, FP-1)
Douglas OA-4 Dolphin (Y1C-26, C-26, FP-2)
Douglas OA-4A Dolphin (Y1C-26A, C-26A)
Douglas OA-4B Dolphin (C-26B)
Douglas OA-4C Dolphin
Douglas YOA-5 (YB-11, YO-44)
Douglas PD-1
Douglas P2D-1
Douglas XP3D-1
Douglas XP3D-2
Douglas P-48
Douglas XP-70
Douglas P-70
Douglas P-70A
Douglas P-70B
Douglas TP-70B
Douglas RD Dolphin (USCG)
Douglas XRD-1 (RD-1) Dolphin
Douglas RD-2 Dolphin
Douglas RD-2 Dolphin, Roosevelt
Douglas RD-3 Dolphin
Douglas RD-4 Dolphin
Douglas R2D-1
Douglas R3D-1
Douglas R3D-2
Douglas R3D-3
Douglas R4D-1 Skytrain
Douglas R4D-2 (R4D-2F) Skytrain
Douglas R4D-3 Skytrain
Douglas R4D-4 Skytrain
Douglas R4D-4R Skytrain
Douglas R4D-4Q Skytrain
Douglas R4D-5 (C-47H) Skytrain
Douglas R4D-5E Skytrain
Douglas R4D-5L (LC-47H) Skytrain
Douglas R4D-5Q (EC-47H) Skytrain
Douglas R4D-5R (TC-47H) Skytrain
Douglas R4D-5S (SC-47H) Skytrain
Douglas R4D-5T Skytrain
Douglas R4D-5Z (VC-47H) Skytrain
Douglas R4D-6 (C-47J) Skytrain
Douglas R4D-6E Skytrain
Douglas R4D-6L (LC-47J) Skytrain
Douglas R4D-6Q (EC-47J) Skytrain
Douglas R4D-6R (TC-47J) Skytrain
Douglas R4D-6S (SC-47J) Skytrain
Douglas R4D-6T Skytrain
Douglas R4D-6Z (VC-47J) Skytrain
Douglas R4D-7 (TC-47K) Skytrain
Douglas R4D-8 (C-117D)
Douglas R4D-8L (LC-117D)
Douglas R4D-8T (TC-117D)

Douglas R4D-8X
Douglas R4D-8Z (VC-117D)
Douglas R5D-1 Skymaster
Douglas R5D-1C Skymaster
Douglas R5D-1F (VC-54N) Skymaster
Douglas R5D-2 (C-54P) Skymaster
Douglas R5D-2-2 Skymaster (US NRL Radar Laboratory Aircraft)
Douglas R5D-2F (VC-54P) Skymaster
Douglas R5D-3 (C-54Q) Skymaster
Douglas R5D-3Z (VC-54Q) Skymaster
Douglas R5D-3P (RC-54V) Skymaster
Douglas R5D-4 (C-54R) Skymaster
Douglas R5D-4R (C-54R) Skymaster
Douglas R5D-5 (C-54S) Skymaster
Douglas R5D-5R (VC-54T) Skymaster
Douglas R5D-5Z (VC-54S) Skymaster
Douglas R6D-1 (C-118B) Liftmaster
Douglas R6D-1Z (VC-118B) Liftmaster
Douglas SBD-1 Dauntless
Douglas SBD-1P Dauntless
Douglas SBD-2 Dauntless
Douglas SBD-2P Dauntless
Douglas SBD-3 Dauntless
Douglas SBD-3A Dauntless (See: A-24)
Douglas SBD-3P Dauntless
Douglas SBD-4 Dauntless
Douglas SBD-4A Dauntless (See: A-24A)
Douglas SBD-4P Dauntless
Douglas SBD-5 Dauntless
Douglas SBD-5A Dauntless
Douglas XSBD-6 Dauntless
Douglas SBD-6 Dauntless
Douglas SBD-6 Dauntless, NASM
Douglas SBD-6A Dauntless
Douglas XSB2D-1
Douglas SDW-1 (See: Dayton Wright)
Douglas Sinbad
Douglas XT2D-1
Douglas T2D-1
Douglas T2D-2 (See: P2D-1)
Douglas XT3D-1
Douglas XT3D-2
Douglas XT-30
Douglas XTBD-1
Douglas TBD-1 Devastator
Douglas TBD-1A Devastator
Douglas XTB2D-1 Skypirate
Douglas TB2D-1 Skypirate
Douglas World Cruiser (DWC)
Douglas World Cruiser (DWC) "Chicago," NASM
Douglas X-3 Stiletto
Douglas X-21A (See: Northrop)
Douglas 7A (See: Northrop)
Douglas 7B
Douglas 8A (See: Northrop)
Douglas 1211-J (1951 Heavy Bomber)

Douhéret (France)
Douhéret Helicopter (1922)

Doutre (Société des Appareils d'Aviation Doutre, SAAD) (France)
Doutre 1911 Biplane (Modified Farman)
Doutre 1912 Biplane
Doutre 1913 Tandem Biplane
Doutre 1914 Military Biplane

Downer Aircraft Industries Inc (Alexandria, MN) *(See: Bellanca Cruisemaster)*

Downey, H. C. (USA)
Downey Bimonoplane (1922)

Doyle Aero Corp (Baltimore, MD)
Doyle O-2 Oriole
Doyle O-2 Oriole Special

Doyle, Richard H. (Mt. Pleasant, IL)
Doyle Moon Maid (Homebuilt)

Doyn Aircraft Inc (Wichita, KS)
Doyn Beech Travelair Conversions
Doyn Cessna 150 Conversions
Doyn Cessna 172 Conversions
Doyn Cessna 175 Conversions
Doyn Cessna Cardinal Conversions
Doyn Dart I (Cessna 170 Conversion)
Doyn Dart II (Piper Apache Conversion)

Dragon Fly SRL (Italy)
Dragon Fly HELIOT
Dragon Fly Model 333

Drake Brothers (Thomasville, GA)
Drake Bros. Training Glider

Dresden (Flugtechnischen Verein, Dresden) (Germany)
Dresden 1921 Biplane Training Glider

Drewer, O. H. (India)
Drewer 1908 Glider

Driggers, Willard R. (Washington, DC)
Driggers D2-A

Driggs Aircraft Corp (Dayton, OH)

In February 1927, Ivan H. Driggs established **Driggs Aircraft Corp** in Dayton, OH. In 1933, after moving to Lansing, MI, the assets of the company were purchased by **Skylark Aircraft Corp** of Muskegon, MI, with the intent to continue the manufacture of Driggs designs. Skylark sold the rights to **Phillips Aviation Co** of Van Nuys, CA, which attempted to revise the Skylark design. For all Driggs Aircraft Corp and Skylark designs, see **Driggs**. For the Phillips-redesigned Skylark, see **Phillips**.

Driggs Coupe
Driggs Dart (See: Driggs, Ivan, Dart I)
Driggs Dart II
Driggs Dart II Special
Driggs Skylark
Driggs Skylark Model 3

Driggs, Ivan H. (Dayton, OH)
Driggs (Ivan) Driggs-Johnson D.J.1
Driggs (Ivan) Dart I

Drouhet (France)
Drouhet 1912 Biplane

Druine (Avions Roger Druine) (France)
Druine 1938 Monoplane
Druine Aigle
Druine D-5 Turbi
Druine D-31 Turbulent
Druine D-60 Condor
Druine D-61 Condor
Druine D-62 Condor

Drzewiecki (France)
Drzewiecki (France) 1912 Tandem
Drzewiecki (France) 1914 Tandem

Drzewiecki, Jerzy (Poland)
Drzewiecki (Jerzy):
D.K. Two-Seat Fighter
J.D.1 (S.L.2) Czarny Kot (Black Cat)

J.D.2 Low-winged Monoplane

DSK Aircraft Corp (North Hollywood, CA) (DSK Airmotive Inc, Fort Walton Beach, FL)
DSK DSK-1 Hawk
DSK DSK-2 Golden Hawk

du Breuil *(See: Breuil)*

du Temple, Félix et Louis (France)
du Temple 1874 Monoplane

Dubanhy-Suthey (France)
Dubanhy-Suthey Glider (c.1910)

Dubna Machinebuilding Plant JSC (Russia)
Dubna Z-2 Selena
Dubna Z-6 Duet
Dubna Z-7 Bekas
Dubna Z-8 Stayer
Dubna-1
Dubna-2

Dubno Secondary School (Poland)
Dubno Secondary School 1926 Glider

DuBois, Bernard J. (France)
DuBois-Rioul 1912 Ornithopter

Dufaux Brothers (Armand and Henri) (Switzerland/France)
Dufaux C1 Fighter (1916)
Dufaux Twin-Engined Fighter (1916)
Dufaux 1909 Biplane

Dufour, Jean (France)
Dufour Biplane Glider (1908)
Dufour No.1 Monoplane (1910)
Dufour No.2 Biplane 2-Prop (1910)
Dufour 1911 Two-Seater

Duhany-Suthy (France)
Duhany-Suthy 1909 Biplane Glider

Duigan, John R. (Australia)
Duigan 1910 Farman-Type Biplane
Duigan 1911 Biplane (See: Avro Duigan Biplane)
Duigan 1911 Wright-Type Glider
Duigan 1913 Biplane

Duluth School of Aeronautics (Duluth, MN)
Duluth School of Aeronautics 1930 Glider

Dumas, Alex (Société Générale de Fabrication de l'Aéroplane A Dumas) (France)
Dumas 1910 Monoplane

Dumod Corp (Opa-Locka, FL)
Dumod Infinité I Modified Beech 18)
Dumod Infinité II Modified Beech 18)

Dumont (France)
Dumont Aeroplane

Dumontet et Talon (France)
Dumontet et Talon 1909 Monoplane

Dumoulin (France)
Dumoulin 1904 Tracteur pour la Navigation Aérienne
Dumoulin Saturnian (Aéroplane Gyropendulaire)

Duncan (USA)
Duncan Sport DX-1 (Homebuilt)

Dunham, Erwin J. (Hamburg, NY)
Dunham Coupe

Dunkerque (Ateliers et Chantiers de Dunkerque) (France)
Dunkerque 1910 Wright Biplane (License-Built)

Dunn Aircraft Corp (Clarinda, IA)
Dunn Cruizaire K-5

Dunne, John William (UK)
Dunne D.1 Glider (1907)
Dunne D.3 (1908)
Dunne D.4 (1908)
Dunne D.5 (1910)
Dunne D.6 Monoplane (1910)
Dunne D.7 Autosafety Plane
Dunne D.7*bis* Monoplane (1912)
Dunne D.8 (1912)
Dunne D.9 (1913)

Duparquet (France)
Duparquet 1911/12 Aeroplane

Duperron-Niepce-Fetterer *(See: D.N.F.)*

Dupuy, Roger (France)
Dupuy D-40

Duramold Aircraft Corp *(See: Fairchild 46)*

Durand Assoc (William H Durand) (Omaha, NE)
Durand A-45
Durand XD-85
Durand Mk V

Durant Aircraft Corp (Rex C. "Cliff" Durant) (Oakland, CA)
Durant 1920 Tour Plane (Rebuilt Standard J-1)

Duray-Mathis (A. Duray and H. Mathis) (France)
Duray-Mathis 1909 Double Canard Biplane

Durban Aircraft Corp (South Africa)
Durban Aeriel Mk.II (See: Genair)

Durl-E-Aire (Canton, IL)
Durl-E-Aire BD-1 (Homebuilt)

Duromold Aircraft Corp *(See: Fairchild 46)*

Duruble, Roland (France)
Duruble RD-02
Duruble RD-03 Edelweiss

D.U.S. (Dabroski and Uszaki) (Poland)
D.U.S.III (LKL I) Ptapta (Putt-Putt)

Dussot, Auguste (France)
Dussot 1910 Flying Boat

Dutel (France)
Dutel 1910 Monoplane

Dutheil-Chalmers et Cie (France)
Dutheil-Chalmers 1909 Biplane

Duvernois (France)
Duvernois Monoplane

D.V.L. (Deutschen Versuchsanstalt für Luft-fahrt) (Germany)
D.V.L. DFS 582
D.V.L. DFS 611

DWL *(See: RWD-8)*

DWLKK (Poland)
DWLKK PW-2D Gapa

Dycer Airport (Los Angeles, CA)
Dycer Sport Biplane

Dyke, John (Fairborn, OH)
Dyke JD-1
Dyke JD-2 Delta

Dyle et Bacalan (France) *(See also: Bordelaise)*
Dyle et Bacalan D.B. 10
Dyle et Bacalan D.B. 20 BT3
Dyle et Bacalan D.B. 30
Dyle et Bacalan D.B. 40
Dyle et Bacalan D.B. 70
Dyle et Bacalan D.B. 71
Dyle et Bacalan D.B. 80
Dyle et Bacalan D.B. 81

Dyn'Aero (France)
Dyn'Aero CR.100
Dyn'Aero MCR-10 Banbi (See: Colomban MC-100)

Dyott, G. M. (UK)
Dyott Bomber (1916)
Dyott Monoplane (1913)

Dzialowski Brothers (Stanislaw & Mieczyslaw Dzialowski) (Poland)
Dzialowski Bydgoszczanka 1925 Glider

Dzialowski, Kruger, Dzialowski *(See: D.K.D.)*

E

E. H. Industries Ltd *(See: European Helicopter Industries Ltd)*

E. M. Smith Co *(See: Emsco)*

E-Systems (Greenville, TX)
E-Systems-Grob Egrett

EAA (Experimental Aircraft Association) (Oshkosh, WI)
EAA Aero-Sport
EAA Aqua Glider (1967)
EAA Biplane
EAA Nesmith Tri Cougar

EAC (France) *(See: Jodel)*

Eagle Aircraft Co (USA)
Eagle DW-1

Eagle Aircraft Co (Boise, ID)
Eagle (ID) 300

Eagle Aircraft Pty Ltd (Australia)
Eagle (Australia) 100
Eagle (Australia) 150
Eagle (Australia) 150A
Eagle (Australia) 150B

Eagle's Perch, Inc (Carrollton, VA)
Eagle's Perch Coaxial Helicopter

Earl Aviation Corp, Ltd (Los Angeles, CA)
Earl Aviation Populair
Earl Aviation Populair 1A

Early Bird Aircraft Co (Erie, CO)
Early Bird Jenny (Scale Replica)
Early Bird Spad XIII (Scale Replica)

Early Flight Concepts, General
Materials relating to development of flying machines predating 1914.

Easley, Phil *(See: Backstrom, Al)*

East Anglian Aviation Co, Ltd (UK)
East Anglian Westlake Monoplane (See: Westlake)

Eastbourne Aviation Co (UK)
Eastbourne 1914 Military-Type Tractor Biplane
Eastbourne 1914 Tractor Seaplane (Circuit of Britain)

Eastern Aeroplane Co (Mineola, NY)
Eastern Aeroplane Co Biplane

Eastern Aircraft Division (General Motors Corp) *(See: General Motors)*

Eastern Ultralights, Inc (Chatsworth, NJ)
Eastern Ultralights Snoop

Eastman Aircraft Corp (Detroit, MI)
Eastman Flivver
Eastman Sea Pirate
Eastman Sea Rover (Flying Yacht, Flying Boat)

Eberhart Aeroplane and Motor Co, Inc (Eberhart Steel Products Co) (Buffalo, NY)
Eberhart "FG-1" Comanche
Eberhart G-3 Target Glider
Eberhart Iroquois
Eberhart SE-5E

Ecker, Herman A. (Newburgh, NY)
Ecker Hydro Plane
Ecker Hydro Plane, NASM

Eclipse Aviation (Albuquerque, NM)
Eclipse 500

Edgar Percival Aircraft, Ltd *(See: Percival, Edgar)*

Edgley Aircraft Ltd (UK)
Edgley (Aircraft) EA7 Optica

Edgley Sailplanes Ltd (UK)
Edgley (Sailplanes) EA9 Optimist

Edo Aircraft Corp (College Point, NY)
Edo Model B Flying Boat "Malolo"
Edo XOSE-1

Emigh Aircraft Corp (Emigh Trojan Aircraft Co) (Douglas, AZ)
Emigh Trojan A-2

Empresa Brasileira de Aeronautica SA *(See: EMBRAER)*

Empresa Nacional de Aeronautica *(See: Enaer)*

Emsco (E. M. Smith Co) (Downey, CA)
Emsco B-2 Challenger
Emsco B-3
Emsco B-3, Around the World Flight
Emsco B-3 "City of Tacoma"
Emsco B-3 Experimental

Emsco B-3 "Morelos"
Emsco B-3-A
Emsco B-3-A "Clasina Madge"
Emsco B-3-A "Regele Carolii"
Emsco B-4 Cirrus
Emsco B-5
Emsco B-7
Emsco B-8 "Asbestos"
Emsco B-10
Emsco Mystery Ship

EMZ (See: Myasishchev)

Enaer (Empresa Nacional de Aeronautica) (Chile)
Enaer A-36 Halcón
Enaer Aguila (License Hawker Hunter)
Enaer T-35 Pillán
Enaer T-36 Halcón

Engel, Richard B. (Natick, MA)
Engel T-1 "Little Chief"

Engels, E. R. (Russia)
Engels II

Engineering Aircraft Corp (Stamford, CT)
Engineering Aircraft Corp EAC-1

Engineering and Research Corp *(See: Erco)*

Engineering Division (US Army Air Service, US Army Air Corps) (Wright Field, OH)
Engineering Division:
B-1 (USB-1)
B-2 (USB-2)
B-3 (USB-3, XB-1)
B-4 (USB-4, XB-2)
XB-1A (USXB-1A) (See also: Dayton-Wright)
Bristol Fighter (See: Curtiss Bristol Fighter)
BVL-12 (See: Pomilio)
CO-1
CO-2
CO-5
CO-6
DB-1 (See: Gallaudet)
FVL-8 (See: Pomilio)
G.A.X.
GA-1 (See: Boeing)
GA-2 (See: Boeing)
GL-2 (G-2, Roche Glider)
GL-3 Glider
NBL-1 Barling Bomber
O-1 (See: Curtiss Bristol Fighter)
PW-1 (See: Verville, Alfred)
R-1 (See: Verville, Alfred)
R-3 (See: Verville-Sperry)
TA-4
TP-1
TW-1 (USXT-3, Type XV Training Airplane)
USAC-1
USAC-2
USAO-1 (See: Curtiss Bristol Fighter)
USD-9 (D.H.9) (See also: Dayton Wright)
USD-9A (D.H.9)
VCP R-1 (See: Verville, Alfred)
VCP-R (See: Verville, Alfred)
VCP-1 (See: Verville, Alfred)
VCP-2 (See: Verville, Alfred)

England, Gordon *(See: Bristol)*

Engle Flying Services (USA)
Engle Flying Services HPK

English Electric (UK)

English Electric Manufacturing Co Ltd was incorporated in 1899 and taken over by **Dick, Kerr & Co Ltd** in 1903. In 1918 Dick, Kerr & Co, the **Coventry Ordnance Works**, and three other concerns were incorporated as **English Electric Co Ltd.** (This company's aircraft division was closed between 1926 and 1938.) **English Electric Aviation Ltd** formed in 1959, and the following year amalgamated with **Vickers-Armstrong** and **Bristol** to form **British Aircraft Corp, Ltd (BAC).** In 1964 a final consolidation of BAC eliminated individual company names. All pre-1964 English Electric aircraft are filed under **English Electric.**

English Electric:
Ayr (M.3)
Canberra A.1 (B.3/45)
Canberra B.Mk.1
Canberra B.Mk.2
Canberra B.Mk.2 Martin Pattern
Canberra PR.Mk.3
Canberra T.Mk.4
Canberra B.Mk.5
Canberra B.Mk.6
Canberra B(I).Mk.6
Canberra PR.Mk.7
Canberra B(I).Mk.8
Canberra PR.Mk.9
Canberra T.Mk.11
Canberra B(I).Mk.12
Canberra T.Mk.13
Canberra B.Mk.15
Canberra E.Mk.15
Canberra B.Mk.16
Canberra T.Mk.17
Canberra T.Mk.18
Canberra TT.Mk.18
Canberra T.Mk.19
Canberra B.Mk.20 (Australian)
Canberra T.Mk.22
Canberra B.Mk.52
Canberra B.(I)Mk.56
Canberra B.Mk.62
Canberra B.(I)Mk.66
Canberra PR.Mk.67
Cork (See: Phoenix)
Kingston Mk.I (P.5)
Kingston Mk.II (P.5)
Kingston Mk.III (P.5)
Lightning F.Mk.1
Lightning F.Mk.1A
Lightning F.Mk.2
Lightning F.Mk.2A
Lightning F.Mk.3
Lightning F.Mk.3A
Lightning T.Mk.4
Lightning T.Mk.5
Lightning F.Mk.6
Lightning F.Mk.52
Lightning F.Mk.53
Lightning F.Mk.53K
Lightning T.Mk.54
Lightning T.Mk.55
Lightning T.Mk.55K
P.1A (Lightning Prototype)
P.1B (Lightning Prototype)
P.11 (Lighting Trainer Prototype)
S-1 Wren

Enstrom Corp (R. J. Enstrom Corp) (Menominee, MI)
Enstrom F-28
Enstrom F-28A
Enstrom F-28-F
Enstrom F-28-FP Sentinel
Enstrom F-28-FX
Enstrom Shark 280C
Enstrom T-28
Enstrom TH-28 (480)

Entler Werk Flugzeug Bau GmbH (Germany)
Entler E.II

EÔA *(See: Eesti Ôhusôidu Aktsiaselts)*

EoN (Elliott's of Newbury) (UK)
EoN Olympia

EoN Type 8 Baby (See: Grunau Baby)

EPA Aircraft Co Ltd (UK)
EPA Fieldmaster (See: NDN)
EPA Firemaster (See: Croplease)

Eparvier, H. (France)
Eparvier 1908 Monoplane
Eparvier 1909 Triplane
Eparvier 1910 Monoplane

Epps, Ben T., Sr (Athens, GA)
Epps (Ben T., Sr) 1907 Monoplane
Epps (Ben T., Sr) 1909 Biplane
Epps (Ben T., Sr) 1912 Monoplane
Epps (Ben T., Sr) 1916 Biplane
Epps (Ben T., Sr) 1925 Monoplane
Epps (Ben T., Sr) 1930 Light Biplane

Epps, Ben T., Jr (Athens, GA)
Epps (Ben T., Jr) 1968 Replica of Father's 1912 Monoplane

Equevilly (Marquis d'Equevilly-Montjustin) (France)
Equevilly 1907/08 Multiplane

Erco (Engineering and Research Corp) (Riverdale, MD)
Erco Ercoupe 310
Erco Ercoupe 415
Erco Twin Ercoupe
Erco YO-55
Erco PQ-13
Erco 191-A

Erickson, Alfred A. (Mojave, CA)
Erickson Sport M-10C

Erickson, Louis G. (Springfield, MA)
Erickson Biplane #1 "Ericka" (1909)

Erla Maschinen-Werk GmbH (Germany)
Erla 5A
Erla 5D

Erle L. Bacon Corp *(See: Bacon)*

Ernst, Emil Robert (New Jersey; Germany)
Ernst 1907 Flying Machine

Eshelman (Cheston L. Eshelman Co) (Baltimore, MD)
Eshelman E.F.100 Winglet
Eshelman FW-5 The Wing

Esjay Aero Co (Esjay = SJ = Anthony Stadlman & E. B. Jaeger) (Chicago, IL)
Esjay 1914 Pusher Biplane Seaplane

Eska (Russia)
Eska-1

Esnault-Pelterie *(See: REP)*

Espenlaub Flugzeugbau (Germany)
Espenlaub Rocket-Powered Glider
Espenlaub 5
Espenlaub 11
Espenlaub E.14
Espenlaub E.16

Essen *(See: Raab-Katzenstein)*

Essex Aero Ltd (UK)
Essex Sprite

Essig (Albert C. & Fred G. Essig) (Los Angeles, CA)
Essig Ace

Essort (France)
Essort Aérien (1912 Monoplane)

Établissements Adolphe Bernard *(See: Bernard)*

Établissements Autoplan *(See: Pischoff)*

Établissements Henri Potez *(See: Potez)*

Établissements Louis de Monge *(See: Monge)*

Établissements Surcouf *(See: Astra)*

Étendard *(See: Dassault)*

Etévé *(See: CGNA Type A Transitional)*

Etrich (Austria-Hungary)
Etrich A.I
Etrich A.II
Etrich Gliders
Etrich Glider, 1908
Etrich Limosin (Limousine)
Etrich Monoplane (Horvath)
Etrich Schwalbe (Swallow)
Etrich Sperling (Sparrow)
Etrich Taube (Dove)
Etrich No I
Etrich 70.01 thru 70.02

Etienne et Cie (France)
Etienne 1911 Aeroplane

Études Aéronautiques et Commerciales (France) *(See: Jodel)*

Euler-Flugmaschinen-Werke (Germany)
Euler B.I
Euler B.II
Euler B.III
Euler Biplane Tractor
Euler C
Euler D
Euler D.I
Euler D.II
Euler Dr.I
Euler Dr.II
Euler Dr.III
Euler Dr.IV
Euler Experimental Military Pusher
Euler Gelber Hund (Yellow Dog)
Euler Monoplane Tractor
Euler Quadruplane
Euler Taube (Dove)
Euler Triplane Flying Boat
Euler Triplane Pusher
Euler Triplane Tractor
Euler 1910 Voisin Copy
Euler 1911 Monoplane
Euler 1911 Pusher Biplane

EuroALA, Ltd (Italy) *(See: Arnet Pereyra)*

Eurocopter (France/Germany)
Eurocopter formed gradually from the helicopter divisions of **Aérospatiale** and **Deutsche Aerospace AG (DASA)** between May 1991 and June 1992. Helicopters of Aérospatiale origin are filed under **Aérospatiale.** DASA products carrying Bo designations appear under **Bölkow.** Joint designs and DASA designs redesignated under the "EC" system are listed under **Eurocopter.**
Eurocopter EC-155
Eurocopter Tiger Attack Helicopter

Eurofighter (Germany/Italy/Spain/UK)
Eurofighter European Fighter Aircraft (EFA)

Eurofighter Typhoon

Europa Aircraft (Lakeland, FL & UK)
Europa XS Mono-Wheel
Europa XS Motor Glider
Europa XS Tri-Gear
Europa Turbo XS Motor Glider

European Helicopter Industries Ltd (E. H. Industries Ltd) (UK/Italy)
European Helicopter EH101

Evans Aircraft Co (La Jolla, CA)
Evans (Aircraft) VP (Volksplane = People's Plane)
Evans (Aircraft) VP-1
Evans (Aircraft) VP-2
Evans (Aircraft) WE-1

Evans Glider Co (Los Angeles, CA)
Evans All Steel Glider Primary Type Pt-2

Evans, Maurice Victor (Hollywood, CA)
Evans (M. V.) Model 1 Glider (1929)

Evektor spol sro (Czech Republic)
Evektor EV97 Eurostar

EWR Süd (Entwicklungsring Süd) (Germany)
In February 1959, **Bölkow GmbH**, **Ernst Heinkel Flugzeugbau AG**, and **Messerschmitt AG**, at the suggestion of the German Defense Ministry, established **Entwicklungsring Süd** (**EWR Süd**) as a central design office for the development of a Mach 2 VTOL interceptor. On 31 December 1964, Heinkel withdrew from the project. On 1 July 1965, EWR Süd became a limited liability company and was later acquired by Messerschmitt-Bölkow-Blohm GmbH as a wholly-owned subsidiary.
EWR Süd Hover Testbed
EWR Süd VJ 101
EWR Süd VJ 101 C
EWR Süd VJ 101 D
EWR Süd Wippe (Brink or Critical Point) (Test Rig)

Experimental Aircraft Association *(See: EAA)*

Experimental Aviation, Inc (Santa Monica, CA)
Experimental Aviation Berkut

Experimental Mashinostroiteleny Zavod *(See: Myasishchev)*

Experimental X-Aircraft *(See: Research Aircraft)*

Explorer (USA)
Explorer PG1 Aqua Glider

Extra Flugzeugbau GmbH (Germany)
Extra EA230 (Aero-Club Der Schweiz)
Extra 260
Extra 260, NASM
Extra 300
Extra 300L
Extra 330
Extra 400

Eyerly Aircraft Corp (Salem, OR)
Eyerly Coupe (Wiffle Hen)
Eyerly Monoplane (1937)

Ezekiel Air Ship Manufacturing Co (Pittsburgh, TX)
Ezekiel Air Ship

F

***F + W** (See: Federal Aircraft Factory, Switzerland)*

***F. Hills & Sons, Ltd** (See: Hillson)*

Fabre, Henri (France)
Fabre Test Boat "l'Essor" ("Soaring")
Fabre 1908 Seaplane
Fabre 1909 Trimotor Seaplane
Fabre 1909 Canard Glider
Fabre 1909 Canard Seaplane Model
Fabre 1910 Goeland (Gull) (Canard Seaplane)
Fabre 1911 Goeland (Gull) (Canard Seaplane)
Fabre 1914 Hydro-Glisseur (Wingless Seaplane)

de Fabrègue (France)
de Fabrègue 1907/08 Gliders

***Fabrica Brasiliera de Avioes** (See: Muniz, Col Antonio)*

***Fabrica de Aviones Anahuac SA** (See: Anahuac)*

***Fábrica Materiales Aerospaciales** (See: FMA)*

***Fábrica Militar de Aviones** (See: FMA)*

***Fabrika Aviona Utva** (See: Utva)*

Faccioli, Aristide (Italy)
Faccioli 1909 Monoplane
Faccioli 1909 Triplane

Fahlusch (Germany)
Fahlbusch Drachenflieger (Dragonfly)

Fahlin Aircraft Co (Olaf "Ole" Fahlin) (Marshall, MO)
Fahlin (Swanson-Fahlin) SF-1
Fahlin (Swanson-Fahlin) SF-2 Plymo-Coupe

Fairchild (Farmingdale, NY)

In 1924, Sherman Fairchild established the **Fairchild Aviation Corp** as the parent company for his many aviation interests. In 1930, the **Aviation Corp (AVCO)** purchased Fairchild Aviation and its subsidiaries, initially operating the various companies under their original names. The following year, Sherman Fairchild repurchased Fairchild Aviation Corp and began repurchasing the subordinate companies. In a December 1936 reorganization, Fairchild Aviation Corp divested itself of all aircraft manufacturing interests, placing them under a new **Fairchild Engine and Airplane Co.** (As Fairchild Aviation Corp never returned to aircraft design or production, this Directory will not cover that corporation's subsequent history.)

The original aircraft manufacturing subsidiary of Fairchild Aviation Corp was **Fairchild Airplane Manufacturing Co**; it was created in 1924 to design and build aircraft as platforms for Fairchild's aerial survey cameras. Fairchild Airplane Manufacturing was one of the subsidiaries purchased by AVCO in 1930, but not one of the first companies repurchased by Sherman Fairchild. In 1931 AVCO combined the aircraft company with **Fairchild Engine Co**, forming **American Airplane and Engine Corp.** Fairchild Aviation Corp bought American Airplane and Engine in 1934, renaming the company the **Fairchild Aircraft Manufacturing and Engine Co.**

In the 1936 reorganization that divided Fairchild Aviation Corp assets, Fairchild Aircraft Manufacturing and Engine Co became Fairchild Engine and Airplane Co and took charge of all Fairchild aircraft and engine holdings. Fairchild Engine and Airplane Co became **Fairchild Engine and Airplane Corp** in 1950 and **Fairchild Stratos Corp** in 1961. With the 1964 purchase of **Hiller Aircraft Corp**, Fairchild Stratos was renamed **Fairchild Hiller Corp**, then, again, renamed **Fairchild Industries** after the separation of all Hiller interests in 1973. Although Fairchild Industries closed and sold its military and commercial aircraft manufacturing divisions in 1987, "Fairchild" aircraft continued to be produced through the Swearingen Metro and Fairchild Dornier lines (see below). All aircraft designed by the various Fairchild companies appear under **Fairchild.**

Fairchild created, purchased, and merged with several companies during its history. The following are the most important subsidiaries:

Fairchild Aircraft Ltd was created in 1929 as Fairchild Aviation Corp's Canadian subsidiary. The company ended all aircraft production in 1948.

The **Kreider Reisner Aircraft Co Inc** was formed in 1927. Kreider Reisner became a wholly-owned division of (first) the Fairchild

Airplane Manufacturing Co in 1929, (second) AVCO's American Airplane and Engine Corp (which renamed KR aircraft "Pilgrims") in 1931, and (third) Fairchild Aircraft Manufacturing and Engine Co in 1934. Kreider-Reisner was renamed the **Fairchild Aircraft Corp** in 1935, becoming Fairchild Engine and Airplane Co's principle US aircraft manufacturing subsidiary. Fairchild Aircraft Corp was renamed the **Fairchild Aircraft Division** in 1939, the **Fairchild Aircraft and Missiles Division** in 1961,

Fairchild, Continued

the **Fairchild Stratos Aircraft and Missiles Division** in 1961, the **Aircraft-Missiles Division** in 1965, and the **Aircraft Division** in 1967. With a growing number of aircraft subsidiaries reporting to Fairchild Industries, the Aircraft Division was broken up in a corporate reorganization of the 1970s. While the Kreider Reisner Midget is listed under **Kreider Reisner**, all Kreider Reisner Challenger series aircraft (designated "KR" biplanes by Fairchild) appear under **Fairchild**.

In 1936 **Fairchild Engine and Airplane Co** founded the subsidiary **Duromold Aircraft Corp** to better account for time spent developing the Duromold wood/resin bonding process and the Model 46 aircraft. In 1938, the majority interest in Duromold was bought by a group of investors (including process inventor Col. Virginius E. Clark), who formed the **Clark Aircraft Corp.** Fairchild kept a minority interest in Clark, retaining Duromold as a holding company. In September 1938, Fairchild renamed its Duromold division **Fairchild Airplane Investment Corp**, and Clark created a subsidiary called **Duramold Aircraft Corp** (note the spelling change). In 1938 Duramold was renamed **Molded Aircraft Corp.** In 1939, Fairchild Engine and Airplane Corp bought back a controlling interest in Clark and renamed Molded Aircraft **Duramold Aircraft Manufacturing Corp.** The Duramold and Clark companies disappeared during one of Fairchild's World War II reorganizations. The Clark/Duramold/Duramold F-46 is listed under **Fairchild**.

In 1952 Fairchild licensed the rights to Dutch Fokker's F.27 medium-range airliner. All Dutch F-27s are listed under **Fokker**, with US production and development appearing under **Fairchild**.

In 1953, the USAF transferred production contracts for the **Chase Aircraft Co, Inc** C 123 to Fairchild. The Chase-built XC 123 and XC 123A appear under **Chase**, while Fairchild's C-123 production is listed under **Fairchild**. Special STOL development of the C 123 airframe appears under the **Stroukoff Aircraft Corp.**

In 1954, the **American Helicopter Co, Inc** (founded 1947) became the **Helicopter Division** of Fairchild Engine and Airplane Corp. The division closed by the end of decade. All American Helicopter designs are filed under **American Helicopter**.

In 1964, Fairchild Stratos purchased **Hiller Aircraft Corp**, and both companies were renamed: **Hiller Aircraft Co Inc** become a subsidiary of Fairchild Hiller Corp. In the 1973 reorganization of Fairchild Hiller into Fairchild Industries, Hiller helicopter interests passed to an independent **Hiller Aviation Inc.** All Hiller designed helicopters are listed under **Hiller**.

In 1965, the **Republic Aviation Corp** became **Republic Aviation Division** (also known as Fairchild Republic) of Fairchild Hiller Corp. In 1987, Republic was shut down when Fairchild Industries ceased building commercial and military aircraft. Aircraft designed by Republic before 1965 appear under **Republic**.

Swearingen Aircraft formed in the late 1950s, modifying Beech aircraft for executive transport. In 1965 the company produced its first new design, the Merlin. In 1970 Swearingen began development of the Metro, a joint venture to be marketed by Fairchild Hiller Corp. As a subsidiary of Fairchild Industries, Swearingen became **Swearingen Aviation Corp**, in 1971, **Fairchild Swearingen** in 1981, and **Fairchild Aircraft Corp** in September 1982. When Fairchild Industries closed its aircraft design and production facilities in 1987, Fairchild Aircraft Corp was sold to **GMF Investments, Inc**; GMF continued to operate the company under the Fairchild name. In 1990, Fairchild Aircraft filed for Chapter 11 protection and was purchased by **Fairchild Acquisition Inc** as **Fairchild Aircraft Inc.** Fairchild Aircraft delivered its last aircraft in 2001. Most Swearingen designs are filed under **Swearingen**; the Metro and Expediter can be found under **Fairchild**.

In 1996, Fairchild Acquisition became Fairchild Aerospace. While continuing to operate Fairchild Aircraft, the company also purchased 80% of the stock of Germany's **Dornier Luftfahrt GmbH** (with the remaining 20% of shares held by **Daimler Benz Aerospace**). Dornier's aircraft manufacturing operations were taken over by **Fairchild Dornier Luftfahrt Beteiligungs GmbH.** In 2000, Fairchild Aerospace was renamed Fairchild Dornier Aerospace, with corporate headquarters moved to Germany. Dornier designs predating Fairchild's takeover are listed under **Dornier**. Subsequent designs are found under **Fairchild Dornier**.

Fairchild YA-10 Thunderbolt II
Fairchild YA-10B Thunderbolt II (N/AW, Night/Adverse Weather)
Fairchild Argus
Fairchild XAT-13 Yankee Doodle
Fairchild XAT-14 Gunner
Fairchild XAT-14A Gunner
Fairchild AT-21 Gunner
Fairchild XBQ-3
Fairchild XC-8 (See: XF-1)
Fairchild C-8 (See: YF-1)
Fairchild C-8A (See: F-1A)
Fairchild (American) Y1C-24 (C-24) Pilgrim
Fairchild C-26A/B
Fairchild XC-31 Pilgrim
Fairchild UC-61 Forwarder
Fairchild UC-61A Forwarder
Fairchild UC-61B Forwarder
Fairchild UC-61C Forwarder
Fairchild UC-61D Forwarder
Fairchild UC-61E Forwarder
Fairchild UC-61F Forwarder
Fairchild UC-61G Forwarder
Fairchild UC-61H Forwarder
Fairchild UC-61J Forwarder
Fairchild UC-61K Forwarder
Fairchild XC-82 Packet
Fairchild C-82A Packet
Fairchild EC-82A Packet
Fairchild XC-82B Packet (See: C-119A)

Fairchild C-82N Packet
Fairchild UC-86
Fairchild UC-96
Fairchild JC-119 Flying Boxcar (Satellite Recovery)
Fairchild C-119A (XC-82B) Flying Boxcar
Fairchild C-119B Flying Boxcar
Fairchild C-119C Flying Boxcar
Fairchild YC-119D (C-128A) Flying Boxcar
Fairchild YC-119E (C-128B) Flying Boxcar
Fairchild C-119F Flying Boxcar
Fairchild C-119G Flying Boxcar
Fairchild AC-119G Shadow Gunship
Fairchild YC-119H Skyvan
Fairchild C-119J Flying Boxcar
Fairchild YC-119K Flying Boxcar
Fairchild AC-119K Stinger Gunship
Fairchild C-119L Flying Boxcar
Fairchild RC-119L Flying Boxcar
Fairchild XC-120 Packplane
Fairchild XC-123 Avitruc (See: Chase)
Fairchild XC-123A Avitruc (See: Chase)
Fairchild C-123B Provider
Fairchild VC-123C Provider
Fairchild (Stroukoff) YC-123D Provider (Boundary Layer Control)
Fairchild (Stroukoff) YC-123E Provider (Pantobase)
Fairchild YC-123H Provider
Fairchild C-123J Provider
Fairchild C-123K Provider
Fairchild NC-123K (AC-123K) Provider
Fairchild UC-123K Provider
Fairchild VC-123K Provider
Fairchild (Stroukoff) YC-134 (Boundary Layer Control)
Fairchild (Stroukoff) YC-134A (BLC, Pantobase)
Fairchild XF-1 (F-1, XC-8)
Fairchild YF-1 (F-1, C-8)
Fairchild F-1A, C-8A
Fairchild (Canada) F 11 Husky
Fairchild F-27 Friendship
Fairchild F-27A Friendship (Fokker F.27 Series 200)
Fairchild F-27B Friendship (Fokker F.27 Series 300)
Fairchild F-27E Friendship
Fairchild F-27F Friendship
Fairchild F-27G Friendship
Fairchild F-27J Friendship
Fairchild F-27M Friendship
Fairchild F-27 (M-258) Military Configuration
Fairchild FH-227 Friendship
Fairchild FH-227B Friendship
Fairchild FH-227C Friendship
Fairchild FH-227D Friendship
Fairchild FH-227E Friendship
Fairchild F 47
Fairchild F 78 (M-82) Packet
Fairchild FB-3 (Special Flying Boat Monoplane)
Fairchild FC-1
Fairchild FC-1A
Fairchild FC-2C
Fairchild FC-2L
Fairchild FC-2W
Fairchild FC-2W, NASM
Fairchild FC-2W2
Fairchild FC-2W2 "Stars and Stripes"
Fairchild FC-2W2 "City of New York"
Fairchild GK-1
Fairchild JK-1
Fairchild J2K-1
Fairchild J2K-2
Fairchild XJQ-1 (FC-2)
Fairchild XJQ-2 (XRQ-2, FC-2)
Fairchild XJ2Q-1
Fairchild KR-21 (Challenger C-6)
Fairchild KR-31 (Challenger C-2)
Fairchild KR-34 (Challenger C-4)
Fairchild M-62
Fairchild M-84
Fairchild M-186
Fairchild M-225
Fairchild M-253
Fairchild M-270D
Fairchild M-284
Fairchild M-288
Fairchild (Swearingen) Metro
Fairchild (Swearingen) Metro II
Fairchild (Swearingen) Metro III
Fairchild (Swearingen) Metro IV
Fairchild (Swearingen) Metro 23
Fairchild (Swearingen) Metro Expediter I
Fairchild XNQ-1
Fairchild (American) Pilgrim 100
Fairchild (Pilatus) Porter (Heli-Porter, Turbo-Porter)
Fairchild PT-19
Fairchild PT-19A
Fairchild PT-19B
Fairchild XPT-23
Fairchild PT-23
Fairchild PT-23A
Fairchild PT-26 Cornell
Fairchild PT-26A Cornell
Fairchild PT-26B Cornell
Fairchild XR2K-1 (F 22)
Fairchild XRQ-2 (See: XJQ-2)
Fairchild XR2Q-1 (See: XJ2Q-1)
Fairchild R4Q-1 Packet
Fairchild R4Q-2 (C-119F) Packet
Fairchild (Canada) SBF-1 Helldiver
Fairchild (Canada) SBF-3 Helldiver
Fairchild (Canada) SBF-4E Helldiver
Fairchild SF-340 (See: Saab 340)
Fairchild YT-31
Fairchild T-46 NGT (Next Generation Trainer)
Fairchild AU-23A Peacemaker (Armed Pilatus Turbo-Porter)
Fairchild OV-12A (Turbo-Porter)
Fairchild VZ-5 Fledgling (M-224-1)
Fairchild 21 (FT-1)
Fairchild 22
Fairchild 24
Fairchild 24R40
Fairchild 34-42 Niska
Fairchild 41
Fairchild 42
Fairchild 45 (F-45)
Fairchild 45-80 Sekani Floatplane
Fairchild 46
Fairchild 51
Fairchild 51A
Fairchild 71
Fairchild 71A
Fairchild 71B
Fairchild 71C
Fairchild 71CM
Fairchild 71D
Fairchild Super 71
Fairchild 91 Baby Clipper (942, XA-942A, XA-942B)
Fairchild 92
Fairchild 125
Fairchild 135
Fairchild 140
Fairchild 150

Fairchild, Walter Lowe (Mineola, NY)

Fairchild (Walter Lowe) 1910 Monoplane I
Fairchild (Walter Lowe) 1910 Monoplane II

Fairchild Dornier (Germany, USA)

Fairchild Dornier Envoy 5
Fairchild Dornier Envoy 7
Fairchild Dornier Envoy 9
Fairchild Dornier 328JET
Fairchild Dornier 428JET
Fairchild Dornier 528JET

Fairchild Dornier 728JET
Fairchild Dornier 928JET

Fairey (UK)
Fairey Albacore
Fairey Barracuda Mk.I
Fairey Barracuda Mk.II
Fairey Barracuda Mk.III
Fairey Barracuda Mk.V
Fairey Battle
Fairey Campania
Fairey Delta One (F.D.1)
Fairey Delta Two (F.D.2)
Fairey F.2 Fighter
Fairey Fantôme/Féroce (Ghost/Savage)
The Fairey Fantôme was designed in 1934 by Marcel Lobelle to meet a Belgian Air Force fighter specification. Several aircraft were assembled at **Avions Fairey** in Gosselies, Belgium. The name for Belgian-assembled aircraft was Féroce.
Fairey Fawn
Fairey F.C.1
Fairey Ferret Mk.I
Fairey Ferret Mk.II
Fairey Ferret Mk.III
Fairey Firefly (1930s)
Fairey Firefly (1940s)
Fairey Firefly Mk.I (1940s)
Fairey Firefly T.Mk.1 (1940s)
Fairey Fleetwing
Fairey Flycatcher
Fairey Fox Mk.I
Fairey Fox Mk.II
Fairey Fox Mk.II Trainer
Fairey Fox Mk.VI
Fairey (Belgium) Fox Mk.VII Mono-Fox Kangourou (Kangaroo)
Fairey Fox Mk.VIII
Fairey Fremantle
Fairey Fulmar
Fairey G.4/31
Fairey Gannet
Fairey Gordon
Fairey Gyrodyne
Fairey Hamble Baby
Fairey Hendon
Fairey Jet Gyrodyne
Fairey Long-Range Monoplane
Fairey N.4 "Atalanta"
Fairey N.4 "Titania"
Fairey N.9
Fairey N.10 (Modified as Fairey III, Prototype of IIIA)
Fairey P.4/34 (Fulmar Prototype)
Fairey Pintail (Type 21, Type XXI, Flycatcher)
Fairey Primer (Tipsy Type M)
Fairey Rotodyne
Fairey S.9/30
Fairey Seafox
Fairey Seal
Fairey Spearfish
Fairey Swordfish
The Tipsy family of light sport aircraft was designed by E. O. Tips, manager of the subsidiary company, **Avions Fairey**, located in Gosselies, Belgium. Starting in 1934, examples were built at the Gosselies factory and at the main Fairey plant at Hayes. The S.2 was also built under license by Aero Engines. Production resumed after World War II, and in 1966, design and manufacturing rights were purchased by **Nipper Aircraft**, Derby, England.
Fairey Tipsy B
Fairey Tipsy BC
Fairey Tipsy Belfair
Fairey Tipsy Junior
Fairey Tipsy M (See: Fairey Primer)
Fairey Tipsy Nipper
Fairey Tipsy S.2
Fairey TSR II (Swordfish Prototype)
Fairey Ultra Light Helicopter
Fairey III (See Fairey N.10)
Fairey IIIA
Fairey IIIB
Fairey IIIC
Fairey IIID
Fairey IIIF

Fairey Britten Norman *(See: Britten Norman)*

Falck, William F. "Bill" (Warwick, NY)
Falck "Chester Special"
Falck Special "Rivets"

Falcon Aircraft, Inc (Knoxville, TN)
Falcon F-1 Sport
Falcon F-2 Agricultural

Falcon Airways (Reidsville, NC) *(See: Lefevers)*

Falcon Jet Corp (Teterboro, NJ)
Falcon Jet Corp sells and services French Dassault Mystère/Falcon aircraft. Files for these aircraft are found under **Dassault**.
Falcon Jet Mystère/Falcon (See: Dassault)

Falconar Aircraft Ltd (Canada)
Chris B. Falconar's **Falconar Aircraft Ltd** appears to have started life in the early 1960s as a subsidiary of **Maranda Aircraft Co, Ltd.** Independent by 1970, the company became the **Falconar Aircraft Division** of **Sturgeon Aircraft Ltd (SAL)** around 1972. There is little record of the company after 1975. Along with some original designs, Falconar produced variants of (as well as kits and plans for) several Jodel aircraft. Jodel designs are listed under **Jodel**, with Falconar variations and original designs filed under **Falconar**.
Falconar AMF-S14 (Based on Maranda Super Loisair)
Falconar F.9
Falconar F.10
Falconar F.11
Falconar F.11-3
Falconar F.12
Falconar Type 121 Teal

Fane, Capt. Gerard (UK)
Fane F.1/40

Fantasy Air (Czech Republic)
Fantasy Air Cora

Farcot (France)
Farcot "Libellule" (Dragonfly)

Farge, Juan de la (Argentina)
Farge Picaflor (Hummingbird)

Farman, Henri (Henry) (France)
Farman (Henri):
Flying Fish (H.F.II)
H.F.I (See: Voisin)
H.F.I*bis* (See: Voisin)
H.F.I*bis*, Farman Modifications
H.F.II, Voisin-Built (See: Voisin)
H.F.II, Farman-Built Le Jabiru
H.F.III
H.F.III Production Variants
H.F.2/2 (1911 High Wing Monoplane)
H.F.6 (1911 Type Militaire)
H.F.7
H.F.10 (1911 Type Militaire De Concourse)
H.F.10-1*bis* (1911 Type Militaire De Concourse)
H.F.11
H.F.12
H.F.14

H.F.15
H.F.16 (1913 Type Militaire)
H.F.17
H.F.18 (1913 Schneider Cup)
H.F.19
H.F.20
H.F.20 "Wake Up, England" (Claude Grahame-White)
H.F.21
H.F.22
H.F.22 Hydro (H.F.22*bis*)
H.F.23
H.F.24 (1913 Baby Sesquiplane)
H.F.25 (1913 Touring Plane)
H.F.26
H.F.27
H.F.30
H.F.33
H.F.35
H.F.36
1918 BN2 Designs

Farman, Maurice (France)
Farman (Maurice):
M.F.1 (1909 Biplane)
M.F.2 Type Coupe Michelin (1910)
M.F.7 Longhorn (1913 Type Tourime)
M.F.7*bis*
M.F.7*ter*
M.F.7 Hydro
M.F.8 Hydro
M.F.9 Hydro
M.F.10 Hydro
M.F.11 Shorthorn (1914 Type Militaire)
M.F.12
M.F.5
M.F.6
M.F.6*bis*

Farman Frères (Société Henri et Maurice Farman) (France)
Farman Despatch Biplane
Farman F.3X (F.121) Jabiru (Stork)
Farman F.4S
Farman F.4X Jabiru (Stork)
Farman F.30 C.2
Farman F.30A (F.XXX)
Farman F.30B
Farman F.31 C.2
Farman F.36 (1918 Type F)
Farman F.40 Horace
Farman F.1.40 Horace (F.40*bis*)
Farman F.1.40 Horace Type Ecole
Farman F.1.40*ter* Horace
Farman F.2.40 Horace
Farman F.41 Horace (TypedeBiplace)
Farman F.41*ter* Horace
Farman F.41H
Farman F.42 Horace
Farman F.1.43 Horace
Farman F.44 Horace
Farman F.45 Horace
Farman F.1.46 Horace (F.46*bis*, F.46e)
Farman F.47 Horace
Farman F.48 Horace
Farman F.49 Horace
Farman F.50 (F.40 Horace Series)
Farman F.50 Bn.2 (Far.13 Bn.2)
Farman F.50 P.6 Limousine
Farman F.51 (F.40 Horace Series)
Farman F.52 (F.40 Horace Series)
Farman F.53 (F.40 Horace Series)
Farman F.54 (F.40 Horace Series)
Farman F.55 (F.40 Horace Series)
Farman F.56 (F.40 Horace Series)
Farman F.57 (F.40 Horace Series)
Farman F.58 (F.40 Horace Series)
Farman F.59 (F.40 Horace Series)
Farman F.60 (F.40 Horace Series)
Farman F.60 (Military, F.60 Bn.2)
Farman F.60 Goliath
Farman F.60 Goliath (Ambulance)
Farman F.60 Goliath TriMotor
Farman F.61 (F.40 Horace Series)
Farman F.61 Goliath
Farman F.62 (F.40 Horace Series)
Farman F.63 Bn.4
Farman F.63 Goliath
Farman F.68 Goliath (Polish Military)
Farman F.70 Series
Farman F.70 Limousine, Avion Militaire
Farman F.71 Ecole
Farman F.73
Farman F.80 Ecole
Farman F.90 Limousine (Eight Seater)
Farman F.100
Farman F.110 A.2 Métallique
Farman F.121 Jabiru (See: F.3X)
Farman F.123 Bn3
Farman F.124
Farman F.140 Super-Goliath
Farman F.140 Bn.4 Super-Goliath
Farman F.160
Farman F.161 (A.2)
Farman F.163 (F.63*bis*)
Farman F.165 Bn.4 Hydravion Goliath
Farman F.168
Farman F.169 Goliath
Farman F.170 Jabiru (Stork)
Farman F.171 Jabiru (Stork)
Farman F.180 Oiseau Bleu (Bluebird)
Farman F.190
Farman F.192
Farman F.193
Farman F.194
Farman F.197
Farman F.198
Farman F.199
Farman F.200
Farman F.201
Farman F.202
Farman F.211 Bn4
Farman F.212 Bn4
Farman F.220 Bn4, Bn5
Farman F.220 B (F.2200)
Farman F.221 Bn5
Farman F.222 (F.222.0)
Farman F.222.1
Farman F.222.2
Farman F.223
Farman F.223.0 (F.2230, NC.2230)
Farman F.223.1 (F.2231)
Farman F.223.2 (F.2232)
Farman F.223.3 (F.2233)
Farman F.223.4 (F.2234, NC.2234)
Farman F.224 TT
Farman F.230
Farman F.231
Farman F.233
Farman F.234
Farman F.235
Farman F.250
Farman F.270
Farman F.280
Farman F.291
Farman F.300
Farman F.301
Farman F.302
Farman F.303
Farman F.304
Farman F.305
Farman F.306
Farman F.310
Farman F.339
Farman F.353
Farman F.355
Farman F.356
Farman F.360
Farman F.390
Farman F.393
Farman F.400
Farman F.402
Farman F.403
Farman F.430
Farman F.431
Farman F.432
Farman F.433
Farman F.450
Farman F.451
Farman F.455

Farman F.1001 (F.1000) Stratosphere Aircraft
Farman F.1020
Farman G.L. Hydravion
Farman Hydravions (1921)
Farman Hydravion de Haute Mer
Farman Hydroglisseurs (Air-Boats)
Farman Moustique (Aviette)
Farman Parasol Sport Monoplane (Petit Avion de Tourisme)
Farman Sport Biplane
Farman Torpilleur
Farman 1921 Bn.4
Farman 1922 Glider
Farman 1923 B.2
Farman 1923 Moto Aviette
Farman 1923 Touring Biplane
Farman 1923 Touring Monoplane
Farman-Farman 1933 Racer
Farman-Renault 1933 Racer

Farnborough-Aircraft.com Ltd (UK)
Farnborough-Aircraft.com F1

Farner (Switzerland)
Farner WF.21

Farnham (La Porte, CO)
Farnham 1 Fly Cycle

Farrar, D. F., Jr (USA)
Farrar LSG-1 Bird Flight Machine
Farrar V-1 Flying Wing

Fasig-Turner (Fairfield, OH)
Fasig-Turner Biplane Racer

Fauber, W. H. (France)
Fauber Hydroplane

Faucett (Compania de Aviación Faucett) (Peru)
Faucett F-19

Fauvel, Charles (France)
Fauvel AV 2
Fauvel AV 3
Fauvel AV 10
Fauvel AV 17
Fauvel AV 22
Fauvel AV 221
Fauvel AV 222
Fauvel AV 36
Fauvel AV 361
Fauvel AV 45
Fauvel AV 451
Fauvel AV 46
Fauvel AV 48
Fauvel AV 50
Fauvel AV 60

Fawcett Aviation Pty (Australia)
Fawcett 120

Fawkes, Joseph Wesley (Los Angeles, CA)
Fawkes 1909 Centrifugal Aeroplane

FBA (Franco-British Aviation Co Ltd) (France)
In January 1913, Louis Schreck, of the **French Wright Co**, and Lt. Jean de Conneau (a.k.a. "André Beaumont") established **Franco-British Aviation Co Ltd** (**FBA**), with works in France but capital mainly from British sources. Schreck and Conneau planned to build flying boats from designs by **Société des Hydroaéroplanes Donnet-Lévêque** and **Société Anonyme des Anciens Chantiers Tellier** (**d'Artois**). Later that year FBA absorbed **Société des Hydroaéroplanes Lévêque** (the remains of Donnet-Lévêque after Jérôme Donnet and François Denhaut left in 1913), continuing the development of Donnet-Lévêque aircraft under the FBA name. After World War I, Schreck, as "Constructeur" for FBA, began to affix his own name to the aircraft, although the company's name did not change. In 1935 FBA was acquired by **Société des Avions Bernard** after several of inactive years. Bernard failed later that year. All Schreck and d'Artois designs pre-dating 1913 are filed under **Schreck** and **d'Artois**, respectively. For Donnet-Lévêque designs, see **Donnet-Lévêque**. For all FBA designs, including Schreck's post-1913 work, see **FBA**.
FBA Model A
FBA Model B
FBA Model C
FBA Model H
FBA Model S
FBA 13 HE2
FBA 16 HE2
FBA 17 HE2
FBA 17 HL2
FBA 17 HMT2
FBA 17 HMT4
FBA 19 HMB2
FBA 21 HMT6
FBA 310

Federal Aircraft Corp (San Bernardino, CA)
Federal (CA) CM-1 Lone Eagle
Federal (CA) CM-2
Federal (CA) CM-3
Federal (CA) XPT Trainer

Federal Aircraft Factory (Switzerland)
Federal (Switzerland) C-3603
Federal (Switzerland) C-3604
Federal (Switzerland) C-3605
Federal (Switzerland) N.20

Federal Aircraft Ltd (Canada)
Federal (Canada) AT-20

Federal Aircraft Works (Minneapolis, MN)
Federal (MN) H150B

Fedorov, D. D. (Russia)
Fedorov DF-1

F.E.G. (See: Coler)

Feiro (Feigl and Rotta) (Hungary)
Feiro I

Felix, Charles R. (Hatfield, PA)
Felix Model A Sport Monoplane

Felixstowe (UK)
Felixstowe F.1
Felixstowe F.2
Felixstowe F.2A
Felixstowe F.2C
Felixstowe F.3
Felixstowe F.5
Felixstowe F-5-L (See: Curtiss; also: Naval Aircraft Factory)
Felixstowe Fury (Super Baby)
Felixstowe Porte Baby

Fenn, Otto A. (New York, NY)
Fenn (Otto A.) 1910 Flying Machine

Fenwick, Robert (UK)
Fenwick 1910 Biplane (See: Thompson, William P.)
Fenwick "Mersey" (1912 Pusher Monoplane)

Ferber, Ferdinand (France)
Ferber Biplane
Ferber No 8 Biplane
Ferber 1904 Glider
Ferber 1925 Glider

Ferbois (Société Industrielle des Metaux et du Bois) (France)
Ferbois C.1

Ferguson, Charles J. (California)
Ferguson Glider

Fernandez, A. (Spain, France)
Fernandez 1909 Biplane

Fernandez Yepez, Augustin (Venezuela)
Fernandez Pursuit (1942)

Fernic (Staten Island, NY)
Fernic Monoplane

Fernic Airplane Co (Westfield, NJ)
Fernic T-9 Cruisaire
Fernic T-10

Ferry, Vernon F. (Stockbridge, MA)
Ferry MT2

Fetterman, Fred O. Y. (Brooklyn, NY)
Fetterman 1939 Monoplane

FFA (Flug & Fahrzeugwerk, AG) (Switzerland)
FFA AS202/18A Bravo
FFA D-3801
FFA D-3802
FFA D-3803
FFA Diamant (Diamond) Sailplane
FFA P-1604

***FFM** (See: D.V.L.)*

FFVS (Flygförvaltningens Verkstad I Stockholm) (Sweden)
FFVS J.22

Fiat (Italy)
In 1980, Fiat, an Italian automobile company, entered the aviation field by opening an aero engine facility. In 1914, Fiat opened its first aircraft subsidiary: **Società Italiana Aviazione (SIA)**. SIA's assets were merged back into the parent company in 1918. Designs over the next fifty years included Celestino Rosatelli's bombers (BR- prefix) and fighters (CR- prefix), and Celestino Rosatelli's aircraft (G- prefix). In 1969, Fiat and Aerfer merged to form Aeritalia. Fiat continued to operate under its own name until 1971. SIA and Aerfer designs are filed under the respective companies. Fiat aircraft designed prior to 1971 are listed under **Fiat**. Subsequent designs are found under **Aeritalia**.

Fiat AC2, AC3, AC4 (See: Ansaldo)
Fiat AL
Fiat Am (See: Ansaldo)
Fiat AN1 (A300/4 with AN.1 Motor)
Fiat AN3
Fiat AP (See: Ansaldo)
Fiat APR.2
Fiat ARF (Modified BR)
Fiat ARS
Fiat AS1
Fiat AS1 Idro (Hydro)
Fiat AS2
Fiat A1 Balilla (See: Ansaldo)
Fiat A3, A5 (See: Ansaldo)
Fiat A115 thru A400 (See: Ansaldo)
Fiat B.12P
Fiat BGA
Fiat BR
Fiat BR.1
Fiat BR.2
Fiat BR.3
Fiat BR.4
Fiat BR.20 Cicogna (Stork)
Fiat BR.20A Cicogna (Stork)
Fiat BR.20L Cicogna (Stork)
Fiat BR.20M Cicogna (Stork)
Fiat BRG
Fiat C.4
Fiat C.5B
Fiat C.6
Fiat C.29
Fiat CR
Fiat CR Asso (Ace)
Fiat CR.1
Fiat CR.2
Fiat CR.10
Fiat CR.20
Fiat CR.20*bis*
Fiat CR.20 Idro (Hydro)
Fiat CR.25
Fiat CR.25D
Fiat CR.30
Fiat CR.30B
Fiat CR.32
Fiat CR.32*bis*
Fiat CR.33
Fiat CR.40
Fiat CR.40*bis*
Fiat CR.41
Fiat CR.42 Falco (Hawk)
Fiat CR.42B Falco (Hawk)
Fiat ICR.42 Falco (Hawk)
Fiat F-86K (North American License)
Fiat F-104G (Lockheed License)
Fiat FC.12
Fiat FC.20
Fiat G.2
Fiat G.2/4
Fiat G.5
Fiat G.5/1
Fiat G.5*bis*
Fiat G.8
Fiat G.12
Fiat G.12 C
Fiat G.12 CA
Fiat G.12 CA "Aula Volante" ("Flying Class-Room")
Fiat G.12 GA
Fiat G.12 L
Fiat G.12 LA
Fiat G.12 LB
Fiat G.12 LGA
Fiat G.12 LP
Fiat G.12 RT
Fiat G.12 RT*bis*
Fiat G.12 T
Fiat G.18
Fiat G.18V
Fiat G.46
Fiat G.46-1B
Fiat G.46-2B
Fiat G.46-3A
Fiat G.46-4
Fiat G.46-4A
Fiat G.46-4B
Fiat G.46-5B
Fiat G.49-1
Fiat G.49-2
Fiat G.50
Fiat G.50V
Fiat G.50*bis*
Fiat G.50*bis* A
Fiat G.50*ter*
Fiat G.55 Centuro (Centaur)
Fiat G.55A Centuro (Centaur)
Fiat G.55B Centuro (Centaur)
Fiat G.56
Fiat G.59
Fiat G.59-1A
Fiat G.59B
Fiat G.59-2B
Fiat G.59-4A
Fiat G.59-4B
Fiat G.61
Fiat G.80
Fiat G.80-1B
Fiat G.80-3B
Fiat G.82
Fiat G.91
Fiat G.91A

Fiat G.91 Pan
Fiat G.91R
Fiat G.91R/3
Fiat G.91R/4
Fiat G.91T
Fiat G.91T/1
Fiat G.91T/3
Fiat G.91T/4
Fiat G.91Y
Fiat G.95
Fiat G.95/4 V/STOL
Fiat G.95/6 VTOL
Fiat G.212
Fiat G.212 AV Trainer
Fiat G.212 CA
Fiat G.212 CP Monterosa
Fiat G.212 TP Monviso Freighter
Fiat G.222 V/STOL
Fiat Hover Rig
Fiat MF.4
Fiat MF.5
Fiat MF.6
Fiat/CMASA MF.10
Fiat N 3
Fiat R.2
Fiat R.22
Fiat R.700
Fiat RS.14
Fiat SIA (See: SIA)
Fiat SP.1 (See: SIA)
Fiat SP.2 (See: Pomilio)
Fiat TR.1
Fiat Vampire Mk.52
Fiat Vertol V-44
Fiat 7002 Helicopter
Fiat-Cansa

Fibera (OY Fibera AB) (Finland)
Fibera KK-1 Utu
Fibera KK-1a Utu
Fibera KK-1b Utu
Fibera KK-1c Utu
Fibera KK-1d Utu
Fibera KK-1e Utu

Fiedler, Paul (Germany)
Fiedler 1910 Monoplane

Fierro y Sea (Mexico)
Fierro y Sea Teziutlan Primary Trainer

Fieseler Werke GmbH (Germany)
Fieseler F 1 (See: Raab-Katzenstein)
Fieseler F 2 Tiger
Fieseler F 3 Wespe (Wasp)
Fieseler F 4
Fieseler Fi 5
Fieseler Fi 97
Fieseler Fi 98
Fieseler Fi 99
Fieseler Fi 103
Fieseler Fi 156 Storch (Stork)
Fieseler Fi 158
Fieseler Fi 167
Fieseler Fi 253 Spatz (Sparrow)
Fieseler Fi 256 Storch (Stork)
Fieseler Fi 333
Fieseler Sailplanes & Gliders (See: Kassel)

Fife, Raymond B. "Ray" (Coronado, CA)
Fife Beachy Replica

Fike, William J. (Anchorage, AK)
Fike A (Homebuilt, 1931)
Fike B (Homebuilt, 1935)
Fike C (Homebuilt, 1936)
Fike D (Homebuilt, 1950)
Fike E (Homebuilt, 1953)

Filippi, Antoine P. (France)
Filippi Helicopter Designs

Filper Corp (San Ramon, CA)
Filper Beta 200
Filper Beta 300
Filper Beta 400A

Filter, Walter (Germany)
Filter 1958 Ornithopter

Finley, Thomas M. (St Louis, MO)
Finley 1918 Flying Machine

Firestone Aircraft Co (G & A, Glider & Autogiro) (Akron, OH) *(See also: Pitcairn-Larsen)*
Firestone GA-45
Firestone GA-45D
Firestone GA-50
Firestone XO-61
Firestone XR-9
Firestone XR-9B
Firestone XR-14

First Strike Aviation *(See: Bowdler)*

Fisher Body Division (General Motors Corp) *(See: General Motors)*

Fisher Flying Products, Inc (South Webster, OH)
Fisher Barnstormer (Ultralite)
Fisher Culex
Fisher FP-101
Fisher FP-202 Koala
Fisher FP-303
Fisher FP-404 Classic
Fisher FP-505 Skeeter
Fisher FP-606 Sky Baby
Fisher Super Koala

Fisk, Edwin M. *(See: Catron and Fisk; also International Aircraft Corp)*

Fitzpatrick, James L. G. (Staten Island, NY)
Fitzpatrick 1960 Ornithopter

Fizir, Rudolf (Yugoslavia)
Fizir AF-2 Amphibian
Fizir FN (Fizir Nastaun = Basic Trainer) (Fizir-Wright)
Fizir FP.1 (Fizir Prelazni = Trainer)
Fizir FP.2 (See: Zmaj)
Fizir LAF
Fizir Nebojsa
Fizir-Castor
Fizir-Loraine
Fizir-Maybach Recon Aircraft
Fizir-Vega AFZ

Flagg, Claude (Buffalo, NY)
Flagg Snyder Special
Flagg Flaggship
Flagg Phantom 1 Special Racer

Flaglor, K. (Northbrook, IL)
Flaglor Scooter

Flair *(See: Fletcher)*

Flanders, R. L. Howard (UK)
Flanders B.2 (1912 Biplane)
Flanders F.4 Monoplane
Flanders 1911 Monoplane
Flanders 1912 Monoplane
Flanders 1913 Biplane
Flanders 1913 Monoplane

Fleet Aircraft Ltd (Canada)
In 1928 Reuben Fleet was unable to convince the directors of his **Consolidated Aircraft Corp** to proceed with production of the Model 14 Husky Junior. Fleet purchased the rights to the aircraft and formed **Fleet Aircraft Inc** in 1929, placing orders for production of 100 of the aircraft (now called "Fleets") with Consolidated. That summer Fleet

Aircraft was purchased by Consolidated, which continued production and development. In 1930 Consolidated formed a **Fleet Aircraft of Canada, Ltd** subsidiary, which passed to Canadian control as **Fleet Aircraft Ltd** in 1936. US-produced Fleets are listed under **Consolidated Model 14.** Canadian Fleets are listed under **Fleet (Canada).**

Fleet (Canada) Model 1
Fleet (Canada) Model 2
Fleet (Canada) Model 7B
Fleet (Canada) Model 7B Fawn I
Fleet (Canada) Model 7C Fawn II
Fleet (Canada) Model 7G Fawn I
Fleet (Canada) Model 10
Fleet (Canada) Model 16
Fleet (Canada) Model 16B
Fleet (Canada) Model 16B Finch II
Fleet (Canada) Model 16F (Brewster-Fleet 10, B-1)
Fleet (Canada) Model 16R
Fleet (Canada) Model 21
Fleet (Canada) Model 21M
Fleet (Canada) Model 50 Freighter
Fleet (Canada) Model 50J Freighter
Fleet (Canada) Model 50K Freighter
Fleet (Canada) Model 60K Fort Prototype
Fleet (Canada) Model 60K Fort I
Fleet (Canada) Model 80 Canuck

Fleetcraft Airplane Corp (Lincoln, NE)
Fleetcraft Model A

Fleetwings (Bristol, PA)
Fleetwings A-1 Target Drone
Fleetwings XBQ-1 & XBQ-2A
Fleetwings XBQ-1
Fleetwings XBQ-2A
Fleetwings XBTK-1
Fleetwings BT-12
Fleetwings XBT-12
Fleetwings XPQ-12A
Fleetwings YPQ-12A
Fleetwings Sea Bird F4
Fleetwings Sea Bird F5
Fleetwings Model 23
Fleetwings Model 33

Fletcher Aviation Co (El Monte, CA) *(See: Fletcher Aviation Corp)*

Fletcher Aviation Corp (Wendell S. Fletcher) (Pasadena, CA; Rosemead, CA)
Fletcher XBG-1
Fletcher XBG-2
Fletcher XCQ-1
Fletcher YCQ-1A
Fletcher CQ-1A
Fletcher FD-25 Defender
Fletcher FL-23
Fletcher FBT-2
Fletcher FU-24 Utility
Fletcher PQ-11
Fletcher PQ-11A

Flettner (Germany)
Flettner Fl 184 Gyroplane
Flettner Fl 185 Heligyro
Flettner Fl 265
Flettner Fl 282 Kolibri (Hummingbird)
Flettner Fl 285
Flettner Fl 339
Flettner 1933 Prototype

Flight Designs, Inc (Salinas, CA)
Flight Designs Flightstar
Flight Designs Jet Wing ATV
Flight Designs 440 ST

Flight Dynamics, Inc (Raleigh, NC)
Flight Dynamics Flightsail Mk I

Fliteways (Lynchburg, VA)
Fliteways Special

Floatplanes, General *(See: Seaplanes, General)*

Florencie (France)
Florencie c.1906 Ornithopter

Florida (University of Florida) (Gainesville, FL)
Florida BDG-1

Florine, Nicolas (Belgium)
Florine Helicopter No.1
Florine Helicopter No.2
Florine Helicopter No.33)
Florine Experimental Helicopter

FLS Aerospace Ltd (UK)
FLS Optica (See: Edgley Aircraft)
FLS Sprint (See: Trago Mills SAH-1)

Flug & Fahrzeugwerk, AG *(See: FFA)*

Flugmaschine Rex GmbH *(See: Rex)*

Flugtechniscen Verein, Dresden (F.V.D.) *(See: Dresden)*

Flugwissenschaftliche Forschunganstalt e.V. München *(See: D.V.L.)*

Flugzeug Reparatur und Bau Anstalt *(See: Fruba)*

Flugzeugwerft Lübeck-Travemünde GmbH *(See: Lübeck-Travemünde)*

Flygförvaltningens Verkstad I Stockholm *(See: FFVS)*

Flying Boats, General *(See: Seaplanes, General)*

Flying Wings & Tailless Aircraft, General

Flyray (USA)
Flyray Jet-Powered Homebuilt (1979)

FMA (Fábrica Militar de Aviones, Fábrica Materiales Aerospaciales) (Argentina)
FMA Ae. C.1 (Civil 1)
FMA Ae. C.2 (Civil 2) "Tenga Confianza" (Ae. M.E.1, Militar Entrenamiento 1)
FMA Ae. C.3 (Civil 3)
FMA Ae. C.3G (Civil 3 Gypsy)
FMA Ae. C.4 (Civil 4)
FMA Ae. M.B.1
FMA Ae. M.B.2
FMA Ae. M.O.1
FMA Ae. M.Oe.1
FMA Ae. M.Oe.2
FMA Ae. M.S.1
FMA Ae. T.1
FMA Calquin (Royal Eagle) (I.Ae.24)
FMA Chingolo (I.Ae. 32)
FMA Clen Antú (Sunray) (I.Ae. 34)
FMA Colibrí (Hummingbird) (I.Ae. 31)
FMA Constancia II
FMA Guarani (I.A. 50)
FMA Huanquero (I.A. 35)
FMA I.A. 37 ("Pulqui III") ("Arrow III")
FMA I.A. 38
FMA I.A. 45
FMA I.Ae D.L.22
FMA I.Ae 23
FMA Mañque (I.Ae. 25)

FMA Ñamcú (Eagle-like Bird) (I.Ae.30)
FMA Pampa I.A. 63
FMA Pampa NG I.A. 63
FMA Pucara (Fortress) (I.A. 58)
FMA Pulqui (Arrow) (I.Ae. 27)
FMA Pulqui (Arrow) II (I.Ae. 33)
FMA Ranquel (I.A. 46)
FMA Urubú (Vulture) (I.A. 41)
FMA 20 El Boyero (Cowboy) (IAe 20)
FMA 21

Focke, Heinrich (Germany)
Focke (Heinrich) A 5
Focke (Heinrich) (Cierva) C.19

Focke Achgelis Flugzeugbau GmbH (Germany)
Focke Achgelis Fa 61
Focke Achgelis Fa 223 Drache (Kite)
Focke Achgelis Fa 224
Focke Achgelis Fa 225
Focke Achgelis Fa 236
Focke Achgelis Fa 266 Hornisse (Hornet)
Focke Achgelis Fa 269
Focke Achgelis Fa 284
Focke Achgelis Fa 325
Focke Achgelis Fa 330 Bachstelze (Water Wagtail)
Focke Achgelis Fa 336

Focke-Wulf Flugzeugbau GmbH (Germany)
Focke-Wulf: Company Designations
Focke-Wulf S 1
Focke-Wulf S 2
Focke-Wulf W 4
Focke-Wulf A 5 (See: Focke, Heinrich)
Focke-Wulf A 16 Family
Focke-Wulf A 17 Möwe (Seagull)
Focke-Wulf GL 18
Focke-Wulf F 19a Ente (Duck)
Focke-Wulf A 20 Habicht (Goshawk)
Focke-Wulf A 21
Focke-Wulf GL 22
Focke-Wulf S 24 Kiebitz (Lapwing)
Focke-Wulf A 28 Habicht (Goshawk)
Focke-Wulf A 29 Möwe (Seagull)
Focke-Wulf A 32 Bussard (Buzzard)
Focke-Wulf A 33 Sperber (Sparrow Hawk)
Focke-Wulf A 38 Möwe (Seagull)
Focke-Wulf A 39
Focke-Wulf A 40
Focke-Wulf A 47
Focke-Wulf L84 (See: Albatros)
Focke-Wulf L100 (See: Albatros)
Focke-Wulf L101 (See: Albatros)
Focke-Wulf L102 (See: Albatros; See also: Fw 55)
Focke-Wulf W102 (See: Albatros; See also: Fw 55)
Focke-Wulf Triebflügel (Thrust-Wing)
Focke Wulf Weihe 50 (See: Jacobs Schweyer)

Focke-Wulf: German Air Ministry Designations
Focke-Wulf Fw 43 (A 43) Falke (Falcon)
Focke-Wulf Fw 44 Stieglitz (Goldfinch)
Focke-Wulf Fw 44 F Stieglitz (Goldfinch)
Focke-Wulf Fw 44 J Stieglitz (Goldfinch)
Focke-Wulf Fw 47
Focke-Wulf Fw 55
Focke-Wulf Fw 56 Stösser (Falcon)
Focke-Wulf Fw 57
Focke-Wulf Fw 58 Weihe (Kite)
Focke-Wulf Fw 62
Focke-Wulf Ta 152 Prototypes
Focke-Wulf Ta 152 A
Focke-Wulf Ta 152 B
Focke-Wulf Ta 152 C
Focke-Wulf Ta 152 E
Focke-Wulf Ta 152 H
Focke-Wulf Ta 152 H, NASM
Focke-Wulf Ta 152 S
Focke-Wulf Ta 153
Focke-Wulf Ta 154 Moskito (Mosquito)
Focke-Wulf Fw 159
Focke-Wulf Ta 183
Focke-Wulf Fw 186 (Cierva C.30 Development)
Focke-Wulf Fw 187 Falke (Falcon)
Focke-Wulf Fw 189 Uhu (Owl)
Focke-Wulf Fw 190 Prototypes
Focke-Wulf Fw 190 A
Focke-Wulf Fw 190 B
Focke-Wulf Fw 190 C
Focke-Wulf Fw 190 D
Focke-Wulf Fw 190 D, NASM
Focke-Wulf Fw 190 E
Focke-Wulf Fw 190 F
Focke-Wulf Fw 190 F-8, NASM
Focke-Wulf Fw 190 G
Focke-Wulf Fw 190 H
Focke Wulf Fw 191
Focke Wulf "Fw 198"
Focke-Wulf Fw 200 Condor
Focke-Wulf Fw 200 S-1 Condor "Brandenburg"
Focke-Wulf Fw 200 Kurier (Courier)
Focke-Wulf Fw 206
Focke-Wulf Ta 211 (See: Ta 154)
Focke-Wulf Ta 254
Focke-Wulf Fw 259
Focke-Wulf Fw 261
Focke-Wulf Fw 272
Focke-Wulf Ta 281
Focke-Wulf Ta 283
Focke-Wulf Fw 300
Focke-Wulf Fw 391
Focke-Wulf Ta 400
Focke-Wulf Fw 491

Focke-Wulf: Special Project Designations
Focke-Wulf Night Fighter with 2xHeS 011, Design II
Focke-Wulf P.011.018
Focke-Wulf P.011.025
Focke-Wulf P.011.045
Focke-Wulf P.021.009
Focke-Wulf P.222.001
Focke-Wulf P.222.004
Focke-Wulf P.222.010
Focke-Wulf P.222.018
Focke-Wulf P.0310.025.1006
Focke-Wulf P.0310.224.20
Focke-Wulf P.0310.224.30
Focke-Wulf P.0310.225
Focke-Wulf P.0310.237.2
Focke-Wulf P.0310.251.13
Focke-Wulf P.413.001
Focke-Wulf P.82114

Fogle Brothers (Donald J. & Forrest F. Fogle) (Butler, PA)
Fogle Hornet

Fokker (Germany, Netherlands, USA)
Anthony Fokker built his first aircraft in 1910. He formed **Fokker Aeroplanbau GmbH** and opened his first factory in Germany in 1912. Following World War I Fokker returned to his native Holland and founded **N.V. Nederlandsche Vliegtuigen Fabriek**; the German factory was renamed **Schweriner Industrie-Werke** and continued production of Fokker transports into the 1920s. In 1945 Fokker aircraft were designed and produced by **N.V. Vereenigde Nederlandsche**, in 1946 by **Vliegtuigen Fabrieken Fokker**, and in 1954 by **N.V Koninklijke Nederlandsche Vliegtuigen Fabrieken Fokker**.

Between 1969 and 1980 NVKNV Fokker merged with Germany's **Vereinigte Flugtechnische Werke GmbH** to form **Fokker-VFW N.V.** At the merger's end Fokker became **Fokker BV**, later restructuring as **N.V. Koninklijke Nederlandse Vliegtuig Fabriek Fokker**.

Fokker aircraft were built under license by several companies, but in the USA Anthony Fokker formed his own aircraft manufacturing (and design) concern. He started in 1920 with **Netherlands Aircraft Manufacturing Co of Amsterdam** which was responsible for sales and information about Fokker imports. In 1924 **Atlantic Aircraft Corp** began construction of Fokker designs in New Jersey. In 1925 the company became **Fokker Aircraft Corp. General Motors Corp** acquired Fokker stock in 1929, forming **General Aviation Corp** as a holding company in 1930 and renaming the Fokker company **General Aviation Manufacturing Corp.** Fokker ended his association with the American company in 1931. (General Aviation became part of **North American Aviation, Inc** in 1934.)

Products of the Dutch, German, and US Fokker organizations are listed under **Fokker**. General Aviation aircraft designed after Fokker left the company (and a few Fokker designs) are found under **General Aviation**. License-built Fokker aircraft are listed under individual licensees.

Fokker A.I (See: M 8)
Fokker A.II (See: M 5L)
Fokker A.III (See: M 5K)
Fokker A-2 (F.IV)
Fokker XA-7 (US Model 17)
Fokker AO-1 (Atlantic Model 3)
Fokker B.I
Fokker B.II
Fokker B.III
Fokker B.IIIc
Fokker B.IV (See also: US F-11)
Fokker B.IVa (See also: US F.11a)
Fokker B Series Unarmed Biplanes
Fokker B (See: M 7)
Fokker B.I (See: M 10E)
Fokker B.II (See: M 11)
Fokker B.II (See: M 17K)
Fokker B.III (See: M 16Z)
Fokker XB-8 (Model 16)
Fokker YB-8 (See: YO-27)
Fokker Y1B-8 (See: Y1O-27)
Fokker C.I (V.38)
Fokker C.Ia
Fokker C.I-W
Fokker C.II
Fokker C.III
Fokker C.IV
Fokker C.IVA
Fokker C.IVB
Fokker C.IVC
Fokker C.IV-W
Fokker C.V-A
Fokker C.V-B
Fokker C.V-C
Fokker C.V-D
Fokker C.V-E
Fokker C.V-W
Fokker C.VI
Fokker C.VII-W
Fokker C.VIII
Fokker C.VIII-W
Fokker C.IX
Fokker C.X
Fokker C.XI-W (C.11-W)
Fokker C.XIV-W
Fokker C-2, Civil (Atlantic Model 7)
Fokker C-2, Civil, "America"
Fokker C-2, Military
Fokker C-2, Military, "Bird of Paradise"
Fokker C-2A
Fokker C-2A "Question Mark"
Fokker C-5 (F-10A)
Fokker C-7 (XC-7)
Fokker C-7A
Fokker Y1C-14 (C-14, F-14)
Fokker Y1C-14A
Fokker C-14B
Fokker Y1C-15 (C-15)
Fokker C-16 (F-11A)
Fokker YC-20 (F-32)
Fokker C-31 (F.27) Friendship
Fokker XCO-4 (CO-4)
Fokker CO-4A
Fokker CO-4 Mail (C-4, Atlantic Model 3)
Fokker XCO-8
Fokker D Series Fighters
Fokker D.I (M 18ZF, M 18ZK)
Fokker D.II (M 17ZF, M 17ZK)
Fokker D.III (M 19F, M 19K)
Fokker D.IV (M 20)
Fokker D.V (M 22E)
Fokker D.VI
Fokker D.VII (V.18)
Fokker D.VII, NASM
Fokker D.VIIF (V.24)
Fokker D.VIII (V.26/2, V.28, E.V)
Fokker D.IX (See also: PW-6)
Fokker D.X
Fokker D.XI (See also: PW-7)
Fokker D.XII
Fokker D.XIII
Fokker D.XIV
Fokker D.XV
Fokker D.XVa
Fokker D.XVI
Fokker D.XVII
Fokker D.XIX
Fokker D.XX
Fokker D.XXI (D.21)
Fokker D.XXIII (D.23)
Fokker D.C.I
Fokker DH-4M-2 (Atlantic Model 1)

Several US manufacturers modified and license built the British de Havilland D.H.4 under the designation DH-4. For British-built aircraft, see **de Havilland D.H.4.** For American-built aircraft, see the individual manufacturers, including **Aeromarine**, **Boeing**, **Dayton Wright**, **Fokker (Atlantic)**, **Gallaudet**, **Keystone**, and **Standard.** When the actual manufacturer cannot be determined from the available evidence, the photos and documents have been filed under **Dayton Wright.**

Fokker Dr.I Triplane (F.I)
Fokker Dr.I Triplane (F.I), Red Baron's Aircraft
Fokker E Series Monoplane Fighters
Fokker E.I (M 5K/MG)
Fokker E.II (M 14)
Fokker E.III (M 14)
Fokker E.IV (M 15)
Fokker E.V (See: D.VIII)
Fokker F Series Civil Aircraft
Fokker F.I (V.44)
Fokker F.II (V.45)
Fokker F.IIB
Fokker F.IID
Fokker F.III
Fokker F.III-W
Fokker F.IV (See also: A-2, T-2)
Fokker F.V
Fokker F.VI (See: PW-5)
Fokker F.VII (Atlantic Model 6, F-7)
Fokker F.VIIA
Fokker F.VIIA "Old Glory"
Fokker F.VIIA-3m
Fokker F.VIIA-3m "Josephine Ford"
Fokker F.VIIA-3m/M
Fokker F.VIIA-3m/W
Fokker F.VIIB-3m
Fokker F.VIIB-3m "Southern Cross"
Fokker F.VIIB-3m (Monospar)
Fokker F.VIIB-3m/M

Fokker F.VIIB-3m/W
Fokker F.VIII
Fokker F.VIIIA
Fokker F.IX (Including License-Built Avia Czechoslovakia F.39)
Fokker F-9 (US Model 7)
Fokker F.IX-M (Avia Czechoslovakia F.39 Bomber)
Fokker F-10 Super TriMotor (TriMotor Deluxe, F-X, F-Ten)
Fokker F-10A
Fokker F-10A Bomber ("XLB-3")
Fokker F.XI
Fokker F-11 (US Model 9, B.IV)
Fokker F-11A (US Model 9, B.IVa)
Fokker F-11A Tandem (US Model 9, B.IVa)
Fokker F-11AHB (US Model 9, B.IVa)
Fokker F.XII
Fokker F.XIV
Fokker F.XIV-3m
Fokker F-14 (US Model 14)
Fokker F-14A (US Model 14)
Fokker F-14B (US Model 14)
Fokker F.XVA
Fokker F.XVB
Fokker F.XVIIA
Fokker F.XVIIB
Fokker F.XVIIC
Fokker F.XVIII
Fokker F.XIX
Fokker F.XX
Fokker F.XXII
Fokker F.XXIII
Fokker F.XXIV
Fokker F.25 Promoter
Fokker F.26 Phantom
Fokker F.27 Series 100 Friendship
Fokker F.27 Series 200 Friendship
Fokker F.27 Series 300 Convertiplane
Fokker F.27 Series 400 Convertiplane
Fokker F.27 Series 400M Convertiplane
Fokker F.27 Series 400M Cartographic
Fokker F.27 Series 500 Long Friendship
Fokker F.27 Series 600 Friendship
Fokker F.27 Maritime
Fokker F.28 Fellowship
Fokker F-32 (US Model 12)
Fokker F.33
Fokker F.XXXVI
Fokker F.XXXVII
Fokker F.39 (License-Built, Avia Czechoslovakia) (See: F.IX)
Fokker F.50
Fokker F.50 King Bird
Fokker F.50 Maritime
Fokker F.50 Maritime Enforcer
Fokker F.50 Sentinel
Fokker F.50 Troopship
Fokker F.LVI
Fokker F.60
Fokker F.70 Executive Jet
Fokker F.80
Fokker F.100, General
Fokker F.130
Fokker XFA-1 (US Model 18)
Fokker F.B.II
Fokker FG-1 Glider
Fokker FLB (See: PJ)
Fokker FT-1 (T.II)
Fokker FT-2 (T.II)
Fokker G Series Attack Aircraft
Fokker G.1
Fokker G.1A
Fokker G.1B
Fokker-Hall (Fokker-Huff) H-51 (US Model 11)
Fokker XJA-1 Super Universal
Fokker K.I (See: M 9)
Fokker XLB-2 (Atlantic Model 5)
Fokker M 2
Fokker M 3
Fokker M 3a
Fokker M 4 Stahltaube (Steel Dove)
Fokker M 5K (A.III)
Fokker M 5K/MG (See: E.I)
Fokker M 5L (A.II)
Fokker M 6
Fokker M 7 (B)
Fokker M 8 (A.I)
Fokker M 9 (K-I)
Fokker M 10E (B.I)
Fokker M 10z
Fokker M 11 (B.II)
Fokker M 12
Fokker M 13
Fokker M 14 (See: E-II, E-III)
Fokker M 15 (See: E-IV)
Fokker M 16E Karausche (Carp)
Fokker M 16Z (B.III)
Fokker M 17
Fokker M 17E
Fokker M 17K (Austria-Hungary B.II)
Fokker M 17 Z
Fokker M 17ZF (See: D.II)
Fokker M 17ZK (See: D.II)
Fokker M 18E
Fokker M 18Z
Fokker M 18ZF (See: D.I)
Fokker M 18ZK (See: D.I)
Fokker M 19F (See: D.III)
Fokker M 19K (See: D.III)
Fokker M 20 (See: D.IV)
Fokker M 20Z
Fokker M 21
Fokker M 22
Fokker M 22E (See: D.V)
Fokker M 22Z (M 20Z)
Fokker XO-27 (US Model 16)
Fokker YO-27 (US Model 16)
Fokker Y1O-27 (US Model 16)
Fokker XO-27A (US Model 16)
Fokker P-1 Partner
Fokker PJ-1 (FLB, Flying Life Boat, US Model 15)
Fokker PJ-2 (FLB, Flying Life Boat, US Model 15)
Fokker PW-5 (V-40, F.VI)
Fokker PW-6 (D.IX)
Fokker PW-7
Fokker RA-1 (TA-1)
Fokker RA-2 (TA-2)
Fokker RA-3
Fokker RA-4 (F-10A)
Fokker S.I (See also: TW-4)
Fokker S.II
Fokker S.IIA
Fokker S.III
Fokker S.IIIW
Fokker S-3 (Atlantic Model 2)
Fokker S.IV
Fokker S.IX/1
Fokker S.IX/2
Fokker S.11 Instructor
Fokker S.12 Instructor
Fokker S.13 Universal Trainer
Fokker S.14 Mach Trainer
Fokker S.14/1 Mach Trainer
Fokker S.14/2 Mach Trainer
Fokker Skeeter (US Model 8)
Fokker Spin (Spider) I
Fokker Spin (Spider) II
Fokker Spin (Spider) III
Fokker Spin (Spider) 1st 1912 Variant
Fokker Spin (Spider) 2nd 1912 Variant
Fokker Spin (Spider) 1st 1913 Variant
Fokker Spin (Spider) 2nd 1913 Variant (M 1)
Fokker Super Universal (US Model 8)
Fokker Super Universal "Virginia"
Fokker T.I
Fokker T.II (See: FT)
Fokker T.III
Fokker T.III-W
Fokker T.IV
Fokker T.IVa
Fokker T.V

Fokker T.VIII-L
Fokker T.VIII-W
Fokker T.VIII-W/C
Fokker T.VIII-W/G
Fokker T.VIII-W/M
Fokker T.IX
Fokker T-2 (F.IV)
Fokker T-2 (F.IV), NASM
Fokker TA-1 (See: RA-1)
Fokker TA-2 (See: RA-2)
Fokker TW-4 (S.I)
Fokker Two-Seat Glider (1922)
Fokker Universal (Atlantic Model 4)
Fokker V.1
Fokker V.2
Fokker V.3
Fokker V.4 (See also: Dr.I)
Fokker V.5
Fokker V.6
Fokker V.7
Fokker V.8
Fokker V.9
Fokker V.10
Fokker V.11
Fokker V.12
Fokker V.13/1
Fokker V.13/2
Fokker V.14
Fokker V.15
Fokker V.16
Fokker V.17
Fokker V.18 (See: D.VII)
Fokker V.19
Fokker V.20
Fokker V.21
Fokker V.22
Fokker V.23
Fokker V.24 (See: D.VIIF)
Fokker V.25
Fokker V.26/1
Fokker V.26/2 (See:D.VIII)
Fokker V.27
Fokker V.28 (See: D.VIII)
Fokker V.29
Fokker V.30
Fokker V.31
Fokker V.32
Fokker V.33
Fokker V.34
Fokker V.35
Fokker V.36
Fokker V.37
Fokker V.38 (See: C.I)
Fokker V.39
Fokker V.40 Fighter (See: PW-5)
Fokker V.40 Sportplane
Fokker V.41
Fokker V.42
Fokker V.43
Fokker V.44 (See: F.I)
Fokker VFW 614
Fokker W.1
Fokker W.2
Fokker W.3
Fokker W.4
Fokker (US) Model 13
Fokker-Republic D.XXIV Alliance (VTOL)

Folkerts, Clayton (Davenport, IA)
Folkerts Mono-Special (1930)
Folkerts Parasol (1928)
Folkerts Special SK-2 "Toots" ("The Foo")
Folkerts Speed King SK-3 "Jupiter - Pride of Lemont"
Folkerts SK-4 (Delbert Bush)
Folkerts Whittenbeck Special (Fordon-Newman Special)
Folkerts #1 Monoplane
Folkerts #2 Monoplane
Folkerts #3 Monoplane
Folkerts #4 Monoplane

Folland Aircraft Ltd (UK)
Folland E.28/40
Folland FO.140
Folland Gnat (FO.141/FO.144)
Folland Midge (FO.139)
Folland 43/37

Fongri (Italy)
Fongri Monoplane

Ford (Detroit, MI) *(See also: Stout)*
Ford ACV (Air-Cushion Vehicle)
Ford Air Coach Monoplane
Ford XB-906 Tri-Motor
Ford C-3 Tri-Motor
Ford C-3A Tri-Motor
Ford C-4A Tri-Motor
Ford C-9 Tri-Motor
Ford Flivver (2-Cylinder Ford)
Ford Flivver (3-Cylinder Anzani)
Ford XJR-1 Tri-Motor
Ford JR-2 (RR-2) Tri-Motor
Ford JR-3 (RR-3) Tri-Motor
Ford RR-4 Tri-Motor
Ford RR-5 Tri-Motor
Ford 2-AT Air Pullman, Air Transport
Ford 2-AT Air Pullman, Air Transport "John Wanamaker"
Ford 2-AT Air Pullman, Air Transport "Maiden Dearborn"
Ford 2-AT Air Pullman, Air Transport "Maiden Dearborn" II
Ford 2-AT Air Pullman, Air Transport "Maiden Dearborn" IV
Ford 2-AT Air Pullman, Air Transport "Maiden Detroit"
Ford 2-F (1928 1-Place Open Land Monoplane)
Ford 3-AT Tri-Motor
Ford 4-AT-A Tri-Motor
Ford 4-AT-A Tri-Motor, Modified 4-AT-1
Ford 4-AT-B Tri-Motor
Ford 4-AT-B Tri-Motor "Floyd Bennett"
Ford 4-AT-C Tri-Motor
Ford 4-AT-D Tri-Motor
Ford 4-AT-E Tri-Motor
Ford 4-AT-1 Tri-Motor
Ford 5-AT-A Tri-Motor
Ford 5-AT-B Tri-Motor
Ford 5-AT-B Tri-Motor "City of Columbus"
Ford 5-AT-B Tri-Motor, NASM
Ford 5-AT-C Tri-Motor
Ford 5-AT-CS Tri-Motor Floatplane
Ford 5-AT-D Tri-Motor
Ford 5-AT-DS Tri-Motor Floatplane
Ford 5-AT-E Tri-Motor
Ford 6-AT-A Tri-Motor
Ford 6-AT-AS Tri-Motor
Ford 6-AT-AS Tri-Motor Floatplane
Ford 7-AT
Ford 8-AT
Ford 9-AT-A Tri-Motor (Modified 4-AT-B)
Ford 11-AT Diesel Tri-Motor
Ford 13-A Tri-Motor (Modified 5-AT-D)
Ford 14-A Tri-Motor

Ford, Jonathan Otto (Detroit, MI)
Ford (Jonathan Otto) 1928 High-Wing Monoplane

Ford-Leigh (Alfredo J Leigh) (Chile & USA)
Ford-Leigh Safety Wing (Modified Bird Model A)

Forlanini, Prof. Enrico (Italy)
Forlanini Hydrofoil Flying Boat Design
Forlanini Steam-Powered Flying Screw Device

Forssman, Villehad (Germany)
Forssman Giant Bomber (See: Siemens-Schuckert)
Forssman Giant Transport (See: Brüning)
Forssman Prinz Sigismund Bulldogge (Bulldog) Monoplane

Fortney, Louis (Oakland, CA)
Fortney 3rd Monoplane (1911)

Foss, Al (Rosemead, CA)
Foss Special "Jinny" (See also: Sorenson "Little Mike")

Foster, Henry C. (Wexford, PA)
Foster Airspeed

Foster, Wikner Aircraft Co, Ltd *(See: Wicko)*

Fouga et Cie (France)
Fouga CM 7
Fouga CM 8
Fouga CM 8-R Sylphe (Sylph), Cyclope (Cyclops)
Fouga CM 10
Fouga C 25 S
Fouga CM 71
Fouga CM 88-R Gémeaux (Gemini)
Fouga CM 100
Fouga CM 101 R
Fouga CM 160
Fouga CM 170 Magister (Schoolmaster)
Fouga CM 171 Makalu
Fouga CM 175 Zéphyr (Zephyr)
Fouga CM 191
Fouga CM 242
Fouga C 301 S
Fouga C 310 P
Fouga CM 311 P
Fouga MC Jalor (Landmark, Beacon)
Fouga Maillet (Hammerhead Shark) (See: SFCA)
Fouga 90

Found Brothers Aviation Ltd (Canada)
Found Model FBA-1A
Found Model FBA-2
Found Model 100 Centennial

Fourney Industries (Carlsbad, NM)
Fourney AirCoupe

Fournier, René (France)
Fournier RF-01
Fournier RF-3
Fournier RF-4
Fournier RF-4D
Fournier RF-5 Sperber (Sparrow Hawk)
Fournier RF-5B Sperber (Sparrow Hawk)
Fournier RF-5D Sperber (Sparrow Hawk)
Fournier RF-6
Fournier RF-9 Motor Glider
Fournier RF-10 Motor Sailplane
Fournier SFS-31 Milan (See: Sportavia)

Fowler, R. G. (San Jose, CA)
Fowler 1910 Glider

Fowler-Gage *(See: Gage)*

Fowlie, Danny A. (Marshall, MO)
Fowlie Phantom

Fox, A. F. (USA)
Fox (A. F.) Biplane

Fox, Frederick L. (Washington, DC)
Fox (Frederick L.) Biplane
Fox (Frederick L.) Kite Glider (1907)
Fox (Frederick L.) Monoplane
Fox (Frederick L.) Monoplane Glider

Fox, Theodore (Dansville, NY)
Fox (Theodore) 1928 Monoplane

Foxjet International (Tony Fox) (Minneapolis, MN)
Foxjet ST/Tri-3 (Foxjet-3)
Foxjet ST-600
Foxjet ST-600S

Frame, Augustus J. (Columbus, OH)
Frame 1928 Monoplane

France-Aviation (France)
France-Aviation Denhaut Flying Boat

Francis & Angell (Lansing, MI)
Francis & Angell Midget Racer
Francis & Angell Whistler

Francis Lombardi & C. *(See: Avia - Italy)*

Frankfort Sailplane Co (Joliet, IL)
Frankfort Cinema I
Frankfort Cinema II (Cinema B)
Frankfort XCG-1 Glider
Frankfort XCG-16 Glider (See: General Airborne Transport)
Frankfort XTG-1
Frankfort TG-1A
Frankfort TG-1A, NASM

Frankfurter Flugzeugbau Max Gerner GmbH *(See: Adler)*

Franklin Aircraft Corp (Franklin, PA)
Franklin Sport
Franklin Sport A
Franklin Sport B (Model Bauer 170)
Franklin Sport Model 65
Franklin Sport Model 70
Franklin Sport Model 85
Franklin Sport Model 90

Franklin Glider Corp (Ypsilanti, MI)
Franklin Glider P-S-2
Franklin Glider P-S-2 "Texaco Eaglet"

Franklin, Moses (Grand Junction, CO)
Franklin (Moses) 1910 Monoplane

Franklin, Wallace H. (Ann Arbor, MI)
Franklin (W. H.) 1929 Preliminary Training Glider

Fratelli Nardi *(See: Nardi)*

Frati, Stelio *(See: Aviamilano)*

Frederick Sage & Co, Ltd *(See: Sage Aircraft)*

Free Air Suspension System *(See: Air Cushion Vehicles)*

Free Flight Aviation Pty, Ltd (Australia)
Free Flight Hornet 130S (Hornet S10)

Freedom Master Corp (Satellite Beach, FL)
Freedom Master FM-2 Air Shark I

Freiberger, Ronald (Kokomo, IN)
Freiberger Ron's I Mod.I

Frenarda (Columbus, OH)
Frenarda Flying Boat

Fréres A. Moreau *(See: Moreau)*

Friedrichshafen (Flugzeugbau Friedrichshafen GmbH) (Germany)
Friedrichshafen C.I
Friedrichshafen D (1917 Quadruplane)
Friedrichshafen D.1
Friedrichshafen FF 1
Friedrichshafen FF 2
Friedrichshafen FF 4
Friedrichshafen FF 7
Friedrichshafen FF 8
Friedrichshafen FF 11
Friedrichshafen FF 17
Friedrichshafen FF 19
Friedrichshafen FF 21
Friedrichshafen FF 29
Friedrichshafen FF 31
Friedrichshafen FF 33
Friedrichshafen FF 33b
Friedrichshafen FF 33c
Friedrichshafen FF 33e
Friedrichshafen FF 33f
Friedrichshafen FF 33h
Friedrichshafen FF 33j
Friedrichshafen FF 33s
Friedrichshafen FF 34
Friedrichshafen FF 35
Friedrichshafen FF 37
Friedrichshafen FF 39
Friedrichshafen FF 40
Friedrichshafen FF 41
Friedrichshafen FF 43
Friedrichshafen FF 44
Friedrichshafen FF 48
Friedrichshafen FF 49b
Friedrichshafen FF 49c
Friedrichshafen FF 59a
Friedrichshafen FF 59b
Friedrichshafen FF 59c
Friedrichshafen FF 60
Friedrichshafen FF 62
Friedrichshafen FF 63
Friedrichshafen FF 64
Friedrichshafen G.I
Friedrichshafen G.II
Friedrichshafen G.III
Friedrichshafen G.IIIa (Oef)
Friedrichshafen G.IV
Friedrichshafen G.V
Friedrichshafen N.I
Friedrichshafen 1913 Biplane (140hp N.A.G.)
Friedrichshafen 1913 Monoplane (70hp Argus)

Friesley Aircraft Corp (Gridley, CA)
Friesley Falcon

Frisbie, John J. (Garden City, NY)
Frisbie 1910 Biplane

Froebe, Doug (Canada)
Froebe Ornicopter

Frost, Edward P. (UK)
Frost Ornithopter (1890)
Frost Second Ornithopter (1902)

Fruba (Flugzeug Reparatur und Bau Anstalt) (Austria-Hungary)
Fruba 1918 Biplane Fighter

Fry, Justus W. (Renton, WA)
Fry 1939 Ornamental Airship Design

Frye Corp (Fort Worth, TX)
Frye F-1 Safari

Fuji Hikoki KK (Fuji Aeroplane Co Ltd) (Japan)
Fuji K5Y (See: Yokosuka)

Fuji Heavy Industries Ltd (Japan)
Fuji (Beech) B-45 Mentor
Fuji FA-200 Aero Subaru
Fuji FA-300
Fuji (Bell) UH-1B (HU-1B)
Fuji KM-2
Fuji KM-2B
Fuji (Cessna) L-19E
Fuji LM-1 Nikko (Beech Mentor)
Fuji T-1
Fuji T-1A
Fuji T-1B
Fuji (Bell) 204B
Fuji (Bell) 205B

Fujinawa, Eiichi (Japan)
Fujinawa Orenco Aeroplane (1921)

Fukunaga Hikoki Seisakusho (Fukunaga Aeroplane Manufacturing Works) (Japan)
Fukunaga Tenryu 3 Trainer (1917)
Fukunaga Tenryu 6 Long-Range Racer
Fukunaga Tenryu 7 Trainer (1921)
Fukunaga Tenryu 8 Trainer (1921)
Fukunaga Tenryu 9 Trainer (1922)
Fukunaga Tenryu 10 Passenger Transport (1922)

Fulda (Germany)
Fulda "Albert" Glider
Fulda FVA Glider

Fuller, Hammond and Assoc (Wilbur Fuller & George Hammond) (Glendale, CA)
Fuller-Hammond Monoplane

Fulton *(See: Continental Inc)*

Funk Aircraft Co (Akron, OH)
Howard and Joseph Funk began designing aircraft in 1933. In 1939 they established the **Akron Aircraft Co**, which closed down a year later. They then established the **Funk Aircraft Co** to continue production and development work. The company halted production during World War II, but resumed again in 1945. When the company closed in 1947, the Funk designs passed into private hands. For all Funk Brothers, Akron, and Funk designs, see **Funk**. (These companies are not related to D. D. Funk Aviation Co, Inc of Salinas, KS: see **Funk Aviation**).
Funk (OH) Model A
Funk (OH) Model B
Funk (OH) Model B, Model E Engine
Funk (OH) Model B75L ("Model L")
Funk (OH) Model B75L ("Model L") Glider Conversion
Funk (OH) Model B85C Bee ("F2B")
Funk (OH) Model CG2 (Glider)
Funk (OH) Model F46A
Funk (OH) Model XC

Funk Aviation (D. D. Funk Aviation Co, Inc) (Salinas, KS)
Funk (KS) F-23 Agricultural Aircraft
Funk (KS) F-23A Agricultural Aircraft
Funk (KS) F-23B Agricultural Aircraft

F.V.D. *(Flugtechnischen Verein, Dresden) (See: Dresden)*

FVM (Sweden)
FVM S 18
FVM S 21
FVM J 23
FVM J 24

F.W.A. (Flugzeugwerke Altenrhein AG) (Switzerland)
F.W.A. AS 32T

G

G & A (Glider & Autogiro) (See: Firestone)

G. Elias & Brothers, Inc (See: Elias)

GA (See: Gippsland Aeronautics Pty Ltd)

GAAC (Worthington, OH)
GAAC Waco II (Ultralight)

Gabardini (Società Incremento Aviazione) (Italy)
Giuseppe Gabardini established **Società Incremento Aviazione** (later **Società Anonima Gabardini per l'Incremento dell'Aviazione** and **Aeronautica Gabardini S.A.**) in 1913. On 1 May 1936, the company was renamed **Construzioni Aeronautiche Novaresi S.A. (C.A.N.S.A.)**. C.A.N.S.A. was absorbed by **Fiat** in 1939, but continued to design and produce aircraft under its own name until forced out of business at the end of World War II. For all pre-1936 designs, including the Lictor series, see **Gabardini**. For all C.A.N.S.A.-originated designs, including all those begun after 1939, see **C.A.N.S.A.**

Gabardini G.8
Gabardini Lictor 90
Gabardini 1910 Hydroaeroplane
Gabardini 1912 Monoplane
Gabardini 1912 Monoplane Hydroaeroplane

Gabriel, Antoni (Poland)
Gabriel Slask (Silesia)

Gabriel Brothers (Pawel & Jan Gabriel) (Poland)
Gabriel P 5
Gabriel P 6
Gabriel P 7

GAC (General Airplanes Corp) (Buffalo, NY)
GAC Model 101 Surveyor
GAC Model 102 Aristocrat
GAC Model 102-A Aristocrat
GAC Model 102-C Aristocrat
GAC Model 102-E Aristocrat
GAC Model 107 Mailplane
GAC Model 111-C Cadet
GAC Model 111-D Sportabout
GAC Model 111-E Postman

Gaffney-Haines (John Robert "Gaff" Gaffney) (Cleveland, OH)
Gaffney-Haines Special "Rusty"

Gage, J. (Los Angeles, CA)
Gage 1912 Tractor Biplane (Fowler Gage), NASM
Gage 1912 Twin Tractor

Gaines, Albert B., Jr (New York, NY)
Gaines 1924 Homebuilt

G.A.L. (General Aircraft Ltd) (UK)
In 1929 **Monospar Wing Co, Ltd** was established to experiment with the strong, light-weight wing structures designed by H. J. Stieger of the **Wm. Beardmore and Co Ltd** aviation department. The success of the Monospar experiments led to the 1934 organization of **General Aircraft Ltd**, absorbing Monospar and its patents. In 1949, General Aircraft merged with **Blackburn Aircraft Ltd** to form **Blackburn and General Aircraft Ltd.** In 1959 Blackburn and General was reorganized as a holding company, **The Blackburn Group Ltd**, with two operating companies: **Blackburn Aircraft Ltd** and **Blackburn Engines Ltd.** Monospar and General Aircraft Ltd designs initiated before the Blackburn merger, or having Monospar (ST) or G.A.L. designation, are filed under **G.A.L.** Blackburn and General designs appear under **Blackburn.**

G.A.L. Monospar ST-3
G.A.L. Monospar ST-4
G.A.L. Monospar ST-6
G.A.L. Monospar ST-10
G.A.L. Monospar ST-11
G.A.L. Monospar ST-12
G.A.L. Monospar ST-18 Croydon
G.A.L. Monospar ST-25 Universal
G.A.L. Monospar ST-25 Universal, Pressurization Tests
G.A.L. Monospar 26
G.A.L. 38 "Fleet Follower"
G.A.L. 41 Monospar
G.A.L. 42 (C. W. Aircraft) Cygnet
G.A.L. 42 Cygnet Mk.II
G.A.L. 45 Owlet
G.A.L. 46
G.A.L. 47 "Field Observation Post"
G.A.L. 48 Hotspur Mk.I
G.A.L. 48 Hotspur Mk.II
G.A.L. 48B Twin Hotspur
G.A.L. 49 Hamilcar Mk.I
G.A.L. 56/01 Medium V
G.A.L. 56/02 Medium U
G.A.L. 56/03 Maximum V
G.A.L. 58 Hamilcar Mk.X
G.A.L. 60 Universal Freighter

Gallaudet...
In 1908, Edson F. Gallaudet established the **Gallaudet Engineering Co** as a consulting firm, later reorganizing the company as the **Gallaudet Co.** In 1917, the company was again reorganized as the **Gallaudet Aircraft Corp** of East Greenwich, RI. Gallaudet left the aviation industry during the 1920s, renting out a portion of its factory space to the newly formed **Consolidated Aircraft Corp** until that company moved its plant to Buffalo, NY. For all aircraft designed or built by the Gallaudet company, see **Gallaudet Aircraft Corp.** For all designs by Edson F. Gallaudet before and after the existence of the Gallaudet company, see **Gallaudet, Edison F.**

Gallaudet Aircraft Corp (East Greenwich, RI)
Gallaudet Aircraft A-1 Bullet
Gallaudet Aircraft B (Monoplane Flying Boat)
Gallaudet Aircraft C-1 Military Tractor Biplane
Gallaudet Aircraft C-2 Military Tractor Biplane
Gallaudet Aircraft C-3 Liberty Tourist
Gallaudet Aircraft CO-1
Gallaudet Aircraft D-1 (A59)
Gallaudet Aircraft D-2
Gallaudet Aircraft D-4
Gallaudet Aircraft D-9
Gallaudet Aircraft DB-1
Gallaudet Aircraft DB-1B
Gallaudet Aircraft DH-4
The British de Havilland D.H.4 was modified and built under license in the United States under the designation DH-4 by several manufacturers. For British-built aircraft, see **de Havilland D.H.4.** For American-built aircraft, see the individual manufacturers, including **Aeromarine, Fokker (Atlantic), Boeing, Dayton Wright, Gallaudet, Keystone,** and **Standard.** When the actual manufacturer cannot be determined from the available evidence, the photos and documents have been filed under **Dayton Wright.**
Gallaudet Aircraft:
DH-4B
E-L-2 Chummy Flyabout
K-4
PW-4
1912 Racing Monoplane
1916 Battle Plane
1916 Military Hydro-Biplane
1922 Navy Monoplane

Gallaudet, Edson F. (Norwich, CT)
Gallaudet (E. F.) Biplane (n.d.)
Gallaudet (E. F.) Kite (1897)
Gallaudet (E. F.) Kite Biplane (1899) (Hydro-Glider)
Gallaudet (E. F.) Inverted Gull-Wing Monoplane (1943)
Gallaudet (E. F.) Skyscraper Biplane

Gallinari (Società Industrie Aeromarittime Gallinari) (Italy)
Gallinari M.G.2

Galway
Galway 1918 Airplane

Gammeter, Harry C. (Bratenahl, OH)
Gammeter 1907 Ornithopter

Gandy-Vrang (Alameda, CA)
Gandy-Vrang 1912 Monoplane

Garaux (France)
Garaux "Airbus"

Garber, Ed (USA)
Garber Penguin (Homebuilt)

Garbrick, Lester (Centre Hall, PA)
Garbrick 1928/29 Biplane

Gardey, Pedro (Argentina)
Gardey 1910 Giroplane

Garrison, Peter (Los Angeles, CA)
Garrison OM-1 Melmoth (Homebuilt)

Garrouste (France)
Garrouste 1923 Glider (Flapping Wings)

Garstecki (Poland)
Garstecki Rywal (Rival) Glider

Gary, William P. (Totowa, NJ)
Gary 1909 Flying Machine ("Hoople")
Gary Flying Barrel Number 1
Gary Flying Barrel Number 2
Gary Flying Barrel Number 3

Gasnier, René (France)
Gasnier No.1 Biplane (1908)
Gasnier No.2 Biplane (1908)
Gasnier No.3 Biplane (1908)

Gasser, L. (Fort Bliss, TX)
Gasser Parasol Monoplane

Gassier (France)
Gassier Sylphe (1911 Monoplane)

Gastambide-Levavasseur (France)
Gastambide-Levavasseur No.1 Biplane (1920)
Gastambide-Levavasseur No.2 Biplane (1920)

Gastambide-Mengin *(See: Antoinette)*

Gasuden (Tokyo Gasu Denki Kogyo KK) (Tokyo Gas & Electrical Industry Co Ltd) (Japan)
Gasuden Koken
Gasuden KR-1
Gasuden KR-2
Gasuden TR-1
Gasuden TR-2
Gasuden Model 1 Trainer
Gasuden Model 2 Trainer
Gasuden Model 3 Trainer

Gates Aircraft Corp (New York, NY)
Ivan R. Gates established the Gates Aircraft Corp in 1929 to build Stampe et Vertongen (RSV) 18 and 26 aircraft.
Gates (RSV) 18-100 (Monoplane)
Gates (RSV) 26-100 (Biplane)

Gates Learjet *(See: Learjet)*

Gates-Day Aircraft Corp *(See: New Standard)*

Gault, Lanford, & Perry (J. L. Gault, C. E. Lanford, J. B. Perry) (Greenville, SC)
Gault, Lanford, & Perry:
PLG-1 (See: Perry, Lanford, & Gault)
GLP-3 (See: Johnson, Luther)

Gaunt, J. (UK)
Gaunt 1911 Biplane
Gaunt 1911 Monoplane

Gavilán (El Gavilán SA) (Colombia)
Gavilán 358

Gazda Engineering (Providence, RI)
Gazda Helicospeeder

GCA (Gruppo Costruzioni Aeronautiche) (Italy)
GCA 1 Pedro
GCA 2 Dumbo
GCA 3 Eta Beta

GECC *(See: Ground Effect Craft Corp)*

Gee Bee *(See: Granville Brothers)*

Geest, Waldemar (Germany)
Geest Möwe (Seagull) Monoplane
Geest 1913 Monoplane
Geest 1916 Single-Seat Fighter

Gefa (Germany)
Gefa 1911 Monoplane

Gehde, W. W. (Fond du Lac, WI)
Gehde Sport Biplane (1927)

Gehrlein, Jay and Rod (USA)
Gehrlein GP-1

Gellet (France)
Gellet "le Gelitas" (Mechanical Bird)

Gemini International, Inc (Sparks, NV)
Gemini Prospector (Ultralight)

Gems *(See: Morane-Saulnier)*

Genair (General Aircraft Pty Ltd) (South Africa)
Genair Aeriel Mk.II (Piel Emeraude Variant)

Genairco (General Aircraft Co Ltd) (Australia)
Genairco 1930 Biplane

General Aeronautical Corp (Cincinnati, OH)
General Aeronautical:
CX-12 Nighthawk
Dietz (Vought) Special (1925)

General Aeronautic Service, Inc (San Francisco, CA)
General Aeronautic Service 1929 Primary Training Glider

General Aeroplane Co (Detroit, MI)
General Aeroplane:
Alpha
Beta (Verville-Type Flying Boat)
Gamma L (Verville-Type Landplane)
Gamma S (Verville-Type Seaplane)

General Airborne Transport Inc (Los Angeles, CA)
General Airborne XCG-16

General Aircraft Co Ltd (Australia) *(See: Genairco)*

General Aircraft Corp (Astoria, NY)
General Aircraft Corp was established in 1941, designing the Skyfarer two-seat aircraft. With the American entry into World War II, General Aircraft switched to experimental work and glider production. In 1943 the company granted exclusive license to the Skyfarer to **Grand Rapids Industries, Inc**, which transferred the license to **LeMars Manufacturing Co** in 1944. LeMars' plans to market the aircraft after the war as the LeMars Skycoupe never came to fruition. For all General Aircraft products, see **General Aircraft Corp**.
General Aircraft Corp G1-80 Skyfarer

General Aircraft Corp (Washington, DC) *(See: Helio)*

General Aircraft Ltd *(See: G.A.L.)*

General Aircraft Pty Ltd (South Africa) *(See: Genair)*

General Airplane Service (Sheridan, WY)
General Airplane Service Model II Agricultural Plane

General Airplanes Corp *(See: GAC)*

General Aviation Corp *(See: General Aviation Manufacturing Corp)*

General Aviation Manufacturing Corp (Dundalk, MD) *(See also: Fokker)*
In 1929, the **General Motors Corp** acquired the stock of the **Fokker Aircraft Corp** and created **General Aviation Corp** as a holding company. In 1930, Fokker Aircraft was renamed **General Aviation Manufacturing Corp**, and, in 1931, Anthony Fokker ended his association with the company. In 1933, General Aviation merged with **North American Aviation Inc** (NAA). In 1934, NAA reorganized as a manufacturing company and took over the General Aviation manufacturing equipment. For aircraft designed before Fokker left the company, see **Fokker**. For aircraft designed before the merger with NAA, see **General Aviation**.
General Aviation GA-43 (Pilgrim 150, Clark GA-43)
General Aviation GA-43A
General Aviation GA-43J (Seaplane)
General Aviation GA-50

General Dynamics Corp (San Diego, CA)
In 1953 the **General Dynamics Corp**, a defense conglomerate which had evolved from the Electric Boatyard, acquired **Consolidated Vultee Aircraft Corp (Convair)**. Convair continued to operate as a division of General Dynamics until 1961, when the Convair facilities were split between the **Convair Division**, in San Diego, and the **Fort Worth Division**, which was primarily responsible for airframe manufacture. The Convair and Fort Worth Divisions were recombined as the **Convair Aerospace Division** in 1970, but were split again in 1974 with responsibilities as before. In 1985, General Dynamics acquired the **Cessna Aircraft Co** as a wholly-owned subsidiary, but never absorbed Cessna's facilities into its own divisional structure before selling it off to **Textron Inc** in 1992. The Fort Worth Division, the only airframe manufacturing portion of General Dynamics, was sold to the **Lockheed Corp**, becoming **Lockheed Fort Worth Co** in 1993. Following the merger of Lockheed and Martin Marietta in 1994, the company was renamed **Lockheed Martin Tactical Aircraft Systems (LMTAS)**. For all aircraft produced by the Convair Division of General Dynamics before the 1961 reorganization, see **Convair**.

For all designs from the Convair (San Diego) Division of General Dynamics post-dating 1961, see **General Dynamics**. For all products of Cessna, see **Cessna**. For all product lines developed by the Fort Worth Division of General Dynamics before the 1993 sale to Lockheed, see **General Dynamics**.

General Dynamics:
YF-16 Fighting Falcon
YF-16/CCV (Control Configured Vehicle)
F-16A Fighting Falcon
NF-16A/AFTI (Advanced Fighter Technology Integration)
F-16B Fighting Falcon
F-16C Fighting Falcon
F-16C CAS/BAI (Close Air Support/ Battlefield Air Interdiction)
F-16D Fighting Falcon
F-16J Fighting Falcon (JASDF)
F-16N Fighting Falcon (USN)
F-16XL Fighting Falcon
F-16/79 Fighting Falcon
F-16/101 Fighting Falcon
F-111 Series
F-111A
F-111A/AFTI (Advanced Fighter Technology Integration)
F-111A/TACT (Transonic Aircraft Technologies)
(Grumman) EF-111A Raven
RF-111A
(Grumman) F-111B
F-111C
F-111D
F-111E
F-111F
F-111K
FB-111A
FB-111H
GETOL (Ground Effect Take-Off and Landing) Proposal
Light Intratheater Transport (LIT)
Supersonic Transport Proposal
Turboprop Ground Attack Proposal
Model 603E (VTXTS Proposal)
Model 640 (Twin-Jet Transport)
Model 660 (Short-Range Transport)
Model 680 (Twin-Jet Transport)
Model 1010 (Twin-Jet Transport)

General Motors Corp, Allison Division *(See: Allison Division of General Motors)*

General Motors Corp, Eastern Aircraft Division (Trenton, NJ)

During World War II, the **General Motors Corp** (GM) converted several of its divisions to aircraft production. The **Eastern Aircraft Division** of GM produced versions of the Grumman F4F and TBF as the FM and TBM (respectively). For all Eastern products, see **General Motors (Eastern)**. For products of the Fisher Body division of GM, see **General Motors (Fisher)**.

GM (Eastern) FM-1 Wildcat
GM (Eastern) FM-1 Wildcat, NASM
GM (Eastern) (FM-1) Wildcat V
GM (Eastern) FM-2 Wildcat
GM (Eastern) FM-2 Wildcat, Civil
GM (Eastern) XF2M-1
GM (Eastern) TBM-1 Avenger
GM (Eastern) TBM-1C Avenger
GM (Eastern) TBM-1 Avenger, Civil
GM (Eastern) TBM-3 Avenger
GM (Eastern) TBM-3D Avenger
GM (Eastern) TBM-3E Avenger
GM (Eastern) TBM-3E2 Avenger
GM (Eastern) TBM-3L Avenger
GM (Eastern) TBM-3N Avenger
GM (Eastern) TBM-3S Avenger
GM (Eastern) TBM-3U Avenger
GM (Eastern) TBM-3W Avenger
GM (Eastern) Avenger TR.Mk.II
GM (Eastern) Avenger AS.Mk.3
GM (Eastern) Avenger AS.Mk.3M
GM (Eastern) Avenger TR.Mk.III
GM (Eastern) Avenger AS.Mk.4
GM (Eastern) Avenger AS.Mk.5
GM (Eastern) Avenger AS.Mk.6
GM (Eastern) Avenger, Civil

General Motors Corp, Fisher Body Division (Flint, MI)

The **Fisher Body Division** of General Motors Corp was involved in aircraft production during both World Wars. For all products of the Fisher division of GM, see **General Motors (Fisher)**. For all products of the Eastern Division of GM, see **General Motors (Eastern)**.

GM (Fisher) XB-39
GM (Fisher) XDH-4L

The British de Havilland D.H.4 was modified and built under license in the United States under the designation DH-4 by several manufacturers. For British-built aircraft, see **de Havilland D.H.4**. For American-built aircraft, see the individual manufacturers, including **Aeromarine**, **Fokker (Atlantic)**, **Boeing**, **Dayton Wright**, **Fisher**, **Gallaudet**, **Keystone**, and **Standard**. When the actual manufacturer cannot be determined from the available evidence, the photos and documents have been filed under **Dayton Wright**.

GM (Fisher) DH-4L (Liberty DH-4B)
GM (Fisher) XP-75 Eagle
GM (Fisher) XP-75A Eagle
GM (Fisher) P-75A, Modified Tail

General Western Aero Corp, Ltd (Burbank, CA) *(See also: Air Transport Manufacturing Co Ltd)*
General Western Meteor

George & Jobling
George & Jobling Biplane (1910)

George D. White Co *(See: White, George D.)*

Geraldson, Gerald (Newcastle, CA)
Geraldson 1910 Aeroplane (Patent)

Gerance des Etablissements Morane-Saulnier *(See: Morane-Saulnier)*

Gérard (France)
Gérard 1784 Ornithopter

Gere, George E. "Bud" (Minneapolis, MN)
Gere Sport Biplane (Homebuilt)

Gerhardt, William Frederick (Dayton, OH)
Gerhardt Cycleplane (Human-Powered Aircraft, 1923)

Germania Flugzeugwerke GmbH (Germany)
Germania Type B
Germania Type C (K.D.D.)
Germania C.IV
Germania Type JM

Gerner, Max *(See: Adler)*

Geske, Gerry (Superior, MT)
Geske Mothbat

Gettins, Edwing G. (Los Angeles, CA)
Gettins Shulz-Type Schooling Glider

GEV (Ground Effect Vehicles) *(See: Air Cushion Vehicles)*

Giannoni, Espartaco (Argentina)
Giannoni E.G. No.1 Parasol

Gibon, Theodor (Clarksville, TN)
Gibon 1902/1903 Aeroplane Patents

Gibson Refrigerator Co (Greenville, MI) *(See: Waco CG-4A)*

Gibson, William Wallace (Canada)
Gibson Multiplane (1911)
Gibson Twin Plane (1910)

Gil-Pazó (Arturo González Gil y José Pazó Montes) (Spain)
Gil-Pazó GP.1

Gilbert, Octave (France)
Gilbert Monoplane

Gilkey, Lynn W. (Newcastle, PA)
Gilkey 1933 Sport Monoplane

Gill-Dosh Airplane Co (Baltimore, MD)
Gill-Dosh 1910 Biplane

Gillis Aircraft Corp (Battle Creek, MI)
Gillis Crusader

Gilmore Brothers (Lyman, Sam, & Harrison Gilmore) (Grass Valley, CA)
Gilmore Brothers Monoplane (c.1910)

Gilpin, Charles W. (Los Angeles, CA)
Gilpin 1925 Six-Passenger Biplane

Gilson, Samuel H. (Salt Lake City, UT)
Gilson Aeroplane (Patent, 1910)

Gippsland Aeronautics Pty Ltd (GA) (Australia)
Gippsland GA-8 Airvan
Gippsland GA-200 AG-Trainer
Gippsland GA-200 Fatman

GIRD (Group for the Study of Jet Propulsion, Jet Propulsion Research Institute) (RNII) (Russia)
GIRD RP-2 (Rocket Plane)

Givaudan (France)
Givaudan No.1

Glaab, Leo C. (Euclid Village, OH)
Glaab 1929 Single-Seat Monoplane

Glasair *(See: Stoddard-Hamilton)*

Glasflügel (Germany)
Glasflügel BS-1
Glasflügel H 101 (See: Start + Flug)
Glasflügel H 201 Standard Libelle
Glasflügel H 201B Standard Libelle
Glasflügel 205 Club Libelle
Glasflügel 206 Hornet
Glasflügel H 301 Libelle
Glasflügel 303 Mosquito
Glasflügel 304
Glasflügel 401 Kestrel
Glasflügel 604

Glen Huffman Aircraft Co *(See: Huffman)*

Glenn L. Martin Co *(See: Martin)*

Glenny and Henderson (UK)
Glenny and Henderson H.F.S.2 Gadfly

Glenview Metal Products Co (Delanco, NJ)
Glenview GMP-1 Skyride

Glideoplane Co (Oklahoma City, OK)
Glideoplane 8

Glider Aircraft Corp (Naples, FL)
Glider Aircraft Corp, General
Glider Aircraft Corp Dragonfly Mk.II
Glider Aircraft Corp Hummingbird Mk.II

Gliders and Sailplanes, General

Gliders, Inc (Gliders, Inc Division of Detroit Aircraft Corp) (Orion, MI)
Gliders, Inc:
Michigander (Primary Training Glider)
Pontoon (Primary Training) Glider
PT-1 Primary Trainer ("Detroit Glider/Gull")
PT-2 Primary Trainer

Global Aircraft Corp (Starkville, MS)
Global GT-3 Global Trainer

Globe Aircraft Corp (Fort Worth, TX)
Originally established as the **Bennett Aircraft Corp** to manufacture aircraft using the Duraloid process, the company reorganized in 1941 as the **Globe Aircraft Corp.** After World War II, Globe began production of a light monoplane, but the company went bankrupt and the rights to the Globe Swift were acquired by **Texas Engineering and Manufacturing Co, Inc** (Temco). For all Bennett and Globe designs, including the Swift, see **Globe.** For all Temco designs, see **Temco.**
Globe Swift GC-1A
Globe Swift GC-1B
Globe Swift 125

Gloster Aircraft Co, Ltd (UK)
The **Gloucestershire Aircraft Co, Ltd** was established in June 1917 as a joint venture of the **Aircraft Manufacturing Co** (**Airco**) and **H. H. Martyn and Co Ltd**, which supplied components for Airco. Immediately after World War I, Gloucestershire acquired the rights to the Nieuport Nighthawk and part of the design staff of the **Nieuport and General Aircraft Co**, providing it an independent engineering staff. In 1926, the company was renamed **Gloster Aircraft Co, Ltd**, primarily because international customers had difficulty pronouncing the original spelling. In 1934, Gloster was acquired by **Hawker Aircraft Ltd**, although it continued to operate under its own name while Hawker merged with a succession of other companies to become the **Hawker Siddeley Group** by 1960. In 1961, Gloster merged with **Sir W. G. Armstrong Whitworth Aircraft Ltd** to form **Whitworth Gloster Aircraft Ltd.** In July 1963, the Hawker Siddeley Group reorganized and Whitworth Gloster became part of the **Avro Whitworth Division** of **Hawker Siddeley Aviation.** The remaining Gloster assets were utilized in non-aviation-related sections of the Hawker Siddeley family. For all Gloucestershire and Gloster designs, see **Gloster.** For designs by Hawker and

its successors, see **Hawker** or **Hawker Siddeley**. For all Armstrong Whitworth designs, see **Armstrong Whitworth**. Whitworth Gloster is not known to have produced any new designs.

Gloster AS.31 Survey
Gloster E.28/39
Gloster E.5/42
Gloster E.1/44
Gloster F.5/34
Gloster F.9/37, Bristol Taurus Engine
Gloster F.9/37, RR Peregrine Engine
Gloster F.9/40 (Meteor Prototypes)
Gloster GA.5 (Javelin Prototypes)
Gloster Gambit (Nakajima A1N1/A1N2 Pattern Aircraft)
Gloster Gamecock I
Gloster Gamecock I, Anti-Flutter Test
Gloster Gamecock II
Gloster Gannet
Gloster Gauntlet Prototype (See: Gloster S.S.18, S.S.19)
Gloster Gauntlet Mk.I
Gloster Gauntlet Mk.II
Gloster Gladiator Mk.I
Gloster Gladiator Mk.II
Gloster Sea Gladiator
Gloster Gnatsnapper Mk.I
Gloster Gnatsnapper Mk.II
Gloster Gnatsnapper Mk.III
Gloster Goldfinch
Gloster Goral
Gloster Gorcock
Gloster Goring
Gloster Grebe Mk.I
Gloster Grebe Mk.II
Gloster Grebe Mk.II, 2-Seat Trainer
Gloster Grouse Mk.I
Gloster Grouse Mk.II
Gloster Guan
Gloster Javelin F(AW).Mk.1
Gloster Javelin F(AW).Mk.2 Prototype
Gloster Javelin F(AW).Mk.2
Gloster Javelin T.Mk.3 Prototype
Gloster Javelin T.Mk.3
Gloster Javelin F(AW).Mk.4
Gloster Javelin F(AW).Mk.5
Gloster Javelin F(AW).Mk.6
Gloster Javelin F(AW).Mk.7
Gloster Javelin F(AW).Mk.8
Gloster Javelin F(AW).Mk.9
Gloster Mars I "Bamel"
Gloster Mars II (See: Gloster Sparrowhawk I)
Gloster Mars III (See: Gloster Sparrowhawk II)
Gloster Mars IV (See: Gloster Sparrowhawk III)
Gloster Mars VI Nighthawk
Gloster Mars X Nightjar
Gloster Meteor Prototypes (See: Gloster F.9/40)
Gloster Meteor F.Mk.1
Gloster Meteor F.Mk.1, Martin Baker Ejection Seat Testbed
Gloster Meteor F.Mk.1, Rolls Royce Trent Turboprop Testbed
Gloster Meteor F.Mk.2
Gloster Meteor F.Mk.3
Gloster Meteor F.Mk.4
Gloster Meteor F.Mk.4:
 Metropolitan Vickers Beryl Testbed
 Rolls RoyceAvon Testbed
 Rolls Royce Derwent Reheat Test
 Rolls Royce Nene Testbed
Gloster Meteor FR.Mk.5
Gloster Meteor T.Mk.7
Gloster Meteor F.Mk.8
Gloster Meteor F.Mk.8:
 Armstrong Siddeley Sapphire Test
 Prone Pilot Testbed
 Rolls Royce Soar Testbed
Gloster Meteor (F.Mk.8) Ground Attack Fighter
Gloster Meteor FR.Mk.9
Gloster Meteor PR.Mk.10
Gloster Meteor NF.Mk.11
Gloster Meteor NF.Mk.12
Gloster Meteor NF.Mk.13
Gloster Meteor NF.Mk.14
Gloster (A-W) Meteor PR.Mk.19
Gloster Sparrowhawk I (Mars II)
Gloster Sparrowhawk II (Mars III)
Gloster Sparrowhawk III (Mars IV)
Gloster SS.18 (Gauntlet Prototype)
Gloster SS.18A (Gauntlet Prototype)
Gloster SS.18B (Gauntlet Prototype)
Gloster SS.19 (Gauntlet Prototype)
Gloster SS.19A (Gauntlet Prototype)
Gloster SS.19B (Gauntlet Prototype)
Gloster SS.37 (Gladiator Prototype)
Gloster TC.33 Bomber-Transport
Gloster TSR.38
Gloster I
Gloster II (Racer)
Gloster III (Racer)
Gloster IIIA (Racer)
Gloster IIIB (Racer)
Gloster IV (Racer)
Gloster IVA (Racer)
Gloster IVB (Racer)
Gloster VI (Racer)

Glowinski, Bronislaw (Poland)
Glowinski 1911 Monoplane

Gluhareff, Michael E. *(See: Vought Gluhareff Dart-Shaped Fighter*

Gnosspelius, Oscar T. (UK)
Gnosspelius 1910 Monoplane
Gnosspelius 1911 Hydro Monoplane
Gnosspelius 1912 Hydro Monoplane
Gnosspelius 1919 Ornithopter
Gnosspelius Gull (1923)

Goddard, Norman A. and Hawkins, K. C. *(See: Goddard/Imperial)*

Goddard/Imperial (Norman A. Goddard & Imperial Airport) (San Diego, CA)
Goddard/Imperial 1927 Monoplane "El Encanto" (Dole Race)

Godwin, Francis (Bishop of Hereford) (UK)
Godwin "Gonsales" Birds ("Man in The Moone," 1638)

Goedecker, J. (Germany)
Goedecker Amphibian (1912)
Goedecker B Type (1915)
Goedecker Monoplane (c.1912)
Goedecker Sturmvogel (Storm Bird)

Goffaux, G. (France)
Goffaux Monoplane

Golaev, A. (Russia)
Golaev 1950 Sports Monoplane

Gold's, Willibald (Poland)
Gold's 1910 Monoplane

Golden Eagle Aircraft Corp (Columbus, OH)
The first Golden Eagle aircraft was built by the **R. O. Bone Co** in 1928/29. Bone reorganized in mid-1929, creating the **Golden Eagle Aircraft Corp**; the company failed during the Depression. All Bone and Golden Eagle products are under **Golden Eagle**.
Golden Eagle Chief
Golden Eagle (Bone) P-1 Single-Seat Sport

Golden Eagle (Bone) P-2 Parasol Monoplane (Sport Monoplane 2P)

Goldhammer, Fred A. (Chicago, IL)
Goldhammer Goldie No.1

Goldwing Ltd (Jackson, CA)
Goldwing Gold Duster (Ultralight)
Goldwing Ltd Goldwing (Ultralight)
Goldwing Nexus (Ultralight)

Gonnel, Arthur & Georges (France)
Gonnel 1911 Uniplan ("l'Aérien")

Gonsales, Domingo *(See: Godwin, Francis)*

Gonzales, Robert (San Francisco, CA)
Gonzales 1912 Biplane

Goodall Aero Co (Chicago, IL)
Goodall Penguin
Goodall Triplane

Goodyear Aircraft Corp (Akron, OH)
Goodyear COIN (Counter-Insurgency Aircraft Proposal)
Goodyear Convoplane
Goodyear FG Corsair, Civil Conversion
Goodyear FG-1 Corsair
Goodyear FG-1A Corsair
Goodyear FG-1D Corsair
Goodyear NFG-1D Corsair
Goodyear FG-3 Corsair
Goodyear (FG) Corsair Mk.IV
Goodyear F2G, Civil Conversion
Goodyear XF2G-1
Goodyear F2G-1
Goodyear F2G-1D
Goodyear F2G-2
Goodyear GA-1 Duck
Goodyear GA-2 Duck
Goodyear GA-2B Duck
Goodyear GA-5
Goodyear GA-13
Goodyear GA-16
Goodyear GA-17
Goodyear GA-22 Drake
Goodyear GA-22A Drake
Goodyear GA-400R Gizmo
Goodyear GA-400T Gizmo
Goodyear Inflatoplane GA-33
Goodyear Inflatoplane GA-447
Goodyear Inflatoplane GA-466
Goodyear Inflatoplane GA-468

Göppingen (Sportflugzeugbau Göppingen Martin Schempp) (Germany)
In 1935, several gliding enthusiasts led by Martin Schempp established **Sportflugzeugbau Göppingen Martin Schempp (Göppingen)** in Göppingen, Germany. In 1938, Wolf Hirth joined Göppingen and the company became **Sportflugzeugbau Schempp-Hirth OHG (Schempp-Hirth)**. In 1940 Hirth established **Wolf Hirth GmbH (Hirth)** as Schempp-Hirth turned more toward subcontract work for other segments of the German aviation industry. Both companies stopped work with the end of World War II, but Hirth began to produce Göppingen designs again in 1951. Schempp-Hirth started producing new designs in 1962. For all Göppingen designs entering production between 1935 and 1938, see **Göppingen**. For all Schempp-Hirth and Hirth designs entering production after 1938, see **Schempp-Hirth**.
Göppingen Gö 1 (Schempp-Hirth Wolf)
Göppingen Gö 3 Minimoa (Sailplane)
Göppingen Gö 4 II (Sailplane)
Göppingen Gö 5 (Modified Hütter 17)
Göppingen Gö 9
Göppingen H-17 (See: Hütter)

Gordon Aircraft Co (Frank Gordon) (Red Bluff, CA)
Gordon A-3 Monoplane (1929)

Gordon and Stryker *(See: Gordon Aircraft Co)*

Gorelov, S. N. (Russia)
Gorelov RG-1

Gossamer Albatross, Condor, and Penguin *(See: MacCready)*

Gotch & Brundage (Gus Gotch & Tom Brundage) (Los Angeles, CA)
Gotch & Brundage Special (1932)

Gotha (Germany)
Gotha 1922 Sailplane

Gotha (Aircraft Dept, Gothaer Waggonfabrik AG) (Germany)
Gothaer Waggonfabrik AG opened an aircraft construction department in 1913, closing it after the end of World War I in 1919. The aircraft department reopened in 1933, closing again following the end of World War II. For clarity, all Gotha products from the first period of operation are listed as first, followed by products from the second period. All are listed as **Gotha**.

Gotha (1913-1919)
Gotha Büchner Biplane
Gotha Büchner Seaplane
Gotha (Ursinus) G.I
Gotha G.II
Gotha G.III
Gotha G.IV
Gotha G.V
Gotha GL.VII (Production)
Gotha GL.VIII
Gotha LD 1
Gotha LD 1a
Gotha LD 2
Gotha LD 3
Gotha LD 5
Gotha LD 6
Gotha LD 6a
Gotha LD 7
Gotha LE 2
Gotha LE 3
Gotha LE 4
Gotha WD 1
Gotha WD 2
Gotha WD 3
Gotha WD 5
Gotha WD 7
Gotha WD 8
Gotha WD 9
Gotha WD 11
Gotha WD 12
Gotha WD 13
Gotha WD 14
Gotha WD 15
Gotha WD 20
Gotha WD 22
Gotha WD 27
Gotha 1913 Biplane Seaplane

Gotha (1933-1945)
Gotha Go 145 A
Gotha Go 145 B
Gotha Go 146
Gotha Go 147
Gotha Go 149
Gotha Go 150
Gotha Go 229 (Horten IX)
Gotha Go 242 A
Gotha Go 242 B
Gotha Go 244
Gotha Go 345

Gotha P.60 A
Gotha P.60 B
Gotha P.60 C

Gould Aviation Co (New York, NY)
Gould O-1 Navy Glider (1942)

Goupil, Alexandre (France)
Goupil 1883 Aeroplane (See also: Curtiss Goupil "Dalton's Duck")

Goupy (Aéroplanes Goupy) (France)
Goupy 1908 No.1 Triplane (Voisin)
Goupy 1909 Biplane (No.2)
Goupy 1911 Biplane (Type Militaire)
Goupy 1912 Hydroaeroplane
Goupy 1913 Type B Hydro
Goupy 1914 Type A
Goupy 1914 Type B

Gourdou-Leseurre (Charles Edouard Pierre Gourdou and Jean Adolfe Leseurre) (Ateliers et Chantiers de la Loire) (France)
Gourdou-Leseurre:
Type a (GL a, GL C1)
Type b (GL b, GL b C1, GL 2 C1)
GL CAP 2 (Type G)
GL 21 C1 (GL B2)
GL 22 ET1 (GL B3)
GL 22 ET2 (GL B5)
GL 23 C1 (GL B4)
GL 23 TS (LGL 23 TS)
GL 24 (GL B6)
GL 32 C1 (See: Loire)
GL 482 C1
GL 633 B1
1923 Retractable Gear Racer
1924 Single-Seat Pursuit
1924 Two-Seat Night Pursuit

Government Aircraft Factory (Australia)
Government Aircraft Factory (Australia):
N-22B Nomad
N-22B Searchmaster L
Jindivik Mk.2
Pika

Grace-N-Air (Bozeman, MT)
Grace-N-Air Vertical Lift Vehicle

Grade, Hans (Germany)
Grade 1909 Monoplane
Grade 1910 Monoplane
Grade 1911 Monoplane
Grade 1912 Monoplane (Schwalbe)
Grade 1914 Military Monoplane

Grahame-White, Claude (Grahame-White Aviation Co, Ltd) (UK)
Grahame-White:
Baby (1910) (See: Burgess)
Experimental Trainer (1917)
GWE.IV Ganymede Trimotor
GWE.6 Bantam (Express Airmail)
GWE.7 Aero-limosine (1918)
GWE.IX Ganymede Bi-motor
Instructional (1919)
New Baby (1911 School Biplane)
Type 6 Military Biplane (1914)
Type 7 Popular (1913)
Type 7c Popular (1913)
Type 8 Hydro-Biplane (1914)
Type 9 Tractor Monoplane (1912)
Type 10 Charabanc (Aerobus) (Type X 5-Seat Passenger Carrier)
Type 11 Warplane (1914)
Type 13 Scout (Circuit Seaplane)
Type 14 Lizzie (Flying Teatray)
Type XV Boxkite(Bi-Rudder 'Bus)
Type XVIII Biplane (1915)
Type XIX (License-Built Breguet V)
Type 20 Experimental Biplane
1913 Pusher Biplane

***Grain** (See: Port Victoria)*

GRAL (Groupement Rouennais d'Aviation Légère) (France)
GRAL 7 1934 Glider
GRAL 8 1934 Glider

***Grand Rapids Industries, Inc** (See: General Aircraft Corp)*

Grandin (France)
Grandin Glider

Granger Brothers (UK)
Granger Brothers Linnet

Grant, Rudolph R. (Norfolk, VA)
Grant-Morse Aerostable
Grant Monoplane

Granville Brothers Aircraft Inc (Gee Bee) (Springfield, MA)
In 1929, the Granville brothers (Ed, Mark, Robert, Tom, and Zantford) established the **Granville Brothers Aircraft Inc** (**Gee Bee**) around Zantford's existing aircraft repair company. After creating several original and controversial aircraft, the company went bankrupt in 1933. While four brothers dispersed in the aircraft industry, Zantford joined with Howell "Pete" Miller and Donald de Lackner, two Gee Bee engineers, to form **Granville, Miller, and De Lackner**, an engineering consulting firm. Zantford died in 1934 and the company declined and folded. For all Granville Brothers aircraft, see **Granville**. For designs by Granville, Miller, and De Lackner, see **Granville, Miller, and De Lackner**.
Granville A ("Gee Bee")
Granville B
Granville C Sportster
Granville C-4 Fourster
Granville C-8 Eightster
Granville D Sportster
Granville E Sportster
Granville Q-1 Ascender
Granville R-1 Super Sportster
Granville R-2 Super Sportster
Granville R-1/2 Super Sportster "Intestinal Fortitude"
Granville X Sportster
Granville Y Senior Sportster (Y-A, YW)
Granville Z Super Sportster "City of Springfield"

Granville, Miller, and De Lackner (GMD) (Springfield, MA)
After the failure of **Granville Brothers Aircraft, Inc** (**Gee Bee**), Zantford Granville joined Howell "Pete" Miller and Donald de Lackner (two of the company's engineers) to form the engineering consulting firm **Granville, Miller, and de Lackner**. The company failed following Granville's death in 1934. For Granville Brothers aircraft, see **Granville**. For all Granville, Miller, and de Lackner designs, see **Granville, Miller, and de Lackner**.
Granville, Miller, and De Lackner:
Aeromobile
Model R-5 International Racer
Model R-6H "QED"/"Conquistador del Cielo"
Vest-Pocket Pursuit

Gray & Talifero (Earl T. Gray & Birney Talifero) (Los Angeles, CA)
Gray & Talifero 1928 4-Place Open Land Biplane

Gray Goose Airways, Inc (Denver, CO)
Gray Goose 1931 Ornithopter

***Grays Harbor Airways Co** (See: Santa Ana Aircraft Co)*

Great Lakes...
In 1929 the Cleveland banking group which had acquired the local assets of the **Glenn L. Martin Co** established **Great Lakes Aircraft Corp** to operate those facilities. Great Lakes became a subsidiary of **Detroit Aircraft Corp** in 1933, but continued to operate under its own name until forced to close in 1934. In 1963, Andrew Oldfield of Cleveland designed a scaled-down version of the Great Lakes 2T-1; he established **Great Lakes Aircraft Co** in 1964 to market the design. The company also updated the original (full-scale) design and began to offer spare parts late in the 1960s. The company opened factories in Enid, OK, and Wichita, KS, in 1974, but closed after 1988. For all products of the original Great Lakes Corp, see **Great Lakes (1929)**. For products of the later Great Lakes Co, including the updated 2T-1, see **Great Lakes (1964)**.

Great Lakes Aircraft Corp (1929) (Cleveland, OH)
Great Lakes (1929):
XBG-1
BG-1
XB2G-1
"Martin74" (Commercial Martin 74)
XSG-1
TG-1
TG-2 (TE-1)
XTBG-1
Model 1 Transport
2T-1, Straight-Wing Prototypes
2T-1
2T-1 Special, Menasco Engine
2T-1 Special, Radial Engine
2T-1A
2T-1A Special
2T-1E Prototype (Cirrus Derby)
2T-1E
2T-1E Special
2T-2 Racer
4A-1 Amphibian

Great Lakes Aircraft Co (1964) (Enid, OK)
Great Lakes (1964):
Model 2T-1A-2
Model X2T-1T (Turboprop)

Green Sky Adventures Inc (Orwell, OH)
Green Sky Zippy Sport

Green, (Greene?) William W. (Niles, MI)
Green (William) 1910/11 Biplane

Greenough, J. J. (USA)
Greenough 1879 Aerobat Patent

Greenwood-Yates (Allan D. Greenwood & George Yates) (Portland, OR)
Greenwood-Yates:
Geodesic-Construction Monoplane
Bi-Craft (1939)

Greer & Robbins (USA)
Greer & Robbins 1911 Parasol Monoplane

Greer, W. M. (New York, NY)
Greer (W.M.) 1917 All-Steel Aircraft

Grega, John W. (Bedford, OH)
Grega GN-1 Aircamper (Homebuilt)

Gregg Aircraft Manufacturing Co (Pueblo, CO)
Gregg Rocket A-75

Gregor
Gregor 5-Seat Aeroplane (c.1910)

Gregor, Michael (New York, NY; Canada)
Gregor (Michael) FDB-1 (See: CCF)
Gregor (Michael) GR-1 (See: Continental, NY)

Gresci, Victor (Argentina)
Gresci Helicopter (1933)

Gressier Aviation Co (New York, NY)
Gressier 1912 Canard Biplane

Gribovsky, V. K (Russia)
Gribovsky G-4
Gribovsky G-5
Gribovsky G-11 (Glider)
Gribovsky G-27
Gribovsky G-28 (TI-28)
Gribovsky G-29

Grigorovich, D. P. (Russia)
Grigorovich DI-3
Grigorovich E-2 (DG-55)
Grigorovich I-Zet
Grigorovich I-ZW
Grigorovich I-1
Grigorovich I-2
Grigorovich I-2*bis*
Grigorovich IP-1
Grigorovich M-5
Grigorovich M-16
Grigorovich M-22
Grigorovich M-23
Grigorovich M-24
Grigorovich MR-2
Grigorovich MR-3
Grigorovich MR-3*bis*
Grigorovich MRL-1
Grigorovich MT-1
Grigorovich MU-2
Grigorovich MUR-1 (MUN-2)
Grigorovich MUR-2
Grigorovich PL-1 (SUVP)
Grigorovich ROM-1
Grigorovich ROM-2
Grigorovich Stal'-MAI
Grigorovich TB-3
Grigorovich TB-5 (TsKB-8)

Grinnell Aeroplane Co (Grinnell, IA)
Grinnell 1915 Tractor Biplane

Grob (Burkhart Grob Flugzeugbau; Grob Werke GmbH & Co, KG) (Germany)
Grob G 102 Astir Club III
Grob G 102 Astir Club IIIB
Grob G 102 Astir CS Jeans
Grob G 102 Astir CS (Club Standard)
Grob G 102 Astir CS-77 (Club Standard)
Grob G 102 Astir Standard II
Grob G 102 Astir Standard III
Grob G 103 Twin Astir
Grob G 103 Acro, Twin II
Grob G 103A Twin II
Grob G 103C Twin III SL (Self-Launch)
Grob G 103C Twin III Acro
Grob G 103T Twin Astir
Grob G 104 Speed Astir II
Grob G 104 Speed Astir IIB
Grob G 109 (Powered Sailplane)
Grob G 109B (Powered Sailplane)

Grob G 115 Heron
Grob GF 200
Grob GF 250
Grob GF 300
Grob GF 350
Grob G 500 Egrett
Grob G 520 Egrett/Strato 1
Grob G 520T Egrett
Grob G 850 Strato 2C
Grob Vigilant T.Mk.1 (RAF G 109B)
Grob Viking T.Mk.1 (RAF G 103A Twin II Acro)

Grokhovsky, P. I. (Russia)
Grokhovsky G-31
Grokhovsky G-37
Grokhovsky G-38
Grokhovsky Inflatable Glider (1936)

Groppius, le E. (Russia)
Groppius GAZ No.5

Gross, Frank R. (USA)
Gross Sky Ghost

Ground Effect Craft Corp (CECC) (Westport, CT)
Ground Effect Craft Corp Flarecraft

Ground Effect Vehicles *(See: Air Cushion Vehicles)*

Group Genesis (USA)
Group Genesis Genesis 1
Group Genesis Genesis 2

Groupement Rouennais d'Aviation Légère *(See: GRAL)*

Grumman (Bethpage, NY)

In December 1929 Leroy Grumman, William Schwendler, and Leon "Jake" Swirbul organized **Grumman Aircraft Engineering Corp**. In 1969, after the company had diversified into a wide variety of non-aviation industries, the company was renamed the **Grumman Corp**, with aviation-related work being centered in the **Grumman Aerospace Corp** subsidiary. In 1973, the company's Gulfstream civil aviation program was merged with **American Aviation Corp** to form **Grumman American Aviation Corp**, while military work continued at Grumman Aerospace. In 1978 Grumman sold its interests in Grumman American to **American Jet Industries (AJI)**, which then purchased the American Aviation interests as well, forming **Gulfstream American Corp** (later **Gulfstream Aerospace Corp**). In 1994, Grumman was acquired by the **Northrop Corp**, during which process Grumman consolidated its holdings and ceased independent airframe design and manufacture. In May 1994 the merger of the two companies was complete and the new company named **Northrop Grumman Corp**. For all Grumman and Grumman Aerospace designs predating 1994 and all Grumman civil designs originating before 1973, see **Grumman**. For all designs post-dating 1994, see **Northrop Grumman**. For all American Aviation designs predating 1973, see **American Aviation**. For all subsequent development of American designs predating 1978 and new designs originating at Grumman American after 1973, see **Grumman American**. For development of all Grumman American designs postdating 1978, see **Gulfstream Aerospace**.

Grumman (AF) XTB3F-1 Guardian
Grumman (AF) XTB3F-1S Guardian
Grumman (AF) XTB3F-2S Guardian
Grumman AF-2S Guardian
Grumman AF-2W Guardian
Grumman (A-6) YA2F-1 Intruder
Grumman A-6A (A2F-1) Intruder
Grumman A-6A (A2F-1) Intruder, Circulation Control Wing
Grumman EA-6A (A2F-1H) Prowler
Grumman A-6B Intruder
Grumman EA-6B Prowler
Grumman A-6C Intruder
Grumman KA-6D Intruder
Grumman A-6E Intruder
Grumman A-6E Intruder, NASM
Grumman A-6F Intruder
Grumman A-6G Intruder
Grumman SA-16A (HU-16A) Albatross
Grumman SA-16A (HU-16A) Albatross, Triphibian Landing Gear
Grumman SA-16B (HU-16B) Albatross
Grumman SA-16B (HU-16B) Albatross, Smithsonian
Grumman SA-16B/ASW (SHU-16B) Albatross
Grumman C-2A Greyhound
Grumman VC-4A Academe (USCG)
Grumman TC-4B (T-41A) Academe
Grumman TC-4C Academe
Grumman XCH-4 Hydrofoil
Grumman XCH-6 Sea Wings Hydrofoil
Grumman "Denison" Hydrofoil
Grumman "Dolphin" Hydrofoil
Grumman E-2A (W2F-1) Hawkeye
Grumman E-2B Hawkeye
Grumman E-2C Hawkeye
Grumman XFF-1
Grumman FF-1
Grumman FF-2
Grumman XF2F-1
Grumman F2F-1
Grumman XF3F-1
Grumman F3F-1
Grumman XF3F-2
Grumman F3F-2
Grumman XF3F-3
Grumman F3F-3
Grumman XF4F-2 Wildcat
Grumman XF4F-3 Wildcat
Grumman F4F-3 Wildcat
Grumman F4F-3A Wildcat
Grumman F4F-3S Wildcat Seaplane ("Wild Catfish")
Grumman XF4F-4 Wildcat
Grumman F4F-4 Wildcat
Grumman XF4F-5 Wildcat
Grumman XF4F-6 Wildcat
Grumman F4F-7 Wildcat
Grumman XF4F-8 Wildcat
Grumman Martlet Mk.I (G-36, G-36A)
Grumman Martlet Mk.II (G-36B)
Grumman (F4F) Wildcat Mk.IV
Grumman XF5F-1 Skyrocket
Grumman XF6F-1 Hellcat
Grumman XF6F-2 Hellcat
Grumman XF6F-3 Hellcat
Grumman F6F-3 Hellcat
Grumman F6F-3 Hellcat, NASM
Grumman F6F-3E Hellcat
Grumman F6F-3K Hellcat (Drone)
Grumman F6F-3N Hellcat
Grumman F6F-3P Hellcat
Grumman XF6F-4 Hellcat
Grumman F6F-5 Hellcat
Grumman NF6F-5 Hellcat
Grumman F6F-5E Hellcat
Grumman F6F-5K Hellcat (Drone)
Grumman F6F-5N Hellcat
Grumman F6F-5P Hellcat
Grumman XF6F-6 Hellcat
Grumman (F6F) Hellcat Mk.I
Grumman (F6F) Hellcat F.Mk.II
Grumman (F6F) Hellcat FR.Mk.II
Grumman (F6F) Hellcat NF.Mk.II

Grumman XF7F-1 Tigercat
Grumman F7F-1 Tigercat
Grumman F7F-1N Tigercat
Grumman F7F-2 Tigercat
Grumman F7F-2D Tigercat
Grumman F7F-2N Tigercat
Grumman F7F-3 Tigercat
Grumman F7F-3N Tigercat
Grumman F7F-3P Tigercat
Grumman F7F-4N Tigercat
Grumman (F7F) Tigercat Mk.I
Grumman XF8F-1 Bearcat
Grumman F8F-1 Bearcat
Grumman F8F-1 Bearcat, Blue Angels
Grumman F8F-1B (F8F-1C) Bearcat
Grumman F8F-1D Bearcat
Grumman F8F-1N Bearcat
Grumman XF8F-2 Bearcat
Grumman F8F-2 Bearcat
Grumman F8F-2 Bearcat "Conquest I," NASM
Grumman F8F-2N Bearcat
Grumman XF9F-2 Panther
Grumman F9F-2 Panther
Grumman F9F-2B Panther
Grumman F9F-2P Panther
Grumman F9F-3 Panther
Grumman XF9F-4 Panther
Grumman F9F-4 Panther
Grumman F9F-5 Panther
Grumman F9F-5 Panther, Blue Angels
Grumman F9F-5P Panther
Grumman XF9F-6 (XF-9F) Cougar
Grumman F9F-6 (F-9F) Cougar
Grumman F9F-6 (F-9F) Cougar, Blue Angels
Grumman F9F-6 (F-9F) Cougar, NASM
Grumman F9F-6P Cougar
Grumman F9F-7 (F-9H) Cougar
Grumman F9F-8 (F-9J) Cougar
Grumman F9F-8 (F-9J) Cougar, Blue Angels
Grumman F9F-8B (AF-9J) Cougar
Grumman F9F-8P (RF-9J) Cougar
Grumman F9F-8T (TF-9J) Cougar
Grumman F9F-8T, Blue Angels
Grumman (F9F) QF-9J Cougar
Grumman XF10F-1 Jaguar
Grumman (F11F) Sapphire Cougar
Grumman F9F-9 (F11F Tiger Prototype)
Grumman F11F-1 (F-11A) Tiger
Grumman F11F-1 (F-11A) Tiger, Blue Angels
Grumman F11F-1F Super Tiger
Grumman F-14A Tomcat
Grumman F-14A Tomcat, Iran
Grumman F-14B (F-14A+) Tomcat
Grumman F-14D Super Tomcat
Grumman G-15 (Argentina J2F)
Grumman G-20(Argentina JF)
Grumman G-21 Goose
Grumman G-21 Goose, NASM
Grumman G-21A (Goose)
Grumman G-21A Retractable Float
Grumman G-21A (Goose), Peru
Grumman G-21B (Goose)
Grumman G-22 "Gulfhawk II"
Grumman G-22 "Gulfhawk II," NASM
Grumman G-23 (CCF) Goblin
Grumman G-32 "Gulfhawk III"
Grumman G-32A "The Red Ship"
Grumman G-37 (F3F-2 Export)
Grumman XG-44 (Widgeon)
Grumman G-44 (Widgeon)
Grumman G-44 (Widgeon) Reengined
Grumman G-44A (Widgeon)
Grumman G-44A (Widgeon) Reengined
Grumman G-58A "Gulfhawk 4"
Grumman G-58A "The Red Ship"
Grumman G-63 Kitten I
Grumman G-65 Tadpole
Grumman G-72 Kitten II
Grumman G-73 Mallard
Grumman G-73 Turbomallard
Grumman G-111 Albatross
Grumman G-159 Gulfstream I
Grumman G-159A Turboliner
Grumman G-164 Ag-Cat Prototypes
Grumman G-164 Ag-Cat
Grumman G-164A Super Ag-Cat A
Grumman G-262 Albatross (Japan)
Grumman G-698 V/STOL (SEMA-X)
Grumman G-1159 Gulfstream II
Grumman GG-1 (FF-1/SF-1 Demonstrator)
Grumman XJF-1 Duck
Grumman JF-1 Duck
Grumman JF-2 Duck (USCG)
Grumman JF-3 Duck
Grumman J2F-1 Duck
Grumman J2F-2 Duck
Grumman J2F-2A Duck
Grumman J2F-3 Duck
Grumman J2F-4 Duck
Grumman J2F-5 Duck
Grumman J2F-6 Duck
Grumman J2F-6 Duck, Mexico
Grumman XJ3F-1 Goose
Grumman J4F-1 Widgeon (USCG)
Grumman J4F-2 Widgeon
Grumman J4F-2 "Petulant Porpoise," NASM
Grumman (J4F) Widgeon Mk.I
Grumman JRF-1 Goose
Grumman JRF-1A Goose
Grumman JRF-2 Goose (USCG)
Grumman JRF-3 Goose (USCG)
Grumman JRF-4 Goose
Grumman JRF-5 Goose
Grumman JRF-5 Goose, Hydroski
Grumman JRF-5G Goose (USCG)
Grumman (JRF) Goose Mk.I
Grumman (JRF) Goose Mk.IA
Grumman (JRF) Goose Mk.II
Grumman OA-9 Goose
Grumman OA-12 Duck
Grumman OA-12A Duck
Grumman OA-12B Duck
Grumman OA-13 Goose
Grumman OA-14 Widgeon
Grumman XP-50
Grumman PG(H)-1 "Flagstaff"
Grumman XSF-1
Grumman SF-1
Grumman XSF-2
Grumman (S-2) XS2F-1 Tracker
Grumman (S-2A) S2F-1 Tracker
Grumman (S-2A) CS2F-1 Tracker (Canada)
Grumman (S-2A) S2F-1E Tracker
Grumman TS-2A (S2F-1T) Tracker
Grumman US-2A Tracker
Grumman S-2B (S2F-1S) Tracker
Grumman US-2B Tracker
Grumman US-2B Tracker, NASM
Grumman (S-2C) S2F-2 Tracker
Grumman RS-2C (S2F-2P) Tracker
Grumman US-2C (S2F-2U) Tracker
Grumman (S-2D) S2F-3 Tracker
Grumman ES-2D Tracker
Grumman S-2E (S2F-3S) Tracker
Grumman S-2F (S2F-1S1) Tracker
Grumman S-2G Tracker
Grumman S-2N Tracker
Grumman (Agusta) S211A (JPATS)
Grumman XSBF-1
Grumman TF-1 (C-1A) Trader
Grumman XTBF-1 Avenger
Grumman TBF-1 Avenger
Grumman TBF-1 Avenger, NASM
Grumman TBF-1C Avenger
Grumman XTBF-3 Avenger
Grumman (TBF-1) Avenger Mk.I
Grumman (UF) XJR2F-1 Pelican (Albatross)
Grumman UF-1 (HU-16C) Albatross
Grumman UF-1G Albatross (USCG)
Grumman UF-2 (HU-16D) Albatross
Grumman UF-2G (HU-16E) Albatross
Grumman (V-1) OF-1 Mohawk
Grumman YOV-1A (YAO-1A) Mohawk
Grumman OV-1A (AO-1A) Mohawk

Grumman OV-1B (AO-1B) Mohawk
Grumman OV-1C (AO-1C) Mohawk
Grumman YOV-1D Mohawk
Grumman OV-1D Mohawk
Grumman WF Tracer Aerodynamic Prototype (Modified TF-1)
Grumman WF-2 (E-1B) Tracer
Grumman X-29A
Grumman X-29A, NASM
Grumman Model 698 (VTOL)
Grumman/Beech Model 730 (VTXTS)
Grumman/BoeingQ/STOL (QuietSTOL)

Grumman American Aviation Corp (Bethpage, NY)

In 1973 **Grumman Corp** merged its Gulfstream civil aviation program with **American Aviation Corp** to form **Grumman American Aviation Corp**. In 1978 Grumman sold its interests in Grumman American to **American Jet Industries (AJI)**. AJI then purchased American Aviation's interests and established **Gulfstream American Corp**. For Grumman civil designs originating before 1973, see **Grumman**. For American Aviation designs originating before 1973, see **American Aviation**. For Grumman American developments of Grumman and American designs, see **Grumman American**. For all development after 1978, see **Gulfstream Aerospace**.

Grumman American AA-1B Trainer
Grumman American AA-5B Tiger
Grumman American TR-2

Grunau (Segel Flugschule Grunau) (Germany)

Grunau Baby 2 (Glider)
Grunau Baby 2B (Glider)
Grunau Baby 3B
Grunau H2PL "Lore" ("Truck")
Grunau "Musterle" ("Model" or "Example")
Grunau 9 (Glider)

Gruneven, Richard Van *(See: Van's Aircraft Inc)*

Gruppo Costruzioni Aeronautiche *(See: GCA)*

Grushin, P. D. (Russia)

Grushin BB-MAI
Grushin Gr-1
Grushin MAI-3 (Sh-Tandem)

Grzeszcyk, Szczepan (Poland)

Grzeszcyk SG-3*bis* Glider
Grzeszcyk SG-3/34 Glider
Grzeszcyk SG-3/35 Glider
Grzeszcyk SG-7 Glider
Grzeszcyk SG-21 Lwów Glider
Grzeszcyk SG-28 Glider
Grzeszcyk SG-38 Glider

Grzeszczyk & Kocjan (Szczepan Grzeszczyk & Antoni Kocjan) (Poland)

Grzeszczyk & Kocjan Mewa (Gull)

Grzmilas, Tadeusz (Poland)

Grzmilas Orkan I (Whirlwind I)
Grzmilas Orkan II (Whirlwind II)

GST *(See: Beriev GST)*

Guangzhou Orlando Helicopters, Ltd (China)

Guangzhou Panda (License-Built Orlando OHA-S-55 Bearcat)

Gudeirdual

Gudeirdual Parachute (Rotor Wing Descent Device, 1947)

Gudkov, M. I. (Russia)

Gudkov Gu-1
Gudkov Gu-82
Gudkov K-37

Guep, Friedrich (Russia)

Guep G-1 (Giup-1)

Guerchais (Avions Guerchais) (France)

Guerchais Light Aircraft (1928)
Guerchais "Stratosphere" (1933)
Guerchais Type 5 (1929)
Guerchais Type 9
Guerchais Type 12 (1931)
Guerchais Type 50 (1932)

Guérin, Jacques (France)

Guérin Varivol Type 1

Guillebeaud (France)

Guillebeaud 1909 Airplane

Guillemin *(See: Blériot Aéronautique Guillemin)*

Gulfstream Aerospace Corp (Savannah, GA)

In 1979 **American Jet Industries (AJI)** purchased **Grumman Corp**'s interests in its **Grumman American Aviation Corp** subsidiary. After purchasing the remaining **American Aviation** interests in the company, AJI formed **Gulfstream American Corp** (later **Gulfstream Aerospace Corp**) from the assets. Gulfstream was purchased by the **Chrysler Corp** in 1985, and by **Allen Paulson** and **Forstmann Little and Co** in 1990. In 1992 Forstmann Little and Co bought out Paulson to become sole owners. For all designs predating 1978, see **Grumman American**, or **Grumman**. For all designs post-dating 1978, see **Gulfstream Aerospace**.

Gulfstream Aerospace:
C-20A Gulfstream III
Gulfstream I Commuter(GAC-159C)
Gulfstream IIB (ASC-300 UPGrade)
Gulfstream III
Gulfstream IV
Gulfstream IV-SP
Gulfstream V
Gulfstream SRA-4
Hustler 500 (See: American Jet)
Peregrine
Gulfstream Aerospace/Sukhoi SSBJ (Supersonic Business Jet)

Gulfstream American *(See: Gulfstream Aerospace)*

Gundersen, Hans (Norway)

Gundersen Airplane (1911 Patent)

Guritzer Van Nes (Austria)

Guritzer Van Nes 1929 Touring Aircraft

Gusty *(See: Limbach)*

Guthier, Roy E. (Chicago, IL)

Guthier 1928 Monoplane

Guttman, Bob (USA)

Guttman Eaglet 1A

Guyon-Cellier-Jaugey (France)

Guyon-Cellier-Jaugey Biplane No.1

Gwinn Aircar Co (Buffalo, NY)

Gwinn Aircar Model I
Gwinn Aircar Model II

Gyrocopter Co, Ltd *(See: Berliner, Emile)*

Gyrodyne Co of America (New York, NY)

Gyrodyne Company of America was established in 1946, purchasing assets of **Helicopters, Inc** (**Bendix Helicopter**) in 1949. Gyrodyne continued development of the Helicopters Inc Model J as the GCA-2. For all Gyrodyne products, see **Gyrodyne**; for pre-1949 development of the GCA-2, see **Bendix Helicopter Model J**.

Gyrodyne GCA-2 Coaxial (Modified Bendix J-2)
Gyrodyne GCA-2A Helidyne Prototype
Gyrodyne GCA-2C GCA Coaxial
Gyrodyne GCA-3 Helidyne
Gyrodyne GCA-5 Gyrodyne
Gyrodyne GCA-7 Helidyne
Gyrodyne GCA-8 Gyroliner
Gyrodyne GCA-9 Helidyne Transport
Gyrodyne XRON-1

Gyronautics, Inc (Ringoes, NJ)

Gyronautics Gyronaut Mk.I

Gzerwinski (Poland)

Gzerwinski S.G.21 Glider
Gzerwinski S.G.28 Glider

H

***H. F. Dorna Co** (See: Dorna)*

***H. W. Wright Co** (See: Wright, H. W.)*

***Habicht de Hentzen** (See: Hentzen)*

Hackenberg, Herman Joseph (East Rutherford, NJ)
Hackenberg 1930 Training Glider

Hadley, C. O. & Blood (Mineola, NY)
Hadley & Blood 1911 Biplane

Haefelin (Germany)
Haefelin 1912 Monoplane

Haessler-Villinger (Helmut Haessler & Franz Villinger) (Germany)
Haessler-Villinger Pedal-Powered Glider

Häfeli (Switzerland)
Häfeli D.H.3
Häfeli D.H.5

Hafner, Raoul (UK)
Hafner Cierva C-30 Autogyro
Hafner R.I Helicopter
Hafner 2nd Model Helicopter
Hafner A.R.III
Hafner A.R.IV (See: Short-Hafner)
Hafner P.D.6 Helicopter
Hafner Rotabuggy
Hafner Rotachute (Rotor Chute)
Hafner Rotatank

Hagfors (ASL Hagfors Aero AB) (Sweden)
Hagfors Opus 280 (See: ARV ARV-1)

Hagofass, Ralph (San Francisco, CA)
Hagofass Glider

Häher (Germany)
Häher 1932 Sport Monoplane

Haig, Larry (USA)
Haig Minibat

***HAL** (See: Hindustan Aeronautics Ltd)*

Halberstadt (Halberstädter FlugzeugWerke GmbH) (Germany)
Halberstadt A.I Taube
Halberstadt B
Halberstadt B.I
Halberstadt B.II
Halberstadt B.III
Halberstadt C.I
Halberstadt C.II
Halberstadt C.III
Halberstadt C.V
Halberstadt C.VII
Halberstadt C.VIII
Halberstadt C.IX
Halberstadt CL.II
Halberstadt CL.IV
Halberstadt CL.IV, NASM
Halberstadt CLS.I
Halberstadt CLS.II
Halberstadt CLS.X
Halberstadt D.I
Halberstadt D.II
Halberstadt D.III
Halberstadt D.IV
Halberstadt D.V
Halberstadt G.I

Hales, George (Long Beach, CA)
Hales Low Wing Model 100 (1932)
Hales Low Wing Model 101 (1932)

Hall, Bert (Lindsay, OK)
Hall (Bert) Model A (1929 2PLM)

Hall, Ernest C. (Warren, OH)
Hall (Ernest C.) 1910 Blériot XI-Type

Hall, J. L. (UK)
Hall (J. L.) Tractor Biplane

***Hall, Randolph Fordham** (See: Cunningham-Hall Aircraft Corp)*

Hall, Robert "Bob" (Springfield Aircraft, Inc) (Springfield, MA)
Hall (Bob) Bulldog
Hall (Bob) Cicada

Hall, Stanley (USA)
Hall (Stanley) Cherokee II
Hall (Stanley) Cherokee RM (John Ree & Terry Miller)
Hall (Stanley) Ibex

Hall, Theodore P. "Ted" (T. P. Hall Engineering Corp) (San Diego, CA)
Hall (T. P.) Convaircar (See: Convair Models 116 and 118)

Hall-Aluminum (Bristol, PA)
Charles Ward Hall Inc formed in 1916 to research the use of aluminum in aircraft design. In 1927 the company became **Hall-Aluminum Aircraft Corp**, which it remained until absorbed by **Consolidated Aircraft Corp** in 1940. Hall's designs are listed under **Hall-Aluminum**.

Hall-Aluminum Air Yacht
Hall-Aluminum Amphibian Fighter
Hall-Aluminum F4C-1 (See: Curtiss)
Hall-Aluminum XFH-1
Hall-Aluminum Monoped
Hall-Aluminum XPH-1
Hall-Aluminum PH-1
Hall-Aluminum PH-2
Hall-Aluminum PH-3
Hall-Aluminum XP2H-1
Hall-Aluminum XPTBH-2

Hallas, Charles (Northville, MI)
Hallas 1929 Monoplane

Halle (Flugzeugwerk Halle GmbH) (Germany)
Halle Fh 104 Hallore

Haller-Hirth Sailplane Corp (Pittsburgh, PA)
Haller-Hirth Haller Hawk
Haller-Hirth Haller Junior Hawk
Haller-Hirth Haller Sparrow Hawk

Halligan Aircraft Co (Beardstown, IL)
The Halligan Plane

Halpin & Huf (Richard Halpin & Tom Huf) (Hatboro, PA)
Halpin & Huf "H & H Special"

Halpin Development Co (Cincinnati, OH) *(See: Metal Aircraft Co)*

Halpin-Graichen *(See: Metal Aircraft Corp)*

Halton *(See: RAF College Halton)*

Halton Aero Club (UK)
Halton HAC-1 Mayfly
Halton HAC-2 Minus
Halton HAC-3 Meteor

Hamble River, Luke and Co (UK)
Hamble River, Luke and Co H.L.1

Hamburger Flugzeugbau GmbH *(See: Blohm und Voss; See also: HFB)*

Hamilton Aero Manufacturing Co (Seattle, WA) *(See: Hamilton, Thomas Foster)*

Hamilton Aero Manufacturing Co Ltd (Canada) *(See: Hamilton, Thomas Foster)*

Hamilton Aerospace (San Antonio, TX)
Hamilton Aerospace (TX) HX-1 Pusher
Hamilton Aerospace (TX) HXT-2 Primary Trainer
Hamilton Aerospace (TX) HX-321 Cross-Country Kit

Hamilton Aircraft Co (Tucson, AZ)
Hamilton (AZ) T-28-R1 Nomair (Military Version)
Hamilton (AZ) T-28-R2 Nomair (T-28A Conversion)
Hamilton (AZ) Westwind Little Liner (Beech D-18 Conversion)
Hamilton (AZ) Westwind II Std (Beech D-18 Conversion)
Hamilton (AZ) Westwind III (Beech D-18 Conversion)
Hamilton (AZ) Westwind IV (Beech D-18 Conversion)

Hamilton Helicopter Inc (Baltimore, MD)
Hamilton (MD) Helicopter

Hamilton Metalplane Co (Milwaukee, WI)
Hamilton Metalplane C-89
Hamilton Metalplane H-18 "Maiden Milwaukee"
Hamilton Metalplane H-19 Silver Eagle
Hamilton Metalplane H-20 Silver Swan
Hamilton Metalplane H-21 Silver Streak
Hamilton Metalplane H-22 Silver Sea-Dan
Hamilton Metalplane H-43 Landplane
Hamilton Metalplane H-44 Seaplane
Hamilton Metalplane H-45 (Wasp)
Hamilton Metalplane H-47 (Hornet)

Hamilton, Al C. (Detroit, MI) *(See: Issoudin)*

Hamilton, Frank (Washington, DC)
Hamilton (Frank) 1893 Airship

Hamilton, Thomas Foster (Seattle, WA; Milwaukee, WI; Canada) *(See also: Hamilton Metalplane)*
Hamilton (Thomas), General
Hamilton (Thomas) 1908 Glider
Hamilton (Thomas) 1910 Aeroplane
Hamilton (Thomas) 1911 Aeroplane
Hamilton (Thomas) 1913 Biplane
Hamilton (Thomas) 1915 Flying Boat
Hamilton (Thomas) 1915 Tractor Biplane
Hamilton (Thomas) 1916 Tractor Biplane (Canada)

Hamilton, Walter (Mineola, NY)
Hamilton (Walter) 1911 Monoplane

Hammond Aircraft Corp (Ann Arbor, MI)
Hammond JH-1 (Y-1 Drone)
Hammond Y-M-95A (Y-1)
Hammond Y-1-M-125A (Y-1)
Hammond Y-1
Hammond Y-125
Hammond Y-150 (Y-1S)
Hammond 100 (Parks P-IH)

Hampshire Aero Club (UK)
Hampshire Aero Club B.H.1 (V. H. Bellamy & R. J. Hilborn) Halcyon

Hampstead Plains Aviation Co *(See: Moisant Monoplane Co)*

Handasyde Aircraft Co (UK)
Handasyde Glider
Handasyde H.2 Monoplane

Handley Page (UK)
Handley Page B (H.P.2) (See: Thompson, William P.)
Handley Page (Reading) Basic Trainer H.P.R.2
Handley Page Bluebird
Handley Page Clive Mk.I (Chitral)
Handley Page Clive Mk.II
Handley Page D
Handley Page E "Yellow Peril" (E/50, H.P.5)
Handley Page F
Handley Page G (G/100, H.P.7)
Handley Page H.P.R.3 Herald
Handley Page H.P.R.5 Herald
Handley Page H.P.R.7 Dart Herald
Handley Page H.P.19 Hanley
Handley Page H.P.21
Handley Page H.P.31 Harrow
Handley Page H.P.32 Hamlet
Handley Page H.P.38
Handley Page H.P.39 Gugnunc
Handley Page H.P.42
Handley Page H.P.43
Handley Page H.P.45

Handley Page H.P.47 (G.4/31)
Handley Page H.P.51
Handley Page H.P.54 Harrow I
Handley Page H.P.54 Harrow II
Handley Page H.P.88
Handley Page H.P.137 Jetstream 1
Handley Page H.P.137 Jetstream 3M
Handley Page H.P.137 Jetstream 200
Handley Page Halifax Prototype
Handley Page Halifax B.Mk.I
Handley Page Halifax B.Mk.II
Handley Page Halifax G.R.Mk.II
Handley Page Halifax A.Mk.III
Handley Page Halifax B.Mk.III
Handley Page Halifax B.Mk.IV
Handley Page Halifax A.Mk.V
Handley Page Halifax B.Mk.V
Handley Page Halifax G.R.Mk.V
Handley Page Halifax B.Mk.VI
Handley Page Halifax G.R.Mk.VI
Handley Page Halifax A.Mk.VII
Handley Page Halifax B.Mk.VII
Handley Page Halifax C.Mk.VIII
Handley Page Halifax A.Mk.IX
Handley Page Halton
Handley Page Hamilton
Handley Page Hampden Prototype
Handley Page Hampden B.Mk.I
Handley Page Hampden T.B.Mk.I
Handley Page Hampden B.Mk.II
Handley Page Hampstead
Handley Page Handcross
Handley Page Hanley
Handley Page Hare
Handley Page Hastings Prototype
Handley Page Hastings C.Mk.1
Handley Page Hastings C.Mk.1A
Handley Page Hastings Met.Mk.1
Handley Page Hastings C.Mk.2
Handley Page Hastings C.Mk.4
Handley Page Hastings T.Mk.5
Handley Page Hendon
Handley Page Hereford
Handley Page Hermes 1
Handley Page Hermes 2
Handley Page Hermes 4
Handley Page Hermes 5
Handley Page Heyford I
Handley Page Heyford IA
Handley Page Heyford II
Handley Page Heyford III
Handley Page Hinaidi Prototype
Handley Page Hinaidi I
Handley Page Hinaidi II
Handley Page Hyderabad
Handley Page L (L/200, H.P.8)
Handley Page Manx
Handley Page Marathon 1
Handley Page Marathon 1A
Handley Page Marathon 1B
Handley Page Marathon 1C
Handley Page Marathon T.Mk.11
Handley Page O/100
Handley Page O/400 (See also: Standard)
Handley Page O/400 "Langley" (See: Standard)
Handley Page R/200
Handley Page Slotted Monoplane X.4B
Handley Page V/1500
Handley Page Victor Prototype
Handley Page Victor B.Mk.1
Handley Page Victor K.Mk.1
Handley Page Victor B.Mk.1A
Handley Page Victor K.Mk.1A
Handley Page Victor B(K).Mk.1A
Handley Page Victor B.Mk.2
Handley Page Victor B./S.R.Mk.2
Handley Page Victor K.Mk.1A
Handley Page Victor, Civil Airliner
Handley Page W.8 (H.P.18)
Handley Page W.8a (H.P.18)
Handley Page W.8b (H.P.18)
Handley Page W.8c (H.P.18)
Handley Page W.9
Handley Page W.10

Hanes, Arnold L. (Los Angeles, CA)
Hanes (Arnold L.) Special "Hornet"

Hanes, John W. (USA)
Hanes (John W.) Homebuilt

Hang Gliders, General

Hannoversche Waggonfabrik AG (Hannover, Hannoveraner, Hawa) (Germany)
Hannover C.II
Hannover C.IV
Hannover CL.II
Hannover CL.IIIa
Hannover CL.IIIb
Hannover CL.V
Hannover F 3
Hannover F 6
Hannover F 10
Hannover F 12
Hannover Grief (Condor) Sailplane
Hannover Vampyr (Vampire)
Hannover Zschach R-Plane Flying Boat Project
Hannover 1921 Sailplane

Hanriot (Avions Hanriot & Co) (France)
French pioneer **René Hanriot** produced several aircraft in the years prior to World War I. During the war he created a new company with chief designer Pierre Dupont (designating their aircraft with "HD" prefixes). In 1930 the company became a division of **Société Générale de Aéronautiques (SGA)**; the 1933 breakup of SGA saw Hanriot reorganize as **Compagnie des Avions Hanriot**. In 1936 Hanriot merged with Farman into **SNCAC (Société Nationale de Constructions Aéronautiques du Centre)**, continuing to design aircraft with Farman and Hanriot designations. The Hanriot name disappeared during World War II. All Hanriot designs are found under **Hanriot**. Note that varied prefixes have been seen with several aircraft designations: H, HD, and LH (Lorraine-Hanriot) have all appeared in print.
Hanriot HD 1
Hanriot HD 2
Hanriot HD 3
Hanriot HD 6
Hanriot HD 7
Hanriot HD 9
Hanriot LH 10
Hanriot LH 11
Hanriot H 12 (HD 12, LH 12)
Hanriot LH 13
Hanriot H 14CR
Hanriot H 14S
Hanriot HD 14
Hanriot H 15 (HD 15)
Hanriot H 16
Hanriot HD 17
Hanriot HD 18
Hanriot H 19 Et2 (HD 19 Et2)
Hanriot LH 21 S
Hanriot HD 22 Coupe Deutsch Racer
Hanriot HD 24
Hanriot HD 24 Type T.O.E.
Hanriot H 25 (HD 25)
Hanriot H 26
Hanriot H 27
Hanriot HD 28 S
Hanriot LH 30
Hanriot H 31
Hanriot H 32 (HD 32)
Hanriot H 32 (HD 32)
Hanriot H 33
Hanriot H 34 (HD 34)
Hanriot H 35
Hanriot H 36
Hanriot H 38
Hanriot H 41 (LH 41 Racer)
Hanriot H 43

Hanriot H 46 (Styx)
Hanriot HD 54
Hanriot LH 61
Hanriot-Biche H 110
Hanriot LH 130
Hanriot H 131
Hanriot H 161
Hanriot H 170
Hanriot H 175
Hanriot H 180
Hanriot H 181
Hanriot H 182
Hanriot H 195
Hanriot H 220
Hanriot H 230
Hanriot H 232
Hanriot H 320
Hanriot H 410
Hanriot LH 420
Hanriot H 431 (LH 431)
Hanriot LH 437
Hanriot H 510
Hanriot H 600
Hanriot 1907 Monoplane
Hanriot 1909 Monoplane
Hanriot 1910 A-1 Monoplane
Hanriot 1910 II Monoplane
Hanriot 1910 A-3 Monoplane "Libellule" ("Dragonfly")
Hanriot 1911 Monoplane
Hanriot 1912 Monoplane

Hansa-Brandenburg *(See: Brandenburg)*

Hanschke, Carl (Germany)
Hanschke 1921 Helicopter

Hanuschke, Bruno (Germany)
Hanuschke 1913 Monoplane

Harbin Aircraft Manufacturing Corp (China)
Harbin H-5 (Hong-5) *Beagle* (Il-28)
Harbin HJ-5 (Hongjiao-5) *Mascot* (Il-28U)
Harbin HZ-5 (Hongzhen-5) *Beagle* (Il-28R)
Harbin PS-5
Harbin Y-11 (Yun-11)
Harbin Y-12 (Yun-12)
Harbin Y-12I (Y-11T1)
Harbin Z-5 (Zhi-5) (Mil Mi-4)
Harbin Z-6 (Zhi-6)
Harbin Z-9 (Zhi-9) Haitun (Dolphin)

Hargrave (USA)
Hargrave (USA) Multiplane Hydro

Hargrave, Lawrence (Australia)
Hargrave (Australia) 1888 Compressed-Air Ornithopter

Harlan Werke GmbH (Germany)
Harlan 70 PS Eindecker (Monoplane)
Harlan 100 PS Military Eindecker
Harlan 125 PS Military Eindecker

Harlow Aircraft Co (Max B. Harlow) (Alhambra, CA)
Note that the letters in Harlow designations stand for Pasadena Junior College (PJC-), Pasadena College (PC-), or Pasadena City College (PCC-).
Harlow UC-80 ((PJC-2)
Harlow PJC-1
Harlow PJC-2
Harlow PJC-4
Harlow PC-5
Harlow PC-5A
Harlow PC-6
Harlow PCC-10 (See also: ATLAS H-10)

Harper Aircraft Co (J. L. Harper) (Cleveland, OH)
Harper Low Wing Monoplane

Harriman Aeromobile Co (J. Emery Harriman, Jr) (Boston, MA)
Harriman (Aeromobile Co):
Aerocar (1910)
Aeromobile (Two Linked Aerocars)
Flying-Machine (1904)
Hydro Aerocar "SeaLandAir" (1912)

Harriman, H. E. (Mineola, NY)
Harriman (H. E.) Multi-Wing (1919)

Harris, Richard H. (Atlanta, GA)
Harris Parasol (1928)

Harrison, John W. (St Louis, MO)
Harrison 1911 Aircraft Patent

Hartman, Arthur J. (Burlington, IA)
Hartman (Arthur J.) 1910 Monoplane

Hartman, Emiel (UK)
Hartman (Emiel) 1959 Ornithopter

Hartwig Industries (San Antonio, TX)
Hartwig Little 'Copter

Hartzell Walnut Propeller Co (Piqua, OH)
Hartzell C-2
Hartzell FC-1
Hartzell FC-2
Hartzell MR-1

Harvard Aeronautical Society (Cambridge, MA)
In 1910 the **Harvard Aeronautical Society** became the first university aeronautical club to design, build, and fly an original aircraft. The aircraft was largely designed, and later flown, by James V. Martin. For all aircraft built by the Harvard Aeronautical Society, see **Harvard**. For later work by James V. Martin, see **Queen** and **Martin, J. V.**
Harvard No.1 (1910)
Harvard 1910 Glider

HarvEd Aircraft Co (Racine, WI)
HarvEd 1928 Monoplane

Hastings (Reed Hastings Aeronautics Co) (Hastings, NE)
Hastings JCH-1

Hatfield, Milton (Elkhart, IN)
Hatfield "Little Bird" Flying Wing

Hatz (Versailles, KY)
Hatz CB-1

Haufe, Walter H. (Neenah, WI)
Haufe HA-S-3 Hobby
Haufe Buzzer 2
Haufe Dale Hawk 2
Haufe Hawk 4

HAWA *(See: Hannoversche Waggonfabrik AG)*

Hawk Industries (Ernie Hawk) (Yucca Valley, CA)
Hawk Gafhawk (General Aviation Freight) 125
Hawk Minihawk

Hawker (UK)
H.G. Hawker Engineering Co, Ltd formed in 1920 succeeding **Sopwith Co.** The name **Hawker Aircraft Ltd** was adopted in 1933. **Hawker-Siddeley Aircraft Company, Ltd** was formed as a public holding company in 1935 with **Hawker Aircraft Ltd** as a subsidiary. Hawker-

Siddeley became the **Hawker Siddeley Group, Ltd** in 1948 with no change to Hawker Aircraft Ltd. In 1959 **Hawker Siddeley Aviation Ltd** was created by the group to control the many aircraft organizations. In July 1963 Hawker Siddeley Aviation reorganized its subsidiary companies into three divisions, one of which was the **Hawker Blackburn Division**. Hawker Siddeley Aviation centralized its organization in 1965, eliminating the three divisions. **British Aerospace** formed in April 1977, absorbing Hawker Siddeley Aviation Limited. Most Hawker aircraft are listed as **Hawker**, with later designs also appearing in the general files of **Hawker Siddeley** and **British Aerospace**.

Hawker Audax
Hawker Cygnet
Hawker Danecock
Hawker Dantorp
Hawker Demon
Hawker Duicker
Hawker F.20/27 Interceptor
Hawker F.9/35
Hawker F.18/37 Centaurus
Hawker Fury (Biplane)
Hawker Fury (F.2/43)
Hawker Hart
Hawker Hardy
Hawker Harrier (Biplane)
Hawker Hartbees (Hartbee, Hartebeeste)
Hawker Hawfinch
Hawker Hector
Hawker Hedgehog
Hawker Henley
Hawker Heron
Hawker Hind
Hawker Hoopoe
Hawker Horizon (See: Raytheon)
Hawker Hornbill
Hawker Hornet
Hawker Horsley
Hawker Hotspur
Hawker Hunter
Hawker Hurricane
Hawker Hurricane Mk.I
Hawker Hurricane Mk.II
Hawker Hurricane Mk.IIc
Hawker Hurricane Mk.IIc, NASM
Hawker Hurricane Mk.IV
Hawker Hurricane Mk.V
Hawker Hurricane Mk.X
Hawker N.7/46 (Sea Hawk Prototype)
Hawker Nimrod
Hawker Osprey
Hawker P.1005 (B.11/41)
Hawker P.1040
Hawker P.1052
Hawker P.1081
Hawker P.1127
Hawker P.V.3 (F.7/30)
Hawker P.V.4 (G.4/31)
Hawker Sea Fury
Hawker Sea Hawk
Hawker Sea Hurricane, Hurricat
Hawker Tempest Mk.I
Hawker Tempest Mk.II
Hawker Tempest Mk.V
Hawker Tempest Mk.VI
Hawker Tomtit
Hawker Tornado
Hawker Turret Demon
Hawker Typhoon
Hawker Typhoon Mk.I
Hawker Typhoon Mk.IA
Hawker Typhoon Mk.IB
Hawker Woodcock

Hawker Siddeley (UK)

H.G. Hawker Engineering Co, Ltd formed in 1920 succeeding **Sopwith Co.** The name **Hawker Aircraft Ltd** was adopted in 1933. **Hawker-Siddeley Aircraft Company, Ltd** was formed as a public holding company in 1935 with **Hawker Aircraft Ltd** as a subsidiary. Hawker-Siddeley became the **Hawker Siddeley Group, Ltd** in 1948 with no change to Hawker Aircraft Ltd. In 1959 **Hawker Siddeley Aviation Ltd** was created by the group for control of the many aircraft organizations. In July 1963 Hawker Siddeley Aviation reorganized its subsidiary companies into three divisions, one of which was the **Hawker Blackburn Division.** Hawker Siddeley Aviation centralized its organization in 1965, eliminating the three divisions. **British Aerospace** formed in April 1977, absorbing Hawker Siddeley Aviation Limited. Most Hawker aircraft are listed as **Hawker**, with later designs also appearing in the files of **Hawker Siddeley** and **British Aerospace**.

Hawker Siddeley Andover C.Mk.1
Hawker Siddeley Andover CC.Mk.2
Hawker Siddeley Argosy
Hawker Siddeley Dominie T.Mk.1 (HS.125)
Hawker Siddeley Harrier GR.Mk.1
Hawker Siddeley Harrier T.Mk.2
Hawker Siddeley Harrier GR.Mk.3
Hawker Siddeley Harrier GR.Mk.7 (Harrier II Plus)
Hawker Siddeley (Harrier) AV-8A
Hawker Siddeley (Harrier) AV-8S (VA.1) Matador (Spain)
Hawker Siddeley (Harrier) TAV-8S (VAE.1) Matador (Spain)
Hawker Siddeley HS.121 Trident 1
Hawker Siddeley HS.121 Trident 2
Hawker Siddeley HS.121 Trident 3
Hawker Siddeley HS.125 Family

Based on **de Havilland**'s initial design for the D.H.125, the HS.125 was developed and produced by **Hawker Siddeley. Beechcraft Hawker Corp**, a US-based **Beech Aircraft Corp** subsidiary, was formed in 1970 to help market, assemble, and help develop the aircraft as the BH 125. Beechcraft Hawker continued in this role after Hawker Siddeley was absorbed by **British Aerospace** in 1977. In June 1993 **Raytheon Co** purchased **British Aerospace's Corporate Jet Division**, merging this with **Beech Aircraft Corp** into a new **Raytheon Aircraft Co** in 1994. Early HS.125/BH 125 versions, through the 700-series, are listed under **Hawker Siddeley**. Versions from the 800- through 1000-series are listed under **British Aerospace**. The Horizon and all subsequent developments of the 125 are listed under **Raytheon**.

Hawker Siddeley HS.125 (Beechcraft Hawker BH 125)
Hawker Siddeley HS.125-1 (D.H.125 Series 1)
Hawker Siddeley HS.125 Series 2
Hawker Siddeley HS.125 Series 3
Hawker Siddeley HS.125-400 (Beechcraft Hawker BH 125-400)
Hawker Siddeley HS.125-600 (Beechcraft Hawker BH 125-600)
Hawker Siddeley HS.125-700 (Beechcraft Hawker BH 125-700)
Hawker Siddeley HS.141 V/STOL
Hawker Siddeley HS.681
Hawker Siddeley HS.748
Hawker Siddeley HS.748 Series 1
Hawker Siddeley HS.748 Series 2
Hawker Siddeley HS.748 Coastguarder
Hawker Siddeley Kestrel FGA.Mk.1
Hawker Siddeley (Kestrel) XV-6A
Hawker Siddeley (Kestrel) XV-6A, NASM
Hawker Siddeley Nimrod MR.Mk.1/2

Hawker Siddeley Nimrod AEW.Mk.3
Hawker Siddeley Sea Harrier FRS.Mk.1
Hawker Siddeley Sea Harrier FRS.Mk.2

Hawks (Frank) HM-1 "Time Flies"
(See: Miller, Howell)

Hawley *(See: Bowlus)*
Hay Aircraft Co (Los Angeles, CA)
Hay Commercial Supersonic Transport

Hayden Bushmaster 2000 Trimotor
(See: Stout Bushmaster 2000)

HCA (Helicopter Corp of America) (Long Island, NY)
HCA (de Bothezat) 1940 Helicopter

Heafey & Heafey Air Ambulance
(See: Stinson SM-8A)

Heath Aircraft Corp (Heath Airplane Co) (Chicago, IL)
Heath Baby Bullet
Heath Cannon Ball Special
Heath Favorite (1923)
Heath Feather Biplane (1914)
Heath Flying Boat
Heath High Wing Monoplane
Heath Center Wing Monoplane
Heath Parasol (1927)
Heath Parasol (1929)
Heath Super Parasol
Heath Sport Plane
Heath Standard
Heath Super Standard
Heath Super Standard Biplane
Heath Tomboy (1926)
Heath 1908 Light Plane

Hecox, L. R. (Bozeman, MT)
Hecox LH1

Hees (France)
Hees 1923 Monoplane Glider

Hegy, Raymond C. (Hartford, WI)
Hegy Chupirosa

Hegy & Zunker (Raymond C. Hegy & Norman W. Zunker) (Hartford, WI)
Hegy & Zunker Special (1929)

Heidenrich (Germany)
Heidenrich 1910 Monoplane

Heine (Steffen Hielaler) (Germany)
Heine 1910 Monocoupe

Heinemann, Gernot W. (Bellingham, WA)
Heinemann GH-1 Sport (Modified Heath Super Parasol)
Heinemann Mosquito
Heinkel (Germany)
Before the assignment of aircraft manufacturers classifications by the Reich Air Ministry in 1934, Heinkel aircraft were designated by a works number preceded by either "HE" ("Heinkel Eindecker" - monoplane), or "HD" ("Heinkel Doppeldecker" - biplane). Beginning with He 46, all Heinkel aircraft used the Air Ministry's designation for Heinkel, "He". Heinkel aircraft are listed here in numerical order.
Heinkel HE 1
Heinkel HE 2
Heinkel HE 3
Heinkel HE 4
Heinkel HE 5
Heinkel HE 5E
Heinkel HE 6
Heinkel HE 7
Heinkel HE 8
Heinkel HE 9
Heinkel HE 10
Heinkel HE 12
Heinkel HD 14
Heinkel HD 15
Heinkel HD 16
Heinkel HD 17
Heinkel HE 18
Heinkel HD 19
Heinkel HD 20
Heinkel HD 21
Heinkel HD 22
Heinkel HD 23
Heinkel HD 24
Heinkel HD 29
Heinkel HE 31
Heinkel HD 32
Heinkel HD 33
Heinkel HD 35
Heinkel HD 36
Heinkel HD 37
Heinkel HD 38
Heinkel HD 39 (He 39)
Heinkel HD 40 (He 40)
Heinkel HD 42
Heinkel HD 43
Heinkel HD 44
Heinkel HD 45
Heinkel He 46
Heinkel He 46 E
Heinkel He 46 F
Heinkel He 51
Heinkel HD 55
Heinkel He 57 Heron
Heinkel He 58
Heinkel He 59
Heinkel He 59 B
Heinkel He 60
Heinkel He 60 a
Heinkel He 60 A
Heinkel He 60 C
Heinkel He 60 D
Heinkel He 63 (Prototype)
Heinkel He 63 L
Heinkel He 64
Heinkel He 64 B
Heinkel He 64 C
Heinkel He 66 (He 50 for China)
Heinkel He 70 Blitz (Lightning) Prototypes
Heinkel He 70 Blitz (Lightning)
Heinkel He 70 A Blitz (Lightning)
Heinkel He 70 B Blitz (Lightning)
Heinkel He 70 D Blitz (Lightning)
Heinkel He 70 D-0 Blitz (Lightning)
Heinkel He 70 E Blitz (Lightning)
Heinkel He 70 G Blitz (Lightning)
Heinkel He 70 K Blitz (Lightning)
Heinkel He 71
Heinkel He 71 B
Heinkel He 72 Kadett (Cadet)
Heinkel He 72B-3 Edelkadett (Senior Cadet)
Heinkel He 74
Heinkel He 100
Heinkel He 100 D
Heinkel He 100 D ("He 113")
Heinkel He 100 V Types (Prototypes)
Heinkel He 111 A
Heinkel He 111 B
Heinkel He 111 C
Heinkel He 111 D
Heinkel He 111 E
Heinkel He 111 F
Heinkel He 111 G
Heinkel He 111 H
Heinkel "He 111 K"
Heinkel He 111 L
Heinkel He 111 P
Heinkel He 111 V Types (Prototypes)
Heinkel He 111 Z (Zwilling, Twin)
Heinkel He 112
Heinkel He 112 B

Heinkel He 112 V Types (Prototypes)
Heinkel "He 113" (See: He 100D)
Heinkel He 114
Heinkel He 115
Heinkel He 115 V Types (Prototypes)
Heinkel He 116
Heinkel He 118
Heinkel He 118 V Types (Prototypes)
Heinkel He 119
Heinkel He 119 V Types (Prototypes)
Heinkel He 162 Spatz (Sparrow), Volksjäger (People's Fighter)
Heinkel He 162 A, NASM
Heinkel He 170
Heinkel He 172
Heinkel He 176
Heinkel He 177 Greif (Griffin)
Heinkel He 178
Heinkel He 219 Uhu (Owl)
Heinkel He 219 Uhu (Owl), NASM
Heinkel He 270
Heinkel He 274
Heinkel He 277 V1 Greif (Griffin)
Heinkel He 280
Heinkel He 343
Heinkel He 535
Heinkel Lerche II (Lark II)
Heinkel Magister CM.170 (See: Fouga Magister)
Heinkel P.1076
Heinkel P.1077 "Julia"
Heinkel P.1078
Heinkel P.1079
Heinkel P.1080

Heinonen, Juhani (Finland)
Heinonen HK-1
Heinonen HK-2

Heinrich (Germany)
Heinrich (Germany) Biplane
Heinrich (Germany) Motor Plane
Heinrich (Germany) Rocket Plane
Heinrich (Germany) Tailless Rocket Plane
Heinrich (Germany) Ultra Light Biplane

Heinrich Aeroplane Co, Inc (Baldwin, NY)
Heinrich (Aeroplane Co):
1909 Monoplane
1911 Model A Monoplane
1914 Model B Military Monoplane
1914 Model D Monoplane
1915 Model C Monoplane
1915 Model E Military Tractor
1915 Model E2 Military Tractor Scout (See: Victor Aircraft Corp)

Heinrich Brothers (Albert S. & Arthur O.) *(See: Heinrich Aeroplane Co, Inc)*

Heintz, Chris *(See: Zenair)*

Heit Parasol *(See: Krupup, Kay)*
Heitmann, Heinrich (Germany)
Heitmann Type I Glider
Heitmann Type II Glider
Heitmann Type III Glider
Heitmann Type IV Glider
Heitmann Type V Glider
Heitmann Type VI Glider

Helicopter Corp of America *(See: HCA)*

Helicopter Engineering Research Corp *(See: Jovair)*

Helicopter Technik Wagner *(See: Wagner)*

Helicopters, Inc *(See: Bendix Helicopter)*

Helicopters, General

Helio Aircraft Corp (Norwood, MA)
Helio H-250 Courier Mark II (Caballero)
Helio H-295 Super Courier
Helio HT-295 Trigear Courier
Helio H-391 Courier
Helio H-392 Strato-Courier
Helio H-395 Super Courier
Helio H-500 Twin Courier
Helio H-550A Stallion
Helio H-580 Twin Courier
Helio H-600 Stallion
Helio H-634
Helio (Koppen-Bollinger, K-B) HelioPlane
Helio Helioplane 2
Helio Helioplane Four (1950)
Helio Hi-Vision Courier
Helio YL-24 (H-391)
Helio U-5A Twin Courier (H-580)
Helio U-10A (L-28A) Courier (H-395)
Helio U-10B Courier (H-395)
Helio U-10D Courier (H-395)
Helio AU-24A (L-28) Stallion (H-550A)

Heliopolis Air Works (Egypt)
Heliopolis Gomhuria (Bücher Bu 181D Development)

Helitec *(See: Aviation Specialties)*

Hellen-Herremans-Lansart (Germany)
Hellen-Herremans-Lansart 1923 Glider
Helton Manufacturing Co (Mesa, AZ; Tracy, CA)
Helton Lark 95

Helwan Air Works (Egypt)
Helwan Al-Kahira (Hispano HA-200)
Helwan HA-300 (Hispano HA-300)

Heminger Brothers (Joel R. & Harold Heminger) (Akron, OH)
Heminger Rebuilt D.H.6

Hempstead Airplane Co (Hempstead, NY)
Hempstead Skylark (See: Dietz-Schriber)

Henderson Aircraft Co (UK) *(See: Hendy)*

Hendrickson, O. J. (Middletown, NJ)
Hendrickson Glider
Hendrickson 1910 Biplane

Hendy Aircraft Co (UK)
Hendy 281 Hobo
Hendy 302
Hendy 321 Hilo
Hendy 3308 Heck

Henlai, Ernesto F. (Argentina)
Henlai Competition Sailplane
Henlai 140 hp
Henlai 275
Henlai 200/52 Flying Wing

Hennesy, Gerald (Washington, DC)
Hennesy Monoplane

Henschel (Germany)
Henschel Hs 121
Henschel Hs 122
Henschel Hs 123
Henschel Hs 124
Henschel Hs 125
Henschel Hs 126
Henschel Hs 128
Henschel Hs 129
Henschel Hs 130
Henschel Hs 132

Henson, W. S. (UK)
Henson Aerial Steam Carriage (1842)

(Habicht de) Hentzen (Germany)
(Habicht de) Hentzen 1924 Glider

Hérard (France)
Hérard 1888 Paddle-Wheel Flying Machine

Herff, A. P. (Boerne, TX)
Herff Monoplane (1920s)

Hergt (F. D. Hergt) (Germany)
Hergt 1918 Monoplane

Hermann, Otto (Canastota, NY)
Hermann 1909 Biplane

Hermanspann, Fred (USA)
Hermanspann Chinook S

Herrick, Gerardus Post (Convertoplane Corp) (New York, NY)
Herrick HV-1 Vertaplane
Herrick HV-2A Vertaplane
Herrick HV-2A Vertaplane, NASM
Herrick HV-3 Vertaplane
Herrick HC-6D Convertoplane

Herring, Augustus (St Joseph, MO)
Herring Aeroplane
Herring 1894 Lilienthal-Type Glider
Herring-Burgess Biplane No.1 Flying Fish (Kingbird, Flying Dragon)
Herring-Burgess Biplane No.2 Flying Fish (Kingbird, Flying Dragon)
Herring-Burgess Biplane No.3 Flying Fish (Kingbird, Flying Dragon)
Herring-Curtiss Aeroplane (See: Curtiss-Herring #1 "Reims Racer")

Herse, Jan *(See: RWD-13)*

Herzog, R. D. (Harvard, NE)
Herzog (R. D.) Meteor
Herzog (R. D.) No. 2

Hess Aircraft Co (Detroit, MI)
Hess H-1 Blue Bird Training
Hess H-2 Blue Bird
Hess 2PH Blue Bird Special

Heston Aircraft Co, Ltd (UK)
Heston A.2/45 A.O.P.
Heston T.1/37
Heston Youngman-Baynes High Lift Research Monoplane
Heston Type 1 Phoenix
Heston Type 5 Racer

HFB (Hamburger Flugzeugbau GmbH) (Germany)
Hamburger Flugzeugbau GmbH was established in July 1933 as a subsidiary of Blohm und Voss shipyards. Company aircraft were designated with "Ha" prefixes. In September 1937 the company was renamed **Blohm und Voss**, with new production carrying "Bv" designations. Blohm und Voss ceased aircraft production at the end of WWII. **Hamburger Flugzeugbau GmbH** (**HFB**) was reestablished in 1955 as a Blohm und Voss subsidiary, building aircraft under the "HFB" designation. The company merged with **Messerschmitt-Bölkow** in May 1969 to form **Messerschmitt-Bölkow-Blohm** (**MBB**). Aircraft built before 1945 (using "Ha" or "Bv" designations) are filed under **Blohm und Voss**. Post-WWII production is filed under **HFB**.
HFB 320 Hansa

Hi-Shear Corp *(See: Wing Aircraft Co)*

Hicks, Chester W. (Waltham, MA)
Hicks (Chester W.) 1910 Glider

Hicks, Lewis (Heaton, NC)
Hicks (Lewis) Homebuilt Monoplane

Higgins Industries Inc (New Orleans, LA)
Higgins (Industries) EB-1 (Enea Bossi)

Higgins, Harry (UK)
Higgins (Harry) Monoplane

High Speed Civil Transport (HSCT) *(See: NASA)*

Higher Planes, Inc (Kansas) *(See: Mitchell Aircraft Corp)*

Highlander Aircraft Corp (USA)
Highlander ARV-1 (See: ARV)

Hikoki Kenkyusho *(See: Nakajima Hikoki KK)*

Hill, Klaus (USA) *(See: US Aviation)*

Hiller (Palo Alto, CA)
The **Aircraft Division of Hiller Industries** was formed in 1942 to develop Stanley Hiller's rotor designs. In 1947, it was established separately as **United Helicopters, Inc**, reverting in 1950 to **Hiller Aircraft Corp**. A 1964 merger with Fairchild resulted in the formation of **Hiller Aircraft Co Inc**, a subsidiary of **Fairchild Hiller Corp**. The 1973 reorganization of **Fairchild Industries** dropped the Hiller name and sold the rights to support H-12s and manufacture new products to a new company, **Hiller Aviation Inc**. This company became **Hiller Helicopters** (a subsidiary of **Rogerson Aircraft Corp**) in 1984, and was later renamed **Rogerson Hiller Corp**. The company was repurchased by in 1994, becoming the **Hiller Aircraft Co**. All Hiller designed helicopters are listed under **Hiller**. [Note that the following listings add the term "Model" to help distinguish company designations from the many similar military designations.]
Hiller YOH-5 (YHO-5, Model 1100)
Hiller YH-23 Raven
Hiller H-23A (Model UH-12A) Raven
Hiller H-23B (Model UH-12B, OH-23B) Raven
Hiller H-23C (OH-23C) Raven
Hiller H-23D (OH-23D) Raven
Hiller H-23E (Model 12E) Raven
Hiller H-23F (Model 12E-4, OH-23F) Raven
Hiller H-23G (OH-23G) Raven
Hiller YH-32 (Model HJ-1 Hornet)
Hiller YH-32A (Sally, 3-Seat)
Hiller Model XH-44 Hiller-Copter
Hiller Model XH-44 Hiller-Copter, NASM
Hiller Model HJ-1 (Model J-1) Hornet
Hiller HOE-1 (Model HJ-1 Hornet)
Hiller HOE-1 (Model HJ-1 Hornet), NASM
Hiller HTE-1 (Model UH-12A)
Hiller HTE-2 (Model UH-12B)
Hiller Model J-5
Hiller XROE-1 Rotorcycle
Hiller YROE-1 Rotorcycle
Hiller YROE-1 Rotorcycle, NASM
Hiller STORC (Self-Ferrying Trans Ocean Rotary-Wing Crane)
Hiller Model UH-4 Commuter
Hiller Model UH-5
Hiller Model UH-12E-4 (E-4)
Hiller Model UH-12L-4 (L-4, SL-4)
Hiller VZ-1 Pawnee (YHO-1E, Flying Platform)

Hiller VZ-1 Pawnee (YHO-1E, Flying Platform), NASM
Hiller Model X-2-235
Hiller X-18 Propelloplane
Hiller Model 360
Hiller Model Ten99
Hiller Model 1100 (FH-1100)

Hillson (F. Hills & Sons, Ltd) (UK)
Hillson Bi-Mono Slip-Wing Aeroplane
Hillson Helvellyn
Hillson Praga (Praga E.114)

Hilton and Brown (Ralph W. Hilton & Harvey W. Brown) (Wichita, KS)
Hilton and Brown Model 1000

HIMAT (Highly Maneuverable Aircraft Technology) (NASA) (USA)
HIMAT (Highly Maneuverable Aircraft Technology), NASM

Hindustan Aeronautics Ltd (HAL) (India)
Hindustan Advanced Light Helicopter (ALH)
Hindustan Ajeet
Hindustan H.T.2

Hino, Kumazo (Japan)
Hino Number 1 (1910)
Hino Number 2 (1911)
Hino Number 3 (1912)
Hino Number 4 Kamikaze-Go (1912)

Hiro Naval Arsenal (Hiro Kaigun Kosho, Hirosho) (Japan)
Hiro B3Y (Navy Type 92 Carrier Attack)
Hiro F.5 Felixstowe Flying Boat
Hiro G2H (Navy Type 95 Land-Based Attack Aircraft)
Hiro H1H (Navy Type 15 Flying Boat)
Hiro H2H (Navy Type 89 Flying Boat)
Hiro H3H (Navy Type 90-1 Flying Boat)
Hiro H4H (Navy Type 91 Flying Boat)
Hiro R-3 Experimental Flying Boat

Hirsch, René (France)
Hirsch Maerch 100

Hirtenberger Patronen Zündhutchen und Metallwarenfabrik AG (Austria)
Hirtenberg HA.11
Hirtenberg HV.12
Hirtenberg HS LO A.130
Hirtenberg HS 9

Hirth (Wolff Hirth GmbH) (Germany) *(See also: Schempp-Hirth)*
Hirth LO-150 (See: Burgfalke)
Hirth Messerschmitt Glider
Hirth Moazagotl

Hise (Detroit, MI)
Hise Trimotor Monoplane

Hispano Aviación SA (Sociedad la Hispano-Suiza SA) (Spain)
Hispano E.30
Hispano E.34
Hispano HA-43
Hispano HA-100 (E-12) (Me 100) Triana
Hispano HA-132-L Chirri
Hispano HA-200 (HA-120) (E-14) (Me 200) Saeta (Arrow)
Hispano HA-220 Supersaeta
Hispano HA-230 Saeta Voyager
Hispano HA-231 Guión (Royal Standard)
Hispano HA-300
Hispano HA-300P (HA-23P) Glider
Hispano HA-1109J Buchón
Hispano HA-1109K Buchón
Hispano HA-1109M Buchón
Hispano HA-1110J
Hispano HA-1110K
Hispano HA-1110M
Hispano HA-1112J
Hispano HA-1112K
Hispano HA-1112M
Hispano HS-42
Hispano Nieuport-Delage 52 C.1 (See: Nieuport)

Historical Aircraft Corp (St Paul, MN)
Historical Aircraft Corp:
F4U Corsair
P-40C Tomahawk
P-51D Mustang
PT-16/PT-20
PZL

Hitachi Kokuki K.K. (Hitachi Aircraft Co Ltd) *(See: Gasuden)*

HOAC *(See: Hoffmann Flugzeugbau)*

Hockaday Aircraft Corp (Burbank, CA)
Hockaday CV-139 Comet

Hocke (Holcker) (Austria-Hungary)
Hocke 1917 Experimental Biplane

Hocker-Denien *(See: Denien, Ralph)*

Hodek, Vincenc (Czechoslovakia)
Hodek HK 101

Hodkinson (Los Angeles, CA)
Hodkinson HT-1 Trimotor Transport

Hoff, C. D. (Chicago, IL)
Hoff 1929 Cabin Monoplane

Hoffelmann, Charles D. (Mineral Wells, TX)
Hoffelmann CH-1 Schatzie (Little Treasure)

Hoffman, Earl (Carnegie, PA)
Hoffman (Earl) Homebuilt

Hoffman, John A. (San Francisco, CA)
Hoffman (John A.) 1911 Aircraft

Hoffmann Flugzeugbau Friesach GmbH (Wolf Hoffmann) (Austria) *(See also: Diamond)*
Hoffmann (Flugzeugbau):
DV 20 Katana Xtreme
DV 22 Speed Katana
H 36 Dimona (Diamond)
H 36 Dimona (Diamond) Mk.II
HK 36 Super Dimona (Diamond) (H 36D)
HK 36R Super Dimona (Diamond)
H 38 Observer
H 40
LF 2 Turbo
LF 2000 Turbo

Hoffmann, Raoul J. (St Petersburg, FL)
Hoffmann (Raoul J.) Flying Wing

Hogan-Moyer Aircraft Corp (Robert J. Hogan, Jr & J. G. Moyer) (Syracuse, NY)
Hogan-Moyer 1928 Special

Holcker *(See: Hocke)*

Holeka, W. T. (USA)
Holeka Type 2 Monoplane (1924)

Holl Aircraft Factory (Parkesburg, PA)
Holl Model 1 Monoplane (1928)

Holland & Holland (UK)
Holland & Holland 1910 Monoplane

Hollandair TB (Netherlands)
Hollandair HA-001 Libel (Dragonfly)

Hollander (Italy)
Hollander A.H.1 (Homebuilt)

Holleville, Roger (France)
Holleville RH-1 Bambi

Hollmann Aircraft (Martin Hollmann) (Cupertino, CA)
Hollmann HA-2M Sportster Gyroplane
Hollmann HA-2M Sportster

Holste, Max *(See: Max Holste)*

Homebuilt, General

Honda Motor Co (Japan)
Honda UA-5

Hongdu Aviation Industry Corp, Ltd *(See: Nanchang)*

Hönningstad (Norway)
Hönningstad Model 5A

Honroth, Edward P. "Ed" (Northfield, OH)
Honroth Special Midget Racer

Hooper, H. G. (USA)
Hooper All-Metal Airplanes (1918)

Hooten Aircraft Co (Springfield, IL)
Hooten H.T.1 Biplane (1929)

Hooton, Gordon B. (USA)
Hooton Monoplane

Hopfner (Flugzeugbau Hopfner) (Austria)
Hopfner H.S.528
Hopfner H.S.829
Hopfner H.V.3
Hopfner H.V.428
Hopfner H.V.6/28

Hopkins & Mead (San Diego, CA)
Hopkins & Mead Monoplane

Hopkins, Robert S. (Reidsville, NC) *(See: Lefevers)*

Hoppi-Copters Inc (American Hoppi-Copters, Inc) (Seattle, WA)
Hoppi-Copters Model 101 (1946)
Hoppi-Copters Model 102 (1947)
Hoppi-Copters Model 103 (1947)
Hoppi-Copters Model 105B (1950)
Hoppi-Copters 1945 Strap-On

Horace Keane Aeroplanes Inc *(See: Keane Aeroplanes, Inc)*

Horace Keane Aircraft Corp *(See: Keane Aircraft Corp)*

Horne, James A. (Portland, OR)
Horne 1911 Aircraft Patent

Horner, Paul G. (Los Angeles, CA)
Horner 1929 Monoplane

Horten (Walter & Reimar) (Germany; Argentina)
Horten H Xb (Piernifero II)
Horten Ho 15 C
Horten Ho 33
Horten I. Ae. 37
Horten I. Ae. 38
Horten Parabola
Horten PUL 10
Horten I Glider (1931)
Horten II (1935)
Horten II (1935), NASM
Horten III (108-250)
Horten III-a
Horten III-b
Horten III-c
Horten III-d
Horten III-e
Horten III-f
Horten III-f, NASM
Horten III-h
Horten III-h, NASM
Horten IV
Horten IV-B
Horten V
Horten VI
Horten VI, NASM
Horten VII (8-226)
Horten VIII
Horten IX (Ho 8-229, "Go 229")
Horten IX (Ho 8-229, "Go 229"), NASM
Horten X
Horten XI
Horten XII
Horten XIII
Horten XIV
Horten XV
Horten 18

Hoshino, Yonezo (Japan)
Hoshino 1914 Aeroplane

Hosler, Russell "Curley" (Huntington, IN)
Hosler B Racer

Hotspur *(See: General Aircraft)*

Houstee, Ricardo (Memphis, TN)
Houstee Monoplane

Hovercraft *(See: Air Cushion Vehicles)*

Hovermarine Corp (Pittsburgh, PA)
Hovermarine HM.2 Mk.4

Hovey, R. W. (Canyon Country, CA)
Hovey Beta Bird
Hovey Whing Ding

Howard, Benjamin O'Dell "Benny" (Houston, TX)
Howard (Benny) 1923 Flyabout Biplane ("DGA-1")
Howard (Benny) 15 Case Standard Modification ("DGA-2")
Howard (Benny) DGA-3 "Pete"
Howard (Benny) DGA-4 "Mike"
Howard (Benny) DGA-5 "Ike"
Howard (Benny) DGA-6 "Mister Mulligan"

Howard Aero Manufacturing Division, Business Aircraft Corp (Howard Aero Inc, D. U. Howard) (San Antonio, TX)
Howard Tri-Motor Travel Air (Modified Beechcraft Travel Air)
Howard 250 (Modified Lockheed Lodestar)
Howard 350
Howard 500 Super Ventura (Modified Lockheed Ventura)

Howard Aircraft Corp (Chicago, IL)
Howard Cargo Carrier
Howard DGA-7
Howard DGA-8
Howard DGA-9
Howard DGA-11
Howard DGA-12
Howard DGA-15
Howard DGA-15P
Howard DGA-15W
Howard DGA-18
Howard DGA-120
Howard DGA-125 Trainer
Howard DGA-160
Howard GH-1 Nightingale
Howard GH-2 Nightingale
Howard GH-3 Nightingale
Howard NH-1

Howell, Phil (Christiansburg, VA)
(See: Mignet)

HPA *(See: Human-Powered Aircraft)*

HSCT *(See: NASA)*

Huabei Machinery Plant (China)
Huabei Y-5 (Yun-5) *Colt* (AN-2)

Hubbard, Gardner G. (Boston, MA)
Hubbard 1910 Monoplane

Hudson-Wright (USA)
Hudson-Wright Model B
Hudson-Wright Model J

Huf, Tom *(See: Halpin & Huf)*

Huff-Daland (Ogdensburg, NY)
In 1920 Thomas Henri Huff and Elliot Daland founded the **Ogdensburg Aeroway Corp** in Ogdensburg, NY. By 1922 the firm had been renamed **Huff, Daland & Co, Inc**, a name held until July 1925 when the company moved to Bristol, PA, as **Huff-Daland Airplanes, Inc**. This became **Keystone Aircraft Corp** in March 1927. All Huff-Daland designs which entered production before March 1927 are listed under **Huff-Daland**; designs which subsequently entered production are listed under **Keystone**.
Huff-Daland AT-1
Huff-Daland AT-2 Dog Ship (Panther)
Huff-Daland XB-1 Super Cyclops (See: Keystone)
Huff-Daland Duster Model 1
Huff-Daland XHB-1 Cyclops
Huff-Daland XHB-3
Huff-Daland HD-1 Early Bird
Huff-Daland HD-1A Early Bird
Huff-Daland HD-1B Early Bird
Huff-Daland HD-4 Bridget
Huff-Daland HD-7 Dizzy Dog
Huff-Daland HD-8 Plover
Huff-Daland HD-8A Petrel
Huff-Daland HD-8C Petrel
Huff-Daland HD-9
Huff-Daland HD-9A
Huff-Daland HD-19L
Huff-Daland HD-19P
Huff-Daland HD-27 Petrel
Huff-Daland HD-31 Big Duster (Petrel 31 Duster)
Huff-Daland HN-1 (HD-19)
Huff-Daland HN-2 Pelican
Huff-Daland HO-1 (HD-19A)
Huff-Daland XLB-1 Pegasus
Huff-Daland LB-1 Pegasus
Huff-Daland XLB-3
Huff-Daland XLB-3A (See: Keystone)
Huff-Daland Pelican (Civil)
Huff-Daland Petrel (Duster)
Huff-Daland Petrel (Duster), NASM
Huff-Daland Petrel 4
Huff-Daland Petrel 5
Huff-Daland TA-2 (HD-4)
Huff-Daland TA-6 (HD-24)
Huff-Daland TW-5
Huff-Daland TW-5C

Huffaker, E. C. (Washington, DC)
Huffaker 1896 Glider
Huffaker 1897 Soaring Bird Model Glider

Hüffer (Germany)
Hüffer Cirrus Sailplane

Huffman (Glen Huffman Aircraft Co) (Detroit, MI)
Huffman 1928 Biplane

Hughes (Culver City, CA)
In 1934 **Hughes Tool Co** began design work on its first aircraft. During 1936 the aviation staff was reorganized as **Hughes Aircraft Co.** In 1953 Hughes Aircraft Co separated from Hughes Tool, leaving behind a **Hughes Helicopter Division**. Hughes Helicopters moved to **Summa Corp** in 1972, to **Hughes Corp** (as **Hughes Helicopter Inc**) in 1981, and finally to **McDonnell Douglas Corp** in 1984 (where it was renamed **McDonnell Douglas Helicopter Co** in 1985). (Hughes Aircraft Co is still active in missiles and avionics.) All aircraft designed by the various Hughes divisions and companies are listed together under **Hughes.** New helicopters developed after the Hughes model 530 are listed under **McDonnell Douglas**.
Hughes D-2
Hughes XF-11 (R-11)
Hughes Feederliner Project
Hughes H-1
Hughes H-1, NASM
Hughes YHO-2
Hughes H-4 (HK-1) Hercules ("Spruce Goose")
Hughes YOH-6 (YHO-6, LOH, Loach)
Hughes H-6 Nightfox
Hughes MH-6A Cayuse
Hughes OH-6A Cayuse
Hughes OH-6A Cayuse Notar (No-Tail-Rotor) Testbed
Hughes EH-6B Cayuse
Hughes MH-6B Cayuse
Hughes OH-6B Cayuse
Hughes AH-6C Cayuse
Hughes MH-6C Cayuse
Hughes OH-6C Cayuse
Hughes OH-6D Cayuse (AHIP)
Hughes EH-6E Cayuse
Hughes MH-6E Cayuse
Hughes AH-6F Cayuse
Hughes AH-6G Cayuse
Hughes MH-6H Cayuse
Hughes AH-6J Cayuse
Hughes MH-6J Cayuse
Hughes XH-17 Flying Crane
Hughes XH-28 Flying Crane
Hughes TH-55A
Hughes YAH-64A Apache
Hughes AH-64 Apache (USMC)
Hughes AH-64 Sea-Going Apache
Hughes AH-64 Petan (Cobra)
Hughes AH-64A Apache
Hughes GAH-64A Apache
Hughes JAH-64A Apache
Hughes AH-64B Apache
Hughes AH-64C Apache
Hughes AH-64D Longbow Apache
Hughes WAH-64D Longbow Apache (UK)
Hughes Hot Cycle Rotorwing
Hughes N-5
Hughes XV-9A
Hughes Winged, Jet, Short-Haul Helicopter Transport
Hughes 200 Family
Hughes 269 Family
Hughes 300 Family
Hughes 369 Family
Hughes 369D
Hughes 500
Hughes 500D
Hughes 500E Olympian (Executive)
Hughes 500M-D Defender
Hughes 500U (Utility)
Hughes 520N (Notar)
Hughes 530 Family
Hughes 530N (Notar)
Hughes 600 And Subsequent (See: McDonnell Douglas)

Hugo, Adolph B., Jr (Tulsa, OK)
Hugo Hu-Go Craft

Hull, Floyd V. (Republic, PA)
Hull Redstone (Curtiss/Hull)

Human-Powered Aircraft, General

Humber, Ltd (UK)
Humber 1910 Monoplane (License Blériot XI)
Humber 1911 Biplane

Humphrey, Jack (UK)
Humphrey 1909 Biplane Hydroplane

Hunt, Ulys H. (Kansas City, MO)
Hunt Experimental (1928)

Hunt & Rettig (Ralph V. Hunt & William Rettig) (Kansas City, MO)
Hunt & Rettig 1928 Special Monoplane

Hunting Aircraft, Ltd (UK)
In 1954, **Percival Aircraft, Ltd**, which had been operating under the **Hunting Group of Companies**, became **Hunting Percival Aircraft, Ltd**; the name was changed to **Hunting Aircraft, Ltd** in 1957. After September 1960 controlling interest passed to **British Aircraft Corp**, although research and production continued under the Hunting name for several years. On 1 January 1964, Hunting was completely merged into **British Aircraft Corp (Operating) Ltd**, and use of the Hunting name was suspended. Nearly all Hunting and Hunting Percival designs were developments of Percival projects; they are filed under **Percival**. Designs begun after 1957 are listed as **Hunting**.
Hunting H.126 (ER.189D) Jet-Flap Research Aircraft

Hunting Firecracker Aircraft Ltd (UK)
Hunting Firecracker Firecracker

Hunting Percival Aircraft, Ltd
(See: Percival Aircraft Ltd)

Huntingdon *(See: Huntington Aircraft Co, Stratford, CT)*
Huntington Aircraft Co (USA)
Huntington (USA) 1915 Tractor

Huntington Aircraft Co (Stratford, CT)
Huntington (CT) Chummy
Huntington (CT) H12 Chum
Huntington (CT) Governor

Huntington Airplane Co (Huntington, WV)
Huntington (WV) 4-place Biplane

Huntington, Dwight W., Jr (Hempstead, NY)
Huntington (Dwight) 1921 Aircraft

Huntington, Howard (Hollis, NY)
Huntington (Howard) 1914 Multiplane

Hurd, E. P. (Detroit, MI)
Hurd HM-1 Monoplane

Hurel-Dubois (Avions Hurel-Dubois, SA) (France)
Hurel-Dubois HD.10
Hurel-Dubois Hurel-Bertin HB.11
Hurel-Dubois Hurel-Bertin HB.12
Hurel-Dubois HD.130
Hurel-Dubois HD.150
Hurel-Dubois HD.31
Hurel-Dubois HD.32
Hurel-Dubois HD.321
Hurel-Dubois HD.322
Hurel-Dubois HD.332
Hurel-Dubois HD.34
Hurel-Dubois HD.37
Hurel-Dubois HD.45

"Hurlburt Hurricane" (Cleveland, OH)
The "Hurlburt Hurricane" was owned by aviation enthusiasts Marge Logan and "Duke" Caldwell. The racer was named for Marge Hurlburt, who was killed in an air show crash on 4 July 1947.
"Hurlburt Hurricane"

Hutchinson Aircraft Co, Inc (Bay St Louis, MS)
Hutchinson Ag Master

Huth, Dr Fritz (Germany)
Huth (Fritz) 1912 Eindecker

Huth, W, (Germany)
Huth (W) Tiefdecker

Hütter, Dr. Ulrich (Germany)
Hütter 17 Sailplane (See also: Schempp-Hirth)
Hütter 17 b Sailplane

Hydrofoils, General

I

IAR (Industria Aeronautica Romana) (Romania)
IAR 2.L
IAR 11 (CV-11)
IAR 12
IAR 14
IAR 15
IAR 16
IAR 21
IAR 22
IAR 23
IAR 24
IAR 37
IAR 80
IAR 81

IAR Brasov (Intreprinderea Aeronautici Romanesc) (Romania)
IAR Brasov IS-28B2 Twin Lark
IAR Brasov IS-28M2 (IAR-34) Motor Lark
IAR Brasov IS-29D2 Lark
IAR Brasov IS-30 Twin Lark
IAR Brasov IS-32

Ibbs, Chester (USA)
Ibbs Flyin' Dutchman "Jezebel"
Ibbs Sportsman

Iberavia SA (Spain)
Iberavia IP-2
Iberavia I-11 Peque
Iberavia I-115

ICA Brasov (Intreprinderea de Constructii Aeronautice) *(See: IAR Brasov)*

ICAR (Intreprindere Pentru Constructii Aeronautice Române) (Romania)
ICAR IAR-36 (Messerschmitt M 36 Derivative)

Icarus (Greek Legend) *(See: Daedalus)*

Icasate-Larios, Félix (Argentina)
Icasate-Larios 1844 Ornithopter

Ichimori, Yoshinori (Japan)
Ichimori Skylark Sport-Plane
Ichimori 1919 Monococque Airplane

ICX Aviation Inc *(See: Yakovlev Yak-40)*

Ideal Aeroplane & Supply Co, Inc (New York, NY)
Ideal Primary Glider, 1931

Iga, Ujihiro (Japan)
Iga Maitsuru-Go Airplane (1911)

III *(See: Iniziative Industriali Italiane SpA)*

Ikarus (Yugoslavia)
Ikarus IK-2
Ikarus Meteor 57
Ikarus Type 451
Ikarus-Mickl

Ilyushin (Russia)
In 1938, Sergei Vladimirovich Ilyushin resigned his post as director of the Soviet Aviation Ministry to form his own design bureau: **OKB Ilyushin.** In 1977, the bureau was renamed **AKI S. V. Ilyushin.** In 1992, **AAI (Aviation Association Ilyushin)** was formed to act as the agent for all Ilyushin aircraft.
Ilyushin DB-3 (TsKB-3, TsKB-30)
Ilyushin DB-4 (TsKB-56)
Ilyushin I-21 (TsKB-32)
Ilyushin Il-2 Shturmovik, *Bark*
Ilyushin Il-4 *Bob* (DB-3F)
Ilyushin Il-8
Ilyushin Il-10 Shturmovik, *Beast*
Ilyushin Il-12 *Coach*
Ilyushin Il-14 *Crate*
Ilyushin Il-16
Ilyushin Il-18 *Clam, Coot*
Ilyushin Il-20
Ilyushin Il-22
Ilyushin Il-28 *Beagle, Mascot*
Ilyushin Il-30
Ilyushin Il-32
Ilyushin Il-38 *May*
Ilyushin Il-40
Ilyushin Il-46
Ilyushin Il-54 *Blowlamp*
Ilyushin Il-62 *Classic*
Ilyushin Il-76 *Candid*
Ilyushin Il-86 *Camber*
Ilyushin Il-96
Ilyushin Il-114
Ilyushin "Moskva" ("Moscow") (Civil DB-3)
Ilyushin TsKB-26
Ilyushin TsKB-57 (Bsh-2)

IMAM *(See: Meridionali; See also: Aerfer)*

IMPA SA (Industrias Metalurgicas y Plasticas Argentinas SA) (Argentina)
IMPA Chorlito (Curlew)
IMPA RR-11
IMPA RR-12D
IMPA RR-13D
IMPA Tu-Sa (Turismo Serie A)

Imperial Airport *(See: Goddard/Imperial)*

Inagaki, Yasuji (Japan)
Inagaki 1917 Tractor Biplane

Inav Ltd (Innovative Aviation) (Oshkosh, WI)
In July 1985, the British firm **Aviation Composites** established **Inav Ltd (Innovative Aviation, Ltd)** to purchase the assets of **Monnett Experimental Aircraft, Inc (MEA)**. For designs originating at MEA before the 1985 purchase, see **Monnett**. For later designs, see **Inav**.
Inav Mercury (Homebuilt)

Indian Aircraft Co (Glendale, CA)
Indian Aircraft Navaho

Indraéro (Société Indraéro) (France)
Indraéro Aéro 101
Indraéro Aéro 110

Industri Pesawat Terbang Nusantara *(See: IPTN)*

Indùstria Aeronáutica Neiva *(See: Neiva)*

Industria Aeronautica Romana *(See: IAR)*

Industrias Aeronáuticas y Mecánicas del Estado (Argentina) *(See: FMA)*

Industrias Cardoen Ltda *(See: Cardoen)*

Industrias Metalurgicas y Plasticas Argentinas SA *(See: IMPA SA)*

Industrie Meccaniche e Aeronautiche Meridionali *(See: Meridionali; also: Aerfer)*

Ingg. Fratelli Nardi *(See: Nardi)*

Iniziative Industriali Italiane SpA (III) (Italian Innovative Industries) (Italy) *(See also: Meteor SpA)*
Iniziative Industriali Italiane:
Raw Arrow 750 Era
Sky Arrow 450 T
Sky Arrow 480 T
Sky Arrow 500 TF
Sky Arrow 650 A/914 Exocet
Sky Arrow 650 Era
Sky Arrow 650 LMT
Sky Arrow 650 SP
Sky Arrow 650 T
Sky Arrow 650 TC
Sky Arrow 650 TCN
Sky Arrow 1200 LC
Sky Arrow 1310 SP
Sky Arrow 1450 A
Sky Arrow 1450 L
Speed Arrow 700 T

Inland Aviation Co (Kansas City, MO)
Inland Sport Monoplane
Inland S-300-DF
Inland R-400 Sportster Monoplane
Inland W-500 Super Sport Monoplane

Innovative Aviation *(See: Inav, Ltd)*

Institute of Science and Technology (Philippines)
Institute of Science and Technology:
XL-14 Maya
XL-15 Tagak
L-17

Instituto Aerotécnico *(See: FMA)*

Instituto de Pesquisas e Desenvolvimento (IPD) *(See: Centro Tecnico de Aeronautica)*

Instituto de Pesquisas Tecnologicas (Technical Research Institute) (IPT) (Brazil)
Instituto de Pesquisas Tecnologicas:
IPT-0 Bichinho
IPT-1 Gafanhoto
IPT-2 Aratinga
IPT-3 Saracura
IPT-4 Planalto
IPT-5 Jaragua
IPT-7 Junior
IPT-8
IPT-9
IPT-10 Junior
IPT-11 Bachão
IPT-12 Caboré
IPT-13
IPT-14 Marreco
IPT-17

Instituto di Aeronautica Politecnico di Milano (Italy)
Instituto di Aeronautica Politecnico di Milano Preti PM-280 Tartuca (Tortoise)

Instruments de Précision MDG *(See: MDG)*

Instytut Lotnictwea (Poland)
Instytut Lotnictwea I-23
Instytut Lotnictwea IS-2
Instytut Lotnictwea Kobra 2000

INTA *(See: AISA)*

Intreprinderea de Constructii Aeronautice) *(See: IAR Brasov)*

Integrated Systems Aero Engineering *(See: ISAE)*

Inter-Air (International Aircraft Manufacturing Inc) (Alexandria, MN) *(See: Bellanca Cruisemaster)*

Intercity Airlines Co (Canada)
Intercity (Sznycer-Gottlieb) SG VI-D

Intermountain Manufacturing Co, Inc (Afton, WY)
Intermountain New Call-Air B-1 Turboprop Conversion

International Aero Construction Co (Woodhaven, NY)
International Aero Construction 1911 Monoplane

International Aircraft Corp (Co) (Venice, CA; Cincinnati, OH) *(See also: Catron and Fisk)*
International Aircraft CF-12
International Aircraft CF-13
International Aircraft F-17 Sportsman
International Aircraft F-17H
International Aircraft F-17M
International Aircraft F-17W
International Aircraft F-18 Air Coach

International Aircraft Manufacturing Inc (Inter-Air) (Alexandria, MN) *(See: Bellanca Cruisemaster)*

International Airship Co (Paterson, NJ)
International Airship Ochoaplane

International Aviation Corp (Cleveland, OH)
International Aviation Duckling

Interplane Ltd (Czech Republic)
Interplane Griffon

Interplane Skyboy
Interplane Skyboy S

Interstate Aircraft & Engineering Corp (El Segundo, CA)
Interstate BQ-4 (TDR-1)
Interstate BQ-6 (TD3R-1)
Interstate XL-6 Envoy (XO-63)
Interstate L-6 Envoy (O-63)
Interstate L-8 (S-1A)
Interstate S-1 Cadet Family
Interstate XTDR-1 (TI-7)
Interstate XTDR-1 (TI-7), NASM
Interstate XTD2R-1

Intreprindere Pentru Constructii Aeronautice Române *(See: ICAR)*

Invincible Metal Furniture Co, Aircraft Div (Manitowoc, WI)
Invincible Model D-D Monoplane
Invincible 1929 Cabin Monoplane

IPD *(See: Centro Tecnico de Aeronautica)*

IPT *(See: Instituto de Pesquisas Tecnologicas)*

IPTN (Industri Pesawat Terbang Nusantara) (Indonesia)
IPTN CN-235 (See: Airtech)
IPTN N-250
IPTN N-270

Irbitis (K. Irbitis) (Latvia)
Irbitis I-1
Irbitis I-2
Irbitis I-3
Irbitis I-4
Irbitis I-5
Irbitis I-6
Irbitis I-7
Irbitis I-8
Irbitis I-9
Irbitis I-11
Irbitis I-12
Irbitis I-14
Irbitis I-15a
Irbitis I-15b
Irbitis I-16
Irbitis I-17
Irbitis I-18

Ireland Aircraft Inc (Ireland Aircraft Co) (Garden City, NY)
Ireland Comet (Curtiss Oriole Fuselage with Ireland High-Lift Wings)
Ireland Meteor
Ireland N-2-B (N2b)
Ireland N-2-C (N2c)
Ireland Privateer
Ireland Privateer P1
Ireland Privateer P2a
Ireland Privateer P3b
Ireland Proposed Flying Boat for Submarines

Irish Aircraft Corp (Sandusky, OH)
Irish (Aircraft Corp) 3-A Aristocrat

Irish, W. E. (USA)
Irish (W. E.) 1905 Feathered-Wing Flying Machines

Irvine, James C. (San Francisco, CA)
Irvine 1909 Aerocycloid

Irwin Aircraft Co (Corning, CA; Sacramento, CA)
Irwin Meteorplane (1912)
Irwin Meteorplane C-C-1
Irwin Meteorplane F-A-1
Irwin Meteorplane LW-3
Irwin Meteorplane M-T (1916)
Irwin Meteorplane M-T-2
Irwin Meteorplane S-P-1

Isaac, Archibald C. J. (Pittsfield, MA)
Isaac 1926 Homebuilt Lightplane
Isacco (Italy, Spain, France, UK)
Isacco Helicogyre No.1 (1926)
Isacco Helicogyre No.2
Isacco Helicogyre No.3 (1929)

Isacson, Sigurd (Sweden)
Isacson Asymmetrical Fighter

ISAE (Integrated Systems Aero Engineering) Logan, UT)
ISAE Omega II (Kitplane)

Ishibashi, Katsunami (Japan)
Ishibashi SPAD XIII Racing Aircraft

Ishida Aerospace Research (Fort Worth, TX)
Ishida TW-68 Tiltwing

Ishikawajima (KK Ishikawajima Hikoki Seisakusho) (Japan)
Ishikawajima Ki-4 (Type 94 Recon)
Ishikawajima R-1 (CM-1)
Ishikawajima R-2
Ishikawajima R-3
Ishikawajima R-5
Ishikawajima T-1
Ishikawajima T-2
Ishikawajima T-3
Ishikawajima Type 91 Fighter

Island Aircraft (UK) *(See: ARV)*

Island Aircraft Corp (Merritt Island, FL)
Island X-199 Spectra

Isobe, Onokichi (Japan)
Isobe 1910 Seaplane
Isobe Number 2 Airplane (1910)
Isobe Rumpler Taube (1915)

Ison, Wayne (Manchester, IN) *(See: PDQ Aircraft Products)*

Israel Aircraft Industries (IAI) (Israel)
Israel Aircraft Industries:
  Arava 201 STOL
  B101
  Commodore 1121 (Jet Commander)
  Commodore 1123 (Jet Commander)
  F-23 Kfir
  Kfir (Young Lion)
  Sea Scan
  Westwind (1124)

Israel, Gordon (Chicago, IL)
Israel (Gordon) "Redhead" Racer

Issoire *(See: Siren)*

Issoudin Aircraft Manufacturing Corp (Detroit, MI)
Issoudin Amphibian (Hamilton)

Itoh (Itoh Hikoki Kenkyusho) (Japan)
Itoh Emi 1
Itoh Emi 2
Itoh Emi 3
Itoh Emi 6 (Fujimara) "Tsubame-Go" ("Swallow")
Itoh Emi 11
Itoh Emi 12
Itoh Emi 13
Itoh Emi 14
Itoh Emi 16 "Fuji-Go"
Itoh Emi 17 Tsurubane No.3
Itoh Emi 18
Itoh Emi 19 "Akira-Go"
Itoh Emi 20 "Oguri-Go"
Itoh Emi 22 "Yamagatakinen-Go"
Itoh Emi 23 Bulldog
Itoh Emi 24 "Akita-Go"
Itoh Emi 25
Itoh Emi 28

Itoh Emi 29 "Taikpku-Go"
Itoh Emi 30
Itoh Emi 31
Itoh Emi 50
Itoh Emi Tsurubane ("Swallow") No.1
Itoh Emi Tsurubane ("Swallow") No.2

Itoh (Itoh Chu Kohku Seibi Kbuhiki Kaisha) (Japan)
Itoh EMI 6 (Fujiwara "Tsubane-Go": "Swallow")
Itoh EMI 9 Trainer
Itoh N-58 Cygnet
Itoh N-62 Eaglet

ITS (Instytut Techniki Szybownictwa: Institute of Gliding Technique) (Poland)
ITS Jaskólka (Swallow)
ITS Wróbel (Sparrow)
ITS-II/a (CWJ 2) Glider
ITS-II/32 (CWJ 2) Glider
ITS-IV
ITS-IVB
ITS-V
ITS-7 Drozd (Thrush)
ITS-8

Ittner-Nürnberg *(See: Nürnberg)*

Ivanov, V. A. (Russia)
Ivanov 1924 Monoplane

Ivensen, P. A. (Russia)
Ivensen I-2

IVL (Finnish National Aircraft Factory) (Finland)
IVL A-22

Izaki, Shozo (Japan)
Izaki Number 2 Sempu-Go Airplane

J

J. Loring Fábrica de Aeroplanos *(See: Loring)*

Jach, Franciszek (Poland)
Jach Bimbus (Bimbo)
Jach Zabus (Froggy)
Jach Zabus 2 (Froggy 2)

Jackson, Arthur Rex (UK)
Jackson 1946 Ornithopter

Jacobs Aircraft Engine Co (Pottstown, PA)
Jacobs Aircraft Engine Model 104 Gyrodyne

Jacobs Schweyer (Hans Jacobs) (Germany)
Jacobs Schweyer Weihe (Kite)
Jacobs Schweyer (Focke Wulf) Weihe (Kite) 50

Jacobs Brothers (H. W. & Frank Jacobs) (Atchison, KS)
Jacobs Bros Multiplane (1911)

Jacquelin (France)
Jacquelin Helicopter (1910)

Jacques Albert de Lailhacar *(See: J.A.L)*

Jacuzzi Brothers (Berkeley, CA)
Jacuzzi Bros 1920 Open-Cockpit Monoplane
Jacuzzi Bros 1920 7-Seat Cabin Monoplane

Jaffe Aircraft Corp (San Antonio, TX)
Jaffe/Swearingen SA-32T

Jamieson Aircraft Co (Jamieson Corp, Charles M. Jamieson) (De Land, FL)
Jamieson D-120
Jamieson D-300
Jamieson J-1 Jupiter
Jamieson J-2-L-1 Jupiter
Jamieson Take 1 (J)

J.A.L (Jacques Albert de Lailhacar) (Spain)
J.A.L. 1910 Monoplane

Jannus Brothers (Anthony "Tony" & Roger W. Jannus) (Baltimore, MD)
Jannus Bros 1914 Flying Boat
Jannus Bros 1915 Flying Boat (Exposition Model)

Janowski, Jaroslaw *(See: Marko-Elektronik)*

Jansson, Ben *(See: BJ)*

J.A.P. (J. A. Prestwich & Co Ltd) (UK)
J.A.P.-Harding 1910 Monoplane

Japan Aeroplane Manufacturing Co *(See: Nihon Kokuki Seizo Kabushiki Kaisha)*

Japan Aeroplane Manufacturing Works *(See: Tachibana)*

Japan Small Aeroplane Co, Ltd *(See: Nippon Kogata Hikoki KK)*

Jarty (France)
Jarty 1935 Aviette

Jatho (Hannoverische Flugzeugwerke GmbH Jatho) (Germany)
Jatho 1909 Biplane (No.3)
Jatho 1910 Monoplane
Jatho 1911 Canard Monoplane (No.4)
Jatho 1911 Monoplane (No.8)

Javelin Aircraft Co (Wichita, KS)
Javelin Wichawk

Jaworski, Wiktor (Poland)
Jaworski (Wiktor) W.J.3 Glider

Jaworski, Wladyslaw *(See: Czerwinski & Jaworski)*

Jayhawk Aircraft Corp (Wichita, KS)
Jayhawk Mars #2

Jeannin (Emile Jeannin Flugzeugbau GmbH) (Germany)
After leaving the **Aviatik** works, engineer Emile Jeannin established **Emile Jeannin Flugzeugbau GmbH** at Johannisthal to build steel-frame monoplanes. The company produced Taube-type aircraft during the early period of World War I, but in 1915 the company was liquidated, with the assets acquired by **National Flugzeug Werke GmbH** (**NFW**) in that June. For all aircraft designed by Jeannin, see **Jeannin.** For all work post-dating the acquisition by NFW, see **NFW**.
Jeannin 1910 Monoplane
Jeannin 1911 Monoplane
Jeannin 1912 Stahltaube
Jeannin 1914 Taube
Jeannin 1915 Biplane

Jeffair Corp (Renton, WA)
Jeffair Barracuda (Homebuilt)

Jenks, Theodore (Los Angeles, CA)
Jenks TJ-2

Jennings, R. C. (Jennings Machine Works) (Uniontown, PA)
Jennings (R. C.) 1911 Monoplane
Jennings (R. C.) 1919 Sportplane

Jennings, W. H. (USA)
Jennings (W. H.) D-9 Monoplane

Jensen, Martin (Charleston, SC)
Jensen (Martin) Trainer 2

***Jensen, Tor** (See: Advanced Soaring Concepts)*

***Jensen, Volmer S.** (See: Volmer Aircraft)*

Jensen Aircraft Corp (Charleston, SC)
Jensen Aircraft Model LW

Jensen Helo (Jensen Helicopter Co Inc) (Tonakset, WA)
Jensen Helo Model 21 "Silver Beetle"

Jerme (France)
Jerme 1909 Biplane

***Jettoplane Aeronautical Research Works** (See: Nishi, Hiro Hiso)*

Jezzi, Leo (UK)
Jezzi 1912 Biplane

Jintoneo
Jintoneo Monoplane

Joachimczyk, Alfred Marceli (Poland)
Joachimczyk 1911 Multiplane

Jodel (Société des Avions Jodel) (Edouard Joly & Jean Delemontez) (France)
Prevented from completing any of their early aircraft designs by the start of World War II, Edouard Joly and Jean Delemontez flew their first homebuilt aircraft in early 1948. The two called their company "Jodel," a name derived from their two last names, and called their first successful aircraft the D.9 Bébé. Jodel designs proved popular, with many aircraft built from kits or plans, and others built under license in several countries. France's Pierre Robin developed a separate line of DR (for Delemontez Robin) aircraft at **Centre Est Aeronautique** (**CEA**, later **Avions Pierre Robin**). Charles Olliver created a similar line as **Colliver**. In Australia, **Frank Rogers** created several highly modified aircraft, and in Canada **Chris Falconar** developed his own variations. All original Jodel designs are listed under **Jodel**, with derived aircraft listed under **Colliver**, **Falconar**, **Robin**, and (**Frank**) **Rogers**.
Jodel D.9 Bébé (Bébé Jodel)
Jodel D.91 Bébé (Bébé Jodel)
Jodel D.92 Bébé (Bébé Jodel)
Jodel D.93 Bébé (Bébé Jodel)
Jodel D.94 Bébé (Bébé Jodel)
Jodel D.95 Bébé (Bébé Jodel)
Jodel D.97 Bébé (Bébé Jodel)
Jodel D.98 Bébé (Bébé Jodel)
Jodel D.99 Bébé (Bébé Jodel)
Jodel D.991 Bébé (Bébé Jodel)
Jodel D.10
Jodel D.11 Club
Jodel D.11A
Jodel D.11B
Jodel D.11C
Jodel D.111
Jodel D.112
Jodel D.117 Grand Tourisme
Jodel D.119 Popuplane
Jodel D.1190S Compostela
Jodel D.12
Jodel D.120
Jodel D.127
Jodel D.13
Jodel D.14
Jodel D.140
Jodel D.140A Mousquetaire (Musketeer)
Jodel D.140AC Mousquetaire (Musketeer) III
Jodel D.140B
Jodel D.140C Mousquetaire (Musketeer) III
Jodel D.140E Mousquetaire (Musketeer) IV
Jodel D.140E1
Jodel D.140R Abeille (Bee)
Jodel D.15
Jodel D.150 Mascaret (Tidal Wave)
Jodel D.150A
Jodel D.1050M Excellence
Jodel D.16
Jodel D.160
Jodel D.160A
Jodel D.18
Jodel D.19
Jodel D.20 Jubile (Jubilee)
Jodel DC.01 (Delemontez-Couchy)

***Joe Ben Lievre Aircraft Corp** (See: Lievre)*

Johansen Aircraft Co (Los Angeles, CA)
Johansen JA-2

John, Herbert F. (Bath, NY)
John's Multiplane (1920)

Johns, Ray (Morristown, PA)
Johns Air Sport
Johns Green Dragon (1940)
Johns Tornado JV
Johns Warrior (1951)
Johns X-3 (Rason Warrior)

***Johnson Aircraft Corp (R. S. Johnson Aircraft Corp) (TX)** (See also: Regent)*
Johnson (TX) Bullet 185 (Texas Bullet)
Johnson (TX) Bullet 205
Johnson (TX) Rocket Prototype
Johnson (TX) Rocket 185

***Johnson Airplane Division, Johnson Airplane & Supply Co (Dayton, OH)** (See also: Driggs)*
Johnson (OH) Bumblebee Monoplane
Johnson (OH) Canary
Johnson (OH) Twin 60

Johnson Brothers (Louis, Harry & Julius Johnson) (Terre Haute, IN)
Johnson Bros 1911 Monoplane

Johnson, Bill (Hillsboro, OR)
Johnson (Bill) Mini Coupe

Johnson, Clarence M. (Detroit, MI)
Johnson (Clarence M.) 1928 2-Place High-Wing Monoplane

***Johnson, E. R.** (See also: Driggs)*
Johnson (E. R.) Omnivator (Helicopter)

Johnson, Kenneth B. (Reynoldsburg, OH)
Johnson (Kenneth B.) Ornithopters

Johnson, Luther C. "Luke" (Johnson Flying Service, Inc) (Greenwood, SC)
Johnson (Luther) "Betty Jo" (GLP 3; Gault, Lanford, & Perry)
Johnson (Luther) Special Midget

Johnson, Richard B. (Chicago, IL)
Johnson (Richard B.) Uniplane

Johnson, Richard H. (USA)
Johnson (Richard H.) Adastra

Johnson, Roy (Seattle, WA)
Johnson (Roy) 1928 Parasol Cabin Monoplane

Johnson, Rufus S. (Dallas, TX)
Johnson (Rufus S.) 1927 Biplane

Johnston-Vaughan Aircraft Co (A. C. Johnston & R. B. Vaughan) (Akron, OH)
Johnston-Vaughan 1928 Special

Jona (Ing. Alberto Jona Studio di Consulenza Aeronautica) (Italy)
Jona J.6

Jones (Ben) Aircraft Co (Schenectady, NY)
Jones (Ben) S-125 (Licensed New Standard)

Jones, L. J. R. (Australia)
Jones (L. J. R.) 4-Seat, High-Wing Monoplane (1930)

Jones, Stanley (Mount Zion, IL)
Jones (Stanley) 10-A Gyrocopter

Jorch, Hans (Germany)
Jorch 1909 Biplane

Jordan, L. F. *(See: 20th Century Aerial Navigation Co)*

Joubert, Jean (France)
Joubert J.4 Avionette (1933)

Jourdain (France)
Jourdain Monoplane (c.1910)

Jourdan (France)
Jourdan 1911 Monoplane

J.O.V. Helicopter Co *(See: Jovair)*

Jovair (Helicopter Engineering Research Corp; Jovair Corp) (Philadelphia, PA)
In 1947 D. K. Jovanovich and F. J. Kozloski left **Piasecki Helicopter Corp** to establish the **Helicopter Engineering Research Corp** in Philadelphia, PA. In 1948 they renamed the company **J.O.V. Helicopter Co** but sold out to **McCulloch Motors Corp** before the end of the year and moved all the company's assets to California, where J.O.V. became the **Airplane Division of McCulloch Motors**. Jovanovich and Kozloski left McCulloch when the Airplane Division was discontinued and formed **Jovair Corp** to continue the development of their helicopters. For all products of Helicopter Engineering Research, J.O.V., and Jovair, see **Jovair**. For all designs originated by the Airplane Division of McCulloch, see **McCulloch**.
Jovair J-2 (Autogiro)
Jovair (JOV) 3
Jovair 4-A
Jovair 4-E Sedan

Juge (France)
Juge 1907 Ornithopter

Juiseux, Comte de (France)
Juiseux Flying Bicycle

Junkers (Junkers Flugzeug Werke AG) (Germany)
In 1913 Dr. Hugo Junkers established **Junkers Motoren Werke** (**Jumo**) in Magdeburg to build marine diesel engines and, later, aircraft engines. In 1915, Junkers formed a new company to build aircraft, but was forced to merge with Anthony Fokker, creating **Junkers-Fokker-Werke AG** (**JFA**). When Fokker withdrew in 1918 to restart his own company in the Netherlands, JFA was renamed **Junkers Flugzeug Werke AG** (still **JFA**). JFA continued to operate, despite Versailles Treaty restrictions, building civil aircraft at the main plant in Dessau, while shifting military work to subsidiary companies in Sweden (**AB Flygindustri**) and Russia. In 1933 the company was nationalized by the Nazi government, and Hugo Junkers was forcibly retired. The combined Junkers aircraft and engine group (**Junkers Flugzeug-und-Motoren Werke AG**, **JFM**) continued to operate throughout the Nazi era. Soviet forces overran most of Junkers' physical plant at the end of WWII, absorbing the factories into the Soviet aviation industry and leaving only the JFM plant at Munich to carry on the name. This last plant was acquired by **Messerschmitt** in 1965. For all products of JFA and its subsidiaries, see **Junkers**. Products of AB Flygindustri, interfiled with main-plant Junkers products, are identified as Junkers (Sweden).
Junkers J 1
Junkers J 2 (E.I)
Junkers J 3
Junkers J 4 (J.I)
Junkers J 7
Junkers J 8
Junkers J 9 (D.I)
Junkers J 10 (CL.I)
Junkers J 10 (Civil Modifications)
Junkers J 11
Junkers J 13 "Annelise" (F 13 Prototype)
Junkers F 13
Junkers F 13a
Junkers F 13a (Junkers-Larsen JL-6)
Junkers F 13b
Junkers F 13c
Junkers F 13d
Junkers F 13f
Junkers F 13g
Junkers F 13h
Junkers F 13k
Junkers J 15
Junkers K 16
Junkers T 19
Junkers A 20L
Junkers A 20W
Junkers J 20 (Ju 20) (Russian Production)
Junkers H 21
Junkers J 21 (Russian Production)
Junkers H 22
Junkers J 22 (Russian Production)
Junkers G 23
Junkers G 23a
Junkers G 23b
Junkers G 23c
Junkers G 23d
Junkers G 23e
Junkers T 23
Junkers G 24
Junkers G 24a
Junkers G 24b
Junkers G 24d
Junkers G 24g

Junkers G 24h
Junkers F 24k
Junkers T 26D (Biplane)
Junkers T 26E (Monoplane)
Junkers T 29
Junkers K 30 (Sweden) (JuG-1 Russia)
Junkers G 31 Prototype
Junkers G 31d
Junkers G 31f
Junkers G 31g
Junkers G 31h
Junkers G 31j
Junkers R Flying Boat
Junkers R.I
Junkers W 33 Prototype
Junkers W 33
Junkers W 33b
Junkers W 33b "Bremen"
Junkers W 33b "Bremen," NASM
Junkers W 33b "Europa"
Junkers W 33c
Junkers W 33d
Junkers W 33f
Junkers W 33h
Junkers J 33 L
Junkers W 34 Prototype
Junkers W 34f
Junkers W 34g
Junkers W 34h
Junkers A 35
Junkers (Sweden) K 37
Junkers G 38a
Junkers G 38b
Junkers G 38c
Junkers G 38d
Junkers K 39
Junkers K 43
Junkers Ju 46 f
Junkers Ju 46 h
Junkers (Sweden) K 47
Junkers Ju 49
Junkers A 50 Junior
Junkers Ju 52 b
Junkers Ju 52 c
Junkers Ju 52/3m, NASM (See: CASA 352L)
Junkers Ju 52/3mb
Junkers Ju 52/3mc
Junkers Ju 52/3mf
Junkers Ju 52/3mg Civil
Junkers Ju 52/3mg Military
Junkers Ju 52/3mh
Junkers (Sweden) K 53
Junkers (Sweden) K 47
Junkers 59
Junkers Ju 60 a
Junkers Ju 60 b
Junkers Ju EF 61 (High Altitude Project)
Junkers Ju 86 V (Prototypes)
Junkers Ju 86 A
Junkers Ju 86 B
Junkers Ju 86 D
Junkers Ju 86 E
Junkers Ju 86 F
Junkers Ju 86 G
Junkers Ju 86 K
Junkers Ju 86 P
Junkers Ju 86 R
Junkers Ju 86 Z
Junkers Ju 87 V Types Stuka (Prototypes)
Junkers Ju 87 A Stuka
Junkers Ju 87 B Stuka
Junkers Ju 87 C Stuka
Junkers Ju 87 D Stuka
Junkers Ju 87 G Stuka
Junkers Ju 87 H Stuka
Junkers Ju 87 R Stuka
Junkers Ju 88 A
Junkers Ju 88 B
Junkers Ju 88 C
Junkers Ju 88 D
Junkers Ju 88 G
Junkers Ju 88 P
Junkers Ju 88 R
Junkers Ju 88 S
Junkers Ju 88 T
Junkers Ju 89
Junkers Ju 90 V1 (Prototype)
Junkers Ju 90 B
Junkers Ju EF 126 Elli (Jet Ground Attack Project)
Junkers Ju EF 128 (Jet Fighter)
Junkers Ju 160 V1 (Prototype)
Junkers Ju 160 A
Junkers Ju 160 D
Junkers Ju 188 Prototype (Ju 88 V27)
Junkers Ju 188 A
Junkers Ju 188 C
Junkers Ju 188 D
Junkers Ju 188 E
Junkers Ju 188 F
Junkers Ju 188 G
Junkers Ju 188 H
Junkers Ju 188 R
Junkers Ju 188 S
Junkers Ju 188 T
Junkers Ju 248 (See: Messerschmitt Me 263)
Junkers Ju 252 V1 (Prototype)
Junkers Ju 252
Junkers Ju 252 A
Junkers Ju 287 V Types (Prototypes)
Junkers Ju 288 V1 (Prototype)
Junkers Ju 288 A Prototypes
Junkers Ju 288 B Prototypes
Junkers Ju 288 C Prototypes
Junkers Ju 290 V1 (Prototype)
Junkers Ju 290 V7 Development A/C
Junkers Ju 290 A
Junkers Ju 352 V1 Herkules (Hercules)
Junkers Ju 352 A Herkules (Hercules)
Junkers Ju 388 J Störtebeker
Junkers Ju 388 K
Junkers Ju 388 K, NASM
Junkers Ju 388 L
Junkers Ju 390 Family
Junkers Ju 488
Junkers Ju 635
Junkers 4-Engine, 2-Hull Flying Boat

***Junkers-Larsen** (See: Junkers F 13a, Junkers-Larsen JL-6)*

Jurca, Marcel (France)
Jurca MJ-2 Tempête (Tempest)
Jurca MJ-7 Gnatsum (Mustang spelled backwards; 2/3 Mustang)
Jurca MJ-77 3/4 Mustang
Jurca MJ-8 3/4 Fw 190
Jurca MJ-9 3/4 Bf 109
Jurca MJ-10 3/4 Spitfire
Jurca MJ-11 3/4 P-47 Thunderbolt

K

K. Irbitis *(See: Irbitis)*

K. J. Barnett Rotorcraft Co *(See: Barnett Rotorcraft)*

K-B *(See: Helio)*

Kaess Aircraft Engineering Corp (Newark, NJ)
Kaess CL-1 Seaplane

Kahn, David (USA)
Kahn Piranha

Kahnt (Germany)
Kahnt Falke (Falcon)

Kaiser (Bristol, PA)
Kaiser Helicopter
Kaiser Tailless Airplane

Kaiser, Rudolf (Germany) *(See: Schleicher)*

Kaiser-Fleetwings *(See: Fleetwings)*

Kaiser-Hammond (Bristol, PA)
Kaiser-Hammond Y "Family Plane"

Kaiserliche Werft (Germany)
Kaiserliche Werft (Kiel) KW

Kaishiki *(See: Rinji Gunyo Kikyu Kenkyu Kai)*

Kaje *(See: Kjeller Flyvemaskinfabrikk)*

Kalinin (Konstantin Alekseyevich Kalinin) (Russia)
Kalinin K-1
Kalinin K-2
Kalinin K-3
Kalinin K-4
Kalinin K-5
Kalinin K-6*bis*
Kalinin K-7
Kalinin K-9
Kalinin K-10
Kalinin K-12 Firebird
Kalinin K-13

Kalkert (Germany)
Kalkert Ka 430

Kaman Aircraft Corp (Bloomfield, CT)
Kaman UH-2A (HU2K-1) Seasprite
Kaman UH-2A Tomahawk
Kaman UH-2A/YJ85 Jet-Augmented Seasprite
Kaman UH-2B Seasprite
Kaman UH-2C Seasprite
Kaman HH-2C Seasprite
Kaman HH-2D Seasprite
Kaman SH-2D/F LAMPS Seasprite
Kaman SH-2D LAMPS Seasprite
Kaman SH-2F LAMPS Seasprite
Kaman YSH-2G LAMPS Seasprite
Kaman HH-43A (H-43A) Huskie
Kaman HH-43B (H-43B) Huskie
Kaman UH-43C (HOK-1) Huskie
Kaman OH-43D (HUK-1) Huskie
Kaman TH-43E (HTK-1)
Kaman HH-43F Huskie
Kaman HTK-1 (See TH-43E)
Kaman HTK-1K Drone
Kaman K-16B V/STOL Amphibian
Kaman K-17
Kaman K-125
Kaman K-190
Kaman K-225
Kaman K-240
Kaman K-250
Kaman K-700
Kaman K-750 (US-Built SA-341 Gazelle)
Kaman K-1125 Huskie III
Kaman MMIRA (Multi-Mission Intermeshing Rotor Aircraft)
Kaman ROMAR (Rotorcraft for Mars)
Kaman Rotochute
Kaman SAVER (Stowable Aircrew Vehicle Escape Rotoseat)

Kaminskas, Rim (Los Angeles, CA)
Kaminskas Jungster II
Kaminskas Papoose Jungster I (8/10th Scale Bü 133)

Kamov (N. I. Kamov) (Russia)
Kamov A-7 (7-EA)
Kamov A-7*bis*
Kamov A-7-3A
Kamov Ka-8
Kamov Ka-10
Kamov Ka-15 *Hen*
Kamov Ka-18 *Hog*
Kamov Ka-22 Vintokryl' *Hoop*
Kamov Ka-25 *Hormone*
Kamov Ka-25K *Hormone*
Kamov Ka-26 *Hoodlum*
Kamov Ka-27 *Helix*
Kamov Ka-32 *Helix* (Civil Ka-27)
Kamov Ka-50 *Hokum*

Kamov-Skrzhinsky (N. I. Kamov & N. K. Skrzhinsky) (Russia)
Kamov-Skrzhinsky KASKR
Kamov-Skrzhinsky KASKR-2

Kansas City Aircraft Corp (Kansas City Aero Manufacturing Corp) (Kansas City MO)
Kansas City Cabin (1926)

Kapferer, Henri (France)
Kapferer 1907 Voisin
Kapferer-Astra (Kapferer-Paulhan): No.1

No.2
No.3

Kappa 77 AS (Czech Republic)
Kappa KP-2U Sova (Owl)

Kaproni Bulgarski *(See: Caproni Bulgaria)*

Kardos, Hugo (Bronx, NY)
Kardos Ornithopter

Karhumäki (Finland)
Karhumäki Karhu 48

Kari-Keen Aircraft (Sioux City, IA)
Kari-Keen 60 (Coupe 60)
Kari-Keen 90
Kari-Keen 90-B (Sioux Coupe Jr)
Kari-Keen 90-C (Sioux Coupe Jr)

Karpinski, Adam (Poland)
Karpinski S.L.1 Akar
Karpinski S.L.1 Akar II

Kasmar, Maxwell (USA)
Kasmar Rollicopter

Kasper, Witold (Seattle, WA)
Kasper Bekas N
Kasper Bekas 1-A
Kasper BKB-1

Kassel (Segelflugzeugbau Kassel) (Germany)
Kassel Anhanger
Kassel "Austria"
Kassel Doppelsitzer (Two-Seater)
Kassel "Professor"
Kassel "Schloss Mainberg"
Kassel "Westpreussen"
Kassel "Wien" (Vienna)
Kassel 12
Kassel 12a
Kassel 17
Kassel 20
Kassel 25
Kassel 28
Kassel 28 "Elida"

Kauffman, Charles H. (Newark, NJ)
Kauffman (Charles) Model A Biplane

Kauffman, W. N. (Davenport, IA)
Kauffman (W. N.) Biplane Flying Boat

Kauffmann (Germany)
Kauffmann No.1
Kauffmann No.3

Kauffold (USA)
Kauffold Special Skybaby

Kawanishi Kokuki KK (Kawanishi Aircraft Co Ltd) (Japan)
Kawanishi E7K1 *Alf* (Navy Type 94 Recon Seaplane)
Kawanishi E8K1 (Experimental 8-Shi Recon Seaplane) (Type P)
Kawanishi E10K1 (Experimental 9-Shi Night Recon Seaplane) (Type T)
Kawanishi E11K1 (Experimental 11-Shi Special Recon Seaplane)
Kawanishi E13K1 (Experimental 12-Shi 3-Seat Recon Seaplane)
Kawanishi E15K1 Shiun, *Norm* (Navy Type 2 High-Speed Recon Seaplane)
Kawanishi H3K1 (Navy Type 90-2 Recon Seaplane)
Kawanishi H6K *Mavis* (Navy Type 97 Flying Boat)
Kawanishi H8K *Emily* (Navy Type 2 Flying Boat Model 12)
Kawanishi K-1 Mail Carrying Aircraft
Kawanishi K-2
Kawanishi K-3
Kawanishi K-5 Mail Carrying Seaplane
Kawanishi K-6 "Harukaze" ("Spring Breeze")
Kawanishi K-7A
Kawanishi K-7B
Kawanishi K-8A
Kawanishi K-8B
Kawanishi K-10
Kawanishi K-11 Experimental Carrier Fighter
Kawanishi K-12 "Sakura" ("Cherry Blossom")
Kawanishi K6K1 (Experimental 11-Shi Intermediate Seaplane Trainer)
Kawanishi K8K1 (Navy Type 0 Primary Seaplane Trainer)
Kawanishi Navy Type 94 Transport
Kawanishi Navy Type 96 Transport
Kawanishi N1K1 Kyofu (Mighty Wind), *Rex* (Navy Seaplane Fighter)
Kawanishi N1K1 Kyofu (Mighty Wind), *Rex* (Navy Seaplane Fighter), NASM
Kawanishi N1K1-J Shiden (Violet Lightning), *George* (Navy Interceptor)
Kawanishi N1K2-J Shiden Kai, *George* (Navy Interceptor)
Kawanishi N1K2-J Shiden Kai, *George* (Navy Interceptor), NASM

Kawasaki Kokuki KK (Kawasaki Aircraft Co Ltd) (Japan)
Kawasaki A-6 (Modified KDA-6)
Kawasaki Army Type Otsu 1 (Kawasaki-Salmson 2 A2)
Kawasaki Army Type 87 Heavy Bomber (Dornier Do N)
Kawasaki C-1
Kawasaki Experimental Carrier Recon Aircraft
Kawasaki Experimental Giyu No.3 Flying Boat
Kawasaki KAG-3 Ram Wing Air-Cushion Vehicle
Kawasaki KAL-1
Kawasaki KDA-2 (Army Type 88-2 Recon Aircraft)
Kawasaki KDA-3 (Army Experimental Fighter)
Kawasaki KDA-5 (Army Type 92 Fighter)
Kawasaki KDC-2 (Experimental Transport)
Kawasaki KDC-5 (C-5)
Kawasaki KH-4
Kawasaki Ki-3 (Army Type 93 Light Bomber)
Kawasaki Ki-5 (Experimental Fighter)
Kawasaki Ki-10 *Perry* (Army Type 95 Fighter)
Kawasaki Ki-28 *Bob* (Army Experimental Fighter)
Kawasaki Ki-32 *Mary* (Army Type 98 Light Bomber)
Kawasaki Ki-45 Toryu (Dragon Slayer), *Nick* (Army Type 2, 2-Engine Fighter)
Kawasaki Ki-45 Toryu, NASM
Kawasaki Ki-48 *Lily* (Army Type 99 Twin Engine Light Bomber)
Kawasaki Ki-56 (Army Type 1 Freight Transport)
Kawasaki Ki-60
Kawasaki Ki-61 Hien (Swallow), *Tony* (Army Type 3 Fighter)
Kawasaki Ki-64 (Army Experimental High-Speed Fighter)
Kawasaki Ki-66 (Army Twin Engine Dive Bomber)
Kawasaki Ki-78 (Army Experimental High-Speed Research Plane)
Kawasaki Ki-96 (Army Twin Engine Experimental Heavy Fighter)
Kawasaki Ki-100 (Army Type 5 Fighter)
Kawasaki Ki-102 *Randy* (Army Type 4 Assault Plane)
Kawasaki Ki-108 (Army Experimental High-Altitude Fighter)

Kawasaki Ki-119 (Army Experimental Light Bomber)
Kawasaki KAL-1
Kawasaki KAL-2
Kawasaki KAT-1
Kawasaki (Dornier Do D) Komet (Comet)
Kawasaki (Boeing-Vertol) KV 107
Kawasaki (Lockheed) P-2J
Kawasaki (Lockheed) T-33
Kawasaki XT-4
Kawasaki (Dornier Do J) Wal (Whale)
Kawasaki (Hughes) 500

Kay Gyroplanes Ltd (UK)
Kay Gyroplane
Kay 331 Gyroplane

Kay Brothers (Eugene & Frank Kay) (Meridian, MS)
Kay Tee 95

Kayaba (KK Kayaba Seisakusho: Kayaba Industrial Co Ltd) (Japan)
Kayaba Heliplane
Kayaba Ka-1
Kayaba Ka-1Kai
Kayaba Ka-2
Kayaba Ku-3
Kayaba Ku-4

Kay-Bee (See: Helio)

Kazan Aviation Institute (Russia)
Kazan KAI-6
Kazan KAI-8
Kazan KAI-9
Kazan KAI-11
Kazan KAI-12
Kazan KAI-13
Kazan KAI-14
Kazan KAI-17
Kazan KAI-19

KB MAI (Russia) *(See also: Aviatika)*
KB MAI 910

Keane (Horace Keane Aeroplanes Inc) (New York, NY)
Keane 4-Cylinder Ford Powered Biplane (1919)
Keane Aeroplanes Ace K-1 (1920)

Keane (Horace Keane Aircraft Corp) (New York, NY)
Keane Aircraft Ace Monoplane (1937)

Kearney, Horace (St Louis, MO)
Kearney-Korn 1911 Farman Type

Keesling, Robert (Anderson, IN)
Keesling Bob-O-Link

Kegel de Kassel *(See: Kassel)*

Keith Rider (West Los Angeles, CA)
Keith Rider R-1
Keith Rider R-1 "San Francisco I"
Keith Rider R-2 "Bumble Bee"
Keith Rider R-2 "San Francisco II"
Keith Rider R-3 "Marcoux-Bromberg Special"
Keith Rider R-3 "Gilmore the Record Breaker"
Keith Rider R-4
Keith Rider R-5 "Elmendorf Special"
Keith Rider R-5 "Jackrabbit"
Keith Rider R-6
Keith Rider R-6 "Eight Ball"

Keith-Weiss (A. Keith & José Weiss) (UK)
Keith-Weiss 1912 Aviette
Keith-Weiss 1916 Monoplane

Keleher, James J. (Fremont, CA)
Keleher Lark

Kellett Autogiro Corp (Philadelphia, PA)
Kellett ASP (Airborne Support Platform)
Kellett YG-1 Autogiro
Kellett YG-1A Autogiro
Kellett YG-1B Autogiro
Kellett K-2 Autogiro
Kellett K-3 Autogiro
Kellett K-4 Autogiro
Kellett KD-1 Autogiro
Kellett KD-1A Autogiro
Kellett KD-1B Autogiro
Kellett KH-2
Kellett KH-15 "Stable Mable"
Kellett KH-17 Autogiro
Kellett K1-X Autogiro
Kellett XO-60 Autogiro
Kellett XO-60 Autogiro, NASM
Kellett YO-60 Autogiro
Kellett XR-2 Autogiro
Kellett XR-3 Autogiro
Kellett XR-8
Kellett XR-8, NASM
Kellett XR-8A
Kellett XR-10 (XH-10)

Kelley, John (Eldorado, CA)
Kelley High Wing Monoplane

Kelly, Dudley R. (Versailles, KY)
Kelly-D (Homebuilt)

Kellner-Béchereau (France)
Kellner-Béchereau Biplace de Tourisme
Kellner-Béchereau E.1
Kellner-Béchereau 28VD

Kemph, James J. (Washington, DC)
Kemph Orniplane

Ken Brock Mfg *(See: Brock)*

Ken S. Coward & Associates *(See: Bee Aviation Associates, Inc)*

Kendle-Nelson (F. J. Kendle & C. E. Nelson) (Chicago, IL)
Kendle-Nelson K-N-1 Monoplane

Kendrick Aeroplane Co, Inc *(See: Alexandria Aircraft Corp)*

Kenilworth Aircraft Club *(See: Driggers, Willard R.)*

Kennedy (Chessborough J. H. Mackenzie-Kennedy) (UK)
Kennedy Giant
Kennedy No.1 (1912 Pusher Biplane)

Kensinger, Ned (Fort Worth, TX)
Kensinger KF Racer
Kensinger Tater Chip Racer

Kentucky Airplane Co (Owensburg, KY)
Kentucky Cardinal

Keppel, Thomas, Robert & Jesse Jr (St Louis, MO)
Keppel Aeroplane (1912)

Kerchone *(See: Aman & Kerchone)*

Ketner Air Coach Co (Moline, IL)
Ketner 1926 Air Coach

Keystone Aircraft Corp (Bristol, PA)
In 1920 Thomas Henri Huff and Elliot Daland founded the **Ogdensburg Aeroway Corp** in Ogdensburg, NY. By 1922 the firm had been renamed **Huff, Daland & Co, Inc**, a name held until July 1925 when the company moved to Bristol, PA, as **Huff-Daland Airplanes, Inc**. This became **Keystone Aircraft Corp** in March 1927. **Loening Aero-**

nautical Engineering Corp became a Keystone operating division in October 1928, though references to the Loening name tapered off by 1930, at which time Keystone became a division of the **Curtiss-Wright Corp**. Keystone was closed in 1936. All Huff-Daland designs which entered production before March 1927 are listed under **Huff-Daland**; designs which subsequently entered production are listed under **Keystone**. All Loening designs appear under **Loening**.

Keystone Air Yacht (See: Loening)
Keystone XB-1 Super Cyclops
Keystone XB-1B Super Cyclops
Keystone B-3A (LB-10A)
Keystone Y1B-4 (LB-13)
Keystone B-4A
Keystone Y1B-5 (LB-14)
Keystone B-5A
Keystone Y1B-6 (LB-13, B-3A)
Keystone B-6A
Keystone Cabin Amphibian (See: Loening)
Keystone Canadian Duster (See: Huff-Daland)
Keystone XHL-1 Air Ambulance (See: Loening)
Keystone XHB-3 (See: Huff-Daland)
Keystone HB Monoplane
Keystone K-47 Pathfinder
Keystone K-47 Pathfinder "American Legion"
Keystone K-47 Pathfinder "Santa Maria"
Keystone K-47 Pathfinder "Sign Carrier I"
Keystone K-55 Pronto
Keystone K-78 Patrician
Keystone K-81
Keystone K-84 Commuter (See: Loening)
Keystone K-85 (See: Loening C-4C)
Keystone LB-1 (See: Huff-Daland)
Keystone XLB-3 (See: Huff-Daland)
Keystone XLB-3A
Keystone XLB-5
Keystone LB-5
Keystone LB-5A
Keystone XLB-6 Panther
Keystone LB-6 Panther
Keystone XLB-7
Keystone LB-7
Keystone LB-8
Keystone LB-9
Keystone LB-10
Keystone LB-11
Keystone LB-11A
Keystone LB-12
Keystone LB-13 (See: Y1B-4 & Y1B-6)
Keystone LB-14 (See: Y1B-5)
Keystone XNK-1 Pup
Keystone NK-1 Pup
Keystone XOK-1
Keystone XO2L-1 (See: Loening)
Keystone XO-10 (See: Loening)
Keystone XO-15 (XP-531)
Keystone XO-37
Keystone OA-2 (See: Loening)
Keystone PK-1
Keystone Pelican (See: Huff-Daland)

Kharkov Aviation Institute (Russia)
Kharkov KhAI-1
Kharkov KhAI-17
Kharkov KhAI-18
Kharkov KhAI-19
Kharkov KhAI-20
Kharkov KhAI-22A
Kharkov KhAI-27

Khioni-Konek Gorbunok (Russia)
Khioni-Konek Gorbunok Kh-4 (Anatra VKh Anadva)
Khioni-Konek Gorbunok Kh-5

Kiceniuk, Taras, Jr (Palomar Mountain, CA)
Kiceniuk Icarus Hang-Glider

Kieff (Germany)
Kieff Primary Glider

Kyle-Smith Aircraft Co (Wheeling, WV)
Kyle-Smith C.3 Pusher Biplane
Kyle-Smith 1918/19 Pusher

Kimball, Asa (USA)
Kimball (Asa) K8 Baby Plane

Kimball, Wilbur M. (New York)
Kimball (Wilbur) Helicopter (1908)
Kimball (Wilbur) N.Y.2
Kimball (Wilbur) 8-Prop Pusher Biplane (1909)

Kimball, Wiley (Severn, MD)
Kimball (Wiley) Monoplane

Kimura, Dr Hidomasa (Japan)
Kimura (Hidomasa) N-52 Lightplane

Kindree, W. P. (Detroit, MI)
Kindree Sky Car

King, Wesley J. (Cleveland, OH)
King King-Bird (1928)

King's Engineering Fellowship (Orange City, IA)
King's Engineering Fellowship Angel

Kingsford Smith Aviation Service Pty Ltd (Australia)
Kingsford Smith EP.9 Conversion (Edgar Percival)
Kingsford Smith KS.3 Cropmaster (See: Yeoman)
Kingsford Smith PL-7 Tanker

Kinman, Duane (USA)
Kinman Super Simple

Kinner Airplane & Motor Co (Glendale, CA)
Kinner Airster
Kinner Canary
Kinner Coupe Monoplane
Kinner Courier "Spirit of Ether"
Kinner Envoy C-7
Kinner Invader
Kinner K-1
Kinner Playboy R-1
Kinner Playboy R-5
Kinner Sedan
Kinner Sportster B-1
Kinner Sportster K-100
Kinner Sportwing B-2
Kinner Sportwing B-2-R (See also: Timm 160 Sportwing)
Kinner XRK-1 (Navy Envoy C-7)

Kippers, Harold M. (Mukwonago, WI)
Kippers K-1 (Homebuilt)

Kirby (USA)
Kirby Cadet Glider
Kirby Gull Sailplane
Kirby Kite Glider

Kirkham, Charles B. (Kirkham Products Co) (Savona, NY)
Kirkham Airboat
Kirkham Pusher Biplane (1911)
Kirkham (Kirkham-Packard) 1926 Schneider Cup Racer
Kirkham Tractor Biplane (1912)
Kirkham Vespa

Kiser, Daniel (Milwaukee, WI)
Kiser Airliner (1926 7-Place Landplane)

Kitfox *(See: Sky Star Aircraft Corp)*

Kitty Hawk *(See: Viking)*

Kjeller Flyvemaskinfabrikk (Norway)
Kjeller Kaje I
Kjeller Kaje II

Klampher, G. F. (Wichita, KS)
Klampher 1930 Monoplane

Klassen, J. H. (San Francisco, CA)
Klassen Monoplane

Kleckler & Zimmerman (USA)
Kleckler & Zimmerman 7-Motor Bomber

Klein, Gustav (Germany)
Klein 2-Seat Monoplane

Klemm Leichtflugzeugbau GmbH (Germany)
Klemm K 2
Klemm C 5
Klemm L 20
Klemm L 21
Klemm L 25
Klemm L 25a
Klemm L 25e (L 25E)
Klemm L 25 I
Klemm L 25 IA
Klemm Kl 25
Klemm Vl 25
Klemm Wl 25
Klemm II a L 25
Klemm L 26
Klemm L 26 Hé
Klemm L 26 I
Klemm L 26 IIa
Klemm L 26 V
Klemm Kl 26
Klemm Vl 26
Klemm Kl 30
Klemm Kl 31
Klemm Kl 32
Klemm Kl 33
Klemm Kl 35
Klemm Wl 35
Klemm Kl 25
Klemm Kl 105

Klemperer (Germany)
Klemperer Glider
Klemperer 1922 Monoplane Canard Glider

Kloeren, Theodore (Philadelphia, PA)
Kloeren Convertiplane

Klosterlein (Germany)
Klosterlein Kondor II

Knepper (Paul H. Knepper Aircraft) (Lehighton, PA)
Knepper KA-1 Crusader
Knepper KAC-4 Crusader
Knepper KAC-5 Crusader

Knight Twister Aircraft Corp *(See: Payne)*

Knoll Aircraft Co (Wichita, KS)
Knoll KN-1
Knoll KN-2
Knoll KN-3

Knoll Brayton Aeronautical Corp (Felix Knoll & Flint Brayton) (Norwich, CT)
Knoll Brayton Sachem (SA-3)

Knoller, Richard (Knoller Flugzeugbau) (Austria-Hungary)
Knoller B.I
Knoller C.I
Knoller C.II
Knoller 30.50 (Prototype)

Koberg, Leslie R. *(See: Bushby M-II Mustang II)*

Kocherigin (Russia)
Kocherigin DI-6 (Tskb-11)
Kocherigin SR (R-9, LBSh, TsKB-27)
Kocherigin-Gurevich TSh-3 (TsKB-24)
Kocherigin-Itskovich LR-1 (TsKB-1)
Kocherigin-Miroshnichenko LB-2LD (B1)

Kochyerigin *(See: Kocherigin)*

Kœchlin (Aéroplanes P. Kœchlin) (France)
Paul Kœchlin (Koechlin) designed and flew his first aircraft in 1908. Later that year he joined with Emil de Pischoff to form **Ateliers de Pischoff-Kœchlin.** In 1910 the company dissolved into **Aéroplanes P. Kœchlin** and **Établissements Autoplan.** Kœchlin's joint designs are listed under **Pischoff-Kœchlin.** The individual designs of Kœchlin and de Pischoff are under **Kœchlin** or **Pischoff.**
Kœchlin Type A Monoplane
Kœchlin Type B Monoplane
Kœchlin Type C Monoplane
Kœchlin Pivot 1911 Type Militaire Monoplane
Kœchlin Pivot 1911 Monoplane
Kœchlin No.1 (1908 Tandem Biplane)
Kœchlin 1909 Monoplane

Kocjan, Antoni (Poland)
Kocjan Czajka (Lapwing) I Glider
Kocjan Czajka (Lapwing) II Glider
Kocjan Komar (Gnat) Glider
Kocjan Mewa (See: Grzeszczyk & Kocjan)
Kocjan Orlik I (Eaglet)
Kocjan Orlik II (Eaglet)
Kocjan Orlik III (Eaglet) (Olympic Orlik)
Kocjan Sokól (Falcon) Glider
Kocjan Sroka (Magpie) Glider
Kocjan Wrona (Crow) Glider

Kokusai (Kokusai Kokuki Kabushiki Kaisha: International Aircraft Co Ltd) (Japan)
Kokusai HK-1
Kokusai Ki-59 (Army Type 1 Transport) *Theresa*
Kokusai Ki-76 *Stella*
Kokusai Ki-86 *Cypress*
Kokusai Ki-105 Ohtori (Phoenix)
Kokusai Ku-7 Manazuru (Crane)
Kokusai Ku-8-II (Army Type 4 Large Transport Glider) *Gander*

Kolb Co Inc (Phoenixville, PA)
Kolb Flyer
Kolb Ultrastar

Kolpakov-Miroshnichenko (Russia)
Kolpakov-Miroshnichenko LB-2LD (B-1)

KOMTA (Russia)
KOMTA 1920 Twin-Engined Triplane
KOMTA 1921 Twin-Engined Triplane

Kondor Flugzeugwerke GmbH (Germany)
Kondor B.1
Kondor D.I
Kondor D.II
Kondor D.6
Kondor D.7
Kondor E.IIIa
Kondor Eindecker (Monoplane)
Kondor Taube (Dove) III
Kondor W.1
Kondor W.2C

Koolhoven Vliegtuigen (Netherlands)
Frederick "Fritz" Koolhoven designed his first aircraft in 1910. He worked briefly with **Deperdussin,** then with **British Deperdussin Syndicate, Ltd.** In 1913 he become a designer for **Sir W.**

G. **Armstrong, Whitworth & Co Ltd**, moving in 1917 to **British Aerial Transport**. In 1923 he returned to the Netherlands as engineer for **Nationale Vliegtuig Industrie (NVI)**. He left NVI in 1926 to set up his own firm, which became **N.V. Koolhoven Vliegtuigen** in 1934. The Koolhoven plant was destroyed by bombing during WWII; Koolhoven died in 1946. All of his aircraft designs after 1913 carried "F.K." designations. They can be found under **Armstrong Whitworth** (F.K.1 thru 10), **B.A.T.** (F.K.22 thru 28), **NVI** (F.K. 30 thru 33), and **Koolhoven** (F.K. 34 thru 58). Some designations were reused; for example, there were two F.K.1 designs. The B.A.T. F.K.23 Bantam was also manufactured by NVI as the F.K.23A.

Koolhoven F.K.34
Koolhoven F.K.35
Koolhoven F.K.37
Koolhoven F.K.38
Koolhoven F.K.39
Koolhoven F.K.40
Koolhoven F.K.41
Koolhoven F.K.42
Koolhoven F.K.43
Koolhoven F.K.45
Koolhoven F.K.46
Koolhoven F.K.47
Koolhoven F.K.48
Koolhoven F.K.49
Koolhoven F.K.50
Koolhoven F.K.51
Koolhoven F.K.52
Koolhoven F.K.53
Koolhoven F.K.54
Koolhoven F.K.55
Koolhoven F.K.56
Koolhoven F.K.57
Koolhoven F.K.58

Koppen, Otto C. (Waban, MA)
Koppen Puritan

Koppen-Bollinger *(See: Helio)*

Korn Brothers (Milton H. & Edward A.) (Shelby County, OH)
Korn Brothers 1910 Monoplane (Benoist Type XII)

Korolev, S. P. (Russia)
Korolev SK-4

Korolov (Russia)
Korolov RP-318-1 Rocket Plane

Korsa, A. (Switzerland)
Korsa I
Korsa T-2

de Korvin, Chevalier (France)
de Korvin 1907 8-Wing Multiplane
de Korvin 1911 Monoplane

Kosellek (France)
Kosellek G 1 Glider (1934)

Kowalski, John (Pittburgh, PA)
Kowalski Biplane (C.1910)
Kowalski "The Bird"

Kozlowski, Wladyslaw & Jerzy
Kozlowski WK 1 Jutrzenka (Dawn)
Kozlowski WK 3

Kraft Super Fli *(See: P. K. Plans Inc)*

Kramme & Zeuthen *(See: Skandinavisk Aero Industri)*

Krapish, Alexander P. (Lowell, MA)
Krapish K-4

Krarup, Kay (Valley Stream, NY)
Krarup Heath Parasol

Kratz Corp (St Louis, MO)
Kratz Serial 1
Kratz Model 1 TB Special

Krekel, Paul (Germany)
Krekel Grille (Cricket) Sailplane

Kreider-Reisner (Hagerstown, MD)

The **Kreider-Reisner Aircraft Co Inc** was founded in 1925 by A. H. Kreider and L. E. Reisner. The company built a Midget racer in 1926 and began production of the Challenger series in 1927. When Kreider-Reisner became a wholly-owned division of the **Fairchild Airplane Manufacturing Corp** in April 1929, the Challenger series was redesignated "KR Biplanes." Kreider-Reisner was renamed the **Fairchild Aircraft Corp** in 1935. The Midget is filed under **Kreider-Reisner**; the Challenger series is filed under **Fairchild**.

Kreider-Reisner Midget

Kress, Wilhelm (Austria)
Kress Drachenflieger (Dragonfly)
Kress Ornithopter (1888)

Kreutzer (Joseph K. Kreutzer Inc) (Los Angeles, CA)
Kreutzer K-1 Air Coach Prototype
Kreutzer K-2 Air Coach
Kreutzer K-3 Air Coach
Kreutzer TM-4
Kreutzer K-5 Air Coach
Kreutzer T-6 Air Coach

Krieger (Germany)
Krieger Parasol
Krieger Taube (Dove)

Kronfeld, Ltd *(See: British Aircraft - B.A.C.)*

Krouse, Frank J. (Denver, CO)
Krouse 1911 Aeroplane Proposal

Krutchkoff, Andre (USA)
Krutchkoff SHP-1 (Modified Schreder HP-14)

Krylov (Russia)
Krylov ASK
Krylov R-2

Krzakala, Wiktor (Poland)
Krzakala 1927/28 Gliders

Kubicki, Jan (Poland)
Kubicki Ikub I

Kucfir, Konrad (Poland)
Kucfir VTOL Project
Kucfir Pirat (Pirate) Glider

Kucher Airplane Corp (Andrew A. Kucher) (Dover, DE)
Kucher Club

Kugisho (Japan)

The Naval Air Arsenal at Yokosuka was known by a variety of names and acronyms from its foundation in 1914 through World War II. Although its WWII aircraft are most commonly listed under **Yokosuka**, during most of the war years, the arsenal was officially known as **Kugisho**, an acronym for **Kaigun Koku-Gijutso-Sho** (Naval Technical Air Arsenal). All aircraft built by the arsenal are filed under **Kugisho**.

Kugisho B4Y1 *Jean* (Navy Type 96 Carrier Attack Bomber)
Kugisho D4Y1 Suisei (Comet), *Judy* (Navy Carrier Bomber Model 11)
Kugisho D4Y2 Suisei (Comet), *Judy* (Navy Carrier Bomber Model 12)

Kugisho D4Y3 Suisei (Comet), *Judy* (Navy Carrier Bomber Model 33)
Kugisho E14Y1 *Glen* (Navy Type 0 Small Recon Seaplane)
Kugisho K1Y2 Type 13 (Navy Seaplane Trainer)
Kugisho K4Y1 Type 90 (Navy Seaplane Trainer)
Kugisho MXY7 Model 11 Ohka (Cherry Blossom) Baka (Crazy)
Kugisho MXY7 Model 11 Ohka (Cherry Blossom) Baka (Crazy), NASM
Kugisho MXY7 Model 22 Ohka (Cherry Blossom) Baka (Crazy)
Kugisho MXY7 Model 22 Ohka (Cherry Blossom) Baka (Crazy), NASM
Kugisho MXY7 Model 43 K-1 Kai Wakazakura Trainer
Kugisho MXY7 Ohka (Cherry Blossom) K-1 Trainer
Kugisho MXY8 Akigusa (Navy Experimental Glider)
Kugisho P1Y1 Model 11 Ginga (Milky Way), *Frances*
Kugisho P1Y1 Model 11 Ginga (Milky Way), *Frances*, NASM
Kugisho P1Y2 Model 16 Ginga (Milky Way), *Frances*
Kugisho Ro-Go Ko-Gata (Recon Seaplane)
Kugisho R2Y-1 Keiun (Beautiful Cloud) (Navy Experimental 18-Shi Recon Plane)

Kühlstein (Germany)
Kühlstein Torpedo Eindecker

Kuhnert (USA)
Kuhnert Photo Plane

Kuibyshev Aviation Institute (Russia)
Kuibyshev Sverchok-1 (Cricket 1)
Kuibyshev VIHR-1

Kungl (Sweden) *(See: FFVS J22)*

Kunicke, George (Bronx, NY)
Kunicke Flying Machine (1910)

Kupfer-Smolensk (M. A. Kupfer) (Russia)
Kupfer-Smolensk 1953 Helicopter

Kurbala (Russia)
Kurbala KG-1 Heavy Glider

Kurz-Kash (C. A. Kurz) (USA)
Kurz-Kash Mailplane

Kuznetsov, V. A. (Russia)
Kuznetsov A-6 Autogyro
Kuznetsov A-8 Autogyro (See: Tsagi A-8 Autogyro)
Kuznetsov A-13 Autogyro
Kuznetsov A-14 Autogyro

Kuznetsov-Mil' (V. A. Kuznetsov & M. L. Mil') (Russia)
Kuznetsov-Mil' A-15 Autogyro

KW *(See: Kaiserliche Werft, Kiel)*

Kyle Smith Aircraft Co (Huntington, WV)
Kyle Smith Aircraft Co, General

Kyushu (Japan)
Kyushu J7W1 Shinden (Magnificent Lightning)
Kyushu J7W1 Shinden (Magnificent Lightning), NASM
Kyushu K9W1 Momiji (Maple) *Cypress*
Kyushu K10W1
Kyushu K11W1 Shiragiku (White Chrysanthemum)
Kyushu Q1W1 Tokai (Eastern Sea) *Lorna*

KZ *(See: Skandinavisk Aero Industri)*

L

La Salle Aircraft Corp (Ottowa, IL)
La Salle F.2

Labaudieet et Puthet (France)
Labaudieet et Puthet 1908 Biplane Glider Series
Labaudieet et Puthet La Comète (1909 Monoplane)

Laboratory Eiffel *(See: Eiffel, Gustaf)*

Lac St-Jean (Avionnerie Lac St-Jean Inc) (Canada)
Lac St-Jean Cyclone

LACAB *(See: Ateliers de Constructions Aéronautiques Belges)*

Lacaille et Lemaire (France)
Lacaille et Lemaire 1910 Monoplane Glider

Lachassagne, Adolf (France)
Lachassagne 1912 Tandem Monoplane
Lachassagne 1923 Tandem Monoplane
Lachassagne 1925 Avionette
Lachassagne 1929 AL 3 Avionette
Lachassagne 1929 AL 5 Avionette

Lacina, Peter W. (Canada)
Lacina (Peter W.) 1929 Flying Machine

Ladougne, Emile (France)
Ladougne Colombe (Pigeon), 3-Cylinder Anzani
Ladougne Colombe (Pigeon), 6-Cylinder Anzani

Lafayette Airplane Works (Los Angeles, CA)
Lafayette Rebuilt Fiske Standard
Lafayette Model T

LaGG *(See: Lavochkin)*

de Lailhacar, Jacques Albert *(See: J.A.L)*

Laird (E. M. Laird Airplane Co) (Emil Matthew "Matty" Laird) (Chicago, IL)
Laird ("Matty"):
Anzani Stunt Biplane (1915)
Baby Biplane (1912)
C-6 Special (1927)
LC-A
LC-B Commercial
LC-C
LC-D Speedwing "Continental Comet"
LC-D Speedwing Jr
LC-D Speedwing "Solution"
LC-D Speedwing "Super Solution"
LC-E Sesquiplane
LC-H3
LC-R Speedwing
Limousine
LTR-14 Pesco Special Meteor (See: Turner-Laird)
Model S Commercial
Model S Sport
Swallow
Twin-Motor Cabin Air Liner

Laird Aircraft Co (Charles L. Laird) (Wichita, KS)
Laird (Charles) Whippoorwill

Laister Sailplanes Inc (South El Monte, CA)
Laister LP-15 Nugget
Laister LP-46
Laister LP-49

Laister, Jack (Highland Park, MI)
Laister LA-1
Laister "Yankee Doodle" (Lawrence Tech Sailplane)

Laister-Kauffmann Aircraft Corp (St Louis, MO)
Laister-Kauffmann LK-10A
Laister-Kauffmann CG-4A (See also: Waco)
Laister-Kauffmann CG-7 (See: Bowlus)
Laister-Kauffmann XCG-10
Laister-Kauffmann XCG-10A Trojan Horse
Laister-Kauffmann YCG-10A (G-10A) Trojan Horse
Laister-Kauffmann XTG-4
Laister-Kauffmann TG-4A

Lake Aero Corp (Christopher J. Lake) (Bridgeport, CT)
Lake (Aero) Stable Airplane (Even-Keel Airplane) (1919)

Lake Aircraft (Sanford, ME)
Colonial Aircraft Corp was established in 1946 by David B. Thurston and Herbert P. Lindblad. The company went bankrupt in 1959; the assets were acquired by the **Lake Aircraft Corp** of Sanford, ME, which was established by Lindblad in 1960. In 1962, the company was taken over by **Consolidated Aeronautics, Inc.** Consolidated Aeronautics established **Aerofab Inc** at Sanford to produce Lake designs which were then marketed through the **Lake Aircraft Division of Consolidated Aeronautics.** For all Colonial designs predating the 1959 bankruptcy, see **Colonial.** For post 1959 development, see **Lake Aircraft.**

Lake (Aircraft) LA4/200 Buccaneer
Lake (Aircraft) LA4/200EP
Lake (Aircraft) LA250 Renegade
Lake (Aircraft) LA270 Turbo Renegade
Lake (Aircraft) Seawolf

Lakes Flying Co (UK)
Lakes Hydro-Monoplane (1913)
Lakes Sea Bird (1912) (See also: Avro Duigan Biplane)
Lakes Water Bird (1911) (See: Avro Curtiss-Type Biplane)
Lakes Water Hen (1912)

Lambert (Robertson, MO)
Lambert Aircraft Engine Corp assumed the assets of the **Velie Motor Corp** in 1929 following the death of Willard Velie. The Velie company's **Mono Aircraft Division** becoming the **Mono Aircraft Corp**. Lambert and Mono went into receivership in 1931, emerging in 1932 as **Lambert Engine and Machine Co** and **Monocoupe Corp** respectively. In July 1934 the two companies were joined under the umbrella of the **Lambert Aircraft Corp**, with Lambert Engine producing aircraft engines and Monocoupe producing aircraft under its own name. Monocoupe was dissolved in 1940 and its assets transferred to an independent company, **Monocoupe Aeroplane and Engine Corp**. For all aircraft designed by Monocoupe and its direct predecessors, see **Monocoupe**. For aircraft developed by Lambert outside of Monocoupe's operations, see **Lambert**.
Lambert Model 1344 (Lambert-Breese X, Breese-Dallas 1, Michigan Model 1)

Lamboley, Francois X. (New York, NY)
Lamboley 1876 Aircraft Patent

Lamburth, Cassius E. (San Francisco, CA)
Lamburth (Cassius E.) 1910 Aeroplane Patent

Laminières (See: Collin de Laminières)

Lampich (Hungary)
Lampich D-2 Pajtás
Lampich L-1
Lampich L-2 Roma
Lampich L-4 Bohóc
Lampich BL-5
Lampich BL-6
Lampich L-9
Lampich NL XXI
Lampich NL XXII

Lamson Aircraft Co, Inc (Yakima, WA)
Central-Lamson Corp was established by Robert Lamson in 1953 to manufacture his Air Tractor agricultural aircraft (which was to be marketed through **Central Aircraft, Inc** of Yakima, WA). In 1955 the company was renamed **Lamson Aircraft Co, Inc**. Towards the end of 1955, the company suspended production. Although Lamson intended to resume work in 1957, it does not appear to have done so. For all Central-Lamson and Lamson products, see **Lamson**.
Lamson (Aircraft Co) L-101 Air Tractor

Lamson, Charles H. (Portland, ME)
Lamson (Charles H.) 1896 Lilienthal-Type Glider

Lamson, Robert (Yakima, WA)
Lamson (Robert) Alcor

Lancair International Inc (Redmond, WA)
Lancair ES
Lancair Super ES
Lancair IV
Lancair IV-P
Lancair 200 Lancer
Lancair 235
Lancair Columbia 300
Lancair 320
Lancair 360 (320/360)

Lancashire Aircraft Co (See: Percival, Edgar)

Lanchester, F. W. (UK)
Lanchester 1897 Cantilever Pusher Monoplane
Lanchester 1907 Cantilever Pusher Monoplane

Landes Frères (France)
Landes Frères 1922 Monoplane Glider

Landes-Breguet (France)
Landes-Breguet 1925 Monoplane Glider

Landes-Derouin (France)
Landes-Derouin Oiseau Bleu (Blue Bird) 1923 Glider

Landgraf Helicopter Co (Fred Landgraf) (Los Angeles, CA)
Landgraf H-2

Lane Aircraft Co (Dallas, TX)
Lane Riviera Amphibian (SIAI-Marchetti FN-333) (See also: Nardi)

Lane, Dick (Fulton, NY)
Lane (Dick) "Flycycle"

Lanford, C. E. (Greenville, SC)
Lanford PLG-1 (See: Perry, Lanford, & Gault)
Lanford GLP-3 (See: Johnson, Luther)

Langley, Samuel P. (Washington, DC)
Langley Aerodrome A (Great Aerodrome, Man-Carrying Aerodrome)
Langley Aerodrome A, Curtiss 1914 Rebuild
Langley Aerodrome A, NASM
Langley Aerodrome No 0 (1891)
Langley Aerodrome No 1 (1891)
Langley Aerodrome No 2 (1892)
Langley Aerodrome No 3 (1892)
Langley Aerodrome No 4 (1895)
Langley Aerodrome No 5 (1895-96)
Langley Aerodrome No 6 (1895-96)
Langley Clockwork Model
Langley Gliding Model Aerodromes (1895)
Langley Ladder Kite (1896)
Langley Model Aerodromes, General
Langley Model Aerodrome No 4 (1895)
Langley Model Aerodrome No 11
Langley Model Aerodrome No 13
Langley Model Aerodrome No 14
Langley Model Aerodrome No 15
Langley Model Aerodrome No 19
Langley Model Aerodrome No 20
Langley Model Aerodrome No 21
Langley Model Aerodrome No 22
Langley Model Aerodrome No 23
Langley Model Aerodrome No 24
Langley Model Aerodrome No 25
Langley Model Aerodrome No 26
Langley Model Aerodrome No 27
Langley Model Aerodrome No 28
Langley Model Aerodrome No 30
Langley Model Aerodrome No 31
Langley Proposed Man-Carrying Aerodrome (1898-99)

Langley "Quarter-Size" Aerodrome (1900-01)
Langley "Rubber-Pull" Model Aerodrome (1895-96)
Langley Whirling Arm (1888-90)

Langley Aviation Corp (New York, NY)
Langley Aircraft Corp XNL-1 (2-4-90)
Langley Aircraft Corp 2-4-90

Lanier Aircraft Corp (Edward M. Lanier) (Marlton, NJ)
Lanier (Edward M.):
Paraplane I Vacu-Jet
Paraplane II
Paraplane Commuter 110
Paraplane 120
Paraplane 130
Paraplane 160
Paraplane 180
Paraplane 443
Paraplane Executive
Paraplane Gazelle
Paraplane Le Bird
Paraplane Sportster
Paraplane Super Commuter

Lanier, Edward H. (University of Miami, FL)
Lanier (Edward H.) XL-1
Lanier (Edward H.) XL-2
Lanier (Edward H.) XL-3
Lanier (Edward H.) XL-4
Lanier (Edward H.) XL-5 Vacuplane
Lanier (Edward H.) XL-6 Vacuplane

Lansing, Arthur D. (Youngstown, OH)
Lansing Sport Monoplane (1929)

Lanzius Aircraft Co (Brooklyn, NY)
Lanzius L I
Lanzius L II

Lark Aircraft Co (Mesa, AZ) *(See: Helton)*

Larkin *(See: LASCO)*

Larnaudi (France)
Larnaudi Single-Engined Biplane Flyingboat (1918)
Larnaudi Twin-Engined Triplane Flyingboat (1918)

Laron Aviation Technologies, Inc (Portales, NM)
Laron Shadow
Laron Star Streak
Laron Streak Shadow

Larribe (France)
Larribe 1912 Aviette

Larsen *(See: Junkers F 13a, Junkers-Larsen JL-6)*

Larsen, Herman (Keasbey, NJ)
Larsen (Herman) 2-C

Larsen, John G. (Chicago, IL)
Larsen "Flying Suit" (1909)

Larson Aero Development (Concord, CA)
Larson D-1 Duster
Larson F-12 Baby Biplane

LASA (Lockheed-Azcarate SA) (Mexico)
In the late-1950s, the **Lockheed Aircraft Corp** established **Lockheed-Azcarate SA** in Mexico to produce a small utility transport designed to the specifications of Mexican General Juan Azcarate.
LASA 60

Lasco (Larkin Aircraft Supply Co) (Australia)
Lasco Lasconder
Lasco Lascoter
Lasco Lascowl

Lasher, C. W. (Winter Springs, FL)
Lasher Renegade I

Laskowitz, I. B. (Brooklyn, NY)
Laskowitz L15C
Laskowitz L15H
Laskowitz L180C
Laskowitz L180H

Lasley, John (Beloit, WI)
Lasley Sport

Lataste (France)
Lataste le Gyroscope

Latécoère (France)
Latécoère LAT N
Latécoère 3 Postal
Latécoère LAT 4
Latécoère LAT 6
Latécoère LAT 7
Latécoère LAT 8
Latécoère LAT 14
Latécoère LAT T.14 Postal
Latécoère LAT 15
Latécoère LAT 16
Latécoère LAT 17
Latécoère LAT 17.1.J
Latécoère LAT 17.1.R
Latécoère LAT 17.3.J
Latécoère LAT 17.3.R
Latécoère LAT 17.4.R
Latécoère LAT 18
Latécoère LAT 19 BN2
Latécoère LAT 20
Latécoère LAT 21
Latécoère LAT 21*bis*
Latécoère LAT 21*ter*
Latécoère LAT 23
Latécoère LAT 25
Latécoère LAT 25.1.R
Latécoère LAT 25.2.R
Latécoère LAT 25.3.R
Latécoère LAT 25.4.J
Latécoère LAT 25.5.R
Latécoère LAT 26
Latécoère LAT 26.2.R
Latécoère LAT 26.3
Latécoère LAT 26.6.R
Latécoère LAT 28.1
Latécoère LAT 28.2
Latécoère LAT 28-3 (Floatplane)
Latécoère LAT 28-5 (Floatplane)
Latécoère LAT 29-0
Latécoère LAT 29-8
Latécoère LAT 300
Latécoère LAT 300 "Croix du Sud" ("Southern Cross")
Latécoère LAT 301
Latécoère LAT 302
Latécoère LAT 32-2
Latécoère LAT 32-3
Latécoère LAT 35
Latécoère LAT 35-0
Latécoère LAT 38
Latécoère LAT 38-0
Latécoère LAT 38-1
Latécoère LAT 44-0
Latécoère LAT 49-0
Latécoère LAT 49-1
Latécoère LAT 500
Latécoère LAT 501
Latécoère LAT 520
Latécoère LAT 521 "Lieutenant de Vaisseau Paris"
Latécoère LAT 522 "Ville de Saint-Pierre"
Latécoère LAT 550
Latécoère LAT 570
Latécoère LAT 582
Latécoère LAT 611 ("Achernar")

Latécoère LAT 631

Latham, Hubert (France)
Latham (Hubert) 110 (1932)

Latham, Jean (France)
Latham E 5
Latham HB 3
Latham HB 5
Latham L-1 Schneider Trophy Racer
Latham L2
Latham 42
Latham 43 HB3
Latham 45
Latham 47
Latham 1920 3-Engined Flying Boat
Latham 1923 1-Engined Hydro
Latham 1928 Hydro

Lather Commercial Aircraft Co, Inc (Portland, OR)
Lather C-2

Laubenthal *(See: Grunau)*

Lauchin, George (USA)
Lauchin 1928 Convertiplane

Läuchli, Walter (Switzerland)
Läuchli (Walter) Monoplane Design

Launoy & Bienvenu (France)
Launoy & Bienvenu 1784 Helicopter Design

Lavell Aircraft Co (Newton, PA) *(See: Fleetwings)*

Lavelli, Carlos Argentino (Argentina)
Lavelli Halcon (Falcon)

Laverda SpA (Italy)
Laverda F.8 Falco (Falcon) (See: Aviamilano)

Laville, André (France/Russia)
Laville DI-4
Laville ZIG-1 (Zavod PS-89)

Lavochkin (Russia)
Lavochkin K-37
Lavochkin La-5
Lavochkin La-7 (La-120) *Fin*
Lavochkin La-9 (La-130) *Fritz*
Lavochkin La-11 (La-140) *Fang*
Lavochkin La-15 (La-174) *Fantail*
Lavochkin La-15UTI (La-180)
Lavochkin La-126
Lavochkin La-138
Lavochkin La-150
Lavochkin La-152
Lavochkin La-154
Lavochkin La-156
Lavochkin La-160
Lavochkin La-168
Lavochkin La-174TK
Lavochkin La-176D
Lavochkin La-190
Lavochkin La-200
Lavochkin La-250 Anaconda
Lavochkin LaGG-1 (I-22)
Lavochkin LaGG-3 (I-301)
Lavochkin LL (Lavochkin Lyushin) Single Seat Fighter

Lavoisier, Pierre (France)
Lavoisier 1946 Monoplane
Lavoisier 1948 Monoplane

Lawrence Sperry Aircraft Co *(See: Sperry)*

Lawrence-Lewis Co (Chicago, IL)
Lawrence-Lewis A-1 Flying Boat
Lawrence-Lewis B-1 Flying Boat
Lawrence-Lewis B-2 Flying Boat

Lawrenz, Lewis (Germany)
Lawrenz 1912 Taube

Lawson, Alfred (Green Bay, WI)
After founding and editing two pioneer aviation magazines - *Fly* (1908) and *Aircraft* (1910) - Alfred Lawson (1869-1954) founded a series of aircraft companies. Established in 1917, The **Lawson Aircraft Corp** of Green Bay, Wisconsin, built two military trainer prototypes and designed several fighter aircraft. The **Lawson Airplane Co** of Milwaukee (1918) built the C-2 Airliner and the L-4 Midnight Airliner. A design similar to the C-2 but powered by pusher engines, the C-1 Airliner, was not built. A separate company, **Lawson Airline Transportation Co** was formed to run the projected national airline system that Lawson planned. **Lawson Aircraft Co** of Plainfield, NJ, (1926) began construction of the 100 Passenger Airliner, which was not completed.
Lawson (Alfred) Battler No.1 Armored Fighter
Lawson (Alfred) C-1 Airliner
Lawson (Alfred) C-2 Airliner
Lawson (Alfred) Ground Hog I Armored Trench Fighter
Lawson (Alfred) L-4 Midnight Airliner
Lawson (Alfred) M.T. One Primary Training Tractor
Lawson (Alfred) M.T. Two
Lawson (Alfred) 100 Passenger Airliner

Lawson, I. N.
Lawson (I. N.) G1 Glider

Lay Brothers Aircraft (Helena, MT)
Lay Brothers SL-4

Lay, Charles E. (Blue Ash, OH)
Lay (Charles E.) Model 1 Gyropter
Lay (Charles E.) A-2A Gyropter

Lazarev, A. A. (Russia)
Lazarev KhAI-3

le Bris, Joseph Marie (France)
le Bris 1857 Glider "l'Albatros Artificiel"
le Bris 1870 Glider "l'Albatros Artificiel"

le Clère *(See: Clère)*

le Prieur, Yves (France)
le Prieur Yvonette Glider (1910)

Leader's International Inc (South Hill, VA)
Leader's International AM-DSII

Leaman, Thomas P. (Ithaca, NY)
Leaman 1918 Training Biplane

The Lear Aircraft Corp (A. J. Lear & C. C. Lear) (Pratt, KS)
Lear (Aircraft Corp) 1929 Biplane

Lear, Inc (Santa Monica, CA)
Lear Learstar

Learjet (Wichita, KS)
In 1953 **Lear Inc** began rebuilding Lockheed Lodestars as Learstars. (Learstar construction was taken over in 1957 by **Pacific Airmotive Corp** which formed **PacAero Engineering Corp** for the conversions.) In 1960 William Lear formed a new company, **Swiss American Aviation Corp**, to produce business jets. The company was renamed **Lear Jet Corp** in 1962,

Gates Learjet Corp in 1970, and **Learjet Corp** in 1987. Lear's Lodestar conversions are listed under **Lear**; Lear business jets are listed under **Learjet**.
Learjet C-21A
Learjet 20 Series
Learjet 23
Learjet 23, NASM
Learjet 24 Series
Learjet 25 Series
Learjet 28 Longhorn
Learjet 29 Longhorn
Learjet 30 Series
Learjet 31 Series
Learjet 35 Series
Learjet 36 Series
Learjet 45 Series
Learjet 50 Series
Learjet 54 Longhorn Series
Learjet 55 Longhorn Series
Learjet 56 Longhorn Series
Learjet 60 Series
Learjet Lear Fan 2100 (Futura)

Leathers, Ward (USA)
Leathers 1918 Tandem Triplane

Lecomte, Alfred (Ateliers Aéronautiques de l'Est) (AAE) (France)
Lecomte (Alfred) Monoplane

Lecomte (H. Lecomte & Cie) (Henri Lecomte) (France)
Lecomte (Henri):
1912 Monoplane, Blériot Type
1912 Monoplane, Diamond X-Section Fuselage
1912 Monoplane, Two-Seat

Lecoq & Monteiro-Allaud (France)
Lecoq & Monteiro-Allaud 1911 Monoplane

Leduc, René (René Leduc Fils) (France)
Leduc 0.10
Leduc 0.16
Leduc 0.21
Leduc 0.22
Leduc RL-16

Ledyard Blake (USA)
Ledyard Blake Biplane

Lee, Cedric (Cedric Lee Co) (UK)
Lee (Cedric) 1914 Monoplane
Lee (Cedric) 1914 Pterygoid Glider

Lee-Richards (Cedric Lee & G. Tilghman Richards) (UK)
Lee-Richards "Flying Saucer" Annular Monoplane (1913)

Lefevers, W. L. (Reidsville, NC)
Lefevers FS-1 Falcon Special
Lefevers Falcon Special II

Lefort (France)
Lefort 1923 Triplane Glider

Leg-Air Corp (Garry LeGare) (Medford, OR)
Leg-Air Sea Hawker

Léger, M. (Monaco)
Léger 1905 Helicopter

Leidorf, Robert (Cleveland, OH)
Leidorf 1910 Aeroplane Patent

Leighnor, William C. (Hutchinson, KS)
Leighnor W-4 Mirage

Leinweber-Curtiss (Chicago, IL)
Leinweber-Curtiss Convertiplane

Leitzen, Hans Martin (Germany)
Hans Martin Leitzen No.231

Lejeune, Louis (France)
Lejeune No.1 (1909) Biplane

Lemâitre & Cie (France)
Lemâitre 1911 Biplane

LeMars Manufacturing Co *(See: General Aircraft Corp)*

Lemont, Harold E., Jr (East Providence, RI)
Lemont Helicopter (1947)

Lenert Aircraft Co (Pentwater, MI)
Lenert All Metal Biplane
Lenert All Metal B
Lenert All Metal C
Lenert All Metal 1930 Model

Lenin Gradets (Russia)
Lenin Gradets 1962 Ultralight

Lenning, G. C. (New York, NY)
Lenning 1909 Biplane Floatplane

Leonard, George W. (Santa Ana, CA)
Leonard 1927 Special

Leonardo da Vinci *(See: da Vinci)*

LePère US Aircraft Corp (LUSAC) (Detroit, MI)
LePère O-11 Triplane
LePère 1
LePère 11 (C-11, C II)
LePère 21

Lepicier, Evangeliste (Brooklyn, NY; Canada)
Lepicier Model A Monoplane (1929)

Les Avions Simplex (France) *(See: Arnoux)*

Lesh, L. J. (Canada)
Lesh 1907 Towed Gliders

Lesseps *(See: de Lesseps)*

Lestage, Arthur-Austin *(See: de Lestage)*

Let Kunovice National Corp (Czechoslovakia)
Let L 200 Morava
Let L 200A Morava
Let L 200D Morava
Let L 410 UPV-E Turbolet
Let L 420
Let L 430
Let L 610

Let-Mont sro (Czech Republic)
Let-Mont Tulák (Rambler)

Letadla Prikryl-Blecha *(See: Ardea)*

Letadla Továrny Aero *(See: Aero, Prague)*

Letecké Opravny Kbely *(See: LOK)*

Letord (Avions E. Letord) (France)
Letord LET.1 A.2
Letord LET.2 A.3
Letord LET.3 B.3
Letord LET.4
Letord LET.5 A.3
Letord LET.6
Letord LET.7 B.3
Letord LET.9 BN.2

Letov (Military Aircraft Works; Vojenská Tovarna na Letadla) (Czechoslovakia)
Letov C.2
Letov L.115 Delfin (Dolphin)
Letov MK-1 Kocour
Letov Š.A.
Letov Š 1
Letov Š 2
Letov Š 3
Letov Š 4
Letov Š 4a
Letov Š 5
Letov Š 6
Letov Š 7
Letov Š 8
Letov Š 9
Letov Š 10 (License Phönix 276)
Letov Š 12
Letov Š 13
Letov Š 14
Letov Š 16
Letov Š 16 B
Letov Š 16 J
Letov Š 16 L
Letov Š 16 T
Letov Š 16r
Letov Š 116
Letov Š 216
Letov Š 316
Letov Š 416
Letov Š 516
Letov Š 616
Letov Š 716
Letov Š 816
Letov Š 916
Letov Š 18
Letov Š 118
Letov Š 218
Letov Š 19
Letov Š 19 W
Letov Š 20
Letov Š 20 J
Letov Š 20 L
Letov Š 20 M
Letov Š 21
Letov Š 22
Letov Š 25
Letov Š 27
Letov Š 28
Letov Š 128
Letov Š 228 E
Letov Š 328
Letov Š 328 F
Letov Š 328 N
Letov Š 328 V
Letov Š 428
Letov Š 528
Letov Š 31
Letov Š 31a
Letov Š 131
Letov Š 231
Letov Š 331
Letov Š 431
Letov Š 32
Letov Š 33
Letov Š 39
Letov Š 139
Letov Š 239
Letov Š 50
Letov Š 50 D
Letov ŠA 20
Letov ŠH 1
Letov Šm 1
Letov Šm a1
Letov ŠM 1
Letov ŠM 2
Letov ŠM 4

Letov Air Ltd (Czech Republic)
Letov Air LK-2M Sluka
Letov Air ST-4 Azték

Letur, François (France)
Letur 1852 Flying Machine

Levasseur, Pierre (P. Levasseur Constructions Aéronautiques) (France)
Levasseur Abril
Levasseur Allege
Levasseur ATI (TK)
Levasseur ATI
Levasseur LB 2 (See: Levy-Biche)
Levasseur 1 (T 03)
Levasseur 2
Levasseur 3
Levasseur 4
Levasseur 5
Levasseur 6
Levasseur 7
Levasseur 8 "l'Oiseau Blanc" ("The White Bird")
Levasseur 9
Levasseur 10 (101)
Levasseur 107
Levasseur 108
Levasseur 11
Levasseur 12
Levasseur PL 12
Levasseur 14
Levasseur 15
Levasseur 154
Levasseur 200
Levasseur 201
Levasseur 400

Levavasseur, Léon (France) *(See also: Antoinette)*
Levavasseur Aéroplane de Villotrans

Levavasseur & Gastambide *(See: Gastambide-Levavasseur)*

Lévêque *(See: Donnet-Lévêque; also: FBA)*

Levier and Associates (La Canada, CA)
Levier Cosmic Wind "Little Toni"
Levier Cosmic Wind "Little Toni," "French Quarter Special"
Levier Cosmic Wind "Minnow"
Levier Cosmic Wind "Slick"

Lévy (Hydravions Georges Lévy) (France)
Lévy Type GL.40 Flying Boat (1917)
Lévy Type R Flying Boat

Lévy-Besson (France)
Lévy-Besson L.B. Flying Boat

Lévy-Biche (Constructions Aéronautiques J. Lévy) (France)
Lévy-Biche LB 2 Shipboard Fighter
Lévy-Biche 4-Ho2

Lévy-Gaillat (France)
Lévy-Gaillat Biplane

Lévy-Lepen (France)
Lévy-Lepen HB.2

Lewis (USA)
Lewis 1917 Flying Boat

Leyat, Marcel (France)
Leyat 1928 l'Autodrome
Leyat l'Hélica (Converted AR)
Leyat l'Hélicat (Converted LeO 20)

LFG Roland (Luftfahrzeug Gesellschaft) (Germany)
LFG Roland C.II Walfisch (Whale)
LFG Roland C.III
LFG Roland C.V
LFG Roland C.VIII
LFG Roland D.I Haifisch (Shark)
LFG Roland D.II
LFG Roland D.IIa
LFG Roland D.III
LFG Roland D.IV
LFG Roland D.V
LFG Roland D.VI
LFG Roland D.VII

LFG Roland D.IX
LFG Roland D.XIII
LFG Roland D.XIV
LFG Roland D.XV
LFG Roland D.XVI
LFG Roland D.XVII
LFG Roland G.I
LFG Roland Pfeil (Arrow) Biplane
LFG Roland R.I
LFG Roland Stahl Taube (Steel Dove)
LFG W
LFG WD
LFG V.3a Susanne
LFG V.8 Bärbel
LFG V.13 Strela
LFG V.14 Binz
LFG V.15
LFG V.16 Rostock
LFG V.17 Schule
LFG V.18 Sassnitz
LFG Stralsund V.19 Putbus
LFG V.20 Arkona
LFG Stralsund V.39
LFG Stralsund V.40
LFG Stralsund V.42
LFG Stralsund V.52
LFG Stralsund V.58
LFG Stralsund V.59
LFG Stralsund V.60
LFG Stralsund V.61

Libanski, Edmund (Poland)
Libanski Jaskolka (Swallow)

Libby Auto Radiator Co (Kansas City, MO)
Libby TG-1 Monoplane

Liberty Aircraft Sales and Manufacturing Co (St Louis, MO)
Liberty Liberty Bell Biplane
Liberty P-2
Liberty Parasol Monoplane

Libossart, Pablo (Argentina)
Libossart Pellegrini (1914 Monoplane)

Lievre (Joe Ben Lievre Aircraft Corp) (San Antonio, TX)
Lievre Primary Glider Model 100

Lifting Bodies, General

Liga Obrony Powietrznej i Przeciwgazowej *(See: L.O.P.P.)*

Lignel (France)
Lignel 44 Taupin
Lignel 46 Coach

Lignes Aériennes Latécoère *(See: Latécoère)*

Light Aero, Inc (Boise, ID) *(See also: Avid)*
Light Aero Avid Flyer

Light's American Sportscopter Inc (Taiwan) *(See: American Sportscopter)*

Ligreau (France)
Ligreau 1924 Mono Aviette

Likosiak, Casimir (Chicago, IL)
Likosiak Homebuilt Biplane (1926)

Lilienthal, Gustav (Germany)
Lilienthal (Gustav) 1914 Ornithopter

Lilienthal, Otto (Germany)
Lilienthal (Otto) 1891 Glider
Lilienthal (Otto) 1892 Glider
Lilienthal (Otto) 1893 Glider
Lilienthal (Otto) 1893 Powered Glider
Lilienthal (Otto) 1894 Glider
Lilienthal (Otto) 1894 Sturmfügelmodell (Storm Flyer)
Lilienthal (Otto) 1895 Biplane Glider
Lilienthal (Otto) 1896 Powered Glider Fledermausfugelapparat (Bat Apparatus)
Lilienthal (Otto) 1894-96 Glider Normal Segelapparat (Sailing Apparatus)

Lillie Co (Canada)
Lillie 1913 Tractor Biplane

Limbach, Gustave (Belgium)
Limbach Gusty Mk.I

Lincoln Aircraft Co, Inc (Lincoln Standard Aircraft Co) (Lincoln, NE)
Lincoln (Aircraft):
A. P. (All-Purpose) Cabin Airplane
Bumble Bee Monoplane
Page Air Coach
Page LP-3 (New Swallow)
Playboy
P.T. Family
Sport Plane (See also: Swanson SS-3)
Standard Cruiser
Standard J-1 (See: Standard J-1)
Standard L-S-3
Standard L-S-5
Standard Raceabout
Standard Speedster
Standard Tourabout

Lincoln, Garland E. (North Hollywood, Ca)
Lincoln (G. E.) LF-1
Lincoln (G. E.) LF-1 as "Nieuport 28"

Ling-Temco-Vought *(See: Vought)*

Linke-Hoffman Werke (Germany)
Linke-Hoffman R.I
Linke-Hoffman R.II

Lioré-Witzig-Dutilleul *(See: Witzig-Lioré-Dutilleul)*

Lioré et Olivier (France)
Lioré le F. Lioré (1911 Twin-Tractor Monoplane)
Lioré le Balsan
Lioré le Plaisant (Pleasant)
Lioré le Protin Contal
Lioré le Lioré Zens (See: Zens)
Lioré et Olivier (Cierva) LeO C.L.10 Autogiro
Lioré et Olivier LeO C.30 Autogiro (Cierva License)
Lioré et Olivier LeO C.302 Autogiro (Cierva License)
Lioré et Olivier LeO C.34 Autogiro (Cierva License)
Lioré et Olivier LeO 3
Lioré et Olivier LeO 4
Lioré et Olivier LeO 4/1
Lioré et Olivier LeO 5S2
Lioré et Olivier LeO 6
Lioré et Olivier LeO 7
Lioré et Olivier LeO 8 (CAn 2)
Lioré et Olivier LeO 9
Lioré et Olivier LeO H.10
Lioré et Olivier LeO 12 Bn.2
Lioré et Olivier LeO 12 Bn.3
Lioré et Olivier LeO H 13 B
Lioré et Olivier LeO H.133
Lioré et Olivier LeO H.134
Lioré et Olivier LeO H 135 B3
Lioré et Olivier LeO H.15
Lioré et Olivier LeO H.180
Lioré et Olivier LeO H.190
Lioré et Olivier LeO H.191
Lioré et Olivier LeO H.192 (H.196)
Lioré et Olivier LeO H.193
Lioré et Olivier LeO H.194
Lioré et Olivier LeO H.197 S
Lioré et Olivier LeO H.198
Lioré et Olivier LeO H.199
Lioré et Olivier LeO 20 Bn.2

Lioré et Olivier LeO 20 Bn.3 (12 Bn.3)
Lioré et Olivier LeO 203
Lioré et Olivier LeO 206
Lioré et Olivier LeO 208
Lioré et Olivier LeO 21
Lioré et Olivier LeO 212
Lioré et Olivier LeO 213
Lioré et Olivier LeO 213N
Lioré et Olivier LeO 215
Lioré et Olivier LeO H.22
Lioré et Olivier LeO H.23
Lioré et Olivier LeO H.240
Lioré et Olivier LeO H.24-2 (H.242)
Lioré et Olivier LeO H.24-2/1 (H.242-1)
Lioré et Olivier LeO H.24-6 (H.246)
Lioré et Olivier LeO H.24-6/1 (H.246-1)
Lioré et Olivier LeO 25
Lioré et Olivier LeO 250 BN.4
Lioré et Olivier LeO H.252
Lioré et Olivier LeO 254
Lioré et Olivier LeO H.255
Lioré et Olivier LeO 256
Lioré et Olivier LeO 257
Lioré et Olivier LeO 259
Lioré et Olivier LeO H.27
Lioré et Olivier LeO 30
Lioré et Olivier LeO 300
Lioré et Olivier LeO H.43
Lioré et Olivier LeO 45
Lioré et Olivier LeO H.46
Lioré et Olivier LeO H.47
Lioré et Olivier LeO H.470
Lioré et Olivier LeO H.49
Lioré et Olivier LeO 50

Lioré-Nieuport *(See: Loire)*

Lioré-Gourdou-Leseurre *(See: Loire)*

Lipczynski, Kazimierz (Poland)
Lipczynski 1925 Monoplane

Lipkowski, Jo'sef (Poland)
Lipkowski 1905 Helicopter

Lipnur (Indonesia) *(See: Angkatan)*

Lippisch, Alexander (Germany)
Lippisch Baby
Lippisch Grüne Post (Green Post)
Lippisch Delta I
Lippisch Delta II "Hermann Köhl"
Lippisch Delta III
Lippisch Delta IV
Lippisch DM-1
Lippisch Hangwind
Lippisch ILA
Lippisch Li P-10
Lippisch Li P-11
Lippisch Li P-12
Lippisch Li P-13
Lippisch Li P-14
Lippisch Li P-15
Lippisch Liliput 65
Lippisch Pegasus (Primary Glider)
Lippisch Professor
Lippisch Prüfling (Examination)
Lippisch Prüfling (Examination), Motorized
Lippisch Steinmann (Stoneman) Experiment (64)
Lippisch Steinmann (Stoneman) V 3
Lippisch Storch (Stork)
Lippisch Wien (See: Kassel)
Lippisch Zögling (Pupil)

List, James A. (Granville, IA)
List Doohickey Model A

Lisunov (Russia)
Lisunov Li-2 *Cab* (PS-84)

Little Wing Autogyros, Inc (Mayflower, AR)
Little Wing Roto-Pup

Liverpool University (UK)
Liverpool Liverpuffin (Human-Powered Aircraft)

Lizette *(See: Ludington)*

LKL (Lubelski Klub Lotniczy, Lublinian Aviation Club) (Poland)
LKL I (See: D.U.S.III)
LKL II
LKL IV
LKL V

Lloyd (Ungarische Lloyd Flugzeug und Motorenfabrik AG) (Austria-Hungary)
Lloyd C.I (Series 41)
Lloyd C.II (Series 42)
Lloyd C.III (Series 43)
Lloyd C.IV (Series 44)
Lloyd C.V (Series 46)
Lloyd D.I (Series 45)
Lloyd LS 1 (20M, 40.01)
Lloyd LS 2 (40.02)
Lloyd 40.03
Lloyd 40.04
Lloyd 40.05
Lloyd 40.06 (KF 1, Kampflugseug 1, Combat Flyer 1)
Lloyd 40.07
Lloyd 40.08
Lloyd 40.09
Lloyd 40.10
Lloyd 40.11
Lloyd 40.12
Lloyd 40.13
Lloyd 40.14
Lloyd 40.15
Lloyd 40.16
Lloyd 40.17, 40.18, 40.19, & 40.20

LMAASA *(See: FMA Pampa)*

Lobet-de-Rouvray (France, Australia)
Lobet-de-Rouvray Ganagoble 05

Locke (UK)
Locke 1927 Avionette

Lockheed (Burbank, CA)

In 1912, the Loughead brothers (Allan and Malcolm), in partnership with Max Mamlock, established the **Alco Hydro-Aeroplane Co** to construct an aircraft of Allan's design. Following the crash of the first item, the company became dormant and the brothers bought out the other stockholders. In 1916 the brothers established the **Loughead Aircraft Manufacturing Co.** The company reorganized in 1919, but continuing financial problems forced its liquidation in 1921. In 1926 the brothers established the **Lockheed Aircraft Co.** The majority stockholder sold his shares to the **Detroit Aircraft Corp** in 1929 and Lockheed became a division of that company until 1931 when it was forced into receivership by the Depression. The company locked its doors in June 1932.

In 1930, Allan and Malcolm formed **Loughead Brothers Aircraft Corp** in Glendale, CA; the company failed in 1934. Allan organized **Alcor Aircraft Corp** in 1937, but the crash of its sole design in 1938 forced it out of business and the Loughead brothers left the aviation field.

Five days after the **Lockheed Division of Detroit Aircraft** failed, Walter Varney, Mr. and Mrs. Cyril Chappellet, R. C. Walker, and Thomas Fortune Ryan III formed the **Lockheed Aircraft Corp** and purchased the assets of the defunct Lockheed Co. In August 1937, Lockheed created a wholly-owned subsidiary, the **AiRover**

Co, renamed **Vega Airplane Co** in 1938. In 1941, Lockheed and Vega merged, with Vega assets being transferred to the **Vega Aircraft Corp** until 1943, when Vega was fully absorbed into Lockheed.

In 1977 the company was renamed **Lockheed Corp** to reflect the diversification of its interests. In 1993 Lockheed acquired the **Forth Worth Division of General Dynamics** as the **Lockheed Fort Worth Division**. In 1994 Lockheed merged with Martin Marietta, forming the **Lockheed Martin Corp** in 1995.

For designs originating with Alco or Loughead Aircraft Mfg Co, see **Lough-ead.** For all products of Loughead Brothers or Alcor, see Loughead Brothers or **Alcor**. For all products of the Lockheed Aircraft Co and Lockheed Aircraft Corp prior to 1995, see **Lock-heed.** [Production while Lockheed was a subsidiary of Detroit is interfiled with Lockheed, but are noted as **Lock-heed (Detroit).**] For products of Lock-heed's AiRover and Vega subsidiaries predating the 1941 merger, see **Vega**; production postdating that merger is interfiled with **Lockheed.** For all designs subsequent to 1995, see **Lockheed Martin.**

Lockheed A-11 (A-12/F-12 Cover Story)
Lockheed A-12MD (Blackbird)
Lockheed A-12T (Blackbird)
Lockheed A-28 Hudson
Lockheed A-28A Hudson
Lockheed A-29 Hudson
Lockheed A-29A Hudson
Lockheed A-29B Hudson
Lockheed (A-28/A-29) Hudson Mk.I (Model B14L)
Lockheed (A-28/A-29) Hudson Mk.II (Model 414)
Lockheed (A-28/A-29) Hudson Mk.III (Model 414)
Lockheed (A-28/A-29) Hudson Mk.IIIA (A-29A)
Lockheed (A-28/A-29) Hudson Mk.IV (Model B14S)
Lockheed (A-28/A-29) Hudson Mk.IVA (A-28)
Lockheed (A-28/A-29) Hudson Mk.V
Lockheed AT-18 Hudson
Lockheed AT-18A Hudson
Lockheed AT-18A Hudson, Civil Use
Lockheed B-34 Ventura
Lockheed B-34A Ventura
Lockheed B-34B Ventura
Lockheed B-37 Ventura
Lockheed (B-34/B-37) Ventura Mk.I (Model 37)
Lockheed Ventura I "Ventillation" (Mod 49 Engine Testbed)
Lockheed (B-34/B-37) Ventura Mk.II (Model 37)
Lockheed (B-34/B-37) Ventura Mk.IIA (Model 137)
Lockheed C-5A Galaxy
Lockheed C-5B Galaxy
Lockheed C-5C Galaxy
Lockheed C-5M Galaxy
Lockheed (Detroit) Y1C-12
Lockheed (Detroit) Y1C-17
Lockheed Y1C-23
Lockheed Y1C-25
Lockheed XC-35
Lockheed XC-35, NASM
Lockheed Y1C-36 (C-36)
Lockheed UC-36A
Lockheed UC-36B
Lockheed UC-36C
Lockheed Y1C-37
Lockheed C-40
Lockheed C-40A
Lockheed C-40B
Lockheed UC-40D
Lockheed Lodestar Mk.I (Model 18)
Lockheed Lodestar Mk.IA (RAF C-59)
Lockheed Lodestar Mk.II (RAF C-60)
Lockheed C-56 Lodestar
Lockheed C-56A Lodestar
Lockheed C-56B Lodestar
Lockheed C-56C Lodestar
Lockheed C-56D Lodestar
Lockheed C-56E Lodestar
Lockheed C-57 Lodestar
Lockheed C-57B Lodestar
Lockheed C-57C Lodestar
Lockheed C-57D Lodestar
Lockheed C-59 Lodestar
Lockheed (C-60) Lodestar, Civil
Lockheed C-60 Lodestar
Lockheed C-60A Lodestar
Lockheed XC-60B Lodestar
Lockheed C-66
Lockheed C-69 Constellation
Lockheed C-69B Constellation
Lockheed C-69C Constellation
Lockheed XC-69E Constellation
Lockheed UC-85
Lockheed UC-101
Lockheed C-111
Lockheed C-121A Constellation
Lockheed VC-121A Constellation
Lockheed VC-121A Constellation "Columbine"/"Columbine II"
Lockheed VC-121B Constellation
Lockheed C-121C Super Constellation
Lockheed RC-121C (EC-121C) Warning Star
Lockheed VC-121C Super Constellation
Lockheed RC-121D (EC-121D) Warning Star
Lockheed VC-121E Super Constellation
Lockheed YC-121F Super Constellation
Lockheed C-121G Super Constellation
Lockheed EC-121H Super Constellation
Lockheed EC-121J Super Constellation
Lockheed NC-121J Super Constellation
Lockheed JC-121K Super Constellation
Lockheed NC-121K Super Constellation
Lockheed EC-121P Super Constellation
Lockheed EC-121Q Super Constellation
Lockheed EC-121R Super Constellation
Lockheed EC-121S Super Constellation
Lockheed EC-121S Coronet Solo
Lockheed EC-121T Super Constellation
Lockheed YC-130 Hercules
Lockheed C-130A Hercules
Lockheed C-130A Hercules, NASM
Lockheed C-130A Hercules, USMC Aerial Refueling Tests
Lockheed AC-130A Spectre (Gunship)
Lockheed DC-130A (GC-130A) Hercules
Lockheed JC-130A Hercules
Lockheed NC-130A Hercules
Lockheed RC-130A Hercules
Lockheed C-130B Hercules
Lockheed HC-130B (R8V-1G) Hercules
Lockheed JC-130B Hercules
Lockheed KC-130B Hercules
Lockheed NC-130B (C-130C) Hercules
Lockheed RC-130B Hercules
Lockheed VC-130B Hercules
Lockheed WC-130B Hercules
Lockheed C-130D Hercules
Lockheed C-130E Hercules
Lockheed AC-130E Spectre (Gunship)
Lockheed DC-130E Hercules
Lockheed EC-130E ABCCC
Lockheed EC-130E Volant Solo/ Commando Solo
Lockheed HC-130E Hercules
Lockheed JC-130E Hercules
Lockheed MC-130E Combat Talon I
Lockheed C-130F (GV-1U) Hercules
Lockheed KC-130F (GV-1) Hercules
Lockheed KC-130F (GV-1) Hercules "Fat Albert" (Blue Angels)
Lockheed LC-130F Hercules
Lockheed C-130G (GV-2U) Hercules
Lockheed EC-130G Hercules

Lockheed C-130H Hercules
Lockheed AC-130H Spectre
Lockheed DC-130H Hercules
Lockheed EC-130H Compass Call
Lockheed HC-130H Hercules, USAF
Lockheed HC-130H Hercules, USCG
Lockheed KC-130H Hercules
Lockheed MC-130H Combat Talon II
Lockheed NC-130H (JHC-130MP) Hercules
Lockheed PC-130H (C-130H-MP) Hercules
Lockheed VC-130H Hercules
Lockheed WC-130H Hercules
Lockheed C-130J Hercules II
Lockheed C-130J Hercules II, RAF
Lockheed C-130K Hercules C.1
Lockheed C-130K Hercules W.2
Lockheed C-130K Hercules C.3
Lockheed HC-130N Combat Shadow
Lockheed HC-130P (MC-130P) Combat Shadow
Lockheed EC-130Q Hercules
Lockheed KC-130R Hercules
Lockheed LC-130R Hercules
Lockheed RC-130S Hercules
Lockheed C-130SS (Stretch/STOL) Hercules
Lockheed C-130T Hercules
Lockheed KC-130T Hercules
Lockheed AC-130U Spectre
Lockheed (C-130) Hercules on Water (HOW)
Lockheed C-140A Jetstar
Lockheed C-140B Jetstar
Lockheed VC-140B Jetstar
Lockheed C-141A Starlifter
Lockheed C-141A Starlifter, NASA Airborne IR Observatory
Lockheed C-141B Starlifter
Lockheed CL-475 (Rigid Rotor Proof-of-Concept Vehicle)
Lockheed CL-475 (Rigid Rotor Proof-of-Concept Vehicle), NASM
Lockheed FO-1 Lightning
Lockheed XFV-1 (Pogo Stick)
Lockheed F-1 (See: Loughead F-1)
Lockheed F-4 Lightning
Lockheed F-4A Lightning
Lockheed F-5A Lightning
Lockheed F-5B Lightning
Lockheed F-5C Lightning
Lockheed XF-5D Lightning
Lockheed F-5E Lightning
Lockheed F-5F Lightning
Lockheed F-5G Lightning
Lockheed F-5G Lightning, Civil
Lockheed F-5G Lightning, Honeywell Test Aircraft
Lockheed YF-12A (Blackbird)
Lockheed F-14 Shooting Star (See: Lockheed F-80)
Lockheed F-22 Lightning 2 (See: Lockheed Martin F-22)
Lockheed (F-80) XP-80 Shooting Star "Lulu Belle"
Lockheed (F-80) XP-80 Shooting Star "Lulu Belle," NASM
Lockheed (F-80) XFP-80 (XF-14) Shooting Star
Lockheed (F-80) XP-80A Shooting Star
Lockheed (F-80) YP-80A Shooting Star
Lockheed (F-80) P-80A Shooting Star
Lockheed (F-80) P-80A Shooting Star:
Aerial Tow Trials
Carrier Suitability Trials
F-94C Rocket Testbed
Ramjet Testbed
Lockheed (F-80) XFP-80A Shooting Star
Lockheed DF-80A Shooting Star
Lockheed EF-80A Shooting Star
Lockheed QF-80A Shooting Star
Lockheed RF-80A (FP-80A, F-14A) Shooting Star
Lockheed (F-80) XP-80B Shooting Star
Lockheed (F-80) P-80B Shooting Star
Lockheed F-80B Shooting Star (GAM-63 Guidance System Test)
Lockheed F-80C Shooting Star
Lockheed NF-80C Shooting Star
Lockheed QF-80C Shooting Star
Lockheed RF-80C Shooting Star
Lockheed QF-80F Shooting Star
Lockheed (F-80) XP-80R Shooting Star
Lockheed XF-90
Lockheed YF-94 Starfire
Lockheed F-94A Starfire
Lockheed YF-94B Starfire
Lockheed F-94B Starfire
Lockheed F-94B Starfire Bomarc Guidance System Testbed
Lockheed F-94B Starfire M-61 Vulcan Cannon Testbed
Lockheed YF-94C (YF-97A) Starfire
Lockheed F-94C (F-97A) Starfire
Lockheed DF-94C Starfire
Lockheed YF-94D Starfire
Lockheed XF-104 Starfighter
Lockheed YF-104A Starfighter
Lockheed YF-104A Starfighter, NASA 818, NASM
Lockheed F-104A Starfighter
Lockheed F-104A Starfighter, CF-104 Pattern Aircraft
Lockheed NF-104A Starfighter
Lockheed QF-104A Starfighter
Lockheed F-104B Starfighter
Lockheed F-104C Starfighter
Lockheed F-104D Starfighter
Lockheed F-104DJ Starfighter
Lockheed F-104F Starfighter
Lockheed F-104G Starfighter
Lockheed RF-104G Starfighter
Lockheed TF-104G Starfighter
Lockheed F-104J Starfighter
Lockheed NF-104N Starfighter, NASA
Lockheed F-104S Starfighter
Lockheed F-117A Nighthawk (Stealth Fighter)
Lockheed XH-51A Aerogyro
Lockheed XH-51A Aerogyro, Compound
Lockheed XH-51N Aerogyro
Lockheed AH-56A Cheyenne
Lockheed Have Blue (Stealth Technology Demonstrator)
Lockheed JO-1
Lockheed JO-2
Lockheed XJO-3
Lockheed Jetstar Prototype (CL-329)
Lockheed Jetstar (L-1329)
Lockheed Jetstar (L-1329) NASA General Purpose Airborne Simulator
Lockheed Jetstar II (L-2329)
Lockheed Jetstar 731 (Airesearch Modification)
Lockheed L-193 (First Generation Jetliner Proposal)
Lockheed L-500 Galaxy (Civil C-5 Proposal)
Lockheed L-1011-1 Tristar
Lockheed L-1011-100 Tristar
Lockheed L-1011-200 Tristar
Lockheed L-1011-300 Tristar
Lockheed L-1011-500 Tristar
Lockheed L-2000 (SST Proposal)
Lockheed YO-3A
Lockheed YO-3A, NASM
Lockheed PV-1 Ventura
Lockheed (PV-1) Ventura (RCAF)
Lockheed PV-2 Harpoon
Lockheed PV-2C Harpoon
Lockheed PV-2D Harpoon
Lockheed PV-3 Ventura
Lockheed (PV) Ventura GR.V
Lockheed (PV) Ventura TT.5
Lockheed XP2V-1 Neptune
Lockheed P2V-1 Neptune
Lockheed P2V-1 Neptune "The Turtle" ("Truculent Turtle")
Lockheed P2V-2 Neptune
Lockheed P2V-2 Neptune, Photo Mapping Modification
Lockheed P2V-2N Neptune
Lockheed P2V-2S Neptune

Lockheed P2V-3 Neptune
Lockheed P2V-3B Neptune
Lockheed XP2V-3C Neptune
Lockheed P2V-3C Neptune
Lockheed P2V-3W Neptune
Lockheed P2V-3Z Neptune
Lockheed P2V-4 (P-2D) Neptune
Lockheed P2V-5 Neptune
Lockheed P2V-5F (P-2E) Neptune
Lockheed P2V-5FD (DP-2E) Neptune
Lockheed P2V-5FE (EP-2E) Neptune
Lockheed P2V-5FS (SP-2E) Neptune
Lockheed (P2V-5F) AP-2E Neptune
Lockheed (P2V-5F) OP-2E Neptune
Lockheed P2V-6 (P-2F) Neptune
Lockheed P2V-6B (P2V-6M, MP-2F) Neptune
Lockheed P2V-6F (P-2G) Neptune
Lockheed P2V-6T (TP-2F) Neptune
Lockheed P2V-7 (P-2H) Neptune
Lockheed P2V-7B Neptune
Lockheed P2V-7LP (LP-2H) Neptune
Lockheed P2V-7S (SP-2H) Neptune
Lockheed P2V-7S (SP-2H) Neptune, NASM
Lockheed P2V-7U (RB-69A) Neptune
Lockheed (P2V-7) AP-2H Neptune
Lockheed (P2V-7) DP-2H Neptune
Lockheed (P2V-7) EP-2H Neptune
Lockheed (P2V-7) NP-2H Neptune
Lockheed (P2V-7*kai*) P-2J Neptune
Lockheed (P2V-7*kai*) UP-2J Neptune
Lockheed (P2V, P-2) Neptune MR.1
Lockheed (P2V, P-2) Neptune MR.2
Lockheed (P2V, P-2) Neptune MR.3
Lockheed (P2V, P-2) Neptune MR.4
Lockheed (YP-3A) YP3V-1 Orion
Lockheed (P-3A) P3V-1 Orion
Lockheed EP-3A Aries
Lockheed NP-3A Orion
Lockheed RP-3A Orion
Lockheed VP-3A Orion
Lockheed WP-3A Orion
Lockheed P-3B Orion
Lockheed EP-3B Aries
Lockheed YP-3C Orion
Lockheed P-3C Orion
Lockheed P-3C Orion Outlaw Hunter/ Oasis
Lockheed EP-3C Orion
Lockheed UP-3C (NP-3C) Orion
Lockheed P-3D Orion
Lockheed RP-3D Orion (Project Magnet)
Lockheed UP-3D Orion
Lockheed WP-3D Orion
Lockheed EP-3E Aries II
Lockheed P-3F Orion
Lockheed P-3H Orion II
Lockheed EP-3J Orion
Lockheed P-3P Orion
Lockheed P-3T Orion
Lockheed P-3W Orion
Lockheed P-3 AEW Sentinel Prototype
Lockheed P-3 AEW Sentinel
Lockheed P-3 Blue Sentinel (US Customs Service)
Lockheed (P-3) CP-140 Aurora (RCAF)
Lockheed (P-3) CP-140A Arcturus (RCAF)
Lockheed P-7A (P-3G) LRAACA
Lockheed (Detroit) YP-24
Lockheed XP-38 Lightning
Lockheed YP-38 Lightning
Lockheed P-38 Lightning
Lockheed P-38 Lightning, Asymmetric Cockpit Testbed
Lockheed XP-38A Lightning
Lockheed P-38D Lightning
Lockheed P-38E Lightning
Lockheed P-38E Lightning Seaplane Testbed
Lockheed P-38E Lightning Swordfish (Air Foil Testbed)
Lockheed P-38F Lightning
Lockheed P-38G Lightning
Lockheed P-38G Lightning Twin Cannon Testbed
Lockheed P-38H Lightning
Lockheed P-38H Lightning Droop Snoot Prototype
Lockheed P-38J Lightning
Lockheed P-38J Lightning, NASM
Lockheed P-38J Lightning "Yippee" (5000th Production P-38)
Lockheed P-38J Lightning Droop Snoot
Lockheed P-38J Lightning Pathfinder
Lockheed XP-38K Lightning
Lockheed P-38K Lightning
Lockheed P-38L Lightning
Lockheed P-38L Lightning Droop Snoot
Lockheed P-38L Lightning Night Fighter Prototype
Lockheed P-38L Lightning Pathfinder
Lockheed TP-38L Lightning
Lockheed P-38M Lightning
Lockheed (P-38) P-322 Lightning
Lockheed (P-38) Lightning Mk.I (Model 322)
Lockheed XP-49
Lockheed XP-58 Chain Lightning
Lockheed PBO-1 Hudson
Lockheed "Plainview" AG(EH)-1 Hydrofoil
Lockheed Q-Star
Lockheed QT-2PC
Lockheed XRO-1
Lockheed XR2O-1
Lockheed XR3O-1 (USCG Electra)
Lockheed R3O-2 (USN Electra Junior)
Lockheed XR4O-1
Lockheed XR5O-1 Lodestar
Lockheed R5O-1 Lodestar
Lockheed R5O-2 Lodestar
Lockheed R5O-3 Lodestar
Lockheed R5O-4 Lodestar
Lockheed R5O-5 Lodestar
Lockheed R5O-6 Lodestar
Lockheed XR6O-1 (XR6V-1) Constitution
Lockheed R7V-1 (C-121J) Super Constellation
Lockheed R7V-1 (C-121J) Super Constellation (Blue Angels)
Lockheed R7V-1P (C-121J) Super Constellation "Phoenix 6"
Lockheed R7V-2 (C-121K) Super Constellation
Lockheed YS-3A Viking
Lockheed S-3A Viking
Lockheed ES-3A Viking
Lockheed KS-3A Viking
Lockheed US-3A Viking (Carrier On-Board Delivery)
Lockheed S-3B Viking
Lockheed SR-71A (Blackbird)
Lockheed SR-71A (Blackbird), NASM
Lockheed SR-71A (Blackbird) LASRE (Linear Aerospike Test)
Lockheed SR-71B (Blackbird)
Lockheed SR-71C (Blackbird)
Lockheed TV-1 (TO-1) Shooting Star
Lockheed TV-2 (TO-2, T-33B) Shooting Star
Lockheed TV-2D (DT-33B) Shooting Star
Lockheed TV-2KD (DT-33C) Shooting Star
Lockheed T2V Seastar Prototype
Lockheed T2V-1 (T-1A) Seastar
Lockheed T-33A (TP-80C, TF-80C) Shooting Star
Lockheed T-33A Shooting Star, NASM
Lockheed T-33A Shooting Star, Acrojets Aircraft
Lockheed T-33A Shooting Star, Thunderbirds Aircraft
Lockheed T-33A Shooting Star, Civil
Lockheed AT-33A Shooting Star
Lockheed DT-33A Shooting Star
Lockheed NT-33A Shooting Star
Lockheed QT-33A Shooting Star
Lockheed RT-33A Shooting Star
Lockheed (T-33A) Silver Star 1 (RCAF)

Lockheed U-2 Prototype
Lockheed U-2A
Lockheed U-2B
Lockheed U-2C
Lockheed U-2C NASM
Lockheed U-2CT
Lockheed U-2D
Lockheed U-2E
Lockheed U-2F
Lockheed U-2G
Lockheed U-2ER (NASA ER-2)
Lockheed U-2R (TR-1A)
Lockheed U-2RT (TR-1B)
Lockheed U-2S
Lockheed U-2ST
Lockheed XV-4A Hummingbird
Lockheed XV-4B Hummingbird II
Lockheed WV Super Constellation Prototype
Lockheed WV-1 (PO-1W) Super Constellation
Lockheed WV-2 (EC-121K) Warning Star
Lockheed WV-2E (EC-121L) Warning Star
Lockheed WV-2Q (EC-121M) Warning Star
Lockheed WV-3 (WC-121N) Super Constellation
Lockheed W2V-1 Warning Star
Lockheed Model 1 Vega
Lockheed Model 1 Vega "Golden Eagle"
Lockheed Model 1 Vega "Miss Teanek"
Lockheed Model 1 Vega, Wilkins/ Eielson Arctic Expedition
Lockheed Model 2 Vega
Lockheed Model 2A Vega
Lockheed Model 2D Vega
Lockheed Model 3 Air Express
Lockheed Model 4 Explorer
Lockheed Model 5 Vega
Lockheed Model 5 Vega "Century of Progress"
Lockheed Model 5 Vega "Winnie Mae" (1st Post Vega)
Lockheed Model 5 Vega "Yankee Doodle"
Lockheed Model 5 Vega Special (Nichols)
Lockheed Model 5A Vega Executive
Lockheed Model 5A Vega Executive "Miss Silvertown"
Lockheed Model 5B Vega
Lockheed Model 5B Vega, Earhart Aircraft, NR7952
Lockheed Model 5B Vega, Earhart Aircraft, NR7952, NASM
Lockheed Model 5B Vega "Winnie Mae" (2nd Post Vega)
Lockheed Model 5B Vega "Winnie Mae" (2nd Post Vega), NASM
Lockheed Model 5C Vega
Lockheed Model 5C Vega High-Speed Special
Lockheed Model 5C Vega Special
Lockheed Model 5C Vega Special (6th Earhart Vega)
Lockheed (Model 5) Vega (1928 National Air Races, X7430)
Lockheed (Model 5) Vega Special "Detroit News"
Lockheed (Detroit) (Model 5) Vega DL-1
Lockheed (Detroit) (Model 5) Vega DL-1B
Lockheed (Detroit) (Model 5) Vega DL-1 Special
Lockheed Model 7 Explorer
Lockheed Model 8 Sirius
Lockheed Model 8 Sirius "Tingmissartoq," NASM
Lockheed Model 8-A Sirius
Lockheed Model 8-C Sirius
Lockheed Model 8-D Altair
Lockheed Model 8-D "Anzac"/ "Lady Southern Cross"
Lockheed Model 8-D Altair Special
Lockheed Model 8-E/8-F Altair
Lockheed (Airover) Model 8-G Altair "Flying Test Stand"
Lockheed (Detroit) (Model 8) Sirius DL-2
Lockheed (Detroit) (Model 8) Altair DL-2A
Lockheed Model 9 Orion
Lockheed Model 9 Orion Special
Lockheed Model 9-A Orion Special "Spirit of Fun"
Lockheed Model 9-B Orion
Lockheed Model 9-C Orion Special
Lockheed Model 9-D Orion
Lockheed Model 9-D Orion "Early Bird" (*Detroit News*)
Lockheed Model 9-D Orion Special (Ingalls Aircraft)
Lockheed Model 9-E Orion
Lockheed Model 9-E Orion-Explorer (Post/Rogers Aircraft)
Lockheed Model 9-F Orion
Lockheed Model 10-A Electra Prototype
Lockheed Model 10-A Electra
Lockheed Model 10-B Electra
Lockheed Model 10-C Electra
Lockheed Model 10-E Electra
Lockheed Model 10-E Electra (Earhart Aircraft, NR16020)
Lockheed Model 12 Electra Junior Prototype
Lockheed Model 12-A Electra Junior
Lockheed Model 12-A Electra Junior, Hot Air Deicing Testbed
Lockheed Model 12-A Electra Junior Tricycle Landing Gear Test
Lockheed Model 12-B Electra Junior
Lockheed Model 12-25 Electra Junior
Lockheed Model 12-26 Electra Junior
Lockheed Model 212 Electra Junior
Lockheed Model 14-H Super Electra
Lockheed Model 14-N Super Electra
Lockheed Model 14-N2 Super Electra (H. Hughes, NX18973)
Lockheed Model 14-W Super Electra
Lockheed Model 14-08 Super Electra
Lockheed Model B14S Hudson (Sperry Gyroscope Aircraft)
Lockheed 18 Lodestar Prototype
Lockheed Model 18-07 Lodestar
Lockheed Model 18-08 Lodestar
Lockheed Model 18-10 Lodestar
Lockheed Model 18-40 Lodestar
Lockheed Model 18-50 Lodestar
Lockheed Model 33 Little Dipper
Lockheed Model 34 Big Dipper
Lockheed Model 144 Excalibre
Lockheed Model 49 Constellation Prototype
Lockheed Model 049 Constellation
Lockheed Model 149 Constellation
Lockheed Model 649 Constellation
Lockheed Model 749 Constellation
Lockheed Model 1049 Super Constellation Prototype
Lockheed Model 1049 Prototype Turboprop Testbed
Lockheed Model 1049 Super Constellation
Lockheed Model 1049C Super Constellation
Lockheed Model 1049D Super Constellation
Lockheed Model 1049E Super Constellation
Lockheed Model 1049G Super Constellation
Lockheed Model 1049H Super Constellation
Lockheed Model 1249 Super Constellation (Turboprop)
Lockheed Model 1649 Starliner
Lockheed Model 75 Saturn
Lockheed Model 82 Hercules Family
Lockheed Model 282A (GL-201) Hercules

Lockheed Model 382 (L-100) Hercules
Lockheed Model 382B (L-100) Hercules
Lockheed Model 382E (L-100-20) Hercules
Lockheed Model 382E (L-100-20) HTTB (High Technology Test Bed)
Lockheed Model 382F (L-100-20) Hercules
Lockheed Model 382G (L-100-30) Hercules
Lockheed Model 382G (L-100-30MP) Hercules, Maritime Patrol
Lockheed Model 382J (L-100J) Hercules II
Lockheed (Model 82) GL-207 (Civil Airfreighter Proposal)
Lockheed Model 86 (Rigid Rotor)
Lockheed Model 286 (Rigid Rotor)
Lockheed Model 188A Electra II
Lockheed Model 188C Electra II
Lockheed Model 188C Electra II, Point Mugu Tracking Aircraft
Lockheed Model 188C Electra II (NCAR Instrument Aircraft)
Lockheed (Aermacchi) MB 339 T-Bird II (JPATS Proposal)
Lockheed (Dassault/Dornier) Alpha Jet (VTXTS Proposal)

Lockheed Martin (Bethesda, MD)

In 1994 **Lockheed Corp** merged with **Martin Marietta Corp**. In 1995 the facilities of the merged companies were reorganized as **Lockheed Martin Corp**. For all products of Lockheed or Martin Marietta designed prior to the 1994 merger, see **Lockheed** or **Martin Marietta**. For all products designed after 1994 see **Lockheed Martin**.

Lockheed Martin:
YF-22 Lightning 2 (ATF) (Model 1132)
F-22A Raptor (ATF) (Configuration 645)
F-22B Raptor (ATF)
Joint Strike Fighter (JSF)

***Lockheed Martin Aircraft Argentina SA (LMAASA)** (See: FMA Pampa)*

Lockheed Martin Alenia Tactical Transport Systems (LMATTS) (Marietta, GA)

Lockheed Martin Alenia:
C-27A Spartan
C-27J Spartan

***Lockheed-Azcarate SA** (See: LASA)*

Lockspeiser, David (UK)

Lockspeiser LDA-01 (Light Defense Aircraft)

Locomotive Terminal Improvement Co (Barrington, IL)

Locomotive "Chicago Javelin"

Loening Aeronautical Engineering Corp (New York, NY)

In 1920 Thomas Henri Huff and Elliot Daland founded the Ogdensburg Aeroway Corp in Ogdensburg, NY. By 1922 the firm had been renamed Huff, Daland & Co, Inc. In July 1925 the company moved to Bristol, PA, as Huff-Daland Airplanes, Inc. This became Keystone Aircraft Corp in March 1927. Loening Aeronautical Engineering Corp became a Keystone operating division in 1928, though references to the Loening name tapered off by 1930. All Huff-Daland designs which entered production before March 1927 are listed under Huff-Daland; designs which subsequently entered production are listed under Keystone. All Loening designs appear under Loening.

Loening (Corp) Aeroboat (See: Queen)
Loening (Corp) Air Yacht (Cabin Amphibian, Duck, CW, C2, etc.)
Loening (Corp) C-4C (Model K-85)
Loening (Corp) XCOA-1
Loening (Corp) COA-1
Loening (Corp) Commuter (Model 219, 1930 Air Yacht)
Loening (Corp) XFL
Loening (Corp) Flying Yacht (Model 23) (1921)
Loening (Corp) XHL-1
Loening (Corp) LS
Loening (Corp) M-2 Kitten
Loening (Corp) M-8
Loening (Corp) M-8-0 (M-80)
Loening (Corp) M-8-1 (M-81)
Loening (Corp) M-8-2 (M-82)
Loening (Corp) M-8-3 (M-83)
Loening (Corp) Monowheel Amphibian (Moth)
Loening (Corp) OL-1
Loening (Corp) OL-2
Loening (Corp) OL-3
Loening (Corp) OL-4
Loening (Corp) OL-5
Loening (Corp) OL-6
Loening (Corp) XOL-7
Loening (Corp) XOL-8
Loening (Corp) OL-8
Loening (Corp) OL-8A
Loening (Corp) OL-9
Loening (Corp) XO2L-1
Loening (Corp) XO2L-2
Loening (Corp) XO-10 (XOA-1A)
Loening (Corp) OA-1A
Loening (Corp) OA-1A, Pan American Good Will Tour
Loening (Corp) OA-1A, Pan American Good Will Tour, NASM
Loening (Corp) OA-1B
Loening (Corp) OA-1C
Loening (Corp) OA-2
Loening (Corp) PA-1
Loening (Corp) PW-2
Loening (Corp) PW-2A
Loening (Corp) PW-2B
Loening (Corp) R-4
Loening (Corp) S-1 Air Yacht

Loening (Grover Loening Aircraft Co) (Garden City, NY)

Loening (Grover) Duckling
Loening (Grover) Monoduck
Loening (Grover) XSL-1 (Design No. 87)
Loening (Grover) XSL-2
Loening (Grover) XS2L-1

Lofland Amphibian Corp (Detroit, MI)

Lofland L.A.G.4
Lofland 2-Place Amphibian

Lohner (Lohnerwerke GmbH) (Austria-Hungary)

Lohner B.I Series 11 Pfeilflieger (Arrow Flyer)
Lohner B.II Series 12 (Type C) Gebirgsflieger (Mountain Flyer)
Lohner B.III Series 13 Pfeilflieger (Arrow Flyer) (Type D)
Lohner B.III Series 14 (Type E)
Lohner B.IV Series 15 (Type G)
Lohner B.V Series 16 (Type H)
Lohner B.VI Series 16.1 (Type H2)
Lohner B.VII Series 17 (Type J)
Lohner B.VII Series 17.3
Lohner B.VII Series 17.8
Lohner C.I Series 18 (Type Jc)
Lohner C.I Series 18.5 (Type Jcr)
Lohner C.I Series 114 & 214 (See: Aviatik)
Lohner C.II Series 19 & 119 (See: Knoller)

Lohner C.II Series 112 (10.19, Type AB)
Lohner D.I Series 111 (10.20, Type AA)
Lohner D.I Series 115 & 315 (See: Aviatik)
Lohner Dr.I 111.04
Lohner Type E Flying Boat
Lohner Type L Flying Boat
Lohner Type L Flying Boat
Lohner Type M Flying Boat
Lohner Parasol
Lohner Type R Flying Boat
Lohner Type S Flying Boat
Lohner Type T Flying Boat
Lohner 10.01 to 10.03 Type E Pfeilfliegers (Arrow Flyers)
Lohner 10.04 Type D Pfeilflieger (Arrow Flyer)
Lohner 10.05 Pfeilflieger (Arrow Flyer)
Lohner 10.06 Pfeilflieger (Arrow Flyer) "Aspern"
Lohner 10.07 Gebirgsflieger (Mountain Flyer)
Lohner 10.08 to 10.11 Taubes (Doves)
Lohner 10.10 (2nd Aircraft) Pfeilflieger (Arrow Flyer)
Lohner 10.12 Gebirgsflieger (Mountain Flyer) Type C
Lohner 10.13 Gebirgsflieger (Mountain Flyer) Type C
Lohner 10.14 Pfeilflieger (Arrow Flyer) Type D
Lohner 10.15 Taube (Dove) (Schichtpreis Eindecker) (Simple Monoplane)
Lohner 10.16 Pfeilflieger (Arrow Flyer)
Lohner 10.17 Type J Pfeilflieger (Arrow Flyer)
Lohner 10.18 Type Jc
Lohner 10.21 Type U
Lohner 10.21, New
Lohner 10.22 Type F
Lohner 10.23 Type AC ("Express I")
Lohner 10.28 Type J
Lohner-Etrich Renn Gebirgs (Mountain Racers) (E-1)
Lohner-Etrich Taube (Dove) Family
Lohner-Umlauff von Frankwell 1910 Biplane

LOK (Letecké Opravny Kbely) (Aircraft Repair Works Kbely) (Czech Republic)
LOK Family Air

Loire (Atelier & Chantier de la Loire) (France)
Loire 10
Loire 11
Loire 132 C1
Loire 18
Loire 20
Loire 21
Loire 24
Loire 30
Loire 43 C1
Loire 45 C1
Loire 46 C1
Loire 50
Loire 60
Loire 70
Loire 70 1
Loire 102 Trans-Atlantic Flyingboat
Loire-Gourdou-Leseurre LGL 32 C1
Loire-Gourdou-Leseurre LGL 321 C1
Loire-Gourdou-Leseurre LGL 33 C1
Loire-Gourdou-Leseurre LGL 34 C1
Loire-Gourdou-Leseurre LGL 35 C1
Loire Nieuport 10
Loire Nieuport 102
Loire Nieuport 130
Loire Nieuport 140
Loire Nieuport 160
Loire Nieuport 161
Loire Nieuport 200 (20)
Loire Nieuport 210 C1 (21)
Loire Nieuport 225
Loire Nieuport 250 (25)
Loire Nieuport 30
Loire Nieuport 40 Family

Lombardi *(See: Avia, Italy)*

London and Provincial (L&P) Aviation (Airplane) Co (UK)
London and Provincial 1914 Biplane (License-built Caudron G.3)
London and Provincial 1917 Trainer (Developed from Caudron G.3)
London and Provincial 1919 Tractor Biplane

Long, David E. "Dave" (Lock Haven, PA)
Long (Dave) Midget Mustang
Long (Dave) Midget Mustang "Mammy"
Long (Dave) LA-1 "P-Shooter"

Long Aircraft (Les Long) (Cornelius, OR)
Long (Les) Hi-Low

Longren, Albin Kasper (Topeka, KS; Torrance, CA)
Longren, Albin Kasper (KS)
Longren #1 Pusher, "Topeka I" ("Dixie Flyer") (1910/11)
Longren #2 Pusher (1912)
Longren #3 Tractor (1914/15)

Longren Aeroplane Co (1916-17) (KS)
Longren Bus (1916)
Longren Model G (1916)
Longren Model H (1916/17)
Longren Model H-2 (1917)

Longren Aircraft Corp (1919-24) (KS)
Longren AK Training Biplane (D-2, Fibre Sport Airplane)
Longren H-2 (Rebuilt)

Longren Aircraft Inc (1933-37) (KS)
Longren NL-13

Longren Aircraft Co (1939-44) (CA)
Longren (1939-1944) - No Aircraft Designed or Produced

Longren Aircraft Co (1959-60) (CA)
Longren Centaur (Rebuilt Military L-13)

Loose, Chester "Chet" (Davenport, IA)
Loose (Chet) Special NR10545
Loose (Chet) Special NR13686
Loose (Chet) Special NX64573

Loose, George H. (San Francisco, CA)
Loose (George) 1909 Monoplane
Loose (George) 1910 Demoiselle Type Monoplane
Loose (George) 1911 7th Demoiselle Type Monoplane
Loose (George) 1911 Biplane

Loose, Walter M. (Reading, PA)
Loose (Walter) 2-Place Open Land Monoplane (1934)

L.O.P.P. (Liga Obrony Powietrznej i Przeciwgazowej: League of Air and Anti-gas Defense) (Poland)
L.O.P.P. Ikas (Icarus)

LoPresti Flight Concepts Inc (Seattle, WA)
LoPresti (Piper) Swiftfire
LoPresti (Piper) Swiftfury

Loring (J. Loring Fábrica de Aeroplanos) (Spain)
Loring R.I

Loring E.II "La Pepa"
Loring R.III
Loring 7-Engined Aircraft

Lorraine-Hanriot *(See: Hanriot)*

Lossé, Emile (France)
Lossé 1910 Gyroscopic Aerial Machine Patent

Lotnicze Warsztaty Doswiadczaine *(See: LWD)*

Loudy, Flavius E. (Mamaroneck, NY)
Loudy Amphibian
Loudy Rotorplane
Loudy L.S.I Scout
Loudy L.II Corps d'Armee
Loudy V Two-Seater Pursuit

Loughead (Loughead Aircraft Manufacturing Co) (Santa Barbara, CA)
In 1912 the Loughead brothers (Allan and Malcolm), partnered with Max Mamlock, established the **Alco Hydro-Aeroplane Co** to construct an aircraft of Allan's design. Following the crash of the first aircraft, the company went dormant, with the brothers buying out the other stockholders. In 1916 the brothers established the **Loughead Aircraft Manufacturing Co.** The company reorganized in 1919, but continuing financial problems forced liquidation in 1921. In 1926 the brothers established the **Lockheed Aircraft Co**, which was purchased by **Detroit Aircraft Corp** in 1929. In 1930 they established **Loughead Brothers Aircraft Corp**, which failed in 1934. Allan Loughead organized **Alcor Aircraft Corp** in 1937 to complete designs begun by Loughead Brothers, but the sole prototype crashed in 1938, forcing the company out of business. For designs originating from Alco or Loughead Aircraft Manufacturing Co, see **Loughead.** For all designs of Loughead Brothers, see **Loughead Brothers.** For Alcor designs, see **Alcor.** For products of Lockheed or its successors, see **Lockheed** or **Lockheed Martin.**
Loughead Model F-1 (Flying Boat)
Loughead Model F-1A (Landplane)
Loughead (Alco) Model G
Loughead 1920 Hydroskimmer
Loughead Model S-1 Sport Biplane

Loughead Brothers Aircraft Corp (Glendale, CA)
Following the sale of the **Lockheed Aircraft Corp** to the **Detroit Aircraft Corp** in 1929, Allan and Malcolm Loughead established the **Loughead Brothers Aircraft Corp.** The company failed in 1934 following the crash of its sole product. Allan Loughead established the **Alcor Aircraft Corp** in Oakland, CA in 1937 to continue the work begun at Loughead Brothers. For all products of Loughead Brothers, see **Loughead Brothers.** For all products of Alcor, see **Alcor.** For all products of Lockheed and its successors, see **Lockheed.**
Loughead Brothers "Olympic" DUO-4

Lovaux Ltd (UK)
Lovaux Optica Scout (See: Edgley Aircraft)
Lovaux Scoutmaster
Lovaux Sprint (See: Trago Mills SAH-1)

Lovejoy, Kenneth F. (Dravosburgh, PA)
Lovejoy 1951 Aircraft Patent

Loving, Neal (Yellow Springs, OH)
Loving Glider NLG-1
Loving Glider S-1
Loving-Wayne Glider S-2
Loving-Wayne WR-1
Loving Roadable WR-2
Loving Roadable WR-3

Lowe, F. Harold *(See: Northern Aerial Transport Co)*

Lowe, Willard, & Fowler Engineering Co *(See: LWF)*

LTG (Luft Torpedo Gesellschaft, Johannisthal) (Torp) (Germany)
LTG SD.1

LTV *(See: Vought)*

LTV Electrosystems (Dallas, TX)
LTV Electrosystems L450F

Lualdi & Cie Spa (Italy)
Lualdi ES53
Lualdi L.55
Lualdi L.57
Lualdi L.59

Lübeck-Travemünde (Flugzeugwerft Lübeck-Travemünde GmbH) (Germany)
Lübeck-Travemünde F 1
Lübeck-Travemünde F 2
Lübeck-Travemünde F 4
Lübeck-Travemünde Single-Seat Seaplane

Lubelski Klub Lotniczy *(See: LKL)*
Lublin (Poland)
Lublin R-I
Lublin R-II
Lublin R-III
Lublin R-IV
Lublin R-VI
Lublin R-VIII
Lublin R-VIII*bis*
Lublin R-IX
Lublin R-X
Lublin R-XI
Lublin R-XII
Lublin R-XIII
Lublin R-XIV
Lublin R-XVI
Lublin R-XVII
Lublin R-XVIII
Lublin R-XIX
Lublin R-XX
Lublin I Glider
Lublin II Glider

Lublinian Aviation Club *(See: LKL)*

Ludington Co (Ludington Exhibition Co) (Philadelphia, PA)
Ludington JN Tractor Biplane
Ludington Lizette

Ludlow, Israel (New York, NY)
Ludlow 1905 Flying Machine

Luft Torpedo Gesellschaft *(See: LTG)*

Luft-Verkehrs Gesellschaft *(See: LVG)*

Luftfahrzeug Gesellschaft *(See: LFG Roland)*

Luftfahrtverein *(See: Raab-Katzenstein)*

Luków Secondary School (Poland)
Luków Secondary School 1927 Glider

Lundgren, Earl (Youngstown, OH)
Lundgren Monoplane (1911)

***LUSAC** (See: LePère LUSAC)*

Luscombe Manufacturing Corp (Montebello, CA)
Luscombe Model 1 (Phantom)
Luscombe Model 4 (Luscombe 90)
Luscombe Model 8 (Luscombe 50)
Luscombe Model 8A (Luscombe 65)
Luscombe Model 8B
Luscombe Model 8C Silvaire
Luscombe Model 8C Silvaire Deluxe
Luscombe Model 8C Silvaire
Luscombe Model 8D Silvaire Trainer
Luscombe Model 8E Silvaire Deluxe
Luscombe Model 8F Silvaire 90
Luscombe Model 10 Silvaire
Luscombe Model 11A Silvaire Sedan

Luyties, Otto (Baltimore, MD)
Luyties 1908 Helicopter

Luzzatti-Stiavelli (Italy)
Luzzatti-Stiavelli Pursuit

LVG (Luft-Verkehrs GmbH) (Germany)
LVG B.I
LVG B.II
LVG B.III
LVG Blériot Type Renneindecker (Racing Monoplane)
LVG C.I
LVG C.II
LVG C.III
LVG C.IV
LVG C.V
LVG C.VI
LVG C.VIII
LVG D.II (D 12)
LVG D.III
LVG D.IV
LVG D.V
LVG D.VI
LVG D 10
LVG E.I
LVG G.I
LVG G.III
LVG V59
LVG 1915

LWD (Lotnicze Warsztaty Doswiadczaine) (Poland)
LWD Junak-1 (Cadet-1)
LWD Junak-2 (Cadet-2)
LWD Junak-3 (Cadet-3)
LWD Szpak
LWD Zak-3
LWD Zaraw (Crane) STOL
LWD Zuch (Daredevil)

LWF (Lowe, Willard, & Fowler Engineering Co) (New York, NY)
LWF (de Havilland) DH-4
LWF (Douglas) DT-2
LWF F
LWF G
LWF G-2
LWF G-3
LWF H Owl
LWF (Curtiss) HS-2L
LWF L Butterfly
LWF T-3 Transport
LWF Twin DH-4
LWF V
LWF V-1
LWF V-2
LWF V-3
LWF VH-1

Lwów (Aviation Assoc of Technical University Students in Lwów) (Poland)
Lwów Monoplane of the "Three"

L.W.S. (Lubelska Wytwórnia Samolotów: Lublinian Aeroplane Plant Co Ltd) (Poland)
L.W.S. AKc
L.W.S. 2
L.W.S. 3 Mewa (Gull)
L.W.S. 4
L.W.S. 6 Zubr (Bison)

Lykoshin-Nikhel'son (V. A. Lykoshin N. G. Nikhel'son) (Russia)
Lykoshin-Nikhel'son LN

Lyons, J. W. (Moline, IL)
Lyons Aerodrome

M

***M. B. Arpin & Co** (See: Arpin)*

M-B (Boutard) (Germany)

M-B Taube (Dove)

M-W-Z Aircraft (Paul Mollenhauer, Richard J Waller, & Julien E. Zerrien) (Chicago, IL)

M-W-Z W-LB-5D

Macchi (Aeronautica Macchi SpA) (Italy)

In 1912, Giulio Macchi established **Nieuport-Macchi SA** in Varese to produce license-built Nieuport aircraft for Italian use. In 1913, Nieuport-Macchi began to manufacture original designs, particularly flying boats and seaplanes. In 1928, the company was renamed **Aeronautica Macchi SpA** (**Macchi**). The company continued to produce aircraft through both World Wars and rebuilt following heavy plant damage during World War II. In January 1981, the company reorganized the variety of interests it had accumulated, transforming Macchi into a holding company and transferring airframe manufacturing to a newly-created subsidiary, **Aermacchi SpA**. In 1983, **Aeritalia** (later **Alenia**) acquired 25% of Macchi, although there was no effect on the operations of its subsidiaries. In early 1997, Aermacchi bought and merged with **SIAI-Marchetti**, concentrating production operations at Aermacchi's plant in Varese. For all products of Nieuport-Macchi, Macchi, and Aermacchi, see **Macchi**. For all Nieuport designs, see **Nieuport**. For all products of SIAI-Marchetti predating the 1997 merger, see **SIAI-Marchetti**.

Macchi AL 60B
Macchi AL 60C
Macchi AL 60D
Macchi AL 60F
Macchi L.1
Macchi L.2
Macchi L.3 (M.3)
Macchi M.4
Macchi M.5
Macchi M.7
Macchi M.7ter
Macchi M.8
Macchi M.9
Macchi M.9bis
Macchi M.9ter
Macchi M.12
Macchi M.14
Macchi M.15
Macchi M.16
Macchi M.16, USN Trials
Macchi M.17
Macchi M.18
Macchi M.19
Macchi M.24
Macchi M.24bis
Macchi M.24ter
Macchi M.26
Macchi M.33
Macchi M.34
Macchi M.37
Macchi M.38
Macchi M.39
Macchi M.40
Macchi M.41
Macchi M.41bis
Macchi M.52
Macchi M.52bis
Macho M.67
Macchi M.70
Macchi M.71
Macchi M.C.72
Macchi M.C.73
Macchi M.C.77
Macchi M.C.94
Macchi M.C.94 Anfibio (Amphibian)
Macchi M.C.99
Macchi M.C.100
Macchi M.C.200 Saetta (Arrow) Prototype
Macchi M.C.200 Saetta (Arrow)
Macchi M.C.202 Folgore Prototype
Macchi M.C.202 Folgore (Thunderbolt)
Macchi M.C.202 Folgore, NASM
Macchi M.C.205 Veltro (Greyhound)
Macchi M.C.205N Orione (Orion)
Macchi M.B.308
Macchi M.B.320
Macchi M.B.323
Macchi M.B.326
Macchi M.B.326C
Macchi M.B.326G (EMBRAER AT-26 Xavante)
Macchi M.B.326H
Macchi M.B.326K
Macchi M.B.326M (Atlas Impala I)
Macchi M.B.335 (See: Aeritalia AM30)
Macchi M.B.339 Pan, Frecce Tricolori
Macchi M.B.339A
Macchi M.B.339B
Macchi M.B.339C
Macchi M.B.339K Veltro 2
Macchi M.290TP RediGO (Ready to GO)
Macchi M.416 (Fokker S.11)
Macchi SF-260 Family
Macchi-Lockheed AL-60 Prototype
Macchi-Lockheed AL-60A

MacCready, Paul (Simi Valley, CA)

MacCready Gossamer Albatross
MacCready Gossamer Albatross, NASM
MacCready Gossamer Condor

MacCready Gossamer Condor, NASM
MacCready Gossamer Penguin
MacCready Quetzalcoatlus Northropi (QN)
MacCready Solar Challenger
MacCready Solar Challenger, NASM

MacDonald Aircraft Co (Robert A. MacDonald) (Concord, CA)
MacDonald (Aircraft) S-21

MacDonald, Vance T. (Detroit, MI)
MacDonald (Vance T.) Sportflight A

MacFie, Robert F. (UK)
MacFie Circuit
MacFie Empress

Madon, Georges (France)
Madon Tailless (1923)

Maestranza Central de Aviación (MCA) (Chile)
Maestranza Central de Aviación:
H.F.XX-02 Trainer
Ticicolo Experimental

MÁG (Ungarische Allgemeine Maschinenfabrik AG, Magyar Altalános Gepgyár RT) (Austria-Hungary)
MÁG-Fokker Triplane 90.01
MÁG-Fokker Triplane 90.02
MÁG-Fokker Triplane 90.03
MÁG-Fokker Triplane 90.04
MÁG-Fokker 90.05

Magaldi-Gallinari *(See: Gallinari M.G.2)*

Magnan, Dr. (France)
Magnan M-2 (Glider)

Magni, Piero (Piero Magni Aviazione) (Italy)
Magni Bi-Vittoria (Victory)
Magni Jona (Alberto Jona) J.6
Magni Jona (Alberto Jona) J.6/S
Magni PM.3-4 Vale (Farewell)
Magni PM.4-1 Supervale (Farewell)
Magni Vale (Farewell)
Magni Vittoria (Victory)

Magyar Altalános Gepgyár RT *(See: MÁG)*

Mahoney-Ryan Aircraft Corp *(See: Ryan)*

MAI (Moscow Aviation Institute) (Russia)
MAI-53
MAI-60
MAI-62
MAI Kvant

Maillet, André *(See: SFCA)*

Maire, M. G. (Switzerland)
Maire 1909 Biplane

Mais, J. E. (Chicago, IL)
Mais 1910 Pusher Biplane

Maiss, Ulrich (Germany)
Maiss Bayern II (Bavaria)

Maiwurm, Paul (New York, NY)
Maiwurm 1929 No.1 Flyworm
Maiwurm 1931 Cyclonic-Rocket

Makarov, Y. V. (Russia)
Makarov ESKA-1 (Surface Skimmer)

Makhonine (France)
Makhonine 123 (1947 Variable-Area Monoplane)
Makhonine 1931 Variable-Area Monoplane

Makina ve Kimya Endustri Kurumu *(See: MKEK)*

Malinowski, Stefan (Poland)
Malinowski Dziaba Glider (1923)
Malinowski Stemal III (1922 Variable Camber-Wing)
Malinowski Stemal VII (1927)

Maliszewski, Roman B. (Milwaukee, WI)
Maliszewski 1928 Monoplane

Mallet, Maurice *(See: Zodiac)*

Man-pi [Manshu (or Mansyu) Hikoki Seizo: Manchurian Aeroplane Manufacturing Co Ltd] (Manchukuo) *(See also: Manko)*
Man-pi Ki-79
Man-pi Ki-98

Man-Powered Aircraft *(See: Human-Powered Aircraft)*

Manchoulas (Belgium)
Manchoulas Type BAC Glider (1934)

Manchurian Airways Co *(See: Manko)*

Manfred Weiss *(See: Weiss Manfred Repiülögép-és Motorgyár)*

Manhattan Aero Co (New York, NY)
Manhattan 1913 Aeroplane (Owned by Charles Wald)

Manko [Manshu (or Mansyu) Koku KK: Manchurian Airways Co] (Manchukuo)
In 1932 the Japanese and their puppet government in Manchukuo established **Manshu [Mansyu] Koku KK** (Manchurian Airways Co Ltd or **Manko**). By 1934, the company was producing Fokker and de Havilland airliners under license at Mukden. From this experience Manko began designing original aircraft, leading to Japanese Army contracts to produce **Kawasaki** and **Nakajima** designs under license, in addition to Manko aircraft. In 1938, airframe production was transferred to a new company, **Manshu [Mansyu] Hikoki Seizo KK** (Manchurian Aeroplane Manufacturing Co Ltd or **Man-pi**). Man-pi continued to produce Manko designs and license-built airframes until the collapse of Manchukuo at the end of World War II. License-built designs are listed under their original design firms. Original Manko designs are under **Manko**. Later Man-pi designs appear under **Man-pi**.
Manko MT-1 Hayabusa (Peregrine Falcon)
Manko MT-2

Mann, Egerton & Co, Ltd (U.K.)
Mann Egerton Type B (Modified Short 184 Seaplane)
Mann Egerton H-2 Seaboat Scout

Manshu (Mansyu) Hikoki Seizo *(See: Man-pi)*

Manshu (Mansyu) Koku KK *(See: Manko)*

Manta Aircraft Corp (Los Angeles, CA)
Manta Long Range Fighter

Manta Products (Oakland, CA)
Manta Fledgling (Fledge) IA
Manta Fledgling (Fledge) IB

Manta Fledgling (Fledge) IIA
Manta Fledgling (Fledge) IIB
Manta Fledge 3

Marchetti Motor Patents, Inc (Paul Marchetti) (San Francisco, CA)
Marchetti M2

Mantelli, Col. Adriano (Argentina)
Mantelli AM-10
Mantelli AM-11 Albatross

Manzolini (Conte Ettore Manzolini di Campoleone) (Italy)
Manzolini Libellula (Dragonfly) II

Marais (France)
Marais 1923 Monoplane Glider

Maranda Aircraft Co, Ltd (Canada)
In 1957, Bernard Maranda acquired rights for R.A.14 Loisirs and R.A.17 light monoplanes from **Roger Adam** of France, marketing the modified designs as Maranda Loisairs. For the original French designs, see **Adam**. For the Canadian derivatives, see **Maranda**.
Maranda Loisair RA14BM1
Maranda Loisair BM3
Maranda Hawk BM4 (Béarn Minicab)
Maranda Falcon BM5
Maranda Lark BM6.

de Marçay (Edmond de Marçay & Cie; Construction Aéronautique Edmond de Marçay) (France)
de Marçay "Passepartout" ("Master Key")
de Marçay 1 C1
de Marçay 2 C1
de Marçay 4 C1

de Marçay-Moonen (E. de Marçay-Moonen) (France)
de Marçay-Moonen 1912 Folding Wing Monoplane

Marchetti, Alessandro (Italy)
Italian engineer Alessandro Marchetti designed several aircraft before becoming the chief engineer for **Società Idrovolanti Alta Italia** (**SIAI** or **Savoia**) in 1922. The company was more commonly called **Savoia Marchetti** after he joined the firm. In 1936, the company was renamed **Società Italiana Aeroplani Idrovolanti** (still **SIAI** or **Savoia-Marchetti**). The common name later became **SIAI-Marchetti**. For all Marchetti designs before joining SIAI, see **Pomilio** and **Marchetti**; for all subsequent work see **SIAI** and **Savoia-Marchetti**
Marchetti M.V.T. (Marchetti-Vickers-Terni) Scout

Marco *(See: Marko-Elektronik Co)*

Marcoux-Bromberg Special *(See: Keith Rider R-3 "Marcoux-Bromberg Special")*

Marichal, Edmundo (Argentina)
Marichal Estudiantil (Scholar)

Marine Aircraft Engineering Co (Detroit, MI)
Marine Model 1927 Flying Boat

Marinens Flyvebaatfabrik (Norway)
Marinens Flyvebaatfabrik W 33 (Brandenburg W 33 Variant)

Marion Aircraft Corp (Marion, OH)
Marion Cabin Monoplane (1929)

Mark (Stahlwerk Mark AG) (Germany)
Mark R III/22
Mark R IV/23
Mark R V/23
Mark 1923 Type Rieseler (Giant) Sport

Märkische Flugzeug-Werke (Germany)
Märkische D.I

Marko-Elektronik Co (Poland)
Marko-Elektronik (Janowski) J-5

Marquardt Aircraft Co (Van Nuys, CA)
Marquardt M-14 Whirlajet

Marquart, Ed (Riverside, CA)
Marquart MA-4 Lancer
Marquart MA-5 Charger

Marquette Aeroplane Co (Indianapolis, IN)
Marquette 1910 Biplane

Marriott, Frederick (UK, USA)
Marriott Aerial Steam Carriage (1842)
Marriott Avitor (c.1863)

Mars (Germany)
Mars Monoplane
Mars 1912 Biplane

Marshall Aircraft Co (Marshall, MO)
Marshall (Aircraft) Montagne

Marshall Flying School, Inc (Marshall, MO)
Marshall (Flying School) Research (Flyer)

Marske Aircraft Corp (Jim Marske) (Michigan City, IN)
Marske Monarch (Glider)
Marske Pioneer IA (Sailplane)
Marske Pioneer II (Sailplane)

Martens, Arthur (Germany)
Martens Motor Glider "Max"

Martin Aircraft Works (UK) *(See: Martin-Baker)*

Martin Marietta Corp (Bethesda, MD)
In 1961, **The Martin Co** merged with **American-Marietta Co** to form the **Martin Marietta Corp.** Martin had already ended its airframe production to concentrate on missiles and electronics, so the Martin facilities continued as the aerospace component of the new firm. In 1994, Martin Marietta merged with the **Lockheed Corp**, forming **Lockheed Martin Corp** in 1995. For pre-1961 Martin designs see **Martin, Glenn L.** and **Wright-Martin**. For the lifting bodies designed by the Martin Divisions of the Martin Marietta Corp, see **Martin Marietta**. For products of Lockheed Martin see **Lockheed Martin**.
Martin Marietta X-24A (Lifting Body)
Martin Marietta X-24B (Lifting Body)
Martin Marietta X-24C (Lifting Body)

Martin, Edward *(See: Martin-Boyd)*

Martin, Glenn L. (Los Angeles, CA; Cleveland, OH; Baltimore, MD)
In 1912, Glenn Martin established the **Glenn L. Martin Co** in Los Angeles to build aircraft of his own design. In 1916, the **Martin Co** merged with the **Wright Aircraft Co**, **Simplex Automobile Co**, **Wright Flying Field, Inc**, and the **General Aeronautic Co of America** to form the **Wright-Martin Aircraft**

Corp. Martin left the company in 1917, forming a new **Glenn L. Martin Co** in Cleveland. In 1929, the company moved to Baltimore. In 1957 the company was renamed **The Martin Co.** In December 1960, Martin ended airframe production, concentrating on missiles and electronics. The following year, Martin merged with the **American-Marietta Co** to form the **Martin Marietta Corp**, with Martin as its aerospace component. In 1995, Martin Marietta merged with the **Lockheed Corp**, forming **Lockheed Martin Corp**.

For designs originating with the various Glenn L. Martin companies, see **Martin**. Lifting bodies of the Martin Divisions of the Martin Marietta Corp are listed under **Martin Marietta**. Products of Lockheed Martin and Wright-Martin are under **Lockheed Martin** and **Wright-Martin**. (Glenn L. Martin's companies should not be confused with the unrelated **James V. Martin Aircraft Co** or with other individuals sharing the Martin name.)

Martin (AM) XBTM-1 Mauler (210)
Martin AM-1 Mauler (210A)
Martin AM-1Q Mauler (210B)
Martin (BM) XT5M-1 (77)
Martin BM-1 (125)
Martin BM-2 (129)
Martin (B-10) XB-907 (139)
Martin (B-10) XB-907A (139)
Martin XB-10 (139)
Martin YB-10 (139A)
Martin YB-10A (139A)
Martin B-10B (139)
Martin B-10M
Martin YB-12 (139B)
Martin YB-12A (139B)
Martin XB-14 (139B)
Martin XB-16 (145)
Martin B-26 Marauder (179)
Martin B-26A Marauder (179A)
Martin B-26B Marauder (179B)
Martin B-26B Marauder (179B) "Flak Bait," NASM
Martin TB-26B (AT-23A) Marauder (179B)
Martin B-26C Marauder (179C)
Martin TB-26C (AT-23B) Marauder (179C)
Martin XB-26E Marauder (179E)
Martin B-26F Marauder (179F)
Martin B-26G Marauder (179G)
Martin XB-26H Marauder "Middle River Stump Jumper"
Martin (B-26) Marauder Mk.I (179A)
Martin (B-26) Marauder Mk.IA (179B)
Martin (B-26) Marauder Mk.II (179C)
Martin (B-26) Marauder Mk.III (179F/G)
Martin B-27 Proposal (182)
Martin B-33 Super Marauder Proposal (189)
Martin XB-33 Super Marauder (190)
Martin XB-48 (223)
Martin XB-51 (234)
Martin B-57 Canberra Combat Transport Proposal
Martin B-57A Canberra (272A)
Martin EB-57A Canberra (272R)
Martin RB-57A Canberra (272R)
Martin B-57B Canberra (272B)
Martin EB-57B Canberra (272B)
Martin EB-57B Canberra (272B), NASM
Martin JB-57B Canberra (272B)
Martin NB-57B Canberra (272B)
Martin B-57C Canberra (272C)
Martin RB-57D Canberra (272D)
Martin B-57E Canberra (272E)
Martin EB-57E Canberra (272E)
Martin RB-57F (WB-57F) Canberra (272P)
Martin B-57G Canberra
Martin COIN (Counter Insurgency Aircraft) Proposal
Martin F
Martin JM-1 Marauder (179C)
Martin JM-2 Marauder (179G)
Martin (JRM) XPB2M-1 Mars (170)
Martin (JRM) XPB2M-1R Mars (170)
Martin JRM-1 Mars (170A)
Martin JRM-2 Mars (170B)
Martin MB-1 (GMB-G)
Martin MB-1 (GMB-G), US Perimeter Flight (1919)
Martin MB-1 (GMB-G), US Perimeter Flight (1919), NASM
Martin (MB-1) GMB-M Night Bomber
Martin (MB-1) GMB-TA Transatlantic Aircraft
Martin (MB-1) GMC, CA Cannon-Equipped Aircraft
Martin (MB-1) GMP, GMT, 12P Transport Aircraft
Martin (MB-1) MP Mailplane
Martin (MB-1) MP Mailplane, Modified to MB-1 Status
Martin MB-2
Martin (MB-2) NBS-1 (License-Built)
Martin MBT Torpedo Bomber
Martin MO-1 (57)
Martin M2O-1 (60)
Martin MS-1 (63)
Martin MT Torpedo Bomber
Martin MT Torpedo Bomber, Racing Modification
Martin N2M-1 (NT-1, 67)
Martin XO3M-1 (83)
Martin PM-1 (117)
Martin PM-1B (117, Brazil)
Martin PM-2 (122)
Martin XP2M-1 (119)
Martin XP2M-2 (119)
Martin P3M-1 (120)
Martin P3M-2 (120)
Martin XP4M-1 Mercator (219)
Martin P4M-1 Mercator (219A)
Martin P4M-1Q Mercator (219B)
Martin XP5M-1 Marlin (237)
Martin P5M-1 (P-5A) Marlin (237A)
Martin P5M-1 (P-5A) Marlin, P5M-2 Development Aircraft
Martin P5M-1G Marlin (237A, USCG)
Martin P5M-1S (SP-5A) Marlin (237A)
Martin P5M-1T (TP-5A) Marlin (237A)
Martin P5M-2 (P-5B) Marlin (237B)
Martin P5M-2G Marlin (237B, USCG)
Martin P5M-2S (SP-5B) Marlin (237B)
Martin XP6M-1 Seamaster (275)
Martin YP6M-1 Seamaster (275A)
Martin P6M-2 Seamaster (275B)
Martin (P6M) Seamistress (Civil P6M Proposal)
Martin XPBM-1 Mariner (162)
Martin PBM-1 Mariner (162)
Martin XPBM-2 Mariner (162)
Martin PBM-3 Mariner (162B)
Martin PBM-3C Mariner (162C)
Martin PBM-3D Mariner (162D)
Martin PBM-3R Mariner (162B)
Martin PBM-3S Mariner (162C)
Martin PBM-5 Mariner (162F)
Martin XPBM-5A Mariner (162G)
Martin PBM-5A Mariner (162G)
Martin PBM-5A Mariner, NASM
Martin PBM-5G Mariner (USCG)
Martin PBM-5S Mariner
Martin (PBM) Mariner GR.Mk.I (RAF)
Martin R
Martin RM-1G (VC-3A, 404C; USCG)
Martin S
Martin SC-1 (69)
Martin SC-2 (69)
Martin T
Martin T Hydro (AeroYacht, Great Lakes Tourer)
Martin T3M-1 (73)
Martin (T3M-2) XSC-6
Martin (T3M-2) SC-6
Martin T3M-2 (73)
Martin XT3M-3
Martin XT3M-4
Martin XT4M-1 (74)
Martin T4M-1 (74)
Martin XT6M-1 (118)

Martin TA
Martin TA Hydro
Martin TA Hydro, William Boeing Long-Wing Special
Martin TT
Martin 66 (NBL-1, MNM-1 Mail Plane)
Martin 68 Liberty Mail Plane
Martin 70 Commercial
Martin 121 Civil Amphibian Proposal
Martin 123 (139 Prototype)
Martin 124 (Export T4M Proposal)
Martin 130 Clipper
Martin 133
Martin 133A
Martin 133B
Martin 133C
Martin 139 (See also: Martin B-10, B-12)
Martin 139W (Export B-10)
Martin 146
Martin 156 Russia Clipper
Martin 162A Tadpole Clipper (PBM Hull Test Vehicle)
Martin 162A Tadpole Clipper, NASM
Martin 163 (250,000 lb. Flying Boat Proposal)
Martin 166 (Export B-10)
Martin 167 Prototype (XA-22)
Martin 167, France
Martin 167-B3 Maryland Mk.I (RAF)
Martin 167-B4 Maryland Mk.II (RAF)
Martin 170 (Civil JRM Mars Proposal)
Martin 187 Baltimore Prototype
Martin 187-B1 Baltimore Mk.I (RAF)
Martin 187-B1 Baltimore Mk.II (RAF)
Martin 187-B1 Baltimore Mk.III (RAF)
Martin 187-B1 Baltimore Mk.IIIA (A-30)
Martin 187-B3 Baltimore Mk.IV (A-30A)
Martin 187-B3 Baltimore Mk.V (A-30A)
Martin 187-B3 Baltimore Mk.V, USN Transonic Research
Martin 187-B4 Baltimore GR.Mk.VI (A-30C)
Martin 193 (Flying Boat Proposal)
Martin 202 (2-O-2), 1st Prototype
Martin 202 (2-O-2), 2nd Prototype
Martin 202 (2-O-2)
Martin 202A (2-O-2)
Martin 228 Civil Transport Proposal
Martin 242 Assault Seaplane Proposal
Martin 259 Seaplane XB-51 Proposal
Martin 270 (M-270, P6M Hull Testbed)
Martin 303 (3-O-3)
Martin 404 (4-O-4) Prototype
Martin 404 (4-O-4)
Martin 1909 Pusher
Martin 1911 Pusher
Martin 1912 Headless Pusher
Martin 1912 Semi-Headless Pusher
Martin 1912 Semi-Headless Pusher (Catalina Flight)
Martin 1913 "Armored" Tractor Biplane
Martin 1914 "Aeroplane Destroyer"
Martin 1914 Special Biplane (Lincoln Beachey Special)

Martin, James Vernon (Elmyra, OH; Dayton, OH; Garden City, NY)

J. V. Martin began designing aircraft as a member of the **Harvard Aeronautical Society** in 1910. In 1911, he directed the construction of a new biplane by the **Queen Aeroplane Co.** He traveled widely before establishing the **Martin Aeroplane Co** in 1917. In 1920 he moved the company to Dayton, OH, (renaming it **Martin Enterprises**) and offered free use of his aeronautical patents to the aviation industry. Two years later he moved again and renamed the company the **Martin Aeroplane Factory**; the plant was later taken over by the **Kirkham Products Co.** For designs predating establishment of the various Martin companies, see **Harvard** or **Queen**. For his later work, see **Martin, James V.**

Martin (J. V.) "Double Tractor" Bomber (1917)
Martin (J. V.) K.III Kitten
Martin (J. V.) K.III Kitten, NASM
Martin (J. V.) K.IV (USN KF-1)

Martin, William H. (Canton, OH)

Martin (W. H.) 1908 Glider
Martin (W. H.) 1908 Glider, NASM

Martin-Baker Aircraft Co, Ltd (UK)

James Martin (not to be confused with American J. V. Martin) established the **Martin Aircraft Works** in 1929. In 1934, Martin joined with Capt. Valentine H. Baker and Francis Francis to form a limited company, renamed **Martin-Baker Aircraft Co, Ltd.** After designing a series of prototype aircraft through the end of World War II, the company changed its concentration to aircraft systems, particularly ejection seats. For all aircraft designs originating from Martin Aircraft Works or Martin-Baker, see **Martin-Baker**.

Martin-Baker M.B.1
Martin-Baker M.B.2
Martin-Baker M.B.3
Martin-Baker M.B.5

Martin-Boyd (Edward Martin & Millard Boyd) (Santa Ana, CA)

Martin-Boyd Monoplane

Martin-Handasyde (UK)

In 1908 George H. Handasyde and H. P. Martin formed a partnership to design and build aircraft. After operating as **Martin-Handasyde** for several years they formally incorporated as **Martinsyde, Ltd** in 1914. For all aircraft developed before the formal incorporation of Martinsyde, see **Martin-Handasyde.** For aircraft designed after incorporation, see **Martinsyde.**

Martin-Handasyde No.1 Monoplane
Martin-Handasyde No.2 Monoplane
Martin-Handasyde No.3 Monoplane
Martin-Handasyde 4.B Dragonfly (1911)
Martin-Handasyde 1911 Monoplane
Martin-Handasyde 1912 Monoplane
Martin-Handasyde 1912 Military Monoplane
Martin-Handasyde 1912 Military Trials Monoplane
Martin-Handasyde 1913 Monoplane
Martin-Handasyde 1914 Transatlantic Monoplane
Martin-Handasyde 1914 Pusher Biplane Racer

Martinsyde, Ltd (UK)

After several years operating as **Martin-Handasyde**, George H. Handasyde and H. P. Martin formally incorporated as **Martinsyde, Ltd** in 1914. The company declined with the end of World War I and went into receivership in May 1921, its assets acquired by the **Aircraft Disposal Co (ADC)**. Handasyde went on to form **Handasyde Aircraft Co, Ltd.** For all products of the Martin and Handasyde partnership before formal incorporation, see **Martin-Handasyde.** For all products of Martinsyde, see **Martinsyde**. For subsequent work by George Handasyde, see **Handasyde.** For "Martinsyde" aircraft designed by ADC, see **ADC.**

Martinsyde Type A Mk.I
Martinsyde Type A Mk.II
Martinsyde Type AS
Martinsyde F.1

Above: Aircraft names and designations are as varied as the designs themselves. Orville and Wilbur Wright's called their first powered aircraft the Flyer. As later versions of the Flyer were produced, the Wright Brothers began describing this aircraft as the 1903 Flyer. Many secondary sources refer to the Kitty Hawk Flyer, a name not used by the Wrights.

Below: Brazilian aviation pioneer Alberto Santos-Dumont sequentially numbered his lighter-than-air aircraft 1 through 14. The No.14 was flown without its gas bag as the 14*bis*, Santos-Dumont's first heavier-than-air design.

Above: French pioneer aviator Louis Blériot's first eleven aircraft were also numbered sequentially. The highly successful No. XI then became the prototype of the Type XI family, with hundreds of aircraft produced. Note that, although Blériot would later spell his name as "Bleriot," the company name always used an "é."

Below: Italian Gianni Caproni's first biplane was built in 1910. Several years later, as part of a formal designation system, he began calling this aircraft the Ca.1.

Right, top: When Geoffrey de Havilland formed his own aircraft company in 1920, he had already built a reputation as one of Britain's finest designers. Although his D.H.1 (shown here) through D.H.18 were designed and built at the Aircraft Manufacturing Co, Ltd (later known as "Airco"), the aircraft were commonly known as de Havillands.

Right, centre: Hugo Junkers designated his first all-metal aircraft the J 1. It was delivered to the German military in late 1915, where its military service designation was E.I (for "Eindecker [Monoplane] No. I").

Right, bottom: The Junkers J 4 biplane was first produced in early 1917. In German military use, it was designated J.I using the new J-series designations for armored attack aircraft. Note that two similar designations – J 1 and J.I – describe two completely different aircraft.

Junkers
Junkers

BRE 19
Nº 107
Nº 1073
4

Left, top: The punctuation of French World War I designations has been well recognized for its inconsistencies. This Farman F.50 Bn.2 has also been noted in reports as the F.50 Bn 2, the F 50 Bn 2, and the Far.13 Bn.2. (Note that the suffix identifies this aircraft as a 2-seat night bomber.)
Left, center: Many aircraft were alternately designated with Roman and Arabic numerals. The Breguet 19 B.2 (2-seat day bomber) appeared in many reports and company brochures as the Breguet XIX B.2. The abbreviated "Bre." prefix (Bre.19 B.2) was also used in place of the full company name to avoid confusion with other companies' model 19s.
Left, bottom: The US Navy purchased ten NC (for Navy/Curtiss) Boats; the first four were built by Curtiss, and the last six by the Naval Aircraft Factory. Initially, each aircraft carried its hull number as part of its designation. Seen here, the NC-4 was the fourth aircraft built. In 1922, the surviving NC Boats received the common designation P2N.
Above: The O-2 was the second major observation type in the US Army's post-1924 designation system. Subsequent variants were identified by letter suffixes: this aircraft is one of 143 Douglas O-2H aircraft. The final variant was the O-2K, which would later be modified and redesignated BT-1 (Basic Trainer 1).
Below: In 1927, a young airmail pilot named Charles Lindbergh hired Ryan Airlines to design and build a special long-range, single-seat monoplane. As the aircraft was engineered for a flight from New York to Paris, the company designated it NYP; Lindbergh chose the name "Spirit of St Louis" in honor of his financial backers.

Left, top: The Glenn L. Martin Co built three Model 130 (a.k.a. M-130) flying boats, delivering all to Pan American Airways in 1935/36. As with its other flying boats, Pan Am named the aircraft "Clippers;" "China Clipper," the best known of the three, served with "Philippine Clipper" and (shown here) "Hawaiian Clipper."

Left, center: The J-2 Cub was introduced by the Taylor Aircraft Co in 1936. About 300 were built before the company was renamed Piper Aircraft Corp in 1937. Piper built more than 350 more before introducing the improved J-3 Cub (seen here) at the end of the year.

Left, bottom: Waco - the Weaver Aircraft Co - created a unique designation system in 1930. A basic model letter (reflecting the fuselage style) was linked to prefixes for (first) engine type and (second) wing style. A suffix indicated the last digit of the model year. Thus, this Jacobs L-4 engine (code letter Y) with a "K" style wing on a 1937 Model S would be designated YKS-7. (Similar aircraft, built with a Jacobs L-5 engines were designated ZKS-7.)

Right, top: Britain's radical Mayo Composite Plane, the brainchild of Imperial Airways' Major Robert Mayo, made its first flight in 1938. The Short Brothers S.21 flying boat "Maia" carried the S.20 floatplane "Mercury" to altitude, launching the smaller aircraft and its cargo for high-speed, long-distance mail delivery.

Right, center: Frederick "Fritz" Koolhoven's aircraft carried F.K.-prefixed designations. The sequence of numbers was unbroken from the F.K.1 through the F.K.58, even though Koolhoven designed aircraft for four companies in two nations between 1913 and 1946. The F.K.56, seen here, was built in the Netherlands by N. V. Koolhoven Vliegtuigen shortly before World War II.

Right, bottom: As the "Arsenal of Democracy," the US produced a somewhat confusing array of designations to identify its aircraft. The Douglas DB-7 (Douglas Bomber, Model 7) was designed as an light bomber for export. In British service the aircraft was known as the Boston (in the light bomber role) or Havoc (in the night fighter and night intruder role). The US Army Air Corps based its unrelated Havoc designations on the projected mission: attack aircraft were A-20s (such as this early A-20A.), pursuit night fighters were P-70s, photo reconnaissance aircraft were F-3s, and the projected heavy observation aircraft would have been the O-53. In US Navy hands, the aircraft was flown as the BD (bomber, Douglas).

Above: Japanese World War II aircraft designations seemed complicated to Western eyes. The Imperial Navy knew this aircraft as the Type 0 Carrier Fighter Model 52 (with Type 0 derived from the Japanese calendar year 2600 (1940), the year the type was first accepted for production. The official name Reisen was abbreviated from Rei Sentoki (Zero Fighter). The short designation was A6M5, interpreted as Carrier Fighter (A), sixth design submitted by Mitsubishi, fifth variant of the type. Western intelligence simplified all of this by assigning easily remembered male names to all Japanese fighters, calling the A6M the *Zeke*.

Below: Between 1922 and 1962, US Navy designations were based on the aircraft's manufacturer. This Vought Corsair was designated F4U-1: "U" was the letter denoting a Vought aircraft, "F4" identified it as the fourth fighter type built by Vought for the Navy, and "-1" as the first variant of the type.

Above: Needing more Corsairs than Vought could produce, the Navy opened production lines with two other companies. As the third fighter type Brewster delivered to the Navy, this Corsair was designated F3A (with the "A" denoting Brewster). Similar aircraft built by Goodyear were designated FG (not "F1G" - Navy designations dropped the "1" from the first of each aircraft type delivered by each manufacturer).
Below: Although the Stearman Aircraft Co became the Wichita Division of the Boeing Aircraft Co in 1941, the division's designs continued to be called Stearman aircraft. While US Navy designations used the company letter "B" for all other Boeing designs, Stearman aircraft continued to use "S" designations. Thus, this version of the Army PT-13D primary trainer carried the somewhat confusing Navy designation N2S-5. (With "N" denoting a trainer, this was the fifth variant of the second Navy trainer design built by Stearman.)

Left, top: Most British military aircraft have been named, rather than designated, as they enter production. Variants of the basic design are generally identified by "mark" numbers. This Supermarine Spitfire is an HF. Mark VII, with the HF denoting a high-altitude fighter. (The mark number could also be correctly presented as "HF. Mk.VII" or "HF.VII.")

Left, center: The Potez 63 Series illustrates a designation style common to many French manufacturers: each subtype was distinguished by a numeric suffix. No aircraft was actually designated Potez 63; instead, the first variant was the 630, followed by the 631, 632, etc. (For clarity, some contemporary documents list these variants as 63.0, 63.1, 63.2, or 63-0, 63-1, 63-2, etc.) This eleventh subtype usually appears as the Potez 63/11.

Left, bottom: Beginning in 1933, Germany numbered all new aircraft sequentially, regardless of designer or manufacturer. The manufacturer was identified by a two-letter company prefix (with a capitalized first letter and lower case second letter, followed by a space - not a hyphen or period). The Messerschmitt 109, originally built by the Bayerische Flugzeugwerke, carried the Bf 109 designation throughout its career, even though the company was renamed Messerschmitt AG in 1938. The "Me" prefix would apply only to later designs. The aircraft shown is a captured Bf 109 F, despite the erroneous designation marked on the print by US evaluators.

Right, upper: The North American Aviation Twin Mustang was originally designated P-82, denoting the 82nd aircraft design in the Army Air Forces' "pursuit" series. In 1948, the new US Air Force changed from pursuit to fighter designations, and this XP-82 became the XF-82. (The "X" prefix marked the aircraft as an experimental prototype.)
Right, lower: As World War II drew to a close, Boeing began production of a highly modified Superfortress variant, the B-29D. To emphasize the advances of the design in shrinking post-War budgets, the type was redesignated B-50. This B-50A was modified for weather reconnaissance duties, but never received the "WB-50" designation used on later variants.

Above: In the late-1940s, North American developed a rocket-firing air defense version of the F-86 Sabre known as the F-95. Downplaying the newness of the design in budgets filled with new designs, the aircraft was redesignated F-86D to emphasize its Sabre heritage.

Below: When Chase Aircraft proved unable to continue development of the C-123 Avitruc, US Air Force planners transferred production contracts to Fairchild. Although prototypes and developmental airframes are considered Chase aircraft in most documentation, all subsequent production aircraft (such as this C-123B Provider) are attributed to Fairchild.

Above: The United Aircraft Corp's Sikorsky Aircraft Division developed its S-58 as an antisubmarine helicopter for the US Navy. In this role, it was known as the HSS Seabat, for helicopter, antisub, first design (no number), Sikorsky. Later development for Marine Corps use resulted in the HUS Seahorse, designating it as a helicopter, utility, first design, Sikorsky. The US Army bought the type as the H-34 Choctaw (following the Army practice of naming its aircraft after Indian tribes). In 1962, introduction of the joint service designation system resulted in all surviving US military S-58s being designated as variants of the H-34. In the UK, Westland undertook licensed production of the type as well as development of advanced, jet-powered versions in its Wessex series of helicopters.

Below: In 1959, the US Army revised the designation system used for its helicopters. Under the new system, the Bell YH-40 Iroquois (the 40th helicopter design, with the "Y" prefix denoting a developmental airframe) became the YHU-1 (first of the new "helicopter, utility" series). The "HU-1" designation would lead to the aircraft's popular, but unofficial, name "Huey." Under the 1962 joint designation system, the Huey became the H-1; with a utility mission prefix, the HU-1 became the UH-1.

Above: Lacking firm intelligence data on Soviet aircraft, NATO (the North Atlantic Treaty Organization) created their own series of code names. Names beginning with "C" marked cargo or transport aircraft: the Antonov An-124, seen here, was known in Western circles as the *Condor*.

Below: The Boeing 707 was one of the Jet Age's premier airliners. Each variant was produced with features specific to its user's needs, and each user was identified by a two-character code. This 707 is a 300B series aircraft (707-300B) built for Pan American (coded 21), or simply a 707-321B. Note that Boeing's airliners never carried a "B" prefix; the unofficial "B.707" and "B-707" designations were created for clarity by secondary sources.

Right, top: This short-fuselage, medium-range version of the 707 was initially designated 707-020, then briefly 717-020, and finally 720. Although this 720B was flown by Pan American, it was originally built for Lufthansa. Therefore, the designation was Boeing 720-030B (using Lufthansa's code number "30" instead of Pan Am's code "21").

Right, center: The original 717 was Boeing's designation aircraft for delivered to the USAF as C-135s. This C-135B would later be modified for reconnaissance duties as an RC-135S. (Note that the lower case "s" in "C-135s" denotes plural; the upper case "S" in "RC-135S" denotes the variant that preceded the RC-135T.)

Right, bottom: In a convenient alignment of two unrelated designation systems, the Douglas DC-9 was purchased by the US military as the C-9. (Shown is a USAF C-9A Nightingale hospital aircraft.) Following a merger with McDonnell, the new McDonnell Douglas Corp changed some DC-9 designations for marketing and publicity purposes. Thus the DC-9-80 through the DC-9-88 were advertised as the MD-80 through MD-88. Documents filed with the US government continued to use the "DC" designations until the introduction of the MD-90. When McDonnell Douglas was bought out by Boeing in 1997, the new owner performed a similar redesignation, marketing the MD-95 as the Boeing 717-200. (Note that the 717-200 was unrelated to the earlier Boeing 720 and C-135 designs mentioned above.)

PAN AM
N783PA
PAN AMERICAN
MIAMI INTER
MATS
U.S. AIR FORCE
MAC
22583
MILITARY AIRLIFT COMMAND
U.S. AIR FORCE

Above: Only a few D.H.125 business jets were produced before de Havilland was absorbed by Hawker Siddeley in 1963; development of the aircraft continued with the designation H.S.125. In 1970, US marketing of the British-built aircraft would come under Beechcraft Hawker Corp (a Beech Aircraft Corp subsidiary), with the type sold as the BH 125. Hawker Siddeley was absorbed into British Aerospace in 1977, and the type continued development and production as the BAe 125. In 1994, Raytheon Co merged British Aerospace's Corporate Jet Division with Beech Aircraft Corp to form the Raytheon Aircraft Co. Under the latest corporate structure, the BAe 125-800XP (shown here) is marketed as the Raytheon Hawker 800XP.

Below: The F-16 Fighting Falcon was originally designed and produced by General Dynamics in the early 1970s. The type was still in production in the 1990s when General Dynamics was bought out by Lockheed. Lockheed subsequently merged with Martin as Lockheed Martin. Although all three companies produced variants of the fighter, this *Directory* lists all F-16s under General Dynamics, the original design firm. Shown is a YF-16.

Martinsyde F.2 Raymor (Fred Raynham & C. W. F. Morgan)
Martinsyde F.3
Martinsyde F.4 Buzzard
Martinsyde F.4A
Martinsyde F.6
Martinsyde G.100
Martinsyde G.102 Elephant
Martinsyde S.1
Martinsyde Semiquaver
Martinsyde Semiquaver, Alula Wing Modification

Maruoka, Katsura (Japan)
Maruoka Human-Powered Screw-Wing Machine (1903)

Maryland Pressed Steel Corp *(See: Bellanca)*

Mas (France)
Mas 1910 Monoplane

Masak, Peter (USA)
Masak Scimitar I
Masak Scimitar II

Masaryk Aviation League *(See: Zlin)*

Masarykova Letecká Liga *(See: Zlin)*

Mason, Monty G. (USA)
Mason Meteor M (Rebuilt Gotch & Brundage Special)
Mason Meteor "Texas Sky Ranger" (Rebuilt Vance Flying Wing)

Masquito Aircraft NV (Belgium)
Masquito M58 Ultralight Helicopter
Masquito M80 Ultralight Helicopter

Massachusetts Institute of Technology *(See: MIT)*

Massia, Count (France)
Massia 1882 Glider (See also: Biot)

Massillon Aero Corp (Massillon, OH)
Massillon Starwing G4

Massy (France)
Massy 1923 Biplane Glider

Mathewson, E. Linn (Denver, CO)
Mathewson Flyer (1911)

Mathieu-Russell (USA)
Mathieu-Russell Model A-1

Matra (Société des Engines Matra) (France)
Matra-Cantinieau M.C.101
Matra-Cantinieau Bamby

Matsui Akira *(See: Tokorozawa Koshiki-2)*

Mattley Airplane & Motor Co, Inc (San Francisco, CA)
Mattley FP-1 (Fliverplane No.1)
Mattley FP-2 (Fliverplane No.2)

Mattullath, Hugo (Detroit, MI)
Mattullath 1899 Twin-Hull Multiplane

Mauboussin (Avions Mauboussin) (France)
Mauboussin:
M.10
M.11
M.112 Corsaire (Corsair)
M.12 Zodiac
M.120 Zodiac
M.121-35 Corsaire (Corsair) Major
M.122 Corsaire (Corsair)
M.122 Corsaire (Corsair) Major
M.123 Corsaire (Corsair)
M.124 Corsaire (Corsair) Major
M.125 Corsaire (Corsair)
M.127 Corsaire (Corsair)
M.129 Corsaire (Corsair)
M.129-48 Corsaire Trainer

Maule Air (Maule Aircraft Corp; Maule Air, Inc) (Moultrie, GA)
In 1963, Belford D. Maule formed the **B. D. Maule Co**, later renamed **Maule Aircraft Corp**, to build and market his M-4 Rocket. Maule Aircraft declared bankruptcy in late 1984 and Maule formed **Maule Air, Inc** to continue production. For all B. D. Maule designs predating the creation of Maule Aircraft, see **Maule, Belford D**. For all subsequent work, see **Maule Air**.
Maule Air M-1
Maule Air M-2
Maule Air M-3
Maule Air Bee Dee M-4
Maule Air M-4 Astro-Rocket
Maule Air M-4 Jetasen
Maule Air M-4 Rocket
Maule Air M-4 Strata-Rocket
Maule Air M-5 Lunar Rocket
Maule Air M-6 Super Rocket
Maule Air M-7 Super Rocket
Maule Air MX-7 Star Rocket
Maule Air M-8-235

Maule, Belford D. (Jackson, MI)
Belford Maule began designing aircraft in his twenties. In 1963 he formed **B. D. Maule Co**, later renamed **Maule Aircraft Corp**, to build and market his M-4 Rocket. For all Maule designs predating incorporation, see **Maule, Belford D**. For later designs, see **Maule Air**.
Maule (B. D.) 1932 Monoplane
Maule (B. D.) 1944 Ornithopter-Assisted Glider

Maupin (Jim Maupin Ltd) (Tehachapi, CA)
Maupin Carbon Dragon
Maupin Windrose
Maupin Woodstock

Maupin-Lanteri (Lan B. Maupin & Bernard P. Lanteri) (Pittsburg, California)
Maupin-Lanteri Black Diamond
Maupin-Lanteri Black Diamond, NASM

Maurice Farman *(See: Farman)*

Mavag (Magyar Allami Vas, -Acel Es Gepgyar) (Hungary)
Mavag Heja

Maverick Manufacturing Inc (Port Orchard, WA)
Maverick Manufacturing Maverick

Max Holste (Société des Avions Max Holste) (France)
Avions Max Holste was established in postwar France around 1947. In 1956, the company was restructured as **Société Nouvelle des Avions Max Holste**. The **Cessna Aircraft Corp** acquired 49% of the company in February 1960, resulting in complete cross-licensing arrangements between the two companies. By 1962, however, most of the Max Holste designs had gone out of production and the company was renamed **Reims Aviation SA**, concentrating primarily on manufacture of Cessna designs in France and subcontract work for other segments of the French aviation industry. In 1989 Cessna sold its holdings in Reims to the **Compagnie Française Chauffor Investissement (CFCI)**. By 1993, Reims was operating

under the supervision of the **Comité Interministérial de Restructuration Industrielle** (**CIRI**) and was seeking other investors. For all designs originating from Max Holste prior to its acquisition by Cessna, see **Max Holste**. Production by Holste and Reims after the acquisition was primarily subcontract or license work; see the original design company (**Cessna**, etc.).

Max Holste:
M.H.20
M.H.52
M.H.53
M.H.152 Broussard
M.H.1521 Broussard
M.H.250 SuperBroussardPrototype
M.H.260 Super Broussard
M.H.262 Frégate (Frigatebird)

Max Oertz Werkes *(See: Oertz)*

Max Plan *(See: Plan, Max)*

Maxair Aircraft Corp (Glen Rock, PA)
Maxair Drifter (Ultralight)
Maxair Hummer (Ultralight)

Maxim, Sir Hiram Stephen (UK)
Maxim 1892 Gyroscopically Controlled Dynamic Biplane
Maxim 1894 Flying Machine
Maxim 1897 Helicopter (See: Pilcher)
Maxim 1909/10 Biplane

Maximum Safety Airplane Co (Wilmington, CA)
Maximum Safety M-1
Maximum Safety M-2 Trainer
Maximum Safety M-3
Maximum Safety M-4L

Mayabb, S. E. (Grand Island, NE)
Mayabb 1929 Monoplane

Mayhew, Nathaniel L. (Beaumont, TX)
Mayhew Aerodrome (1910 Patent)

Maykemper (Germany)
Maykemper Aeromobil

Mayo Co *(See: Simplex Aircraft Co)*

MBB (Messerschmitt-Bölkow-Blohm) *(See: Bölkow, Boeing-Vertol, and MBB/Kawasaki)*

MBB/Kawasaki (Germany)
MBB/Kawasaki BK-117

MCA *(Maestranza Central de Aviación)*

McAvoy, James M. (Georgia Tech, Atlanta, GA)
McAvoy MPA-1 (Man-Powered Aircraft 1)

McCabe, Dewey (Lexington, NE)
McCabe (Dewey) 1915 Biplane

McCabe, Ira Emmet (Lexington, NE)
McCabe (Ira) Baby 1919 Biplane
McCabe (Ira) Lexington 1914 Biplane
McCabe (Ira) Glider No.1 (1911)
McCabe (Ira) Glider No.2 (1911)

McCallum, DeWitt Clinton (Los Angeles, CA)
McCallum Aerodrome (1911 Patent)

McCandless Aviation Ltd (Rex McCandless) (UK)
McCandless M-2 Gyroplane
McCandless M-4 Gyroplane

McCarley, Charles E. (Hueytown, AL)
McCarley Mini-Mac (Homebuilt)

McCarthy
McCarthy Air Scout

McClary, Earl E. (Huntington Park, CA)
McClary Monoplane (1931)
McClary Power Glider "A" (1929)

McConnell, Leonard Charles (San Jose, CA)
McConnell 1929 Monoplane

McCook Aircraft Corp (McCook, NE)
McCook Commercial (1928)

McCormick-Romme (Harold F. McCormick & William S. Romme) (Cicero, IL)
McCormick-Romme Umbrellaplane

McCrae, Alexander J. (Hempstead, NY)
McCrae Dart

McCulloch (Airplane Division, McCulloch Motors Corp) (Los Angeles, CA)
In 1948, **McCulloch Motors Corp** acquired the assets and staff of **J.O.V. Helicopter Co** of Philadelphia, PA. McCulloch moved the company to California and absorbed it as the **Airplane Division of McCulloch Motors**. When McCulloch received no orders for aircraft and the division was shut down, the staff formed **Jovair Corp** to continue the development of the original J.O.V. products. For products of J.O.V. and Jovair, see **Jovair**. For aircraft developed at McCulloch, see **McCulloch**.
McCulloch MC-4
McCulloch MC-4C (YH-30)
McCulloch MC-4E

McCurdy, J. A. D. *(See: Aerial Experiment Association; also Canadian Aerodrome)*

McCurdy-Willard *(See: Queen-McCurdy)*

McDaniel, Arthur (Toledo, OH)
McDaniel (Arthur) Model 2 (1928)

McDaniel (Taylor) Gliders, Inc (Washington, DC)
McDaniel (Taylor) Pneumatic ("Rubber") Glider No.1
McDaniel (Taylor) Pneumatic ("Rubber") Glider No.2

McDonnell (St Louis, MO)
In 1928, after several years working with various aircraft companies, James S. McDonnell, Constantine Zakhartchenko, and James Cowling formed **J. S. McDonnell Jr & Associates**. A lack of funding and the Depression doomed the company, leading McDonnell to rejoin existing companies. In 1938, however, he again set out on his own, establishing **McDonnell Aircraft Corp** in 1939, with its office in St Louis, MO. In 1942, the company acquired an interest in **Platt-LePage Aircraft Co** of Eddystone, PA, acquiring Platt-LePage outright in 1944, and liquidating it in 1946. In 1966, the company was renamed the **McDonnell Co**, a result of diversification production and design efforts. At the same time, McDonnell acquired control of the **Douglas Aircraft Co** and, in 1967, the two companies merged to form the

McDonnell Douglas Corp (MDC), with **McDonnell Aircraft Co (MCAir)** and **Douglas Aircraft Co (DAC)** as operating divisions. Designs of McDonnell companies prior to the 1967 merger are listed under **McDonnell.** Subsequent designs are listed under **McDonnell Douglas.**

McDonnell Doodlebug (Model 1)
McDonnell (FH) XFD-1 Phantom
McDonnell FH-1 (FD-1) Phantom
McDonnell FH-1 (FD-1) Phantom, NASM
McDonnell XF2H-1 (XF2D-1) Banshee
McDonnell F2H-1 Banshee
McDonnell F2H-2 Banshee
McDonnell F2H-2B Banshee
McDonnell F2H-2N Banshee
McDonnell F2H-2P Banshee
McDonnell F2H-3 (F-2C) Banshee
McDonnell F2H-3 (F-2C) Banshee, Canada
McDonnell F2H-4 (F-2D) Banshee
McDonnell XF3H-1 Demon
McDonnell F3H-1 Demon
McDonnell F3H-1N Demon
McDonnell F3H-2 (F-3B) Demon
McDonnell F3H-2M (MF-3B) Demon
McDonnell F3H-2N (F-3C) Demon
McDonnell F3H-2P Demon
McDonnell (F-4) AH-1
McDonnell (F-4) F4H Mockup
McDonnell (F-4) YF4H-1 Phantom II
McDonnell (F-4) YF4H-1 Phantom II Project Top Flight
McDonnell F-4A (F4H-1F) Phantom II
McDonnell F-4A (F4H-1F) "Sageburner," NASM
McDonnell F-4A (F4H-1F) "Skyburner"
McDonnell F-4B (F4H-1) Phantom II
McDonnell F-4B (F4H-1) Phantom II, Project High Jump
McDonnell F-4B (F4H-1) Phantom II, USAF Evaluation
McDonnell NF-4B Phantom II
McDonnell QF-4B Phantom II
McDonnell RF-4B Phantom II
McDonnell F-4C (F-110A) Phantom II
McDonnell YRF-4C Phantom II
McDonnell RF-4C Phantom II
McDonnell F-4D Phantom II
McDonnell YF-4E Phantom II
McDonnell YF-4E Phantom II, PACT/CCV
McDonnell F-4E Phantom II
McDonnell F-4E Phantom II, Project Agile Eagle
McDonnell F-4E Phantom II, Thunderbirds
McDonnell F-4E(F) Phantom II Single-Seat Proposal
McDonnell F-4EJ Phantom II (Japan)
McDonnell RF-4EJ Phantom II (Japan)
McDonnell F-4F Phantom II, Germany
McDonnell F-4G Phantom II, Navy
McDonnell F-4G Phantom II Wild Weasel
McDonnell YF-4J Phantom II
McDonnell F-4J Phantom II
McDonnell F-4J Phantom II, Blue Angels
McDonnell YF-4K Phantom FG.Mk.1 (Royal Navy)
McDonnell F-4K Phantom FG.1 (Royal Navy)
McDonnell YF-4M Phantom FRG.Mk.2 (RAF)
McDonnell F-4M Phantom FRG.Mk.2 (RAF)
McDonnell F-4N Phantom II Prototype
McDonnell F-4N Phantom II
McDonnell F-4S Phantom II
McDonnell F-4S Phantom II, NASM
McDonnell F-4T Phantom II
McDonnell F-4X Phantom II
McDonnell F-4(FVS) Phantom II (Variable Geometry)
McDonnell XF-85 (XP-85) Goblin
McDonnell XF-88 Voodoo
McDonnell XF-88A Voodoo
McDonnell XF-88B Voodoo
McDonnell F-101A Voodoo
McDonnell JF-101A Voodoo (Operation Fire Wall)
McDonnell NF-101A Voodoo (J79 Testbed)
McDonnell YRF-101A Voodoo
McDonnell RF-101A Voodoo
McDonnell RF-101A Voodoo, Operation Sun Run
McDonnell F-101B Prototype Voodoo (NF-101B)
McDonnell F-101B Voodoo
McDonnell CF-101B Voodoo (RCAF)
McDonnell EF-101B Voodoo (RCAF)
McDonnell F-101C Voodoo
McDonnell RF-101C Voodoo
McDonnell RF-101C Voodoo, NASM
McDonnell F-101D Voodoo
McDonnell F-101F Voodoo
McDonnell CF-101F Voodoo (RCAF)
McDonnell RF-101G Voodoo
McDonnell RF-101H Voodoo
McDonnell H-1 (Helicopter Proposal)
McDonnell XH-20 Little Henry
McDonnell XH-35 (See: McDonnell XV-1)
McDonnell XHCH-1 (Model 86 Cargo Helicopter Proposal)
McDonnell XHJD-1 Whirlaway
McDonnell XHJD-1 Whirlaway, NASM
McDonnell XHRH-1 (Model 78 Assault Helicopter Proposal)
McDonnell XP-67 Bat
McDonnell XV-1 Convertiplane (XL-25)
McDonnell XV-1 Convertiplane (XL-25), NASM
McDonnell Model 18 (Tandem Engine Fighter Proposal)
McDonnell Model 60 (Interceptor Proposal)
McDonnell Model 82 (USN Class HO Convertiplane) (See also: McDonnell XV-1)
McDonnell Model 110 (Interceptor Proposal)
McDonnell Model 119A (UCX Entry)
McDonnell Model 120
McDonnell Model 153A Missileer
McDonnell Model 188E (Breguet 941 License Proposal)
McDonnell Model 188H (Breguet 941 License Proposal)
McDonnell Model 220 (119B)

McDonnell Douglas Corp (St Louis, MO)

In 1966, the **McDonnell Aircraft Corp** was renamed the **McDonnell Co.** At the same time, McDonnell acquired control of the **Douglas Aircraft Co.** In 1967, the two companies merged to form the **McDonnell Douglas Corp (MDC)**, with **McDonnell Aircraft Co (MCAir)** and **Douglas Aircraft Co (DAC)** operating divisions. In 1984 McDonnell Douglas purchased **Hughes Helicopters Inc**, which was renamed **McDonnell Douglas Helicopter Co (MDHC)** in 1985. In 1992 MDC reorganized, with DAC becoming an independent civil transport manufacturer, and the remaining subsidiaries (including MCAir and MDHC) forming **McDonnell Douglas Aerospace (MDA)** to concentrate on military and government contracts. MDHC was briefly offered for sale, but was reintegrated into the MDA structure by 1995. In 1997, the **Boeing Co** purchased MDC, with all operations placed under the Boeing name.

McDonnell company designs predating the 1967 merger are listed under **McDonnell**; Douglas designs predating the merger are under **Douglas**. Designs postdating the merger are under **McDonnell Douglas**. For Hughes designs predating the 1984 purchase (up to and including the Hughes 530) see **Hughes**; subsequent designs are listed under **McDonnell Douglas Helicopters**. Products developed after the 1997 purchase by Boeing are listed under **Boeing**.

McDonnell Douglas:
- AST (Advanced SST Proposal)
- KC-10A Extender Prototype
- KC-10A Extender
- YC-15 (Advanced Medium STOL Transport)
- C-17A Globemaster III Prototype
- C-17A Globemaster III
- DC-9 and prior (See: Douglas)
- DC-10 Twin Proposal
- DC-10 Series 10
- DC-10 Series 10CF
- DC-10 Series 15
- DC-10 Series 20 (DC-10 Series 40 Prototype)
- DC-10 Series 30
- DC-10 Series 30CF
- DC-10 Series 30F
- DC-10 Series 40
- DC-10 Series 40D
- DC-10 Orbis
- KDC-10 (Dutch Tanker)
- F-15 RPRV
- YF-15A Eagle
- F-15A Eagle
- F-15A Streak Eagle
- F-15B (TF-15A) Eagle
- NF-15B Agile Eagle
- NF-15B Agile Eagle (S/MTD Demonstrator)
- YF-15B Eagle (F-15E Strike Eagle Prototype)
- F-15C Eagle
- F-15D Eagle
- F-15DJ Eagle (Japan)
- F-15E Strike Eagle
- F-15I Ra'am (Israel)
- F-15J Eagle (Japan)
- F-15S (F-15XP) Strike Eagle (Saudi Arabia)
- F/A-18A Hornet Prototypes
- F/A-18A Hornet
- F/A-18A Hornet, Blue Angels
- F/A-18A Hornet, High Angle of Attack Research
- AF-18A Hornet (RAAF)
- CF-18A (CF-188A) Hornet (RCAF)
- EF-18A (C.15) Hornet (Spain)
- F/A-18B (TF-18A) Hornet Prototype
- F/A-18B (TF-18A) Hornet
- F/A-18B, Blue Angels
- F/A-18C Hornet
- F/A-18D Hornet
- F/A-18D Night Attack Hornet
- F/A-18E Super Hornet
- F/A-18F Super Hornet
- F/A-18F Super Hornet C[2]W
- MD-10 (Advanced DC-10 Freighter)
- (MD-11) MD-100 (Proposal)
- MD-11 Prototype
- MD-11
- MD-11CF
- MD-11ER
- MD-11F
- MD-12
- MD-17 (Civil C-17 Proposal)
- MD-80 - MD-87 (See: Douglas DC-9)
- MD-88 (DC-9-88)
- MD-90-30
- MD-90-30ER
- MD-90-40
- MD-91X (MD-80 UHB Refit Proposal)
- MD-95 (Boeing 717-200)
- MD-100 (See: MD-11)
- MD-2001 (NASP Proposal)
- MD-XX (Wide-Body Proposal)
- T-45A Goshawk (VTXTS Program)
- T-45C Goshawk
- AV-8A Harrier (See: Hawker Siddeley)
- TAV-8A Harrier (See: Hawker Siddeley)
- YAV-8B Harrier II
- AV-8B Harrier II
- EAV-8B (VA.2) Matador II (Spain)
- TAV-8B Harrier II
- X-36
- Model 265 Vectored Lift Fighter
- 1968 "Space Glider" Proposal

McDonnell Douglas-Fokker MDF-100 (See: Fokker)

McDonnell Douglas-General Dynamics A-12

McDonnell Douglas Helicopter Corp (Tucson, AZ)

In 1984, **Hughes Helicopter Inc** was purchased by the **McDonnell Douglas Corp** and, in 1985, renamed **McDonnell Douglas Helicopter Co (MDHC)**. As part of a 1992 reorganization, MDHC became part of **McDonnell Douglas Aerospace (MDA)**, but shortly afterwards was offered for sale. In 1995, MDHC was reintegrated into the MDA structure. In 1997 the **Boeing Co** purchased McDonnell Douglas, integrating all operations under the Boeing name. For all products of Hughes Helicopter, up to and including the Hughes 530, see **Hughes**; for all subsequent designs by MDHC, see **McDonnell Douglas Helicopter**. Later designs are listed under **Boeing**.

McDonnell Douglas Helicopter
- MD 500-Series (See: Hughes)
- MD 600N (MD 630N)
- MD 900 Explorer (MDX)
- MD 900 Combat Explorer
- McDonnell Douglas Helicopter-Bell LHX

McDowell, Walter E. (Maywood, CA)

McDowell 1930 Monoplane

McFarland Aircraft Co (Spring Valley, CA)

McFarland MPT-3 (Glider)

McFarland PT-4 (Glider)

McGaffey Airplane Development Co (Inglewood, CA)

McGaffey Aviate

McGee, Don (Saginaw, MI)

McGee 1915 Biplane

McKinnie Aircraft Co (Fargo, ND)

McKinnie 165

McKinnon Enterprises Inc (Sandy, OR)

McKinnon G-21C Goose (Modified Grumman G-21A)

McKinnon G-21D Turboprop Goose (Modified Grumman G-21A)

McKinnon G-21G TurboGoose (Modified Grumman G-21A)

McKinnon Super Widgeon (Modified Grumman G-44)

McMullen Aircraft Co (Tampa, FL)

McMullen Airliner

McMullen Experimental MC-1 Biplane

McPherson, John B., IV (Abington, PA)

McPherson Model 1 Monoplane

McWorter, John E. (St Louis, MO)

McWorter Autoplane (1919)

MDB Flugtechnik AG (Switzerland)

MDB MD-3-160 (Developed Dätwyler MD-3)

MDC Max Dätwyler AG (See: Dätwyler, Max)

MDG (Instruments de Précision MDG) (France)
MDG LD-45
MDG LD-46
MDG LD-261 Midgy-Club

MDM (Poland)
MDM-1 Fox

Mead Gliders (Chicago, IL)
Mead Challenger (Contest Glider)
Mead Rhon Ranger RR-1 (Primary Training Glider)

Medwecki, Józef (Poland)
Medwecki HL 2
Medwecki M 9
Medwecki O.2

Medwecki (Józef) & Nowakowski (Zygmunt) *(See: Samolot)*

M.E.E.T. *(See: M.I.I.T.)*

Mefford, George *(See: Pollitt & Mefford)*

Meger, Mike (Marinette, WI)
Meger Heli-Star

Megone, W. B. (UK)
Megone 1912 Biplane

Melberg, Raymond (Denver, CO)
Melberg (Raymond) MG-3

Melberg, Richard (Spokane, WA)
Melberg (Richard) 1930 Glider

Melberg-Greenameier-Ward (Raymond Melberg, Conrad Greenameier, & Rowan C. Ward) (Denver, CO)
Melberg-Greenameier-Ward MGW-1

Melone, Pasco M.
Melone Parasol Monoplane

Melsheimer, Frank (USA)
Melsheimer FM-1

Melton, C. C. (Kansas City, MO)
Melton Baby Biplane Houperzine (Homebuilt)

Menefee Airways Inc (New Orleans, LA)
Menefee Crescent
Menefee Special (See: Wedell-Williams Model 44/92)

Mengle, Leon A. (Escondido, CA)
Mengle M-2

Menzimer, Verne (Oceanside, CA)
Menzimer Cavalier SA 102.5

Mepler (France)
Mepler 1922 Biplane Glider

Merckle Flugzeugbau GmbH (Germany)
Merckle Flugzeugbau GmbH was established in the 1950s to conduct rotorcraft research under contract to the Federal German government. In 1968 the company was acquired by **Dornier GmbH.**
Merckle SM-67

Mercury Air (Bradford, PA)
Mercury Air Shoestring

Mercury Aerial Service *(See: Aerial Service Corp)*

Mercury Aircraft Corp *(See: Aerial Service Corp)*

Mercury Aircraft Inc *(See: Aerial Service Corp)*

Mercury Flying Corp (New York, NY)
Mercury Racer (Al Williams' 1929 Schneider Entry)

Meridionali (Industrie Meccaniche e Aeronautiche Meridionali; IMAM) (Italy)
Officine Ferroviarie Meridionali began producing aircraft in 1923. In 1934, **Societa Anonima Industrie Aeronautiche Romeo** was created to assume Meridionali's aeronautical activities. In 1936, the company name changed to **Industrie Meccaniche e Aeronautiche Meridionali**, a part of the Breda group. **Aerfer** (later **Aerfer-Industrie Aerospaziali Meridionali SpA**) was established by the 1955 merger of **Officine di Pomigliano per Costruzioni Aeronautiche e Ferroviarie** and Meridionali. For all pre-1955 designs, see **Meridionali**; for post-1955 designs, see **Aerfer.**
Meridionali Ro 1
Meridionali Ro 1*bis*
Meridionali Ro 5
Meridionali Ro 10
Meridionali Ro 25
Meridionali Ro 26
Meridionali Ro 30
Meridionali Ro 37
Meridionali Ro 37*bis*
Meridionali Ro 41
Meridionali Ro 41B
Meridionali Ro 43
Meridionali Ro 44
Meridionali Ro 45
Meridionali Ro 51
Meridionali Ro 57
Meridionali Ro 58
Meridionali Ro 63

Merkel Airplane Co (Wichita, KS)
Merkel Mark II

Merrian, H. Warren (UK)
Merrian Glider

Merrill Aircraft Co (Albert A. Merrill) (Pasadena, CA)
Merrill 1914 Tractor Biplane
Merrill 1926 Variable Incidence Biplane (C.I.T.9)
Merrill 1930 Variable Incidence Glider
Merrill 1931 Variable Incidence Biplane
Merrill 1932 Variable Incidence Monoplane

Mersey Aeroplane Co (UK)
Mersey 1912 Monoplane

Merx (Germany)
Merx 1912 Multiplane

Messer, Glenn E. (Birmingham, AL)
Messer 1926 Biplane

Messerschmitt (Germany)
From the end of World War I to 1922, Willy Messerschmitt and Frederich Harth collaborated on a series of gliders. In 1923, Messerschmitt established his own company, **Flugzeugbau Messerschmitt**, which he incorporated in 1926 as **Messerschmitt Flugzeugbau GmbH.** Under pressure from the Bavarian government in 1927, Messerschmitt and **Bayerische Flugzeugwerke AG** (**BFW**, established 1926) pooled resources under an agreement by which BFW provided manufacturing

facilities and Messerschmitt acted as the design and development office, retaining all design and patent rights. In 1931 BFW went into receivership but as an independent company Messerschmitt was unaffected; by 1933 Messerschmitt had amassed enough capital to bring BFW out of receivership under his direction, effectively merging the two companies. In 1939 BFW and Messerschmitt reorganized as Messerschmitt AG. After World War II the company switched to sub-contract work and production of license-built designs. In June 1968, the company merged with **Bölkow GmbH** to form **Messerschmitt-Bölkow GmbH (MB)**, which merged with **Hamburger Flugzeugbau GmbH** in 1969 to form **Messerschmitt-Bölkow-Blohm GmbH (MBB)**. For BFW designs originating outside of Messerschmitt, see **BFW** (1926). For BFW designs originating with Messerschmitt or designs postdating 1933, see **Messerschmitt**. For products of MB and MBB, see **Bölkow**.

Messerschmitt (Harth) S 5
Messerschmitt (Harth) S 6
Messerschmitt (Harth) S 7
Messerschmitt (Harth) S 8
Messerschmitt S 9
Messerschmitt (Harth) S 10
Messerschmitt (Harth) S 12
Messerschmitt (Harth) S 13
Messerschmitt S 14
Messerschmitt S 15
Messerschmitt S 16a Bubi (Baby)
Messerschmitt S 16b Betty
Messerschmitt M 17 Ello
Messerschmitt (BFW) M 18 (Prototype)
Messerschmitt (BFW) M 18a
Messerschmitt (BFW) M 18b
Messerschmitt (BFW) M 18c
Messerschmitt (BFW) M 18d
Messerschmitt (BFW) M 18w
Messerschmitt (BFW) M 19
Messerschmitt (BFW) M 20 (Prototype)
Messerschmitt (BFW) M 20a
Messerschmitt (BFW) M 20b
Messerschmitt (BFW) M 21a
Messerschmitt (BFW) M 21b
Messerschmitt (BFW) M 22
Messerschmitt (BFW) M 23a
Messerschmitt (BFW) M 23b
Messerschmitt (BFW) M 23c
Messerschmitt (BFW) M 24a
Messerschmitt (BFW) M 24b
Messerschmitt (BFW) M 25
Messerschmitt (BFW) M 26
Messerschmitt (BFW) M 27a
Messerschmitt (BFW) M 27b
Messerschmitt (BFW) M 28
Messerschmitt (BFW) M 29a
Messerschmitt (BFW) M 31
Messerschmitt (BFW) M 32
Messerschmitt (BFW) M 33
Messerschmitt (BFW) M 35a
Messerschmitt (BFW) M 35b
Messerschmitt (BFW) M 36
Messerschmitt Bf 108 A (M 37) Taifun (Typhoon)
Messerschmitt Bf 108 B Taifun (Typhoon)
Messerschmitt Bf 108 C Taifun (Typhoon)
Messerschmitt Bf 108 D Taifun (Typhoon)
Messerschmitt (Bf 108) XC-44 Taifun (Typhoon)
Messerschmitt Bf 109 Prototypes
Messerschmitt Bf 109 B
Messerschmitt Bf 109 C
Messerschmitt Bf 109 D (Dora)
Messerschmitt Bf 109 E (Emil)
Messerschmitt Bf 109 F (Friedrich)
Messerschmitt Bf 109 G (Gustav)
Messerschmitt Bf 109 G, NASM
Messerschmitt Bf 109 H
Messerschmitt Bf 109 J
Messerschmitt Bf 109 K
Messerschmitt Bf 109 T (Träger; Carrier Version)
Messerschmitt Bf 109 Z (Zwilling; Twin Bf 109)
Messerschmitt Bf 110 Prototypes, Bf 110 V1
Messerschmitt Bf 110 A
Messerschmitt Bf 110 B
Messerschmitt Bf 110 C
Messerschmitt Bf 110 D
Messerschmitt Bf 110 E
Messerschmitt Bf 110 F
Messerschmitt Bf 110 G
Messerschmitt Bf 110 H
Messerschmitt Bf 161
Messerschmitt Bf 162 Jaguar
Messerschmitt Bf 163 (1935 STOL Design, P.1051)
Messerschmitt Me 163 A Komet (Comet)
Messerschmitt Me 163 B Komet Prototype
Messerschmitt Me 163 B Komet
Messerschmitt Me 163 B Komet, NASM
Messerschmitt Me 163 C Komet
Messerschmitt Me 163 D Komet (See: Me 263)
Messerschmitt Me 200 (See: Hispano HA-200 Saeta)
Messerschmitt Bf 208
Messerschmitt Me 209 (Bf 109 R; Speed Record Aircraft)
Messerschmitt Me 209 A (High-Altitude Fighter)
Messerschmitt Me 210 Prototypes
Messerschmitt Me 210 A
Messerschmitt Me 210 B
Messerschmitt Me 210 C
Messerschmitt Me 210 D
Messerschmitt Me 261
Messerschmitt Me 262 Prototypes
Messerschmitt Me 262 A-1a Schwalbe (Swallow)
Messerschmitt Me 262 A-1a, NASM
Messerschmitt Me 262 A-1a/U-1 Schwalbe (Swallow)
Messerschmitt Me 262 A-1a/U-3 Schwalbe (Swallow)
Messerschmitt Me 262 A-1a/U-4 Schwalbe (Swallow)
Messerschmitt Me 262 A-2a Sturmvogel (Storm-Bird)
Messerschmitt Me 262 A-2a/U-2 Sturmvogel
Messerschmitt Me 262 B-1a Schwalbe (Swallow)
Messerschmitt Me 262 B-1a/U-1 Schwalbe (Swallow)
Messerschmitt Me 262 B-2 (HeS 011A Night Fighter Proposal)
Messerschmitt Me 263 (Me 163 D, Ju 248)
Messerschmitt Me 264
Messerschmitt Me 309
Messerschmitt Me 310 Prototype (Me 210 V13)
Messerschmitt Me 321 A Gigant (Giant)
Messerschmitt Me 321 B Gigant
Messerschmitt Me 323 D Gigant
Messerschmitt Me 323 E Gigant (Giant)
Messerschmitt Me 328 A
Messerschmitt Me 328 B
Messerschmitt Me 329
Messerschmitt Me 410 Hornisse (Hornet) Prototypes
Messerschmitt Me 410 A Hornisse
Messerschmitt Me 410 A Hornisse, NASM
Messerschmitt Me 410 B Hornisse
Messerschmitt Me 609 (Twin Me 309)
Messerschmitt P.1078
Messerschmitt P.1101
Messerschmitt P.1104
Messerschmitt P.1106
Messerschmitt P.1107
Messerschmitt P.1110
Messerschmitt P.1111
Messerschmitt P.1112

Messerschmitt P.1114

Messerschmitt-Bölkow-Blohm *(See: Bölkow; also: Boeing-Vertol; also MBB/Kawasaki)*

Metal Aircraft Corp (Cincinnati, OH)
Metal Flamingo G-MT-6
Metal Flamingo G-1 (Prototype)
Metal Flamingo G-2
Metal Flamingo G-2H
Metal Flamingo G-2W

Meteor SpA Costruzioni Aeronautiche ed Elettroniche (Italy) *(See also: Iniziative Industriali Italiane)*
Meteor Gufo
Meteor Mirach 20 Pelican

Meteore *(See: SPCA)*

Methvin *(See: Aircraft Methvin Inc)*

Mexico *(See: TNCA)*

Meyer Aircraft (George Meyer Aircraft) (Corpus Christi, TX)
Meyer (George) Little Toot

Meyer, Gus, Jr (Goleta, CA)
Meyer (Gus) C-9 Monoplane

Meyer, Ray (Quincy, IL)
Meyer (Ray) Wren Sport

Meyers, Charles W. (Columbus Junction, IA)
Meyers (C. W.) Commercial (1920)
Meyers (C. W.) Midget (1926)
Meyers (C. W.) Sport (1924)

Meyers Aircraft Co (Tecumseh, MI)
Allen H. Meyers established **Meyers Aircraft Co** in 1936. The company produced a well-received series of training and light aircraft but failed to achieve prominence. In July 1965 the company was acquired by **Aero Commander, Inc** (a subsidiary of **Rockwell-Standard**) as the **Tecumseh Division of Aero Commander**. In 1967 Aero Commander sold the drawings, tooling, and jigs for the Meyers designs. **Interceptor Corp** acquired much of this material and marketed a pressurized turbine-powered version of the Meyers 200 as the Interceptor 400. At the same time the **Meyers Aircraft Manufacturing Co** of Denver, CO attempted to reintroduce the reciprocating-engined Meyers 200D. By 1992 a reformed Meyers Aircraft in Deerfield Beach, FL had obtained the rights and was again attempting to market the aircraft. For all designs originating with Meyers Aircraft, see **Meyers**. For the Interceptor redesign of the Meyers 200, see **Interceptor**. For all Aero-Commander designs, see **Aero-Commander**.
Meyers (Aircraft Co) OTW
Meyers (Aircraft Co) 125
Meyers (Aircraft Co) 145
Meyers (Aircraft Co) 165
Meyers (Aircraft Co) 200A
Meyers (Aircraft Co) 200B
Meyers (Aircraft Co) 200C
Meyers (Aircraft Co) 200D
Meyers (Aircraft Co) (Aero-Commander) AC-200
Meyers (Aircraft Co) 400 (See: Interceptor 400)

Meysenburg, Richard (San Diego, CA)
Meysenburg 1929 Glider

M.G.H. Airplane Co (William Monahan, Henry W. Gastman, & Behrend H. Hullen) (Auburn, CA)
M.G.H. L.M.1

M.G.I. (Motorized Gliders of Iowa, Inc) (Clear Lake, IA)
M.G.I. Teratorn TA

Miami Aircraft Co (Miami, OK)
Miami (Aircraft Co) A-1 (1-A) Special Biplane

Miami Aircraft Corp (Hialeah, FL)
Miami (Aircraft Corp) Maid (Model B)

Micco Aircraft Co (Ft Pierce, FL)
Micco SP20

Michel, André (France)
Michel "Flying Auto"

Michigan Aircraft Co (Grand Rapids, MI)
Michigan 1918 Flying Boat

Michigan Aircraft Co (1933/34) (Detroit, MI) *(See: Lambert Model 1344)*

Microturbo SA (France)
Microturbo Microjet 200

Mid-Continent Aircraft Co (Tulsa, OK) *(See also: Spartan Aircraft Co)*
Mid-Continent Maiden Tulsa

Mid-West Airplane Co (Omaha, NE) *(See: Durand)*

Midland Aircraft Co (Los Angeles, CA)
Midland 2-Seat Light Plane (1940)

Midwest Aircraft Co (Wyandotte, MI)
Midwest Light Plane
Midwest Mercury (1945)

Midwest MU-1/MI-1 *(See: Steinhauser)*

Midwest Microlites *(See: Waspair)*

MiG (Russia)
In 1939 P. A. Voronin and P. V. Dementyev convinced Stalin to establish a new aircraft design team at OKB Polikarpov under the management of Anushavan "Artyom" Ivanovich Mikoyan. Mikoyan agreed on the condition that Mikhail Iosifovich Guryevich be his assistant. In December the new team became autonomous and was named **OKB-155** (Opytno konstruktorskoe byuro, Experimental Design Bureau). In keeping with Soviet practice, OKB-155 was generally known after its leaders as **OKB Mikoyan i Guryevich (MiG)**, with Guryevich's name included at Mikoyan's insistence. Following the collapse of the Soviet Union in 1992, the bureau became **Moskovskii Mashinostroitelnyy Zavod Imieni A. I. Mikoyana** (Moscow Machine Plant named for A. I. Mikoyan). In 1993 it was renamed **ANPK MiG** (Aviatsionnyi Nauchno-Proizvodstvenniy Kompleks - ANPK MiG Imieni A. I. Mikoyana; Aviation Scientific-Production Complex "MiG" named for A. I. Mikoyan). In 1995, ANPK MiG merged with the Moscow Aircraft Production Organization **(MAPO)** to form **MAPO MiG**. In January 1996, a presidential decree joined MAPO MiG with the **Kamov Helicopter Co** and other aviation-related companies to form **VPK MAPO** (Voyenno Promyshlennyi Komplex MAPO; Military

Industrial Group, Moscow Aircraft Production Organization (MiG MAPO)). For all designs originating with OKB MiG, ANPK MiG see **MiG**. For all MAPO MiG and MiG MAPO designs having "MiG" designations, see **MiG**. For all Kamov designs see **Kamov**.

MiG I-7U
MiG I-75
MiG I-210
MiG I-220 (A)
MiG I-221 (2A)
MiG I-222 (3A)
MiG I-224 (4A)
MiG I-225 (5A)
MiG I-250 (N)
MiG I-270 (Zh)
MiG I-320 (R-1)
MiG I-350 (M)
MiG I-360 (SM-2)
MiG I-370
MiG I-380
MiG I-410
MiG I-420
MiG MiG-1 (I-200)
MiG MiG-3
MiG MiG-3U (I-230, D)
MiG MiG-5
MiG MiG-8 Utka (Duck)
MiG MiG-9 *Fargo* (I-301, FS)
MiG MiG-9 UTI (I-301T, FT-1)
MiG MiG-9 UTI (FT-2)
MiG MiG-9L (FK)
MiG MiG-9M (I-308, FR)
MiG MiG-15 *Fagot-A* (I-310, S, SU, SV)
MiG MiG-15 UTI *Midget* (ST)
MiG MiG-15*bis* *Fagot-B* (SD)
MiG MiG-15*bis* *Fagot-B* (SD), NASM
MiG MiG-15*bis* (ISh)
MiG MiG-15*bis* (SYe, LL)
MiG MiG-15*bis* Burlaki
MiG MiG-15*bis*P (SP-1)
MiG MiG-15*bis*R (SR)
MiG MiG-15*bis*S (SD-UPB)
MiG MiG-17 *Fresco-A* (I-330, SI)
MiG MiG-17F *Fresco-C*
MiG MiG-17P *Fresco-B*
MiG MiG-17PF *Fresco-D*
MiG MiG-17PFU *Fresco-E*
MiG MiG-17R (SR-2)
MiG MiG-19 *Farmer-A*
MiG MiG-19P *Farmer-B*
MiG MiG-19PM *Farmer-E*
MiG MiG-19R
MiG MiG-19S *Farmer-D*
MiG MiG-19SU
MiG MiG-19SV *Farmer-C*
MiG MiG-21 *Fishbed* (Ye-5)
MiG MiG-21 *Fishbed-B* (Ye-6)
MiG MiG-21F *Fishbed-C*
MiG MiG-21F-13 *Fishbed-E*
MiG MiG-21F-13 *Fishbed-E*, NASM
MiG MiG-21FL *Fishbed-D*
MiG MiG-21I Analog
MiG MiG-21K
MiG MiG-21M *Fishbed-J*
MiG MiG-21MF *Fishbed-J*
MiG MiG-21MT
MiG MiG-21P
MiG MiG-21PD *Fishbed-G*
MiG MiG-21PF *Fishbed-D*
MiG MiG-21PFM *Fishbed-F*
MiG MiG-21PFS *Fishbed-F*
MiG MiG-21R *Fishbed-H*
MiG MiG-21S
MiG MiG-21SM *Fishbed-J*
MiG MiG-21SMT *Fishbed-K*
MiG MiG-21U *Mongol-A*
MiG MiG-21UM *Mongol-B*
MiG MiG-21US *Mongol-B*
MiG MiG-21Ye
MiG MiG-21*bis* *Fishbed-L*
MiG MiG-21*bis*-SAU *Fishbed-N*
MiG MiG-23 *Flogger* Prototype (23-11)
MiG MiG-23B *Flogger-F*
MiG MiG-23BK *Flogger-H*
MiG MiG-23BM *Flogger-F*
MiG MiG-23BN *Flogger-F*
MiG MiG-23M *Flogger-B*
MiG MiG-23MF *Flogger-E*
MiG MiG-23ML *Flogger-G*
MiG MiG-23MLD *Flogger-K*
MiG MiG-23MS *Flogger-E*
MiG MiG-23P *Flogger-G*
MiG MiG-23PD STOL Prototype *Faithless*
MiG MiG-23S *Flogger-A*
MiG MiG-23UB *Flogger-C*
MiG MiG-25BM *Foxbat-F*
MiG MiG-25MP (See: MiG Ye-155MP)
MiG MiG-25P *Foxbat-A*
MiG MiG-25PD *Foxbat-E*
MiG MiG-25PDS *Foxbat-E*
MiG MiG-25PU *Foxbat-C*
MiG MiG-25R *Foxbat-B*
MiG MiG-25RB *Foxbat-B*
MiG MiG-25RBF *Foxbat-D*
MiG MiG-25RBK *Foxbat-D*
MiG MiG-25RBS *Foxbat-D*
MiG MiG-25RBSh *Foxbat-D*
MiG MiG-25RBT *Foxbat-B*
MiG MiG-25RBV *Foxbat-B*
MiG MiG-25RU *Foxbat-C*
MiG MiG-25 Civil Transport Derivative
MiG MiG-27 *Flogger-D*
MiG MiG-27D *Flogger-J1*
MiG MiG-27K *Flogger-J2*
MiG MiG-27L *Flogger-J1*
MiG MiG-27M *Flogger-J1*
MiG MiG-29 *Fulcrum-A*
MiG MiG-29K *Fulcrum-D*
MiG MiG-29KVP
MiG MiG-29KU
MiG MiG-29M *Fulcrum-E*
MiG MiG-29S *Fulcrum-C*
MiG MiG-29SE *Fulcrum-C*
MiG MiG-29UB *Fulcrum-B*
MiG MiG-31 *Foxhound-A*
MiG MiG-31M *Foxhound-B*
MiG MiG-33
MiG MiG-35
MiG MiG 1-42 (MFI)
MiG MiG 18-50
MiG MiG-110
MiG MiG-AT
MiG MiG-AS
MiG MiG-ATS
MiG SVB
MiG TA4
MiG Ye-1
MiG Ye-2
MiG Ye-2A *Faceplate*
MiG Ye-4
MiG Ye-5 (See: MiG-21)
MiG Ye-6 (See: MiG-21)
MiG Ye-6V
MiG Ye-8
MiG Ye-50
MiG Ye-66A
MiG Ye-133
MiG Ye-150
MiG Ye-152
MiG Ye-152A *Flipper*
MiG Ye-152M
MiG Ye-155M (Ye-266M)
MiG Ye-155MP (MiG-25MP, MiG-31)
MiG Ye-155P
MiG Ye-155R
MiG Ye-166
MiG Ye-266M (See: MiG Ye-155M)

Mignet, Henri (France)

Mignet HM.14 Pou du Ciel (Flying Flea)
Mignet (Crosley) HM.14 Pou du Ciel (Flying Flea), NASM
Mignet HM.280 Pou du Ciel (Flying Flea)
Mignet HM.290 Pou du Ciel (Flying Flea)
Mignet HM.293 Pou du Ciel (Flying Flea)
Mignet HM.350 Pou du Ciel (Flying Flea)
Mignet HM.360 Pou du Ciel (Flying Flea)
Mignet HM.380 Pou du Ciel (Flying Flea)
Mignet HM.390 Pou du Ciel (Flying Flea)

Miguet (France)

Miguet 1923 Lilienthal Type Monoplane Glider

M.I.I.T. (Russia)
M.I.I.T. "Rote Presnia"

Mikhal'kevich-Sharapov (Russia)
Mikhal'kevich-Sharapov MSh

Mikhel'son (Russia)
Mikhel'son NU-4
Mikhel'son U-3

Mikhel'son-Korvin-Shishmarev (Russia)
Mikhel'son-Korvin-Shishmarev MK-1 Rybka (Little Fish)

Mil (Russia)
In 1947, Mikhail Leontyevich Mil established a design bureau (Opytno konstruktorskoe byuro, OKB) to develop rotorcraft. After Soviet practice it was called **OKB Mil**. After the collapse of the Soviet Union in 1992, the OKB was renamed **Moskovskii Vertoletnay Zavod Imieni M. L. Milya** (**MVZ Mil**; Moscow Helicopter Plant named for M. L. Mil). In 1994 **Kazan Helicopters**, which had been part of the Mil establishment, became a separate joint stock company. For all OKB Mil and MVZ Mil designs and all Kazan designs with Mi- designations, see **Mil**. For all other Kazan designs see **Kazan**.

Mil Mi-1 *Hare* Prototype
Mil Mi-1NKh *Hare*
Mil Mi-1S *Hare*
Mil Mi-1U *Hare* (Mi-1T, Mi-1UT)
Mil Mi-2 *Hoplite* Prototype
Mil Mi-2 *Hoplite* (See: WSK-PZL V-2)
Mil Mi-3 (See: Mil Mi-1S)
Mil Mi-4 *Hound-A*
Mil Mi-4 *Hound-B*
Mil Mi-4 *Hound-C*
Mil Mi-4L *Hound*
Mil Mi-4P *Hound*
Mil Mi-4S *Hound*
Mil Mi-6AYaSh *Hook-D*
Mil Mi-6P *Hook*
Mil Mi-6T *Hook-A*
Mil Mi-6VKP *Hook-B*
Mil Mi-8 Prototype *Hip-A*
Mil Mi-8 *Hip-B*
Mil Mi-8K
Mil Mi-8MT *Hip-H*, Mi-17)
Mil Mi-8AMT(Sh) (Mi-17)
Mil Mi-8P *Hip-B*
Mil Mi-8PPA *Hip-K*
Mil Mi-8PS *Hip-C*
Mil Mi-8R
Mil Mi-8S *Hip-C*
Mil Mi-8SMV *Hip-J*
Mil Mi-8T *Hip-C*
Mil Mi-8AT *Hip-C*
Mil Mi-8BT
Mil Mi-8TB *Hip-E*
Mil Mi-8TBK *Hip-F*
Mil Mi-8TG
Mil Mi-8TM
Mil Mi-8TP
Mil Mi-8TV *Hip-C*
Mil Mi-8TZ
Mil Mi-8VZPU *Hip-D*
Mil Mi-9 *Hip-G*
Mil Mi-10 *Harke-A*
Mil Mi-10K *Harke-B*
Mil Mi-12 *Homer*
Mil Mi-14BT *Haze-B*
Mil Mi-14P
Mil Mi-14PL *Haze-A*
Mil Mi-14PLM *Haze-A*
Mil Mi-14PS *Haze-C*
Mil Mi-17 *Hip-H*
Mil Mi-17 (See also: Mi-8MT, Mi-8AMT(Sh))
Mil Mi-17-1
Mil Mi-17-1V
Mil Mi-17-1VA
Mil Mi-17P (*Hip-K derivative*, Mi-8MTPB)
Mil (Kazan) Mi-17MD Night
Mil (Kazan) Mi-17KF
Mil Mi-18
Mil Mi-19
Mil Mi-22 *Hook-C*
Mil Mi-24A *Hind-A*
Mil Mi-24A *Hind-B*
Mil Mi-24BMT
Mil Mi-24D *Hind-D*
Mil Mi-24K *Hind-G2*
Mil Mi-24P *Hind-F*
Mil Mi-24PS
Mil Mi-24RKR *Hind-G1* (Mi-24RCh)
Mil Mi-24U *Hind-C*
Mil Mi-24V *Hind-E* (Mi-24W)
Mil Mi-24VM
Mil Mi-24VP
Mil Mi-25 *Hind-D*
Mil Mi-26 *Halo* Prototype
Mil Mi-26 *Halo*
Mil Mi-26A
Mil Mi-26M
Mil Mi-26MS
Mil Mi-26P *Halo*
Mil Mi-26T *Halo*
Mil Mi-26TM
Mil Mi-26TS
Mil Mi-26TZ
Mil Mi-27
Mil Mi-28 *Havoc*
Mil Mi-28N *Havoc*
Mil Mi-30 Tilt-Wing
Mil Mi-32
Mil Mi-34 *Hermit*
Mil Mi-34A *Hermit*
Mil Mi-34S *Hermit*
Mil Mi-34VAZ *Hermit*
Mil Mi-35 *Hind-E*
Mil Mi-35M *Hind-E*
Mil Mi-35P *Hind-E*
Mil Mi-38
Mil Mi-40
Mil Mi-52
Mil Mi-54
Mil Mi-58
Mil V-12

Miles, F. G. (F. G. Miles Ltd) (See: Miles)

Miles (UK)
The Miles brothers, F. G. and George, began to operate a small, unlicensed flying business near Shoreham in 1916. By 1926 the business had grown and been organized as **Southern Aircraft Ltd.** By 1929 the pair, now joined by F.G. Miles' wife, Blossom, had begun to design their own aircraft, although they lacked the facilities for quantity production. In 1932, they reached an agreement with **Phillips & Powis Ltd** at Woodley and formed **Phillips & Powis Aircraft, Ltd** to manufacture their designs. In 1936, Rolls-Royce Ltd purchased a significant interest in the company, which it held until 1941, allowing the three Miles to purchase Rolls' interest. They renamed the firm **Miles Aircraft Ltd** and continued to operate it until 1946 when the cancellation of wartime contracts caused the company to fail financially; the assets were purchased by **Handley Page Ltd.** F. G. and Blossom Miles established **F. G. Miles Ltd** in 1947, being rejoined by George Miles in 1951. In 1960 the company was acquired by **British Executive and General Aviation Ltd** (**Beagle**) as **Beagle-Miles Ltd.** In 1962 Beagle-Miles was merged with **Beagle-Auster Aircraft Ltd** to create **Beagle Aircraft Ltd**, which carries on design work originated in both firms. For all designs originating with the Miles brothers, Southern Aircraft, Phillips & Powis Aircraft, Miles Aircraft, or F. G.

Miles Ltd, see **Miles**. For designs originating from Miles, but developed by Handley Page, Beagle-Miles, or Beagle, see **Handley Page** or **Beagle**.
Miles Metal Martlet
Miles Southern Martlet
Miles M.1 Satyr
Miles M.2 Hawk
Miles M.2A Hawk
Miles M.2B Hawk
Miles M.2C Hawk
Miles M.2D Hawk
Miles M.2E Hawk Speed Six
Miles M.2F Hawk Major
Miles M.2G Hawk Major
Miles M.2H Hawk Major
Miles M.2L Hawk Speed Six
Miles M.2M Hawk Major
Miles M.2P Hawk Major
Miles M.2R Hawk Major de Luxe (3L)
Miles M.2S Hawk Major
Miles M.2T Hawk Major
Miles M.2U Hawk Speed Six
Miles M.2W Hawk Trainer
Miles M.2X Hawk Trainer
Miles M.2Y Hawk Trainer
Miles M.3 Falcon Prototype
Miles M.3A Falcon Major
Miles M.3B Falcon Six
Miles M.3B Falcon Six, RAE Taperwing Testbed
Miles M.3C Falcon Six
Miles M.3D Falcon Six
Miles M.3E Falcon Six
Miles M.4 Merlin
Miles M.5 Sparrowhawk
Miles M.5 Sparrowhawk (RAE Testbed Aircraft)
Miles M.6 Hawcon
Miles M.7 Nighthawk
Miles M.8 Peregrine
Miles M.8 B.L.S. Peregrine (Boundary Layer Suction)
Miles M.9 Kestrel
Miles M.9A Master I
Miles M.10 (Proposal)
Miles M.11 Whitney Straight Prototype
Miles M.11 B.L.S. Whitney Straight
Miles M.11A Whitney Straight
Miles M.11B Whitney Straight
Miles M.11C Whitney Straight
Miles M.12 Mohawk
Miles M.13 Hobby
Miles M.14 Hawk Trainer III
Miles M.14 Magister
Miles M.14 Magister, M.18 Wing Testbed
Miles M.14 Magister, Trailer Wing Trials
Miles M.14A Magister
Miles M.14B Magister
Miles M.15 (T.1/37 Trainer)
Miles M.16 Mentor
Miles M.17 Monarch
Miles M.18 Trainer Prototype
Miles M.18 Trainer Prototype, Nose Wheel Tests
Miles M.18 Mk.2 Trainer
Miles M.18 Mk.3 Trainer
Miles M.18HL Trainer (RAE High-Lift Testbed)
Miles M.19 Master II
Miles M.20
Miles M.21 (Trainer Proposal)
Miles M.22 (Fighter Proposal)
Miles M.23 (Fighter Proposal)
Miles M.24 Master Fighter
Miles M.25 Martinet
Miles M.26 (See: Miles X. Series)
Miles M.27 Master III
Miles M.28 Mercury
Miles M.29 (Trainer Proposal)
Miles M.30 X Minor
Miles M.31 Master IV (Proposal)
Miles M.32 (Powered Glider Proposal)
Miles M.33 Monitor Prototype
Miles M.33 Monitor
Miles M.34 (Attack Aircraft Proposal)
Miles M.35 Libellula (Dragonfly)
Miles M.36 Montrose (Proposal)
Miles M.37 Martinet Trainer
Miles M.38 Messenger Mk.I (RAF)
Miles M.38 Messenger Mk.2A
Miles M.38 Messenger Mk.2B
Miles M.38 Messenger Mk.2C
Miles M.38 Messenger Mk.4
Miles M.39 Libellula (Dragonfly) (Bomber Proposal)
Miles M.39B Libellula (Dragonfly) (M.39 5/8th Testbed)
Miles M.40 (Transport Proposal)
Miles M.41 (Transport Proposal)
Miles M.42 Libellula (Dragonfly) (Attack Fighter Proposal)
Miles M.43 Libellula (Dragonfly) (Attack Fighter Proposal)
Miles M.44 (Fighter Proposal)
Miles M.45 (Trainer Proposal)
Miles M.46 (Engine Testbed Proposal)
Miles M.47 (Pilotless Aircraft Proposal)
Miles M.48 Messenger Mk.3
Miles M.49 (Pilotless Aircraft Proposal)
Miles M.50 Queen Martinet
Miles M.51 Minerva (Proposal)
Miles M.52 Supersonic Project
Miles M.53 (Trainer Proposal)
Miles M.54 (Transport Proposal)
Miles M.55 Marlborough (Proposal)
Miles M.56 (Transport Proposal)
Miles M.57 Aerovan Mk.I
Miles M.57 Aerovan Mk.II
Miles M.57 Aerovan Mk.III
Miles M.57 Aerovan Mk.IV
Miles M.57 Aerovan Mk.V
Miles M.57 Aerovan Mk.VI
Miles M.58 (Fighter Proposal)
Miles M.59 (Transport Proposal)
Miles M.60 Marathon Prototypes
Miles M.60 Marathon (See: Handley Page Marathon)
Miles M.61 (Freighter Proposal)
Miles M.62 (Freighter Proposal)
Miles M.63 Libellula (Dragonfly) (Mail Plane Proposal)
Miles M.64 (L.R.5)
Miles M.65 Gemini Mk.1 (Prototype)
Miles M.65 Gemini Mk.1A
Miles M.65 Gemini Mk.1B
Miles M.65 Gemini Mk.2 (Prototype)
Miles M.65 Gemini Mk.3 (Prototype)
Miles M.65 Gemini Mk.3A
Miles M.65 Gemini Mk.3B
Miles M.65 Gemini Mk.3C
Miles M.65 Gemini Mk.4
Miles M.65 Gemini Mk.7
Miles M.65 Gemini Mk.8
Miles M.66 (Liaison Aircraft Proposal)
Miles M.67 (Transport Proposal)
Miles M.68 Boxcar
Miles M.69 Marathon Mk.II
Miles M.70 (Advanced Trainer Proposal)
Miles M.71 Merchantman
Miles M.72 Aerovan (Proposal)
Miles M.73 (Transport Proposal)
Miles M.74 (Light Aircraft Proposal)
Miles M.75 Aries
Miles M.76 (Kendall K.1) Durestos Glider Wing
Miles M.77 SparrowJet
Miles M.100 Student
Miles X.2 (Transport Proposal)
Miles X.3 (Transport Proposal)
Miles X.7 (Transport Proposal)
Miles X.9 (Transport Proposal)
Miles X.10 (Transport Proposal)
Miles X.11 (Transport Proposal)
Miles X.12 (Bomber Proposal)
Miles X.13 (Transport Proposal)
Miles X.14 (Empire Transport Proposal)

Miles & Atwood (Lee Miles & Leon Atwood) (California)
Miles & Atwood Special (Racer)

***Military Aircraft Corp** (See: Miller, Howell)*

Military Aircraft Works (Czechoslovakia) *(See: Letov)*

Mill Basin Aircraft, Inc (Brooklyn, NY)
Mill Basin Super-Gull 1W

Miller Aircraft Corp *(See: Miller, Howell)*

Miller Brothers (Irvine, PA)
Miller Bros. 1911 Biplane

Miller Corp (John J. Miller) (New Brunswick, NJ)
Miller Corp MCA-1 Amphibian Biplane

Miller, Charles W. *(See: Witteman Bros.)*

Miller, Franz (Italy)
Miller (Franz) 1909 Monoplane

Miller, Guy *(See: Miller-Ybarra)*

Miller, Henry (Farmington, MI)
Miller (Henry) Parasol (1928)

Miller, Howell W. (East Hartford, CT)
Howell Miller assisted the Granville Brothers on the design of their "GeeBee" racing aircraft. He went on to start a series of short-lived companies to market his original designs, including the **Military Aircraft Corp**, **Miller Aircraft Corp**, and **Aerovel Aircraft Corp** (all of Springfield, MA), the **American Armament Corp** of New York City, and the **Summit Aeronautical Corp** of Bendix, NJ. All Howell Miller designs are filed under **Miller, Howell.**
Miller (Howell) Dasch Trainer
Miller (Howell) HM-1
Miller (Howell) HM-1 "Time Flies"
Miller (Howell) HM-4 Aerovel
Miller (Howell) HM-5
Miller (Howell) MAC-1 (Rebuilt "Time Flies")
Miller (Howell) VPB
Miller (Howell) VPO
Miller (Howell) VPP
Miller (Howell) VPT
Miller (Howell) Zeta

Miller, Jim W. (Marble Falls, TX)
Miller (Jim) JM-1
Miller (Jim) JM-2
Miller (Jim) Special "Little Gem"

Miller, John J. *(See: Miller Corp)*

Miller, M. (La Porte, TX)
Miller (M.) M-6 Stove Bolt (Racer)

Miller, Merle B. (Savannah, GA)
Miller (Merle B.) Red Bare-Un

Miller, W. (UK)
Miller (W.) 1843 Ornithopter

Miller, W. Terry (Furlong, PA)
Miller (W. Terry) Tern I (Sailplane)
Miller (W. Terry) Tern II (Sailplane)
Miller (W. Terry) WM-2 (Powered Sailplane)

Miller, Wilson W. (Oneida, NY)
Miller (Wilson) 1927 Monoplane

Miller-Ybarra (Guy Miller & Demis Ybarra) (Pittsburgh, PA)
Miller-Ybarra Model 1 Monoplane

Millet-Lagarde (Société Millet-Lagarde) (France)
Millet-Lagarde ML.10

Millicer, Henry K. *(See: Victa)*

Milliken, William (Old Town, ME)
Milliken M-1

Mills-Fulford (UK)
Mills-Fulford 1910 Monoplane

Miloš Bondy a Spol *(See: Avia Czechoslovakia)*

Milutinovic, Eng. Prof. (Yugoslavia)
Milutinovic Type 215

Minerva Aircraft Corp (Akron, OH)
Minerva 1929 Sport Monoplane

Minicab *(See: Ord-Hume)*

Mira, Virgilio (Argentina)
Mira Golondrina I (Swallow)
Mira Golondrina II (Swallow)
Mira Golondrina III (Swallow)
Mira Golondrina IV (Swallow)

Mirage *(See: Dassault)*

Mirage Aircraft Corp (Klamath Falls, OR)
Mirage Celerity (Homebuilt)

Mississippi State University *(See: MSU)*

Mistel (German WWII Composite Attack Aircraft)
Mistel Prototype (Bf 109 F + DFS 230 Composite)
Mistel 1 (Bf 109 F + Ju 88A-4 Composite)
Mistel 2 (Fw 190 F-8 + Ju 88 G-1 Composite)
Mistel S 2 (Fw 190 F-8 + Ju 88 G-1 Composite, Trainer)
Mistel 3A (Fw 190 A-8 + Ju 88 A-4 Composite)
Mistel S 3A (Fw 190 A-8 + Ju 88 A-4 Composite, Trainer)
Mistel 3B (Fw 190 A-8 + Ju 88 H-4 Composite)
Mistel 3C (Fw 190 F-8 + Ju 88 G-10 Composite)
Mistel S 3C (Fw 190 F-8 + Ju 88 G-10 Composite, Trainer)
Mistel 5 (Jet Aircraft Composite Projects)

Mistral *(See: Armella-Senemaud; also: SNCASE)*

MIT (Massachusetts Institute of Technology) (Cambridge, MA)
MIT Burd (1972 Human-Powered)
MIT Chrysalis (1980)
MIT Eagle (1987 Human-Powered)
MIT Monarch (1984 Human-Powered)
MIT No.9 Glider (#409)
MIT 1922 Glider

Mitchell Aircraft Corp (Porterville, CA)
In 1976 Don S. Mitchell established **Mitchell Aircraft Corp** to build and market ultralight kits from his own designs. For all Mitchell designs predating 1976, see **Mitchell, Don S.** For post-1976 Mitchell designs, see **Mitchell Aircraft Corp.** For post 1997 development of the line, see **AmeriPlanes.**
Mitchell (Aircraft) A-10 Silver Eagle
Mitchell (Aircraft) AG-38
Mitchell (Aircraft) P-38 Lightning
Mitchell (Aircraft) Super Wing U-2
Mitchell (Aircraft) Super Wing U-2, NASM
Mitchell (Aircraft) Wing B-10
Mitchell (Aircraft) Wing T-10

Mitchell, Don S. (Don S. Mitchell Co) (San Lorenzo, CA)
Mitchell (Don) Nimbus I
Mitchell (Don) Nimbus II
Mitchell (Don) Nimbus III
Mitchell (Don) Super Osprey 280

Mitchell-Prizeman (C. G. B. Mitchell & R. W. Prizeman) (UK)
Mitchell-Prizeman Scamp

Mitchell-Proctor Aircraft (UK)
Mitchell-Proctor Kittiwake I

Mitrovich, Milenko (Yugoslavia)
Mitrovich M.M.S.3

Mitsubishi (Japan)
In 1920 the **Mitsubishi Nainenki Seizo KK** (Mitsubishi Internal Combustion Manufacturing Co Ltd) began operation in Nagoya, Japan, building licensed versions of European aircraft and engines as well as original aircraft. In 1928 the company was renamed **Mitsubishi Kokuki KK** (Mitsubishi Aircraft Co Ltd); it had become one of Japan's major aircraft manufacturers, with separate divisions building Army, Navy, and civil aircraft. The company suspended its aviation-related operations in 1945 after World War II. At the conclusion of the 1952 Peace Treaty, it resumed work as **Mitsubishi Jukogyo KK** (Mitsubishi Heavy Industries Ltd). For all Mitsubishi products, see **Mitsubishi**.
Mitsubishi:
Army Type 87 Light Bomber (2MB1)
Army Type 92 Recon Aircraft (2MR8)
A5M1, Navy Exp 9-Shi Single Seat Fighter (Ka-14)
A5M1, Navy Type 96 Carrier Fighter Model 1)
A5M2a (Navy Type 96 Carrier Fighter Model 2-1)
A5M2b (Navy Type 96 Carrier Fighter Model 2-2)
A5M3a (Navy Type 96 Carrier Fighter Model 2-3)
A5M4 *Claude* (Navy Type 96 Carrier Fighter 24/34)
A5M4-K *Claude* (Navy Type 2 Training Fighter)
A6M1 Reisen (Navy 12-Shi Exp Carrier Fighter)
A6M2 Reisen (Navy 12-Shi Exp Carrier Fighter)
A6M2a Reisen *Zeke* (Navy Type 0 Carrier Fighter 11)
A6M2b Reisen *Zeke* (Navy Type 0 Carrier Fighter 21)
A6M2-K Reisen *Zeke* (Navy Type 0 Training Fighter 11)
A6M2-N Reisen *Rufe* (Navy Type 2 Fighter Seaplane)
A6M3 Reisen *Hamp* (*Hap*, *Zeke*) (Navy 0 Carrier Fighter 32)
A6M3 Reisen *Zeke* (Navy Type 0 Carrier Fighter 22)
A6M3a Reisen *Zeke* (Navy 0 Carrier Fighter 22-Koh)
A6M4 Reisen *Zeke* (Navy Type 0 Carrier Fighter 42)
A6M5 Reisen *Zeke* (Navy Type 0 Carrier Fighter 52)
A6M5 Reisen *Zeke* (Navy Type 0 Carrier Fighter 52), NASM
A6M5b Reisen *Zeke* (Navy 0 Carrier Fighter 52-Otsu)
A6M5c Reisen *Zeke* (Navy 0 Carrier Fighter 52-Hei)
A6M5d-S Reisen *Zeke* (Night Fighter Modification)
A6M5-K Reisen *Zeke* (Navy 0 Training Fighter 22)
A6M7 Reisen *Zeke* (Navy 0 Carrier Fighter 62/63)
A6M7 Reisen *Zeke*, NASM
A6M8c Reisen *Zeke* (Navy 0 Carrier Fighter 54-Hei)
A7M Reppu *Sam* (Navy Ko-Type Carrier Fighter)
A7M1 Reppu *Sam* (Navy 17-Shi Exp Carrier Fighter)
A7M2 Reppu *Sam* (Navy Ko-Type Carrier Fighter)
B1M1 (Navy Type 13-1 Carrier Attack Bomber, 2MT1)
B1M2 Tora (Navy Type 13-2 Carrier Attack Bomber, 2MT5)
B1M3 (Navy Type 13-3 Carrier Attack Bomber, Model 2MT2)
(B1M) Navy Type 13 Carrier Attack Bomber 2MT4 Ohtori)
B2M1 (Navy Type 89-1 Carrier Attack Aircraft, 3MR4)
B2M2 (Navy Type 89-2 Carrier Attack Aircraft, 3MR4)
B4M1 (Navy 9-Shi Exp Carrier Attack Aircraft)
B5M1 *Mabel* (*Kate 61*) (Navy Type 97 Carrier Bomber 2)
Civil Type 89 General Purpose Aircraft
Civil Type 90 General Purpose Aircraft
C1M (Navy Type 10 Carrier Recon Aircraft)
C1M1 (Navy Type 10 Carrier Recon Aircraft, 2MR1)
C1M2 (Navy Type 10 Carrier Recon Aircraft, 2MR2/3/4)
C5M1 *Babs* (Navy Type 98 Recon Aircraft 1)
C5M2 *Babs* (Navy Type 98 Recon Aircraft 2)
D3M1 (Navy 11-Shi Experimental Carrier Bomber)
Experimental Short Range Recon Aircraft (2MR7)
Experimental Type R Flying-Boat (R-2)
Experimental Special-Purpose Carrier Recon Aircraft (2MR5)
Experimental 7-Shi Carrier Attack Aircraft (3MT10)
Experimental 7-Shi Carrier Fighter (1MF10)
F1M2 *Pete* (Navy Type 0 Observation Seaplane 11)
F-1
XF-2
F-2A
F-2B
F3B1 Trainer
G1M1 (Navy 8-Shi Special Recon Aircraft, Ka-9)
G1M1 (Navy Type 93 Attack Bomber)
G3M1 (Navy 9-Shi Experimental Medium Bomber, Ka-15)
G3M1 *Nell* (Navy Type 96 Medium Bomber 11)
G3M1-L *Nell* (Navy Type 96 Transport 11)
G3M2 *Nell* (Navy Type 96 Medium Bomber 21)
G3M2 *Nell 21* Civil Conversions, "Nippon"
G3M2 *Nell* (Navy Type 96 Medium Bomber 22)
G3M2 *Nell* 22, Civil Conversions
G3M3 *Nell* (Navy Type 96 Medium Bomber 23)
G4M1 Navy Experimental 12-Shi Attack Bomber
G4M1 *Betty* (Navy Type 1 Land Attack Bomber 11)
G4M1 *Betty* (Navy Type 1 Land Attack Bomber Model 12)
G4M2 *Betty* (Navy Type 1 Land Attack Bomber Model 22)
G4M2a *Betty* (Navy Type 1 Land Attack Bomber 24)
G4M2b *Betty* (Navy Type 1 Land Attack Bomber 25)
G4M2c *Betty* (Navy Type 1 Land Attack Bomber 26)

G4M2e *Betty* (Navy Type 1 Land Attack Bomber 24J)
G4M3 *Betty* (Navy Type 1 Land Attack Bomber 34)
G4M3 *Betty* (Navy Type 1 Land Attack Bomber 34), NASM
G6M1 (Navy Type 1 Wingtip Convoy Fighter)
G6M1-K (Navy Type 1 Large Land Trainer)
G6M1-L2 (Navy Type 1 Transport)
Hanriot 28 Trainer
Hato-Type Survey Aircraft
Hayabusa-Type (Peregrine Falcon) Fighter (1MF2)
Hibari-Type (Skylark) Trainer
Hinazuru-Type Transport (Mitsubishi-Airspeed Envoy)
J2M1 Raiden *Jack* (Navy Experimental 14-Shi Interceptor)
J2M2 Raiden *Jack* (Navy 14-Shi Interceptor 11)
J2M3 Raiden *Jack* (Navy 14-Shi Interceptor 21)
J2M3a Raiden *Jack* (Navy 14-Shi Interceptor 21-Koh)
J2M4 Raiden *Jack* (Navy 14-Shi Interceptor 34)
J2M5 Raiden *Jack* (Navy 14-Shi Interceptor 33)
J2M6 Raiden *Jack* (Navy 14-Shi Interceptor 31)
J4M1 Senden *Luke* (Navy 17-Shi Exp Otsu Interceptor)
J8M1 Shusui (Navy 19-Shi Exp Rocket-Interceptor)
J8M2 Shusui (Navy 19-Shi Rocket Interceptor)
Ka-8 (Experimental 8-Shi 2-Seat Fighter)
Ka-14 (Experimental 9-Shi 1-Seat Fighter)
K3M1 (Navy Type 90 Crew Trainer; Ka-2)
K3M2 *Pine* (Navy Type 90-1 Crew Trainer)
K3M3 *Pine* (Navy Type 90-2 Crew Trainer)
K3M3-L (Navy Type 90-2 Crew Trainer)
K6M1 (Navy 11-Shi Experimental Advanced Trainer Seaplane)
K7M1 (Navy 11-Shi Experimental Crew Trainer)
Karigane I (Wild Goose)
Karigane II (Wild Goose)
Ki 1 (Army Trainer; Hanriot HD-14)
Ki-1-I (Army Type 93-1 Heavy Bomber)
Ki-1-II (Army Type 93-2 Heavy Bomber)
Ki-2-I *Louise* (*Loise*) (Army Type 93-1 2-Engine Light Bomber)
Ki-2-II *Louise* (*Loise*) (Army Type 93-2 2-Engine Light Bomber)
Ki-7 (Experimental Crew Tainer)
Ki-14
Ki-15-I *Babs* (Army Type 97 Command Recon Aircraft 1)
Ki-15-II *Babs* (Army Type 97 Command Recon Aircraft 2)
Ki-18 (Experimental Fighter)
Ki-20 (Army Type 92 Heavy Bomber)
Ki-21-I *Sally* (*Jane*) (Army Type 97 Heavy Bomber 1)
Ki-21-Ia *Sally* (Army Type 97 Heavy Bomber 1-Koh)
Ki-21-Ib *Sally* (Army Type 97 Heavy Bomber 1-Otsu)
Ki-21-Ic *Sally* (Army Type 97 Heavy Bomber 1-Hei)
Ki-21-IIa *Sally* (Army Type 97 Heavy Bomber 2-Koh)
Ki-21-IIb *Sally* (Army Type 97 Heavy Bomber 2-Otsu)
Ki-30 *Ann* (Army Type 97 Light Bomber)
Ki-33 *Tina* (Experimental Fighter)
Ki-35
Ki-39
Ki-40
Ki-42
Ki-46-I *Dinah* (Army Type 100 Command Recon Aircraft 1)
Ki-46-II *Dinah* (Army Type 100 Command Recon Aircraft 2)
Ki-46-II Kai *Dinah* (Army Type 100 Operations Trainer)
Ki-46-III *Dinah* (Army Type 100 Command Recon Aircraft 3)
Ki-46-III Kai *Dinah* (Army Type 100 Air Defense Fighter)
Ki-46-IV *Dinah* (Army Type 100 Command Recon Aircraft 4)
Ki-47
Ki-50
Ki-51 *Sonia* (Army Type 99 Assault Aircraft)
Ki-57-I *Topsy* (Army Type 100 Transport 1)
Ki-57-II *Topsy* (Army Type 100 Transport 2)
Ki-67-I Hiryu (Flying Dragon) *Peggy* (Army Type 4 Heavy Bomber 1)
Ki-67-I Kai Hiryu (Improved Flying Dragon) *Peggy* (Army Type 4 Special Attack Aircraft 1)
Ki-69
Ki-73
Ki-83
Ki-90
Ki-95
Ki-97
Ki-99
Ki-103
Ki-109
Ki-112
Ki-118
Ki-200 Shusui (Sword Stroke)
Ko 1 (Army Trainer, Nieuport 81-E2)
L4M1 *Topsy* (Navy Type 0 Transport 1)
LR-1
MC-1
MC-20-I *Topsy*
MC-20-II *Topsy*
MC-21 *Sally*
MH2000
MS-1
MU-2A
MU-2B
MU-2C (See: Mitsubishi LR-1)
MU-2D
MU-2E (See: Mitsubishi MU-2S)
MU-2F
MU-2G
MU-2J
MU-2JF
MU-2K
MU-2K (JASDF)
MU-2L
MU-2M
MU-2N
MU-2P
MU-2S
(MU-2) Marquise
(MU-2) Solitaire
MU-300 Diamond I
MU-300 Diamond IA
MU-300 Diamond II (See: Beech Beechjet 400A)
Navy Type 10 Carrier Fighter (1MF1-5)
Navy Type 10 Carrier Torpedo Aircraft (1MT1N)
Navy Type 93 Land-Based Attack Aircraft (3MT5, 3MT5A)
Ohtori-Type (Phoenix) Communications Aircraft, *Eve*
Type R Transport Flyingboat
R-1.2 Trainer
R-2.2 Trainer
R-4 Survey Aircraft
RP-1
T-1.2
XT-2
T-2
T-2 CCV (Control Configured Vehicle)
T-2A

T-2A (Blue Impulse Aircraft)
Taka-Type (Hawk) Experimental Carrier Fighter (1MF9)
Tobi-Type (Final Effort) Recon Aircraft (2MR1)
Tombo-Type Trainer (Dragonfly) (2MS1)
Washi-Type (Eagle Eyes) Light Bomber (2MB2)

Mix, Arthur (Chicago, IL)
Mix Flying Arrow (1924 Dayton Bicycle Trophy Race)

MKEK (Makina ve Kimya Endustri Kurumu) (Turkey)
MKEK 1 (See: THK-15)
MKEK 2 (See: THK-16)
MKEK 4 Ugur (Luck)
MKEK 5 (See: THK-5)
MKEK 5A (See: THK-5A)
MKEK 6 (See: THK-14)
MKEK 7 (See: THK-2)

M. L. Aviation Co (UK)
M. L. Utility ("Flying Mattress")

Möbius-Pocher (C. Möbius & Pocher) (Germany)
Möbius-Pocher 1922 Monoplane Glider
Möbius-Pocher 1929 Biplane Glider

Moczarski (Ludwik), Idzkowski (Jan), and Ploszajski (Jerzy) (Poland)
Moczarski, Idzkowski, and Ploszajski MIP Smyk (Brat)

Mohawk Aircraft Corp (Minneapolis, MN)
Mohawk (M-1) Pinto Prototype
Mohawk (Model M-1) MLV Pinto
Mohawk Model M-1-C New Pinto
Mohawk Model M-1-C Redskin
Mohawk Model M-2-C
Mohawk XPT-7
Mohawk Spurwing
Mohawk Twin-Engine

Mohlar, Capt W. Hoseas "Bill" *(See: North Star)*

Mohme Aero Engineering Corp (New Brunswick, NJ)
Mohme 1927 Monoplane

Mohn, Albert B. "Al" (Grand Rapids, MI)
Mohn Special Racer (1948 Goodyear Race)

Mohr, Fred W. (Riceville, IA)
Mohr Experimental Monoplane

Moineau, Rene (France)
Moineau Pusher
Moineau C-1 Monoplane

Moinicken, Christ (Aberdeen, SD)
Moinicken 1924 Special

Moisant Monoplane Co (New York, NY)
In 1910 John and Alfred Moisant formed the **International Aviators** as a touring exhibition troupe. In January 1911, shortly after John's death, the company became **Moisant International Aviators**. Its role was expanded to include the management of the Moisant Long Island flying school and the production of licensed versions of French aircraft designs. By the end of the year Moisant had acquired the **Hempstead Plains Aviation Co**, planning to establish a large plant near College Park, MD. Financial difficulties ended Moisant's expansion plans, and the company, then called **Moisant Monoplane Co**, disappeared in 1914. For aircraft built by Moisant or Hempstead Plains, see **Moisant Monoplane Co.** Earlier aircraft designed independently by John Moisant, see **Moisant, John B.**
Moisant (Co):
1911 Monoplane (Blériot Type)
1911 Biplane (Farman Type)
1914 Bluebird Monoplane

Moisant, John B. (France/USA)
American **John Moisant** to learn to fly in France in 1909. He designed several monoplanes before returning to the USA to form an aviation school with his brothers. In 1910 he and his brother Alfred formed the **International Aviators**, a touring aviation troupe. Shortly after the start of the first tour in December 1910, John was thrown from his aircraft and killed. For aircraft designed by John Moisant, see **Moisant, John B.** For subsequent family aircraft designs see **Moisant Monoplane Co.**
Moisant (John) 1910 Monoplane
Moisant (John) 1910 Monoplane Aluminio-plane

Moiseyenko *(See: Moiseenko)*

Moiseenko, V. L. (Russia)
Moiseenko P-1 (2U-B3)

Molded Aircraft Corp *(See: Fairchild 46)*

Molino OY *(See: Eiri Avion OY)*

Moller International (Paul S. Moller) (Davis, CA)
Moller Merlin 100
Moller Merlin 150
Moller Merlin 200 (200XR) Aerobot
Moller Merlin 300
Moller Merlin 400 Aerobot
Moller XM-4 Aerobot
Moller XM-5 Commuter
Moller XM-6
Moller XM-7

Möller, H. G. (Germany)
Möller Stomo 3
Möller Stromer
Möller Stürmer S5

Monaghan, Richard (USA)
Monaghan Osprey

Monarch Aircraft Corp (Riverside, IL)
Monarch Light Commercial
Monarch Model 2

Moncassin (France)
Moncassin 1918 Single-Engined Biplane Flyingboat
Moncassin 1918 Twin-Engined Biplane Flyingboat

Mong, Ralph E. (Tulsa, OK)
Mong Sport (Homebuilt)

de Monge (Vicomte Louis de Monge de Franceau; Établissements Louis de Monge) (Belgium/France) *(See also: Buscaylet-de Monge)*
de Monge M.101 C2 (A2)
de Monge 1914 Experimental Parasol
de Monge 1918 Experimental Biplane
de Monge 1921 Deutsch Cup Racer (5.1)
de Monge 1921 Lumiére (Light)
de Monge 1922 Sailplane
de Monge 1923 Experimental, Twin-Anzani
de Monge 1923 Sailplane
de Monge 1924 Experimental, Single-Hispano

Moniot (Avions Automobiles Philippe Moniot) (France)
Moniot APM-20-1 Lionceau

Monnett Experimental Aircraft Inc (Oshkosh, WI)
John Monnett established **Monnett Experimental Aircraft Inc (MEA)** to design and market homebuilt aircraft kits. In July 1985 the company assets were purchased by **INAV Ltd (Innovative Aviation)**, a wholly owned subsidiary of the British company **Aviation Composites**. For designs originating from MEA, see **Monnett**. For all designs begun after the July 1985 purchase, see **INAV**.
Monnett Monerai-P (Homebuilt)
Monnett Monerai-S (Homebuilt)
Monnett Monex
Monnett Moni (Homebuilt)
Monnett Moni, NASM
Monnett Sonerai (Homebuilt)
Monnett Sonerai II (Homebuilt)
Monnett Sonerai IIL (Homebuilt)
Monnett Sonerai IILT (Homebuilt)

***Mono Aircraft Co** (See: Monocoupe)*

Mono-Twin Aircraft Corp (Beverly Hills, CA)
Mono-Twin MT-25

Mono-Van Aircraft, Inc (Toledo, OH)
Mono-Van 1929 Monoplane

Monocoupe (Moline, IL)
Originally called **Central States Aero Co (CSA), Central States Airplane Co** was established in 1927 to build Don Luscombe's "Monocoupe." In January 1928, the company became the **Mono Aircraft Division** of **Velie Motor Corp.** Following Willard Velie's death in March 1929, the Velie interests were sold to **Allied Aviation Industries**, a holding company. By May, these interests were divided into two separate companies: the **Lambert Aircraft Engine Corp** and the **Mono Aircraft Co** of Moline, IL. Both companies passed into receivership in 1931, reemerging in 1932 as the **Lambert Engine and Machine Co** and the **Monocoupe Corp.** In July 1934 the two companies joined under the newly-formed **Lambert Aircraft Corp**, with Monocoupe continuing to operate under its own name. In 1940 the company was dissolved and its assets passed to the **Monocoupe Aeroplane and Engine Corp** (transferring operations to Orlando, FL).
In September 1941, Monocoupe acquired the **Bristol Aircraft Corp** of Bristol, VA and its Canadian subsidiary **Bristol Aircraft Products Ltd.** The operations of these three companies were combined under the **Universal Molded Products Corp**, with Monocoupe forming a separate division of the company. Aircraft production halted during World War II, resuming briefly in 1948-1950 under the name **Monocoupe Airplane and Engine Corp.** In 1955, the corporate assets were acquired by a West Virginia aviation group, which reorganized the company as **Monocoupe Aircraft of Florida, Inc** and transferred operations to Melbourne, FL. For all products of Central States, Mono, and Monocoupe in their various incarnations, see **Monocoupe**. Products developed by Lambert outside of Monocoupe's operations are listed under **Lambert**.
Monocoupe Meteor
Monocoupe Monocoach Model H (Twin Monocoach)
Monocoupe Monocoach Model H Special (Twin Monocoach)
Monocoupe Monocoach Model 201
Monocoupe Monocoach Model 275
Monocoupe Monocoupe Prototype
Monocoupe Monocoupe Model 5
Monocoupe Monocoupe Model 22
Monocoupe Monocoupe Model 70
Monocoupe Monocoupe Model 70, NASM
Monocoupe Monocoupe Model 70V
Monocoupe Monocoupe Model 90 (Mono) ("Model 500")
Monocoupe Monocoupe Model 90 Deluxe
Monocoupe Monocoupe Model 90-A
Monocoupe Monocoupe Model 90-A "Little Mulligan"
Monocoupe Monocoupe Model 90-AF
Monocoupe Monocoupe Model 90-AL
Monocoupe Monocoupe Model 90J
Monocoupe Monocoupe Model 110
Monocoupe Monocoupe Model 110 Special (Clipped-Wing)
Monocoupe Monocoupe Model 110 Special "Little Butch," NASM
Monocoupe Monocoupe Model 113
Monocoupe Monocoupe Model 113 Special
Monocoupe Monocoupe Model 125
Monocoupe Monocoupe Model D-145
Monocoupe Monocoupe D-145, Lindbergh Aircraft (NR211)
Monocoupe Monoprep
Monocoupe Monoprep 90
Monocoupe Monoprep 218
Monocoupe Monosport Model 1 (Monosport 249)
Monocoupe Monosport 2
Monocoupe Monosport D
Monocoupe Monosport G (See also: Culver Dart-G)
Monocoupe Monosport G Deluxe

***Monoplans Borel-Morane** (See: Borel)*

***Monospar Aircraft Ltd** (See: G.A.L.)*

***Monospar Wing Co Ltd** (See: G.A.L.)*

Monsted-Vincent Aeronautical Inc (New Orleans, LA)
Monsted-Vincent MV-1 Starflight

Montagne (France)
Montagne 1923 Monoplane Glider

Montaigne, Olaf (Sweden)
Montaigne Wingless Aircraft (1932)

Monte-Copter Inc (Seattle, WA)
Monte-Copter Model 15 Triphibian

Montee Aircraft Co (James, Ralph, Kenneth, & Harold Montee (Santa Monica, CA)
Montee Biplane
Montee MR1 Monoplane
Montee Thornton Monoplane

Montel, Pierre & Philippe Jean (France)
Montel Quasar 200

Montgolfier (France)
Montgolfier Demoiselle-Type Parasol

Montgomery, John Joseph (Santa Clara, CA)
Montgomery (John) 1884 Glider
Montgomery (John) 1884 Glider Replica (for *Gallant Journey*)
Montgomery (John) "Santa Clara" 1905 Tandem Glider
Montgomery (John) 1905 Tandem Monoplane
Montgomery (John) 1911 Glider "Evergreen"

Montgomery (John) 1911 Glider, NASM
Montgomery (John) 1911 Glider Replica (Lockheed Replica)

Montgomery, Samuel (Stockton, CA)
Montgomery (Samuel) Spider Flying Machine (1910 Patent)

Montreal Aircraft Industries, Ltd
(See: Reid Aircraft Co)

Moody, Preston Tugman (Lacrosse, WA)
Moody Flying Machine (1911 Patent)

Mooney Aircraft Corp (Wichita, KS)
In 1948 Albert Mooney, former Chief Designer for **Culver Aircraft Corp**, established **Mooney Aircraft Corp.** In February 1969, the company declared bankruptcy and was acquired by **American Electronic Laboratories (AEL)**. By November 1969 the company was again bankrupt and its assets were acquired by **Butler Aviation InterNational, Inc.** In July 1970, Butler merged the Mooney assets with **Ted Smith Aerostar**, which it had previously acquired, to form **Aerostar Aircraft Corp.** Aerostar production was suspended in early 1972. In October 1973 Aerostar was acquired by **Republic Steel Corp** of Columbus, OH, which renamed the company **Mooney Aircraft Corp.** After Republic Steel was taken over by the **LTV Corp** in 1984, Republic's aviation interests were disposed of, with Mooney going to a group of private investors operating as **Mooney Holding Corp.** In spring 1985, the company was acquired by **Lake Amphibian Inc** (30%) and a partnership led by M. Alexandre Couvelaire and M. Michel Seydoux (70%), continuing to operate as the **Mooney Aircraft Corp** of Kervile, TX. For all designs originating with Mooney, including those built under Aerostar nameplates, see **Mooney.** For products of LTV and Lake, see **LTV** or **Lake.**

Mooney A2-A Cadet (Redesigned Ercoupe)
Mooney Flivver (1931)
Mooney M10 Cadet (Redesigned Ercoupe)
Mooney M18 Mite
Mooney M18C Mite (Wee Scotsman)
Mooney M18C Mite, NASM
Mooney M18C55 Mite
Mooney M18L Mite
Mooney M18LA Mite (Wee Scotsman)
Mooney M19 Cub-Killer
Mooney M20 Mark 20
Mooney M20A Mark 20A
Mooney M20B Mark 21
Mooney M20C Mark 21
Mooney M20C Ranger
Mooney M20D Master
Mooney M20E Chaparral
Mooney M20E Super 21
Mooney M20F Executive (Aerostar 220)
Mooney M20G Statesman
Mooney M20J Allegro
Mooney M20J AT (Advanced Trainer)
Mooney M20J MSE
Mooney M20J MSE Limited
Mooney M20J 201
Mooney M20J 201 AT (Advanced Trainer)
Mooney M20J 201 LM (Lean Machine)
Mooney M20J 201 MSE
Mooney M20J 201 MSE Limited
Mooney M20J 201 SE (Special Edition)
Mooney M20J 205
Mooney M20K Encore
Mooney M20K 231
Mooney M20K 252
Mooney M20K 252 TSE
Mooney M20L PFM
Mooney M20M TKS
Mooney M20M TLS
Mooney M20M Bravo TLS
Mooney M20R Ovation
Mooney M20R Ovation TKS
Mooney M20R Ovation 2
Mooney M20S Eagle
Mooney M22 (Twin)
Mooney M22 Mustang

Mooney, Albert (USA)
Mooney (Albert) Flivver (1931)

Moore, Arlen C. (USA)
Moore (Arlen C.) SS-1

Moore, R. S. (Washington, DC)
Moore (R. S.) 1912 Monoplane

Moore, Virgil B. (Los Angeles, CA)
Moore (Virgil) Road-Runner

Moore-Brabazon, Col J. T. C. (Brabazon, Lord of Tara) (UK)
(See also: Short; also Voisin)
Moore-Brabazon 1908 Biplane Glider

Morane-Saulnier (France)
In 1910, Léon Morane, Gabriel Borel, and Raymond Saulnier established **Société Anonyme des Aéroplanes Morane-Borel-Saulnier.** By mid-1911, Borel had left to found his own company and his place had been taken by Morane's brother Robert. Saulnier and the two Moranes reorganized and established **Société Anonyme des Aéroplanes Morane-Saulnier** in October 1911. This company avoided the late-1930s nationalization of the French aviation industry. After the company filed for bankruptcy in November 1962, control passed to **Établissements Henri Potez** in January 1963. Potez reorganized the company as **Société d'Exploitation des Établissements Morane-Saulnier (SEEMS).** In May 1965, SEEMS was taken over by **Sud-Aviation**, which reorganized the company as **Gerance des Établissements Morane-Saulnier (GEMS).** In 1966, GEMS became a full subsidiary of Sud-Aviation, which reorganized its light aircraft operations around Morane-Saulnier under the name **Société de Construction d'Avions de Tourisme et d'Affaires (SOCATA).**

For all products of Morane-Saulnier, SEEMS, and GEMS, as well as SOCATA products having Morane-Saulnier (M.S.) designations, see **Morane-Saulnier.** For SOCATA developments of the Rallye series postdating 1966 and which did not receive M.S. designations, see **SOCATA.** For products of Établissements Henri Potez and Sud-Aviation, see **Potez** or **Sud-Aviation**, respectively. For early aircraft designed by Saulnier, see **Saulnier, Raymond.**

Morane-Saulnier:
- Model A (Mo.S.11)
- Model A Racer
- Model (A) PP (Pau-Paris Aircraft)
- Model AC (Mo.S.23)
- Model AE
- Model AF (Mo.S.28)
- Model AI (Mo.S.27, 29, & 30)
- Model AI, Musee de l'Air
- Model AI (Mo.S.27)
- Model AI (Mo.S.30)
- Model AN (Mo.S.31)
- Model ANB (Mo.S.31)
- Model ANL (Mo.S.32)
- Model ANR (Mo.S.33)

Model ANS (Mo.S.34)
Model AR (Mo.S.35)
Model AS
Model AU
Model AV
Model BB (Mo.S.7)
Model BH (Mo.S.8)
Model BI
Model C
Model F
Model G (Mo.S.2)
Model (G) WR (Russian Model G)
Model G Seaplane
Model G (Mo.S.19, Parasol Prototype)
Model H (Mo.S.1)
Model H (Mo.S.1)
Model I (Mo.S.6)
Model L (Mo.S.3)
Model LA (Mo.S.4)
Model M (Mo.S.13)
Model N (Mo.S.5)
Model N Racer
Model O Racer
Model P (Mo.S.21)
Model P (Mo.S.26)
Model S (Mo.S.10)
Model T (Mo.S.25)
Model TRK (Mo.S.9)
Model U
Model V (Mo.S.22)
M.S.35 EP2
M.S.43
M.S.45
M.S.50
M.S.51 ET2
M.S.101
M.S.121 C1
M.S.122
M.S.129
M.S.130
M.S.131
M.S.137
M.S.139
M.S.147
M.S.148
M.S.149
M.S.152
M.S.180
M.S.181
M.S.185
M.S.221 C1
M.S.222 C1
M.S.223 C1
M.S.224 C1
M.S.225 C1
M.S.226
M.S.227
M.S.230
M.S.234
M.S.238
M.S.260
M.S.275
M.S.275 C1S
M.S.275-2
M.S.276
M.S.277
M.S.278
M.S.300
M.S.301
M.S.315
M.S.317
M.S.325
M.S.332
M.S.340
M.S.341
M.S.342
M.S.343
M.S.350
M.S.405
M.S.406
M.S.430
M.S.471
M.S.472
M.S.474
M.S.475
M.S.479
M.S.500 Criquet (Locust)
M.S.500 Criquet (Locust), NASM
M.S.502 Criquet (Locust)
M.S.560
M.S.561
M.S.563
M.S.571
M.S.700 Pétrel (Petrel)
M.S.730 Alcyon (Halcyon)
M.S.733 Alcyon (Halcyon)
M.S.755 Fleuret (Foil/Sword)
M.S.760 Paris
M.S.880
M.S.880A
M.S.880B Rallye (Rally) Club
M.S.880B Rallye (Rally) 100T
M.S.881 Rallye (Rally) Club
M.S.885 Super Rallye (Rally)
M.S.886 Super Rallye (Rally)
M.S.887 Rallye (Rally) 125
M.S.890 Rallye Commodore
M.S.892 Rallye Commodore 150
M.S.893 Rallye Commodore 180
M.S.894 Rallye Minerva 220 (See also: Waco)
1912 Hydrobiplane
1912 Type Militaire (Monoplane)
1913 Monoplane (Three Seat)

Moravan Narodni Podnik *(See: Zlin)*

Moreau Brothers (Fréres A. Moreau) (France)
Moreau (Brothers) Aérostable (1912)

Moreau, Jean-Marie (France)
Moreau (Jean-Marie) JM 10 (JMM 10) (See also: Delanne D.II)

Moreau-Berthaud *(See: Berthaud)*

Moreland Aircraft Inc (El Segundo, CA)
Moreland M-1
Moreland S-3

Morisson and Nawrot (Józef Morisson and Józef Nawrot) (Poland)
Morisson and Nawrot:
Ostrovia I (Moryson I)
Ostrovia II (Moryson II)
Moryson III

Morok (New York, NY)
Morok 1910 Monoplane

Morrisey (Santa Ana, CA)
In 1949 William J. Morrisey established **Morrisey Aircraft Co** to exploit his two-place monoplane. The company recapitalized to take the aircraft into production and became **Morrisey Aviation Inc**. In 1959, Morrisey sold the rights to the aircraft to **Shinn Engineering Inc**. Shinn ceased production in 1962 and the type certificates reverted to Morrisey. In 1967 Morrisey sold them to George Varga, who established **Varga Aircraft Corp**, with Morrisey acting as designer and engineering test pilot. When Varga ended production in 1982, Morrisey formed **The Morrisey Co** to build kitplanes of his design. In 1985, the company was renamed **Morrisey Aircraft Co**; Morrisey reacquired the type certificates to his earlier designs in 1986. For all Morrisey designs, including aircraft built by Shinn and Varga, see **Morrisey**.
Morrisey Bravo I Model OM-1
Morrisey Bravo II Model OM-1-2
Morrisey 1000C Nifty
Morrisey 2000C
Morrisey 2150
Morrisey (Shinn) 2150A Kachina
Morrisey (Varga) 2150TG
Morrisey (Varga) 2180A

Morrisey (Varga) 2180TG

Morrison, Lester A. (Fort Wayne, IN)
Morrison A-1 (1928)

Morita, Shinzo (Japan)
Morita 1911 Monoplane

Morrow Aircraft Corp (San Bernadino, CA)
Morrow (Aircraft Corp) Model 1-L Victory Trainer

Morrow, J. E. (Hammondsport, NY)
Morrow (J. E.) 1905 Helicopter

Morse, Robert
Morse Hang Glider

Morton Brothers (G. R. & C. L. Morton) *(See:McCook Aircraft Corp)*

Moryson, Józef *(See: Morisson and Nawrot)*

Moskalev, A. S. (Russia)
Moskalev SAM-2 (Mu-3)
Moskalev SAM-5
Moskalev SAM-5-2*bis*
Moskalev SAM-7 Sigma
Moskalev SAM-9 Strela
Moskalev SAM-10
Moskalev SAM-10*bis*
Moskalev SAM-11
Moskalev SAM-11*bis*
Moskalev SAM-13
Moskalev SAM-25

Moss Brothers Aircraft Ltd *(See: Mosscraft)*

Mosscraft (Moss Brothers Aircraft Ltd) (UK)
Mosscraft MA-1

Moswey (Switzerland)
Moswey 3

Moth Aircraft Corp (St Louis, MO), Moth Manufacturing Corp (Lowell, MA)
Moth American Moth
Moth Gypsy Moth

Motor Gliders Inc (Dayton, OH)
Motor Gliders Buzzard

Motorized Gliders of Iowa, Inc *(See: M.G.I.)*

Moudy, Loenard E. (Visalia, CA)
Moudy 1928 Monoplane

Mouillard, Louis-Pierre (France, Algeria, Egypt)
Mouillard 1856 Glider (1st Glider)
Mouillard 2nd Glider
Mouillard 1865 Glider (3rd Glider)
Mouillard 1878 Glider (4th Glider)
Mouillard 1892/96 Glider

Moulin (France)
Moulin 1912 Aviette

Moundsville Aircraft Corp (Moundsville, WV)
Moundsville X2LC Biplane (1928)
Moundsville 1928 Lone Eagle
Moundsville 1929 Lone Eagle

Mozhaiskii, Aleksander Fedorovich (Russia)
Mozhaiskii 1880s Experimental Aircraft

MPA (Man-Powered Aircraft) *(See: Human-Powered Aircraft)*

Mráz (Továrna Letadel inz J. Mráz) (Czechoslovakia)
Mráz M-1A Sokol (Falcon)
Mráz M-1B Sokol (Falcon)
Mráz M-1C Sokol (Falcon)
Mráz M-1D Sokol (Falcon)
Mráz M-1E Sokol (Falcon)
Mráz M-2 Skaut
Mráz M-3 Bonzo
Mráz XLD 40 Mir
Mráz L 40 Meta-Sokol

Mrozinski (Germany)
Mrozinski 1912 Monoplane

MSU (Mississippi State U., Starkville, MS)
MSU XAZ-1 Marvelette
MSU XV-11 Marvel

Mudry (Avions Mudry et Cie, Auguste Mudry) (France)
Mudry CAP 10 (See: CAARP)
Mudry CAP 20 (See: CAARP)

Mueller (USA)
Mueller 5 Windsock

Muessig, Otto G. (Milwaukie, OR)
Muessig P-2 ("Music M-100")

Müller (Flugzeugbau Gebruder Müller) (Germany)
Müller G.M.G.II, Opel Rocket Testbed

Mulliners Aeroplanes (UK)
Mulliners 1910 Monoplane
Mulliners 1911 Biplane Glider

Mullins Manufacturing Co (Salem, OH)
Mullins (Modisette) H-S "Hot Shot"
Mullins (Modisette) "Robot" Bomber

Multiplane, Ltd *(See: Jacobs Bros)*

Mummert, Harvey C. (Hammondsport, NY)
Mummert Cootie
Mummert Light plane (Monoplane Racer)
Mummert Sportplane

München (Akademy Fliegerclub München) (Germany)
München 1922 Glider

Muniz, Col Antonio (Casmuniz, Cassio Muniz S.A) (Brazil)
Muniz Casmuniz M 5
Muniz Casmuniz M 7
Muniz Casmuniz M 9
Muniz Casmuniz M-11 (CNNA HL-1)
Muniz Casmuniz 52

Munk, Max M. (Washington. DC)
Munk M-B-1

Muntz *(See: Baynes)*

Munter (UK)
Munter I Biplane (1912)
Munter II Biplane (1913)

Muraszew & Tomaszewski (Aleksander Muraszew & Kazimierz Tomaszewski) (Poland)
Muraszew & Tomaszewski M.T.1 Glider

Murchio Aircraft Corp (Thomas A. Murchio) (Paterson, NJ)
Murchio M-3

Muré (France)
Muré 1934 Glider

Mureaux (France)
In 1918 **Ateliers des Mureaux** was established to build licensed versions

of other manufacturers' aircraft. In 1930, Mureaux merged with **Ateliers de Construction du Nord de la France (ANF)**, a rail stock, to form **Ateliers de Construction du Nord de la France et des Mureaux (ANF-Mureaux)**. In 1934, ANF-Mureaux joined **Blériot** and **Farman** to form the **Union Corporative Aéronautique (U.C.A.)** in order to pool government production orders. With the 1936 nationalization of the French aeronautical industry ANF-Mureaux became part of the **Société Nationale de Constructions Aéronautiques du Nord (SNCAN)**. For Mureaux and ANF-Mureaux products predating the nationalization of the industry, see **Mureaux**. For SNCAN designs, see **SNCAN**.

Mureaux 1 C1
Mureaux (Brunet) 3 C2
Mureaux (Brunet) 3 R2
Mureaux (Brunet) 4 C2
Mureaux 110 R2
Mureaux 111 R2
Mureaux 112 R2
Mureaux 112 GR
Mureaux 113 R2
Mureaux 113 Cn2
Mureaux 114 Cn2
Mureaux 115 R2
Mureaux 116
Mureaux 117 R2
Mureaux 118
Mureaux 119
Mureaux 120 Rn3
Mureaux 121 Rn3
Mureaux 130 AR
Mureaux 130 A2
Mureaux 131 A2
Mureaux 140 T
Mureaux 160 T
Mureaux 170 C1
Mureaux 180 C2
Mureaux 190 C1
Mureaux 200 A3
Mureaux MB.411

Mutgatroyd (USA)
Murgatroyd 1910 Biplane

Murphy Aircraft Co (Canada)
Murphy (Aircraft Co) Maverick
Murphy (Aircraft Co) Rebel
Murphy (Aircraft Co) Renegade
Murphy (Aircraft Co) Renegade Spirit
Murphy (Aircraft Co) Renegade II

Murphy, Michael (USA)
Murphy (Michael) Upside Down Lander (Modified Taylor E-2)

Murray & Yarton Airplane Manufacturing Co (Moran, KS)
Murray & Yarton Sport (1930)

Murray, J. J. (Denver, CO)
Murray (J. J.) 1912 Monoplane

Murray, W. Roland (Glendale, CA)
Murray (W. Roland) M7

Myasishchev (Russia)
Myasishchev DVB-102
Myasishchev M-4 (MYa-4) *Bison-A*
Myasishchev M-4 (MYa-4) *Bison-B*
Myasishchev M-4 (MYa-4) *Bison-C*
Myasishchev M-50 (MYa-50) *Bounder*
Myasishchev M-50 (MYa-50) *Bounder*
Myasishchev M-52
Myasishchev M-52D Doubler
Myasishchev M-56
Myasishchev Nuclear-Powered Aircraft
Myasishchev VB-109

Myers Brothers (Arthur F. Myers) (Omaha, NE)
Myers Brothers 1928 Monoplane

Myers, George Francis (Jackson Heights, NY)
Myers Annular Multiplane "Flying Donut"
Myers Model MH-2 (1925 Helicopter)
Myers 1935 Aircraft
Myers 1954 Convertible Aircraft

***Mystere** (See: Dassault)*

N

NAC (National Aircraft Corp) (King of Prussia, PA)
NAC Dream (Ultralight)

NACA (National Advisory Committee for Aeronautics) (USA)
NACA 110-T Turboprop VTOL

Nagler, Bruno (White Plains, NY)
Austrian Bruno Nagler moved to the US following World War II, forming **Nagler Helicopter Co** Inc in 1955. The company became **Nagler Vertigyro Co Inc** in 1965, **Vertidynamics Corp** in 1970, and **Nagler Aircraft Co** in 1971. All products of Nagler's American companies are listed under **Nagler**.
Nagler NH-160 Helicopter
Nagler VG-1 Vertigyro
Nagler VG-2
Nagler VG-2C
Nagler VG-2P

Nagler-Rolz Flugzeugbau GmbH (Bruno Nagler & Franz Rolz) (Germany)
Nagler-Rolz NR.54 V1
Nagler-Rolz NR.54 V2
Nagler-Rolz NR.54 V2, NASM
Nagler-Rolz NR.55

Naglo Bootswerft (Germany)
Naglo Quadruplane

Nakajima Hikoki KK (Hikoki Kenkyusho) (Japan)
Nakajima:
A1N (Navy Type 3 Carrier Fighter)
A2N (Navy Type 90 Carrier Fighter) (NY)
A3N (Navy Type 90 Training Fighter)
A4N (Navy Type 95 Carrier Fighter) (YM)
A6M2-N *Rufe* (Navy Type 2 Fighter Seaplane) (AS-1)
Army Model 2 Ground Taxiing Trainer
Army Type 91 Fighter (NC)
"Army Type 97 Fighter Seaplane" *Adam*
AT-2
"AT-27 Twin-engined Fighter" *Gus*
B3N (Navy Experimental 7-Shi Carrier Attack Aircraft) (Y3B)
B4N (Navy Experimental 9-Shi Carrier Attack Bomber) (Q)
B5N *Kate* (Navy Type 97 Carrier Attack Bomber)
B6N Tenzan, *Jill* (Navy Carrier Attack Bomber)
B6N Tenzan, *Jill* (Navy Carrier Attack Bomber), NASM
B-6 Biplane
Bulldog Fighter
C2N (Navy Fokker Recon Plane)
C3N (Navy Experimental 10-Shi Carrier Recon Plane) (S)
C6N Saiun (Painted Cloud), *Myrt*
C6N Saiun (Painted Cloud), *Myrt*, NASM
D2N (Navy Experimental 8-Shi Carrier Bomber)
D3N (Navy Experimental 11-Shi Carrier Bomber) (DB)
E1Y (Navy Type 14 Recon Seaplane)
E2N (Navy Type 15 Recon Seaplane)
E4N (Navy Type 90-2 Recon Seaplane)
E8N *Dave* (Navy Type 95 Recon Seaplane) (MS)
E12N (Navy Experimental 12-Shi 2-Seat Recon Seaplane)
Fishery Seaplane
G5N Shinzan (Heart of Mountains), *Liz*
G8N Renzan (Mountain Range), *Rita* (Navy Experimental 18-Shi Attack Bomber)
G10N Fugaku (Mt Fuji) (Navy Experimental Super Heavy Bomber)
J1N *Irving* (Navy Type 2 Recon)
J1N1-S Gekko (Moonlight) *Irving* (Navy Night Fighter)
J1N1-S Gekko (Moonlight) *Irving* (Navy Night Fighter), NASM
J5N Tenrai (Heavenly Thunder) (Navy Experimental 18-Shi Otsu Type Interceptor)
Ki-4 (Army Type 94 Recon)
Ki-6 (Army Type 95-2 Trainer)
Ki-8 (Army Experimental Two-Seat Fighter) (DF)
Ki-11 (Army Experimental Fighter) (PA)
Ki-12 (Army Experimental Fighter)
Ki-13 (Army Experimental Direct Cooperation Plane)
Ki-16 (Army Experimental Transport)
Ki-19 (Army Experimental Heavy Bomber)
Ki-27 *Nate* (*Abdul*) (Army Type 97 Fighter)
Ki-31(Army Experimental Light Bomber)
Ki-34 *Thora* (Army Type 97 Transport)
Ki-37 (Army Experimental Twin-engined Fighter)
Ki-41 (Army Experimental High-speed Transport)
Ki-43 Hayabusa (Peregrine Falcon), *Oscar* (*Jim*) (Army Type 1 Fighter)
Ki-43 Hayabusa (Peregrine Falcon), *Oscar* (*Jim*) (Army Type 1 Fighter), NASM

Ki-44 Shoki (Devil Queller), *Tojo (John)* (Army Type 2 Single-Seat Fighter)
Ki-49 Donryu (Storm Dragon), *Helen* (Army Type 100 Heavy Bomber)
Ki-52 (Army Experimental Dive Bomber)
Ki-58 (Army Experimental Escort Fighter)
Ki-62 (Army Experimental Fighter)
Ki-63 (Army Experimental Fighter)
Ki-68 (Army Experimental Long-range Bomber)
Ki-75 (Army Experimental 2-engine High-altitude Night Fighter)
Ki-80 (Army Experimental Multi-Seat Fighter)
Ki-82 (Army Experimental Bomber)
Ki-84 Hayate (Gale), *Frank* (Army Type 4 Fighter)
Ki-87 (Army Experimental High-altitude Fighter)
Ki-101 (Army Experimental Twin-engined Night Fighter)
Ki-113 (Army Experimental Fighter)
Ki-115 Tsurugi (Sabre) (Army Special Attacker)
Ki-115 Tsurugi (Sabre) (Army Special Attacker), NASM
Ki-117 (Army Experimental Fighter)
Ki-201 Karyu (Fire Dragon) (Army Experimental Fighter Attacker)
Ki-230 (Army Experimental Special Attacker)
Kikka (Orange Blossom) (Navy Special Attacker)
Kikka (Orange Blossom) (Navy Special Attacker), NASM
Ko 2 Trainer
Ko 3 Fighter/Trainer
Ko 4 Fighter
K1Y (Navy Type 13 Trainer)
L1N *Thora* (Navy Type 97 Transport)
L2D *Tabby* (Navy Type 0 Transport)
LB-2 Experimental Long-range Attack Aircraft
N-35 Tokubaku (Navy Experimental 7-Shi Carrier Bomber)
N-36 Transport
NAF-1 (Navy Experimental 6-Shi Carrier Two-Seat Fighter)
NAF-2 (Navy Experimental 8-Shi Carrier Two-Seat Fighter)
Navy Experimental 6-Shi Carrier Attack Bomber
Navy Experimental 6-Shi Carrier Bomber
Navy Experimental 6-Shi Special Bomber
Navy Experimental 7-Shi Carrier Fighter (NK1F)
Navy Experimental 7-Shi Recon Seaplane
Navy Experimental 7-Shi Special Bomber
Navy Experimental 9-Shi Single-Seat Fighter (PA-kai)
P-1 Mail Carrier
Y3B (Navy Experimental 7-Shi Carrier Attack Bomber)
Yokosho Ro-Go Ko-Gata (Navy Recon Seaplane)
Type 1 Biplane
Type 3 Biplane
Type 4 Biplane
Type 5 Biplane
Type 5 Trainer
Type 6 Biplane
Type 7 Biplane
Nakajima-Avro 504 Trainer
Nakajima-Breguet Recon Seaplane
Nakajima-Douglas DC-2 *Tess*
Nakajima-Fokker Ambulance Aircraft
Nakajima-Fokker Super Universal Transport
Nakajima Hansa Recon Seaplane

NAL (India) *(See: National Aerospace Laboratories)*

Naleszkiewicz & Nowotny (Jaroslaw Naleszkiewicz & Adam Nowotny (Poland)
Naleszkiewicz & Nowotny:
NN 1 Glider
NN 2 Glider
NN 2*bis* Glider

Naleszkiewicz, Jaroslaw (Poland)
Naleszkiewicz J.N.1 Zabus II (Froggy II) Glider

NAM (National Aero Manufacturing Corp) (Philippines) *(See: PADC)*

Namèche et Wagnon (France)
Namèche et Wagnon 1934 Glider

Namco *(See: Nihon Kokuki Seizo Kabushiki Kaisha)*

Nanchang (China)
Nanchang CJ-5 (Yak-18)
Nanchang CJ-6 (Chujiao-6)
Nanchang Haiyan (Petrel)
Nanchang K-8 Karakorum
Nanchang J-6 (See: Shenyang)
Nanchang N-5A
Nanchang Q-5 (Qiang-5) (A-5, "F-6*bis*," "F-9") *Fantan*
Nanchang Q-5M (A-5M)
Nanchang Y5 (See: Shijiazhuang)

Narahara, Sanji (Japan)
Narahara Number 1 (1910)
Narahara Number 2 (1911)
Narahara Number 3 (1911)
Narahara Number 4 Ohtori-Go

Nardi SA Per Costruzioni Aeronautiche (Ingg. Fratelli Nardi) (Italy)
Nardi FN.305
Nardi FN.310
Nardi FN.315
Nardi FN.333 Riviera (See also: Lane)

NASA (National Aeronautics and Space Administration) (USA)
NASA AD-1 (Ames/Dryden-1)
NASA High Speed Civil Transport (HSCT)
NASA Hyper 3
NASA Hypersonic Transport Studies
NASA M2-F1 Lifting Body
NASA M2-F1 Lifting Body, NASM
NASA Pathfinder (Solar-Powered Research Aircraft)
NASA Quiet Short Haul Research Aircraft (QSRA, QSTOL)
NASA X-30 NASP (See: North American Rockwell)

Nash Aircraft Co. (Russell Nash) (Chapin, IL)
Nash Model N (1928)

Nash-Kelvinator Corp (Detroit, MI)
Nash-Kelvinator R-6 (See: Sikorsky)

National Advisory Committee for Aeronautics *(See: NACA)*

National Aero Manufacturing Corp (NAM) (Philippines) *(See: PADC)*

National Aeronautics and Space Administration *(See: NASA)*

National Aeronautics Co (USA)
National Aeronautics Cassutt Special IIIM (See: Cassutt)

National Aerospace Laboratories (NAL) (India)
National Aerospace Labs Hansa
National Aerospace Labs Saras

National Aircraft Builders Assoc (Berkeley, CA) *(See: Peters, Alvia)*

National Aircraft Corp (Beaverton, OR)
National Aircraft (OR) N-5-300

National Aircraft Corp (NAC) (Burbank, CA)
National Aircraft (CA) NA-75

National Aircraft Corp (Hagerstown, MD) *(See: Custer Channel Wing Corp)*

National Aircraft Corp (King of Prussia, PA) *(See: NAC)*

National Airplane & Motor Co (Billings, MT)
National Airplane & Motor Blue Bird

National Airways System (Lomax, IL)
National Airways System:
Air-King
Air-King "City of Peoria"
Air-King 2
Air-King 27
Air-King 27 Special
Air-King 28

National Flugzeug-Werke, GmbH *(See: NFW)*

National Glider Association (NGA) (Detroit. MI)
National Glider Association Primary Training Glider

National Research Associates (College Park, MD)
National Research Associates Aqua Gem Air Cushion Boat

National Research Council of Canada *(See: NRC)*

National Security *(See: Security Aircraft Corp)*

Nationale Vliegtuig Industrie *(See: NVI)*

Naugle (Latrobe, PA)
Naugle N-6
Naugle N-12 Mercury
Naugle 80 Mercury

Naval Aircraft Factory (NAF) (Philadelphia, PA)
Naval Aircraft Factory Pre-1923 Designations
Naval Aircraft Factory:
BS-1
Canard Glider (1921)
DH-4B
DT-2
DT-4
DT-5
F-5L (PN-5)
F-6L
H-16 (C-1) (See also: Curtiss)
HS-3 (See also: Curtiss)
M-81 (See also: Loening)
Mercury Racer (See: Mercury Flying Corp)
MF (See also: Curtiss)
N-1
NC-1 thru NC-4 (See: Curtiss)
NC-5
NC-6
NC-7
NC-8
NC-9
NC-10
NM-1
NO-1
NO-2
PT-1
PT-2
SA-1
SA-2
"Model T" (Sea Sled)
TB Flyingboat
TF Series
TG-1
TG-2
TG-3
TG-4
TG-5
TR-1
TR-3
TS-1 (See also: Curtiss)
TS-2
TS-3
UO-2
VE-7 (See: Vought)

Naval Aircraft Factory Post-1922 Designations
Naval Aircraft Factory:
XBN-1
XFN-1
XLRN-1
LR2N
N2N-1
XN3N-1
N3N-1
XN3N-2
XN3N-3
N3N-3
N3N-3, NASM
N4N
XN5N-1
XOSN-1 (XO2N-1, XO2N-2)
OS2N-1 Kingfisher (See also: Vought OS2U-3)
PN-7 (F-7L)
PN-8
PN-9
PN-10
PN-11
PN-12
XP4N-1 (XP2N-1)
XP4N-2 (P2N-1)
PBN-1 Nomad
SBN-1 (See also: Brewster XSBA-1)
SON-1 Seagull (See also: Curtiss SOC-3)
XTN-1
XT2N-1

Navion *(See: North American L-17)*

NDN Aircraft Ltd (N. D. Norman) (UK)
NDN Fieldmaster
NDN Freelance
NDN NDN-1 Firecracker (See: Hunting Firecracker)
NDN NDN-1T Turbo Firecracker

Neale, John V. (UK)
Neale Pup
Neale 6 Monoplane
Neale 7 Biplane

Nebraska Aircraft Corp *(See: Lincoln)*

Neico Aviation (Santa Paula, CA) *(See: Lancair)*

Neilson Steel Aircraft Corp (Thomas Scott Neilson & D. S. Neilson) (Berkley, CA)
Neilson Coupe Monoplane (1929)

Neiva (Sociedade Aeronáutica Neiva, Indùstria Aeronáutica Neiva, Sociedade Construtora Aeronáutica Neiva, Ltda) (Brazil)
Neiva B-2 Monitor
Neiva BN-1
Neiva Campeiro

Neiva Carajà
Neiva Lanceiro
Neiva Minuano
Neiva Paulistinha-56-C
Neiva IPD 5901 Regente
Neiva Seneca III
Neiva Universal
Neiva Universal T-25

Nelsch, William (St Louis, MO)
Nelsch Flying Flivver

Nelson Aircraft Corp (Nelson Specialty Corp, Ted Nelson) (San Fernando, CA, Pittsburgh, PA)
Nelson (Aircraft) BB-1 Dragonfly
Nelson (Aircraft) Bowlus Bumblebee
Nelson (Aircraft) PG-185 Hummingbird
Nelson (Aircraft) PG-185B Hummingbird NASM
Nelson (Aircraft) Terra-Plane

Nelson, Erik Sigfrid (San Francisco, CA)
Nelson (Erik) 1930 Helicopter

Nelson/Driscoll (Nels J. Nelson & Benjamin B. Driscoll) (New Britain, CT)
Nelson/Driscoll Fleetwing

Neman, I. G. (Russia)
Neman KhAI-1
Neman KhAI-5 (R-10)
Neman KhAI-6 (SFR)

Nemeth Helicopter Co (Chicago, IL)
Nemeth (Helicopter Co) "Pauletta"

Nemeth, Stephen Paul (Chicago, IL)
Nemeth (Stephen Paul) Circular Compound Wing Airplane
Nemeth (Stephen Paul) Rotary Wing Airplane (Cyclogiro)

Nemett *(See: Nemeth)*

Nemuth *(See: Nemeth)*

Neoteric USA Inc (Terre Haute, IN)
Neoteric Neova I Hovercraft
Neoteric Neova II Hovercraft

Nessler (France)
Nessler 1923 Biplane Glider
Nessler 1925 Avionette
Nessler 1928 Avionette

Netherlands Helicopter Industry N.V. (NHI) (Netherlands)
Netherlands Helicopter H-3 Kolibrie (Hummingbird)

Neukom, Albert (Switzerland)
Neukom 9-meter Sailplane (See also: Pfenninger)

New England Air Transport Co (Providence, RI)
New England F2WG (1929)

New England Aircraft Co *[See: Miller (Howell) HM-1 "Time Flies"]*

New Piper Aircraft Inc *(See: Piper Aircraft Corp)*

New Standard Aircraft Corp (Paterson, NJ)
Charles Healy Day designed his first aircraft in 1909, becoming a designer for the new **Glenn L. Martin Co** in 1911. During 1912/13 Day managed his own tractor biplane shop, returning to Martin in mid-1913, and moving to **Sloane Aeroplane Co** in 1914. Sloane became part of **Standard Aero Corp** in 1916; it was there that Day designed the Standard J-1. Standard closed in 1919. In 1927 Day and Ivan Gates formed **Gates-Day Aircraft Corp**, which was incorporated as **New Standard Aircraft Corp** in 1928. Day resigned in 1931, independently designing one further aircraft, which is listed under **Day**. His designs for Martin, Sloane, and Standard are found under those companies. Gates-Day and New Standard designs are under **New Standard**.
New Standard D-24 (G-D-24)
New Standard D-25
New Standard D-25A
New Standard D-25K
New Standard D-26
New Standard D-26A
New Standard D-27
New Standard D-27A
New Standard D-28
New Standard D-29A
New Standard D-29B
New Standard D-29P
New Standard D-29S
New Standard D-30
New Standard D-31
New Standard D-32
New Standard D-33
New Standard NT-1
New Standard NT-2

New York Aero Construction Co (Newark, NJ)
New York Aero Construction 1917 Twin-Motored Seaplane

New York State Education Department (Albany, NY)
New York GT Ground Trainer Glider

New Zealand Aerospace Industries Ltd (New Zealand)
New Zealand Aerospace Airtourer (See: AESL; also Victa)

Newbury *(See: EoN)*

Newhouse, Richard A., & Son (Rocky Hill, NJ)
Newhouse NS-1 Parasol

NFW (National Flugzeug-Werke, GmbH) (Germany)
NFW B.I
NFW E.I
NFW E.II
NFW Experimental Monoplane

NGA *(See: National Glider Association)*

NHI *(See: Netherlands Helicopter Industry N.V.)*

Niagara Amphibious Aircraft Corp *(See: Amphibious Aircraft Corp)*

Nicholas Airplane Co (Marshall, MO)
Nicholas J-1 (Rebuilt Standard J-1)

Nicholas-Beazley Airplane Co, Inc (Marshall, MO)
In 1927 the **Nicholas-Beazley Aircraft Co, Inc** hired Walter H. Barling and began marketing their NB-3 as a Barling product. In 1929 Barling formed his own **Barling Aircraft Corp.** Nicholas-Beazley products, including their Barling designs, are listed under **Nicholas-Beazley**. The earlier Barling Bomber (manufactured by **Witteman-Lewis** as the NBL-1) is filed under **Engineering Division**, while Barling's later designs are included in the **Barling Aircraft Corp** section.
Nicholas-Beazley: Birdwing

NB-2
(Barling) NB-3
(Barling) NB-3G
NB-4
NB-8-G Trainer
NB-9
NBPG (Power Glider)
Plymo-Coupe (See: Fahlin)
"Pobjoy Special" Racer
SF-1 (See: Fahlin)

Nicholson, H. G., Jr (Tonawanda, NY)
Nicholson KN-1 Junior

Nicollier (Avions H. Nicollier) (France)
Nicollier HN.433 Menestrel
Nicollier HN.434 Super Menestrel
Nicollier HN.500 Bengali
Nicollier HN.600 Week-End
Nicollier HN.700 Menestral II

Nielsen & Winther (Denmark)
Nielsen & Winther Type Aa N&W 1 (Nieuport Fighter Type)

Neimi, Leonard *(See: Arlington Sisu)*

Nieuport (Société Anonyme des Établis-sements Nieuport, Nieuport-Delage) (France) *(See also: Loire)*
Nieuport (France):
BM (See: Tellier T.5)
Bomber with 160hp Renault
Experimental Triplane #1
Experimental Triplane #2
Fighter with 150hp Hispano-Suiza
Fighter with 230hp le Rhône 9L
Fighter with 250hp Clerget
Fighter w/275hp Lorraine-Dietrich 8Bd
Fighter w/300hp Hispano-Suiza 8Fb
Recon Aircraft w/150hp Hispano-Suiza
Recon Aircraft w/150hp le Rhône
Recon Aircraft W/180hp Lorraine-Dietrich
Recon Aircraft w/200hp Hispano-Suiza
S (See: Tellier T.6)
TM (See: Tellier T.8)
2-Seat Fighter w/370hp Lorraine-Dietrich 12Da
2-Seater w/120hp Hispano-Suiza
I
II Family
IIIA
IV Family
4R 450 (See: Tellier "Vonna")
VI Family
Niner (Russian 1-Seat 10 Variant)
10 Monoplane
X Sesquiplane
11 Monoplane
11 (XI) Sesquiplane
12 (XII)
12*bis* C2
13 (XIII)
14
15
16
17
17*bis*
18
19
20
20*bis*
21
22 m (29*bis*)
23
24
24*bis*
25
27
28 C1
28 C1, NASM
29 C1
29 Racer
30 B2
30T
30T.1
30T.2
31 Sesquiplane
32 RH
33
37 C1
37 Racer
Nieuport-Delage Ni-D 38
Nieuport-Delage Ni-D 39
Nieuport-Delage Ni-D 390
Nieuport-Delage Ni-D 391
Nieuport-Delage Ni-D 391[2]
Nieuport-Delage Ni-D 391[3]
Nieuport-Delage Ni-D 393
42 C1
42 C2
43
44
450 (1929/30 Schneider Racer)
46
48 C1
481
50
52 C1
Nieuport-Delage Ni-D 540
Nieuport (France) 580
Nieuport-Delage Ni-D 590
Nieuport (France) 62 C1
Nieuport-Delage Ni-D 640
Nieuport-Delage Ni-D 641
72 C1
740
80
81
82
83
941 (Tailless)
121
122 Ca
122 C1
123
124
125
1912-13 Sesquiplane Floatplane
1914 Gordon-Bennett Sesquiplane
1916 Biplane with 150hp Clerget
1916 Twin-Fuselage Aircraft
1917 Mono/Sesqui-plane Fighter
1920 Sesquiplane Racer, 1921/22
Dunne D.8 Flying Wing

Nieuport, UK (The Nieuport & General Aircraft Co Ltd) (UK)
Nieuport (UK) B.N.1
Nieuport (UK) Goshawk
Nieuport (UK) L.C.3 Empire Type
Nieuport (UK) L.C.1 Nighthawk
Nieuport (UK) London
Nieuport (UK) LS.1
Nieuport (UK) LS.2
Nieuport (UK) Nightjar
Nieuport (UK) Mark V Seaplane
Nieuport (UK) Mark VI Seaplane
Nieuport (UK) Nieuhawk (Mailplane)

Nieuport-Delage *(See: Nieuport)*

Nieuport-Macchi *(See: Macchi)*

Nihon Hikoki KK (Nihon Aeroplane Co, Ltd) (Japan)
Nihon (Hikoki KK) Nippi NH-1 Hibari (Skylark)
Nihon (Hikoki KK) K8Nil (12-Shi Experimental Seaplane Trainer)
Nihon (Hikoki KK) 7LP1 (13-Shi Experimental Small Transport)

Nihon Kokuki Seizo Kabushiki Kaisha (Nihon Aeroplane Mfg Co; NAMCO) (Japan)
Nihon (Kokuki Seizo) YS-11

Nihon University (Japan)
Nihon University Human-Powered Aircraft:
Linnet

Linnet II
Stork I
Stork B

Nikitin, V. V. (Russia)
Nikitin NV-1
Nikitin NV-2
Nikitin NV-2*bis* (UTI-5)
Nikitin NV-4
Nikitin NV-5 (U-5)
Nikitin NV-6 (UTI-6)
Nikitin 1950 Monoplane

Nikitin-Shevchenko (V. V. Nikitin & V. V. Shevchenko) (Russia)
Nikitin-Shevchenko IS-1
Nikitin-Shevchenko IS-2

Nikol, Jerzy (Poland)
Nikol A-2

Niles Aircraft Corp (Niles, MI)
Niles (Aircraft) Williams Monoplane

Niles, Joe (Ephrata, WA)
Niles 1928 Monoplane

Nimmo, Rodney (Hollywood, CA)
Nimmo NAC-SB-1 (Nimmo Aircraft Co - Szekely Biplane)
Nimmo Special Racer "Pfttt" (Argander Special)

Nipper Aircraft *(See: Fairey Tipsy Nipper)*

Nippon Hikoki Seisakusho *(See: Tachibana)*

Nippon Kogata Hikoki KK (Japan Small Aeroplane Co, Ltd) (Japan)
Nippon Hachi (Bee) Motor Glider

Nishi, Hiro Hisa (Japan)
Nishi Jettoplane

Nishida, Matsuzo (Japan)
Nishida 1923 Sport Aeroplane

Nixon, John F. (Naugatuck, NY)
Nixon 1924 Special

Noble, Richard (UK)
Noble ARV Super2 (See: ARV)
Noble Farnborough F1 (See: Farnborough-Aircraft.com)

Noel (France)
Noel 1914 Reconnaissance Aircraft

Nolan Aircraft Co (USA)
Nolan Monoplane (Morane-Borel Type)

Noni, Gerardo J. (Argentina)
Noni Biplane
Noni Monoplane

Noonan-Wiseman *(See: Wiseman-Peters)*

Noorduyn Aviation Ltd (Canada)
Noorduyn YC-64 (C-64, UC-64) Norseman
Noorduyn YC-64 Norseman, NASM
Noorduyn UC-64A (C-64A) Norseman
Noorduyn UC-64B (C-64B) Norseman
Noorduyn Harvard Mk.IIB (AT-16)
Noorduyn JA-1 Norseman
Noorduyn Norseman, British Military
Noorduyn Norseman, Civil

Noran Aircraft Co (San Francisco, CA)
Noran Bat P-1 Monoplane (1929)

Nord *(See: SNCAN)*

Nordflug (Germany)
Nordflug Möwe (Seagull) 1923 Glider

Nordska-Phoenix (Sweden)
Nordska-Phoenix Tummerliten
Nordska-Phoenix 1923 Type Louis Monoplane Glider
Nordska-Phoenix 1923 School Biplane
Nordska-Phoenix J.21
Nordska-Phoenix J.23
Nordska-Phoenix S.1
Nordska-Phoenix S.21

Norman Aeroplane Co Ltd (N. D. Norman) (UK)
Norman (Aeroplane) Firecracker (See: Hunting Firecracker)

Norman Thompson Flight Co (UK)
Norman Thompson N.1B
Norman Thompson N.T.2B
Norman Thompson N.T.4A

Norman, H. J. (New York, NY)
Norman (H. J.) Glider (Lifeboat)

Norris Aircraft Manufacturer (R. V. Norris) (San Francisco, CA)
Norris #1

North American Aviation, Inc (Los Angeles, CA)
North American Aviation Inc was incorporated in 1928. The company name remained unchanged until a 1967 merger with **Rockwell-Standard Corp** created **North American Rockwell Corp.** In 1973 the company became **Rockwell International Corp** with the North American name remaining in the **North American Aerospace Operations** division. Most North American and Rockwell aircraft are listed under **North American.** (See also **Commander.**)

North American:
XAJ-1 Savage (NA-146)
AJ-1 (A-2A) Savage
AJ-2 (A-2B) Savage (NA-184)
AJ-2P Savage (NA-175, NA-183)
XA2J-1 (NA-163)
XA3J-1 Vigilante (NA-233)
A3J-1 (A-5A) Vigilante
A3J-2 (A-5B) Vigilante
A3J-3P (RA-5C) Vigilante
A-27 (NA-44, NA-69)
A-36A Apache (NA-97)
AT-6 Texan
AT-6A (T-6A) Texan
AT-6B Texan (NA-84)
AT-6C (T-6C) Texan (NA-88)
AT-6D (T-6D) Texan
XAT-6E Texan
AT-6F (T-6F) Texan (NA-121)
T-6G Texan
LT-6G Texan (NA-168)
T-6H Texan (See: T-6G)
T-6J Texan (NA-186) (See: CCF Harvard 4)
AT-16 (See: Noorduyn Harvard)
B-1 Lancer (See: Rockwell)
XB-21 Dragon (NA-39)
B-25 Mitchell (NA-62)
B-25A Mitchell (NA-62A)
B-25B Mitchell (NA-62B)
B-25C Mitchell
TB-25C (AT-24C) Mitchell
B-25D Mitchell (NA-87, NA-100)
TB-25D (AT-24A) Mitchell
XB-25E Mitchell (NA-94)
XB-25F Mitchell
XB-25G Mitchell
B-25G Mitchell (NA-96)
TB-25G (AT-24B) Mitchell
B-25H Mitchell (NA-98)
B-25J Mitchell (NA-108)
B-25J Mitchell, NASM
CB-25J Mitchell
TB-25J (AT-24D) Mitchell

VB-25J Mitchell
TB-25K Mitchell
TB-25L Mitchell
TB-25M Mitchell
TB-25N Mitchell
XB-28 (NA-63)
XB-28A (NA-67)
XB-45 Tornado (NA-130)
B-45A Tornado (NA-147)
DB-45A Tornado
JB-45A Tornado
TB-45A Tornado
B-45B Tornado
B-45C Tornado (NA-153)
DB-45C Tornado
JB-45C Tornado
RB-45C Tornado (NA-153)
XB-70A (RS-70) Valkyrie
BC-1 (NA-36)
BC-1A (NA-55)
BC-1B
BC-1I (NA-36)
BC-2 (NA-54)
BT-9 Yale (NA-19)
BT-9A Yale (NA-19A)
BT-9B Yale
BT-9C Yale (NA-29)
BT-9D Yale (NA-26)
Y1BT-10 (NA-29, NA-30)
BT-14 Yale (NA-58)
BT-14A Yale
XFJ-1 Fury (NA-134)
FJ-1 Fury (NA-141)
XFJ-2 Fury (NA-179)
FJ-2 Fury (NA-181)
XFJ-2B Fury (NA-185)
FJ-3 (F-1C) Fury
FJ-3D (DF-1C) Fury
FJ-3D2 (DF-1D) Fury
FJ-3M (MF-1C) Fury
FJ-4 (F-1E) Fury
FJ-4B (AF-1E) Fury
FJ-4F Fury
F-6B Mustang (P-51A)
F-6C Mustang (P-51B/C)
F-6D (RF-51D) Mustang (P-51D)
TF-6D (TRF-51D) Mustang (P-51D)
F-6K (RF-51K) Mustang (P-51K)
F-10 Mitchell
XP-86 Sabre (NA-140)
P-86A (F-86A) Sabre
(P-86A) F-86A Sabre, NASM
DF-86A Sabre
RF-86A Sabre
P-86B Sabre (NA-152) (Built as F-86A-5)
F-86C Sabre (See: F-93)
YF-86D (YF-95A) Sabre (NA-164)
F-86D Sabre (Sabre Dog) (NA-165, -173, -177, etc)
F-86E Sabre
QF-86E Sabre
F-86F Sabre
RF-86F Sabre
TF-86F Sabre
F-86G Sabre (See: F-86D; Built as F-86D-20)
YF-86H Sabre
F-86H Sabre (NA-187, NA-203)
F-86J Sabre (See also: Canadair)
YF-86K Sabre (NA-205)
F-86K Sabre
F-86L Sabre
YF-93A (NA-157)
F-93A
YF-95A Sabre (See: YF-86D)
F-100 Interceptor (NA-211)
YF-100 (YF-100A) Super Sabre
YQF-100 Super Sabre
QF-100 Super Sabre
F-100A Super Sabre (NA-192)
RF-100A Super Sabre
F-100B Super Sabre (See: F-107)
F-100C Super Sabre
F-100D Super Sabre
F-100D Super Sabre, NASM
F-100F Super Sabre
QF-100F Super Sabre
NF-100F Super Sabre
YF-107A (NA-212)
YF-108A Rapier (NA-257)
Harvard Mk.I
Harvard Mk.II
Harvard Mk.IIA
Harvard Mk.IIB (See Noorduyn AT-16 Harvard)
Harvard Mk.III
Harvard Mk.IV (See: CCF)
QL-17 Navion
L-17A (U-18A) Navion (NA-154)
(Ryan) L-17B (U-18B) Navion
L-17C (U-18C) Navion
L-17D Super Navion
Mitchell Mk.I (UK)
Mitchell Mk.II (UK)
Mitchell Mk.III (UK)
Mustang Mk.I (UK) (NA-73, NA-83)
Mustang Mk.IA (UK) (NA-91)
Mustang Mk.II (UK)
Mustang Mk.III (UK)
Mustang Mk.IV (UK)
NA-16
NA-16-1A (NA-32)
NA-16-1E (NA-61)
NA-16-1G (NA-43)
NA-16-1GV (NA-45)
NA-16-2A (NA-42)
NA-16-2H (NA-20)
NA-16-2H (NA-27)
NA-16-2K (NA-33)
NA-16-3 (NA-71)
NA-16-3C (NA-48)
NA-16-4 (NA-41, NA-46, NA-56)
NA-16-4M (NA-31, NA-38)
NA-16-4P (NA-34)
NA-16-4R (NA-37)
NA-16-4RW (NA-47)
NA-16-5
NA-18
NA-21
NA-22
NA-35
NA-40
NA-40A
NA-40B
NA-44 (See also: A-27)
NA-50
NA-50A (See: P-64)
NA-53
NA-57
NA-64 (See also: Yale)
NA-70
NA-73X
NA-80
NA-86
NA-89
NA-158
NA-236 LRI (Long Range Interceptor)
NA-237 FBX (Fighter Bomber Experimental)
NA-239 SAC Bomber
NA-295 (Val)
Navion (NA-143, -145)
NJ-1 (NA-28)
NJ-2 (NA-28)
XO-47 (GA-15)
O-47A (NA-25, NA-60)
O-47A (NA-25, NA-60), NASM
O-47B (NA-51)
XP-51 Mustang (NA-73)
P-51 (F-6A) Mustang (NA-91)
P-51A Mustang (NA-99)
XP-51B (XP-78) Mustang (NA-101)
P-51B (F-51B) Mustang
P-51C Mustang
P-51C Mustang "Excalibur III," NASM
P-51D (F-51D) Mustang
P-51D (F-51D) Mustang, NASM
TP-51D Mustang
XP-51F Mustang (NA-105)
XP-51G Mustang (NA-105)
P-51H (F-51H) Mustang
XP-51J Mustang (NA-105)
P-51K (F-51K) Mustang
P-51L Mustang (NA-129)
P-51M Mustang (NA-124)
P-64 (NA-50A, NA-68)
XP-82 Twin Mustang (NA-120)
XP-82A Twin Mustang

P-82B (F-82B) Twin Mustang
P-82B "Betty Joe"
P-82C (F-82C) Twin Mustang
P-82D (F-82D) Twin Mustang
P-82E (F-82E) Twin Mustang
P-82F (F-82F) Twin Mustang
P-82G (F-82G) Twin Mustang
P-82H (F-82H) Twin Mustang
PBJ-1C Mitchell
PBJ-1D Mitchell
PBJ-1G Mitchell
PBJ-1H Mitchell
PBJ-1J Mitchell
Sabre (UK) (See: F-86, Fuji, & Canadair)
Sabre Bat
Sabreliner (Sabre Commander)
SNJ-1 Texan (NA-52)
SNJ-2 Texan (NA-65, NA-79)
SNJ-3 Texan
SNJ-4 Texan (NA-88)
SNJ-4 Texan, NASM
SNJ-5 Texan (NA-88)
SNJ-5C Texan (NA-88)
SNJ-6 Texan (NA-121)
SNJ-7 Texan
SNJ-7B Texan
SNJ-8 Texan (NA-198)
XSN2J-1 (NA-142)
Supersonic Transport Proposal
T2J-1 (T-2A) Buckeye
XT2J-2 (YT-2B) Buckeye (NA-280)
T2J-2 (T-2B) Buckeye
T-2C Buckeye
T-2X Buckeye
XT-28 (XBT-28) Trojan (NA-159)
T-28A Trojan
T-28B Trojan
T-28BD Trojan
T-28C Trojan
T-28D Trojan
AT-28D Trojan
YAT-28E Trojan (NA-284)
XT-30
T-39 Sabreliner Prototype (UTX)
T-39A (CT-39A) Sabreliner
T-39B Sabreliner
T-39D (T3J-1) Sabreliner
T-39E Sabreliner
T-39F Sabreliner
YOV-10A Bronco (NA-300)
OV-10A Bronco
OV-10B Bronco
OV-10C Bronco
YOV-10D Bronco
OV-10D Bronco
OV-10E Bronco
OV-10F Bronco
X-15
X-15, NASM
X-15A-2
Yale (NA-64)

North American Rockwell:
XFV-12A (See: Rockwell XFV-12A)
X-30 (See: Rockwell X-30)
X-31 (See: Rockwell-MBB X-31)

North Star Aircraft Corp (St Cloud, MN)

North Star Liberty Bell "The Spirit of St Cloud"
North Star M6 Liberty Bell

Northern Aerial Transport Co (UK)

Northern (Lowe) H.L.(M).9 Marlburian

Northern Aircraft Inc (Alexandria, MN) (See: Bellanca Cruisemaster)

Northrop (USA)

After proving himself as an aircraft designer for Loughead, Davis-Douglas, and Lockheed, John K. "Jack" Northrop formed his own **Avion Corp** in 1927. In 1929 he reorganized under **United Aircraft and Transport Corp (UAT)** as the **Northrop Aircraft Corp**. Following a 1932 split with UAT, he formed the **Northrop Corp**, a partially owned subsidiary of **Douglas Aircraft**. In 1937, the Northrop Corp became the **Douglas El Segundo Division**. Jack Northrop remained with Douglas until 1939, when he formed **Northrop Aircraft Inc**. This company was renamed the **Northrop Corp** in 1959. A 1994 purchase of the **Grumman Corp** resulted in the creation of the **Northrop Grumman Corp**. Jack Northrop's designs for other companies are found under those companies. All designs of Avion and the various Northrop companies appear under Northrop. Aircraft designed by the Grumman Corp appear under **Grumman**.

Northrop YA-9
Northrop YA-13 (2-C)
Northrop XA-16 (Gamma 2C)
Northrop A-17 (8A)
Northrop A-17A
Northrop A-17AS
Northrop A-33
Northrop Alpha 2
Northrop Alpha 2 Floatplane
Northrop Alpha 3
Northrop Alpha 3 Floatplane
Northrop Alpha 4
Northrop Alpha 4, NASM
Northrop XBT-1
Northrop BT-1
Northrop BT-1S (Douglas Civil Designation 1-X)
Northrop XBT-2
Northrop B-2 Spirit (Stealth Bomber)
Northrop XB-35
Northrop YB-35
Northrop YB-49
Northrop YRB-49
Northrop Beta 3
Northrop C-19 Alpha
Northrop C-125 Raider
Northrop Delta 1
Northrop Delta 1A
Northrop Delta 1B
Northrop Delta 1D
Northrop Delta 1E (Gamma)
Northrop Delta II
Northrop (Avion) Experimental No.1 (1929 Flying Wing)
Northrop XFT-1
Northrop XFT-2
Northrop F2T-1 (P-61)
Northrop YF-5A (N-156F)
Northrop F-5A Freedom Fighter
Northrop F-5B Freedom Fighter
Northrop F-5C Freedom Fighter
Northrop F-5D Freedom Fighter
Northrop F-5E Tiger II
Northrop F-5F Tiger II
Northrop RF-5E Tigereye
Northrop F-15 Reporter
Northrop YF-17
Northrop F-18L Hornet
Northrop F-20 (F-5G) Tigershark
Northrop XF-89 Scorpion
Northrop YF-89A Scorpion
Northrop F-89A Scorpion
Northrop DF-89A Scorpion
Northrop F-89B Scorpion
Northrop DF-89B Scorpion
Northrop F-89C Scorpion
Northrop YF-89D Scorpion
Northrop F-89D Scorpion
Northrop YF-89E Scorpion
Northrop F-89F Scorpion
Northrop F-89G Scorpion
Northrop F-89H Scorpion
Northrop F-89J Scorpion
Northrop Gamma 2A "Sky Chief"
Northrop Gamma 2B "Polar Star," NASM
Northrop Gamma 2D
Northrop Gamma 2E
Northrop Gamma 2G

Northrop Gamma 2J
Northrop Gamma 2L
Northrop Gamma 5A
Northrop Gamma 5B
Northrop Gamma 8A A-17A Prototype
Northrop Gamma 8A Nomad (Export A-17A)
Northrop Gamma 8A-2
Northrop Gamma 8A-3N
Northrop Gamma 8A-3P
Northrop Gamma 8A-4
Northrop Gamma 8A-5
Northrop HL-10 Lifting Body
Northrop JB
Northrop JB-1
Northrop JB-10
Northrop M2-F1 Lifting Body (See: NASA)
Northrop M2-F2 Lifting Body
Northrop M2-F3 Lifting Body
Northrop M2-F3 Lifting Body, NASM
Northrop MX-324
Northrop MX-334
Northrop N-1A
Northrop N-1M Jeep
Northrop N-1M Jeep, NASM
Northrop N-3PB
Northrop N-9M Flying Wing
Northrop N-23 (N-32) Pioneer
Northrop N-102 Fang
Northrop XP-56 Black Bullet
Northrop XP-56 Black Bullet, NASM
Northrop XP-61 Black Widow
Northrop YP-61 Black Widow
Northrop P-61A Black Widow
Northrop P-61B Black Widow
Northrop P-61C Black Widow
Northrop P-61C Black Widow, NASM
Northrop XP-61E Black Widow
Northrop XP-79 Flying Ram
Northrop P-530 Cobra
Northrop XP-948 (Model 3A)
Northrop RT-1 (Delta)
Northrop YT-38 Talon
Northrop T-38A Talon
Northrop Tacit Blue
Northrop X-4 Bantam
Northrop X-21A (LFC, Laminar Flow Control Wing)
Northrop-McDonnell Douglas YF-23

Northrop Grumman (USA)
Northrop Grumman Greyhound 21

North Star Aerial Service Corp (Rochester, NY)
North Star 1927 3-Place Monoplane

Northwest Aircraft & Motor Corp (Seattle, WA) *(See also: Clifford)*
Northwest Model A
Northwest Training Monoplane

Northwestern *(See: Waco)*

Norton, F. H. (USA)
Norton Fighter
Norton N.16 Night Bomber

Nourse & Leighton (W. L. Nourse & V. Leighton) (Kansas City, MO)
Nourse 1929 Monoplane

Noury Aircraft Ltd (Canada)
Noury Tourer
Noury T-65 Noranda

Nowakowski, Zygmunt *(See: Samolot)*

Nowotny, Adam (Poland)
Nowotny N-y *4bis*

Nozawa Koku Kenkyusho (Nozawa Aviation Research Institute) (Japan)
Nozawa X-1 Light Aeroplane
Nozawa Z-1 Light Aeroplane

NRC (National Research Council of Canada)
NRC Tailless Glider

Nu-Day Aircraft Co (Springfield, IL)
Nu-Day Messenger

Nuclear Powered Aircraft, General

Numancia *(See: Dornier)*

Nürnberg (Ittner-Nürnberg) (Germany)
Nürnberg D-10 Sailplane (1922)
Nürnberg D-14 (1921)

Nuvoli Ing (Italy)
Nuvoli N.3
Nuvoli N.3S
Nuvoli N.5R

Nuwaco Aircraft Co (Littleton, Co)
Nuwaco T-10 Taper Wing

NVI (Nationale Vliegtuig Industrie) (Netherlands) *(See also: Koolhoven)*
NVI F.K.30
NVI F.K.31
NVI F.K.32
NVI F.K.33

O

O. E. Szekely Corp *(See: Szekely)*

O'Meara, J. K. "Jack" *(See: Wild, Horace B.)*

Oakland Airmotive *(See: Bay Aviation Services)*

OAW (Ostdeutsche Albatroswerke) *(See: Albatros)*

Oberlerchner, Josef (Austria)
Oberlerchner Mg-23 Sailplane
Oberlerchner Mg-23SL Sailplane

Obre (France)
Obre Biplane
Obre Biplane #3
Obre Monoplane

Occidental Aircraft Corp (Washington, DC)
Occidental Super STOL Model 100

Ocenášek, Ludvík (Austria-Hungary)
Ocenášek 1910 Monoplane

Odier-Bessiere (France)
Odier-Bessiere Clinogyre

Odier-Vendôme *(See: Vendôme)*

Oeffag (Österreichische Flugzeugfabric AG) (Austria)
Oeffag Type AF (50.08 - 50.12)
Oeffag Type BF (50.13)
Oeffag C.I
Oeffag C.II
Oeffag Type CF (50.14)
Oeffag Type K
Oeffag 50.01
Oeffag 50.02
Oeffag 50.03
Oeffag 50.04
Oeffag 50.05
Oeffag 50.06
Oeffag 50.07
Oeffag 50.15

Œhmichen (Oehmichen), Étienne (France)
Œhmichen Helicopter #1
Œhmichen Helicopter #2
Œhmichen Helicopter #3
Œhmichen 1936 Helicostat
Œhmichen 1938 Helicostat

Oertz (Max Oertz Werkes) (Germany)
Oertz Biplane I
Oertz Monoplane
Oertz W 4
Oertz W 5
Oertz W 6 Flugschoner
Oertz W 8

Officine Sommese Aeronautica *(See: OSA)*

Oficina de Mantencao e Recuperacao de Avioes Ltda *(See: Muniz, Col Antonio)*

Oficinas Gerais de Material Aeronáutico *(See: OGMA)*

Ogawa, Saburo (Japan)
Ogawa Number 1 (1917)
Ogawa Number 2 (1918)
Ogawa Number 3 Taxiing Trainer (1918)
Ogawa Number 5 Trainer (1921)

Ogden Aeronautical Corp (Inglewood, CA)
Ogden Osprey

Ogdensburg Aeroway Corp *(See: Huff-Daland)*

ÖGL (Estonia)
ÖGL PN-3

OGMA (Oficinas Gerais de Material Aeronáutico) (Portugal)
OGMA products are modifications, repairs, and license-built versions of other aircraft

Oguri, Tsunetaro (Japan)
Oguri-Curtiss Jenny Trainer (1919)
Oguri Number 2 Trainer (1920)

Ohio Aero Manufacturing Corp (Youngstown, OH)
Ohio Aero Cabin
Ohio Airmaster Coupe
Ohio Youngstown Youngster

Ohio Valley School of Aeronautics (Moundsville, WV)
Ohio Valley School of Aeronautics 1928 Sport Monoplane

Ohm (Richard) & Stoppelbein (Gordon) (Rochester, NY)
Ohm & Stoppelbein Special

Ojea, Emilio Alvarez (Argentina)
Ojea El Argentino (The Argentine) Cargo Glider
Ojea Revised Mignet Pou-du-Ciel (Flying Flea)

Okay Airplane Co Inc (Okay, OK)
Okay SK-1

OKB MiG *(See: MiG)*

OKB Mil *(See: Mil)*

Oldershaw, Vernon (USA)
Oldershaw O-2
Oldershaw O-3

Oldfield Aircraft Co (Barney Oldfield) (Cleveland, OH)
Oldfield Baby Great Lakes

Olesen Aircraft Inc (Bridgeport, CT)
Olesen Hydro Glider 01 (1930)

Oliver, Leo E. (Sanger, CA)
Oliver 1940/41 Transport Patents

Olliver, Charles *(See: Colliver)*

Ollo-Shprangel (M. Ollo & V. Shprangel) (Russia)
Ollo-Shprangel 1950 Monoplane

Olmstead, Dr. Charles M. (Le Roy, NY)
Olmstead 1910 Pusher
Olmstead 1910 Pusher, NASM

Olympian (Quonset Point, RI)
Olympian ZB-1 (Human-Powered)

Omega Aircraft Corp (New Bedford, MA)
Omega BS-12 (SB-12) Helicopter

Omega II *(See: ISAE)*

On Mark Engineering Co (Van Nuys, CA)
On Mark B-26K (A-26A) Counter-Invader
On Mark Marketeer (B-26)
On Mark Marksman (B-26)
On Mark Guppy Family (See: Aero Spacelines)

Ong Aircraft Corp (Kansas City, MO)
Ong M-32W

Opel (Germany)
Opel-Rekord
Opel-Sander RAK.1

d'Oplinter, Jean de Wouters (Belgium)
d'Oplinter 1934 twin-Engined Pusher

Optica Industries Ltd (UK)
Optica Industries Optica (See: Edgley Aircraft)

Option Air Reno (Reno, NV)
Option Air Acapella 100L
Option Air Acapella 100LR
Option Air Acapella 200L
Option Air Acapella 200S

O.R.B. (Italy)
O.R.B. Flying Boat

Orca Aircraft Ltd (UK)
Orca SAH-1 (See: Trago Mills)

Ord-Hume, Arthur W. J. G. (UK)
Ord-Hume Minicab GY-20R

Orenco (Ordnance Engineering Corp) (New York, NY)
Orenco Type A
Orenco Type B
Orenco Type C
Orenco Type C2
Orenco Type C3
Orenco Type C4
Orenco Type D
Orenco Type D2 (PW-3)
Orenco Type E
Orenco Type E2
Orenco Type F Tourister
Orenco Type F2 Tourister II
Orenco Type H2
Orenco Type H3
Orenco Type I Sport Boat
Orenco Type P-19 The Sociable
Orenco Type PW-3

Orlogsvaerftet (Danish Royal Dockyards) (Denmark)
Orlogsvaerftet:
FB.III Maager (Seagull)
HM.1 (Brandenburg W 33 Variant)
HM.2 (Heinkel He 8 Variant)

Orme, Harry (Washington, DC)
Orme 1912 Biplane

Ornithopters, General

Ort, Daniel J. (Detroit, MI)
Ort Sport Monoplane

Ortego, Leo (Alexandria, LA)
Ortego Helicopter

Ös, Ludwig, Jr
Ös Helicopter

OSA (Officine Sommese Aeronautica) (Italy)
OSA 135

Osprey Aircraft (Sacramento, CA)
Osprey I (X-28A Air Skimmer)
Osprey Pereira GP-3 Osprey II
Osprey Pereira GP-4 (Homebuilt)

Ostdeutsche Albatroswerke *(See: Albatros)*

Österreichische Flugzeugfabric AG *(See: Oeffag)*

Otto (Flugmaschinen-Werke Gustav Otto) (Germany)
Otto B
Otto B.I
Otto C.I
Otto C.II
Otto Pusher Biplane
Otto 1911 Biplane
Otto 1912 Monoplane
Otto 1912 Monoplane Racer
Otto 1913 Biplane Seaplane (100hp Argus)
Otto 1913 Monoplane Racer

Ottowa (Canada)
Ottowa MPA (Man-Powered Aircraft)

Ouest-Aviation *(See: SNCASO)*

Overcashier Aircraft Manufacturing Co (Detroit, MI)
Overcashier (Curtiss) JN-4D Type
Overcashier Model O.H.11 Cabin Monoplane (Model A.O.)
Overcashier O-12

Overland Airways Inc (Omaha, NE)
Overland Model L
Overland Sport 60

Owen & Bates (George A. Owen & George A. Bates) (Hartford, CT)
Owen & Bates 1911 Aerial Machine

Owl, George (Gardena, CA)
Owl OR65-2 Formula I Racer

Ozaki, Yukiteru (Japan)
Ozaki 1917 Tractor Biplane
Ozaki Soga-Go Airplane (1917)

P

P. K. Plans, Inc (Vista, CA)
P. K. Plans Kraft Super Fli

P. Levasseur Constructions Aéronautiques *(See: Levasseur)*

P. M. Co (UK)
P. M. Caudron-Type Biplane, 1914

Pac-Aero*(See: Pacific Airmotive Corp)*

Pacer Aircraft Corp (Perth Amboy, NJ)
Pacer Aircraft Corp Pacer

Pacific & Western Aviation (Australia)
Pacific & Western Air Tourer (See: Victa)

Pacific Aero Products Co (Seattle, WA)
Pacific Aero Products:
Seaplane Model I
Seaplane Model II

Pacific Aeronautical Industries (San Francisco, CA)
Pacific Aeronautical Model A

Pacific AeroSystem Inc (San Diego, CA)
Pacific AeroSystem Sky Arrow

Pacific Aircraft Co (Brea, CA)
Pacific Aircraft (Brea, CA):
J-30 (Low-wing Junkers-Type)
WT2 (Low-wing Junkers-Type)

Pacific Aircraft Co (Ken Coward) (La Jolla, CA)
Pacific Aircraft (La Jolla, CA) D-8 Sailplane

Pacific Airmotive Corp (Pac-Aero) (Burbank, CA)
Pacific Airmotive 580 (See: Allison Division of General Motors)
Pacific Airmotive Learstar
Pacific Airmotive PAC-1
Pacific Airmotive Tradewind

Pacific Airplane & Supply Co (Venice, CA)
Pacific Hawk
Pacific Standard C-1

Pack & Associates (Nashville, TN)
Pack Johnny Reb Racer

Packard Motor Car Co (Detroit, MI)
Packard 1920 Model A Biplane

Packard-Kirkham *(See: Kirkham)*

Packard-LePère *(See: LePère LUSAC)*

PADC (Philippine Aerospace Development Corp) (Philippines)
PADC Bo 105 (See: Bölkow/MBB)
PADC Islander (See: Britten Norman BN-2A)
PADC Lancair (See: Lancair)
PADC S.211 (See: SIAI-Marchetti)
PADC SF.260 (See: SIAI-Marchetti)
PADC Utility Monoplane

Paix & Dersig
Paix & Dersig Monoplane

Pakistan Aeronautical Complex (Pakistan)
Pakistan Aeronautical Complex:
K-8 (See: Nanchang)
Mushshak (Licensed Saab Safari)
Super Mushshak (Licensed Saab Safari)

Palmgren, David A. *(See: American Aeroplane Co)*

Pallissard & Co (Cicero, Il) *(See also: Partridge & Keller)*
Pallissard Partridge & Keller Biplane

Palmer, Joe Collier *(See: Southwestern Engineering & Mapping)*

Palmer, Paul J. (USA)
Palmer Pusher-Tractor Biplane

PAMA (Constructions de Planeurs à Moteur Auxiliare) (France)
PAMA Light Monoplane

Panavia (International)
Panavia Tornado

Pander (H. Pander & Zonen) (Netherlands)
Pander Type D Avionette
Pander DF Monoplane
Pander EC.60
Pander EF.85
Pander EG.100
Pander Type P
Pander Postjager
Pander 1925 Moto Aviette

Panstwowe Zaklady Lotnicze *(See: PZL)*

Papa 51 Ltd, Co (Nampa, ID)
Papa 51 Thunder Mustang

Papin & Rouilly (France)
Papin & Rouilly Gyroptère (1915)

PAR (Brazil) *(See: Centro Tecnico de Aeronautica)*

PAR (Parks Alumni Racers) (Ferguson, MO)
PAR Special (See also: Trefethen TRW Special)

Para-Copter Corp (Lindenhurst, NY)
Para-Copter M-7LPC Gyroplane

Paramount Aircraft Corp (Saginaw, MI)
Paramount Cabinaire
Paramount Sportster 120

Paraplane Corp (Pennsauken, NJ)
Paraplane (Ultralight)

Paraplane Commuter 110 & 120 *(See: Lanier)*

Parisano Aerial Navigation Co of America Inc (New York, NY)
Parisano Paraplane

Park Bros (Evan & Elwood) (Beloit, KS)
Park Bros Park-1

Parker (C. W. Parker Co) (Leavenworth KS)
Parker (C. W.) 1910 Airplane Factory

Parker, Calvin Y. (Coolidge, AZ)
Parker (C. Y.) Teenie (Homebuilt)
Parker (C. Y.) Teenie Two
Parker (C. Y.) Tin Wind (Homebuilt)

Parker, Francis Frederick (Oakland, CA)
Parker (Francis):
Canard Pusher Biplane (1913)
Firecracker
Firecracker 2
Tractor Biplane 1
Tractor Biplane 2
Tractor Biplane 3

Parker, John *(See: American Air Racing, Inc)*

Parker, Raymon H. (Los Angeles, CA)
Parker (Raymon H.) Tiny Mite
Parker (Raymon H.) RP-9 T-Bird

Parker, Raymond L. (Lakewood, Co)
Parker (Raymond L.) "Sperry Messenger" (Homebuilt)

Parker, W. L. (USA)
Parker (W. L.) Ranger

Parker, Will D. "Billy" (Fort Collins, Co) *(See also: Dewey Airplane Co)*
Parker (Will D.) Pusher

Parker, Willard (Willard Parker Aircraft Corp) (Cleveland, OH)
Parker (Willard) Pal

Parks Aircraft Inc (Oliver L. Parks) (St Louis, MO)
Parks Arrow
Parks P-1
Parks P-2 Speedster
Parks P-2A Speedster
Parks P-3
Parks P-4

Parks-Miller (Erie, CO)
Parks-Miller Sparrow

Parnall (UK)
The firm of **Parnall & Sons, Ltd** had long been active in Bristol as manufacturers of weights and measures and shop fittings. In 1898, the firm was acquired by **W & T Avery**, although the family name was retained. During World War I, the firm was enlisted to build aircraft for several other manufacturers, including Sopwith, Avro, and Short. Parnall also built 2 of its own designs, the Scout and the Panther. After the war, the company returned to the shop fittings business. The managing director of Parnall, George Parnall, resigned in 1919 and formed his own firm, **George Parnall & Co** in 1920. Starting in the 1930s, the Air Ministry again awarded contracts to **Parnall & Sons**, which continued to build airframes and components through 1965. George Parnall & Co built components and its own designs. In 1935 the company was sold to the **Hendy Aircraft Co** and **Nash & Thompson Ltd**, and was renamed **Parnall Aircraft Ltd**, which continued in the aircraft industry through 1945. In this finding aid, initials indicate which of the three Parnall companies built particular aircraft: PS for Parnall & Sons, GP for George Parnall, and PA for Parnall Aircraft.
Parnall Autogiro (Cierva C.10) (GP)
Parnall Elf (GP)
Parnall Hamble Baby Convert (PS)
Parnall Heck 2C (PA)
Parnall Imp (GP)
Parnall Panther (PS)
Parnall Parasol (GP)
Parnall Perch (GP)
Parnall Peto (GP)
Parnall Pike (GP)
Parnall Pipit (GP)
Parnall Pixie I (GP)
Parnall Pixie II (GP)
Parnall Pixie III Biplane (GP)
Parnall Plover (GP)
Parnall Possum (GP)
Parnall Puffin (GP)
Parnall Type 382 (Parnall Heck Mk.III) (PA)

Parsons Corp (Traverse City, MI)
Parsons XV-11 Marvel (See: MSU)

Parsons-Jocelyn (Lindsay Parsons & Rodney Jocelyn) (Ambler PA)
Parsons-Jocelyn PJ-260

Partenavia Costruzioni Aeronautiche SpA (Italy)
Partenavia P48B Astore (Goshawk)
Partenavia P52 Tigrotto
Partenavia P53 Aeroscooter
Partenavia P55 Tornado
Partenavia P57 Fachiro
Partenavia P59 Jolly
Partenavia P64 Oscar
Partenavia P66C Charlie
Partenavia P66D Delta
Partenavia P68 Observer
Partenavia P68B Victor
Partenavia P68C
Partenavia P70 Alpha
Partenavia P86 Mosquito
Partenavia Spartacus

Partridge (Elmer L.) & Keller (Henry C. "Pop") *(Cicero, IL)*
(See also: Pallissard)
Partridge & Keller:
#1 (1913)
1913 Tractor Trainer
1914 Biplane "Looper" (Stinson)

Pasadena City College, Pasadena Junior College *(See: Harlow)*

Pashinin, M. N. (Russia)
Pashinin I-21

Pashley Brothers (Cecil Lawrence Pashley & Eric Clowes Pashley) (UK)
Pashley Brothers 1914 Biplane

Pasotti Legnami Spa (Italy)
Pasotti F.6 Airone
Pasotti F.9 Sparviero (Sparrow)

Pasped Aircraft Co (Glendale, CA)
Pasped Skylark
Pasped Skyrider

Passat, M. B. (UK)
Passat Seagull (1912 Folding-Wing Monoplane)

Patchen (Marvin Patchen Inc) (Ramona, CA)
Patchen TSC-1 T-Boat (Teal Flying Boat) (See: Thurston)
Patchen TSC-1A1 Teal Amphibian
Patchen TSC-2 Explorer
Patchen TSC-2 Observer

Patco *(See: Philippine Aviation Corp)*

Paterson, Compton (UK/South Africa)
Paterson 1911 Biplane
Paterson 1914 Biplane

Paton, Robert (Carrington, ND)
Paton Aeroplane (1910)

Patterson-Francis (USA)
Patterson-Francis 1913 Flying Boat

Paukner, Arthur E. (Wauwatosa, WI)
Paukner Experimental (1928 Monoplane)

Paul H. Knepper Aircraft *(See: Knepper)*

Paul Schmidt *(See: Schmidt)*

Paul Strasburg Flying Service *(See: Strasburg)*

Paulhan, Louis (Societé Anonyme d'Aviation Paulhan) (France)
Paulhan Biplane No.1 (1911)
Paulhan Biplane No.2 (1911)
Paulhan Biplane No.3 (1911)
Paulhan Duck (1911)
Paulhan Triplane (1911)
Paulhan-Curtiss Flying Boat (1912)
Paulhan-Curtiss Flying Boat "Daily Mail Waterplane"
Paulhan-Curtiss Triad (1913)
Paulhan-Houry Model Aeroplane
Paulhan-Tatin Torpille (Torpedo) (1911)

Paulic, Rudy (Seattle, WA)
Paulic XT-3B

Paup Aircraft Inc (Carroll, IA)
Paup P-Craft (Ultralight)

Pavin (France)
Pavin 1923 Parasol Monoplane

Pavlov, A. N. (Russia)
Pavlov Orenburgskii-Osoaviakhim

Payen, (Payen-Aviation Delta) (France)
Payen Pa 49/B Katy
Payen Pa 112 Fléchair
Payen Pa 225
Payen Pa 266
Payen Pa 350 CD

Payne (V. W. Payne Engineering Co) (Cicero, IL; Escondido, CA)
Payne Knight Twister
Payne Knight Twister Imperial
Payne Knight Twister Junior JR-75
Payne Knight Twister KT-95
Payne MC-7 Monoplane

Pazmany Aircraft Corp (Ladislao Pazmany) (San Diego, CA)
Pazmany PL-1
Pazmany PL-1B Chieh-Shou (Chinese-Built PL-1)
Pazmany PL-2
Pazmany PL-2/LT-200 (Indonesian-Built PL-2)
Pazmany PL-4A
Pazmany PL-4B
Pazmany PL-6
Pazmany PL-7

PDQ Aircraft Products (Elkhart, IN)
PDQ-1
PDQ-2

Pearse (New Zealand)
Pearse 1908 Monoplane

Pearson-Williams (Venice, LA)
Pearson-Williams PW-1 "Mr. Smoothie"

Peekay Aircraft Corp (Marion, MA)
Peekay Helicopter

Peel Glider Boat Corp (College Point, NY)
Peel Glider Boat Z-1

Pegna (Pegna Rossi Bastianelli) (Italy)
Pegna Cabin Monoplane

Peitz (USA)
Peitz Model 101 Monoplane

Pellarini, Luigi (Italy)
Pellarini A.E.R.1 (See: Aeronova)
Pellarini PL-5C Aerauto

Pelzner, Willy (Germany)
Pelzner Hang Glider

Pemberton Billing Ltd (UK)
Noel Pemberton Billing founded his first aviation business in 1909, in South Fambridge, Essex, England. The firm failed within a year, but a new company, **Pemberton Billing, Ltd** was started at Woolston, Southampton in 1912. The firm was renamed **Supermarine Aviation Works Ltd** in 1916.
Pemberton Billing P.B.1
Pemberton Billing P.B.2
Pemberton Billing P.B.9 Scout
Pemberton Billing P.B.23E Push-Project
Pemberton Billing P.B.25 Scout
Pemberton Billing P.B.29E Zeppelin Destroyer
Pemberton Billing P.B.31E Nighthawk

Pénaud, Alphonse (France)
Pénaud Hélicoptère (1871)
Pénaud Planaphore (1872)
Pénaud & Gauchot Aéroplane Amphibie (1876)

Penfield Brothers (Thomas & John Penfield) (Paradise, CA)
Penfield 1926 Aircraft

Penguin
Penguins are, as the name implies, aircraft incapable of flight, and were used at flight schools during World War I to train student pilots in ground handling. Usually, smaller wings would be placed on a regular aircraft (See: Blériot XI Pingouin), but

occasionally a simple design might be fabricated from "the ground up."
Penguin Glider Towing Trainer

Penhoët (Société des Chantiers et Ateliers de la Loire-Penhoët) (France) *(See also: Wibault-Penhoët)*
Penhoët RP Flying Boat

Pennsylvania Aircraft Syndicate (Philadelphia, PA) *(See also: Wilford Aircraft Corp)*
Pennsylvania XOZ-1
Pennsylvania Wilford Gyroplane (WRK Gyroplane)

Penta Aeronautic (Italy)
Penta Super Cardinal

Penetcost, Horace T. (Seattle, WA) *(See also: Hoppi-Copters Inc)*
Penetcost E III Hoppi-Copter
Penetcost E III Hoppi-Copter, NASM

Percival Aircraft, Ltd (UK)
In 1932, Australian Edgar W. Percival formed the **Percival Aircraft Co** in England. The company reorganized in 1937 as **Percival Aircraft, Ltd.** After operating for some time as a member of the **Hunting Group of Companies**, the company was renamed **Hunting Percival Aircraft, Ltd** in 1954; the name changed to **Hunting Aircraft, Ltd** in 1957. After September 1960, controlling interest passed to **British Aircraft Corp**, although research and production continued under the Hunting name for several years. On 1 January 1964, Hunting was completely merged into **British Aircraft Corp (Operating) Ltd**, and use of the Hunting name was suspended. Edgar Percival returned to aviation in 1954, forming **Edgar Percival Aircraft, Ltd** (see Percival, Edgar). Nearly all Hunting and Hunting Percival designs were developments of Percival projects; they are filed under **Percival**. The few designs begun after 1957 are listed as **Hunting**.
Percival Gull Four
Percival Gull Six
Percival Helicopter (P.74)
Percival Jet Provost (P.84)
Percival Mailplane(See: Saunders-Roe/Percival)
Percival Merganser (P.48)
Percival Mew Gull
Percival Pembroke (P.66)
Percival Petrel (Military Q.6 P.16E)
Percival Prentice (P.40, T.Mk.1)
Percival President (P.66)
Percival Prince Mk.1 (P.50)
Percival Prince Mk.2 (P.50)
Percival Prince Mk.3 (P.50)
Percival Prince Mk.3B (P.50)
Percival Proctor Mk.I P.28)
Percival Proctor Mk.III (P.34)
Percival Proctor Mk.IV (P.31, T.9/41) (Preceptor)
Percival Proctor Mk.V
Percival Proctor Mk.VI
Percival Provost (P.56, T.16/48)
Percival Q.4, Q.6
Percival Q.6 (P.16E)
Percival Sea Prince (P.57)
Percival Survey Prince (P.54)
Percival Vega Gull

Percival, Edgar (Edgar Percival Aircraft, Ltd) (UK)
After losing control of **Percival Aircraft, Ltd** to the **Hunting Group of Companies**, Edgar Percival formed **Edgar Pecival Aircraft, Ltd** in 1954. In 1958 Percival sold this company, which then reformed as **Lancashire Aircraft Co** and continued to produce Percival's E.P.9 as the Prospector E.P.9 before going out of business in the 1960s. The products of both companies appear under **Percival, Edgar**.
Percival (Edgar) EP.9 Prospector

Perco (USA)
Perco Wasp C DMZ (Homebuilt Ercoupe 415)

Performance Aircraft, Inc (Olathe, KS)
Performance Aircraft Turbine Legend

Performance Aviation Manufacturing Group (Williamsburg, VA)
Performance Aviation (VA):
PM-100 ILV (Individual Lifting Vehicle)
PAM-100B ILV

Perkins, A. J. E. (Ajep) (UK) *(See: Wittman W-8 Tailwind)*

Perl, Harry N. (USA)
Perl PG-130 Penetrator Sailplane

Perrin (France)
Perrin 1926 Helicopter

Perry, Beadle & Co (E. W. Copeland Perry & F. P. H. Beadle) (UK)
Perry-Beadle 1914 Biplane
Perry-Beadle 1914 Flying Boat
Perry (E. W. C.) 1914 Sponson Wing Floatplane Patent

Perry (J. B.), Lanford (C. E.), & Gault (J. L.) (Greenville, SC)
Perry, Lanford, & Gault PLG-1
Perry, Lanford, & Gault GLP-3 (See: Johnson, Luther)

Perry, Thomas Osborne (Lombard, IL)
Perry Helicopter (1924)

Pertolini (Carlos Colombi Petrolini) (Petrolini Hermanos, Societa Anonima Industrial y Comercial) (Argentina)
Pertolini El Boyero (See: FMA 20)

Pescara de Castellicio, Marquis Raoul Pateras (Spain)
Pescara Helicopter No.1
Pescara Helicopter No.2
Pescara Helicopter No.3
Pescara Helicopter No.4

Peschkes, Fritz (Germany)
Peschkes Glider

Peters, Alvia (National Aircraft Builders Assoc) (Berkeley, CA)
Peters Monoplane (X-845M)
Peters Mono-Ford (10335)
Peters NABA (X-10682)

Peterson, Adolphe C. (Minneapolis, MN)
Peterson (Adolphe C.) Multipowered Aeroplane

Peterson, Albin K. (Washington, DC)
Peterson (Albin) Flivver Plane

Peterson, Max A. (USA)
Peterson (Max A.) J-4 Javelin

Petlyakov, Vladimir Mikhailovitch (Detainee Design Bureau KB-100) (Russia)
Petlyakov Pe-2
Petlyakov Pe-3

Petlyakov Pe-8

Petrie, C. M. (Kremlin, MT)
Petrie V Monoplane

Petróczy-Kármán-Zurovec (PKZ) (Austria)
PKZ-1 Helicopter
PKZ-2 Helicopter

Peugeot-Rossel (France)
Peugeot-Rossel Biplane
Peugeot-Rossel Monoplane

Peyret, Louis (France)
Peyret 1922 Tandem Monoplane Glider
Peyret 1923 Monoplane Glider
Peyret 1924 Moto Aviette
Peyret 1925 Monoplane Glider
Peyret-Mauboussin Scorpion II

Peyton-Sonntag (USA)
Peyton-Sonntag Monoplane

Pfalz Flugzeugwerke GmbH (Germany)
Pfalz A.I
Pfalz C.I
Pfalz D Type Experimental ("D.XI")
Pfalz D Type, Modified E.V
Pfalz D.III
Pfalz D.IIIa
Pfalz D.IV
Pfalz D.VI
Pfalz D.VII
Pfalz D.VIII
Pfalz D.XI (See: D Type Experimental, "D.XI")
Pfalz D.XII
Pfalz D.XII, NASM
Pfalz D.XIV
Pfalz D.XV
Pfalz Dr.I
Pfalz E.I
Pfalz E.II
Pfalz E.III
Pfalz E.IV
Pfalz E.V
Pfalz E.VI

Pfau, Audre M. "Aud" (Holyoke, CO)
Pfau Pedco Special (1948 Goodyear Race)
Pfenninger, W. (Switzerland)
Pfenninger Elfe (Brownie) M Sailplane
Pfenninger Elfe (Brownie) PM-3
Pfenninger Elfe (Brownie) S-1
Pfenninger Elfe (Brownie) S-2
Pfenninger Elfe (Brownie) S-3
Pfenninger Elfe (Brownie) 1
Pfenninger Elfe (Brownie) 2

Pfitzner, Alexander L. (Hammondsport, NY)
Pfitzner Monoplane (1909)

Pfluytmann
Pfluytmann Biplane

Phantom Knight Airplane (Aircraft) Co (Oak Park, IL)
Phantom Knight PKCA

Pheasant Aircraft Co (Memphis, MO)
Pheasant H-10
Pheasant M-1
Pheasant Traveller

Phelps, Leland E. (Toledo, OH)
Phelps Model 1 (Rebuilt Van Valkenburg VM 11)

Philippine Aerospace Development Corp *(See: PADC)*

Philippine Aviation Corp (Patco) (Philippines)
Philippine Aviation Corp Snipe

Phillips & Powis Aircraft Ltd *(See: Miles)*

Phillips Aviation Co (Los Angeles, CA)
Phillips (Aviation Co) Aeroneer 1B
Phillips (Aviation Co) Skylark CT-1
Phillips (Aviation Co) Skylark CT-2
Phillips (Aviation Co) XPT, X-PT (See: Aeroneer 1B)

Phillips, Claude B. (Duluth, MN)
Phillips (Claude B.) MW-1

Phillips, Lt Donald B. (McCook Field, OH)
Phillips (Lt Donald B.) Alouette

Phillips, Edward L. (St Louis, MO)
Phillips (Edward L.) 1940 Sport Plane
Phillips, Horatio (UK)
Phillips (Horatio) Aëroplane (1893)

Phillips, Peter *(See: SpeedTwin Developments)*

Phillips, Robert J. (Nutley, NJ)
Phillips (Robert J.) #90V473 (1930)

Phillips, W. H. (UK)
Phillips (W. H.) Helicopter (1842)

Phinn, Willard J. (Chicago, IL)
Phinn B-7 Phinn Arrow Biplane

Phoenix (Germany)
Phoenix Meteor L 2
Phoenix Meteor L II d

Phoenix Dynamo Manufacturing Co, Ltd (UK)
Phoenix Helicopter Observation Platform
Phoenix P.1 Seaplane
Phoenix P.3 Seaplane
Phoenix P.5 Cork Mk.I
Phoenix P.5 Cork Mk.II
Phoenix P.6 Pulex Patrol Flying Boat
Phoenix P.7 Eclectic Transport Flying Boat

Phoenix OKB (Russia) *(See: Aviacor)*

Phönix Flugzeugwerke AG (Austria-Hungary)
For Phönix production of **Albatros**, **Brandenburg**, **Knoller**, or **UFAG** aircraft, see each individual company.
Phönix C.I
Phönix D.I
Phönix D.II
Phönix D.IIa
Phönix D.III
Phönix Meteor L.2
Phönix Type 13 Reconnaissance Biplane
Phönix 20.01 (öAlb.01)
Phönix 20.02
Phönix 20.03
Phönix 20.04 (Modified Albatros B.I(Ph))
Phönix 20.05 thru 20.07 Series (Modified Albatros B.I(Ph))
Phönix 20.08 & 20.09 Series (Modified Brandenburg C.I)
Phönix 20.10
Phönix 20.11 & 20.12
Phönix 20.13 (Modified Brandenburg C.I (Ph))
Phönix 20.14 (Modified Brandenburg D.I (Ph))
Phönix 20.17
Phönix 20.18 (D.II Prototype)
Phönix 20.19 (Type 10)
Phönix 20.21 (Type 10)
Phönix 20.20 (Modified Brandenburg D.I (Ph))

Phönix 20.23 (Modified D.II)
Phönix 20.24 & 20.25 (Kirste Fighters)
Phönix 20.28 & 20.29
Phönix 350 HP Biplane
Phönix 1917 Monoplane
Phönix 1918 Rotary-Engined Fighter
Phönix 1918 Rotary-Engined Parasol Fighter

Piaggio (Industrie Aeronautiche e Meccaniche Rinaldo Piaggio SpA (Italy)

Piaggio (Pegna) P.B.N2
Piaggio P.6
Piaggio P.6*bis*
Piaggio P.6*ter*
Piaggio P.7 (P.c.7, P.7C)
Piaggio P.8
Piaggio P.9
Piaggio P.10
Piaggio P.16
Piaggio P.23M
Piaggio P.23R
Piaggio P.32-I
Piaggio P.32-II (P.32*bis*)
Piaggio P.50-I
Piaggio P.50-II
Piaggio P.108A
Piaggio P.108B
Piaggio P.108C
Piaggio P.108T
Piaggio P.111
Piaggio P.119
Piaggio P.136
Piaggio P.136-L
Piaggio P.136-L1
Piaggio P.136-L2
Piaggio P.148/P.149
Piaggio P.148
Piaggio P.149
Piaggio P.149D
Piaggio P.150
Piaggio P.166
Piaggio P.166B
Piaggio P.166C
Piaggio P.166M
Piaggio P.166S
Piaggio P.166-DL3
Piaggio P.180
Piaggio-Douglas PD-808

Piaggio-Nardi *(See: Nardi)*

Piasecki Aircraft Corp (Philadelphia, PA)

P-V Engineering Forum was organized in 1941 and incorporated in 1943. In 1946 the company became **Piasecki Helicopter Corp.** In 1955 a second company, **Piasecki Aircraft Corp (PiAC)**, was formed. Piasecki Aircraft remains in business at this writing; Piasecki Helicopter became **Vertol Aircraft Corp** in March 1955, the **Vertol Division of Boeing (Boeing-Vertol)** in 1960, **Boeing Vertol Co** (a Boeing subsidiary) in 1972, and finally, **Boeing Helicopters** in 1985. Aircraft designed by either of the Piasecki companies are listed under **Piasecki**. Aircraft designed by Vertol or Boeing-Vertol are under **Boeing-Vertol**.

Piasecki Airgeep, Airgeep II (See: PA-59H, PA-59K)
Piasecki XH-16 Transporter
Piasecki YH-16 Transporter
Piasecki YH-16A Transporter
Piasecki YH-21 Workhorse
Piasecki H-21A Workhorse
Piasecki H-21B (CH-21B) Workhorse
Piasecki H-21C (CH-21C) Shawnee
Piasecki XH-21D Shawnee
Piasecki H-25A Army Mule
Piasecki XHJP-1 (HUP-1 Prototype)
Piasecki XHRP-X "Dogship" (PV-3)
Piasecki XHRP-1 Rescuer (Flying Banana)
Piasecki XHRP-1 Rescuer (Flying Banana), NASM
Piasecki HRP-1 Rescuer (Flying Banana)
Piasecki HRP-2 Rescuer (Flying Banana)
Piasecki HUP-1 Retriever (See also: XHJP-1)
Piasecki HUP-2 (UH-25B) Retriever
Piasecki HUP-2S Retriever
Piasecki HUP-3 Retriever (UH-25C)
Piasecki HUP-4 Retriever
Piasecki PA-2B Ringwing VTOL
Piasecki PA-4 Sea Bat Drone VTOL
Piasecki (PA-59H) VZ-8P (B) Airgeep II
Piasecki PA-59K Sky-Car
Piasecki (PA-59K) VZ-8P Airgeep
Piasecki PA-59N Seageep
Piasecki PD-18 Commercial Helicopter (Civil HUP)
Piasecki PH-42 (Civilian H-21A)
Piasecki PV-2
Piasecki PV-2, NASM
Piasecki 16H-1 Pathfinder
Piasecki 16H-1A Pathfinder II
Piasecki Model 44

Picchio *(See: Procaer Picchio F.15)*

Pichancourt

Pichancourt 1879 Ornithopter

Pickering, William H. (Boston, MA)

Pickering Wind Tricycle

Picken (H. B. Picken & Associates) (Canada)

Picken Helicon

Piel Aviation SA (Avions Claude Piel) (France)

Piel CP20 Pinocchio
Piel CP30 Emeraude (Emerald)
Piel CP100 Emeraude (Emerald)
Piel CP301 Emeraude (Emerald)
Piel CP301A Emeraude (Emerald)
Piel CP301C Emeraude (Emerald)
Piel CP302 Emeraude (Emerald)
Piel CP303 Emeraude (Emerald)
Piel CP304 Emeraude (Emerald)
Piel CP305 Emeraude (Emerald)
Piel CP306 Emeraude (Emerald)
Piel CP311 Emeraude (Emerald)
Piel CP316 Emeraude (Emerald)
Piel CP320 Super Emeraude
Piel CP320A Super Emeraude
Piel CP321 Super Emeraude
Piel CP323A Super Emeraude
Piel CP323AB Super Emeraud
Piel CP328 Super Emeraude
Piel CP402 Donald
Piel CP500
Piel CP601 Diamant (Diamond)
Piel CP602 Diamant (Diamond)
Piel CP603 Super Diamant
Piel CP604 Super Diamant
Piel CP605 Super Diamant
Piel CP605B Super Diamant
Piel CP607 Super Diamant
Piel CP70 Beryl
Piel CP750 Beryl
Piel CP751 Beryl
Piel CP80 Zef (Zephyr, Breeze)
Piel CP90 Pinocchio
Piel CP150 Onyx
Piel CP1320 Saphir (Sapphire)

Pierce (S. S. Pierce Aeroplane Corp) (Southampton, NY)

Pierce Sport Biplane

Pietenpol, Bernard H. (Spring Valley, MN)

Pietenpol Air Camper
Pietenpol Scout
Pietenpol Scout B-4A
Pietenpol Sky Camper

Pifer, Roy (Mansfield, OH)

Pifer P-1 Skeet

Pigeon Fraser *(See: Albree)*

***Pigeon Hollow Spar Co** (See: Albree)*

Piggott Brothers and Co, Ltd (UK)
Piggott No.1 Biplane
Piggott 1911 Monoplane
Piggott 1912 Military Biplane

***PIK** (See: Eiri)*

Pilatus Flugzeugwerke AG (Switzerland)
Pilatus B-4 Glider (PC-11AF)
Pilatus Britten Norman (See: Britten Norman)
Pilatus P-2
Pilatus P-3
Pilatus P-4
Pilatus PC-6 Porter
Pilatus PC-6/A Turbo-Porter
Pilatus PC-7 Turbo-Trainer
Pilatus PC-8 Twin Porter
Pilatus PC-9 Advanced Turbo-Trainer
Pilatus PC-12 Turbo Transport
Pilatus SB-2 Pelikan (Pelican)
Pilatus UV-20 Chiricahua (PC-6)

Pilcher, Percy Sinclair (UK)
Pilcher Bat Glider 1895
Pilcher Beetle Glider (1895)
Pilcher Gull Glider (1896)
Pilcher Hawk Glider (1896)
Pilcher Helicopter (1897-99)

***Pilgrim** (See: Fairchild)*

Pioneer Aero Trades School, Inc (New York, NY)
Pioneer (Aero Trades School) BS2 Biplane
Pioneer (Aero Trades School) Snipe

Pioneer Aircraft Corp (Long Island, NY)
Pioneer (Aircraft Corp) Type A Sport Plane

Pioneer International Aircraft Corp (Manchester, CT)
Pioneer (International Aircraft) Flightstar

Piotrków Secondary School (Poland)
Piotrków Secondary School 1913 Glider

Piper Aircraft Corp (USA)
In October 1929, William T. Piper, Sr helped convince **Taylor Brothers Aircraft Manufacturing Co** to move from Rochester, NY to Bradford, PA. When the company went bankrupt in August 1930, the assets were purchased by Piper and re-organized as **Taylor Aircraft Co.** In 1935, Piper bought out all Taylor interests, renaming the company **Piper Aircraft Corp** in 1937. Also in 1937, a fire destroyed the Bradford factory and the company moved to Lock Haven, PA. In 1969, control of the company shifted from the Piper family to Chris Craft and Bangor Punta, though the Piper name was retained. Bangor Punta assumed full control in 1977. The company was, in turn, purchased by **Lear Siegler, Inc** in 1984 and **M. Stuart Millar** in 1987. The company filed for bankruptcy in 1991, coming out of Chapter 11 protection when its assets were purchased by **Newco Pac Inc** in July 1995. It was renamed **New Piper Aircraft Inc.** Aircraft built before the 1937 name change, including the J-2 Cub, are listed under **Taylor Aircraft Co.** All others are listed under **Piper**.
Piper Clipper Amphibian (See: Applegate Piper AP-1 Duck Amphibian)
Piper Duck Amphibian (See: Applegate Piper AP-1, Piper P-1 Cub Clipper)
Piper Grasshopper
Civilian J-3 Cubs, along with light planes of several other manufacturers, were used as liaison and reconnaissance aircraft in Army war games in 1940/41. General Innis P. Swift dubbed the aircraft "Grasshoppers," and the nickname came to be associated with all military light aircraft, though it was never an official name for any particular model. (For military Cubs, see: J-3C Cub, 1941 War Games; HE-1; L-4; L-18; L-21; NE-1; and O-59.)
Piper HE-1 (AE-1)
Piper J-2 Cub (See: Taylor J-2 Cub)
Piper J-3 Cub
Piper J-3 Cub, NASM
Piper J-3 Cub, Civilian L-4
Piper J-3 Cub Flitfire
Piper J-3 Cub Modified (Pobjoy)
Piper J-3 Cub Sport
Piper J-3C Cub
Piper J-3C Cub Crop Duster
Piper J-3C Cub Floatplane
Piper J-3C Cub "Little Bear"
Piper J-3C Cub Tricycle Gear Mod
Piper J-3C Cub Modified
Piper J-3C Cub Special
Piper J-3C Cub, 1941 War Games
Piper J-3F Cub
Piper J-3F Cub Crop Duster
Piper J-3F Cub Flitfire
Piper J-3F Cub Floatplane
Piper J-3L Cub
Piper J-3L Cub Flitfire
Piper J-3L Cub Floatplane
Piper J-3P Cub
Piper J-3P Cub Floatplane
Piper J-4 Cub Coupe
Piper J-4A Cub Coupe
Piper J-4E Cub Coupe
Piper J-5 Cub Cruiser
Piper J-5A Cub Cruiser
Piper J-5B Cub Cruiser
Piper J-5C Cub Cruiser
Piper J-5CO
Piper L-4A
Piper L-4B
Piper L-4B, NASM
Piper L-4B Crosswind Landing Gear
Piper L-4B Floatplane
Piper L-4C
Piper L-4D
Piper L-4E
Piper L-4F
Piper L-4G
Piper L-4H Floatplane
Piper L-4J
Piper L-4J Floatplane
Piper YL-14
Piper L-18A
Piper L-18B (PA-11)
Piper L-18C (PA-11)
Piper YL-21
Piper L-21A
Piper TL-21A
Piper L-21B (U-7A)
Piper NE-1
Piper NE-2
Piper YO-59
Piper O-59
Piper O-59A
Piper P-1 Cub Clipper (See also: Applegate Duck Amphibian)
Piper PA-6 Sky Sedan
Piper PA-8 Skycycle
Piper PA-11 Cub Special
Piper PA-11 Cub Special Agricultural Sprayer
Piper PA-11 Cub Special Floatplane
Piper PA-11 Cub Special Tricycle Gear Mod
Piper PA-12 Super Cruiser
Piper PA-12 Super Cruiser "City of Washington," NASM

Piper PA-12S Super Cruiser Floatplane
Piper PA-14 Family Cruiser
Piper PA-15 Vagabond
Piper PA-16 Clipper
Piper PA-16S Clipper Floatplane
Piper PA-17 Vagabond Trainer
Piper PA-18 Super Cub
Piper PA-18 Super Cub, NASM
Piper PA-18 Super Cub Ski Mod
Piper PA-18 Super Cub Whittaker Tandem Landing Gear Mod
Piper PA-18AS Super Cub Floatplane
Piper PA-18T Special
Piper PA-18-A Super Cub Agricultural Airplane
Piper PA-20 Pacer
Piper PA-20-125 Pacer
Piper PA-20S Pacer Floatplane
Piper PA-22 Tri-Pacer
Piper PA-22 Tri-Pacer, Experimental Dual Engine Modification
Piper PA-22-108 Colt
Piper PA-22-125 Tri-Pacer
Piper PA-22-135 Tri-Pacer
Piper PA-22-150 Tri-Pacer
Piper PA-22S-150 Pacer Floatplane
Piper PA-22-160 Tri-Pacer
Piper PA-23 Apache Prototype ("Twin Stinson")
Piper PA-23 Apache
Piper PA-23 Apache, NASM
Piper PA-23 Apache, Wilson Jet Profile Modification
Piper PA-23-150 Apache
Piper PA-23-160 Apache
Piper PA-23-160 Apache F
Piper PA-23-160 Apache G
Piper PA-23-160 Apache G
Piper PA-23-235 Apache
Piper PA-23-235 Apache Super Custom
Piper PA-23-250 Aztec
Piper PA-23-250 Aztec B
Piper PA-23-250 Aztec C
Piper PA-23-250 Aztec E
Piper PA-23-250 Aztec F
Piper PA-23-250 Turbo Aztec F
Piper PA-24 Comanche
Piper PA-24 Comanche B
Piper PA-24-180 Comanche
Piper PA-24-250 Comanche
Piper PA-24-260 Comanche
Piper PA-24-260 Comanche B
Piper PA-24-260 Comanche C
Piper PA-24-400 Comanche
Piper PA-24-600 Turbo Comanche
Piper PA-25-150 Pawnee
Piper PA-25-235 Pawnee
Piper PA-25-235 Pawnee B
Piper PA-25-260 Pawnee D
Piper PA-28-140 Cherokee
Piper PA-28-140 Cherokee B
Piper PA-28-140 Cherokee Cruiser
Piper PA-28-140 Flite Liner
Piper PA-28-151 Cherokee Warrior
Piper PA-28-160 Cherokee Prototype
Piper PA-28-161 Cadet
Piper PA-28-161 Warrior II
Piper PA-28-161 Warrior III
Piper PA-28-180 Cherokee B
Piper PA-28-180 Cherokee C
Piper PA-28-180 Cherokee Challenger
Piper PA-28-180 Cherokee D
Piper PA-28-180 Archer
Piper PA-28R-180 Cherokee Arrow
Piper PA-28-181 Archer II
Piper PA-28-181 Archer III
Piper PA-28R-200 Cherokee Arrow
Piper PA-28-200 Cherokee Arrow II
Piper PA-28-201T Turbo Dakota
Piper PA-28R-201 Cherokee Arrow
Piper PA-28R-201 Cherokee Arrow III
Piper PA-28R-201T Turbo Arrow
Piper PA-28R-201T Turbo Cherokee Arrow III
Piper PA-28RT-201 Arrow IV
Piper PA-28RT-201T Turbo Arrow IV
Piper PA-28-235 Cherokee
Piper PA-28-235 Cherokee B
Piper PA-28-235 Cherokee C
Piper PA-28-235 Cherokee D
Piper PA-28-235 Cherokee Pathfinder
Piper PA-28-236 Dakota
Piper PA-28R-300 Pillon
Piper PA-30 Twin Comanche
Piper PA-30 Twin Comanche B
Piper PA-30 Turbo Twin Comanche
Piper PA-31 Navajo
Piper PA-31 Navajo C
Piper PA-31 Navajo II (Prototype Navajo Chieftain)
Piper PA-31P Navajo
Piper PA-31T Cheyenne
Piper PA-31T Cheyenne I
Piper PA-31T Cheyenne II
Piper PA-31-310 Navajo
Piper PA-31-310 Turbo Navajo B
Piper PA-31-325 Navajo C/R
Piper PA-31-350 Navajo Chieftain (Chieftain after 1979/80)
Piper PA-31P-350 Mojave
Piper PA-31-425 Pressurized Navajo
Piper PA-32-260 Cherokee Six
Piper PA-32-300 Cherokee Six
Piper PA-32-300 Lance
Piper PA-32R-301 Saratoga SP
Piper PA-32-301T Turbo Saratoga
Piper PA-32R-301 Saratoga II HP
Piper PA-32R-301T Saratoga II TC
Piper PA-32R-301T Turbo Saratoga SP
Piper PA-34 Seneca
Piper PA-34-200 Seneca
Piper PA-34-200T Seneca II
Piper PA-34-220T Seneca III
Piper PA-34-220T Seneca IV
Piper PA-34-220T Seneca V
Piper PA-35 Pocono
Piper PA-38-112 Tomahawk II
Piper PA-39 Twin Comanche C/R
Piper PA-42-680 Cheyenne III
Piper PA-42-720 Cheyenne IIIA
Piper PA-42-1000 Cheyenne 400 (Cheyenne 400LS)
Piper PA-44-180 Seminole
Piper PA-44-180T Turbo Seminole
Piper PA-46-350P Malibu Mirage
Piper PA-48 Enforcer
Piper PA-60-290 Aerostar 600A
Piper PA-60-290 Aerostar 601A
Piper PA-60-290P Aerostar 601P
Piper PA-60-290P Aerostar 602P (Sequoya)
Piper PA-60-290T Aerostar 601B
Piper PT-1
Piper PWA-1 Skycoupe
Piper-Stinson 49
Piper TG-8 Glider
Piper UO-1 (U-11A)

Pippart-Noll (Germany)
Pippart-Noll Type 3

Pirie, J. A. (USA)
Pirie Rukh

Pisarenko, V. O. (Russia)
Pisarenko VOP-2

de Pischoff, Emil (France/Austria)
Emil de Pischoff designed and flew his first aircraft in 1907. In 1908 he joined with Paul Kœchlin (Koechlin) to form **Ateliers de Pischoff-Kœchlin.** In 1910 the company dissolved into **Aéroplanes P. Kœchlin** and **Établissements Autoplan.** Kœchlin's joint designs are listed under **Pischoff-Kœchlin.** The individual designs of de Pischoff (including Autoplan) and Kœchlin are under **Pischoff** or **Kœchlin.**
de Pischoff 1907 Biplane Glider

de Pischoff No.1 (1907 Biplane)
de Pischoff Autoplan Family (PWP - Pischoff-Werner-Pfederer)
de Pischoff 1920 Avionette
de Pischoff Estafette (1921 Avionette)

de Pischoff-Kœchlin (Ateliers de Pischoff-Kœchlin) (France)
de Pischoff-Kœchlin 1908 Tandem Wing Monoplane

Pitcairn (Pitcairn-Cierva) (Willow Grove, PA)
In 1924, Harold Frederick Pitcairn (1897-1960) founded two companies: **Pitcairn Aircraft** for the design and construction of aircraft and for research into rotary wing flight, and **Pitcairn Aviation** for flight operations. The two companies were incorporated in 1926, and a third company, **Pitcairn Aeronautics**, took over helicopter and rotary wing research. In 1929, Pitcairn purchased the US rights to Juan de la Cierva Codorníu's autogyro patents, proposing to cross-license Pitcairn rotary wing patents with Cierva's company. Pitcairn Aeronautics was then renamed **Pitcairn-Cierva Autogiro Company of America (PCA)** and, in 1931, the **Autogiro Company of America**. Pitcairn Aircraft was renamed **Pitcairn Autogiro Company, Inc** in 1933. Pitcairn Aircraft, Pitcairn Autogiro, Autogiro Company of America, Pitcairn-Cierva, and Pitcairn Autogiro Company designs are all listed under **Pitcairn**. See also **Pitcairn-Alfaro** and **Pitcairn-Larsen**.
Pitcairn AC-35 Roadable Autogiro
Pitcairn AC-35 Roadable Autogiro, NASM
Pitcairn YG-2 (PA-33) Autogiro
Pitcairn XOP-1 (PCA-2) Autogiro
Pitcairn XOP-2 (PA-34) Autogiro
Pitcairn PA-1 Fleetwing
Pitcairn PA-2 Sesqui-Wing
Pitcairn PA-3 Orowing
Pitcairn PA-3A Orowing II
Pitcairn PA-4 Fleetwing II (Fleetwing Deluxe)
Pitcairn PA-5 Mailwing
Pitcairn PA-5 Mailwing, NASM
Pitcairn PA-5 Sport Mailwing
Pitcairn PA-6 Sport Mailwing
Pitcairn PA-6 Super Mailwing
Pitcairn PA-6B Sport Mailwing
Pitcairn PA-7 Sport Mailwing
Pitcairn PA-7 Super Mailwing
Pitcairn PA-7S Super Mailwing
Pitcairn PA-8 Super Mailwing
Pitcairn PA-16 Cabin Autogiro Study
Pitcairn PA-17 Autogiro Study
Pitcairn PA-18 Autogiro
Pitcairn PA-19 Cabin Autogiro
Pitcairn PA-20 Autogiro
Pitcairn PA-21 Autogiro
Pitcairn PA-22 Cabin Autogiro
Pitcairn PA-23 Autogiro Study
Pitcairn PA-24 Autogiro
Pitcairn PA-25 Design Study
Pitcairn PA-26 Cabin Autogiro Study
Pitcairn PA-27 Design Study
Pitcairn PA-28 Design Study
Pitcairn PA-29 Cabin Autogiro Study
Pitcairn PA-30 Cabin Autogiro Study
Pitcairn PA-31 Ambulance Autogiro Study
Pitcairn PA-32 Autogiro Study
Pitcairn PA-36 Whirl Wing
Pitcairn PA-37 Autogiro Study
Pitcairn PA-38 Autogiro Study
Pitcairn PA-39 (See: Pitcairn-Larsen PA-39)
Pitcairn PA-40 Training Aircraft
Pitcairn PA-41 Autogiro Study
Pitcairn PA-42 Autogiro Study
Pitcairn PA-43 Ambulance Autogiro
Pitcairn PAA-1 Autogiro
Pitcairn PAA-2 Autogiro
Pitcairn PCA-1A Autogiro
Pitcairn PCA-1A Autogiro, NASM
Pitcairn PCA-1B Autogiro
Pitcairn PCA-2 Autogiro
Pitcairn PCA-2-30 Autogiro (See: Pitcairn-Alfaro PCA-2-30)
Pitcairn PCA-3 Autogiro
Pitcairn PCA-4 Autogiro Study

Pitcairn-Alfaro (Cleveland, OH)
The PCA-2-30 was a one of a kind experimental autogiro commissioned by Harold Pitcairn and built by Heraclio Alfaro in 1930. It is not related to the Pitcairn PCA-2.
Pitcairn-Alfaro PCA-2-30

Pitcairn-Cierva *(See: Pitcairn)*

Pitcairn-Larsen Autogiro Co (Willow Grove, PA)
In 1940, production at **Pitcairn Autogiro** had ceased when longtime Pitcairn designer Agnew Larsen formed **Pitcairn-Larsen** to produce the PA-39 Autogiro for the Royal Navy. The PA-39 was a highly-modified, refurbished Pitcairn PA-18.
Pitcairn-Larsen PA-39 Autogiro

Pitts Aviation Enterprises Inc (Homestead, FL; Afton, WY)
Pitts Formula One Racer (Mid-Wing)
Pitts Formula One Racer (Low-Wing)
Pitts S-1 Special
Pitts S-1 Special "Little Stinker," NASM
Pitts S-1C Special
Pitts S-1D Special
Pitts S-1L Special
Pitts S-1S Special
Pitts S-1S Special "Maryann," NASM
Pitts S-1T Special
Pitts S-1T Special, Clipped Wing
Pitts S-2A Special
Pitts S-2C
Pitts S-2S Special
Pitts Samson
Pitts Samson, Steve Wolf Replica

Pittsburgh Metal Airplane Co *(See: Thaden Metal Aircraft Co)*

Pittsburgh School of Aviation (Pittsburgh, PA)
Pittsburgh Mac Jr

Pivot (France) *(See also: Kœchlin)*
Pivot 1910 Monoplane

PKZ *(See: Petroczy-Karman-Zurcvec)*
Plage & Laskiewicz *(See: Lublin)*

Plage, Emil (Poland)
Plage Torpedo Monoplane

Plan, Max (France)
Plan P.F. 204
Plan P.F. 214

Plane-Mobile (Zuck-Whitakaker) (Los Angeles, CA)
Plane-Mobile Roadable I
Plane-Mobile Roadable II (Design)

Planes Ltd (UK)
Planes Ltd 1910 Biplane (See: Thompson, William P.)

Planet Aircraft Ltd (UK)
Planet Satellite

Platt, Haviland Hull (USA)
Platt Cyclogiro

Platt-LePage Aircraft Co (Eddystone, PA)
Platt-LePage XR-1
Platt-LePage XR-1, NASM
Platt-LePage XR-1A
Platt-LePage Rotor-Prop

Plazzeriaud, M. (Switzerland)
Plazzeriaud 1909 Triplane

Pliska, John V. *(See: Coggin & Pliska)*

PLV *(See: Vought, Chauncey)*

Pober (Paul H. Poberezney) (Hales Corners, WI)
Pober Pixie
Pober Sport

Pocard (France)
Pocard 1923 Monoplane Glider

de Poix et de Roig (France)
de Poix et de Roig 1912 Two-Seat Monoplane

Polikarpov (Russia)
Polikarpov BDP (S-1)
Polikarpov BDP-2 (Glider)
Polikarpov DI-2
Polikarpov DIT-1 (DIT-2)
Polikarpov I-1 (IL-400)
Polikarpov I-3
Polikarpov I-5 (VT-12)
Polikarpov I-6
Polikarpov I-15 (TsKB-3)
Polikarpov I-152 (I-15*bis*, TsKB-3*bis*)
Polikarpov I-153 (I-15*ter*)
Polikarpov I-153DM Ramjet Modification
Polikarpov I-16 (TsKB-12)
Polikarpov I-17 (TsKB-15, TsKB-19, TsKB-33)
Polikarpov I-180
Polikarpov I-185
Polikarpov Ivanov
Polikarpov N-1 (ITP)
Polikarpov NB Type T
Polikarpov NPI/SVB (TsKB-44/48, SPB)
Polikarpov P-2
Polikarpov P-5
Polikarpov PL-5
Polikarpov R-1
Polikarpov R-4
Polikarpov R-5
Polikarpov S-1
Polikarpov S-2
Polikarpov TB-2
Polikarpov TIS Type A
Polikarpov U-2 (Po-2)
Polikarpov U-2S (Po-2S)
Polikarpov VI-1 (2I-N1)

Polish Airmen's Assoc (Poland)
Polish Airmen's Assoc Mechanik (Mechanic) Glider

Poll Giant *(See: Brüning Giant Triplane Transport)*

Pollay Brothers (Polly Brothers) (Canada)
Pollay 1914 Biplane

Polliwagen, Inc (Garden Grove, CA)
Polliwagen A
Polliwagen B
Polliwagen C

Polson Ironworks (Canada)
Polson M.F.P. Warplane

Polyteknikkojen Ilmailukerho (PIK) *(See: Eiri)*

Pomilio (Fabrica Aeroplani Ing. O. Pomilio, Italy; Pomilio Brothers Corp, New York, NY)
Pomilio Alpha
Pomilio BVL-12 (Bomber-Victory-Liberty-12)
Pomilio C.1 (PC.1)
Pomilio D (PD)
Pomilio E (PE)
Pomilio F
Pomilio F*bis*
Pomilio FVL-8 (Fighter-Victory-Liberty-8) (PVL-8)
Pomilio G
Pomilio Gamma
Pomilio PY
Pomilio SP.1 (See: SIA)
Pomilio SP.2 (Savoia-Pomilio)
Pomilio SP.3 (Savoia-Pomilio)
Pomilio SP.4 (Savoia-Pomilio)
Pomilio Zeta
Pomilio Zeta*bis*

Ponche et Primard Tubavion *(See: Tubavion)*

Poncelet (Belgium)
In 1925 Poncelet and Demonty joined to form **SABCA**; see each company for appropriate listings.
Poncelet 1923 Sailplane
Poncelet 1924 Moto Aviette

Ponnier (A. Ponnier) (France)
Ponnier Biplane I (Biplace Métallique)
Ponnier D.III Monoplane
Ponnier D.VIII Monoplane
Ponnier F.5
Ponnier L.1
Ponnier M.1
Ponnier M.2
Ponnier P.1
Ponnier 1916 Pusher

Pontkowski-Lichorsik Machine Co *(See: Vought, Chauncey)*

Pontkowski-Lichorsik-Vought *(See: Vought, Chauncey)*

Poop Deck (USA)
Poop Deck 2 (Homebuilt)

Pope, Leon M. (Plymouth, MI)
Pope (Leon) Thunderbird

Pope, Malcolm (USA)
Pope (Malcolm) Rocket Glider (1930)

Popov, M. I. (Russia)
Popov MP-5
Popov MP-6
Popov-1
Popov VII

Popular Mechanics *Magazine (USA)*
Popular Mechanics Teenie Two (See: Parker, Calvin Y.)
Popular Mechanics 1909 Glider

Porokhovshchikov, A. A. (Russia)
Porokhovshchikov IV
Porokhovshchikov 6

Port Victoria (RNAS Experimental Construction Depot, Isle of Grain) (UK)
Port Victoria Grain Griffin
Port Victoria PV.1
Port Victoria PV.2
Port Victoria PV.4
Port Victoria PV.5A
Port Victoria PV.7, PV.8
Port Victoria PV.7 Grain Kitten
Port Victoria PV.8 Eastchurch Kitten
Port Victoria PV.9

Porte *(See: Felixstowe)*

Porterfield Aircraft Corp (Kansas City, MO)

In 1934 Ed Porterfield, ex-President of **American Eagle Aircraft Corp**, established the **Porterfield Aircraft Corp** to manufacture a light aircraft. In March 1939 Roscoe Turner joined the company as vice president. Although the company name was not changed, aircraft where labeled as "Porterfield-Turner Airplanes." The company undertook subcontract work during World War II until August 1943 when Porterfield sold the powered-aircraft interests to the **Columbia Aircraft Corp** and glider subcontract facilities to the **Ward Manufacturing Co.** All Porterfield products are listed under **Porterfield**.

Porterfield Model 35 Flyabout
Porterfield CP-40 Zephyr Prototype
Porterfield CP-40 Zephyr
Porterfield Model 50 Collegiate (CP-50, FP-50, LP-50)
Porterfield Model 65 Collegiate (CP-65, FP-65, LP-65)
Porterfield Model 65, "Sweet P Field"
Porterfield CP-75 Deluxe
Porterfield C-145 (PT-125)

Portsmouth Aviation Ltd (UK)

Portsmouth Aerocar

Posnansky, Hernan (USA)

Posnansky PF-1 White Knight

***Post** (See: Wiley Post Aircraft Corp)*

Potez (Société des Aéroplanes Henry Potez) (France)

In 1916 Henri Potez and Marcel Bloch established **Société d'Études Aéronautiques (SEA)**. In 1918 Potez, Bloch, and Basin formed **Anjou Aéronautique** to produce SEA's first successful aircraft. When contracts were canceled after World War I, Anjou was liquidated. Potez immediately established **Société Anonyme des Aéroplanes Henri Potez** and began to convert SEA airframes to civil use.

In 1933, Potez acquired **Chantiers Aéro-Maritimes de la Seine (CAMS)**, which he continued to operate under its own name. With the January 1937 nationalization of the French aviation industry, Potez became part of **Société Nationale des Constructions Aéronautiques du Nord (SNCAN)**, and CAMS was split between SNCAN and **Société Nationa! de Constructions Aéronautiques de Sud-Est (SNCASE)**. Without aircraft to build, Potez concentrated on engine development; after World War II the company was renamed **Société des Moteurs Potez**. In 1952, Potez returned to aircraft design, and the company was renamed **Société des Avions et Moteurs Henri Potez**. In May 1958 Potez purchased **Air Fouga**, which became **Potez Air Fouga** until absorbed into the main Potez firm in September 1961. The company failed in 1967, and its factories were taken over by **Sud-Aviation**.

For SEA and CAMS products, see **SEA** and **CAMS**. All Potez designs, including those built by SNCAN and SNCASE, appear under **Potez**. Designs by Fouga, Air Fouga, and Potez Air Fouga, are under **Fouga**.

Potez IV (See: SEA IV; see also: Potez VII)
Potez VII (7) (See also: SEA IV)
Potez VIII (8) Family
Potez VIII P Parasol Sailplane
Potez IX (9) Limousine
Potez X Bp2 (10)
Potez X A Colonial
Potez XI Cap2 (11)
Potez XII (12) Racer
Potez XIV (14)
Potez XV A2 (15)
Potez XV S (15 S)
Potez XVI C1 (16)
Potez XVII (17) (Bulgaria)
Potez XVIII (18)
Potez 19 Bp2
Potez 19 BpR3
Potez 21
Potez 22 Bp2
Potez 220 A3
Potez 23 C1
Potez 230 C1
Potez 24
Potez 25
Potez 25 A2
Potez 25 B2
Potez 25 Et2
Potez 25 GR
Potez 25 M (Monoplane)
Potez 25 O (Ocean)
Potez 25 Postal
Potez 25 TOE (Théatres des Opérations Extérieures)
Potez 25/5
Potez 25/55
Potez 25/67
Potez 26 C1
Potez 27 A2
Potez 28 B3
Potez 28 M
Potez 28/2
Potez 29 T7
Potez 29/2
Potez 29/4 T7
Potez 29/5 T7
Potez 29/6
Potez 29/8 T7
Potez 29/11
Potez 30
Potez 31 C-N2
Potez 32 T7
Potez 32/2 T7
Potez 32/3 T7
Potez 32/4 T7
Potez 32/5
Potez 33/1 T7
Potez 33/2 T7
Potez 33/3 T7
Potez 33/4 T7
Potez 34 R2
Potez 35 M B3
Potez 36 T2
Potez 36/3 T2
Potez 36/4 T2
Potez 36/5 T2
Potez 36/6 Tn2
Potez 36/10 T2
Potez 36/11 T2
Potez 36/13 T2
Potez 36/14 T2
Potez 36/15 T2
Potez 36/17 T2
Potez 36/19 T2
Potez 36/21 T2
Potez 36/23 T2
Potez 36/30 Tn2
Potez 36/31 Tn2
Potez 370 R2
Potez 371 Tn2
Potez 38 T11
Potez 39/0 A2
Potez 39/1 A-C2
Potez 39/1*bis* A-C2
Potez 39/2 A-B-CN2 (39/22)
Potez 39/2*bis* A2
Potez 39/3 A-B-CN2
Potez 39/3*bis* A2
Potez 39/10 R2
Potez 400 Colonial
Potez 401 Colonial
Potez 402 Colonial
Potez 403 T10
Potez 41 Bn5

Potez 41 E
Potez 420
Potez 421
Potez 430 T2/3
Potez 431 T3
Potez 432 T3
Potez 434 T2/3
Potez 435 T3
Potez 436 T3
Potez 437 T2
Potez 438 A-T.n.3
Potez 439 T3
Potez 439A T3
Potez 450
Potez 452
Potez 453 C1
Potez 490 R-B2
Potez 50 A2
Potez 501
Potez 502 A2
Potez 503
Potez 506
Potez 51 ET2
Potez 53 Racer
Potez 532 Racer
Potez 533 Racer
Potez 540
Potez 541
Potez 542
Potez 543
Potez 544
Potez 56 E
Potez 560 T3
Potez 561 T3
Potez 566 T3
Potez 567
Potez 568 P3
Potez 580 T3
Potez 582 T4
Potez 584 T3
Potez 585 T3
Potez 586 T4
Potez 59
Potez 600 T2 Sauterelle (Grasshopper)
Potez 60/43
Potez 620 T16
Potez 621 T7
Potez 630
Potez 631
Potez 632
Potez 633
Potez 634 Tn2
Potez 636 C3
Potez 637 A3
Potez 639 A2
Potez 63/11 R3
Potez 63/12
Potez 63/13
Potez 63/16
Potez 650
Potez 661 T14
Potez 662 T14
Potez 670 C2/3
Potez 671 C2/3
Potez 700
Potez 75
Potez 840
Potez-CAMS 110 (See: CAMS)
Potez-CAMS 120 (See: CAMS)
Potez-CAMS 141 (See: CAMS)
Potez-CAMS 160 (See: CAMS)
Potez-CAMS 161 (See: CAMS)
Potez-CAMS 170 (See: CAMS)
Potez-Heinkel CM 191 (See: Fouga)

Potter, James W. (Pasadena, CA)
Potter Experimental Sport

Poulain (France)
Poulain Orange Biplane

Powell, C. H. (Detroit, MI)
Powell (C. H.) P.H. Racer (1925)
Powell (C. H.) P.H.2

Powell, John C. (Middletown, RI)
Powell (J. C.) Acey Deucy (Homebuilt)

Powers, George W. (Des Moines, IA)
Powers (G. W.) 1927 Biplane

Powers, William C. (Mobile, AL)
Powers (W. C.) Helicopter (1862)

Praga (Ceskomoravská-Kolben-Denek SA) (Czechoslovakia)
Praga BH-36
Praga E 36
Praga BH-39 AG
Praga BH-39 G
Praga BH-39 NZ
Praga E 39
Praga E 39 AG
Praga E 39 G
Praga E 39 NZ
Praga E 40
Praga BH-41
Praga E 41
Praga BH-44
Praga E 44
Praga E 45
Praga E 46
Praga E 48
Praga E 49
Praga E 51
Praga E 52
Praga E 60
Praga BH-111
Praga E 114 B
Praga E 114 D
Praga E 115
Praga E 117
Praga E 139
Praga E 140
Praga E 141
Praga E 144
Praga E 210
Praga E 211
Praga E 214
Praga E 215
Praga E 240
Praga E 241
Praga E 244
Praga E 341
Praga E 344

Pratt, Read & Co (Deep River, CT)
Pratt Read LBE-1 ("Glomb")
Pratt Read XLNE-1
Pratt Read LNE-1
Pratt Read PR-G1
Pratt Read TG-32

Prauss, Stanislaw (Poland)
Prauss P.S.1

P.R.B. *(See: Bastianelli)*

Preiss, Henry (Canada)
Preiss RHJ-7 (Converted Schreder HP-14)
Preiss RHJ-8
Preiss RHJ-9

Prescott Aeronautical Corp (Wichita, KS)
Prescott Pusher
Prescott Pusher II

Prest Airplane & Motors (Arlington, CA)
Prest Baby Pursuit

Prestwich (J. A. Prestwich & Co Ltd) *(See: J.A.P.)*

Prestwick *(See: Scottish Aviation)*

Preti *(See: Instituto di Aeronautica Politecnico di Milano; See also: Agusta)*

Prewitt Aircraft Co (Clifton Heights, PA)
Prewitt Rotorchute

Pride, Christopher (UK)
Pride 1909 Monoplane

Priesel, Guido (Austria-Hungary)
Priesel KEP (Kampfeinsitzer Priesel) Fighter

Prigent
Prigent Flying Machine (c.1700s)

Prikryl-Blecha *(See: Ardea)*

Princeton University (Princeton, NJ)
Princeton University X-3A GEM Ground Effect Machine
Princeton University X-3B GEM Ground Effect Machine
Princeton University X-3C GEM Ground Effect Machine

Prini-Berthaud *(See: Berthaud)*

Private Explorer, Inc (Grangeville, ID)
Private Explorer

Privateer *(See: Ireland)*

Procaer (Progetti Costruzioni Aeronautiche (Italy)
Procaer F.15/B
Procaer F.400 Cobra
Procaer Picchio F.15

Project Ornithopter, Inc (Canada)
Project Ornithopter Model 1 "Big Flapper"

Pruett/Cowan (D. W. Pruett & Charles Virgil Cowan) (Sacramento, CA)
Pruett/Cowan CAC 1 (Cowan Aviation Co)

Progetti Costruzioni Aeronautiche SpA *(See: Procaer)*

Prop and Wing Degree, Kenilworth Aircraft Club *(See: Driggers, Willard R.)*

Prudden, George Henry (San Diego, CA)
George Henry Prudden founded **Prudden-San Diego Airplane Co** in November 1927 and incorporated **Prudden Aircraft Corp** in August 1928. Prudden left in November 1928, moving to the **Atlanta Aircraft Corp**. In April 1929, Prudden Aircraft Corp became **Solar Aircraft Co** and Prudden-San Diego was liquidated.
Prudden-San Diego:
SE-1 Special Monoplane
Tri-Motor Transport
Prudden-Whitehead Low Wing Tri-Motor (See: Atlanta Aircraft)

Prue, Irving (USA)
Prue 215
Prue IIA
Prue UHP-1
Prue Super Standard

Pruett Pusher (John Pruett) (Crosby, MS)
Pruett Pusher JP-1 (1960)

Pruitt, Kenneth D. (Albuquerque, NM)
Pruitt B-31C

Pterodactyl Ltd (Watsonville, CA)
Pterodactyl Acro-Dactyl
Pterodactyl Ascender
Pterodactyl Ascender II
Pterodactyl Ascender II+
Pterodactyl Ascender II+2
Pterodactyl Fledgling
Pterodactyl Fledgling (Pfledge) Prototype
Pterodactyl Fledgling (Pfledge) OR (Oshkosh Replica)
Pterodactyl Fledgling (Pfledge) X
Pterodactyl Fledgling (Pfledge) 430D
Pterodactyl Light Flyer
Pterodactyl NFL (Not Foot Launchable)
Pterodactyl Ptiger
Pterodactyl Ptraveler
Pterodactyl Ptug

Puget Pacific Planes Inc Wheelair *(See: Wheelair)*

Pulsar *(See: Aero Designs Pulsar)*

Pulawski, Zygmunt (Poland)
Pulawski S.L.3 Glider

Pusterla, Attilio (Bathbeach, NY)
Pusterla Aeroplane (1909)

Putilov, Aleksandr Ivanovich (Russia)
Putilov Stal-2 (Steel)
Putilov Stal-3 (Steel)
Putilov Stal-5 (Steel)
Putilov Stal-11 (Steel)
Putilov Stal-12 (Steel)

Pützer (Alfons Pützer K.G. Flugzugbau) (Germany)
Pützer Elster-B (Magpie-B)
Pützer Moraa

PWS (Podlaska Wytwórnia Samolotów - Podlasian Aircraft 14) (Poland)
PWS Sep (Vulture) Glider
PWS 1
PWS 2
PWS 3
PWS 4
PWS 5a
PWS $5t_2$
PWS 6
PWS 7
PWS 8
PWS 8*bis*
PWS 10
PWS 10M
PWS 11
PWS 11*bis*
PWS 11S.M.
PWS 12
PWS 12*bis*
PWS 12S.P.
PWS 14
PWS 15
PWS 16
PWS 16*bis*
PWS 17
PWS 18
PWS 19
PWS 20
PWS 20*bis*
PWS 20T
PWS 20*ter*
PWS 21
PWS 21*bis*
PWS 23T
PWS 24
PWS 24*bis*
PWS 24T
PWS 24WT
PWS 26
PWS 27
PWS 28
PWS 33 Wyzel (Pointer)
PWS 35 Ogar (Hound)
PWS 36
PWS 37
PWS 40
PWS 41
PWS 42 Sokól (Falcon)
PWS 46 Ciolkosz Bomber Project
PWS 49
PWS 50
PWS 51

PWS 52
PWS 54
PWS 60
PWS 61
PWS 62
PWS 101 Glider
PWS 102 Redkin (Shark) Glider
PWS 103 Glider

PZL (Panstwowe Zaklady Lotnicze: National Aviation Establishments) (Poland)

PZL P.1
PZL L.2
PZL.3
PZL.4
PZL.5
PZL P.6
PZL P.7
PZL P.8
PZL P.9
PZL P.11
PZL.12 (PZL-H)
PZL.16
PZL.18
PZL.19
PZL.22
PZL P.23 Karas (Carp)
PZL P.24
PZL.26
PZL.27
PZL.30
PZL P.37 Los (Elk)
PZL P.38 Wilk (Wolf)
PZL P.42 Karas (Carp)
PZL P.43 Karas (Carp)
PZL.44 Wicher (Gale)
PZL P.45 Sokól (Falcon)
PZL P.46 Sum (Sheat-fish)
PZL P.48 Lampart (Leopard)
PZL P.49 Mis (Teddy Bear)
PZL P.50 Jastrzab (Hawk)
PZL P.62 Dabrowski Fighter Project

PZL Bielsko (Poland)

PZL Bielsko SZD 22C Mucha Standard
PZL Bielsko SZD 24C Foka
PZL Bielsko SZD 30 Pirat (Pirate)
PZL Bielsko SZD 36 Cobra
PZL Bielsko SZD 38A Jantar 1
PZL Bielsko SZD 41 Jantar Standard
PZL Bielsko SZD 42 Jantar 2
PZL Bielsko SZD 42A Jantar 2B
PZL Bielsko SZD 45 Ogar
PZL Bielsko SZD 48-2 Jantar Standard 2
PZL Bielsko SZD 48-3 Jantar Standard 3
PZL Bielsko SZD 50-1 Puchacz
PZL Bielsko SZD 50-3 Puchacz
PZL Bielsko SZD 51-1 Junior
PZL Bielsko SZD 51-2 Junior
PZL Bielsko SZD 55-1
PZL Bielsko SZD 59 Acro

PZL Krosno (Poland)

PZL Krosno KR 03A Puchatek

PZL Mielec (Wytwórmia Sprzetu Komunika-cyjnego PZL Mielec SA) (Poland)

PZL Mielec SBLim-1 (MiG-15UTI)
PZL Mielec SBLim-1A (MiG-15UTI)
PZL Mielec Lim-2 (MiG-15*bis*)
PZL Mielec Lim-5 (Mig-17F)
PZL Mielec Lim-5M
PZL Mielec Lim-5P (Mig-17PF)
PZL Mielec Lim-5R
PZL Mielec Lim-6*bis* (MiG-17)
PZL Mielec Lim-6*bis*R
PZL Mielec Lim-6M
PZL Mielec Lim-6MR
PZL Mielec M4 Tarpan
PZL Mielec M18 Dromader (Dromedary)
PZL Mielec M20 Mewa (Gull)
PZL Mielec M26 Iskierra (Little Spark)
PZL Mielec M28 Skytruck
PZL Mielec M93
PZL Mielec M96 Iryda (Iridium)
PZL Mielec TS-11 Iskra (Spark)
PZL Mielec TS-11*bis* Iskra (Spark)

PZL Swidnik SA (Zygmunta Pulawskiego-PZL Swidnik) (Poland)

PZL Swidnik Kania (Kitty Hawk)
PZL Swidnik PW-5 Smyk
PZL Swidnik V-2 *Hoplite* (Mi-2)
PZL Swidnik W-3 Sokól (Falcon)

PZL Warsawa-Okecie (Poland)

PZL Warsawa-Okecie:
PZL-101 Gawron
PZL-102B Kos (Blackbird)
PZL-104 Wilga 35 (Oriole)
PZL-104 Classic Wilga (Oriole) 80
PZL-105L Flaming (Flamingo)
PZL-106 Kruk
PZL-110 Koliber (Hummingbird)
PZL-111 Koliber (Hummingbird) Senior
PZL-112 Koliber (Hummingbird) Junior
PZL-126P Mrówka (Ant)
PZL-130 Orlik (Spotted Eaglet)
PZL-140 Orlik (Spotted Eaglet)
PZL-240 Pelikan (Pelican)

Q

Qantas (Queensland & Northern Territory Aerial Services, Ltd) (Australia)
Qantas D.H.50J (License-Built)

Quad City Ultralight Aircraft Corp (Moline, IL)
Quad City Challenger
Quad City Challenger II

Quaissard (France)
Quaissard GQ.01 Monogast

Queen Aeroplane Co (New York, NY)
Queen Aeroboat (Queen 1912 Flying Boat) (Loening Design)
Queen Blériot-Type, 1911
Queen Twin Monoplane, 1911 (Wilis McCormick Design)
Queen-Martin Biplane, 1911 (J. V. Martin Design)
Queen McCurdy 1911 Biplane

Queirolo, Luigi (Italy)
Queirolo A.22
Queirolo QR.2
Queirolo QR.2*bis*
Queirolo QR.14 (See: CAT)

Questair Inc (Greensboro, NC)
Questair Spirit
Questair Venture

Quetzalcoatlus Northropi (See: MacCready)

Quick Aeroplane Dusters (Houston, TX)
Quick (TX) Special (Standard J-1 Ag Mod)

Quick Flight (USA)
Quick Flight Quiet Bird

Quick, William Lafayette (Carmichael, AL)
Quick (Wm) 1904 Monoplane

Quick, William & Joseph (Huntsville AL)
Quick (W & J) 1908 Monoplane

Quickie Aircraft Corp (Mojave, CA)
Quickie QAC-1
Quickie Q-2 (QAC-2)
Quickie QAC-3
Quickie Q-200

Quicksilver Enterprises Inc (Temecula, CA)
Quicksilver CT 400
Quicksilver GT 500
Quicksilver MX Sport
Quicksilver MX Sprint

Quiet Short Haul Research Aircraft (QSRA, QSTOL) (See: NASA)

Quikkit Division of Rainbow Flyers, Inc (Dallas, TX)
Quikkit Glass Goose

R

***R. J. Enstrom Corp** (See: Enstrom Corp)*

***R. W. Rambler** (See: Rinehart & Whalen)*

Raab-Katzenstein Flugzeugwerk GmbH (Germany)
Raab-Katzenstein:
Glider
F 1 Tigerschwalbe (Tiger Swallow)
Kl I (K I) Schwalbe (Swallow)
L.U.G.C. VI
RK 2 Pelikan (Pelican)
RK 6
RK 7
RK 8
RK 9 Grasmücke(MeadowWarbler)
RK 12
RK 25 Erka
RK 26
RK 29
RK 32

Radab (Sweden)
Radab Windex 1200C

Rader, Homer John (Fort Worth, TX)
Rader Sport (1928)

Radley, James (UK)
Radley 1911 Monoplane

Radley-England (James Radley & E. C. Gordon England) (UK)
Radley-England Waterplane 1
Radley-England Waterplane 2
Radley-England Shoreham (1913)

Radley-Moorhouse (James Radley & W. B. Rhodes-Moorhouse) (UK)
Radley-Moorhouse 1912 Monoplane

R.A.E. Aero Club (Royal Aircraft Establishment, Farnborough) (UK)
R.A.E. Hurricane
R.A.E. Scarab
R.A.E. Scirocco
R.A.E. Zephyr

***RAF** (See: Royal Aircraft Factory)*

RAF College Halton (Royal Air Force College Halton) (UK)
RAF College Halton Jupiter Human-Powered Aircraft

Rafaelyants, Aram Nazarovich (Russia)
Rafaelyants PR-5
Rafaelyants PR-12
Rafaelyants RAF-1
Rafaelyants RAF-2
Rafaelyants RAF-11

***Rainbow Flyers, Inc** (See: Quikkit)*

Ralls, Earl M. (Sacramento, CA)
Ralls Airship

Ralston, Walter (Beaumont, TX)
Ralston 1st Aircraft
Ralston 2nd Aircraft, 1913 Biplane
Ralston 3rd Aircraft, Kemp Motor Monoplane

Ramsey Aircraft (Minneapolis, MN)
Ramsey Flying Bathtub (Ultralight)

Rand Robinson Engineering Inc (Huntington Beach, CA)
Rand Robinson KR-1
Rand Robinson KR-1B
Rand Robinson KR-2
Rand Robinson KR-3 Amphibian

Ranger Aircraft Corp (Oklahoma City, OK)
Ranger Model A (See: Coffman)
Ranger Model W

Rankin Brothers (Alexander, Charles, & Arch Rankin) (Northville, MI)
Rankin Brothers R-B-M 3 (R-B-M 4)

Rans Inc (Hays, KS)
Rans S-4 Coyote
Rans S-5 Coyote
Rans S-6 Coyote II
Rans S-7 Courier
Rans S-9 Chaos
Rans S-10 Sakota
Rans S-11 Pursuit
Rans S-12 Airaile
Rans S-14 Airaile
Rans S-16 Shekari

Rasmussen, Christian (Argentina)
Rasmussen 1910 Monoplane

***Ratmanoff & Co (France)** (See also: Beer)*
Ratmanoff 1912 Canard (See: Drzewiecki)
Ratmanoff 1913 Monoplane Trainer

Ravaud, Roger (France)
Ravaud Aeroscaphe

Rawdon Brothers Aircraft, Inc (Wichita, KS)
Rawdon T-1

Raytheon Aircraft Co (Wichita, KS)

Raytheon Co's 1980 purchase of **Beech Aircraft Co** left both companies' names unchanged. At the time, **Beechcraft Hawker Corp**, a Beech subsidiary, was aiding in the development and marketing of the Hawker Siddeley HS.125. (This alliance survived both Hawker Siddeley's 1977 absorption into **British Aerospace Corp** and Beech's purchase by **Raytheon.**) In August 1993, Raytheon purchased the **British Aerospace Corporate Jet Division**, which it then named **Raytheon Corporate Jets**. A year later, Beech and Raytheon Corporate Jets merged to from **Raytheon Aircraft Co.** All original Beech designs (and their subsequent development) are listed under **Beech.** HS.125 aircraft through the Series 700 are listed under **Hawker Siddeley**; Series 800 through 1000 appear under **British Aerospace.** Later HS.125 variants (including the Hawker Horizon) and all post-1994 designs are listed under **Raytheon.**

Raytheon Hawker 4000 Horizon
Raytheon Premier I
Raytheon Premier II
Raytheon Premier III
Raytheon T-6A Texan II (See: Beech)

Raz-Mut *(See: St Germain)*

Read, Robert C. (Wilton, CT)
Read 1946 Transport Patent

Rearwin Airplanes, Inc (Kansas City, KS)
Rearwin Cloudster (Models 8090, 8125, 8135)
Rearwin Junior (Models 3000, 4000)
Rearwin Ken-Royce (Model 2000C)
Rearwin Ranger (Skyranger) (Models 165, 175, 185, 190)
Rearwin Speedster (Model 6000)
Rearwin Sportster (Models 7000, 8500, 9000)

Rectenwald, John J. (Pittsburg, PA)
Rectenwald 1911 Aeroplane Pat nt

Red Ball Motor Truck Corp (Indianapolis, IN)
Red Ball Type A Biplane (1927)

Red Bird Aircraft Corp (Bern, KS)
Red Bird Model 28

Redshaw (USA)
Redshaw 1917 Articulated Fuselage Design

Redwing Aircraft Ltd *(See: Robinson Aircraft Co)*

Reece, George D. S. (St Louis, MO)
Reece 1910 Airship Patent

Reed Hastings Aeronautics Co *(See: Hastings)*

Reedley Glider Club (Reedley, CA)
Reedley Glider Club Monoplane Glider

Regent Aircraft Inc (McAllen, TX) *(See also: Johnson Aircraft Corp)*
Regent Rocket
Regent Rocket 260
Regent Rocket 400

Reggiane (Officine Meccaniche Reggiane SA, Caproni) (Italy)
Reggiane CA 405 Procellaria
Reggiane P.32*bis* (See: Piaggio p.32-II)
Reggiane RE 2000 Falco I (Hawk)
Reggiane RE 2001 Falco II (Hawk)
Reggiane RE 2002 Ariete (Ram)
Reggiane RE 2003
Reggiane RE 2004
Reggiane RE 2005 Sagittario (Archer)
Reggiane RE 2006
Reggiane RE 2007

Reichelt, Hermann (Germany)
Reichelt Biplane (1910)
Reichelt Monoplane

Reid (Australia)
Reid (Australia) 1923 Biplane

Reid Aircraft Co (Canada)
Reid (Aircraft) (Curtiss-Reid) Rambler

Reid and Sigrist, Ltd (UK)
Reid and Sigrist Bobsleigh (RS.4)
Reid and Sigrist Desford (RS.3) Trainer
Reid and Sigrist Snargasher (RS.1) Trainer
Reid and Sigrist Trainer Prototype

Reinhard, Gerhard (Germany)
Reinhard Cumulus

Reinke, Ernest (Germany)
Reinke Convertiplane (Flying Auto)

Reissner, Hans von (Germany)
Reissner Ente (Duck)
Reissner 1907 Biplane (Voisin-Built)
Reissner 1912 Canard

Remington-Burnelli *(See: Burnelli)*

Remy, Adrien (France)
Remy Ocean Hydro-Glider

Renard Motor & Airplane Co (Alfred Renard) (Belgium)
Renard Epervier (Sparrowhawk)
Renard R-17
Renard R-30
Renard R-31
Renard R-32
Renard R-33
Renard R-34
Renard R-35
Renard R-36
Renard R-37
Renard R-38
Renard R-40
Renard R-44
Renard R-46

Renard, Charles (France)
Renard (Charles) Helicopter

Renault, Louis (France)
Renault O1 Bomber BN2 (1918)
Renault 1924 Single Seat Monoplane
Renault 1925 Scout

Renaux, A. J. (France)
Renaux 1784 Ornithopter

Renegade *(See: Lasher)*

Rensselaer Polytechnic Institute (RPI) (Troy, NY)
Rensselaer RP-1 Sailplane
Rensselaer RP-2 Glider
Rensselaer RP-3 Sailplane
Rensselaer RP-4

Rentel-Krylov (V. F. Rentel & V. Ya. Krylov) (Russia)
Rentel-Krylov MA-1 (LIG-9)

Rentel-Lisichkin (I. Lisichkin) (Russia)
Rentel-Lisichkin NIAI-1

R.E.P. (Robert Esnault-Pelterie) (France)

R.E.P. Biplane Glider (1904)
R.E.P. C1 Fighter (1918)
R.E.P. D (1911)
R.E.P. E-80 (1912)
R.E.P. F Two-Seater (1911)
R.E.P. F Three-Seater (1912)
R.E.P. I-80 (1913)
R.E.P. K Floatplane (1914)
R.E.P. K-80 (1913)
R.E.P. Monoplan de Cours (1911)
R.E.P. N
R.E.P. Parasol Vision Totale (1914)
R.E.P.1 (1907)
R.E.P.2 Monoplane (1908)
R.E.P.2*bis* (1909)
R.E.P.2*bis* Type B (1910)
R.E.P. Three-Seater (1911)

***Repair (Switzerland)** (See: F.W.A.)*

Replica Plans (Canada)

Replica Plans S.E.5a

Republic (Farmingdale, NY)

Following a 1939 reorganization, the **Seversky Aircraft Corp** was renamed the **Republic Aviation Corp**. In 1965, Republic became the **Republic Aviation Division** of **Fairchild Hiller Corp**. Reorganized again in 1971, the division became the **Fairchild Republic Co**, a subsidiary of **Fairchild Industries**. Designs first put into production by Seversky are listed under **Seversky**. Subsequent production of Seversky and Republic designs appears under **Republic**. Aircraft introduced by Fairchild Republic appear under **Fairchild**.

Republic (Sud) Alouette II
Republic AT-12 (See: Seversky)
Republic EP-1 (See: Seversky)
Republic XF-12 (R-12) Rainbow
Republic XF-84 (XP-84) Thunderjet
Republic XF-84 (XP-84) Thunderjet, NASM
Republic YF-84A (YP-84A) Thunderjet
Republic F-84B (P-84B) Thunderjet
Republic F-84C Thunderjet
Republic F-84D Thunderjet
Republic F-84E Thunderjet
Republic YF-84F (YF-96A) Thunderstreak
Republic F-84F Thunderstreak
Republic YRF-84F Thunderflash
Republic RF-84F Thunderflash
Republic F-84G Thunderjet
Republic XF-84H Thunderjet
Republic YF-84J Thunderjet
Republic RF-84K Thunderjet
Republic XF-91 Thunderceptor
Republic YF-96A Thunderstreak
Republic XF-103
Republic YF-105A Thunderchief
Republic YF-105B Thunderchief
Republic F-105B Thunderchief
Republic F-105D Thunderchief
Republic F-105D Thunderchief, NASM
Republic F-105F Thunderchief
Republic F-105G Thunderchief
Republic XF-108
Republic YP-43 Lancer
Republic P-43 Lancer
Republic P-43A Lancer
Republic P-43B Lancer
Republic P-43C Lancer
Republic P-43D Lancer
Republic P-43E Lancer
Republic XP-44 (AP-4J, AP-4L) Rocket (Warrior)
Republic XP-47B Thunderbolt
Republic P-47B Thunderbolt
Republic P-47C Thunderbolt
Republic P-47D (F-47D) Thunderbolt
Republic P-47D Thunderbolt, NASM
Republic XP-47E Thunderbolt
Republic XP-47F Thunderbolt
Republic P-47G Thunderbolt
Republic TP-47G Thunderbolt
Republic XP-47H Thunderbolt
Republic XP-47J Thunderbolt
Republic XP-47K Thunderbolt
Republic P-47M Thunderbolt
Republic XP-47N Thunderbolt
Republic P-47N (F-47N) Thunderbolt
Republic XP-69
Republic XP-72
Republic RC-2 Airliner
Republic RC-3 Seabee
Republic RC-3 Seabee, NASM

Réquillard (France)

Réquillard 1910 Monoplane

Research Aircraft

Materials in these files relate to the general subject of special research aircraft, including the US "X-planes," built after World War II. Specific aircraft and projects will continue to be filed under designer or manufacturer.

Resnier de Goué (France)

Resnier de Goué 1801 Flying Machine

Retz, Robert R. (Elon College, NC)

Retz R8 Special

***Revelation Aircraft Co** (See: Aviation Engineering School)*

Revolution Helicopter Corp, Inc (Excelsior Springs, MO)

Revolution HelicopterTalon Mini-500
Revolution Helicopter Voyager-500

Rex (Flugmaschine Rex GmbH) (Germany)

Rex 1915 Single-Seat Scout
Rex 1916 Single-Seat Scout
Rex 1917 Single-Seat Scout

Rey (Société des Avions F. J. Rey) (France)

Rey R.1

***RFB** (See: Rhein-Flugzeugbau)*

RFD Co (R. F. Dagnall) (UK)

RFD Primary Glider (Dagling)
RFD Target Glider

***RFW, Inc** (See: White Lightning Aircraft Corp)*

Rhett, Henry P. (Hempstead, NY)

Rhett 1911 Aeroplane Patent

Rhein-Flugzeugbau GmbH (RFB) (Germany)

RFB Fan Ranger (See: Rockwell/MBB)
RFB Fantrainer ATI-2
RFB Fantrainer AWI-2
RFB Fantrainer 400
RFB Fantrainer 600
RFB Fantrainer 800
RFB H40 (See: Hoffmann)
RFB RF-1 Channel-Wing STOL
RFB RW-3 Passat Multoplane
RFB (Lippisch) X-113 Am Aerofoil Boat
RFB X-114 Aerofoilcraft
RFB/Grumman American Fanliner

***Rhein-West-Flug, Fisher u. Companie** (See: Rhein-Flugzeugbau)*

Rhodes, Inc (Pittsburgh, PA)

Rhodes AirRhoCar Hoverscooter

Rhodes Berry (Los Angeles, CA)

Rhodes Berry Silver Sixty

Rhön-Rossitten Gesellschaft E.V. (RRG) (Germany) *(See also: DFS)*
Rhön-Rossitten Gliders

Ricci Brothers (Italy)
Ricci Hydroplane
Ricci R.1 Twin-Hull Flying Boat
Ricci R.2 Gull-Wing Seaplane
Ricci R.3 Triplane Flying Boat
Ricci R.4 Quadriplane Flying Boat
Ricci R.5 Tandem Triplane Flying Boat
Ricci R.6 Triplane
Ricci R.9 Two-Seat Triplane

Rich Airplane Co (N. B. Rich Co, Nelson Bernard Rich) (Springfield, MA)
Rich-Twin 1-X-A

Richard (Belgium)
Richard 1925 Monplane Glider

Richard Warner Aviation Inc (Covington, LA) *(See: Anderson)*

Richard, P. A. (France/Russia)
Richard TOM-1

Richards, G. Tilghman *(See: Lee-Richards)*

Richardson Aeroplane Corp (Holden C. Richardson) (Philadelphia, PA & Washington, DC)
Richardson Seaplane Body
Richardson Tandem Biplane Flying Boat
Richardson Twin Tractor Seaplane (See: Washington Navy Yard)

Richart, Rollant T. (Santa Barbara, CA)
Richart High Wing Monoplane (1931)

Richmond Airways, Inc (Greenridge, NY)
Richmond Sea-Hawk

Richter, Hans (Germany)
Richter Hang Glider
Richter 1921 Biplane Glider
Richter 1929 Glider (Cross-Channel)

Rickman (USA)
Rickman Ornithopter (c.1910)

Ricks, Fred (Yuma, AZ)
Ricks Handley Sport (Ed Handley)

Rider, Keith *(See: Keith Rider)*

Ridgefield Manufacturing Co (Ridgefield, NJ)
Ridgefield XPG-2A

Rieger, August Richard (Chicago, IL)
Rieger 1910 Airship Patent

Rieseler (Kleinflugzeugbau Rieseler) (Germany)
Rieseler R.III
Rieseler R.IV/23

Riffard, Marcel (France)
Riffard 1934 Monoplane Prototype

Riga (Riiga) (Latvia, Russia)
Riga Chaka 1
Riga 1
Riga 50

Riggs, Arlis (San Bernadino, CA)
Riggs (Arlis) Homebuilt Helicopter

Riggs, E. A. "Gus" (Terra Haute, IN)
Riggs (E. A. "Gus") 1913 Biplane

Rikugun Kokugijutsu Kenkyujo (Army Aerotechnical Research Institute (Japan)
Rikugun Ki-93

Riley Aeronautics (Fort Lauderdale, FL)
Riley Rocket (Modified Cessna 310)
Riley Turbo-Executive 400 (Converted de Havilland D.H.104)
Riley Turbo-Rocket (Modified Cessna 310)
Riley 55 (See: Temco)

Riley, Theron G. (Flint, MI)
Riley 1928 Low-Wing Monoplane

Rimailho (France)
Rimailho 1912 Biplane

Rinehart & Whalen (Howard Rinehart & Bernard Whalen) (Hartford, CT)
Rinehart & Whalen R. W. Rambler
Rinehart & Whalen Wren

Rinek (Charles Norvin) Aero Manufacturing Co (Easton, PA)
Rinek No.1 Biplane
Rinek (Voisin) No.2 Biplane

Rinji Gunyo Kikyu Kenkyu Kai (Provisional Military Balloon Research Assoc, PNBRA) (Japan)
Rinji Gunyo Kikyu Kenkyu Kai:
- Army Experimental Model 3 Fighter
- Army Experimental 3-Seat Light Bomber
- Army Type Mo (M. Farman Type) 1913
- Army Type Mo (M. Farman Type) 1914, Converted
- Army Type Mo-4 (H. Farman 4)
- Army Type Mo-5 (M. Farman 5)
- Army Type Mo-6 (M. Farman 6)
- Army Model 2 Ground Taxiing Trainer
- Army Model 3 Ground Taxiing Trainer
- Kaibo Gikai KB Experimental Flying Boat
- Kaishiki No.1
- Kaishiki No.2
- Kaishiki No.3
- Kaishiki No.4
- Kaishiki No.5
- Kaishiki No.6
- Kaishiki No.7
- Kaishiki No.7 Small Aeroplane
- Koshiki-1 Experimental Recon
- Koshiki-2 Experimental Fighter
- Koshiki A-3 Experimental Long-Range Recon Aircraft
- Seishiki-1
- Seishiki-2
- Standard H-3 Trainer (1917)

Ringhoffer-Tatra *(See: Tatra)*

Ritchell, Prof Charles F. (Bridgeport, CT)
Ritchell 1878 Flying Machine

Ritz Aircraft Co (Wartrace, TN)
Ritz Standard Model A (Ultralight)

Riverside Aircraft Corp (Riverside, CA)
Riverside Penguin Ground Trainer

RLU *(See: Breezy Aircraft Co)*

Roamair (Roam Aircraft Corp) (Van Nuys, CA)
Roamair Cadet
Roamair Sport
Roamair Sport Deluxe

Robart, Henri (France)
Robart 1910 Biplane
Robart 1926 Steam-Motor Glider

Robbins and Schiefer (San Diego, CA)
Robbins and Schiefer 1917 Pursuit

***Robert Esnault-Pelterie** (See: R.E.P.)*

Roberts Sport Aircraft (Yakima, WA)
Roberts Sport Aircraft Sceptre I

Roberts, Donald (USA)
Roberts (Donald) Cygnet

Robertson Aircraft (St Louis, MO)
Robertson (St Louis) RO-1504 Rovaire

***Robertson Aircraft Corp (Anglum, MO)** (See: Curtiss)*

Robertson Aircraft Corp (James Robertson) (Renton, WA)
Robertson (Renton) B1-RD

Robertson Aircraft Sales and Service (South Africa)
Robertson (South Africa) Aeriel Mk.II (See: Genair)

Robertson Aircraft Systems (Seattle, WA)
Robertson (Seattle) Ultralights

Robertson Development Corp (Clayton, MO)
Robertson Development SRX-1 Skylark

***Robertson, Donald** (See: Aeronautical & Automobile College, Redhill, UK)*

Robertson, Milton & Russell (Alameda, CA)
Robertson (Alameda) Waterplane

Robey & Co Ltd (UK)
Robey Fighting Machine
Robey Single Seat Scout

Robin, Pierre (Avions Pierre Robin) (France)
In 1957 Pierre Robin formed **Centre Est Aéronautique** (**CEA**) to produce designs derived from **Jodel** aircraft. (DR designations (for Delemontez Robin) reflected the participation of designer Jean Delemontez, one of Jodel's founders.) In 1969, CEA was renamed **Avions Pierre Robin**. When company control passed to **CFCI** (Chaufour Investissement) in 1988, Robin left to form Robin SA, an after-sales support company. All Robin designs, including those produced by CEA, are listed under **Robin**.
Robin ATL
Robin ATL Club (ATL Bijou)
Robin ATL Club Model 88
Robin ATL Club Model 89
Robin ATL Voyage
Robin ATL 2+2
Robin D140R Abeille (Bee) (See: Jodel)
Robin DR.100 (DR.1) Ambassadeur (Ambassador)
Robin DR.105A
Robin DR.1050 Ambassadeur (Ambassador)
Robin DR.1050 Sicile
Robin DR.1050M Excellence(Sky King)
Robin DR.1051 (Sky Queen)
Robin DR.1051M1 Sicile Record
Robin DR.200
Robin DR.220
Robin DR.220 2+2
Robin DR.220/108
Robin DR.220A
Robin DR.221 Dauphin
Robin DR.250 Captaine (Captain)
Robin DR.250/160 Captaine (Captain)
Robin DR.250/180 Captaine (Captain)
Robin DR.253 Regent
Robin DR.300
Robin DR.300/108 2+2
Robin DR.300/108 2+2 Tricycle
Robin DR.300/125 Petit(Little)Prince
Robin DR.300/140 l'Acrobat
Robin DR.300/140 Petit(Little)Prince
Robin DR.300/180 Remorquer
Robin DR.300/180R Remorqueur
Robin DR.315 Cadet
Robin DR.315 Petit(Little)Prince
Robin DR.330
Robin DR.340 Major
Robin DR.360 Chevalier (Knight)
Robin DR.360 Major 160
Robin DR.380 Prince
Robin DR.400 Chevalier (Knight)
Robin DR.400 Dauphin
Robin DR.400 Major 80
Robin DR.400 Remo V6
Robin DR.400 Remo 212
Robin DR.400/V6
Robin DR.400/100 Cadet
Robin DR.400/100 2+2
Robin DR.400/108 2+2
Robin DR.400/120 Dauphin
Robin DR.400/120 Dauphin 2+2
Robin DR.400/120 Dauphin 80
Robin DR.400/12 Petit(Little)Prince
Robin DR.400/125 Dauphin
Robin DR.400/125 Petit (Little) Prince
Robin DR.400/125i
Robin DR.400/140 Major
Robin DR.400/140B Dauphin 4
Robin DR.400/140B Major
Robin DR.400/160 Chevalier (Knight)
Robin DR.400/160 Major
Robin DR.400/160 Major 80
Robin DR.400/180 Regent
Robin DR.400/180R Remorqueur (Remo 180)
Robin DR.400/180RP
Robin DR.400/200R Remo 200
Robin DR.400/2001 President
Robin DR.400GL Regent III
Robin DR.500i President
Robin HR.100
Robin HR.100/4+2
Robin HR.100/180
Robin HR.100/200 Royale
Robin HR.100/210 Safari
Robin HR.100/230TR
Robin HR.100/250TR President
Robin HR.100/285 Tiara
Robin HR.200
Robin HR.200/100 Club
Robin HR.200/120B (Robin 200)
Robin HR.200/140
Robin HR.200/160 Acrobin
Robin HR.400
Robin R.1180 Aiglon (Eaglet)
Robin R.2100
Robin R.2100A
Robin R.2112 Alpha
Robin R.2120U
Robin R.2160 Sport
Robin R.2160 Alpha Sport
Robin R.2160A Acrobatique
Robin R.2160A Rafale
Robin R.2160D
Robin R.2160i
Robin R.2160M
Robin R.3000ATAL
Robin R.3000L
Robin R.3100 (R.3000/100)
Robin R.3120
Robin R.3120 2+2 (R.3000/120)
Robin R.3140 (R.3000/140)
Robin R.3140 Testbed

Robin R.3140E (R.3000/140)
Robin R.3140T
Robin R.3150
Robin R.3160 (R.3000/160)
Robin R.3160GT
Robin R.3160L
Robin R.3160R
Robin R.3160T
Robin R.3170
Robin R.3180
Robin R.3180GT
Robin R.3180GT1
Robin R.3180GT2
Robin R.3180R
Robin R.3180S
Robin R.3180T
Robin X-4

Robinson Aircraft Co (P. G. Robinson) (UK)
Robinson (UK) Redwing
Robinson (UK) Redwing II

Robinson Helicopter Co, Inc (Torrance, CA)
Robinson (Helicopter) R22
Robinson (Helicopter) R22 Alpha
Robinson (Helicopter) R22 Beta
Robinson (Helicopter) R22 IFR Trainer
Robinson (Helicopter) R22 Mariner
Robinson (Helicopter) R22 Police
Robinson (Helicopter) R44 Astro
Robinson (Helicopter) R44 IFR Trainer
Robinson (Helicopter) R44 News
Robinson (Helicopter) R44 Police

Robinson, "Banger Bill" (Pacoima, CA)
Robinson (Bill) Special "Suzie Jayne" (Rebuilt Brown B-1)

Robinson, Hugh A. (Rochester, NY)
Robinson (Hugh A.) 1911 Biplane

Robinson, James Thomas (Los Angeles, CA)
Robinson (James Thomas) Sailaire

Robinson, S. D. *(See: Miami Aircraft Co, OK)*

Robinson, W. C. (Grinnell, IA) *(See also: Grinnell)*
Robinson (W. C.) Monoplane

Robiola, Attilio (Italy)
Robiola 1912 Hydroplane
Robiola 1912/13 Helicopter
Robiola 1914 Twin-Engine

Roche, Jim (Dayton, OH)
Roche 1924 Parasol

Roche-Aviation SA (France)
Roche (France) T.30
Roche (France) T.35
Roche (France) T.39
Roche (France) 107

Roché-Dahse
Roché-Dahse Lightplane

Rocheville Airplane Manufacturing Co (Charles Rocheville) (USA)
Rocheville Arctic Tern
Rocheville Deeble Double Action Engine
Rocheville Modified Fokker D-VII
Rocheville Modified SPAD
Rocheville Special
Rocheville Variable Camber Monoplane

Rock Segelflugzeugbau (Germany)
Rock Krähe

Rock Island Oil and Refining Co, Inc (Wichita, KS)
Rock Island Monarch 26 (A-26 Conversion)
Rock Island Consort 26 (A-26 Conversion)

Rocket Aircraft Corp *(See: Johnson Aircraft Corp)*

Rockwell International (El Segundo, CA) *(See also: North American)*
Rockwell B-1A Lancer
Rockwell B-1B Lancer
Rockwell Commander (See: Commander)
Rockwell Thrush Commander
Rockwell XFV-12A (V/STOL)
Rockwell X-30
Rockwell-MBB Fan Ranger (Rockwell/DASA Ranger 2000) (JPATS)
Rockwell-MBB X-31A

Roe, Alliott Verdon *(See: Avro)*

Roe, Carl (Mineola, NY)
Roe (Carl) 1942 Aircraft Patent

Roesner (Germany)
Roesner 1913 Taube (Dove), Type I
Roesner 1913 Taube (Dove), Type II

Rogalski, Wigura & Drzewieki *(See: RWD)*

Rogers Aeronautical Manufacturing Co, Inc (USA)
Rogers (Aeronautical) Arrow Air Transport
Rogers (Aeronautical) RBX

Rogers Construction Co (1917) (USA)
Rogers (Construction) (Rogers-Day) Air Express
Rogers (Construction) (Rogers-Day) Biplane
Rogers (Construction) 1917 Float Plane

Rogers, Frank (Australia)
Rogers (Frank) Sky Princess (Rebuilt Jodel D.11)

Rogers-Day (Charles H. Day) *(See: Rogers Construction Co)*

Rogerson Hiller Corp *(See: Hiller)*

Rogozarski (Prva Srpska Fabrika Aeroplana Zivojin Rogojarsky) (Yugoslavia)
Rogozarski R-100
Rogozarski SIM X
Rogozarski SIM XII-H (Petit SIM)
Rogozarski SIM XIV-H (Grand SIM)

Rohner, Oscar Arthur (Aurora, CO)
Rohner Model A (Single-Seat Land Monoplane)

Rohr (Fred H.) Aircraft Corp (Chula Vista, CA)
Rohr MO-1

Rohrbach Metall-Flugzeugbau GmbH (Germany)
Rohrbach Ro I Flying Boat
Rohrbach Ro II Roche Flying Boat
Rohrbach Ro III Rotri Flying Boat
Rohrbach Ro IIIa Rodra Flying Boat
Rohrbach Ro IV Ronix "Inverness" Flying Boat
Rohrbach Ro V Rocco Flying Boat
Rohrbach Ro VI Rorex
Rohrbach Ro VII Robbe I Flying Boat
Rohrbach Ro VIIa Robbe IIa Flying Boat
Rohrbach Ro VIII Roland
Rohrbach Ro VIIIa Roland IIa
Rohrbach Ro IX Rofix
Rohrbach Ro X Romar Flying Boat
Rohrbach Ro Xa Romar II Flying Boat

Rohrbach Rofox
Rohrbach Rolas
Rohrbach Rostra Flying Boat
Rohrbach Revolving-Winged Aircraft
Rohrbach-Beardmore "Inflexible" (See: Beardmore)

Rohrig, Bernard F. (San Diego, CA)
Rohrig 1910 Farman-Type Pusher

Roig *(See: Poix et de Roig)*

Roks-Aero Inc *(See: Aeroprogress)*

Roland *(See: LFG Roland)*

Rolladen-Schneider (Germany)
Rolladen-Schneider LS-1
Rolladen-Schneider LS-3
Rolladen-Schneider LS-4
Rolladen-Schneider LS-6
Rolladen-Schneider LS-7
Rolladen-Schneider LS-8

Rolland, Yves (Chicago, IL)
Rolland 1914 Twin-Screw Monoplane

Rollason Aircraft & Engines Co (UK)
Rollason:
D.62 Condor (Redesigned Druine)
D.62A Condor (Redesigned Druine)
D.62B Condor (Redesigned Druine)

Rollé (France)
Rollé 1925 Monoplane Glider

Rolls, Charles (UK)
Rolls Power Glider (1910)

Rolls Royce, Ltd (UK)
Rolls Royce Thrust Measuring Rig (Flying Bedstead)

Roloff-Liposky-Unger *(See: Breezy Aircraft Co)*

Romano (Chantiers Aéronavals E. Romano) (France)
Romano R.3
Romano R.5
Romano R.6
Romano R.7
Romano R.8
Romano R.9
Romano R.10
Romano R.11
Romano R.12
Romano R.13
Romano R.14
Romano R.15
Romano R.16
Romano R.80
Romano R.90

Romeo *(See: Meridionali; See also: Aerfer)*

Ron's I *(See: Freiberger)*

Ronchetti, Razzetti Aviacion SA (RRA) (Italy)
Ronchetti Razzetti J-1 Martin Fierro

ROS-Aeroprogress *(See: Aeroprogress)*

Rose Aeroplane & Motor Co (Chicago, IL)
Rose Parrakeet Model A-1
Rose Parrakeet Model A-2

Rosenman-Rozewski (Poland)
Rosenman-Rozewski 1910 Ornithopter

Ross Aircraft Corp (Orin E. Ross) (Amityville, NY)
Ross Aircraft Corp RS-1 Sportplane
Ross Aircraft Corp RS-2L

Ross, Harland (Los Angeles, CA)
Ross (Harland) RS-1 (See: Ross-Stevens)
Ross (Harland) R-2 Ibis Sailplane
Ross (Harland) R-3 Sailplane
Ross (Harland) RH-3 Sailplane
Ross (Harland) RJ-5
Ross (Harland) RJK-5
Ross (Harland) R-6

Ross-Smith (Australia)
Ross-Smith Cantilever Biplane (1923)

Ross-Stevens (Harland Ross & Harvey Stevens (Los Angeles, CA)
Ross-Stevens RS-1 Zanonia Sailplane

Rossel-Peugeot *(See: Peugeot-Rossel)*

Rotali-Mandelli (France)
Rotali-Mandelli 1934 Light Biplane

Rotary Air Force, Inc (RAF) (Canada)
Rotary Air Force RAF 1000
Rotary Air Force RAF 2000

Rotary Wing Flight *(See: Vertical Flight, General)*

RotaWings Inc (New York, NY)
RotaWings Rota-Aircoach

Rotec Engineering Inc (Duncanville, TX)
Rotec Panther Plus (Ultralight)
Rotec Panther 2 Plus (Ultralight)
Rotec Rally 2B (Ultralight)
Rotec Rally 3 (Ultralight)
Rotec Rally Sport (Ultralight)
Rotec Rally Spraymaster (Ultralight)

Roteron *(See: Thomas, William H.)*

Rotor Plane Co of America (New York, NY)
Rotor Plane General

Rotor-Craft Corp (Glendale, CA)
Rotor-Craft Heli-Jeep
Rotor-Craft RH-1 Pinwheel
Rotor-Craft XR-11 Dragonfly
Rotor-Craft Skyhook
Rotor-Craft X-2

Rotorcraft SA (Pty) Ltd (South Africa)
Rotorcraft Minicopter

Rotorway Aircraft, Inc (Rotorway International) (Tempe, AZ)
Rotorway Exec Helicopter
Rotorway Exec 90 Helicopter
Rotorway RW 133 Scorpion
Rotorway Exec 162F Helicopter

Rotter, Lajos (Hungary)
Rotter Nemere

Rouffaer Aircraft (Oakland, CA)
Rouffaer R-6

de Rouge, Charles (France)
de Rouge l'Élytroplan (1934)

Roulier (France)
Roulier 1917 Flying Boat

Rousch Brothers (Charles W. & Berl Rousch) (Robinson, IL)
Rousch Brothers Single Bay (1927)

Roussel, Jacque (France)
Roussel (Jacque) 10

Roussel, Maurice (France)
Roussel (Maurice) 30

Roux (France)
Roux Ornithopter

Rowland, Charles Obediah (Chicago, IL)
Rowland 1911 Aerial Vessel Patent

Royal Air Force College Halton
(See: RAF College Halton)

Royal Aircraft Corp (Detroit, MI)
Royal (Detroit) Air Trainer

Royal Aircraft Corp (Milwaukee, WI)
Royal (Milwaukee) Gull Amphibian
(See: Piaggio P.136-L, P.136-L2)

Royal Aircraft Corp (Stewart Manor, NY) *(See: Bird Aircraft Corp, NY)*

Royal Aircraft Establishment Aero Club *(See: R.A.E. Aero Club)*

Royal Aircraft Factory (RAF) (UK)
RAF A.E.2
RAF A.E.3
RAF B.E.1
RAF B.E.2
RAF B.E.2a
RAF B.E.2b
RAF B.E.2c
RAF B.E.2d
RAF B.E.2e
RAF B.E.2f
RAF B.E.2g
RAF B.E.3
RAF B.E.4
RAF B.E.5
RAF B.E.6
RAF B.E.7
RAF B.E.8
RAF B.E.8a
RAF B.E.9
RAF B.E.9a
RAF B.E.10
RAF B.E.11
RAF B.E.12
RAF B.E.12a
RAF B.E.12b
RAF B.S.1
RAF B.S.2
RAF C.E.1
RAF F.E.1
RAF F.E.2
RAF F.E.2a
RAF F.E.2b
RAF F.E.2c
RAF F.E.2d
RAF F.E.2e
RAF F.E.2f
RAF F.E.2g
RAF F.E.2h
RAF F.E.3 (A.E.1)
RAF F.E.4
RAF F.E.5
RAF F.E.6
RAF F.E.7
RAF F.E.8
RAF F.E.8, NASM
RAF F.E.9
RAF F.E.10
RAF F.E.12
RAF N.E.1
RAF R.E.1
RAF R.E.2 (H.R.E.2)
RAF R.E.3
RAF R.E.4
RAF R.E.5
RAF R.E.6
RAF R.E.7
RAF R.E.8
RAF R.E.8a
RAF R.E.9
RAF S.E.1
RAF S.E.2
RAF S.E.2 Rebuilt (S.E.2a)
RAF S.E.3
RAF S.E.4
RAF S.E.4a
RAF S.E.5
RAF S.E.5a
RAF S.E.5b
RAF SE-5E (See: Eberhart)
RAF S.E.6
RAF S.E.7
RAF T.E.1

Roxborough Glider Club (Philadelphia, PA)
Roxborough Glider Club Primary Training Glider

RRA *(See: Ronchetti, Razzetti Aviacion SA)*

RRG *(See: Rhön-Rossitten Gesellschaft E.V.)*

Rubberworks (Van Nuys, CA)
Rubberworks Rubber Bandit (Piloted, Rubber-Band Powered)

Rudlicki, Jerzy (Rudlicky) (Poland) *(See also: Lublin)*
Rudlicki Glider No.1
Rudlicki Glider No.2
Rudlicki Glider No.3
Rudlicki Glider No.4
Rudlicki Glider No.5
Rudlicki Glider No.6
Rudlicki Glider No.7
Rudlicki Glider No.8
Rudlicki Glider No.9
Rudlicki R-I Monoplane
Rudlicki R-II Monoplane
Rudlicki R-III Monoplane
Rudlicki R-IV Monoplane
Rudlicki R-V Monoplane

Rummell, H. B. (Findlay, OH)
Rummell Monoplane 1 (1929)

Rumpler-Werke AG (Germany)
Rumpler B.I (Typ 4A)
Rumpler Typ Berlin-Wien Monoplane (1912)
Rumpler C.-Type Aircraft
Rumpler C.I (Typ 5A2)
Rumpler C.IV (Typ 6A5 & 6A7)
Rumpler C.V
Rumpler C.VII
Rumpler C.VIII (Typ 6A8)
Rumpler C.IX
Rumpler C.X (Typ 8C14)
Rumpler D.I (Typ 8D1)
Rumpler Typ DDD Biplane (1914/15)
Rumpler Eggers Monoplane (1910)
Rumpler G.I (Typ 4A15, 5A15)
Rumpler G.II (Typ 5A16)
Rumpler G.III (Typ 5A16 & 6G2)
Rumpler Haefelin Monoplane (1910)
Rumpler Pegelow Monoplane (1910)
Rumpler Plage Biplane (1910)
Rumpler Schudeisky Biplane (1910)
Rumpler Sohlmann Monoplane (1912)
Rumpler Stein Monoplane (1910)
Rumpler Taube (Dove), 1910-12
Rumpler Taube (Dove), 1913
Rumpler Taube (Dove), Military
Rumpler Taube (Dove), 2-Engined
Rumpler Twin-Hull Flying Boat (Transatlantic)
Rumpler Wasser-Taube (Water Dove)
Rumpler Typ 3C Monoplane (1913)
Rumpler Typ 3F Float Monoplane
Rumpler Typ 4A13 Biplane
Rumpler Typ 4A14 Biplane
Rumpler Typ 4B1 Float Biplane
Rumpler Typ 4B11 Float Biplane
Rumpler Typ 4B13 Float Biplane
Rumpler Typ 4B2 Float Biplane
Rumpler Typ 4C Monoplane
Rumpler Typ 4E Flyingboat
Rumpler Typ 5A2 Biplane

Rumpler Typ 5A4 Biplane
Rumpler Typ 5DDD Biplane (1914/15)
Rumpler Typ 6A2
Rumpler Typ 6BI Single-Seat Combat Floatplane
Rumpler Typ 6BII Two-Seat Combat Floatplane
Rumpler Typ 7D Fighter
Rumpler 7D1 Fighter
Rumpler 7D2 Fighter
Rumpler 7D4 Fighter
Rumpler 7D7 Fighter
Rumpler Typ E12 & 4B13 Float Biplanes
Rumpler 1913 Monoplane

Rupert, Walter (Portland, OR)

Rupert Improved V (Rupert Special)

Ruschmeyer Aircraft-Production (Germany)

Ruschmeyer R 90-230 RG

Russell Airways, Inc (Jack Russell) (Fredrick, OK)

Russell Convertible Monoplane (Comet) (Converts to Biplane)

Rust (Hilmar & Roy Rust) (Waring, TX)

Rust Model 1
Rust Model 2

Rutan Aircraft Factory (RAF) (Mojave, CA)

Rutan Model 27 Vari Viggen
Rutan Model 31 Vari-Eze Prototype
Rutan Model 32 Special Performance Vari-Viggen
Rutan Model 33 Vari-Eze
Rutan Model 33 Vari-Eze, NASM
Rutan Model 35 AD-1 Skew Wing (See: NASA)
Rutan Model 40 Defiant
Rutan Model 54 Quickie
Rutan Model 54 Quickie, NASM
Rutan Model 59 Predator
Rutan Model 61 Long-EZ
Rutan Model 68 Amsoil Biplane Racer
Rutan Model 72 Grizzly
Rutan Model 73 NGT (62% Fairchild T-46)
Rutan Model 74 Defiant
Rutan Model 76 Pond Racer (See: Scaled Composites Inc)
Rutan Model 76 Voyager
Rutan Model 76 Voyager, NASM
Rutan Model 77 Solitaire
Rutan Model 81 Catbird
Rutan Model 97 Microlight (See: Scaled Composites Inc)
Rutan Model 115 Starship (See: Scaled Composites)
Rutan Model 120 Predator (See: Scaled Composites Inc)
Rutan Model 133 AT3 (ATTT, Smut) (See: Scaled Composites)
Rutan Model 143 Triumph (NGCT) (See: Scaled Composites)
Rutan Model 143J Triumph (NGCT) (See: Scaled Composites)
Rutan Model 144 CM-44 California Microwave (See: Scaled Composites)
Rutan Model 151 Ares (LATS) (See: Scaled Composites Inc)
Rutan Model 202 Boomerang (See: Scaled Composites Inc)
Rutan Model 324 Scarab RPV (See: Scaled Composites Inc)
Rutan PARLC (US Navy)

R.W. (Stanislaw Rogalski & Stanislaw Wigura) (Poland)

R.W.1

RWD (Stanislaw Rogalski, Stanislaw Wigura & Jerzy Drzewieki) (Poland)

RWD-1
RWD-2
RWD-3
RWD-4
RWD-5
RWD-6
RWD-6*bis* (RWD-13)
RWD-7
RWD-8
RWD-9
RWD-10
RWD-13
RWD-14

Ryan Aeronautical Co (San Diego, CA)

World War I pilot Claude Ryan began his barnstorming career in the early 1920s, forming **Ryan Flying Co** in 1922. His additional work refurbishing and converting surplus aircraft led to the formation of **Ryan Airlines Inc** at San Diego, California, in 1925. In 1927, Ryan left the firm, which his partner moved to St. Louis, Missouri, under the name **B. F. Mahoney Aircraft Corp**; due to the popularity of the Ryan name, the company became **Mahoney-Ryan Aircraft Corp** in 1928 (even though Claude Ryan was no longer associated). Mahoney-Ryan was sold to the **Detroit Aircraft Corp** early in 1929, operating under its own name until later in the year, when Claude Ryan returned and renamed the company the **Ryan Aircraft Corp**. Ryan and Detroit closed their doors in the early 1930s, victims of the Great Depression.

In 1928, after Mahoney moved the firm to St Louis, a group of former Ryan employees remained in San Diego to form the **Ryan Mechanics Monoplane Co**; the new company was renamed the **Federal Aircraft Corp** later in the year.

Claude Ryan formed a new design and manufacturing company, the **Ryan Aeronautical Corp**, in 1934. In 1968, Ryan sold the company to **Teledyne Inc**; it would continue to operate as **Teledyne-Ryan**. All products of the Claude Ryan and Mahoney companies are found under **Ryan**. Products of Ryan Mechanics appear under **Federal Aircraft Corp**. Although Teledyne-Ryan's numerous unmanned aircraft are not included in this listing, that company's Model 410 appears under **Ryan**.

Ryan B Series
Ryan B.1 Brougham
Ryan B.1 M.G.M. Special
Ryan B.1 Special (Pot-Bellied Brougham)
Ryan B.1X Brougham
Ryan B.3 Brougham
Ryan B.5 Brougham
Ryan B.5A Brougham
Ryan B.7 Brougham
Ryan BLC-7 Business STOL
Ryan BlueBird
Ryan Cloudster
Ryan C 1 Foursome
Ryan C 1A Foursome
Ryan CM-1 Lone Eagle (See: Federal Aircraft Corp)
Ryan CM-2 (See: Federal Aircraft Corp)
Ryan CM-3 (See: Federal Aircraft Corp)
Ryan XFR-1 Fireball
Ryan FR-1 Fireball
Ryan FR-1 Fireball, NASM
Ryan XFR-4 Fireball
Ryan XF2R-1 Dark Shark
Ryan Flex Wing
Ryan Flex Wing Air Cargo Delivery System
Ryan L-17 Navion (See: North American)
Ryan M Series

Ryan M-1
Ryan M-2
Ryan M-2C
Ryan M-3C
Ryan Navion (See: North American)
Ryan NR-1
Ryan NYP "Spirit of St Louis"
Ryan NYP "Spirit of St Louis," NASM
Ryan NYP-2
Ryan YO-51 Dragonfly
Ryan Primary Phase Trainer (jet)
Ryan XPT-16
Ryan YPT-16
Ryan XPT-16A
Ryan PT-16A (YPT-16A)
Ryan PT-20
Ryan PT-20, Civil Use
Ryan PT-20A
Ryan PT-20B
Ryan PT-21
Ryan PT-22 Recruit
Ryan PT-22, Civil Use
Ryan PT-22A Recruit
Ryan PT-22C Recruit
Ryan YPT-25
Ryan S-C Series
Ryan S-C
Ryan SCW
Ryan Speedster (See: Parks P-2A)
Ryan S-T Sport Trainer
Ryan ST-A
Ryan ST-A Special
Ryan ST-B
Ryan STK
Ryan STM
Ryan STM-2
Ryan STM-S2
Ryan STM-2E
Ryan STM-2P
Ryan STW
Ryan ST-3
Ryan ST-3S
Ryan Standard
Ryan XV-5A Vertifan
Ryan XV-5B Vertifan
Ryan XV-8A Fleep
Ryan X-1 "Doodle Bug" (Mahoney-Ryan Special)
Ryan X-13 Vertijet
Ryan X-13 Vertijet, NASM
Ryan VZ-3 Vertiplane
Ryan VZ-11 Liftfan V/STOL (See: XV-5)
Ryan Model 410

Rydberg, John T. (Garwood, NJ)
Rydberg 1911 Flying-Machine Aeroplane Patent

Ryder *(See: Keith Ryder)*

Ryl'tsev, S. N. (Russia)
Ryl'tsev Mars

Ryley, L. G. (UK)
Ryley 1914 Glider

Ryson Aviation Corp (San Diego, CA)
Ryson ST-100 Cloudster

Rystedt, Ingemar K. (Dayton, OH)
Rystedt 1930 Airplane Patent

S

S & S (Ambrose Stampo & John Sillivan) (Pittsburgh, PA)
S & S 1928 Monoplane

S-Wing spol sro (Czech Republic)
S-Wing Swing

SAAB *(See: Skandinaviska Aero A.B.)*

Saab (Svenska Aeroplan Aktiebolaget AB) (Sweden)
Saab MFI-15 Safari
Saab MFI-17 Supporter
Saab 17
Saab 17A
Saab 17B
Saab 17BS
Saab 17C
Saab 18A
Saab 18B
Saab 21A
Saab 21R
Saab 22 Svenska
Saab 29 Tunnan (Barrel)
Saab 29 Tunnan (Barrel), NASM
Saab 32 Lansen (Lance)
Saab 35 Draken (Dragon)
Saab 37 Viggen (Thunderbolt)
Saab 39 Gripen
Saab 90 Scandia
Saab 90A Scandia
Saab 90A-2 Scandia
Saab 91A Safir (Sapphire)
Saab 91B Safir (Sapphire)
Saab 91C Safir (Sapphire)
Saab 91D Safir (Sapphire)
Saab 100B Argus
Saab 105
Saab SAR-200
Saab 201
Saab 210
Saab-Fairchild SF-340
Saab SF-340B
Saab 2000

SAAC *(See: Learjet)*

Saalfeld Aircraft Co (San Diego, CA)
Saalfeld Skyscooter

Sääski, Osakeyhtiö (Finland)
Sääski 1929 Biplane

S.A.B. (Société Aérienne Bordelaise) *(See: Bordelaise; also Dyle et Bacalan)*

S.A.B. (Société des Appareils Bêchéreaux) (France)
S.A.B. (Bêchéreau) S.A.B. 1 C1
S.A.B. (Bêchéreau) S.A.B. 4
S.A.B. (Bêchéreau) C.M.18

S.A.B. (Société des Avions Bernard) *(See: Bernard)*

SABCA (Société Anonyme Belge de Constructions Aéronautiques) (Belgium)
SABCA AR-2
SABCA Cambgul
SABCA Demonty-Poncelet Limousine
SABCA RSV-22
SABCA S.2
SABCA S.XI
SABCA S.XII
SABCA S.20
SABCA S.30 Sportplane
SABCA S.40
SABCA S.45*bis*
SABCA S.46
SABCA S.47
SABCA S.48
SABCA S.50

Sablatnig Flugzeugbau GmbH (Germany)
Sablatnig C.I
Sablatnig C.II
Sablatnig C.III
Sablatnig N.I
Sablatnig P.I
Sablatnig P.III
Sablatnig SF 1
Sablatnig SF 2
Sablatnig SF 3
Sablatnig SF 4
Sablatnig SF 5
Sablatnig SF 6
Sablatnig SF 7
Sablatnig SF 8

Sabreliner Corp (St Louis, MO)
Sabreliner (Agusta) SF260
Sabreliner (Socata) Omega JPATS

Sablier (France)
Sablier Type 12 Avionette (1934)
Sablier 1923 Chanute-type Biplane
Sablier 1923 Parasol
Sablier 1925 Monoplane Glider

Sabourault-Boussières-Touya *(See: SBT)*

S.A.C.A.N.A. *(See: Bille)*

Sachem Aeronautical Corp *(See: Knoll Brayton)*

Sack, Arthur (Germany)
Sack AS 6 "Fliegende Bierdeckel" (Flying Beertray)

S.A.D. (Société Anonyme des Appareils d'Aviation Doutre) *(See: Doutre)*

Sadler, Bill (Scottsdale, AZ)
Sadler Vampire

SAFCA *(See: Ponnier M.2)*

Sage Aircraft (Frederick Sage & Co, Ltd) (UK)
Sage Long Distance Passenger Seaplane
Sage Long Distance 2-Seater Land Machine
Sage Long Distance 2-Seater Seaplane
Sage 1-Seat Sporting Land Machine
Sage Single Seat Sporting Seaplane
Sage 6 Passenger Seaplane, Enclosed Type
Sage 2 Passenger Seaplane, Open Type
Sage Training Seaplane
Sage 2-Seater Sociable Land Machine
Sage 2-Seater Sociable Seaplane
Sage 2-Seater Sporting Land Machine
Sage 2-Seater Sporting Seaplane
Sage I
Sage II
Sage III
Sage IV
Sage IVa
Sage IVb
Sage IVc

SAI (Singapore Aircraft Industries)
SAI (Singapore) A-4S-1 Super Skyhawk (Converted McDonnell Douglas A-4S

SAI *(See: Skandinavisk Aero Industri)*

SAI (Società Aeronautica Italiana) (Italy) *(See also: Ambrosini)*
SAI (Italy) SAI.1
SAI (Italy) SAI.2
SAI (Italy) SAI.2S
SAI (Italy) SAI.3
SAI (Italy) SAI.3S
SAI (Italy) SAI.7
SAI (Italy) SAI.10 Grifone
SAI (Italy) SAI.207
SAI (Italy) SAI.403 Dardo
SAI (Italy) S.S.3 (See SCA)
SAI (Italy) S.S.4

SAIC *(See: Shanghai Aviation Industrial Corp)*

Sailplanes, General *(See: Gliders and Sailplanes)*

Saiman (Società Anònima Industrie Meccaniche Aeronautiche Navali) (Italy)
Saiman C.4
Saiman C.10
Saiman LB.4
Saiman 200
Saiman 202
Saiman 202/M
Saiman 204

St Germain, Jean (Canada)
St Germain Raz-Mut

St Hubert (Ecole d'Aviation de St Hubert) (Belgium)
Orta-St-Hubert Monoplane (1929)

St Louis Aircraft Corp (St Louis, MO)
St Louis Cardinal C2-60
St Louis Cardinal C2-85
St Louis Cardinal C2-90
St Louis Cardinal C2-100
St Louis Super Cardinal C2-110
St Louis XCG-5
St Louis XCG-6
St Louis PT-LM-4
St Louis PT-1W
St Louis PT-2
St Louis XPT-15
St Louis YPT-15
St Louis PT-35

St Stephens College (Annandale-on-Hudson, NY)
St Stephens College F-131 Glider

S.A.I.S. *(See: Sloan & Cie)*

Saito, Sotoichi (Japan)
Saito Saigai Aeroplane (1912)

SAL *(See: Sturgeon Air Ltd; See also: Falconar)*

Salino, Pedro (Argentina)
Salino 1914 Monoplane

Salminen, Eino (USA)
Salminen 1918 Monoplane

Salmson (Société de Moteurs Salmson) (France)
Salmson D.6 Cricri (Cricket)
Salmson Phrygane (Phrygian)
Salmson 2 A2
Salmson 3 C1
Salmson 4 Ab2
Salmson 5
Salmson 6
Salmson 7 A2
Salmson 1923 Moto Aviette
Salmson-Moineau S.M.1
Salmson-Moineau S.M.2

Salvay-Stark Aircraft Co *(See: Skyhopper)*

SAML (Società Anonima Meccanica Lombarda) (Italy)
SAML Aviatik B.I (License-built)
SAML S.1
SAML S.2

Samolot *(Russian word for "airplane")*

Samolot (Wielkopolska Wytwórnia Samolotów 'Samolot' Sp Akc: Wielkopolan Aeroplane Plant 'The Aeroplane" Co Ltd) (Poland)
Samolot M.N.2
Samolot M.N.3
Samolot M.N.4
Samolot M.N.5

Samsonov, P. D. (Russia)
Samsonov MBR-5
Samsonov MDR-7

SAN (Société Aéronautique Normande) (France) *(See also: Jodel)*
SAN 101
SAN D140R Abeille (Bee) (See: Jodel)

San Diego Airplane Manufacturing Co *(See: Walsh, Charles F.)*

San Francisco Aero Club (San Francisco, CA)
San Francisco Aero Club:
Glider "Floyd Bennett Jr"
1909 Glider

Sanchez-Besa (Spain/France)
Sanchez 1909 Biplane
Sanchez-Besa Multiplane
Sanchez-Besa Seaplane
Sanchez-Besa Wright Type
Sanchez-Besa 1912/13 Hydro

Sanders Aeroplane Co (UK)
Sanders Teacher
Sanders 1909 Biplane (Type 1 No.1)
Sanders Type 1 No.2 Biplane
Sanders Type 2 No.1 Biplane

Sänger, Dr Eugen Albert (Germany)
Sänger Silbervogel (Silverbird)

Santa Ana Aircraft Co (Santa Ana, CA)
Santa Ana VM-1 Activian

Santoro, Aurelio (Argentina)
Santoro 1917 Monoplane

Santos-Dumont, Alberto (Brazil & France)
Santos-Dumont 14*bis*
Santos-Dumont 15
Santos-Dumont 17
Santos-Dumont 18 Hydroplane
Santos-Dumont 19 Demoiselle
Santos-Dumont 20 Demoiselle
Santos-Dumont 21 Demoiselle
Santos-Dumont 22 Demoiselle

Sargent-Fletcher *(See: Fletcher Aviation Corp)*

Saro *(See: Saunders-Roe)*

Sato-Maeda (Dr Hiroshi Sato & Mr Kenichi Maeda (Japan)
Sato-Maeda SM-OX-1 Human-Powered Aircraft

Saturn Engineering (USA)
Saturn Engineering Taylor Meteor

Saul Aircraft Corp (W. I. Saul) (Carroll, IA)
Saul 1000 Triad

Saulnier, Raymond (France)
Raymond Saulnier worked for a time with Louis Blériot, Gabriel Borel, and Léon Morane. In October 1911, Saulnier and the Morane brothers (Léon and Robert) joined to form the **Société Anonyme des Aéroplanes Morane-Saulnier.** For all Saulnier designs predating the establishment of Morane-Saulnier, see **Saulnier**, Raymond. For all later work, see **Morane-Saulnier**.
Saulnier 1910 Parasol Monoplane, Anzani Engine
Saulnier 1910 Parasol Monoplane, Darracq Engine
Saulnier 1911 Monoplane (E.N.V. Engine)

Saunders-Roe (UK)
Founded in 1830, **S. E. Saunders Ltd** began flying boat production in 1912. In 1928 Sir A. V. Roe purchased a majority interest and changed the company name to **Saunders-Roe Ltd**, or **Saro**. In 1951 Saro took over the **Cierva Autogiro Co, Ltd** and formed a Saunders helicopter division. (Products of this division were built as **Saro-Cierva** until 1952, when the Cierva name was dropped.) In 1959 Saunders-Roe became a **Westland** subsidiary. In October 1966 a reorganized **Westland Helicopters Ltd** formed its **Cowes Division**, dropping the Saunders-Roe name. Saunders-Roe Ltd aircraft are listed as **Saunders-Roe**; some rotary wing products are included in the **Westland** section.

<u>Saunders/Saunders-Roe Named Projects:</u>
Saunders-Roe Duchess
Saunders Kittiwake
Saunders-Roe Rotorcoach

<u>Saunders/Saunders-Roe Projects in Numerical Sequence</u>
Saunders T.1
Saunders A.3 Valkyrie
Saunders A.4 Medina
Saunders-Roe A.7 Severn
Saunders-Roe A.10 Multi-Gun (F.20/27)
Saunders A.14 Metal Hull
Saunders-Roe A.17 Cutty Sark
Saunders-Roe A.19 Cloud
Saunders-Roe A.21 Windhover
Saunders-Roe (Segrave) A.22 Meteor
Saunders-Roe/Percival A.24 Cruiser (See: Spartan, UK)
Saunders-Roe/Percival A.24 Mailplane (See also: Spartan, UK)
Saunders-Roe A.27 London
Saunders-Roe A.27 London Mk.I
Saunders-Roe A.27 London Mk.II
Saunders-Roe A.29 Cloud
Saunders-Roe A.33
Saunders-Roe S.36 Lerwick
Saunders-Roe S.35 Shetland (See: Short)
Saunders-Roe A.37
Saunders-Roe S.38
Saunders-Roe S.38A
Saunders-Roe S.39
Saunders-Roe S.39A
Saunders-Roe S.40
Saunders-Roe S.41
Saunders-Roe S.42
Saunders-Roe S.42A
Saunders-Roe SR.45 Princess
Saunders-Roe SR.45 Princess Landplane
Saunders-Roe SR.53
Saunders-Roe P.121
Saunders-Roe P.154 (F.124T)
Saunders-Roe P.162
Saunders-Roe P.162B
Saunders-Roe SR.177
Saunders-Roe P.192
Saunders-Roe P.501 Skeeter Family (Cierva W.14)
Saunders-Roe P.531 (See also: Westland Scout & Wasp)
Saunders-Roe 1033 (Hiller XROE-1) Rotorcycle
Saunders-Roe SR.A/1 "Squirt"
Saunders-Roe SR.N/1 Hovercraft

<u>Saunders-Roe Projects to Air Ministry Specifications</u>
Saunders-Roe R.3/38
Saunders-Roe R.14/40

Savage, Alfred D. (St Louis, MO)
Savage Glider 1 (1932)

Savary (Société Anonyme des Aéroplanes Robert Savary) (France)
Savary 1910 Racer (Single-Propeller, Race No.19)
Savary 1910 Racer (Double-Propeller, Race No.45)
Savary 1911 Military Biplane (Double-Propeller)
Savary 1911/12 Military Biplane (Single-Propeller)
Savary 1912 Biplane (Double-Propeller)

Savel'ev, Vladimir F. *(See: Zalewski)*

Savieliev, Vladimir *(See: Zalewski)*

Savoia, Umberto (Italy)
Savoia 1909 Farman-Type Biplane

Savoia-Marchetti (Italy) *(See also: American Aeronautical Corp)*
Savoia-Marchetti S.5 thru S.50 (See: SIAI)
Savoia-Marchetti S.51
Savoia-Marchetti S.52
Savoia-Marchetti S.53
Savoia-Marchetti S.55
Savoia-Marchetti S.55A
Savoia-Marchetti S.55C
Savoia-Marchetti S.55M
Savoia-Marchetti S.55P
Savoia-Marchetti S.55X
Savoia-Marchetti S.56
Savoia-Marchetti S.57
Savoia-Marchetti S.58
Savoia-Marchetti S.59
Savoia-Marchetti S.59*bis*
Savoia-Marchetti S.62
Savoia-Marchetti S.62*bis*
Savoia-Marchetti S.62C
Savoia-Marchetti S.62P
Savoia-Marchetti S.63
Savoia-Marchetti S.64
Savoia-Marchetti S.65

Savoia-Marchetti S.66
Savoia-Marchetti S.67
Savoia-Marchetti S.69
Savoia-Marchetti S.71
Savoia-Marchetti S.72
Savoia-Marchetti S.73
Savoia-Marchetti S.74
Savoia-Marchetti SM.75
Savoia-Marchetti SM.78
Savoia-Marchetti SM.79 Sparviero (Sparrowhawk)
Savoia-Marchetti SM.79B Sparviero (Sparrowhawk)
Savoia-Marchetti SM.80
Savoia-Marchetti SM.81 Pipistrello (Bat)
Savoia-Marchetti SM.81B Pipistrello (Bat)
Savoia-Marchetti SM.82 Canguro (Kangaroo)
Savoia-Marchetti SM.83
Savoia-Marchetti SM.84 Civil
Savoia-Marchetti SM.84 Military
Savoia-Marchetti SM.85
Savoia-Marchetti SM.89
Savoia-Marchetti SM.91
Savoia-Marchetti SM.92
Savoia-Marchetti SM.93
Savoia-Marchetti SM.95
Savoia-Pomilio SP.1 (See: SIA)
Savoia-Pomilio SP.2 thru SP.4 (See: Pomilio)

SBT (Sabourault-Boussières-Touya)
SBT 1934 Glider

SCA (Société Commerciale Aéronautique) (France)
SCA SFR 10 (See: Delanne D.II; also: Moreau JM 10)

SCA (Stabilimento Costruzioni Aeronautiche) (Italy)
SCA S.S.1
SCA S.S.2
SCA S.S.3
SCA S.S.4

SCAA *(See: de Lesseps)*

Scaled Composites Inc (Rutan Developed) (Mojave, CA)
Scaled Composites V-Jet II (See: Williams International Corp)
Scaled Composites Model 76 Pond Racer
Scaled Composites Model 97 Microlight
Scaled Composites Model 115 Starship (See: Beech Starship 1 Prototype)
Scaled Composites Model 120 Predator
Scaled Composites Model 133 AT3 (ATTT, Smut)
Scaled Composites Model 143 Triumph (NGCT)
Scaled Composites Model 143J Triumph (NGCT)
Scaled Composites Model 144 CM-44 UAV
Scaled Composites Model 151 Ares (LATS)
Scaled Composites Model 202 Boomerang

SCAN (Société de Constructions Aéro Navales) (France)
SCAN 20 Flying Boat
SCAN 30

Scanlan, Thomas W. (USA)
Scanlan SG-1A

Schapel, Rod *(See: Wing Research)*

Scheibe-Flugzeugbau GmbH (Egon Scheibe) (Germany)
Scheibe Bergfalke (Mountain Falcon) (Mü 13E Development)
Scheibe Bergfalke (Mountain Falcon) 2
Scheibe Bergfalke (Mountain Falcon) 2/55
Scheibe Bergfalke (Mountain Falcon) 2/55, Motorized
Scheibe Bergfalke (Mountain Falcon) 3
Scheibe Bergfalke (Mountain Falcon) 4
Scheibe SF-23 Sperling (Sparrow)
Scheibe SF-24A Motorspatz (Motorized Sparrow)
Scheibe SF-24B Motorspatz (Motorized Sparrow)
Scheibe SF-25 Falke (Falcon)
Scheibe SF-25B Falke (Falcon)
Scheibe SF-25C Falke (Falcon)
Scheibe SF-25E Falke (Falcon)
Scheibe SF-25-2000 Falke (Falcon)
Scheibe SF-26
Scheibe SF-27A
Scheibe SF-27M
Scheibe SF-28A Tandem Falke (Falcon)
Scheibe SFS-31 Milan (See: Sportavia)
Scheibe SF-34
Scheibe Spatz (Sparrow)
Scheibe L-Spatz (Sparrow) 3
Scheibe Venture T.Mk.2 (See: Slingsby T.61E Venture)
Scheibe Zugvogel 3A
Scheibe Zugvogel 3B

Schelde *(See: De Schelde)*

Scheller, Bernhard (Germany)
Scheller 2-Seat Monoplane

Schempp-Hirth (Sportflugzeugbau Schempp-Hirth OHG) (Germany)
In 1935, several gliding enthusiasts led by Martin Schempp established **Sportflugzeugbau Göppingen Martin Schempp (Göppingen)** in Göppingen, Germany. In 1938, Wolf Hirth joined Göppingen and the company became **Sportflugzeugbau Schempp-Hirth OHG (Schempp-Hirth)**. In 1940 Hirth established **Wolf Hirth GmbH (Hirth)** as Schempp-Hirth turned more toward subcontract work for other segments of the German aviation industry. Both companies stopped work with the end of World War II, but Hirth began to produce Göppingen designs again in 1951. Schempp-Hirth started producing new designs in 1962. For all Göppingen designs entering production between 1935 and 1938, see **Göppingen**. For all Schempp-Hirth and Hirth designs entering production after 1938, see **Schempp-Hirth**.
Schempp-Hirth Discus A
Schempp-Hirth Discus B
Schempp-Hirth Discus BM
Schempp-Hirth Doppelraab Glider
Schempp-Hirth Duo Discus
Schempp-Hirth Gö 5 (See: Göppingen Gö 5)
Schempp-Hirth Gö 9 (See: Göppingen Gö 9)
Schempp-Hirth Janus
Schempp-Hirth Janus B
Schempp-Hirth Janus C
Schempp-Hirth Janus CM
Schempp-Hirth Janus CT
Schempp-Hirth Janus T.Mk.1 (RAF Janus C)
Schempp-Hirth Minimoa (See: Göppingen Gö 3)
Schempp-Hirth Wolf (See: Göppingen Gö 1)
Schempp-Hirth Mini-Nimbus
Schempp-Hirth Mini-Nimbus C
Schempp-Hirth Nimbus 2
Schempp-Hirth Nimbus 2B
Schempp-Hirth Nimbus 2C

Schempp-Hirth Nimbus 3/22.9
Schempp-Hirth Nimbus 3/24.5
Schempp-Hirth Nimbus 3T
Schempp-Hirth Nimbus 3D
Schempp-Hirth Nimbus 3DT
Schempp-Hirth Nimbus 3DM
Schempp-Hirth Nimbus 4
Schempp-Hirth Nimbus 4D
Schempp-Hirth Nimbus 4DM
Schempp-Hirth Nimbus 4DT
Schempp-Hirth SHK
Schempp-Hirth Standard Austria
Schempp-Hirth Standard Austria SH-1
Schempp-Hirth Cirrus
Schempp-Hirth Cirrus 81 (Jastreb)
Schempp-Hirth Cirrus 75
Schempp-Hirth Ventus A
Schempp-Hirth Ventus B
Schempp-Hirth Ventus BT
Schempp-Hirth Ventus C
Schempp-Hirth Ventus CT
Schempp-Hirth Ventus 2a
Schempp-Hirth Ventus 2b
Schempp-Hirth Ventus 2c
Schempp-Hirth Ventus 2cT
Schempp-Hirth Ventus 2cM

von Schertel, Freiherr (Germany)
von Schertel 1924 Glider

Scheutzow Helicopter Corp (Webb Scheutzow) (Columbia Station, OH)
Scheutzow Model B Helicopter

Schiefer and Sons (San Diego, CA) *(See: Robbins and Schiefer)*

Schill, Paul (New York, NY)
Schill 1912 Hydro Pusher

Schiller (USA)
Schiller 1912 Monoplane

Schiller and Barros (A. Schiller & A. A. Barros) (Brazil)
Schiller and Barros AB-1
Schiller and Barros AB-2

Schindler Brothers (Rudolf and Wincenty) (Poland)
Schindler Brothers Aquila (1909 Monoplane)

Schindler und Brzesky (Austria-Hungary)
Schindler und Brzesky Tractor Monoplane

Schleicher (Flugzeugbau Alexander Schleicher) (Germany)
Schleicher Condor 4
Schleicher "Frankfurt/M"
Schleicher Grunau Baby 3B (See: Grunau)
Schleicher "Hol's der Teufel"
Schleicher Rhönadler (Rhine Eagle)
Schleicher Rhönbussard (Rhine Buzzard)
Schleicher Rhönsperber (Rhine Sparrow Hawk)
Schleicher Seeadler (Sea Eagle)
Schleicher Sperber (Sparrow Hawk) Junior
Schleicher Sperber (Sparrow Hawk) Senior
Schleicher TG-9 (USAF ASK-21)
Schleicher Valiant T.Mk.1 (RAF ASW-19)
Schleicher Vanguard T.Mk.1 (RAF ASK-21)

<u>Schleicher Numeric Designations:</u>
Schleicher Ka-1
Schleicher Ka-2 Rhönschwalbe (Rhine Sparrow)
Schleicher Ka-3
Schleicher Ka-4 Rhönlerche (Rhine Lark)
Schleicher Ka-6
Schleicher Ka-6B
Schleicher Ka-6BR
Schleicher Ka-6C
Schleicher Ka-6CR
Schleicher Ka-6CRPE
Schleicher Ka-6E
Schleicher Ka-7
Schleicher Ka-8
Schleicher Ka-8B
Schleicher Ka-8C
Schleicher ASW-12
Schleicher ASK-13
Schleicher ASK-14
Schleicher ASW-15
Schleicher ASW-15B
Schleicher ASW-17
Schleicher ASW-19
Schleicher ASW-20
Schleicher ASW-20B
Schleicher ASW-20BL
Schleicher ASW-20C
Schleicher ASW-20CL
Schleicher ASW-20F
Schleicher ASW-20FP
Schleicher ASW-20L
Schleicher ASK-21
Schleicher ASW-22
Schleicher ASW-22B
Schleicher ASW-22BE
Schleicher ASK-23
Schleicher ASW-24
Schleicher ASW-24E
Schleicher ASH-25
Schleicher ASH-25E
Schleicher ASH-26E
Schleicher ASW-27

Schlicht, Charles A. (Phillips, WI)
Schlicht A-100 Parasol

Schmid, H. G. (Switzerland)
Schmid Korsa (Corsica) T I

Schmide (Austria)
Schmide I

Schmidt, Theodore (Seattle, WA)
Schmidt V/STOL Aircraft (1966 Patent)

Schmitt (Ateliers de Constructions Mécaniques & Aéronautiques Paul Schmitt) (France)
Schmitt (Paul) BN 3/4
Schmitt (Paul) C2
Schmitt (Paul) Floatplane
Schmitt (Paul) S.B.R.
Schmitt (Paul) S.F.R.
Schmitt (Paul) Variable Incidence Biplane
Schmitt (Paul) 3
Schmitt (Paul) 6
Schmitt (Paul) 7 (B.R.A.H.)
Schmitt (Paul) 9
Schmitt (Paul) 10 B2
Schmitt (Paul) 1910 Tailless

Schmitt (Maurice), Flostoy, et Rigaud (Roger) (SFR) *(See: SCA)*

Schmitt, Maximilian (Paterson, NJ)
Schmitt (Maximilian) 1915 Monoplane

Schmuck Aircraft Co (Los Angeles, CA)
Schmuck Commercial Sport No.1
Schmuck S.3 Biplane

Schmutzhart, Berthold (Washington, DC)
Schmutzhart SCH-1 (Glider)

Schneider (Établissements Schneider, Service d'Aviation) (France)
Schneider (France) Bpr 3

Schneider (France) Quadrimotor Type Henri-Paul
Schneider (France) 10M

Schneider (Edmund & Harry) (Germany, Australia)
Schneider (Edmund) Grunau Baby (See: Grunau)
Schneider (Edmund & Harry) ES-59 Arrow
Schneider (Edmund & Harry) ES-60
Schneider (Edmund & Harry) ES-60 Series 2 Boomerang
Schneider (Edmund & Harry) ES-60B Super Arrow

Schneider, Franz (Germany)
Schneider (Franz) 1918 Biplane

Schneider, Fred P. (New York, NY)
Schneider (Fred P.) Biplane, c.1908
Schneider (Fred P.) Biplane, c.1910

Schoenfeldt, Bill (Long Beach, CA)
Schoenfeldt Special "Firecracker"

Schreck, Louis (France)
In 1913 aircraft designer and engineer Louis Schreck and Lt. Jean de Conneau (aka "André Beaumont") established **Franco-British Aviation Co Ltd (FBA)** with British capital but with airframe construction to take place in France under Schreck's direction. After World War I, Schreck, as "Constructeur" for FBA, began to affix his own name to FBA aircraft, although the company name did not change. All Schreck designs pre-dating 1913 are filed under **Schreck**, with subsequent work under **FBA**.
Schreck Biplane
Schreck Hydrapsilon (1912)

***Schreck-FBA** (See: FBA)*

Schreder, Richard E. "Dick" (Bryan, OH)
Schreder Airmate HP-7 Sailplane
Schreder RHJ-7 Sailplane
Schreder Airmate HP-8 Sailplane
Schreder RHJ-8 Sailplane
Schreder Airmate HP-9 Sailplane
Schreder RHJ-9 Sailplane
Schreder Airmate HP-10 Sailplane
Schreder Airmate HP-11 Sailplane
Schreder Airmate HP-11A Sailplane
Schreder HP-12 Sailplane
Schreder HP-12A Sailplane
Schreder HP-13 Sailplane
Schreder HP-14 Sailplane
Schreder HP-14C Sailplane
Schreder HP-14T Sailplane
Schreder HP-15 Sailplane
Schreder RS-15 Sailplane
Schreder HP-16 Sailplane
Schreder HP-17 Sailplane
Schreder HP-18 Sailplane
Schreder HP-19 Sailplane
Schreder HP-20 Sailplane

Schroeder, Arlo E. (Newton, KS)
Schroeder Hawk-Pshaw (Schroeder-Meyer Special)

Schroeder, Ernest H. (San Francisco, CA)
Schroeder Cyclogiro

Schroeder-Wentworth Associates (Rudolph W "Shorty" Schroeder & John R Wentworth) (Glencoe, IL)
Schroeder-Wentworth Mercury S.W.A. (1929 Safety Plane)

Schule von Hilvety, Georges C. (Argentina)
Schule von Hilvety 1923 Parasol

Schuler (Germany)
Schuler 1909 Monoplane

Schultz, Art (USA)
Schultz ABC Glider
Schultz Nucleon Glider
Schultz TG-16 (AAF ABC Glider)

Schulze, Gustav (Germany)
Schulze 1908/09 FlugMachine
Schulze 1912 Eindecker I
Schulze 1912 Eindecker II
Schulze 1912 Eindecker III
Schulze 1913 Eindecker
Schulze 1922 Glider

***Schmuck Aircraft** (See: Monarch)*

Schütte-Lanz (Luftfahrzeugbau Schütte-Lanz) (Germany)
Schütte-Lanz C.I
Schütte-Lanz D.I
Schütte-Lanz D.II
Schütte-Lanz D.III
Schütte-Lanz D.IV
Schütte-Lanz D.V
Schütte-Lanz D.VI
Schütte-Lanz D.VII
Schütte-Lanz D.I
Schütte-Lanz Dr.I
Schütte-Lanz G.I
Schütte-Lanz G.II
Schütte-Lanz G.III
Schütte-Lanz G.IV
Schütte-Lanz G.V
Schütte-Lanz Schül (School) R.I
Schütte-Lanz Übersee-Flugzeug

Schwade Flugzeug und Motorenbau (Germany)
Schwade Farman-Type Biplane
Schwade Taube (Dove)
Schwade 1914 Single-Seat Fighter
Schwade 1915 Single-Seat Fighter

Schwarzwald (USA, Germany)
Schwarzwald Mü 13D3 (See: Akadflieg München)

Schweizer Aircraft Corp (Elmira, NY)
Schweizer Company Designations
Schweizer aircraft are organized by model number; each model number has a prefix indicating the number of seats. Many aircraft also have three-letter prefixes, with the first two letters (SG) standing for "Schweizer Glider," and the final letter designating the glider type ("C" for Cargo, "P" for Primary training, "S" for Sailplane, or "M" for Motorized Glider).
Schweizer SGU 1-1 (SGP 1-1)
Schweizer SGU 1-2
Schweizer SGU 1-3
Schweizer SGU 1-6
Schweizer SGU 1-7
Schweizer SGS 2-8
Schweizer SGC 8-10
Schweizer SGC 15-11
Schweizer SGS 2-12
Schweizer SGC 6-14
Schweizer SGC 1-15
Schweizer SGU 1-16
Schweizer SGS 1-17
Schweizer SGS 2-18
Schweizer SGU 1-19
Schweizer SGU 1-19, Powered (SGM 1-19)
Schweizer SGS 1-20 (SGU 1-20)
Schweizer SGS 1-21
Schweizer SGU 2-22
Schweizer SGU 2-22EK, NASM
Schweizer SGS 1-23 Family
Schweizer SGS 1-24
Schweizer SGS 2-25
Schweizer SGS 1-26
Schweizer SGS 1-26A

Schweizer SGS 1-26B
Schweizer SGS 1-26C
Schweizer SGS 1-26D
Schweizer SGS 1-26E
Schweizer 2-27
Schweizer 7-28
Schweizer SGS 1-29
Schweizer 1-30 Motorized Glider
Schweizer 2-31
Schweizer SGS 2-32
Schweizer SGS 2-33
Schweizer SGS 1-34
Schweizer SGS 1-35
Schweizer SGS 1-36 Sprite
Schweizer SGM 2-37 (SA 2-37)
Schweizer SA 2-38A Condor
Schweizer TSC-1A1 Teal Amphibian (See:Patchen)

Schweizer Military Designations
Schweizer LNS-1 (SGS 2-8)
Schweizer TG-2 (SGS 2-8)
Schweizer TG-2A (SGS 2-8)
Schweizer XTG-3 (SGS 2-12)
Schweizer TG-3 (SGS 2-12)
Schweizer TG-3A (SGS 2-12)
Schweizer TG-3 (SGS 1-26E)
Schweizer TG-4A (SGS 2-33)
Schweizer TG-7A (SGM 2-37)
Schweizer X-26A
Schweizer X-27A Tow Target Plane

Schweyer *(See: Jacobs Schweyer)*

SCIM (Société Generale des Constructions Industrielles et Méchaniques) *(See: Borel)*

Scibor-Rylski, Adam (Poland)
Scibor-Rylski SR-3

Scintex (Scintex-Aviation SA) (France) *(See also: Piel)*
Scintex Emeraude (Emerald)
Scintex Super Emeraude (Emerald)

Scott, G. E. (Long Beach, CA)
Scott 1928 3-place Biplane

Scottish Aviation (UK)
Scottish Aviation Bulldog 120
Scottish Aviation (Pioneer) A.4/45 (Prototype)
Scottish Aviation Pioneer CC.Mk.1
Scottish Aviation Prestwick Pioneer
Scottish Aviation Twin Pioneer
Scottish Aviation Twin Pioneer, Modified

Scraggs, Roy B. (Eugene, OR)
Scraggs 1929 Dart

Scrogham, Charles (Weed Patch Hill, IN)
Scrogham Ornithopter

SEA (Société d'Études Aéronautiques) (France)
SEA 1
SEA 4 C2
SEA 4 Airline Conversion (See: Potez VII)
SEA 4 PM
SEA VII (See: Potez VII)
SEA Floatplane

Seahawk Industries Inc *(See: Condor Aircraft)*

Seaplanes, General

Searcy, William N. (Silverton, CO)
Searcy 1911 Flying Machine Patent

Seaton-Karr *(See: Wright, Howard, 1908 Pusher Biplane)*

Seattle Aeromotive Corp (Seattle, WA)
Seattle Transport Monoplane (1930)

Seawind/S.N.A., Inc (Kimberton, PA)
Seawind/S.N.A. Seawind 2000
Seawind/S.N.A. Seawind 2500
Seawind/S.N.A. Seawind 3000

SECAN (Société d'Études et de Constructions Aéro-Navales) (France)
SECAN S.U.C.10 Courlis (Curlew)

SECAT (Société d'Études et de Constructions d'Avions de Tourisme (France)
SECAT LD.45 Biplane
SECAT RG-60 Ultralight Biplane
SECAT RG-75 Cabin Monoplane
SECAT S.4 Monoplane
SECAT S.5 Monoplane

SETCA (France)
SETCA Milan

SECM (Société d'Emboutissage et de Constructions Méchaniques) (France)
In 1916, M. Amiot founded the **Société d'Emboutissage et de Constructions Méchaniques (SECM)** at Colombes, a suburb of Paris, France. This metal pressing and stamping firm built **Morane**, **Sopwith**, and **Breguet** aircraft under license. In 1921 the firm introduced the Type XX series, the first aircraft produced under the SECM name; aircraft of this series are filed under **SECM.** In 1929 SECM amalgamated with **Latham & Cie**, taking over Latham's flying-boat works at Caudebec-en-Caux, France. The combined company was known as **SECM Avions Amiot**; the company's Type 100 series aircraft are filed under **SECM Amiot.** The SECM name was dropped by the mid-1930s, when the company became known as **Avions Amiot.** The Type 300 series aircraft of this period are filed under **Amiot.** With the German occupation of France during World War II, Avions Amiot built Ju 52 aircraft under the control of Junkers. After the war, the former Amiot company was nationalized as **Atelier Aéronautique de Colombes**, and continued to produce Ju 52s under the designation AAC.1; these aircraft are filed under Junkers.
SECM Type 12 BN2
SECM Type XX Family
SECM Type XX Lutèce
SECM Type XXII
SECM Type XXII Modification (E.T.2)
SECM Type XXIII
SECM Type XXIV
SECM Type XXVI

SECM Amiot (SECM Avions Amiot) (France)
In 1916, M. Amiot founded the **Société d'Emboutissage et de Constructions Méchaniques (SECM)** at Colombes, a suburb of Paris, France. This metal pressing and stamping firm built **Morane**, **Sopwith**, and **Breguet** aircraft under license. In 1921 the firm introduced the Type XX series, the first aircraft produced under the SECM name; aircraft of this series are filed under **SECM.** In 1929 SECM amalgamated with **Latham & Cie**, taking over Latham's flying-boat works at Caudebec-en-Caux, France. The combined company was known as **SECM Avions Amiot**; the company's Type 100 series aircraft are filed under **SECM Amiot.** The SECM name was dropped by the mid-1930s, the company then being known as **Avions**

Amiot. The Type 300 series aircraft of this period are filed under **Amiot**. With the German occupation of France during World War II, Avions Amiot built Ju 52 aircraft under the control of **Junkers**. After the war, the former Amiot company was nationalized as **Atelier Aéronautique de Colombes**, and continued to produce Ju 52s under the designation AAC.1; these aircraft are filed under **Junkers**.

SECM Amiot 100
SECM Amiot 101 C1
SECM Amiot 122 B3
SECM Amiot 122 BP3
SECM Amiot 140 M
SECM Amiot 110 C1
SECM Amiot 110 S Hydravion
SECM Amiot 120 BN2
SECM Amiot 120 B3
SECM Amiot 120 S
SECM Amiot 122 BP2
SECM Amiot 122 BP3
SECM Amiot 123 BP3
SECM Amiot 124 BP3
SECM Amiot 130 R2
SECM Amiot 140 M
SECM Amiot 141 M
SECM Amiot 142 M4
SECM Amiot 143 M5
SECM Amiot 143 M5
SECM Amiot 144 M5
SECM Amiot 146
SECM Amiot 147
SECM Amiot 150 BE (150 M)
SECM Amiot 180 BN5

***Secretarea de Marina MX-1 Programa** (See: Dirección General de Reparaciones y Construciones Navales)*

Security (Security National) Aircraft Corp (Long Beach, CA)

Security National Airster S-1
Security National Airster S-1-A
Security National Airster S-1-B

Sedel'nikov-Korvin (A. N. Sedel'nikov V. L. Korvin) (Russia)

Sedel'nikov-Korvin SK

***SEEMS** (See: Morane-Saulnier)*

***Segelflugschule Grunau** (See: Grunau)*

***Segelflugzeugbau Kassel** (See: Kassel)*

***Segrave, Sir Henry** (See: Saunders-Roe A.22)*

Seibel Helicopter Co (Charles M. Seibel) (Wichita, KS)

Seibel YH-24 Sky Hawk
Seibel H-41 (See: Cessna)
Seibel S-3
Seibel S-4

Sellers, Mathew Bacon (Louisville, KY)

Sellers 1908 Quadruplane (Sellers No5)
Sellers 1909 Quadruplane (Sellers No6)
Sellers 1924 Multiplane

Selnaszka, Wladyslow (Poland)

Selnaszka Bozena

Semenaud (France)

Semenaud C1
Semenaud 1918 1-Engine Flying Boat
Semenaud 1918 2-Engine Flying Boat

Semenov-Polikarpov (A. A. Semenov & N. N. Polikarpov) (Russia)

Semenov-Polikarpov PM-1

Semenov-Sutugin-Gorelov (A. A. Semenov, L. I. Sutugin, & S. N. Gorelov) (Russia)

Semenov-Sutugin-Gorelov Tri Druga (Three Friends)

***Sempu Hiko Gakko (Sempu Flying School)** (See: Izaki)*

Sepecat (Société Européene de Production de l'Avion École de Combat et Appui Tactique) (International)

Sepecat Jaguar A
Sepecat Jaguar B (T.Mk.2)
Sepecat Jaguar E
Sepecat Jaguar Export/ International
Sepecat Jaguar Fly-by-Wire (FBW)
Sepecat Jaguar (Hindustan Aeronautics Ltd, License-Built)
Sepecat Jaguar M
Sepecat Jaguar Night Strike
Sepecat Jaguar S (GR.Mk.1)

Sequoia Aircraft Corp (Richmond, VA)

Sequoia Falco (Homebuilt)
Sequoia 300/301 (Homebuilt)

Serrell, Edward Wellman (USA)

Serrell Helicopter (1861)

Sesefsky, S. (Romania)

Sesefsky 1923 Observation Biplane

SET (Societatea pentru Exploatâri Technice) (Romania)

SET.III
SET.XV

Setter, M. B. (USA)

Setter Step Glider

Seversky (USA)

A former Russian aviator, Alexander P. de Seversky founded the **Seversky Aircraft Corp** at Farmingdale, Long Island, NY, in 1931. De Seversky left in a 1939 reorganization, and the company was renamed the **Republic Aviation Corp**. Designs put into production by Seversky are listed under **Seversky**; aircraft put into production after the 1939 reorganization are listed under **Republic**.

Seversky AP-1 (Army Pursuit)
Seversky AP-2 Racer
Seversky AP-4
Seversky AP-7 Racer
Seversky AP-8
Seversky AP-9 Racer
Seversky AT-12 (Model 2PA-204A) Guardsman
Seversky BT-8 (SEV-3XAR)
Seversky EP-1 (European Pursuit)
Seversky EP-1-68
Seversky EP-1-106 (J-9) (Sweden)
Seversky Executive Transport
Seversky NF-1
Seversky P-35
Seversky P-35A
Seversky XP-41
Seversky P-43 (See: Republic)
Seversky XP-44 Warrior (See: Republic)
Seversky SEV-DS "Doolittle Special" Racer
Seversky SEV-S2 (S-2) Racer
Seversky SEV-S2 (S-2) "Drake Bullet" Racer
Seversky SEV-1XP (Experimental Pursuit)
Seversky SEV-2PA (2PA) Convoy Fighter
Seversky SEV-2PA-A (2PA-A)

Seversky SEV-2PA-B3 *Dick* (2PA-B3) (A8V1) (Japan)
Seversky SEV-2PA-L (2PA-L)
Seversky SEV-2PA-204A (2PA-204A) (B-6) (Sweden)
Seversky SEV-2XP
Seversky SEV-3
Seversky SEV-3M (Military)
Seversky SEV-3M-WW (Wright Whirlwind)
Seversky SEV-3XAR
Seversky Super Clipper
Seversky X-BT

SFAN (Société Française d'Aviation Nouvelle) (France)
SFAN 2 (II)
SFAN H 3
SFAN 4
SFAN 5

SFCA (Société Française de Constructions Aéronautiques (France)
SFCA Maillet 20 (201)
SFCA MN (André Maillet & Edmond Nennig) Type A
SFCA Taupin (Beetle)

SFECMAS (Société Française d'Études et de Constructions de Matériels Aéronautiques Spéciaux) (France)
SFECMAS 1402 Gerfaut 1A (Gyrfalcon)

***SFG** (See: Grunau)*

***SFLA** (See: Société Francaise de Locomotion Aerienne)*
***SFR** (See: SCA)*

Sfreddo & Paolina (Jorge Sfreddo & Luis Paolina) (Argentina)
Sfreddo & Paolina SYP-I El Nacional
Sfreddo & Paolina SYP-II

***SGCIM (Société Generale des Constructions Industrielles et Mechaniques)** (See: Borel)*

Shaanxi Transport Aircraft Factory (China)
Shaanxi Y-8 (Yun-8) *Cub* (Antonov An-12BP)

Shackleton-Murray (W. S. Shackleton & C. Lee Murray) (UK)
Shackleton-Murray SM.1

Shanghai Aviation Industrial Corp (SAIC) (China)
Shanghai MD-82 (McDonnell Douglas License)
Shanghai Y-10 (Yun-10)

Shannon & Buente (Harry D. Shannon, Houma, LA & Benjamin E. Buente, Jr, Evansville, IN)
Shannon & Buente SB-1 Special

Shavrov, Vadim Borisovich (Russia)
Shavrov Sh-1
Shavrov Sh-2 Shavruske
Shavrov Sh-3
Shavrov Sh-5
Shavrov ASh-5
Shavrov Sh-7

Shcherbakov, A. Y. (Russia)
Shcherbakov Shche-2 Shchuka (Pike)
Shcherbakov TS-1 STOL

***Shearer, H. A.** (See: Irvine, J. C.)*

Shenstone, B. S. (Canada)
Shenstone Harbinger

Shenyang Aircraft Co (SAC) (China)
Shenyang HU-1 Seagull Motor Glider
Shenyang HU-2 (See: Nanchang)
Shenyang J-1 (F-4) *Fagot* (MiG-15)
Shenyang JJ-1 (FT-4) *Fagot* (MiG-15UTI)
Shenyang J-5 (F-5) *Fresco* (MiG-17F)
Shenyang JJ-5 (MiG-17UTI)
Shenyang J-6 (Jian-6, F-6) *Farmer* (MiG-19)
Shenyang J-6 III
Shenyang JJ-6 (Jianjiao-6, FT-6) *Farmer* (MiG-19UTI)
Shenyang J-7 *Fishbed*(See: Xian)
Shenyang JJ-7 (See: Xian)
Shenyang J-8 (Jian-8, F-8) *Finback-A*
Shenyang J-8 II (F-8 II) *Finback-B*
Shenyang Q-5 *Fantan* (See: Nanchang)

Shepherd, F. R. (Riverbank, CA)
Shepherd Skylark (1927)

Sherpa Aircraft Manufacturing Co (Aloha, OR)
Sherpa Aircraft Sherpa

***Shevchenko-Nikitin** (See: Nikitin-Shevchenko)*

Shigeno, Kiyotake (Japan)
Shigeno Wakadori-Go (Young Bird) Aeroplane (1912)

Shijiazhuang Aircraft Plant (China)
Shijiazhuang Qingting (Dragonfly) W5A
Shijiazhuang Qingting W5B
Shijiazhuang Qingting W6
Shijiazhuang Y5 (An-2)

Shin Meiwa Industry Co, Ltd (Japan)
Shin Meiwa PS-1 (PX-S)
Shin Meiwa SS-2A
Shin Meiwa US-1

***Shinn Engineering Inc** (See: Morrisey)*

Shirato Hikoki Kenkyusho (Japan)
Shirato Asahi-Go Aeroplane
Shirato Anzani Ground Taxi Trainer
Shirato Iwao-Go Aeroplane
Shirato Takeru-Go Aeroplane
Shirato Kaoru-Go (Fragrance)
Shirato 20
Shirato 25 Kuma-Go
Shirato 28 Trainer
Shirato 31 Trainer
Shirato 32 Racer
Shirato 37 Racer
Shirato 38 Trainer
Shirato 40 Trainer

Shishmarev, M. M. (Russia)
Shishmarev R III

Shoemaker, Joseph Clark (Bridgeton, NJ)
Shoemaker (Shoemaker-Chanonhouse) Biplane

Short Brothers (UK)
In November 1908, Hugh Oswald, Albert Eustace, and Horace Leonard Short established **Short Brothers Ltd** to build Wright aircraft under license in England. In 1919 the name was revised to **Short Brothers (Rochester & Bedford) Ltd.** In 1936 Short Brothers convinced the British Air Ministry to establish a new factory in Belfast and established **Short & Harland Ltd,**

jointly owned by Short Brothers and Harland & Wolff, to manage the plant. In November 1947 Short Brothers and Short & Harland merged to form **Short Brothers & Harland Ltd** of Belfast. In June 1977 the company readopted the name **Short Brothers Ltd.** By the 1980s, 100% of the company stock was held by the British Government and, in 1984, the company was renamed **Short Brothers PLC.** In October 1989 the company was purchased by **Bombardier Inc of Canada** to become the **European Group of Bombardier Aerospace**, although the company still operates under the Short Brothers name.

The company name has been known variously as **"Short"** or **"Shorts,"** particularly after World War II. Although these two terms have often been used interchangeably, the term "Short" was generally used by the company when referring to itself up until 1947. With the merger of Short Brothers and Short & Harland in 1947, company literature began to use the term "Shorts" with some consistency (although aircraft would continue to be marked with the term "Short" for another twenty years). For the purposes of this listing, most aircraft designs will be listed as "Short;" the term "Shorts" will be used only on those designs so-listed in company releases (such as the Shorts S.C.5 Britannic).

Beginning with, the 26th aircraft built at the Short Eastchurch and Rochester factories, the company issued constructor's numbers with an "S." prefix. In 1921 the company instituted a series of "Design Index" numbers (using the same "S." prefix), as well as several separate series of design numbers with "P.D.," "S.B.," "S.C.," and "S.D." prefixes. To avoid confusion, the Short/Shorts aircraft are divided into three sections: one for pre-1921 designs, a second for post-1920 designs with Preliminary Design or Design Index numbers, and a third for unnumbered post-1920 designs.

Short, Airplane Designs Through 1921

Short Type A (Admiralty Type 166 Seaplane)
Short Admiralty Type 166 Seaplane
Short Admiralty Type 827 Seaplane
Short Admiralty Type 830 Seaplane
Short Type B (Admiralty Type 178 Seaplane)
Short Biplane No.1 (1909)
Short Biplane No.2 (1909)
Short Biplane No.3 (1909)
Short Bomber, Long Fuselage (1915)
Short Bomber, Short Fuselage (1915)
Short Bomber, Twin-Engined (1915)
Short Type C (Improved Admiralty Type 74 Seaplane)
Short N.1B Shirl
Short N.1B Shirl "Shamrock" (S.538)
Short N.2A Experimental Scout Seaplane No.1 (S.313)
Short N.2A Experimental Scout Seaplane No.2 (S.313)
Short N.2A Experimental Scout Seaplane No.3 (S.364)
Short N.2B (N66 & N67)
Short N.3 Cromarty
Short S.26 Pusher Biplane "The Dud"
Short S.27 Type Pusher Biplane
Short Improved S.27 Type Pusher Biplane
Short S.27 Type Tandem Twin Biplane ("Gnome Sandwich," "The Vacuum Cleaner")
Short S.28 Type Pusher Biplane "Little Willy" (1910)
Short S.29 Type Pusher Biplane
Short S.32 Type Pusher Biplane
Short Rebuilt S.32 Type Pusher Biplane
Short S.33 Type Pusher Float Biplane
Short S.34 Type Pusher Biplane
Short S.35 Type Pusher Biplane
Short S.36 Type Tractor Biplane
Short S.38 Type Pusher Biplane
Short S.39 Type Triple-Twin Biplane
Short S.39 Type Pusher Biplane
Short S.41 Type Tractor Biplane
Short Improved S.41 Type (Admiralty 42)
Short S.45 Type Tractor Biplane
Short S.45 Type Dual-Control Tractor Biplane
Short S.46 Type Twin-Gnome Monoplane "Double-Dirty"
Short S.47 Type Triple-Tractor Biplane "Field Kitchen" (T4)
Short S.63 Type Seaplane Folder (RNAS 81 & 82)
Short S.65 Type Seaplane Folder (RNAS 89 & 90)
Short S.68 Type Seaplane Folder
Short S.69 Type Tractor Seaplane (RNAS 74 - 80)
Short S.80 Type Nile Seaplane
Short S.81 Type Gunbus Seaplane
Short S.82 Type Seaplane Folder (RNAS 119 - 122, 186)
Short S.86 Type Torpedo Seaplane Folder (RNAS 186)
Short S.87 Type Tractor Seaplane (RNAS 135-136)
Short S.89 Type Torpedo Seaplane (Type B, RNAS 178)
Short S.90 Type Tractor Seaplane Folder (RNAS 161 - 166)
Short S.301 Type Seaplane (RNAS 9781 - 9790)
Short S.106 Type (Admiralty Type 184, Short Two-Two-Five)
Short S.106 Type Cut Short (Modified Admiralty Type 184)
Short Modified S.106 Type (Admiralty 184-B) (See: Mann Egerton)
Short Modified S.106 Type, Martin Stabilizer (Admiralty Type 184)
Short Improved S.106 Type (Admiralty Improved Type 184)
Short S.245 (Admiralty Type 184-D Seaplane)
Short Silver Streak (S.543)
Short Sporting Type Seaplane
Short 50hp Gnome Monoplane (Possibly RNAS 14)
Short 310 Type A (310-A) Seaplane
Short 310 Type A4 (310-A4, Type 320) Seaplane
Short 310 Type B (310-B) Seaplane (North Sea Scout)
Short-Wright Flyer (License-Built Wright Model A)
Short-Wright Glider (1909)

Short/Shorts Aircraft 1921-Present (By Preliminary Design or Design Index Number)

Shorts P.D.1 (B.35/46)
Shorts P.D.2 (R.2/48)
Shorts P.D.3 (Solent M.R. proposal)
Shorts P.D.5 (N.114T)
Shorts P.D.6 (S.A.4 Gyron Testbed)
Shorts P.D.7 Rocket Fighter (F.124T)
Shorts P.D.8
Shorts P.D.9 (B.126T)
Shorts P.D.10

Shorts P.D.12
Shorts P.D.13 (NA.39, M.148)
Shorts P.D.14 Executive
Shorts P.D.15 Freighter
Shorts P.D.17 (G.O.R.339)
Shorts P.D.17/3 VTOL Launching Platform
Shorts P.D.19
Shorts P.D.20 Transatlantic
Shorts P.D.21 VTOL Tactical Transport
Shorts P.D.22 SST
Shorts P.D.23 Naval Strike VTOL
Shorts P.D.24 Air-Bus
Shorts P.D.25 Ground Attack VTOL
Shorts P.D.26 Short-Range Feeder-Liner
Shorts P.D.27 Short-Range Feeder-Liner
Shorts P.D.29 Long-Range SST
Shorts P.D.34 VTOL Launcher
Shorts P.D.44 Naval Strike F.R.
Shorts P.D.45 VTOL Low Level Strike
Shorts P.D.46 "Jumping Jet"
Shorts P.D.47 S.C.5 STOL
Shorts P.D.48 (S.C.9)
Shorts P.D.49 Light Strike VTOL
Shorts P.D.50 (S.C.7 VTOL Testbed)
Shorts P.D.51 Skyvan 2
Shorts P.D.52 (5-Seater, based on Baron)
Shorts P.D.53 (S.C.5 with C-141 Wing)
Shorts P.D.55 NATO V/STOL Tactical Transport
Shorts P.D.56 NATO VTOL Fighter
Shorts P.D.59 V/STOL Skytruk Light Transport
Shorts P.D.60 Short-Range Civil Transport
Shorts P.D.62 Skyvan Feeder-Liner
Shorts P.D.64 Light Communication
Shorts P.D.65 Feeder-Liner
Shorts P.D.66 Light Rotorcraft
Shorts P.D.69 Maritime Recce
Shorts P.D.75 200-Seat Air Bus
Shorts P.D.77 Strategic Transport
Shorts P.D.78 Counter Insurgency
Shorts P.D.80 Feeder-Liner
Short S.1 Cockle (Stellite)
Short S.2 (Metal-Hulled Felixstowe F.5, Tin Five)
Short S.3 Springbok Mk.I
Short S.3a Springbok Mk.II
Short S.3b Chamois
Short S.4 Satellite (Tin Kettle, Parker's Iron Balloon)
Short S.5 Singapore Mk.I
Short S.6 Sturgeon
Short S.7 Mussel I
Short S.7 Mussel II
Short S.8 Calcutta (S.8/1)
Short S.8/2 Calcutta (Rangoon) (See also: Breguet 52-1 & 53-0)
Short S.8/8 Rangoon
Short S.9
Short S.10 Gurnard Mk.I
Short S.10 Gurnard Mk.II
Short S.11 Valetta
Short S.12 Singapore II
Short S.12 Singapore IIA
Short S.12 Singapore IIB
Short S.12 Singapore IIC
Short S.14 Sarafand
Short S.19 Singapore III
Short S.15 K.F.1 (Kawanishi H3K Prototype)
Short S.16 Scion Prototype (Alpha)
Short S.16 Scion I
Short S.16 Scion II
Short S.17 Kent
Short S.17/L (L.17) (Scipio Landplane)
Short S.18 (Knuckleduster, R.24/31)
Short S.20/S.21 Mayo Composite
Short S.20 "Mercury" (Mayo Composite, Upper Component)
Short S.21 "Maia" (Mayo Composite, Lower Component)
Short S.22 Scion Senior
Short S.22 Scion Senior, Sunderland Hull Testbed
Short S.23 Empire Boat (C-Class Flying Boat)
Short S.23M (Empire Boat Military Conversion)
Short S.24
Short S.25 Sunderland Prototype (R.2/33)
Short S.25 Sunderland Mk.I
Short S.25 Sunderland Mk.II
Short S.25 Sunderland Mk.III
Short S.25 Sunderland Mk.III Civil Conversions
Short S.25 Sunderland Mk.IIIA
Short S.25 Sunderland Mk.IV (Seaford)
Short S.25 Sunderland Mk.V
Short S.25 Sunderland Mk.V Civil
Short S.25 Sunderland GR.Mk.5
Short S.25 (Conversion) Hythe
Short S.25 (Conversion) Sandringham Mk.1
Short S.25 (Conversion) Sandringham Mk.2
Short S.25 (Conversion) Sandringham Mk.3
Short S.25 (Conversion) Sandringham Mk.4
Short S.25 (Conversion) Sandringham Mk.5 (Plymouth Class)
Short S.25 (Conversion) Sandringham Mk.6
Short S.25 (Conversion) Sandringham Mk.7 (Bermuda Class)
Short S.26 Empire Boat (G-Class Flying Boat)
Short S.27
Short S.28
Short S.29 Stirling Prototype (B.12/36)
Short S.29 Stirling Mk.I
Short S.29 Stirling Mk.II
Short S.29 Stirling Mk.III
Short S.29 Stirling Mk.IV
Short S.29 Stirling Mk.V
Short S.30 Empire Boat (Long-Range C-Class Flying Boat)
Short S.31 Stirling (½-Scale S.29 Aerodynamic Test Article)
Short S.32
Short S.33 Empire Boat (C-Class Flying Boat)
Short S.34 Cannon-Armed Bomber (B.1/39)
Short S.35 Shetland Mk.I
Short S.36 Super Stirling (B.8/41)
Short S.37 "Silver Stirling" (S.29 Civil Conversion)
Short S.38 (S.A.1) Sturgeon Prototype
Short S.38 (S.A.1) Sturgeon PR.Mk.1
Short S.39 (S.A.2) Sturgeon TT.Mk.2
Short S.40 Shetland Mk.II (Civil)
Short S.41 (S.A.3, N.7/46)
Short S.42 (S.A.4) Sperrin
Short S.42 (S.A.4) Sperrin, DeHavilland Gyron Testbed
Short S.43 (S.A.5)
Short S.44 (S.A.6) Sealand Prototype
Short S.44 (S.A.6) Sealand Prototype, Modified
Short S.44 (S.A.6) Sealand Production Airframes
Short S.44 (S.A.6) Sealand Production Airframes, "Nadia"
Short S.44 (S.A.6) Sealand Indian Navy
Short S.45 Seaford
Short S.45A Solent Mk.1
Short S.45A Solent Mk.2

Short S.45A Solent Mk.3
Short S.45A Solent Mk.4
Short S.46 (S.A.7)
Short S.47 (S.A.8) Commercial Flying Boat Project
Short S.48 (S.A.9) Military Glider (X.30/46)
Shorts S.312 Tucano (See: EMBRAER)
Short S.B.1 Aeroisoclinic Glider
Short S.B.2 Sealand 2
Short S.B.3 Sturgeon (ASW Proposal)
Short S.B.4 Sherpa
Short S.B.5 1st Configuration
Short S.B.5 2nd Configuration
Short S.B.5 3rd Configuration
Short S.B.5 4th Configuration
Short S.B.6 (P.D.4) Seamew Prototype
Short S.B.6 Seamew AS.Mk.1
Short S.B.7 Sealand III
Short S.B.8 Helicopter
Short S.B.9 Sturgeon TT.Mk.3
Short S.C.1 (P.D.11)
Short S.C.2 Seamew AS.Mk.2
Short S.C.3 (P.D.16)
Shorts S.C.5 Britannic (Belfast Prototype) (P.D.18)
Shorts S.C.5 Britannic 1
Shorts S.C.5 Britannic 2
Shorts S.C.5 Britannic 3
Shorts S.C.5 Britannic 3A
Shorts S.C.5 Britannic 4
Shorts S.C.5/10 Belfast C.Mk.1
Shorts S.C.5/21 Belfast (P.D.47)
Shorts S.C.7 Skyvan Prototype (P.D.36)
Shorts S.C.7 Skyvan I
Shorts S.C.7 Skyvan Series 2
Shorts S.C.7 Skyvan Series 3
Shorts S.C.7 Skyvan Series 3 Skyliner
Shorts S.C.7 Skyvan Series 3M
Short S.C.8 (P.D.43) (2-Seat S.C.1)
Shorts S.D.3-30 (330) Prototypes
Shorts S.D.3-30 (330)
Shorts 330-UTT (Utility Tactical Transport)
Shorts 330-200
Shorts 330-200 Sherpa
Shorts (330-200) C-23A Sherpa
Shorts (330-200) C-23B Sherpa
Shorts SD3-60 (360) Prototypes
Shorts 360 (SD3-60)
Shorts 360 (SD3-60) ADV (Advanced)
Shorts 360-300
Shorts 360-300F

Short/Shorts Aircraft, 1921-Present (Without Design Index Numbers)
Short Nimbus (Sailplane)
Short-Bristow Crusader (1927 Schneider Trophy)
Short-Hafner A.R.IV

Showa Hikoki Kogyo KK (Japan)
Showa L2D2 (Navy Type 0 Transport) *Tabby* (See also: Nakajima)

SIA (France)
SIA BN2
SIA Coanda Lorraine Bomber

SIA (Societa Italiano Aviazione) *(See also: Fiat, also: Pomilio)*
With the outbreak of World War I, the Italian Fiat company obtained a license to build Farman MF.11s. The company also handled development and production of the Savoia-Pomilio SP.1. In 1916 Fiat formed the subsidiary **SIA** (**Societa Italiano Aviazione**), which was renamed **Fiat Aviazione** in Early 1918. Aircraft designed before 1918 are filed under **SIA**, with later development under **Fiat**.
SIA 5B (License-built Farman MF.11)
SIA 7B1
SIA 7B2
SIA 9B
SIA 1200
SIA SP.1 (Savoia-Pomilio)

S.I.A.G. (Società Industrie Aeromarittime Gallinari) *(See: Gallinari)*

SIAI (Società Idrovolanti Alta Italia) (Italy)
SIAI M.V.T.
SIAI S.5
SIAI S.8
SIAI S.9
SIAI S.12
SIAI S.13
SIAI S.16
SIAI S.16*bis*
SIAI S.16*bis* M
SIAI S.16*ter* "Gennariello"
SIAI S.16R
SIAI S.17
SIAI S.19
SIAI S.21
SIAI S.22
SIAI S.23
SIAI S.23 Passenger Version
SIAI S.24
SIAI S.29
SIAI S.31
SIAI S.43
SIAI S.51 & Subsequent (See: Savoia-Marchetti)

SIAI (Società Italiana Aeroplani Idrovolanti) (See: Savoia-Marchetti)

SIAI-Marchetti (Italy)
SIAI-Marchetti FN-333 Riviera (See: Nardi; also: Lane)
SIAI-Marchetti S.205 (See also: Waco Aircraft Co)
SIAI-Marchetti S.208
SIAI-Marchetti S.210
SIAI-Marchetti S.211
SIAI-Marchetti SA.202 Bravo (See also: Waco Aircraft Co)
SIAI-Marchetti SF.250
SIAI-Marchetti SF.260 Minerva (See also: Waco Aircraft Co)
SIAI-Marchetti SF.600TP Canguro (Kangaroo)
SIAI-Marchetti SH-4 Silvercraft
SIAI-Marchetti SM.101
SIAI-Marchetti SM.102
SIAI-Marchetti SM.1019
SIAI-Marchetti 3V-1 Eolo

SIAT WMD (Siebelwerke ATG GmbH) (Germany)
SIAT 223

Siddeley Deasy Motor Car Co Ltd (UK)
Siddeley Deasy R.T.1
Siddeley Deasy S.R.2 Siskin
Siddeley Deasy Siskin II (See: Armstrong Whitworth)
Siddeley Deasy Sinaia

Sido, Jozef (Poland)
Sido S.1
Sido S.2

Siebel Flugzeugwerke KG (Germany)
Siebel Fh 104 Hallore (See: Halle)
Siebel Si 201
Siebel Si 202 Hummel (Bumble Bee)
Siebel Si 204
Siebel Si 204 A
Siebel Si 204 D

Siebelwerke ATG GmbH *(See: SIAT)*

Siegel, Mieczyslaw (Poland)
Siegel Lightplane (1927)
Siegel MS 1 Glider
Siegel MS 2 Glider
Siegel MS 3 Glider
Siegel MS 8 Glider

Siegrist, Rudolf (Valley City, OH)
Siegrist AS1 Ilse

Siemens, Werner von (Germany)
Siemens 1847 Ornithopter

Siemens-Halske *(See: Siemens-Schuckert)*

Siemens-Schuckert Werke (SSW) (Germany)
Siemens-Schuckert B
Siemens-Schuckert Bulldogg
Siemens-Schuckert Bulldogg, Mercedes
Siemens-Schuckert D.D5
Siemens-Schuckert D.Dr.I
Siemens-Schuckert D.I
Siemens-Schuckert D.Ia
Siemens-Schuckert D.Ib
Siemens-Schuckert D.II
Siemens-Schuckert D.IIe
Siemens-Schuckert D.III
Siemens-Schuckert D.III, Long
Siemens-Schuckert D.III, Short
Siemens-Schuckert D.IV
Siemens-Schuckert D.IVa
Siemens-Schuckert D.V
Siemens-Schuckert D.VI
Siemens-Schuckert E.I
Siemens-Schuckert E.II
Siemens-Schuckert Kann Project (R-Plane)
Siemens-Schuckert L.I (G.III)
Siemens-Schuckert R.VIII
Siemens-Schuckert R.IX
Siemens-Schuckert Taube (Dove) Type
Siemens-Schuckert 1909 Aeroplane
Siemens-Schuckert 1910 Aeroplane
Siemens-Schuckert 1911 Monoplane
Siemens-Schuckert Forssman, First version
Siemens-Schuckert Forssman, Final version
Siemens-Schuckert Steffen R-Planes
Siemens-Schuckert Steffen R.I
Siemens-Schuckert Steffen R.II
Siemens-Schuckert Steffen R.III
Siemens-Schuckert Steffen R.IV
Siemens-Schuckert Steffen R.V
Siemens-Schuckert Steffen R.VI
Siemens-Schuckert Steffen R.VII

Siemens-Waco *(See: Waco 125)*

Sien Nin Hai (China)
Sien Nin Hai Single-Seat Reconnaissance Seaplane
Sien Nin Hai Two-Seat Trainer
Sien Nin Hai Three-Seat Touring Biplane

Sierra *(See: Aircraft Industries, Inc)*

Sigismund von Preussen, Prinz Friedrich (Germany)
Sigismund Bulldogge Monoplane (See: Forssman)
Sigismund 1910 Monoplane
Sigismund 1911 Monoplane
Sigismund 1914 Monoplane

Sikorsky, Igor Ivanovich (Russia)
Igor Ivanovich Sikorsky began designing helicopters and aircraft in 1909. In March 1918, he left Russia and, in time, immigrated to the United States, where he resumed his aviation activities and established **Sikorksy Aero Engineering Corp.** For all aircraft designed by Sikorsky prior to his leaving Russia, see **Sikorsky, Igor Ivanovich.** For all Sikorsky's work in America, see **Sikorsky.**
Sikorsky (Igor) Helicopter No.1 (1908)
Sikorsky (Igor) Helicopter No.2 (1910)
Sikorsky (Igor) S-1
Sikorsky (Igor) S-2
Sikorsky (Igor) S-3
Sikorsky (Igor) S-4
Sikorsky (Igor) S-5
Sikorsky (Igor) S-5A
Sikorsky (Igor) S-6
Sikorsky (Igor) S-6A
Sikorsky (Igor) S-6B
Sikorsky (Igor) S-7
Sikorsky (Igor) S-8
Sikorsky (Igor) S-9
Sikorsky (Igor) S-10, 1913
Sikorsky (Igor) S-10, Hydro
Sikorsky (Igor) S-10A
Sikorsky (Igor) S-10B
Sikorsky (Igor) S-11
Sikorsky (Igor) S-12
Sikorsky (Igor) S-13
Sikorsky (Igor) S-14
Sikorsky (Igor) S-15
Sikorsky (Igor) S-16
Sikorsky (Igor) S-17
Sikorsky (Igor) S-18
Sikorsky (Igor) S-19
Sikorsky (Igor) S-20
Sikorsky (Igor) S-21 Grand "Bol'shoi Bal'tisky" (Great Baltic)
Sikorsky (Igor) S-21 Grand "Russkiy Vityaz" (Russian Knight)
Sikorsky (Igor) S-22 Il'ya Muromets
Sikorsky (Igor) S-22 Il'ya Muromets Beh
Sikorsky (Igor) S-23 Il'ya Muromets Veh
Sikorsky (Igor) S-24G-1 Il'ya Muromets
Sikorsky (Igor) S-25G-2 Il'ya Muromets
Sikorsky (Igor) S-25G-3 Il'ya Muromets
Sikorsky (Igor) S-25G-4 Il'ya Muromets
Sikorsky (Igor) S-26D-1 Il'ya Muromets
Sikorsky (Igor) S-26D-2 Il'ya Muromets
Sikorsky (Igor) S-27E

Sikorsky (Stratford, CT)
In 1923, Russian émigré Igor Sikorsky established the **Sikorsky Aero Engineering Corp.** The company was reorganized as the **Sikorsky Manufacturing Corp** in 1925 and as the **Sikorsky Aviation Corp** in 1928. In 1929 Sikorsky became a subsidiary of the **United Aircraft and Transport Corp** (**UATC**, later **United Aircraft Manufacturing Corp** and **United Aircraft Corp, UATC**), in July 1935 becoming the **Sikorsky Aircraft Division** of UATC. In April 1939, UATC merged Sikorsky and its **Chance Vought Division** to form the **Vought-Sikorsky Aircraft Division** of UATC. In 1943, the two operations were again separated, with helicopter development being centered in the re-established **Sikorsky Aircraft Division** of UATC. In January 1995 UATC recreated **Sikorsky Aircraft Corp** as a subsidiary. For aircraft designed by Igor Sikorsky prior to his immigration to the USA, see **Sikorsky, Igor Ivanovich.** For products of the various forms of the Sikorsky Aircraft Corp, including products of the Vought-Sikorsky Division of UATC which originated in Sikorsky offices, see **Sikorsky (USA).**

Sikorsky (USA) C-6
Sikorsky (USA) C-6A
Sikorsky (USA) Y1C-28
Sikorsky (USA) YSH-3A (YHSS-2) Sea King
Sikorsky (USA) NH-3A Sea King
Sikorsky (USA) RH-3A Sea King
Sikorsky (USA) SH-3A (HSS-2) Sea King
Sikorsky (USA) VH-3A (HSS-2Z) Sea King
Sikorsky (USA) (SH-3A) CHSS-2 Sea King (Canada)
Sikorsky (USA) CH-3B Sea King
Sikorsky (USA) CH-3C Jolly Green Giant
Sikorsky (USA) SH-3D Sea King
Sikorsky (USA) VH-3D Sea King
Sikorsky (USA) CH-3E Jolly Green Giant
Sikorsky (USA) HH-3E Jolly Green Giant
Sikorsky (USA) HH-3F Pelican (USCG)
Sikorsky (USA) SH-3H Sea King
Sikorsky (USA) YH-18A
Sikorsky (USA) YH-18B
Sikorsky (USA) YH-19 Chickasaw
Sikorsky (USA) H-19A (UH-19A) Chickasaw
Sikorsky (USA) SH-19A (HH-19A) Chickasaw
Sikorsky (USA) H-19B (UH-19B) Chickasaw
Sikorsky (USA) SH-19B (HH-19B) Chickasaw
Sikorsky (USA) H-19C (UH-19C) Chickasaw
Sikorsky (USA) H-19D (UH-19D) Chickasaw
Sikorsky (USA) H-34A (CH-34A) Choctaw
Sikorsky (USA) VH-34A Choctaw
Sikorsky (USA) H-34B (CH-34B) Choctaw
Sikorsky (USA) H-34C (CH-34C) Choctaw
Sikorsky (USA) HH-34D Choctaw
Sikorsky (USA) H-34G.I (Germany)
Sikorsky (USA) H-34G.III (Germany)
Sikorsky (USA) YH-37 Mojave
Sikorsky (USA) H-37A (CH-37A) Mojave
Sikorsky (USA) H-37B (CH-37B) Mojave
Sikorsky (USA) XH-39 (S-59)
Sikorsky (USA) HH-52A Seaguard
Sikorsky (USA) CH-53A Sea Stallion
Sikorsky (USA) HH-53B Super Jolly
Sikorsky (USA) HH-53B Super Jolly, Pave Low Testbed
Sikorsky (USA) HH-53C Super Jolly
Sikorsky (USA) CH-53D Sea Stallion
Sikorsky (USA) RH-53D Sea Stallion
Sikorsky (USA) VH-53D Sea Stallion
Sikorsky (USA) CH-53D/G Sea Stallion (Germany)
Sikorsky (USA) CH-53E Super Stallion
Sikorsky (USA) MH-53E Sea Dragon
Sikorsky (USA) MH-53H Pave Low III
Sikorsky (USA) MH-53J Pave Low IIIE
Sikorsky (USA) YCH-54A Tarhe
Sikorsky (USA) CH-54A Tarhe
Sikorsky (USA) CH-54B Tarhe
Sikorsky (USA) XH-59A Advancing Blade Concept Demonstrator
Sikorsky (USA) XH-59B Advancing Blade Concept Demonstrator
Sikorsky (USA) EH-60A (EH-60C) Black Hawk (Quick Fix II)
Sikorsky (USA) HH-60A Credible Hawk
Sikorsky (USA) UH-60A Black Hawk
Sikorsky (USA) EH-60B Black Hawk (SOTAS Testbed)
Sikorsky (USA) SH-60B Sea Hawk
Sikorsky (USA) UH-60C Black Hawk
Sikorsky (USA) HH-60D Night Hawk
Sikorsky (USA) SH-60F Ocean Hawk (CV Hawk)
Sikorsky (USA) HH-60G Pave Hawk
Sikorsky (USA) MH-60G Pave Hawk
Sikorsky (USA) HH-60H Rescue Hawk
Sikorsky (USA) HH-60J Jayhawk
Sikorsky (USA) MH-60K Black Hawk
Sikorsky (USA) UH-60L Black Hawk
Sikorsky (USA) VH-60N (VH-60A) White Hawk
Sikorsky (USA) UH-60P Black Hawk
Sikorsky (USA) UH-60Q Black Hawk
Sikorsky (USA) SH-60R Sea Hawk
Sikorsky (USA) XHJS-1
Sikorsky (USA) HNS-1
Sikorsky (USA) HNS-1, USCG
Sikorsky (USA) HOS-1
Sikorsky (USA) HOS-1, USCG
Sikorsky (USA) HO2S-1
Sikorsky (USA) HO3S-1
Sikorsky (USA) HO3S-1, Autopilot Testbed
Sikorsky (USA) HO3S-1G (USCG)
Sikorsky (USA) HO4S-1
Sikorsky (USA) HO4S-1G (USCG)
Sikorsky (USA) HO4S-2G (USCG)
Sikorsky (USA) HO4S-3 (UH-19F)
Sikorsky (USA) HO4S-3G (HH-19G)
Sikorsky (USA) HO5S-1
Sikorsky (USA) HO5S-1, USCG
Sikorsky (USA) HRS-1
Sikorsky (USA) HRS-2
Sikorsky (USA) HRS-2 ROR (Rocket-On-Rotor) Testbed
Sikorsky (USA) HRS-3 (CH-19E)
Sikorsky (USA) XHR2S-1
Sikorsky (USA) HR2S-1 (CH-37C)
Sikorsky (USA) HR2S-1W
Sikorsky (USA) XHSS-1 (YSH-34G) Seabat
Sikorsky (USA) HSS-1 (SH-34G) Seabat
Sikorsky (USA) HSS-1 (SH-34G) Seabat (France)
Sikorsky (USA) HSS-1 (SH-34G) Seabat (Italy)
Sikorsky (USA) HSS-1F (SH-34H) Seahorse
Sikorsky (USA) HSS-1N (SH-34J) Seabat
Sikorsky (USA) HSS-2 (See: H-3)
Sikorsky (USA) HUS-1 (UH-34D) Seahorse
Sikorsky (USA) HUS-1 (UH-34D) Seahorse, NASM
Sikorsky (USA) HUS-1A (UH-34E) Seahorse
Sikorsky (USA) HUS-1G (HH-34F) Seahorse (USCG)
Sikorsky (USA) HUS-1Z (VH-34D) Seahorse
Sikorsky (USA) JRS-1
Sikorsky (USA) JRS-1, NASM
Sikorsky (USA) Y1OA-8 (OA-8)
Sikorsky (USA) XPS-1
Sikorsky (USA) XPS-2 (XRS-2)
Sikorsky (USA) PS-3 (RS-3)
Sikorsky (USA) XP2S-1
Sikorsky (USA) XPBS-1 (Flying Dreadnaught)
Sikorsky (USA) RS-1
Sikorsky (USA) RS-4 (Impressed S-38A)
Sikorsky (USA) RS-5 (Impressed S-41A)
Sikorsky (USA) XR-4
Sikorsky (USA) XR-4, NASM
Sikorsky (USA) YR-4A
Sikorsky (USA) YR-4B (YH-4B)
Sikorsky (USA) R-4B (H-4B)
Sikorsky (USA) R-4B Civil Conversions
Sikorsky (USA) R-4B Single-Blade Rotor Testbed
Sikorsky (USA) R-4B Tail Rotor Testbeds

Sikorsky (USA) (R-4) Hoverfly Mk.I
Sikorsky (USA) XR-5 (VS-317)
Sikorsky (USA) XR-5 (VS-317), NASM
Sikorsky (USA) YR-5A (YH-5A)
Sikorsky (USA) R-5A (H-5A)
Sikorsky (USA) R-5D (H-5D)
Sikorsky (USA) YR-5E (YH-5E)
Sikorsky (USA) R-5F (H-5F)
Sikorsky (USA) (R-5) H-5G (C)
Sikorsky (USA) (R-5) H-5H
Sikorsky (USA) XR-6
Sikorsky (USA) XR-6A
Sikorsky (USA) YR-6A (YH-6A)
Sikorsky (USA) R-6A (H-6A)
Sikorsky (USA) XSS-2
Sikorsky (USA) S-28
Sikorsky (USA) S-29-A
Sikorsky (USA) S-31
Sikorsky (USA) S-32
Sikorsky (USA) S-33 Messenger
Sikorsky (USA) S-34
Sikorsky (USA) S-35 (Fonck 1926 Transatlantic Attempt)
Sikorsky (USA) S-36A
Sikorsky (USA) S-36B
Sikorsky (USA) S-37 "Ville de Paris"/ "Southern Star"
Sikorsky (USA) S-37-B "Guardian" (Bomber Variant)
Sikorsky (USA) S-38 Amphibion Prototype
Sikorsky (USA) S-38A Amphibion
Sikorsky (USA) S-38AH Amphibion
Sikorsky (USA) S-38B Amphibion
Sikorsky (USA) S-38B Amphibion "'untin Bowler"
Sikorsky (USA) S-38BH Amphibion
Sikorsky (USA) S-38BS Amphibion "Osa's Ark"
Sikorsky (USA) S-38BT Amphibion
Sikorsky (USA) S-38C Amphibion
Sikorsky (USA) S-39 Sport Amphibion, 1st Prototype
Sikorsky (USA) S-39 Sport Amphibion, 2nd Prototype
Sikorsky (USA) S-39A Sport Amphibion
Sikorsky (USA) S-39B Sport Amphibion
Sikorsky (USA) S-39C Sport Amphibion
Sikorsky (USA) S-39CS Sport Amphibion "Spirit of Africa"
Sikorsky (USA) S-40 Clipper
Sikorsky (USA) S-40A Clipper
Sikorsky (USA) S-41 Amphibion Prototype
Sikorsky (USA) S-41A Amphibion
Sikorsky (USA) S-41B Amphibion
Sikorsky (USA) S-42 Clipper
Sikorsky (USA) S-42A Clipper
Sikorsky (USA) S-42B Clipper
Sikorsky (USA) S-43A Amphibion (Baby Clipper)
Sikorsky (USA) S-43B Amphibion (Baby Clipper)
Sikorsky (USA) S-43W Amphibion (Baby Clipper)
Sikorsky (USA) S-43WB Amphibion (Baby Clipper)
Sikorsky (USA) S-43WH Amphibion (H. Hughes Aircraft)
Sikorsky (USA) VS-44A
Sikorsky (USA) S-51
Sikorsky (USA) S-51, Japan
Sikorsky (USA) S-51, Canada
Sikorsky (USA) S-52-1
Sikorsky (USA) S-52-2
Sikorsky (USA) S-52-3
Sikorsky (USA) S-52, Orlando Helicopters Electric-Powered Mod
Sikorsky (USA) S-54
Sikorsky (USA) S-55
Sikorsky (USA) S-55, France
Sikorsky (USA) S-55, RCAF
Sikorsky (USA) S-55, RCN
Sikorsky (USA) S-55-T (Aviation Specialties Turbine Conversion)
Sikorsky (USA) S-56
Sikorsky (USA) S-58, Canada
Sikorsky (USA) S-58B
Sikorsky (USA) S-58C
Sikorsky (USA) S-58T
Sikorsky (USA) S-60 (Skycrane Prototype)
Sikorsky (USA) S-61 Prototype
Sikorsky (USA) S-61A
Sikorsky (USA) S-61A, Rotoprop Testbed
Sikorsky (USA) S-61A, Denmark
Sikorsky (USA) S-61L
Sikorsky (USA) S-61N
Sikorsky (USA) S-61R
Sikorsky (USA) S-62A
Sikorsky (USA) S-64A Skycrane Prototype
Sikorsky (USA) S-64B Skycrane (Proposal)
Sikorsky (USA) S-64E Skycrane
Sikorsky (USA) S-64F Skycrane
Sikorsky (USA) S-65A Sea Stallion
Sikorsky (USA) S-65C Sea Stallion
Sikorsky (USA) S-65Ö Sea Stallion (Austria)
Sikorsky (USA) S-65-300 Compound Helicopter Proposal
Sikorsky (USA) S-66 (AAFSS Proposal)
Sikorsky (USA) S-67 Blackhawk (Gunship Demonstrator)
Sikorsky (USA) S-69 (See: H-59)
Sikorsky (USA) S-70A Black Hawk
Sikorsky (USA) S-70A Black Hawk, Turkey
Sikorsky (USA) S-70B Sea Hawk
Sikorsky (USA) S-70C
Sikorsky (USA) S-72 Rotor Systems Research Aircraft
Sikorsky (USA) S-72X1 RSRA (X-Wing Proposal)
Sikorsky (USA) S-75 Advanced Composite Airframe Program
Sikorsky (USA) S-76A Spirit
Sikorsky (USA) S-76A+ Spirit
Sikorsky (USA) S-76B Spirit
Sikorsky (USA) (S-76B) H-76 Eagle
Sikorsky (USA) (S-76B) H-76 Eagle Fantail Demonstrator
Sikorsky (USA) (S-76B) H-76N
Sikorsky (USA) S-76C Spirit
Sikorsky (USA) S-80E Super Stallion
Sikorsky (USA) S-80M Sea Dragon
Sikorsky (USA) S-92C Helibus
Sikorsky (USA) S-92IU (S-92M) Helibus (Growth Hawk)
Sikorsky (USA) VS-300 Rotor Test Stand
Sikorsky (USA) VS-300, 1st Configuration
Sikorsky (USA) VS-300, 2nd Configuration
Sikorsky (USA) VS-300, 3rd Configuration
Sikorsky (USA) VS-300, 4th Configuration
Sikorsky (USA) VS-300, 4th Configuration, NASM
Sikorsky (USA) VS-316A (See: HNS, R-4)
Sikorsky (USA) VS-316B (See: HOS, R-6)
Sikorsky (USA) VS-327 (See: HO2S, R-5)

Sil'vansky, A. V. (Russia)
Sil'vansky IS (I-220)

Silesia (First Silesian Aircraft Factory) (Poland)
Silesia S-1
Silesia S-2
Silesia S-3
Silesia S-4
Silesia S-10

Silvaire Aircraft Corp *(See: Luscombe)*

Silver-Wing Aircraft Co, Inc (Boulder, CO)
Silver-Wing 1928 Monoplane

Silvercraft SpA (Italy)
Silvercraft SH-4
Silvercraft SH-200

Silverman/Dunayer (Morris H. Silverman & Harold Dunayer (Coney Island, NY)
Silverman/Dunayer 1936 Primary Glider

Silverston (Milwaukee, WI)
Silverston Milwaukee No 2 Pendulum System Ornithopter

SIMB (Société Industrielle des Métaux et du Bois) *(See: Bernard)*

Simonet (Belgium)
Simonet 1924 Moto Aviette
Simonet 1925 Monoplane Glider

Simmering-Graz-Pauker (SGP) (Austria)
Simmering-Graz-Pauker M222 Flamingo

Simmonds Aircraft Ltd (Sir Oliver Simmonds) (UK) *(See also: Spartan Aircraft Ltd)*
Simmonds Avron
Simmonds Spartan

Simmons, Herbert H. (San Diego, CA)
Simmons SP-1 Sport Monoplane

Simplex (Les Avions Simplex) (France) *(See: Arnoux)*

Simplex Aircraft Co (New Haven, CT)
Simplex (CT) Tractor Biplane (Mayo Type A)

Simplex Aircraft Corp (Defiance, OH)
Simplex (OH) Kite
Simplex (OH) Red Arrow K-2-C
Simplex (OH) Red Arrow K-2-S
Simplex (OH) Red Arrow L-2-S
Simplex (OH) Red Arrow R-2-D
Simplex (OH) Red Arrow W-2-S
Simplex (OH) W-5-C
Simplex (OH) 1910 Monoplane

Sindicato Aereo Argentino *(See: Artigala)*

Sindlinger, Fred G. (Puyallup, WA)
Sindlinger HH-1 5/8th Scale Hawker Hurricane

Singapore Aircraft Industries *(See: SAI)*

Sioux Aircraft Corp Kari-Keen Jr *(See: Kari-Keen 90-B & 90-C)*

Sioux City Glider Club (Sioux City, IA)
Sioux City Glider Club PT Glider No.1

SIPA (Société Industrielle Pour l'Aéronautique) (France)
SIPA S.10 (See: Arado Ar 396)
SIPA S.11
SIPA S.12 Advanced Trainer
SIPA S.47
SIPA 200 Minijet
SIPA 300 Trainer
SIPA 300R
SIPA 901
SIPA 1000 Coccinelle (Ladybug)

Sipowicz, Aleksander (Poland)
Sipowicz 1925 Monoplane

Siren Works and Aerodrome (France)
Siren C-30S Edelweiss
Siren PIK 30 (Developed from Eiri Avion PIK-20E)

Sisler Aircraft Corp (Burnsville, MN)
The SF-2A Cygnet was originally designed by Capt. A. M. Sisler and marketed through the **Sisler Aircraft Corp.** In October 1983 the rights were acquired by **Hapi Engines Inc** of Eloy, AZ. All Sisler designs are listed under **Sisler**.
Sisler SF-1 Pipit
Sisler SF-2 Wistler (Homebuilt)
Sisler SF-2A Cygnet (Homebuilt)

Sisu *(See: Arlington Aircraft Co)*

Skandinavisk Aero Industri A/S (SAI) (Denmark)
SAI K.Z.II Kupe
SAI K.Z.II Sport
SAI K.Z.II Trainer
SAI K.Z.III
SAI K.Z.IV Ambulance
SAI K.Z.VII Lark
SAI K.Z.VIII
SAI K.Z.X

Skandinaviska Aero A.B. (SAAB) (Sweden)
Skandinaviska Aero BHT-1 Beauty

Skoda-Kauba Flugzeugbau (Czechoslovakia)
Skoda-Kauba SK-V1
Skoda-Kauba SK-V1a
Skoda-Kauba SK-V2
Skoda-Kauba SK-V3
Skoda-Kauba SK-V4 (SK-257)
Skoda-Kauba SK-V5
Skoda-Kauba SK-V6
Skoda-Kauba SK SL6
Skoda-Kauba SK-V7
Skoda-Kauba SK-V8
Skoda-Kauba SK-P-14

Skraba, Boleslaw (Poland)
Skraba S.T.3

Skrzhinsky, N. K. (Russia)
Skrzhinsky A-12
Skrzhinsky 2-EA
Skrzhinsky 4-EA

Sky Romer Manufacturing Co Inc (Ft Wayne, IN)
Sky Romer 101

Sky Star Aircraft Corp (Nampa, ID)
Sky Star Kitfox Classic IV
Sky Star Kitfox Safari
Sky Star Kitfox Speedster
Sky Star Kitfox Vixen
Sky Star Kitfox Series 5

Skycraft, Inc (Fort Worth, TX)
Skycraft (TX) Skyshark

Skycraft Industries, Inc (Venice, CA)
Skycraft (CA) Model 447

Skyhopper Airplanes, Inc (Encino, CA)
Skyhopper 10
Skyhopper 11
Skyhopper 20

Skylark *(See: Robertson)*

Skylark Manufacturing Co (Venice, CA)
Skylark Skycraft Model 445

Skyote Aeromarine Ltd (Boulder, CO)
Skyote Homebuilt

Skytrader Corp (Kansas City, MO) *(See Also: Dominion)*
Skytrader Scout-STOL
Skytrader ST1700 Conestoga
Skytrader ST1700 MD Evader

Skyway Engineering Co (Carmel, IN)
Skyway AC-35 Autogiro

Skywise Aviation Pty Ltd *(See: Sadler)*

Slavin, J. J. (Los Angeles, CA)
Slavin 1912 Biplane

Slingsby (UK)

Frederick N. Slingsby, a pilot and partner in a small woodworking and furniture shop, began building gliders the early 1930s. In 1934 he formed the engineering firm of **Slingsby, Russell & Brown Ltd** to continue his work. In July 1939 the company reorganized as **Slingsby Sailplanes Ltd**, but plans to open a new factory were curtailed by the outbreak of World War II. The company resumed civil sales after the war, but in 1955 was placed under the **Shaw-Slingsby Trust.** By the early 1960s, the company had diversified into structural woodworking (windows, etc) to supplement glider construction. In 1967, the company was split into **Slingsby Joinery Ltd** and **Slingsby Aircraft Ltd**, both under **Slingsby Aircraft Holdings Ltd.** Following a devastating factory fire in November 1968, the company's fortunes declined until July 1969 when Slingsby Holdings went into receivership. **Vickers Ltd** acquired the assets of the company, establishing **Slingsby Sailplanes Ltd** as a division of the **Vickers Ltd Shipbuilding Group.** In 1975 the company became **Vickers-Slingsby**, a division of **Vickers Ltd Offshore Engineering Group.** By 1980, the company had separated as the **Aircraft Division of Slingsby Engineering Ltd.** Glider production ended in 1982, with the company emphasizing powered aircraft as **Slingsby Aviation Ltd.** By 1989 the company was a subsidiary of the Aerospace and Defence division of **M. L. Holdings PLC.** All Slingsby products, including those built while a subsidiary of other companies, appear under **Slingsby.**

Slingsby T.1 Falcon 1
Slingsby T.2 Falcon 2
Slingsby T.3 Primary Glider
Slingsby T.4 Falcon 3
Slingsby T.5 Grunau Baby 2
Slingsby T.6 Kirby Kite 1
Slingsby T.7 Kirby Cadet TX.Mk.1
Slingsby T.8 Kirby Tutor
Slingsby T.9 King Kite
Slingsby T.10 Kirby Kitten
Slingsby T.11 Kirby Twin
Slingsby T.12 Kirby Gull 1
Slingsby T.13 Petrel 1
Slingsby T.14 Gull 2
Slingsby T.15 Gull 3
Slingsby T.16
Slingsby T.17
Slingsby T.18 Hengist I
Slingsby T.19
Slingsby T.20
Slingsby T.21A
Slingsby T.21B
Slingsby T.21B Sedbergh TX.Mk.1
Slingsby T.21C
Slingsby T.22 Petrel 2
Slingsby T.23 Kite 1A
Slingsby T.24 Falcon 4
Slingsby T.25 Gull 4
Slingsby T.26 Kite 2
Slingsby T.27
Slingsby T.28 (See: Slingsby T.21B)
Slingsby T.29 Motor Tutor
Slingsby T.30 Prefect
Slingsby T.30 Prefect TX.Mk.1 (RAF)
Slingsby T.31 Tandem Tutor
Slingsby T.31B Kirby Cadet TX.Mk.3
Slingsby T.32 Gull 4B
Slingsby T.33
Slingsby T.34A Sky 1
Slingsby T.34B Sky 2
Slingsby T.35
Slingsby T.36
Slingsby T.37 Skylark 1
Slingsby T.38 Grasshopper TX.Mk.1
Slingsby T.39 Target
Slingsby T.40 Hayhow
Slingsby T.41 Skylark 2
Slingsby T.42 Eagle
Slingsby T.43 Skylark 3
Slingsby T.44 Stratoferic
Slingsby T.45 Swallow
Slingsby T.45 Swallow T.Mk.1
Slingsby T.46
Slingsby T.47
Slingsby T.48
Slingsby T.49B Capstan
Slingsby T.49C Powered Capstan
Slingsby T.50 Skylark 4
Slingsby T.51 Dart
Slingsby T.52
Slingsby T.53A
Slingsby T.53B
Slingsby T.53C
Slingsby T.54
Slingsby T.55
Slingsby T.56 S.E.5A Replica
Slingsby T.57 Sopwith Camel F.1 Replica
Slingsby T.58 Rumpler C.IV Replica
Slingsby T.59C (Glasflügel) Kestrel
Slingsby T.59D (Glasflügel) Kestrel 19
Slingsby T.59H (Glasflügel) Kestrel 22
Slingsby T.61A Falke (Falcon)
Slingsby T.61B (Scheibe SF-25B) Falke (Falcon)
Slingsby T.61C (Scheibe SF-25C) Falke (Falcon)
Slingsby T.61E Venture T.Mk.2 (RAF)
Slingsby T.61G Falke (Falcon)
Slingsby T.65A Vega
Slingsby T.65C Sport Vega
Slingsby T.65D Vega
Slingsby T.66 (See: Fairey Tipsy Nipper)
Slingsby T.67 Firefly
Slingsby T.67M Firefly
Slingsby (T.67M) T-3A Firefly

Slinn Aeroplane Co (USA)
Slinn B Monoplane (1910)

Slip Stream Industries, Inc (Wautoma, WI)
Slip Stream Dragonfly
Slip Stream Genesis
Slip Stream Revelation
Slip Stream Scepter
Slip Stream SkyBlaster
Slip Stream SkyQuest

Sloan & Cie (S.A.I.S.) (France)
Sloan (France) Bicurve Biplanes

Sloan, Douglas (Australia)
Sloan (Australia) 1911/12 Biplane

Sloane Aircraft Co, Inc (Plainfield, NJ)
Sloane Exhibition Tractor (Sloane #4)
Sloane Flying Boat (Sloane #2)
Sloane H Tractor Biplane (Sloane #5)
Sloane H-1 Tractor Biplane (Sloane #6)
Sloane H-2 Recon Biplane
Sloane H-3 Trainer

Sloane Military Tractor Biplane (Sloane #3)
Sloane Scout Monoplane (Sloane #1)

SME Aviation Sdn Bhd (Malaysia)
SME MD-3-160 AeroTiga (See: MDB Flugtechnik)

Smith Brothers (USA)
Smith Brothers 43 (Glider)

Smith, Arthur (Fort Wayne, IN)
Smith (Art) Biplane (1911)

Smith, B. L. (Miami, FL)
Smith (B. L.) Arrowhead Safety Plane

Smith, Claude P. (Reidsville, NC) *(See: Lefevers)*

Smith, Donald L. J. (Long Beach, CA)
Smith (Donald L. J.) Experimental Monoplane (1928)

Smith, E. M. *(See: Emsco; American Albatross; Rocheville; Zenith)*

Smith, Ed (Camden, NJ)
Smith (Ed) ES-5 Monoplane (1929)

Smith, Everett M. (Camden, NJ)
Smith-Cirigliano SC-1 Baby Hawk

Smith, Frank (Norco, CA)
Smith (Frank) DSA-1 Miniplane

Smith, Glen A. (Mason City, IA)
Smith (Glen A.) S-2

Smith, J. E. *(See: Sorenson "Little Mike")*

Smith, John W. (Chicago, IL)
Smith (John W.) 1910 Monoplane
Smith (John W.) 1911 Monoplane
Smith (John W.) 1912 Monoplane
Smith (John W.) 1913 Monoplane
Smith (John W.) 1919 Monoplane Seaplane (Patent)

Smith, Kyle *(See: Kyle-Smith Aircraft Co)*

Smith (L. B.) Aircraft Corp (Miami, FL)
Smith (L. B.) Tempo II

Smith, Rexford (Washington, DC)
Smith (Rexford) Biplane No.1
Smith (Rexford) Biplane No.2
Smith (Rexford) Biplane No.3

Smith, Ted *(See: Ted Smith Aircraft Co, Inc)*

Smith, Wayne R. (Richmond, VA)
Smith (Wayne R.) 1928 Biplane

Smolik (Czechoslovakia) *(See also: Letov)*
Smolik Sm 1
Smolik Sm 2
Smolik Sm 3
Smolik Sm 4
Smolik Sm 6
Smolik Sm 9
Smolik Sm 11
Smolik Sm 15

Smolinski, Hank (Point Magu, CA)
Smolinski Ford Pinto Auto-plane (1973)

S.N.A., Inc *(See: Seawind/S.N.A.)*

SNCAC (Société Nationale de Constructions Aéronautiques du Centre) (Aérocentre) (France)
In 1937 France nationalized its military aircraft production factories and organized them geographically into six companies. In central France the factories were organized under **Société Nationale de Constructions Aéronautiques du Centre (SNCAC)**. All their products are listed under **SNCAC**. See also: **SNCAN**, **SNCASE**, **SNCASO**, **SNCAO**, **SNCAM**, and **Arsenal**.
SNCAC NC.150
SNCAC NC.211 Cormoran (Cormorant)
SNCAC NC.701 Martinet (Swift) (Siebel 204 Development)
SNCAC NC.702 Martinet (Swift) (Siebel 204 Development)
SNCAC NC.853
SNCAC NC.856-A
SNCAC NC.856-H
SNCAC NC.856-N
SNCAC NC.900 (Focke-Wulf Fw 190A)
SNCAC NC.1071
SNCAC NC.1080
SNCAC NC.2001 Abeille (Bee)
SNCAC NC.2234
SNCAC NC.3021

SNCAM (Société Nationale de Constructions Aéronautiques du Midi) (France)
In 1937 France nationalized its military aircraft production factories and organized them geographically into six companies. In southern France production was organized under **Société Nationale de Constructions Aéronautiques du Midi (SNCAM)**. SNCAM was absorbed by **SNCASE** during 1941. See also: **SNCAN**, **SNCAC**, **SNCASE**, **SNCASO**, **SNCAO**, and **Arsenal**.
SNCAM D.520 (See: Dewoitine)
SNCAM HD.730

SNCAN (Société Nationale de Constructions Aéronautiques du Nord) (Nord) (France)
In 1937 France nationalized its military aircraft production factories and organized them geographically into six companies. In northern France production organized under **Société Nationale de Constructions Aéronautiques du Nord (SNCAN)**. In 1970 Nord merged with **Sud-Aviation** to form **Aérospatiale**. All pre-1970 designs are listed under **SNCAN**. See also: **SNCASE**, **SNCASO**, **SNCAO**, **SNCAM**, **SNCAC**, and **Arsenal**.
SNCAN 262 (See: Max Holste M.H. 262)
SNCAN 500
SNCAN 750 Norelfe (Elf)
SNCAN C.800 Sailplane (See: Caudron)
SNCAN 1000 (Messerschmitt Bf 108D)
SNCAN 1002 Pingouin (Penguin)
SNCAN 1101 Noralpha
SNCAN 1101 Ramier (Dove)
SNCAN 1200
SNCAN 1201 Norécrin (Jewel Box)
SNCAN 1203 Norécrin (Jewel Box) II
SNCAN 1222
SNCAN 1300 (See: Grunau Baby)
SNCAN 1402 Noroit
SNCAN 1402 Gerfaut (Gyrfalcon) (See: SFECMAS 1402 Gerfaut 1A)
SNCAN 1405 Gerfaut (Gyrfalcon) II
SNCAN 1500 Griffon (Griffin)
SNCAN 1500 Noréclair (Lightning)
SNCAN 1500-02 Griffon (Griffin) II
SNCAN 1601

SNCAN 1700 Norelic
SNCAN 1750 Norelfe (Elf)
SNCAN 2100 Norazur (Azure)
SNCAN 2200
SNCAN 2500 Noratlas (Atlas)
SNCAN 2501 Noratlas (Atlas)
SNCAN 2502 Noratlas (Atlas)
SNCAN 2503 Noratlas (Atlas)
SNCAN 3201
SNCAN 3202
SNCAN 3400

SNCAO (Société Nationale de Constructions Aéronautiques du Ouest) (France)

In 1937 France nationalized its military aircraft production factories and organized them geographically into six companies. In western France production was organized under **Société Nationale de Constructions Aéronautiques du Ouest** (**SNCAO**). SNCAO was absorbed by **SNCASO** during 1941. See also: **SNCAN**, **SNCAC**, **SNCASE**, **SNCASO**, **SNCAM**, and **Arsenal**.

SNCAO LN.140 (See: Loire)

SNCASE (Société Nationale de Constructions Aéronautiques du Sud-Est) (France)

In 1937 France nationalized its military aircraft production factories and organized them geographically into six companies. In southeastern France production organized under **Société Nationale de Constructions Aéronautiques du Sud-Est** (**SNCASE**). During 1941 SNCASE absorbed **SNCAM**. In 1957 SNCASE merged with **Ouest-Aviation** (**SNCAO**) to form **Sud-Aviation**. All pre-1957 SNCASE products are listed under **SNCASE**. See also: **SNCAN**, **SNCASO**, **SNCAO**, **SNCAM**, **SNCAC**, and **Arsenal**.

SNCASE Aquilon (North Wind) (de Havilland D.H. 112 Sea Venom)
SNCASE (Sikorsky) S.55 Éléphant Joyeux (Happy Elephant)
SNCASE SE 100 (See: Lioré et Olivier LeO 50)
SNCASE SE 102
SNCASE SE 103
SNCASE SE 105
SNCASE SE 161 Languedoc (Bloch 161)
SNCASE SE 162 Languedoc (Bloch 162)
SNCASE SE 200
SNCASE SE 210 Caravelle (Caravel) Prototypes
SNCASE SE 210 Caravelle (Caravel) Super B
SNCASE SE 210 Caravelle I
SNCASE SE 210 Caravelle III
SNCASE SE 210 Caravelle VI.N
SNCASE SE 210 Caravelle VI.R
SNCASE SE 210 Caravelle 10.R
SNCASE SE 210 Caravelle 10.BIR
SNCASE SE 210 Caravelle 11.R
SNCASE SE 212 Durandal
SNCASE C.302 (Cierva C.30) (See: Lioré-et-Olivier C.302)
SNCASE SE 315B Lama (Llama)
SNCASE SE 535 Mistral (Cold Northern Wind) (de Havilland 100)
SNCASE SE 1010
SNCASE SE 1210
SNCASE SE 2010 Armagnac
SNCASE SE 2100 Tailless
SNCASE SE 2310
SNCASE SE 2410 Grognard (Grumbler)
SNCASE SE 2415 Grognard
SNCASE SE 3000
SNCASE SE 3101
SNCASE SE 3110
SNCASE SE 3120 Alouette (Skylark) I
SNCASE SE 3130/Sud SA 3180 Alouette (Skylark) II
SNCASE SE 3130 Alouette Gouverneur (Governor)
SNCASE SE 3160/Sud SA 316/Sud SA 319 Alouette (Skylark) III
SNCASE SE 3200 Frelon (Hornet)
SNCASE SE 4000
SNCASE SE 5000 Baroudeur (Fighter)

SNCASO (Société Nationale de Constructions Aéronautiques du Sud-Ouest) (France)

In 1937 France nationalized its military aircraft production factories and organized them geographically into six companies. In southwestern France production organized under **Société Nationale de Constructions Aéronautiques du Sud-Ouest** (**SNCASO**). SNCASO absorbed **SNCAO** in 1941, and became **Ouest-Aviation** in 1956; a 1957 merger with **SNCASE** formed **Sud-Aviation**. All pre-1957 designs are listed as **SNCASO**. See also: **SNCAN**, **SNCASE**, **SNCAO**, **SNCAM**, **SNCAC**, and **Arsenal**.

SNCASO (Bloch) MB 700
SNCASO SO 30 Atar
SNCASO SO 30 Nene
SNCASO SO 30C
SNCASO SO 30N Bellatrix
SNCASO SO 30P
SNCASO SO 30R (Bretagne, Bellatrix)
SNCASO SO 80
SNCASO SO 90
SNCASO SO 94
SNCASO SO 95 Corse II (Corsica)
SNCASO SO 1100 (Giravion, Ariel I)
SNCASO SO 1110 Ariel II
SNCASO SO 1120 Ariel III
SNCASO SO 1220
SNCASO SO 1221 Djinn (Genie)
SNCASO (SO 1221 Djinn) YHO-1
SNCASO SO 1310 Farfadet (Goblin)
SNCASO SO 4000
SNCASO SO 4050 Vautour (Vulture)
SNCASO SO 4050 Vautour II
SNCASO SO 4050 Vautour IIA
SNCASO SO 4050 Vautour IIN
SNCASO SO 4050 Vautour IIRB
SNCASO SO 6000 Triton
SNCASO SO 6020 Espadon (Swordfish)
SNCASO SO 6021 Espadon
SNCASO SO 6025 Espadon
SNCASO SO 6026 Espadon
SNCASO SO 7060 Deauville
SNCASO SO 8000 Norval (Narwhal)
SNCASO SO 9000 Trident
SNCASO SO 9050 Trident II
SNCASO SO M1
SNCASO SO M2

Snead and Co (USA)

Snead XCG-11

SNECMA (Société Nationale d'Études et de Construction de Moteurs d'Aviation) (France)

SNECMA Coléoptère

Snow Aeronautical Corp (Leland Snow) (Olney, TX) *(See also: Air Tractor)*

Snow (TX) S-2B
Snow (TX) S-2C

Snow Aviation International Inc (Columbus, OH)

Snow (OH) SA-204C
Snow (OH) SA-210AT

Snyder, Cloyd *(See: Arup)*

Snyder Motor Gliders Inc (Orval Huff "Bud" Snyder) (Dayton, OH)

Snyder Baby Bomber
Snyder MG-1 Buzzard
Snyder OHS-III

Soar (Society of Aircraft Restoration) (Commack, NY)
Soar ME-109 Replica
Socata (Société de Construction d'Avions de Tourisme et d'Affaires) (France)
In 1966, **Sud-Aviation** reorganized its light aircraft operations around **Gerance des Établishments Morane-Saulnier** (**GEMS**, the remnants of the bankrupt **Morane-Saulnier** company), creating **Société de Construction d'Avions de Tourisme et d'Affaires** (**SOCATA**) as a subsidiary. In January 1970 **Nord Aviation**, **Sud-Aviation**, and **Sereb** merged to form **Aérospatiale**. SOCATA transferred without change, becoming the light aircraft subsidiary of Aérospatiale. For all products of SOCATA see **SOCATA**. For all products of Morane-Saulnier predating the establishment of SOCATA, including all SOCATA production with Morane-Saulnier (M.S.) designations, see **Morane-Saulnier**.
Socata Rallye 100S Sport
Socata Rallye 235G Guerrier (Warrior)
Socata Rallye 235GT
Socata TB9 Tampico Club
Socata TB10 Tobago
Socata TB20 Trinidad
Socata TB21 Trinidad TC
Socata TB30 Epsilon
Socata TB31 Omega
Socata TB360 Tangara (Tanager)
Socata TBM700
Socata TBM60.000 Sarohale

Sociedade Aeronáutica Neiva *(See: Neiva)*

Sociedade Aerotec Ldta (Brazil) *(See: Aerotec)*

Sociedade Avibras Ltda *(See: Avibras)*

Sociedade Construtora Aeronáutica Neiva, Ltda *(See: Neiva)*

Società Aeroplani Livio Agostini *(See: Alaparma)*

Società Anònima Industrie Meccaniche Aeronautiche Navali *(See: Saiman)*

Società Anònima Meccanica Lombarda *(See: SAML)*

Società Idrovolanti Alta Italia *(See: SIAI)*

Società Industrie Aeromarittime Gallinari *(See: Gallinari)*

Società Italiana Brevetti Antoni *(See: Antoni)*

Società Italiano Aviazione *(See: SIA)*

Société Aérienne Bordelaise *(See: Bordelaise; See also: Dyle et Bacalan)*

Société Aéronautique Blanchard *(See: Blanchard)*

Société Aéronautique Normande *(See: SAN)*

Société Anonyme des Aéroplanes Borel *(See: Borel)*

Société Anonyme des Aéroplanes Morane-Saulnier *(See: Morane-Saulnier)*

Société Anonyme des Aéroplanes Robert Savary *(See: Savary)*

Société Anonyme des Appareils d'Aviation Doutre (S.A.D.) *(See: Doutre)*

Société Anonyme des Établissements Aéronautiques Weymann *(See: Weymann)*

Société Anonyme des Établissements Borel *(See: Borel)*

Société Anonyme des Établissements Nieuport *(See: Nieuport)*

Société Anonyme Française de Constructions Aéronautiques *(See: Ponnier M.2)*

Société Armella-Senemaud *(See: Armella-Senemaud)*

Société Commerciale Aéronautique *(See: SCA)*

Société d'Études Aéronautiques *(See: SEA)*

Société d'Études et de Constructions Aéro-Navales *(See: SECAN)*

Société d'Études et de Constructions d'Avions de Tourisme *(See: SECAT)*

Société d'Exploitation des Établissements Morane-Saulnier *(See: Morane-Saulnier)*

Société de Construction d'Appareils Aériens (SCAA) *(See: de Lesseps, Robert)*

Société de Construction d'Avions de Tourisme et d'Affaires *(See: Socata)*

Société de Constructions Aéronautiques Astra *(See: Astra)*

Société de Constructions Aéronautiques et Navales Marcel Besson *(See: Besson)*

Société de Gérance des Avions Bernard *(See: Bernard)*

Société de Moteurs Salmson *(See: Salmson)*

Société des Appareils Bêchéreau *(See: S.A.B.)*

Société des Avions Bernard (S.A.B.) *(See: Bernard)*

Société des Avions Jodel *(See: , Jodel)*

Société des Hydroaéroplanes Lévêque *(See: Donnet-Lévêque; also: FBA)*

Société Européene de Production de l'Avion École de Combat et Appui Tactique *(See: Sepecat)*

Société Française d'Etudes et de Constructions de Matériels Aéronautiques Spéciaux *(See: SFECMAS)*

***Société Française d'Aviation Nouvelle** (See: SFAN)*

Société Française de Locomotion Aérienne (SFLA) (France)
Société Francaise de Locomotion Aérienne 1910 Monoplane

Société Française de Vol à Voile (France)
Société Française de Vol à Voile Sulky (1934 Glider)

***Société Generale des Constructions Industrielles et Méchaniques** (See: Borel)*

***Société Generale des Constructions Industrielles et Mechaniques** (See: Borel)*

***Société Industrielle d'Aviation Latécoère** (See: Latécoère)*

***Société Industrielle des Métaux et du Bois (S.I.M.B)** (See: Bernard; See also: Ferbois)*

***Société les Avions F. J. Rey** (See: Rey)*

***Société Millet-Lagarde** (See: Millet-Lagarde)*

***Société Nationale d'Etude et de Construction de Moteurs d'Aviation** (See: SNECMA)*

***Société Nouvelle des Avions Max Holste** (See: Max Holste)*

***Société Pour la Construction d'Avions Métalliques Avimeta** (See: Avimeta)*

***Société Provencale de Constructions Aéronautiques** (See: SPCA)*

***Société "Zodiac"** (See: Zodiac)*

Soko (Yugoslavia)
Soko G2-A Galeb (Seagull)
Soko J1 Jastreb (Hawk)

Solar Aircraft Co (San Diego, CA)
Solar (CA) MS-1
Solar (CA) MS-2

***Solar Challenger** (See: MacCready Solar Challenger)*

Solar Powered Aircraft Development (SPAD) (UK)
Solar Powered Aircraft Development TO Solar One

Solar-Powered Aircraft, General

Solbrig, Oscar (Wichita, KS)
Solbrig-Benoist 1914 Biplane

Soldenhoff, Alexander (Germany)
Soldenhoff A1 Tailless Aircraft
Soldenhoff A2 Tailless Aircraft
Soldenhoff A3 Tailless Aircraft
Soldenhoff A4 Tailless Aircraft
Soldenhoff A5 Tailless Aircraft

Soltycki (Poland)
Soltycki Nieuport Conversion

Somalvico (Italy)
Somalvico Seaplane

Somerville (Oshkosh, WI)
Somerville (WI) MX.II (Homebuilt)

Somerville, William E. (Coal City, IL)
Somerville (IL) 1911 Biplane
Somerville (IL) 1911 Monoplane

***Sommer (France)** (See also: Deutsche Sommer Werke)*
Sommer Aérobus (Mid-1911 Biplane) (Sommer II Biplane)
Sommer Early-1910 Biplane
Sommer Mid-1910 Biplane
Sommer L (1912 Biplane)
Sommer Late-1911 Monoplane (Sommer II Monoplane, Sommer E)
Sommer 1912 Monoplane (Sommer F)

Sonoda, Takehiko (Japan)
Sonoda 1912 Biplane

Sons, John A. (Humble, TX)
Sons Training Monoplane (1929)

Sopwith (UK)
In late 1911 Thomas Sopwith established **Sopwith Aviation Co Ltd** in Kingston-on-Thames. In September 1920 the company, now named **Sopwith Aviation and Engineering Co, Ltd**, voluntarily liquidated following losses on motorcycle production. Much of Sopwith's staff went on to establish **H. G. Hawker Engineering Co, Ltd.** For all Sopwith products, see **Sopwith**.

Sopwith Admiralty Type 137 Seaplane
Sopwith Admiralty Type 806 Gunbus
Sopwith Admiralty Type 807 Folder Seaplane
Sopwith Admiralty Type 860 Seaplane
Sopwith Admiralty Type 880 ("Greek") Seaplane
Sopwith Antelope
Sopwith Atlantic (Hawker/Grieve Transatlantic Attempt)
Sopwith Baby, Early Production
Sopwith Baby
Sopwith Bat Boat I
Sopwith Bat Boat II
Sopwith Bee
Sopwith Bomber (B.1/B.2)
Sopwith Buffalo
Sopwith Bulldog Mk.I (2.F.R.2, 1-Bay Wings)
Sopwith Bulldog Mk.I (2.F.R.2, 2-Bay Wings)
Sopwith Bulldog Mk.II (2.F.R.2)
Sopwith Type C Seaplane
Sopwith Camel F.1
Sopwith Camel F.1, Airship Launch Experiments
Sopwith Camel F.1, Home Defense Camel
Sopwith Camel F.1, 2-Seat Mod
Sopwith Camel F.1/1 Taper-Wing Camel
Sopwith Camel T.F.1 Trench Fighter
Sopwith Camel 2.F.1 Ship Camel
Sopwith Churchill (Sociable)
Sopwith Cobham Mk.I
Sopwith Cobham Mk.II
Sopwith Cuckoo (T.1)
Sopwith Dolphin (5.F.1) Prototype
Sopwith Dolphin (5.F.1)
Sopwith Dolphin (5.F.1) Night Flyer
Sopwith Dove Single-Seater
Sopwith Dove Two-Seater
Sopwith Dragon 1st Prototype ("Dragonfly Snipe")
Sopwith Dragon 2nd Prototype
Sopwith Dragon
Sopwith Gnu
Sopwith Gordon Bennett Racer (1914)
Sopwith Grasshopper
Sopwith Hippo (3.F.2)
Sopwith L.R.T.Tr. (Long Range Tractor Triplane)
Sopwith Pup
Sopwith Rainbow
Sopwith Rhino
Sopwith Salamander (2.T.F.1/T.F.2)
Sopwith Schneider (RN Seaplane)

Sopwith Scooter (Monoplane No.1)
Sopwith Sigrist Bus
Sopwith SL.T.B.P. (Light Tractor Biplane)
Sopwith Snail (8.F.1)
Sopwith Snail (8.F.1), Monocoque
Sopwith Snapper (R.M.1)
Sopwith Snark
Sopwith Snipe (7.F.1) Prototype
Sopwith Snipe (7.F.1)
Sopwith Snipe (7.F.1), NASM
Sopwith Snipe (7.F.1), Civil Use
Sopwith Sparrow
Sopwith Swallow (Monoplane No.2)
Sopwith Tabloid (S.S.)
Sopwith Tabloid (S.S.), 1914 Schneider Cup Racer
Sopwith Tabloid (S.S.), Scout
Sopwith Triplane, Clerget Triplane
Sopwith Triplane, Hispano-Suiza Triplane
Sopwith Wallaby (Matthews/Kay UK-Australia Attempt)
Sopwith 1½ Strutter Single-Seat Bomber
Sopwith 1½ Strutter Single-Seat Home Defense Fighter
Sopwith 1½ Strutter Two-Seat Fighter
Sopwith 1½ Strutter Two-Seat Observation/Recon (1A.2)
Sopwith 1912 Tractor Biplane (3-Seat, "Hybrid")
Sopwith 1913 Tractor Biplane (3-Seat, "R.G.")
Sopwith 1913 Tractor Seaplane, "Anzani" Tractor
Sopwith 1913 Tractor Seaplane, "Circuit of Britain"
Sopwith 1915 Tractor Biplane (Two-Seat, "Spinning Jenny")
Sopwith 1919 Schneider Cup Racer

Sorenson, Keith (Van Nuys, CA)
Sorenson "Deer Fly"
Sorenson "Little Mike" (See also: Foss "Jinny")

Sorrell Aviation (Sorrell Aircraft Co, Ltd) (Tenino, WA)
Sorrell Golden Condor
Sorrell SNS-1 Negative Stagger Biplane
Sorrell SNS-2 Guppy
Sorrell SNS-6 Hiperbipe (Prototype)
Sorrell SNS-7 Hiperbipe (Production)
Sorrell SNS-8 Hiperlight EXP
Sorrell SNS-9 EXP II
Sorrell SNS-10 Intruder

Southern Aircraft Construction Co (South Africa)
Southern Aeriel Mk.II (See: Genair)

Southern Aircraft Corp (Houston, TX)
Southern (TX) BM-10
Southern (TX) BM-10A
Southern (TX) BM-10C
Southern (TX) Flying Automobile
Southern (TX) Southernaire

Southern Aircraft, Inc (Birmingham, AL)
Southern (AL) Messer Airboss 2 Biplane

Southern Aircraft Ltd *(See: Miles)*

Southern Cross Aviation Ltd (Australia)
Southern Cross SC-1

Southern Eagle (Southern Eagles Aviation Club) (Baltimore, MD)
Southern Eagle Model A

Southern Gyroplane Co (Miami, FL)
Southern Gyroplane

Southhampton University (UK)
Southhampton University Man-Powered Aircraft (SUMPAC)

Southwest Airways Inc (Oklahoma City, OK)
Southwest Airways R.T. No.1

Southwestern Engineering & Mapping Co (Mexia, TX)
Southwestern Palmer P3 Model I

Sowder, Nicholas M. (Beatrice, NE)
Sowder 1911 Dirigible Bi-Parachute

SPAD (France)

In 1913, Armand Deperdussin was jailed on fraud charges, forcing **Établissements A. Deperdussin** (established 1910) into receivership. Louis Béchéreau (Technical Director of the firm) and Louis Blériot (**Blériot Aéronautique** founder) took control of the company, which they renamed **"SPAD"** (erroneously thought to mean "speed" in the then-fashionable international language *Volapuk*). Contemporary company literature soon justified the new name as an acronym for **Société Anonyme pour l'Aviation et ses Derives**. After the war in 1919, Blériot merged the SPAD establishment into Blériot Aéronautique, although designs originating with the SPAD staff continued to receive SPAD designations well into the 1930s. For all Deperdussin designs predating 1913, see **Deperdussin**. For all Blériot designs not having SPAD designations, see **Blériot**. For all designs of both SPAD firms, as well as Blériot designs with SPAD designations, see **SPAD**.

SPAD A
SPAD A.1 (SA.1)
SPAD A.2 (SA.2)
SPAD A.3 (SA.3)
SPAD A.4 (SA.4)
SPAD SB
SPAD SC
SPAD SD
SPAD SE
SPAD SF
SPAD SG
SPAD SH
SPAD SI
SPAD SJ
SPAD SK
SPAD V (S.5)
SPAD VII (S.7)
SPAD 7 CR
SPAD XI A2 (S.11 A2)
SPAD XII (S.12)
SPAD XIII (S.13)
SPAD XIII (S.13) "Smith IV," NASM
SPAD S.13E
SPAD XIV (S.14)
SPAD XV (S.15)
SPAD XVI (S.16)
SPAD XVI (S.16), General Mitchell, NASM
SPAD XVII (S.17)
SPAD XVIII (S.18)
SPAD XIX (S.19)
SPAD XX C2 (S.20)
SPAD S.20*bis* (Racer)
SPAD S.21
SPAD S.22
SPAD S.24
SPAD S.25 (Type Grands Raids)
SPAD S.27
SPAD S.29
SPAD S.30
SPAD S.31
SPAD S.33
SPAD S.34
SPAD S.36
SPAD S.37

SPAD S.40
SPAD S.41
SPAD S.42
SPAD S.45
SPAD S.46
SPAD S.47
SPAD S.48
SPAD S.50
SPAD S.51
SPAD S.510
SPAD S.54
SPAD S.55
SPAD S.56
SPAD S.61
SPAD S.61, Altitude Record Aircraft
SPAD S.66
SPAD S.81 C1
SPAD S.91 C1 Jockey
SPAD S.116
SPAD S.126

Spalding, Reuben Jasper (Rosita, CO)
Spalding 1889 Ornithopter Patent

Sparling, J. N. (St Louis, MO)
Sparling 1910 Biplane
Sparling 1910 Monoplane

Sparrow, W. W. (Healdton, OK)
Sparrow Model 1 (1928 Monoplane)

Spartan Aircraft Co (Tulsa, OK)
Spartan (OK) C-2-60
Spartan (OK) C-3
Spartan (OK) C-3-165
Spartan (OK) C-3-165C
Spartan (OK) C-3-166
Spartan (OK) C-3-225
Spartan (OK) C-4-225
Spartan (OK) C-4-300
Spartan (OK) C-4-301
Spartan (OK) C-5-300
Spartan (OK) C-5-301
Spartan (OK) E
Spartan (OK) F
Spartan (OK) UC-71 Executive
Spartan (OK) Executive, RAF
Spartan (OK) Executive 7J
Spartan (OK) Executive 7W
Spartan (OK) Executive 7W-F
Spartan (OK) Executive 7W-P
Spartan (OK) Executive 7X
Spartan (OK) Executive 8W Zeus
Spartan (OK) Executive 12
Spartan (OK) FBW-1
Spartan (OK) Glider
Spartan (OK) NP-1
Spartan (OK) Model 12

Spartan Aircraft Ltd (UK) *(See also: Simmonds Aircraft Ltd)*
Spartan (UK) A.24 Mailplane (See also: Saunders-Roe)
Spartan (UK) Arrow
Spartan (UK) Clipper
Spartan (UK) Cruiser I (A.24)
Spartan (UK) Cruiser II (A.24)
Spartan (UK) Cruiser III (A.24)
Spartan (UK) Three-Seater Mk.I
Spartan (UK) Three-Seater Mk.II

Spartanburg Aviation Co (Spartanburg, SC) *(See: Storms)*

SPCA (Société Provençale de Constructions Aéronautiques) (France)
SPCA Météore (Meteor) 63 Type Syrie (Syria)
SPCA I type 10E5
SPCA II type 20T3 Bn4
SPCA III type 30M4
SPCA IV type Hermes
SPCA V type 50E7
SPCA VII type 40T
SPCA VII type 41T
SPCA VII type 40T Sanitaire (Hospital)
SPCA VIII type 80 Col^2
SPCA type 81
SPCA type 90 Col.3

Specialized Aircraft (Camarillo, CA)
Specialized Aircraft Tri Turbo-3

Spectrum Aircraft Corp (Van Nuys, CA)
Spectrum SA-550 Spectrum-One (Cessna Skymaster Mod)

SpeedTwin Developments, Ltd (UK)
SpeedTwin E2E

Spencer, Percival H. (Sun Valley, CA)
Spencer Ornithopter (1930s)
Spencer S-12-C Amphibian Air Car
Spencer S-12-E Amphibian Air Car

Spencer-Larsen Aircraft Corp (Amityville, NY)
Spencer-Larsen SL-12C

Sperry (Lawrence Sperry Aircraft Co) (Farmingdale, NY)
Sperry Airboat
Sperry Avro 504K Aerial Torpedo
Sperry Curtiss JN-4 Monoplane
Sperry Curtiss N-9 1917 Aerial Torpedo
Sperry Curtiss Speed Scout 1918 Aerial Torpedo
Sperry Messenger M-1
Sperry Messenger M-1, NASM
Sperry Messenger M-1, Airship Hook-on Testbed
Sperry Messenger M-1, Chamberlin Mod
Sperry Messenger M-1, Dropable Landing Gear Testbed
Sperry Messenger M-1, Gyro Stabilizer Testbed
Sperry Messenger M-1, Variable Camber Wing
Sperry Messenger M-1A
Sperry Messenger MAT Aerial Torpedo
Sperry R-3 (See: Verville-Sperry)
Sperry Sport Plane
Sperry Standard E-1 1920 Aerial Torpedo
Sperry Vickers Virginia Aerial Torpedo
Sperry 1911 Tractor
Sperry 1919 Triplane Bomber Amphibian

Spezio, Tony & Dorothy (Bethany, OK)
Spezio Sport DAL 1

Spinks Industries, Inc (Fort Worth, TX)
Spinks Model 11 Akromaster

Sport Aircraft International (Hillsboro, OR)
Sport Aircraft Mini Coupe
Sport Aircraft Mini Coupe TG

Sport Flight Engineering Inc (Grand Junction, CO)
Sport Flight Sky Pup (Ultralight)

Sportavia (France, Germany)
Sportavia RF-5B (See: Fournier)
Sportavia SFS-31 Milan

Sportina Aviacija (Lithuania)
Sportina LAK 12 Lietuva

Spratt, Dr George, Sr & George, Jr (Coatesville, PA)
Spratt Controllable Wing Amphibian
Spratt 1908 Controllable Wing Glider

Spratt 1934 Experimental
Spratt 1945 Controllable Wing

Springfield Aircraft Co (Springfield, MA; Agawam, MA)
Springfield (Curtiss) JN-4D
Springfield Bulldog (V High Wing Racing)

***Springfield Aircraft, Inc** (See: Hall, Robert "Bob")*

Spyker (Nederlandse Auto U. Vliegtuig Fabrik) (Netherlands)
Spyker V.2 Biplane Trainer
Spyker V.3 Single-Seat Scout
Spyker V.4 Two-Seat Scout

SR-1 Ultra Sport Aviation(SR-1 Enterprises) (St Paul, MN)
SR-1 Ultra Sport SR-1 Hornet

SSS (Edmund Szutkawski, Leszek Szwarc, and Jan Staszek) (Poland)
SSS 1936 1-Seat Tailless Monoplane

***SSW (Siemens-Schuckert Werke)** (See: Siemens-Schuckert)*

***Staaken** (See: Zeppelin-Staaken)*

Stabilaire Inc (Danvers, MA)
Stabilaire Safe Monoplane

Staeckel, Bruno (Germany)
Staeckel Two-Wheeled Monoplane
Staeckel Three-Wheeled Monoplane

***Stahlwerk Mark AG** (See: Mark)*

Stampe et Renard (Constructions Aéronautiques Stampe et Renard) (Belgium)
Stampe (et Renard) SR.7B
Stampe (et Renard) SV.4D

Stampe et Vertongen (Belgium)
Stampe (et Vertongen) RSV (Renard-Stampe-Vertongen)
Stampe (et Vertongen) SV.4
Stampe (et Vertongen) SV.4b
Stampe (et Vertongen) SV.4c
Stampe (et Vertongen) SV.5
Stampe (et Vertongen) SV.7
Stampe (et Vertongen) SV.10
Stampe (et Vertongen) SV.18 (See also: Gates)
Stampe (et Vertongen) SV.22
Stampe (et Vertongen) RSV.22-180
Stampe (et Vertongen) SV.26 (See also: Gates)
Stampe (et Vertongen) RSV.28-180 Type III
Stampe (et Vertongen) SV.46

Standard Aircraft (Aero) Corp (Elizabeth, NJ)
Standard-Built Caproni Ca.5 Night Bomber
Standard D Float Plane

The British de Havilland D.H.4 was modified and built under license in the United States under the designation DH-4 by several manufacturers. For British-built aircraft, see **de Havilland D.H.4**. For American-built aircraft, see the individual manufacturers, including **Aeromarine**, **Atlantic**, **Boeing**, **Dayton Wright**, **Gallaudet**, **Keystone**, and **Standard**. When the actual manufacturer cannot be determined from the available evidence, the photos and documents have been filed under **Dayton Wright**.

Standard DH-4 Liberty Plane
Standard E-1 Pursuit (M Defense Plane)
Standard E-4 Mail Plane
Standard E-5 Triplane
Standard E-7
Standard E-8 Night Bomber
Standard Gnome Rotary Stunting Biplane (1915)
Standard H-3 (See also: Sloane H Series)
Standard H-4-H Training Seaplane
Standard Handley Page O/400
Standard Handley Page O/400 "Langley"
Standard (Curtiss) HS-1-L Flying Boat
Standard (Curtiss) HS-2-L Flying Boat
Standard J-1
Standard J-1, NASM
Standard J-1 (SN-1) Sikorsky Wings
Standard JH-1 Training Seaplane
Standard JR-1B
Standard 1919 Twin-Tractor Floatplane

Stanley Sailplane (USA)
Stanley Sailplane

Star Aircraft Co (Bartlesville, OK)
Star Cavalier
Star Cavalier Model B
Star Cavalier Model C
Star Cavalier Model D
Star Cavalier Model E

Star Flight Aircraft (Liberty, MO)
Star Flight StarFire
Star Flight TriStar
Star Flight TX1000 Ultralight

Star•Kraft Inc (Fort Scott, KS)
Star•Kraft 700

Star-Lite Aircraft Inc (San Antonio, TX)
Star-Lite Aircraft Star-Lite

Starck, André (France)
Starck AS.57
Starck AS.70
Starck AS.71
Starck AS.75
Starck AS.80 Lavadoux

***Starduster** (See: Stolp)*

Stark (Lomita, CA)
Stark (CA) Sport-Aire
Stark (CA) Sport-Aire II (See: Trefethen-Thistle)

Stark Flugzeugbau AG (Germany) (Wilhelm Stark) & Stark Ibérica SA (Spain)
Stark (Wilhelm) Turbulent Prototype (Redesigned Druine)
Stark (Wilhelm) Turbulent-D (Redesigned Druine)

Starling Aircraft Co (Minneapolis, MN)
Starling Aircraft H-11 Biplane
Starling Aircraft H-12 Imperial

***Starling Burgess Co** (See: Burgess)*

Start Aviation Circle (Poland)
Start Aviation Circle K.L.S.1
Start Aviation Circle K.L.S.2
Start Aviation Circle K.L.S.3
Start Aviation Circle ZE-1 Cytrynka (Lemon)

Start + Flug (Germany)
Start + Flug H101 Salto

Stástík, Jan (Austria-Hungary, Czechoslovakia)
Stástík Dreadnaught

States Aircraft Corp (States Aircraft Co) (Chicago, IL)
States (IL) 1929 Monoplane

States Aircraft Corp (Center, TX)
States (TX) B-3
States (TX) Rover 75

Statler, William "Bill" *(See: Foss "Jinny;" see also: Sorenson "Little Mike")*

Stauffer, C. L. (Elkhart, IN)
Stauffer 1931 Gyroplane

Stealth Aircraft, General

Stearman Aircraft Co (Wichita, KS)

In 1927 Lloyd Stearman left **Travel Air Aircraft Co** (est. 1924) and joined with the **Lyle-Holt Aircraft Corp**, Travel Air's West Coast distributor, to form **Stearman Aircraft Co** in Venice, CA. Later that year, the company moved to Wichita, KS. The company became a subsidiary of **United Aicraft and Transport Corp (UATC)** in 1929 and, in September 1931, absorbed UATC's **Northrop Aircraft Corp** subsidiary. With the dissolution of UATC in 1934, Stearman became a subsidiary of the **Boeing Airplane Co** (another former-UATC subsidiary and the parent company of the **Boeing Aircraft Co** of Seattle, WA), becoming the **Stearman Aircraft Division** of Boeing Airplane in 1938. In 1941, the Stearman Division of Boeing Airplane became the **Wichita Division** as part of the pre-war expansion and rationalization of Boeing Airplane's various manufacturing subsidiaries, all of which were joined into a single company in the December 1947 merger of Boeing Airplane and Boeing Aircraft. For all aircraft designed by Stearman predating the 1941 renaming of the division, see **Stearman (1927)**. For aircraft produced at Boeing's Wichita facility not originating with Stearman, see **Boeing**. This company should not be confused with the **Stearman Aircraft Corp** established by Lloyd Stearman in 1968; for designs of this later company, see **Stearman (1968)**.

Stearman (1927) XA-21 (X100)
Stearman (1927) XAT-15 (X120)
Stearman (1927) YBT-3 (6-C)
Stearman (1927) YBT-5 (6-D)
Stearman (1927) XBT-17 (X91)
Stearman (1927) C-1
Stearman (1927) C-1 Modified (Salmson Engine)
Stearman (1927) C-2
Stearman (1927) C-2B
Stearman (1927) C-3B Sport Commercial
Stearman (1927) C-3B Special
Stearman (1927) C-3C (C-3D)
Stearman (1927) C-3MB
Stearman (1927) C-3P
Stearman (1927) C-3R Business Speedster
Stearman (1927) C-3R Business Speedster, Peru
Stearman (1927) CAB-1 Coach
Stearman (1927) LT-1 Light Transport
Stearman (1927) M-2 Speedmail
Stearman (1927) NS-1 (73)
Stearman (1927) N2S-1 Kaydet (A75N1)
Stearman (1927) N2S-2 Kaydet (B75)
Stearman (1927) N2S-3 Kaydet (B75N1)
Stearman (1927) N2S-4 Kaydet (A75N1)
Stearman (1927) N2S-5 Kaydet (E75)
Stearman (1927) N2S-5 Kaydet (E75), NASM
Stearman (1927) XOSS-1 (X85)
Stearman (1927) YPT-9 (6-A)
Stearman (1927) YPT-9B (6-L)
Stearman (1927) YPT-9C
Stearman (1927) PT-13 Kaydet (75)
Stearman (1927) PT-13A Kaydet (A75)
Stearman (1927) PT-13B Kaydet (A75)
Stearman (1927) PT-13D Kaydet (E75)
Stearman (1927) PT-17 Kaydet (A75N1)
Stearman (1927) PT-17A Kaydet (A75N1)
Stearman (1927) PT-17B Kaydet
Stearman (1927) PT-18 Kaydet (A75J1)
Stearman (1927) PT-18A Kaydet (A75J1)
Stearman (1927) PT-27 Kaydet (D75N1)
Stearman (1927) 4-C Junior Speedmail
Stearman (1927) 4-CM Junior Speedmail
Stearman (1927) 4-D Junior Speedmail
Stearman (1927) 4-DM Junior Speedmail
Stearman (1927) 4-E (4-W) Junior Speedmail
Stearman (1927) 4-E Special
Stearman (1927) 4-EM Junior Speedmail
Stearman (1927) 6-A Cloudboy
Stearman (1927) 6-D Cloudboy (XBT-915)
Stearman (1927) 6-F Cloudboy
Stearman (1927) 6-H Cloudboy
Stearman (1927) 6-L Cloudboy
Stearman (1927) 6-P Cloudboy
Stearman (1927) X70 (XPT-943)
Stearman (1927) 73 Family
Stearman (1927) A73B1 (Cuba)
Stearman (1927) 75 Family
Stearman (1927) X75 (Prototype)
Stearman (1927) A75L3
Stearman (1927) A75L5 (China)
Stearman (1927) A75N1 (Peru)
Stearman (1927) 76 Family
Stearman (1927) 76B4
Stearman (1927) 76C (A76C3, B76C3)
Stearman (1927) 76D (76D1, 76D3)
Stearman (1927) 80
Stearman (1927) 81
Stearman (1927) X90

Stearman Aircraft Corp (1968) (Washington, DC)

In 1968 Lloyd Stearman, founder of the original **Stearman Aircraft Co** (est. 1927), Carl Argent, and Norman L. Meyers established the **Stearman Aircraft Corp** to develop new designs. For all products of this company, see **Stearman (1968)**. This Stearman company is not related to the earlier company, which was absorbed by Boeing. For products of the earlier Stearman company, see **Stearman (1927)**.

Stearman (1968) Model MP

Stearman-Hammond Aircraft Co (San Francisco, CA)

In 1934 Lloyd Stearman was requested by the Bureau of Air Commerce to assist the **Hammond Aircraft Corp** of San Francisco, CA in developing the Hammond Model Y, which had been designed to Bureau specifications for a safe and easy-to-use aircraft. In 1936 Stearman joined the company, which was renamed the **Stearman-Hammond Aircraft Corp**, but had left by 1938 when production stopped due to financial difficulties. For all products of Stearman-Hammond, see **Stearman-Hammond**. For other Stearman designs, see **Stearman (1927)** and **Stearman (1968)**.

Stearman-Hammond:
JH-1
Model Y Prototype
Model Y Prototype, NASM
Model Y-1 (Y-125)
Model Y-1S

Stebbins-Geyult (Norwich, CT)
Stebbins-Geyult Model A (1909 Biplane/Triplane)
Stebbins-Geyult Model B (1910 Biplane/Triplane)

Steele, John Thomas (East Syracuse, NY)
Steele 1929 Glider

Steen Aero Lab, Inc (Brighton, CO)
Steen Skybolt

Stefanutti, Sergio *(See: SCA)*

Steinhauser Sailplane Co (Mundelein, IL)
Steinhauser Midwest MI-1
Steinhauser Midwest MU-1

Steinkamp, Carl (Springfield, OH)
Steinkamp Ornithopter

Stemme, Reiner (Germany)
Stemme S 10
Stemme S 10V
Stemme S 10VT
Stemme TG-1 (USAF S 10)

Stemmer, William (Philadelphia, PA)
Stemmer 1911 Aeroplane Patent

Stepanchenok, V. A. (Russia)
Stepanchenok S-1

Stepanich (Rosedale, NY)
Stepanich #1 Highwing 1 PLM
Stepanich #2 Midwing 2 PLM

Stephens (B. Stephens & Son Co) (Woonsocket, RI)
Stephens (B.) 1915 Flying Boat

Stephens, C. L. (Rubidoux, CA)
Stephens (C. L.) Akro

Stephens, James S. (Steco) (Cicero, IL)
Stephens (J.) 1914 Hydroaeroplane

Steward-Davis, Inc (Long Beach, CA)
Steward-Davis Packet (C-82 Conversion)
Steward-Davis Jetpacket 1600 (C-82 Conversion)
Steward-Davis Jetpacket 3400 (Skytruck 3400) (C-82 Conversion)
Steward-Davis STOLmaster
Steward-Davis Super-Catalina (PBY-5A Conversion)

Stewart Aircraft Corp (Donald Stewart) (Salem, OH)
Stewart (Donald) Headwind
Stewart (Donald) Foo Fighter

Stewart, J. H. (USA)
Stewart (J. H.) 1921 Aeroplane Patent

Stewart (W. F. Stewart Co) (Flint, MI)
Stewart (W. F.) M-1 (Monoplane X-1)
Stewart (W. F.) M-2

Stieber, Maurice C. (France)
Stieber 1926 Helicopter

Stieger *(See: G.A.L.)*

Stierlin, Robert (Switzerland)
Stierlin Helicopter #1
Stierlin Helicopter #2 Merlin
Stierlin Helicopter #3 (RS-65)
Stierlin Helicopter #3 Modif. (RS-65B)
Stierlin Helicopter #4 (RS-40)
Stierlin Helicopter #4 Modif. (RS-40A)
Stierlin Helicopter #5 Biplace

Stiles Aircraft Inc (Chicago, IL)
Stiles Dragonfly Monoplane (1927)

Stingray *(See: Ace Aircraft Composite)*

Stinson, Jack (San Antonio, TX)
Stinson (Jack) Aircoupe

Stinson (Northville, MI)

In 1925 Edward A. Stinson, Harvey J, Campbell, and William A. Mara formed the **Stinson Aircraft Syndicate**. The Syndicate formally incorporated as the **Stinson Aircraft Corp** in May 1926. In January 1929 the **Cord Corp** obtained a controlling interest in the company, merging it with **Corman Aircraft Inc** the following year. In November 1934, the **Aviation Corp (AVCO)** acquired Cord and all its subsidiaries, including Stinson. In 1938 AVCO liquidated Stinson, reestablishing it as the **Stinson Aircraft Division**, an operating division of the **Aviation Manufacturing Corp (AMC)**. In a stock swap in October 1939, the **Vultee Aircraft Division** of AMC acquired control of Stinson, retaining it as the **Stinson Division of Vultee Aircraft Inc** when that company was established in November 1939. When Vultee formally merged with **Consolidated Aircraft Corp** in 1943, Stinson became a division of the resulting **Consolidated Vultee Aircraft Corp (Convair)**. In June 1948 Convair closed down the Stinson plant and, in December, sold the Stinson assets and rights to the **Piper Aircraft Corp**. In April 1949 Piper established a Stinson division to market the unsold or uncompleted airframes it had obtained from Convair; the rights to the various Stinson designs were later sold to **Universal Aircraft Industries**. For all designs originating at Stinson, both as an independent company and as a subsidiary of Cord, AMC, Vultee, and Convair, see **Stinson**. For further developments of Stinson designs by Piper, see **Piper**.

Stinson (Aircraft) Model A Amphibian
Stinson (Aircraft) Model A Tri-Motor
Stinson (Aircraft) AT-19 Reliant (V-77)
Stinson (Aircraft) AT-19A Reliant (V-77)
Stinson (Aircraft) AT-19B Reliant (V-77)
Stinson (Aircraft) AT-19 Reliant (V-77), Civil Use
Stinson (Aircraft) (AT-19) Reliant Mk.I (V-77)
Stinson (Aircraft) Model B
Stinson (Aircraft) UC-81 (SR-8B, SR-9E)
Stinson (Aircraft) UC-81A (SR-10G)
Stinson (Aircraft) UC-81B (SR-8E)
Stinson (Aircraft) UC-81C (SR-9C)
Stinson (Aircraft) XC-81D (SR-10F)

Stinson (Aircraft) UC-81E (SR-9F)
Stinson (Aircraft) UC-81F (SR-10F)
Stinson (Aircraft) UC-81G (SR-9D)
Stinson (Aircraft) UC-81H (SR-10E)
Stinson (Aircraft) UC-81J (SR-9E)
Stinson (Aircraft) UC-81K (SR-10C)
Stinson (Aircraft) UC-81L (SR-8C)
Stinson (Aircraft) UC-81M (SR-9EM)
Stinson (Aircraft) UC-81N (SR-9B)
Stinson (Aircraft) C-91(SM-6000)
Stinson (Aircraft) CQ-2 (L-1F aerial target controller)
Stinson (Aircraft) HW-75 Prototype
Stinson (Aircraft) HW-75 ("Stinson 105")
Stinson (Aircraft) HW-80 Model 10 ("Stinson 105")
Stinson (Aircraft) HW-90 Voyager Model 10A
Stinson (Aircraft) HW-90 Voyager Model 10B
Stinson (Aircraft) Model L, SR-5 Reliant Replacement Proposal
Stinson (Aircraft) Model L, Monoplane Trainer Cover Story
Stinson (Aircraft) L-1 (O-49, L-49) Vigilant
Stinson (Aircraft) L-1A (O-49A) Vigilant
Stinson (Aircraft) L-1B (O-49B) Vigilant
Stinson (Aircraft) L-1C Vigilant
Stinson (Aircraft) L-1D Vigilant
Stinson (Aircraft) L-1E Vigilant
Stinson (Aircraft) L-1F Vigilant
Stinson (Aircraft) L-1T Vigilant (Glider Tow Variant)
Stinson (Aircraft) (L-1) Vigilant Mk.IA
Stinson (Aircraft) L-5 Sentinel
Stinson (Aircraft) L-5 Sentinel, NASM
Stinson (Aircraft) L-5A Sentinel
Stinson (Aircraft) L-5B Sentinel
Stinson (Aircraft) L-5C Sentinel
Stinson (Aircraft) L-5E Sentinel
Stinson (Aircraft) L-5E Sentinel, Noise Reduction Testbed
Stinson (Aircraft) L-5G Sentinel
Stinson (Aircraft) L-9 (O-49) (HW-90)
Stinson (Aircraft) L-9A (Provisionally AT-19A) Voyager (HW-90)
Stinson (Aircraft) L-9B (Provisionally AT-19B) Voyager (HW-90)
Stinson (Aircraft) L-12 (SR-5A)
Stinson (Aircraft) L-12A (SM-7B)
Stinson (Aircraft) L-13 (See: Convair)
Stinson (Aircraft) Model M
Stinson (Aircraft) Model O
Stinson (Aircraft) OY-1 Sentinel
Stinson (Aircraft) OY-2 Sentinel
Stinson (Aircraft) YO-54 Voyager (10 Voyager)
Stinson (Aircraft) Model R Junior
Stinson (Aircraft) Model R-2 Junior
Stinson (Aircraft) Model R-3 Junior
Stinson (Aircraft) RQ-1 Reliant
Stinson (Aircraft) XR3Q-1 Reliant
Stinson (Aircraft) Model S (Junior)
Stinson (Aircraft) SB-1 Detroiter Prototype
Stinson (Aircraft) SB-1 Detroiter
Stinson (Aircraft) SM-1 Detroiter
Stinson (Aircraft) SM-1B Detroiter
Stinson (Aircraft) SM-1D Detroiter
Stinson (Aircraft) SM-1DX Detroiter (Packard Diesel Testbed)
Stinson (Aircraft) SM-1F Detroiter
Stinson (Aircraft) SM-2 Junior
Stinson (Aircraft) SM-2AA Junior
Stinson (Aircraft) SM-2AB Junior
Stinson (Aircraft) SM-2AC Junior
Stinson (Aircraft) SM-3
Stinson (Aircraft) SM-4
Stinson (Aircraft) SM-5
Stinson (Aircraft) SM-6A Detroiter
Stinson (Aircraft) SM-6B Detroiter
Stinson (Aircraft) SM-7A Junior
Stinson (Aircraft) SM-7B Junior
Stinson (Aircraft) SM-8A Junior
Stinson (Aircraft) SM-8D Junior (Packard Diesel)
Stinson (Aircraft) SM-6000 Airliner Prototype (Corman 3000)
Stinson (Aircraft) SM-6000 Airliner
Stinson (Aircraft) SM-6000B Airliner (Model T)
Stinson (Aircraft) SR Reliant
Stinson (Aircraft) SR-5 Reliant Special
Stinson (Aircraft) SR-5A Reliant
Stinson (Aircraft) SR-5C Reliant
Stinson (Aircraft) SR-5D Reliant
Stinson (Aircraft) SR-5E Reliant
Stinson (Aircraft) SR-5F Reliant
Stinson (Aircraft) SR-6 Reliant
Stinson (Aircraft) SR-6A Reliant
Stinson (Aircraft) SR-6B Reliant
Stinson (Aircraft) SR-7A Reliant
Stinson (Aircraft) SR-7B Reliant
Stinson (Aircraft) SR-8A Reliant
Stinson (Aircraft) SR-8B Reliant
Stinson (Aircraft) SR-8C Reliant
Stinson (Aircraft) SR-8D Reliant
Stinson (Aircraft) SR-8E Reliant
Stinson (Aircraft) SR-9B Reliant
Stinson (Aircraft) SR-9C Reliant
Stinson (Aircraft) SR-9D Reliant
Stinson (Aircraft) SR-9F Reliant
Stinson (Aircraft) SR-10B Reliant
Stinson (Aircraft) SR-10C Reliant
Stinson (Aircraft) SR-10D Reliant
Stinson (Aircraft) SR-10E Reliant
Stinson (Aircraft) SR-10F Reliant
Stinson (Aircraft) SR-10F Reliant, Human Pick-Up Aircraft, NASM
Stinson (Aircraft) SR-10G Reliant
Stinson (Aircraft) SR-10J Reliant
Stinson (Aircraft) SR-10K Reliant
Stinson (Aircraft) Model T (See: SM-6000B)
Stinson (Aircraft) Model U
Stinson (Aircraft) "105" (See: Stinson HW-75, HW-80)
Stinson (Aircraft) Model 105 (See: Convair L-13)
Stinson (Aircraft) Model 108 Voyager 125 Prototype
Stinson (Aircraft) Model 108 Voyager 150 Prototype
Stinson (Aircraft) Model 108 Voyager 150
Stinson (Aircraft) Model 108-1 Voyager 150
Stinson (Aircraft) Model 108-2 Voyager 165
Stinson (Aircraft) Model 108-3 Voyager 165

***Stipa** (See: Caproni)*
Stits Aircraft Corp (Riverside, CA)
Stits (SA-1) Junior
Stits SA-2 Sky Baby
Stits SA-2 Sky Baby, NASM
Stits SA-3A Playboy (Single-Seat)
Stits SA-3B Playboy (Two-Seat)
Stits-Besler SA-4A Executive
Stits SA-5A Flut-R-Bug (Single-Seat)
Stits SA-5B Flut-R-Bug (Single-Seat)
Stits SA-6A Flut-R-Bug (2-Seat Tandem)
Stits SA-6B Flut-R-Bug (2-Seat Tandem)
Stits SA-6C Flut-R-Bug (2-Seat Side-by-Side)
Stits SA-7A Skycoupe
Stits SA-7B Skycoupe
Stits SA-7D Skycoupe
Stits SA-8A Skeeto
Stits SA-9A Skycoupe
Stits SA-11A Playmate

Stockett, Henry (Bethany, OK)
Stockett 1928 Light Plane

Stoddard-Hamilton Aircraft, Inc (Arlington, WA)
Stoddard-Hamilton:
Glastar
Glasair I FT (Fixed Tricycle)
Glasair I RG (Retractable Gear)
Glasair I TD (Tail Dragger)
Glasair II FT (Fixed Tricycle)
Glasair II RG (Retractable Gear)
Glasair II TD (Tail Dragger)
Glasair II-S FT (Fixed Tricycle)
Glasair II-S RG (Retractable Gear)
Glasair II-S TD (Tail Dragger)
Glasair Super II-S FT (Fixed Tricycle)
Glasair Super II-S RG (Retractable Gear)
Glasair Super II-S TD (Tail Dragger)
Glasair III
Glasair III LP (Lightning Protected)
Glasair III Propjet
Glasair Super III

STOL (Short Take-Off and Landing), General

Stolp Starduster Corp (Riverside, CA)
Stolp SA-100 Starduster
Stolp SA-300 Starduster Too (Modified)
Stolp SA-500 Starlet
Stolp SA-700 Acroduster I
Stolp SA-750 Acroduster Too
Stolp SA-900 V-Star

Stoppelbein, Gordon *(See: Ohm & Stoppelbein)*

Storms Aviation Co (N. E. Storms Aircraft Co) (Asheville, NC)
Storms Flying Flivver Prototype
Storms Flying Flivver
Storms Parasol Sport (Whiz Banger)

Stout (Detroit, MI) *(See also: Ford)*
Stout Amphibian
Stout AS-1 Air Sedan, Metal (Cabin Model Batwing)
Stout AS-1 Air Sedan, Plywood (Cabin Model Batwing)
Stout Batwing (Bat, Vampire Bat)
Stout Bushmaster SBM-1 TriMotor
Stout Bushmaster 2000 TriMotor
Stout Experimental Monoplane
Stout Ornithopter
Stout Skycar (1930)
Stout Skycar (1930), NASM
Stout Skycar (1941)
Stout ST All-Metal Torpedo Plane
Stout SX Types Design Reports
Stout SX-3
Stout SX-4
Stout SX-6
Stout SX-7
Stout Tri-Motor (See: Stout Bushmaster 2000; See also: Ford)
Stout 2-AT Air Pullman (See Ford 2-AT)
Stout 3-AT Tri-Motor (See: Ford 3-AT)
Stout 5-Seat Executive Aircraft

Strack (Germany)
Strack 1913 Monoplane (55hp Hilz)

Strasburg (Paul Strasburg Flying Service) (Detroit, MI)
Strasburg MF Flying Boat

Strat Aircraft Co (Stratford, CT)
Strat Atom
Strat M-21

Straughan Aircraft Corp (Wichita, KS)
Straughan A (Straughan-Holmes Biplane, Wiley Post Model A)
Straughan 2

Strauss, Joseph B. (USA)
Strauss 1926 Biplane

Stringfellow, John (UK)
Stringfellow 1843 Aerial Steamcarriage
Stringfellow 1848 Steam-Powered Monoplane
Stringfellow 1868 Steam-Powered Triplane
Stringfellow 1881 Tandem-Wing Monoplane

Striplin Aircraft Corp (Lancaster, CA)
Striplin FLAC
Striplin Lone Ranger
Striplin Skyranger
Striplin Star Ranger

Strojnik, Alex (USA)
Strojnik S-2

Stroukoff Aircraft Corp (Trenton, NJ)
Chase Aircraft Co, Inc was established in 1943 to produce assault cargo aircraft for military use. In 1953, the company was purchased by **Willys Motors, Inc** of Toledo, OH, a wholly-owned subsidiary of **Kaiser-Frazer Corp.** In June 1953, the USAF canceled production contracts for the Chase C-123 and transferred production to **Fairchild Engine and Aircraft Corp.** Kaiser-Frazer phased out its Chase operations, while the Chase design staff established **Stroukoff Aircraft Corp.** For all Chase aircraft including the XC-123 and XC-123A, see **Chase.** For production C-123s, see **Fairchild.** For post-1953 development work on the C-123 airframe, see **Fairchild.**
Stroukoff MS-8
Stroukoff MS-29

Sturgeon Air Ltd (SAL) (Canada) *(See also: Falconar)*
Sturgeon Air 2/3d Mustang (Jurca M.J.7)
Sturgeon AMF-S14 (See: Falconař)

Sturm, Reinhart
Sturm Airship

Sturtevant Aeroplane Co (Boston, MA)
Sturtevant Model A
Sturtevant Model A Special
Sturtevant Model B
Sturtevant Battleplane
Sturtevant Battleplane, Trimotor
Sturtevant S
Sturtevant S-4 Seaplane
Sturtevant Scout Sesquiplane

Stuttgart (Germany)
Stuttgart (WLV) "Württemberg"
Stuttgart "Brenner" Glider
Stuttgart 1921 Glider
Stuttgart 1922 Glider

Suarez, Pablo (Argentina)
Suarez 1895 Glider

Sud-Aviation, Société Nationale de Constructions Aéronautiques (France)
In 1957 **Ouest-Aviation** merged with **SNCASE** to form **Sud-Aviation.** In 1970 Sud merged with **SNCAN** and **Sereb** to form **Aérospatiale.** Pre-1957 designs are listed under their original firms. Sud designs can be found under **Sud** and **Aérospatiale.**

Sud-Aviation SA 318 Alouette (Lark) II (See: SNCASE SE 3130)
Sud-Aviation SA 316/319 Alouette (Lark) III (See: SNCASE SE 3160)

Sud-Est *(See: SNCASE)*

Sud-Ouest *(See: SNCASO)*

Sukhanov (Russia)
Sukhanov Diskoplan

Sukhoi OKB (Russia)
Sukhoi pre-1949 designs:
Sukhoi ShB
Sukhoi Su-1 (I-330)
Sukhoi Su-2 (ANT-51)
Sukhoi Su-3 (I-360)
Sukhoi Su-4
Sukhoi Su-5 (I-107)
Sukhoi Su-6
Sukhoi Su-7
Sukhoi Su-8 (DDBSh)
Sukhoi Su-9
Sukhoi Su-10
Sukhoi Su-11
Sukhoi Su-12
Sukhoi Su-15 (Samolet 'P')
Sukhoi Su-17

Sukhoi post-1955 designs:
Sukhoi Su-7 *Fitter*
Sukhoi Su-9 *Fishpot*
Sukhoi Su-11 *Fishpot*
Sukhoi Su-12
Sukhoi Su-15 *Flagon*
Sukhoi Su-15 V/STOL *Flagon B*
Sukhoi Su-17 *Fitter*
Sukhoi Su-19 *Fencer*
Sukhoi Su-20 *Fitter*
Sukhoi Su-22 *Fitter*
Sukhoi Su-24 *Fencer*
Sukhoi Su-25 *Frogfoot*
Sukhoi Su-26
Sukhoi Su-27 *Flanker*
Sukhoi Su-29
Sukhoi Su-34
Sukhoi Su-35

Sullivan Aircraft Co (Wichita, KS)
Sullivan (Co) SG-1 Glider

Sullivan Aircraft Manufacturing Corp (Wichita, KS)
Sullivan (Corp) K-3 Crested Harpy

Summit Aeronautical Corp *(See: Miller, Howell)*

Summit Motor Car Co (Seattle, WA)
Summit Angeles Monoplane

Sun Aerospace Group, Inc (Nappanee, IN)
Sun Aerospace Sun Ray 100

Sunbeam Motor Car Co (UK)
Sunbeam Bomber (1917)

Sundorph Aeronautical Corp (Cleveland, OH; Los Angeles, CA)
Sundorph Model 2 (XA-1)

Sundstedt, Hugo (Liberty, NY)
Sundstedt S.N.B.
Sundstedt Twin-Engined Seaplane

Super V *(See: Bay Aviation Services)*

Super 580 Aircraft Inc (Carlsbad, CA) *(See: Allison Division of General Motors)*

Superior Aircraft Co (Culver City, CA)
Superior Satellite

Supermarine Aviation Works Ltd (UK)
Supermarine A.D. Boat
Supermarine A.D. Submarine Patrol Seaplane
Supermarine A.D.1 Navyplane
Supermarine Air Yacht
Supermarine Attacker
Supermarine Channel
Supermarine Commercial Amphibian (Seal Mk.I)
Supermarine E.10/44
Supermarine N.1B Baby
Supermarine Nanok
Supermarine Nighthawk (See: Pemberton Billing)
Supermarine S.4
Supermarine S.5
Supermarine S.6
Supermarine S.6A
Supermarine S.6B
Supermarine Scapa
Supermarine Scarab
Supermarine Scimitar
Supermarine Scylla
Supermarine Sea Eagle
Supermarine Sea King
Supermarine Sea Lion I (S.1)
Supermarine Sea Lion II (S.2) (Sea King II)
Supermarine Sea Lion III (S.3)
Supermarine Sea Otter
Supermarine Sea Urchin
Supermarine Seafang
Supermarine Seafire Mk.I
Supermarine Seafire Mk.IB
Supermarine Seafire Mk.II
Supermarine Seafire F.Mk.IIC
Supermarine Seafire FR.Mk.IIC
Supermarine Seafire LR.Mk.IIC
Supermarine Seafire F.Mk.III
Supermarine Seafire FR.Mk.III
Supermarine Seafire LF.Mk.III
Supermarine Seafire F.Mk.XV
Supermarine Seafire F.Mk.XVII
Supermarine Seafire F.Mk.45
Supermarine Seafire F.Mk.46
Supermarine Seafire F.Mk.47
Supermarine Seagull
Supermarine Seagull ASR.Mk.I
Supermarine Seal Mk.II
Supermarine Seamew
Supermarine Sheldrake
Supermarine Solent
Supermarine Southampton
Supermarine Southampton X
Supermarine Sparrow
Supermarine Sparrow Monoplane
Supermarine Spiteful
Supermarine Spitfire Prototype
Supermarine Spitfire F.Mk.I
Supermarine Spitfire F.Mk.I Floatplane
Supermarine Spitfire PR.Mk.I
Supermarine Spitfire F.Mk.IA
Supermarine Spitfire F.Mk.IB
Supermarine Spitfire F.Mk.II
Supermarine Spitfire F.Mk.IIA
Supermarine Spitfire F.Mk.IIB
Supermarine Spitfire ASR.Mk.IIB (Mk.IIC)
Supermarine Spitfire F.Mk.III
Supermarine Spitfire PR.Mk.III
Supermarine Spitfire F.Mk.IV
Supermarine Spitfire PR.Mk.IV
Supermarine Spitfire F.Mk.V
Supermarine Spitfire F.Mk.VA
Supermarine Spitfire LF.Mk.VA
Supermarine Spitfire F.Mk.VB
Supermarine Spitfire F.Mk.VB Floatplane
Supermarine Spitfire LF.Mk.VB
Supermarine Spitfire F.Mk.VC
Supermarine Spitfire LF.Mk.VC
Supermarine Spitfire F.Mk.VI
Supermarine Spitfire PR.Mk.VI
Supermarine Spitfire F.Mk.VII
Supermarine Spitfire HF.Mk.VII
Supermarine Spitfire HF.Mk.VII, NASM

Supermarine Spitfire PR.Mk.VII
Supermarine Spitfire F.Mk.VIII
Supermarine Spitfire F.Mk.VIII Trainer Conversion (T.Mk.8)
Supermarine Spitfire HF.Mk.VIII
Supermarine Spitfire LF.Mk.VIII
Supermarine Spitfire F.Mk.IX
Supermarine Spitfire HF.Mk.IX
Supermarine Spitfire LF.Mk.IX
Supermarine Spitfire LF.Mk.IX Floatplane
Supermarine Spitfire PR.Mk.9
Supermarine Spitfire T.Mk.9
Supermarine Spitfire PR.Mk.X
Supermarine Spitfire PR.Mk.XI
Supermarine Spitfire F.Mk.XII
Supermarine Spitfire PR.Mk.XIII
Supermarine Spitfire F.Mk.XIV
Supermarine Spitfire FR.Mk.XIV
Supermarine Spitfire FR.Mk.XIVe
Supermarine Spitfire LF.Mk.XVI
Supermarine Spitfire F.Mk.XVIII
Supermarine Spitfire FR.Mk.XVIII
Supermarine Spitfire PR.Mk.XIX
Supermarine Spitfire Mk.20 Series
Supermarine Spitfire Mk.20
Supermarine Spitfire F.Mk.21
Supermarine Spitfire F.Mk.22
Supermarine Spitfire F.Mk.23
Supermarine Spitfire F.Mk.24
Supermarine Stranraer
Supermarine Swan
Supermarine Swift
Supermarine Walrus (N-1, N-2, Seagull Mk.V)
Supermarine Type 179 (Giant)
Supermarine Type 224 (F.7/30, Spitfire)
Supermarine Type 316 - 318 (B.12/36)
Supermarine Type 322 (S.24/37) Dumbo
Supermarine Type 508
Supermarine Type 510
Supermarine Type 525
Supermarine Type 528/535
Supermarine Type 541
Supermarine Type 544
Supermarine Type 545
Supermarine Type 559
Supermarine Type 571

Supersonic & Hypersonic Flight, General

Supersonic & Hypersonic Flight, Hypersonic Transport (HST)
HST, USA, HRA (Hypersonic Research Aircraft) Program
HST, USA, NASP (National AeroSpace Plane, X-30) Program

Supersonic & Hypersonic Flight, Supersonic Transport (SST)
SST, UK/France (See: Concorde)
SST, Russia (See: Tupolev Tu-144 *Charger*)
SST, USA, Next Generation SST
SST, USA, Scar (Supersonic Cruise Aircraft Research)
SST, USA, SCAT (Supersonic Commercial Air Transport)
SST, USA, SCAT 15F

Surcouf (Établissements Surcouf) *(See: Astra)*

Surrey Flying Service, Ltd (UK)
Surrey AL-1

Suzukaze (Fictional Co) (Japan)
Suzukaze 20 SSF *Omar*

SVA *(See: Ansaldo SVA)*

Svenska Aero Aktiebolaget (Sweden)
Svenska Falken (Falcon)
Svenska Jaktfalk (Hunting Hawk)
Svenska Pirat (Pirate)

Svenska Aeroplan Aktiebolaget *(See: Saab)*

Swallow Aeroplane Co (Rockfall, CT)
Swallow (CT) A Ultralight
Swallow (CT) B Microlight

Swallow Aircraft Corp (Swallow Airplane Manufacturing Co) (Wichita, KS)
Swallow (KS) B1 mailplane
Swallow (KS) Model C Coupé
Swallow (KS) Commercial
Swallow (KS) "Dallas Spirit"
Swallow (KS) F-28
Swallow (KS) Flight Commander
Swallow (KS) F28W (F-28-AX)
Swallow (KS) F165
Swallow (KS) G-29
Swallow (KS) LT-65
Swallow (KS) New Swallow
Swallow (KS) OBL
Swallow (KS) Special
Swallow (KS) Standard
Swallow (KS) Super-Swallow
Swallow (KS) TP (Training Plane)
Swallow (KS) TP-K (Training Plane)
Swallow (KS) TP-W (Training Plane)
Swallow (KS) OX-5
Swallow (KS) Hisso
Swallow (KS) J-5
Swallow (KS) F28-AX
Swallow (KS) Sport HA
Swallow (KS) Sport HW
Swallow (KS) Sport HC

Swan, William G. (Atlantic City NJ)
Swan (William) 1931 Steel Pier Rocket Plane

Swan Air Corp (Canada)
Swan (Air Corp) Model 101

Swanson Airplane Co (Swanson Aircraft Co Inc) (Hopewell, VA)
Swanson (Airplane) W-15 Coupe (See also: Fahlin SF-1)
Swanson (Airplane) SS-3 (See also: Lincoln Standard Sport Plane)
Swanson (Airplane) SS-4

Swanson, Carl (Zion, IL)
Swanson (Carl) Flyabout A-12

Swanson, Darwin F. (Murray, IA)
Swanson (Darwin F.) B-4-T

Swanson, Don Frank (Springfield, OR)
Swanson (Don Frank) C-O-2

Swanson, Swen (Williamsburg, VA) *(See also: Swanson Airplane Co, Swanson-Freeman, Swanson-Fahlin)*
Swanson (Swen) 1917 Monoplane
Swanson (Swen) 1919 Biplane

Swanson-Fahlin *(See: Fahlin)*

Swanson-Freeman (Swen Swanson & Edgar Freeman) (Vermillion, SD)
Swanson-Freeman SF-1V

Swearingen (San Antonio, TX)
In 1959, Ed Swearingen formed the **Swearingen Aircraft Co** in San Antonio, TX. The company started out modifying Beech aircraft for executive transport, but by 1965 produced its first new design: the Merlin. In 1970 Swearingen began development of the Metro, a joint venture to be marketed by **Fairchild Hiller Corp**. As a

subsidiary of **Fairchild Industries**, Swearingen became **Swearingen Aviation Corp** in 1971, **Fairchild Swearingen** in 1981, and **Fairchild Aircraft Corp** in 1982. At the same time, Ed Swearingen formed a new, independent **Swearingen Aircraft, Inc.** In 1995, Swearingen Aircraft and **Sino Aerospace International Inc** of Taiwan formed the **Sino Swearingen Co** to jointly develop Swearingen's SJ30-2 design. While mose Swearingen modifications and designs are filed under **Swearingen**, files on the Metro and Expediter appear under **Fairchild.**

Swearingen Excalibur (Modified Beech Twin Bonanza)
Swearingen Merlin I
Swearingen Merlin II
Swearingen Merlin IIA
Swearingen Merlin III
Swearingen Merlin IV
Swearingen Queen Air 800 (Beech Modification)
Swearingen SJ30
Swearingen SJ30-2
Swearingen SX200
Swearingen SX300 (SA-29) (Kit Plane)

Swedenborg, Emanuel (Sweden)
Swedenborg 1714 Flying Machine

Swendet (Switzerland)
Swendet Hang Glider

Swift (Poland)
Swift (Poland) S-1

Swift Aircraft Corp (Wichita, KS)
Swift (KS) Swift
Swift (KS) Model 18
Swift (KS) Model 19 Trainor Training Plane

Swiftfury *(See: Lopnesti Piper Aircraft Engineering Co)*

Swiss American Aircraft Corp *(See: Learjet)*

Sylvaire Manufacturing Ltd (Canada)
Sylvaire Bushmaster (Ultralight)

Synchromotion, Inc *(See: Taylorcraft)*

SZD *(See: PZL Bielsko)*

Szekely Aircraft & Engine Co (Otto E. Szekely) (Holland, MI)
Szekely SR-3 "Flying Dutchman"

Sznycer-Gottlieb (See: Intercity)

Szynkiewicz, Roman (Poland)
Szynkiewicz 1928 Glider

T

T. P. Hall Engineering Corp (See: Hall, Theodore P. "Ted")

Ta-Ho-Ma Airplane and Motor Corp (Elgin, IL)
Ta-Ho-Ma Model 9A Biplane

Tachibana, Ryokan (Japan)
Tachibana Sakamoto Number 6 Biplane (1915)
Tachibana Suzuki Gyro Number 2 Tractor Biplane (1915)
Tachibana Umino Seaplane (1915)

Tachikawa Hikoki KK (Japan)
Originally named **Ishikawajima**, the company was reformed in 1952 as **Shin Tachikawa Kokuki KK** (New Tachikawa Aircraft Co Ltd). In the listing below, only the R-HM is from the Shin Tachikawa period.
Tachikawa Ki-9 *Spruce*
Tachikawa Ki-17 *Cedar*
Tachikawa Ki-36 *Ida*
Tachikawa Ki-54 *Hickory*
Tachikawa Ki-55 *Ida*
Tachikawa Ki-70 *Clara*
Tachikawa Ki-74 *Patsy, Pat*
Tachikawa Ki-77
Tachikawa Ki-94-I
Tachikawa Ki-94-II
Tachikawa R-HM
Tachikawa R-38
Tachikawa R-52
Tachikawa R-53

Tact-Avia Corp (Marina Del Rey, CA)
Tact-Avia Fokker E.III Ultralight
Tact-Avia Morane N Ultralight

***Tadeusz Soltyk** (See: TS)*

Taft Airplane Corp (Elizabeth City, NC)
Taft Kingfisher Flying Boat

Taifun Flugzeugbau GmbH (Germany)
Taifun ME-108

***Tailless Aircraft** (See: Flying Wings)*

Tairov, V. K. (Russia)
Tairov OKO-1
Tairov Ta-1 (OKO-6)
Tairov Ta-3

Takasou, Takayuki (Japan)
Takasou TN-6 Airplane (1917)
Takasou Number 4 Airplane (1914)
Takasou Number 5 Airplane (1915)

***Tallares Nacionales de Construcciones Aeronáuticas** (See: TNCA)*

Tamai, Seitaro (Japan)
Tamai Number 1 Seaplane (1912)
Tamai Number 2 Nippon-Go Seaplane
Tamai NFS Tamai Number 1 Airplane
Tamai NFS Tamai Number 2 Trainer
Tamai NFS Tamai Number 3 Trainer
Tamai Number 5 Trainer
Tamai Number 24 Trainer

Tamamushi (Japan)
Tamamushi Bi-Plane

Tampier, René (France)
Tampier Avion-Automobile
Tampier T.2
Tampier T.4
Tampier T.6 Multiplace de Combat

Taneja Aerospace and Aviation Ltd (India)
Taneja Hansa (See: National Aerospace Labs)
Taneja P68 Observer (See: Partenavia)
Taneja P68B Victor (See: Partenavia)

Tanski, Czeslaw (Poland)
Tanski Flying Machine Proposals (late-1800s)

Tarczynski (Tadeusz) & Stepniewski Wieslaw) (Poland)
Tarczynski & Stepniewski TS-1/34 Promyk (Small Radius) Glider

Taris (France)
Taris Monoplane (1910)

Tarrant (W. G. Tarrant Ltd) (UK)
Tarrant Tabor

Tatin, Victor (France)
Tatin Compressed-Air Aeroplane (1879)
Tatin-Richet Steam Aeroplane (1890)
Tatin-de la Vaulx Monoplane (1907) (See: Clément-Bayard 1908 Twin-Boom Monoplane)

Tatra (Ringhoffer-Tatra Moravsko-Slezska Vozovka AS) (Czechoslovakia)
Tatra T 1
Tatra T 2
Tatra T 3
Tatra T 101
Tatra T 126 (Licensed Avro 626)
Tatra T 131 (Licensed Bücker 131
Tatra T 201
Tatra T 301
Tatra T 401

Taubman Aircraft Co (Akron, OH)
(See: Babcock)

Taylor, Taylor Brothers, Taylor-Young, Taylorcraft (USA, UK)

Clarence Gilbert Taylor founded or participated in several companies bearing his name. Soon after he and his brother Gordon built their first aircraft, they founded the **Arrowing Co** in Newark, NY in 1926. In September 1927, they moved to Rochester, NY, as **Taylor Brothers Aircraft Manufacturing Co**, incorporating in April 1928 as **Taylor Brothers Aircraft Corp**. That month, Gordon Taylor was killed in an aircraft crash. William T. Piper, Sr and other Bradford, PA, businessmen helped the company move to their city in 1929. Following bankruptcy in 1930, the company reorganized as the **Taylor Aircraft Co** with Taylor and Piper as partners. Taylor left the company in 1935. Piper moved the company to Lock Haven, PA, and, in 1937, changed the name to **Piper Aircraft Corp**. In this finding aid, Taylor's earliest designs appear under **Arrowing**; later designs up to the J-3 Cub are listed under **Taylor Brothers, Taylor Aircraft Co**; the J-3 and subsequent Lock Haven aircraft appear under **Piper**.

After leaving Lock Haven, Taylor started a new line of aircraft with **Taylorcraft Aviation Co** in 1936. The company started in Butler, PA, then moved to Alliance, OH, where, in 1937, it incorporated as the **Taylor-Young Airplane Co** (though aircraft were still sold as Taylorcraft designs). In 1939 the corporate name changed to **Taylorcraft Aviation Corp**. Following bankruptcy in 1946, the company reorganized March 1947 as **Taylorcraft, Inc**; production moved to Conway, PA, in 1948. In 1955, the type certificates for Taylorcraft's two-place models were purchased by the **Univair Aircraft Corp**, while Taylorcraft concentrated on production of the Model 20. A new management group took over Taylorcraft in 1958, moving the company to Connellville, PA. With a 1963 bankruptcy, remaining assets were purchased by Univair. In 1968, Taylorcraft was reestablished in Alliance, OH, as **Taylorcraft Aviation Corp**. The company was sold again in 1985, with operations were moved to the former Piper factory in Lock Haven, PA. The company filed for Chapter 11 protection in 1987, and was purchased by **Long World Aviation** in June 1989. The **Aircraft Acquisition Corp** purchased the company at another bankruptcy auction in November 1989. (Aircraft Acquisition also started **Taylor Kits Corp** to sell a kit version of the F-21B.) In January 1991, **East Kent Capital** bought the Taylorcraft assets, renamed the company **Taylorcraft Aircraft**. The Lock Haven plant was closed in August. The company sold in early 1995 to **Synchromotion, Inc** and sold again in late 1995, organizing as **B. Taylorcraft Aerospace, Inc** in Greensboro NC. All products and designs of this string of companies appear under **Taylorcraft**.

C. G. Taylor worked on several other aircraft projects, including his twin-engine Meteor for **Saturn Engineering** and the Taylor Bird, a kit plane marketed by his son under the company name of **Taylor Aero Industries Inc**. These products are filed under their respective companies.

Taylor Brothers, Taylor Aircraft Co (USA)

Taylor A-2 Chummy (See Arrowing A-2 Chummy)
Taylor B-2 Chummy
Taylor C-2 Chummy
Taylor E-2 Cub
Taylor F-2 Cub
Taylor H-2 Cub
Taylor J-2 Cub
Taylor (Piper) J-2 Cub, NASM
Taylor Loomis Special (1925 Monoplane)
Taylor Special (1928 Monoplane)
Taylor Glider

Taylor Aero Industries Inc (Tipp City, OH)

Taylor (Aero Industries) Bird

Taylorcraft (USA)

Taylorcraft Model A
Taylorcraft Model B Family
Taylorcraft BC
Taylorcraft BC-12 Deluxe
Taylorcraft BC-12C
Taylorcraft BC-12D Twosome (Ace)
Taylorcraft BC-12D-65
Taylorcraft BC-12D-85 Deluxe
Taylorcraft BC-65
Taylorcraft BF-12 Deluxe Floatplane
Taylorcraft BF-65
Taylorcraft BL
Taylorcraft BL-55 Floatplane
Taylorcraft BL-65
Taylorcraft DC-65 Tandem Trainer
Taylorcraft DCO-65 Tandem Trainer
Taylorcraft DL-65 Tandem Trainer
Taylorcraft F-19
Taylorcraft F-21
Taylorcraft F-21A
Taylorcraft F-21B
Taylorcraft F-22 Classic
Taylorcraft F-22A Tractor
Taylorcraft F-22B Ranger
Taylorcraft F-22C Trooper
Taylorcraft F-22S Floatplane
Taylorcraft LBT-1
Taylorcraft L-2B
Taylorcraft L-2M
Taylorcraft YO-57
Taylorcraft TG-6
Taylorcraft Model 15
Taylorcraft Model 15A Tourist
Taylorcraft Model 16
Taylorcraft Model 18
Taylorcraft Model 19 Sportsman
Taylorcraft Model 20 Ranchwagon
Taylorcraft Model 20 Seabird
Taylorcraft Model 20 Zephyr 400
Taylorcraft Model 20AG Topper
Taylorcraft Model 47 (See: BC-12D "Model 47")

Taylor, Don (Evansville, IN)
Taylor (Don) Tinker Toy

Taylor, John F. (UK)
Taylor (John) J.T.2 Titch
Taylor (John) Monoplane

Taylor, Molt (Longview, WA)
Taylor (Molt) Aerocar
Taylor (Molt) Coot
Taylor (Molt) Mini-Imp

Taylor, Truman (Honolulu, HA)
Taylor (Truman) 1935 1-seat Biplane

Taylorcraft Aeroplanes Ltd (British Taylorcraft) (UK)

In 1938, A. L. Wykes purchased the British rights to manufacture Taylorcraft designs, founding **Taylorcraft Aeroplanes Ltd**. The name of the company changed to **Auster Aircraft Ltd** in 1946. Auster was purchased, along with Miles Aircraft by the Pressed Steel Company to form **British Executive and General Aircraft, Ltd (Beagle)**. In 1966, Beagle was sold to the British government. Plans to resume production were not realized, and operations were suspended in 1968. All British Taylorcraft and Auster production is listed under **Auster**.

TBM 700 *(See: Socata TBM700)*

TCD Ltd (Tiger Cub Developments Ltd) (UK)
TCD Ltd Sherwood Ranger
TCD Ltd Sherwood Ranger XP
TCD Ltd Tiger Cub

Tcheranovsky *(See: Cheranovsky, B.I.)*

Tchetverikov *(See: Chetverikov)*

TeamTango *(See: DFL Holdings)*

Tebaldi (Ing. Tebaldi) (Italy)
Tebaldi's design for a fighter sesquiplane was built by **Zari Brothers** in 1919. The design rights and prototype were purchased by **Breda**. Despite modifications, the aircraft did not enter production.
Tebaldi Pursuit Sesquiplane

Ted Smith Aircraft Co, Inc (Van Nuys, CA)
In 1966, Ted Smith established **Ted Smith Aircraft Co, Inc.** In 1968 the company was purchased by the **American Cement Co**, which operated it briefly before selling it to **Butler Aviation International, Inc** in late 1969. In July 1970 Butler merged Ted Smith with its **Mooney Aircraft Corp** subsidiary as the **Aerostar Aircraft Corp** in Wichita, KS. In Early 1972, Butler suspended Aerostar production and Smith reacquired rights to the Aerostar design, establishing **Ted R. Smith & Associates** to manufacture the aircraft. In 1976, the company was renamed **Ted Smith Aerostar Corp.** In March 1978 **Piper Aircraft Corp** acquired the company as the **Santa Maria Division** of Piper, incorporating Aerostar designs into the Piper product line as the PA-60. In 1981 Piper closed the Santa Maria facility and moved Aerostar production to Vero Beach, FL until 1984, when it ended production. In 1991, the company reemerged as **Aerostar Aircraft Corp.**
All pre-1991 designs originating with Ted Smith are filed under **Ted Smith.** Aerostar Aircraft Corp designs originating with Mooney are filed under **Mooney.** Aerostar designs begun after 1991 are filed under **Aerostar Aircraft Corp.** Other Piper products are found under **Piper.**
Ted Smith Aerostar 340/400
Ted Smith Aerostar 360
Ted Smith Aerostar 600
Ted Smith Aerostar 601
Ted Smith Aerostar 601P

Teledyne Ryan *(See: Ryan)*

Tellier (Établissements Tellier, A. Tellier & Cie) (France)
Tellier Monoplane (1910)
Tellier T.1 Monoplane Racer (1910)
Tellier T.2
Tellier T.3
Tellier T.4
Tellier T.5 (Nieuport BM)
Tellier T.6 (Nieuport S)
Tellier T.7
Tellier T.8 (Nieuport TM)
Tellier TE.1 (1921 Parasol)
Tellier Vonna (Nieuport 4R 450)
Tellier-Nieuport Flying Boat (See: Nieuport)

Teman Aircraft (Westminster, CA)
Teman Mono-Fly

Temco Aircraft Corp (Dallas, TX)
Texas Engineering and Manufacturing Co, Inc formed in 1947 to produce the Globe Swift. Reformed as **Temco Aircraft Corp**, by the mid-1950s, the firm designed and produced aircraft and target drones. In 1960 the company ended aircraft production, merging to form **Ling-Temco Electronics Inc**, which then merged to form **Ling-Temco-Vought, Inc (LTV)** in 1961. Temco aircraft designed and produced before the mergers are listed as **Temco**; LTV designs are found under **Vought.**
Temco Riley 55
Temco Swift
Temco TT-1 Pinto (Model 51)
Temco YTT-1 Pinto (Model 51)
Temco Turbo-Commuter (DC-3 Turboprop Conversion)
Temco T-35 Buckaroo
Temco T-35A Buckaroo
Temco TE-1A
Temco TE-1B Buckaroo
Temco Model 33 Plebe
Temco Model 58

Temple *(See: Texas)*

Templeton-McMullen (William & Winston Templeton, William McMullen) (Canada)
Templeton-McMullen 1911 Tractor Biplane

Tena, Chris (Hillsboro, OR)
Tena Mini Coupe

Teratorn Aircraft, Inc (Clear Lake, IA)
Teratorn Tierra Ultralight
Teratorn Tierra II Ultralight

Texas Aero Manufacturing Co (George Williams Airplane Mfg Co, Texas Aero Corp) (Temple, TX)
Texas Aero (Temple Monoplane):
C-4
Commercialwing
Commercialwing, 1st Prototype
Commercialwing, 2nd Prototype
Speedwing
Sportsman (Sport)
Trimotor (1929)
1908 Monoplane

Texas A&M (Texas Agricultural and Mechanical College) (College Station, TX)
Texas A&M AG-1

Texas Airplane Manufacturing Co Inc (Dallas, TX) *(See: Carstedt)*

Texas Bullet *(See: Johnson Aircraft Bullet 185)*

Texas Engineering and Manufacturing Co *(See: Temco)*

Texas Helicopter Corp (Irving, TX)
Texas Helicopter M74 Wasp
Texas Helicopter M74A Wasp
Texas Helicopter M79S Wasp II
Texas Helicopter M79T Jet Wasp II

***Teziutlan** (See: TNCA Teziutlan)*

Thaden Metal Aircraft Corp (San Francisco, CA)
Thaden T-1 Argonaut
Thaden T-2
Thaden T-4

Thalman, Harry (Salt Lake City, UT)
Thalman Geodetic
Thalman Geodetic T-4

Theodoresco, J. (France)
Theodoresco Monoplane

Thibault de St André (France)
Thibault 1784 Glider

Theis, Jim (Zumbrota, MN)
Theis Nighthawk Ornithopter

THK (Türk Hava Kurumu Uçak Fabrikasi, Turkish Air League Factory) (Turkey)
THK-1
THK-2
THK-5
THK-5A
THK-11
THK-13
THK-14
THK-15
THK-16

Thomas Brothers Co (Ithaca, NY)
Thomas Brothers Co, which formed in 1909, incorporated as the **Thomas Brothers Aeroplane Co** in 1913. A 1917 bailout by **Morse Chain Co** resulted in formation of **Thomas-Morse Aircraft Corp.** Thomas-Morse became **Consolidated Aircraft Corp** subsidiary in 1929, and was dissolved in 1935. Pre-1935 designs are listed under **Thomas Brothers** and **Thomas-Morse**.
Thomas (Brothers) B-3
Thomas (Brothers) BP
Thomas (Brothers) D-2
Thomas (Brothers) D-2 Seaplane
Thomas (Brothers) D-5
Thomas (Brothers) E
Thomas (Brothers) Flying Boat
Thomas (Brothers) HS
Thomas (Brothers) Metal Flying Boat
Thomas (Brothers) Monoplane
Thomas (Brothers) Nacelle Pusher
Thomas (Brothers) Nacelle Pusher, Single-Place
Thomas (Brothers) Nacelle Pusher, Three-Place
Thomas (Brothers) T-2
Thomas (Brothers) TA
Thomas (Brothers) TA Hydro
Thomas (Brothers) No.1
Thomas (Brothers) 1st Flying Boat
Thomas (Brothers) 2nd Flying Boat
Thomas (Brothers) Model 10A
Thomas (Brothers) Model 10AT

Thomas, Julian P. (New York)
Thomas (Julian) Wind Wagon (1907)

Thomas, William H. (Los Angeles, CA)
Thomas (W. H.) Roteron XM-1 Helicopter

Thomas-Morse Aircraft Corp (Ithaca, NY)
Thomas Brothers Co formed in 1909 and was incorporated as the **Thomas Brothers Aeroplane Co** in 1913. A 1917 bailout by **Morse Chain Co** resulted in formation of **Thomas-Morse Aircraft Corp**, which became a subsidiary of **Consolidated Aircraft Corp** in 1929. Consolidated dissolved Thomas-Morse in 1935. Pre-1935 designs are listed as **Thomas Brothers** and **Thomas-Morse**.
Thomas-Morse MB-1
Thomas-Morse MB-2
Thomas-Morse MB-3
Thomas-Morse MB-3A (See: Boeing)
Thomas-Morse MB-4
Thomas-Morse MB-6
Thomas-Morse MB-7
Thomas-Morse MB-10
Thomas-Morse XO-6
Thomas-Morse "O-6"
Thomas-Morse "O-6B"
Thomas-Morse XO-19
Thomas-Morse O-19
Thomas-Morse O-19A
Thomas-Morse O-19B
Thomas-Morse O-19C
Thomas-Morse O-19E
Thomas-Morse XO-21
Thomas-Morse YO-23
Thomas-Morse Y1O-33
Thomas-Morse XO-932
Thomas-Morse XO-942
Thomas-Morse XP-13 Viper
Thomas-Morse XP-13A Viper
Thomas-Morse R-5 (MB-11, P-268)
Thomas-Morse S-4
Thomas-Morse S-4B
Thomas-Morse S-4C
Thomas-Morse S-4C Post-War, Clipped-Wing Racer
Thomas-Morse S-4E
Thomas-Morse S-5 Prototype (S-4 on Floats)
Thomas-Morse S-5
Thomas-Morse S-6
Thomas-Morse S-7
Thomas-Morse S-9
Thomas-Morse SH-4
Thomas-Morse TM-22
Thomas-Morse TM-23
Thomas-Morse TM-24

***Thomas-Pigeon Aeroplane Co (Boston, MA)** (See: Albree)*

Tomasini (Italy)
Tomasini 1923 Biplane Glider

Thompson, Carl (Suffern, NY)
Thompson (Carl) Special "Mitey Mite"

Thompson, Carl G. (St Louis, MO)
Thompson (Carl G.) Patent (1930)

Thompson, John A. (Tucson, AZ)
Thompson (John A.) Special 101

Thompson, Stephen W. (Dayton, OH)
Thompson (S. W.) Patent (1940)

Thompson, William P. (UK)
Thompson (Wm) Ornithopter (1916)
Thompson (Wm) Pendulum Stability Biplane ("The Scrapheap")

Thomson, Melvin (Quarry, WI)
Thomson Sport I (1928)

***Thöne & Fiala (Austria-Hungary)** (See: Knoller B.I; also Aviatik D.I)*

***Thor, Hank** (See: BJ)*

Thorp Aircraft Co (John W. Thorp) (Pacoima, CA) (Thorp Engineering Co; Burbank, CA)
Thorp T-11 Sky Scooter
Thorp T-111
Thorp T-18 Tiger
Thorp T-211 AeroSport

Thuau (France)
Thuau Monoplane (1910)

Thulin (Enoch Thulins Aeroplanfabrik) (Sweden)
Thulin D
Thulin FA-2
Thulin K
Thulin N

Thunderbird Aircraft Inc (Glendale, CA)
Thunderbird Commercial
Thunderbird W-14

Thunderwings (Scottsdale, AZ)
Thunderwings Scale Spitfire

Thurston Aircraft Corp (David B. Thurston) (Sanford, ME)
Thurston TSC-1 T-Boat (Teal Flying Boat)
Thurston TSC-1A1 Teal Amphibian (See: Patchen)
Thurston TSC-2 Explorer (See: Patchen)

Thurston Monoplane *(See: Aerial Engineering Corp)*

Thurston, Dr. A. P. *(See: Maxim 1909/10 Biplane)*

Tianjin *(See: Shenyang)*

Tientsin *(See: Shenyang)*

Tierra Aircraft, Inc *(See: Teratorn Aircraft, Inc)*

Tiger Cub Developments Ltd (UK) *(See: TCD Ltd)*

Tijuana Aircraft Co (Mexico) *(See: Cia Aeronautica Constructora y de Transport SA de Tijuana)*

Tikhonravov, N. K. (Russia)
Tikhonravov I-302

Tilbury-Fundy (Bloomington, IL)
Tilbury-Fundy Flash

Timm Aerocraft *(See: Wally Timm Aerocraft)*

Timm Aircraft Corp (Glendale & Van Nuys, CA)
Timm Cabin Biplane (1928)
Timm C-185
Timm K-90 Collegiate
Timm N2T-1 Tutor
Timm XN2T-2
Timm Pacific Hawk
Timm PT-175-K
Timm PT-220-C (Civil Conversion of N2T-1 Tutor)
Timm S-160-K (PT-160-K) Aeromold
Timm Tractor Biplane (1916)
Timm T-18
Timm T-800
Timm T-840
Timm Model 160 Sportwing (Kinner Sportwing B-2-R)
Timm-Hughes Pursuit Racer (See: Hughes H-1)

Timson-Albree (Swanpscott, MA)
Timson-Albree Monoplane (1913)

Tips & Smith Inc (J. C. Tips) (Houston, TX)
Tips & Smith United Biplane (1927)

Tipsy *(See: Fairey)*

Tipton, G. W. (Oklahoma City, OK)
Tipton Three-Place (1928)

Titoff (Russia)
Titoff Inflatible Glider

TL Ultralight (Czech Republic)
TL-32 Typhoon
TL-96 Star
TL-132 Condor
TL-232 Condor Plus
TL-532 Fresh

TNCA (Tallares Nacionales de Construcciones Aeronauticas) (Mexico)
TNCA MWT-1
TNCA Teziutlan
TNCA 3-E-130
TNCA 4-E-131 Quetzalcoatl

Toczolowski (Henryk) & Wulw (Poland)
Toczolowski & Wulw TW 12

Todd, Edgar (Pueblo, CO)
Todd Monoplane
Todd 1930 Racer

Todd, Edward (Madison, WI)
Ed Todd Special

Tokaev, G. A. (Russia)
Tokaev Utka (Duck) (See: MiG-8)

Tokorozawa (Army Tokorozawa Aviation School) (Japan)
Tokorozawa Koshiki-2

Tokyo Gasu Denki Kogyo KK (Tokyo Gas & Electrical Industry Co Ltd) *(See: Gasuden)*

Tokyo Hikoki Seisakusho (Tokyo Aeroplane Manufacturing Works) *(See: Narahara)*

Tolstykh, I. P. (Russia)
Tolstykh IT-2
Tolstykh IT-6

Tomashevich (Russia)
Tomashevich I-110
Tomashevich Pegas (Pegasus)

Toneray (USA)
Toneray Helicopter

Topa Aircraft Co (Oxnard, CA)
Topa Scout

Topeka Aeronautical Service *(See: Corben)*

Torp *(See: LTG)*

Torra (Jaime & Francisco) (Argentina)
Torra Glider
Torra 1914 Monoplane

Toussaint (France)
Toussaint Multi-Wing Aircraft

Továrna Letadel inz J. Mráz *(See: Mráz)*

Towle (Towle Aircraft Co Inc; Towle Marine Aircraft Engineering Co) (Detroit, MI)
Towle Amphibian
Towle TA-2 Amphibion
Towle TA-3 Amphibion
Towle WC Amphibian (World Cruiser)

Townsend K. R. (Tulsa, OK)
Townsend A-1 Special (See: Loose, Chet, Special NX64573

Toyo Koku KK (Japan)
Toyo TT-10

Tradewind *(See: Pacific Airmotive Corp)*

Tradewind Turbines (Amarillo, TX)
Tradewind Turbines Propjet Bonanza (Beech A36 Conversion)

Trager-Bierens (Kempes Trager & John Bierens) (USA)
Trager-Bierens T-3 Alibi

Trago Mills Ltd (UK) *(See: Orca)*

Train (Établissements E. Train) (France)
Train Monoplane (1911)

Transall (Transporter Allianz) (International)
Transall C-160A
Transall C-160D
Transall C-160Z

Transcendental Aircraft Corp (Glen Riddle, PA)
Transcendental 1-G Convertiplane
Transcendental 2 Convertiplane

Transland Aircraft (Torrance, CA)
Transland AG-2

Trapeznikov-Krasnikov (G. V. Trapeznikov & A. V. Krasnikov) (Russia)
Trapeznikov-Krasnikov STI-1

Travel Air Manufacturing Co (Wichita, KS)
The **Travel Air Manufacturing Co** was established by Walter Innes, Lloyd Stearman, Walter Beech, and Clyde Cessna in January 1925. Eventually Cessna and Stearman left the company to found their own firms and, in 1929, Travel Air was acquired by the **Curtiss-Wright Corp**. In 1930 Travel Air merged with Curtiss-Wright's **Curtiss-Robertson Airplane Manufacturing Co** as the **Curtiss-Wright Airplane Co** and Beech left to found his own company. For all products of Travel Air developed before the 1930 merger, see **Travel Air**. For all products of the Curtiss-Wright Airplane Co, including those marketed with the "Travel Air" name, see **Curtiss**.
Travel Air Model B
Travel Air Model R "Mystery Ship"
Travel Air Model R, Chevrolair Engine
Travel Air Model R, Barnes Aircraft
Travel Air Model R, Davis Aircraft
Travel Air Model R, Mussolini Aircraft
Travel Air Model R, Shell Oil Aircraft
Travel Air Model R, Texaco Aircraft
Travel Air Model 12 (See: Curtiss CW-12)
Travel Air Model 14 (See: Curtiss CW-14)
Travel Air Model 15 (See: Curtiss CW-15)
Travel Air Model 16 (See: Curtiss CW-16)
Travel Air 1000
Travel Air 2000 ("OX-5 Travel Air")
Travel Air 2000, Bessler Steam Engine Testbed
Travel Air 2000, Reengined Aircraft
Travel Air 2000-T (Tank 115 hp Engine)
Travel Air S-2000
Travel Air 3000, "Hisso Travel Air"
Travel Air 3000, Reengined Aircraft
Travel Air 4000 (Model 4)
Travel Air A-4000
Travel Air B-4000
Travel Air B9-4000
Travel Air C-4000
Travel Air D-4000
Travel Air E-4000
Travel Air T-4000 (Tanager Travel Air)
Travel Air W-4000
Travel Air (4000) 4-D
Travel Air (4000) D-4-D
Travel Air (4000) D-4-D Pepsi Skywriter
Travel Air (4000) A-4-E
Travel Air (4000) 4-P
Travel Air 5000
Travel Air 5000 "Oklahoma" (1926 Dole Race)
Travel Air 5000 "Woolaroc" (1926 Dole Race)
Travel Air 6000
Travel Air A-6000-A
Travel Air 6000-B
Travel Air 6000-B Special
Travel Air (6000) 6-A
Travel Air (6000) 6-B
Travel Air 7000
Travel Air 8000
Travel Air 9000
Travel Air 10
Travel Air 10-B
Travel Air 10-D
Travel Air 11

Travis, James (Cascade, MT)
Travis Aerodrome (1911)

Trefethen-Thistle (Al Trefethen & A. Thistle) (Lomita, CA)
Trefethen-Thistle Sport Aire (See: Stark)
Trefethen-Thistle Sport Aire II
Trefethen TRW Special

Trekker Aircraft Corp Royal Gull *(See: Piaggio P.136)*

Trella Aircraft Co (F. Jay Trella) (Detroit, MI)
Trella T-17 Biplane
Trella T-19 Biplane
Trella T-21 Special
Trella T-100
Trella T-101
Trella T-102
Trella T-103
Trella T-104 Biplane
Trella T-105
Trella T-106 Pusher

Tremaine Aircraft (W. D. Tremaine) (San Diego, CA)
Tremaine Cabin Monoplane
Tremaine HB Racer
Tremaine Hummingbird
Tremaine WT2 Junkers-Type Monoplane (See: Pacific Aircraft)

Tri-R Technologies (Oxnard, CA)
Tri-R Kis

Tridair (Costa Mesa, CA)
Tridair 206L-3ST (Converted Bell 206 Long Ranger)
Tridair 206L-4ST Gemini (Converted Bell 206 Long Ranger)

Trier and Martin Ltd (M. Trier and George Martin) *(See: Martin-Handasyde 4.B Dragonfly)*

Trimmer, Gilbert G. (Mineola, NY)
Trimmer B.P.1 Biplane (1929)

Trimmer-Purchase (Gilbert G. Trimmer & Louis S. Purchase) (Mineola, NY)
Trimmer-Purchase T-2 Trimcraft

Trinks, O. (Germany)
Trinks Monoplane (c.1913)

Trompenburg (Nederlandse Auto u. Vliegtuig Fabrik, Trompenburg) (Netherlands)
Trompenburg Spyker

Troy Air Service (Hatchville, MA)
Troy N.M.J-1A (Nelson M. Jones)

Trump, Frederic L. (Minneapolis, MN)
Trump Sport (1928)

TS (Tadeusz Soltyk) (Poland)
TS-8 Bies

TsAGI (Tsentral'nyi Aerogidrodinamicheskii Institut) (Central Aero-Hydrodynamics Institute) (Russia)
TsAGI A-8 Autogyro

Tsybin (P. V.) (Russia)
Tsybin Ts-25

Tsybin (P. V.) & Kolyesnikov (D. N.) (Russia)
Tsybin & Kolyesnikov KTs-20 Glider

Tubavion (Charles Ponche & Primard) (France)
Tubavion 1912

Tubular Aircraft Products Co, Inc (USA) *(See: Thorp T-211 AeroSport)*

Tufts University (Medford, MA)
Tufts Glider

Tulacz, Piotr (Poland)
Tulacz M.1 Glider

Tulsa Mitchell Wing, Inc (Tulsa, OK) *(See: Mitchell Aircraft Corp)*

Tunison Manufacturing Co (Los Angeles, CA)
Tunison Scout

Tunmer *(See: Dutel)*

Tupolev (Russia)

In 1921, Andrei Nikolayevich Tupolev was placed in charge of the Aviation and Hydrodynamics Department (**AGO, Aviatsii i Gidrodinamiki Otdel**) of the Soviet Union's Central Aero-Hydrodynamic Institute (**TsAGI, Tsentralnoya Aero- Gidrodinamiki Institut**). By 1925, AGO had expanded to include construction and was renamed **AGOS (Aviatsii i Gidrodinamiki Otdel, Stroitel'stvo**; Aviation and Hydrodynamic Office, Construction) and Tupolev was directing a number of design teams. In Dec 1930, Tupolev was confirmed as Chief Constructor and Director of AGOS; three years later AGOS was expanded into **KOSOS (Konstruktorskii Otdel Opitnogo Samolyetostroyeniya**; Design Department for Experimental Aircraft Construction). In 1937, Tupolev was tried during one of Stalin's purges and imprisoned; in 1938 the imprisoned aeronautical engineers were organized into **TsKB-29 (Tsentral'noe Konstrukorskoe Byuro**; Central Construction Bureau 29) with Tupolev in *de facto* charge. Tupolev was released in July 1941, remaining in control of TsKB-29 so that, by the end of World War II, Tupolev's group was considered an **OKB (Opytno Konstruktorskoe Byuro**; Experimental Design Bureau) tracing its history back to AGO. Operations continued as **OKB Tupolev**, surviving Tupolev's death in 1972, until the collapse of the Soviet Union in 1990, when the OKB was reorganized as **ANTK Tupolev (Aviatsionnyi Nauchno-Tekhnicheskiy Kompleks Imieni A. N. Tupoleva**; Aviation Scientific-Technical Complex Named for A. N. Tupolev). With no new military orders, in 1995 ANTK Tupolev reorganized as a joint stock Company, **Tupolev JSC**.
For all designs originating from Tupolev's bureaus and its successors see **Tupolev**. For designs begun under Tupolev's direction, but which were developed and/or produced by another design bureau, see the appropriate bureau.

Tupolev ANT-1
Tupolev ANT-2
Tupolev ANT-8 (MDR-2)
Tupolev ANT-9
Tupolev (ANT-9) PS-9
Tupolev ANT-11
Tupolev ANT-14
Tupolev ANT-18
Tupolev ANT-20 "Maksim Gorkii" (MG)
Tupolev ANT-20*bis* (PS-124)
Tupolev ANT-20V
Tupolev ANT-25 (RD, CAHI-25)
Tupolev ANT-28
Tupolev ANT-32
Tupolev ANT-35
Tupolev ANT-35*bis* (PS-35)
Tupolev ANT-38
Tupolev ANT-43
Tupolev ANT-45
Tupolev ANT-47
Tupolev ANT-48
Tupolev ANT-49
Tupolev ANT-50
Tupolev ANT-51
Tupolev ANT-51*bis*
Tupolev ANT-53
Tupolev C-Prop
Tupolev DB-1 (ANT-36, RD-VV)
Tupolev DB-2 (ANT-37)
Tupolev DB-2B (ANT-37*bis*)
Tupolev DB-2D (ANT-37*bis*)
Tupolev DI-8 (ANT-46)
Tupolev DIP (ANT-29)
Tupolev (I-4) ANT-5 Prototype
Tupolev I-4
Tupolev I-4 *bis*
Tupolev I-4 Zvyeno (I-4Z)
Tupolev I-5 (ANT-12)
Tupolev I-8 (ANT-13) Zhokei (Jockey)
Tupolev I-12 1st Prototype (ANT-23)
Tupolev I-12 2nd Prototype (ANT-23*bis*)
Tupolev (I-14) ANT-31
Tupolev (I-14) ANT-31*bis*
Tupolev I-14
Tupolev MDR-4 (ANT-27)
Tupolev MI-3 (ANT-21)
Tupolev MI-3D (ANT-21*bis*)
Tupolev MK-1 (MK, ANT-22)
Tupolev (MTB-2) ANT-44
Tupolev (MTB-2) ANT-44*bis* (ANT-44D)
Tupolev PB (Model 57)
Tupolev R-3 (ANT-3)
Tupolev (R-6) ANT-7 First Prototype
Tupolev (R-6) ANT-7 Second Prototype
Tupolev R-6
Tupolev R-6 Limuzin
Tupolev R-6a
Tupolev KR-6
Tupolev MR-6 (KR-6P)
Tupolev (R-6) P-6
Tupolev (R-6) MP-6
Tupolev (R-6) PS-7
Tupolev R-7 (ANT-10)
Tupolev (SB) ANT-40 No.2
Tupolev (SB) PS-40 2M-100A
Tupolev (SB) PS-40U
Tupolev (SB) PS-41
Tupolev (SB) PS-41 2M-103U
Tupolev (SB) PS-41*bis*

Tupolev SB 2HS (ANT-40 No.1)
Tupolev SB 2M-100
Tupolev SB 2M-100A
Tupolev SB 2M-103
Tupolev SB 2M-104
Tupolev SB 2M-105
Tupolev SB 2M-106
Tupolev SB 2RTs (ANT-40 2RTs)
Tupolev SB 1938-39
Tupolev SB-3/2M-103
Tupolev SB-3B/2M-103
Tupolev SB-3*bis*
Tupolev SB/TSH (SB/3K)
Tupolev SB*bis* 2M-103
Tupolev SB*bis*-3 2M-103
Tupolev SDB (Model 63)
Tupolev SK-1 (ANT-30)
Tupolev T-1 (ANT-41)
Tupolev (TB-1) ANT-4 Prototype
Tupolev (TB-1) ANT-4 Second Aircraft
Tupolev (TB-1) ANT-4*bis*
Tupolev (TB-1) G-1
Tupolev TB-1
Tupolev TB-1 "Strana Sovyetov"
Tupolev TB-1P (TB-1A)
Tupolev TB-1T
Tupolev (TB-3) ANT-6 Prototype
Tupolev (TB-3) ANT-6/AM-34R "Aviaarktika"
Tupolev (TB-3) ANT-6/M-34RD
Tupolev (TB-3) G-2
Tupolev TB-3
Tupolev TB-3 *Zvyeno* Parasite Fighter Experiments
Tupolev TB-3/AM-34FRN
Tupolev TB-3/AM-34TK
Tupolev TB-3/M-17F
Tupolev TB-3/M-34R
Tupolev TB-3/M-34RN
Tupolev TB-3D
Tupolev TB-4 (ANT-16)
Tupolev TB-6 (ANT-26)
Tupolev TB-7 (ANT-42) (See also: Petlyakov Pe-8)
Tupolev Triton
Tupolev TSh-B (ANT-17)
Tupolev Tu-1 (Model 63)
Tupolev Tu-2, First Prototype (Model 58, Aeroplane 103)
Tupolev Tu-2 Second Prototype (Model 59, Aeroplane 103U)
Tupolev Tu-2 M-82 Engine (Model 60, Aeroplane 103V)
Tupolev Tu-2 Paravan
Tupolev Tu-2 Production (Model 60, Aeroplane 103VS)
Tupolev (Tu-2) UBT (UBT-2)
Tupolev Tu-2/ASh-83 ("Tu-2M")
Tupolev Tu-2D, Model 62
Tupolev Tu-2D, Model 67
Tupolev Tu-2DB (Model 65)
Tupolev Tu-2K
Tupolev Tu-2S (Model 61)
Tupolev Tu-2S (Model 61) Freighter
Tupolev Tu-2Sh
Tupolev Tu-2T
Tupolev Tu-4 *Bull* (B-4)
Tupolev Tu-4 *Bull* Tanker
Tupolev Tu-4K *Bull*
Tupolev Tu-4LL *Bull*
Tupolev Tu-4R *Bull*
Tupolev Tu-4T *Bull*
Tupolev Tu-6 *Bat*
Tupolev Tu-8 *Bat* (Model 69)
Tupolev Tu-10 *Bat* (Model 68)
Tupolev Tu-12 *Cart* (Model 70)
Tupolev Tu-12 Jet Bomber (Model 77)
Tupolev Tu-14 Prototype (Model 81)
Tupolev Tu-14 *Bosun*, Model 89R
Tupolev Tu-14 *Bosun*, Model 89T
Tupolev Tu-14 *Bosun*, Model 93
Tupolev Tu-16 *Badger* (Model 88)
Tupolev (Tu-16) M-16 (Target Drone)
Tupolev Tu-16 Tsyklon
Tupolev Tu-16A (Model 88)
Tupolev Tu-16K-10 *Badger-C* (Model 88)
Tupolev Tu-16K-10-26 *Badger-C Mod*
Tupolev Tu-16K-11-16 *Badger-G*
Tupolev Tu-16K-26 *Badger-G Mod*
Tupolev Tu-16KRM
Tupolev Tu-16KS *Badger-B* (Model 88)
Tupolev Tu-16LL
Tupolev Tu-16M *Badger* (Model 88)
Tupolev Tu-16N
Tupolev Tu-16P *Badger-F*
Tupolev Tu-16P *Badger-L*
Tupolev Tu-16P Mod *Badger-J*
Tupolev Tu-16PP *Badger-H*
Tupolev Tu-16R *Badger-E* (Model 92)
Tupolev Tu-16RM *Badger-J*
Tupolev Tu-16S *Badger-G Mod*
Tupolev Tu-16Sh
Tupolev Tu-16T
Tupolev Tu-16Ye *Badger-D*
Tupolev Tu-16Z
Tupolev Tu-22, Model 82
Tupolev Tu-22, Model 105 (Tu-105)
Tupolev Tu-22B *Blinder-A* (*Bullshot*, *Beauty*) (Model 105A)
Tupolev Tu-22K *Blinder-B*
Tupolev Tu-22KD *Blinder-B*
Tupolev Tu-22KP
Tupolev Tu-22M-0 *Backfire-A* (Tu-22KM)
Tupolev Tu-22M-1
Tupolev Tu-22M-2 *Backfire-B*
Tupolev Tu-22M-3 *Backfire-C*
Tupolev Tu-22M-R
Tupolev Tu-22P
Tupolev Tu-22PD
Tupolev Tu-22R *Blinder-C*
Tupolev Tu-22RD *Blinder-C*
Tupolev Tu-22RDK *Blinder-C*
Tupolev Tu-22RDM *Blinder-C*
Tupolev Tu-22U *Blinder-D*
Tupolev Tu-24P
Tupolev Tu-24R
Tupolev Tu-24S
Tupolev Tu-24Skh
Tupolev Tu-24ST
Tupolev Tu-24T
Tupolev Tu-24V
Tupolev Tu-34
Tupolev Tu-95/I
Tupolev Tu-95/II
Tupolev Tu-95 *Bear* (*Type 40*)
Tupolev Tu-95I
Tupolev Tu-95K-20 *Bear-B*
Tupolev Tu-95K-22 *Bear-G*
Tupolev Tu-95KD *Bear-B*
Tupolev Tu-95KM *Bear-C*
Tupolev Tu-95LL
Tupolev Tu-95M *Bear-A*
Tupolev Tu-95M Mod
Tupolev Tu-95M-5
Tupolev Tu-95M-55
Tupolev Tu-95MR *Bear-E (Bear-C)*
Tupolev Tu-95MS *Bear-H*
Tupolev Tu-95MS-6 *Bear-H*
Tupolev Tu-95N
Tupolev Tu-95RTs *Bear-D*
Tupolev Tu-95SM-20 (Tu-95K)
Tupolev Tu-95U
Tupolev Tu-96
Tupolev Tu-98 *Backfin*
Tupolev Tu-104A *Camel*
Tupolev Tu-104B *Camel*
Tupolev Tu-110 *Cooker*
Tupolev Tu-114 Rossiya (Russia) Prototype
Tupolev Tu-114
Tupolev Tu-114D *Cleat*
Tupolev Tu-116
Tupolev Tu-124 *Cookpot*
Tupolev Tu-124K *Cookpot*
Tupolev Tu-124V *Cookpot*
Tupolev Tu-126 *Moss*
Tupolev Tu-128 Prototype *Fiddler-A* (Tu-28-80)
Tupolev Tu-128 *Fiddler-A*
Tupolev Tu-128A
Tupolev Tu-128B *Fiddler-B*
Tupolev Tu-128M

Tupolev Tu-128UT
Tupolev Tu-130
Tupolev Tu-134 *Crusty* (Model 124A)
Tupolev Tu-134 IMARK
Tupolev Tu-134A *Crusty*
Tupolev Tu-134B *Crusty*
Tupolev Tu-134BSh *Crusty*
Tupolev Tu-134BU *Crusty*
Tupolev Tu-134LL *Crusty*
Tupolev Tu-134Sh *Crusty*
Tupolev Tu-134SKh *Crusty*
Tupolev Tu-134UBL *Crusty*
Tupolev Tu-142 *Bear-F*
Tupolev Tu-142 *Bear-F, Mod I*
Tupolev Tu-142M *Bear-F Mod III*
Tupolev Tu-142M-Z *Bear-F Mod II*
Tupolev Tu-142M-Z *Bear-F Mod IV*
Tupolev Tu-142MK
Tupolev Tu-142MLL
Tupolev Tu-142MP
Tupolev Tu-142MR *Bear-J*
Tupolev Tu-144 Prototype *Charger-A*
Tupolev Tu-144 *Charger-B*
Tupolev Tu-144D
Tupolev Tu-144LL (NASA/Tupolev Testbed)
Tupolev Tu-154 Prototype *Careless*
Tupolev Tu-154A *Careless*
Tupolev Tu-154B *Careless*
Tupolev Tu-154M *Careless*
Tupolev Tu-154M2 *Careless*
Tupolev Tu-154S *Careless*
Tupolev Tu-156 (Model 156S)
Tupolev Tu-156M
Tupolev Tu-160 *Blackjack* (Model 70)
Tupolev Tu-160 SK *Blackjack*
Tupolev Tu-204-22
Tupolev Tu-204-100
Tupolev Tu-204-120
Tupolev Tu-204-122
Tupolev Tu-204-200
Tupolev Tu-204-220
Tupolev Tu-204-230
Tupolev Tu-204-400
Tupolev Tu-204C
Tupolev Tu-214
Tupolev Tu-216
Tupolev Tu-224
Tupolev Tu-234 (Tu-204-300)
Tupolev Tu-244 (SST)
Tupolev Tu-304
Tupolev Tu-324
Tupolev Tu-330
Tupolev Tu-334-100
Tupolev Tu-334-100D
Tupolev Tu-334-200
Tupolev Tu-334 Propfan
Tupolev Tu-336
Tupolev Tu-404
Tupolev Tu-414
Tupolev Model 56
Tupolev Model 64
Tupolev Model 66
Tupolev Model 71
Tupolev Model 72 ("Tu-8")
Tupolev Model 73 ("Tu-14")
Tupolev Model 75 *Cart*
Tupolev Model 78
Tupolev Model 79
Tupolev Model 80
Tupolev Model 83
Tupolev Model 84
Tupolev Model 85
Tupolev Model 86
Tupolev Model 87
Tupolev Model 88
Tupolev Model 90
Tupolev Model 91 *Boot* (Tu-91)
Tupolev Model 94
Tupolev Model 99
Tupolev Model 101
Tupolev Model 102
Tupolev Model 103
Tupolev Model 106
Tupolev Model 107
Tupolev Model 108
Tupolev Model 111
Tupolev Model 115
Tupolev Model 117
Tupolev Model 119 (Nuclear Aircraft)
Tupolev Model 121
Tupolev Model 125
Tupolev Model 134
Tupolev Model 135
Tupolev Model 148
Tupolev Model 155
Tupolev Model 156
Tupolev Model 304
Tupolev Model 306
Tupolev Project 509

Turbay, Alfredo (Argentina)
Turbay T-1 Tucan (Toucan) (X-1)
Turbay T-1B
Turbay T-2
Turbay T-3A
Turbay T-4
Turbay Two-Seat Glider

Türk Hava *(See: THK)*

Turner, Barry (Culver City, CA)
Turner (Barry) Rota 1 Autogyro

Turner Aircraft Inc (E. L. Turner) (Grandview, TX)
Turner (Aircraft) T-40
Turner (Aircraft) T-40A
Turner (Aircraft) Super T-40A
Turner (Aircraft) T-40B
Turner (Aircraft) T-40C

Turner Educational Development Enterprises (Stratford, CT)
(See: Turner Aircraft, Inc)

Turner-Hodson (John E. Turner & Charles F. Hodson) (UK)
Turner-Hodson Ornithopter (1930)

Turner-Laird (Roscoe Turner, Emil Matthew Laird) (USA)
The Turner-Laird RT-14 Meteor, also known as the Turner Racer, was designed by Roscoe Turner with engineering assistance by Prof. Howard Barlow of the University of Minnesota. It was built by the **Lawrence W. Brown Aircraft Co** of California. Found to be too heavy upon completion in 1936, the aircraft was shipped to Chicago to be re-designed and rebuilt by **Emil Matthew (Matty) Laird**, who called it the Laird LTR-14 Pesco Special Meteor.
Turner-Laird RT-14 Meteor
Turner-Laird RT-14 Meteor, NASM

Turolf Eklund *(See: Eklund)*

Tuscar Metals Inc (Cleveland, OH)
Tuscar Tail-Less HA 70-71

Tusco (Tulsa Manufacturing Co) (Galveston, TX)
Tusco Navion D

20th Century Aerial Navigation Co (Chicago, IL)
20th Century Jordan's Improved Gyro Finback Mon-o-plane
20th Century Jordan's Oscillating Wing Airplanes
20th Century Jordan's Parachute Machine

Twentieth Century Aircraft Corp (Amsterdam, NY)
Twentieth Century:
2-place Open Cockpit Monoplane
3-place Cabin Monoplane
6-place Cabin Monoplane

TWI (Germany)
TWI Taifun 17EII

Twining Aeroplane Co (E. W Twining) (UK)
Twining (Aeroplane) 1910 Glider
Twining (Aeroplane) Biplane No. 1
Twining (Aeroplane) Biplane No. 2

Twining, H. L. (USA)
Twining (H. L.) Ornithopter

Twining, Pry (San Francisco, CA)
Twining (Pry) Monoplane (1910)

Twining-Eaton (H. L. Twining; Warren & S. F. Eaton) (Los Angeles, CA)
Twining-Eaton Monoplane (1911)

U

Udet Flugzeugbau GmbH (Germany)

In 1923 Ernst Udet and E. Scheuermann (ex-director of **Bayerische Flugzeug Werke**), established **Udet Flugzeugbau GmbH in Munich**. In 1926 the assets of the company were acquired by the newly-established **Bayerische Flugzeugwerke AG** (not directly related to the earlier BFW). For Udet products up to the U 12, see **Udet**; for later developments, see **BFW (1926)**.

Udet U 7 Kolibri (Hummingbird)
Udet U 8
Udet U 10
Udet U 10a
Udet U 11 Kondor (Condor)
Udet U 12 Flamingo

Ueda, Shotaro (Japan)

Ueda Hiryu-Go Aeroplane (1909 Biplane)
Ueda 1908 Biplane Glider

UEP (Ultra Efficient Products Inc) (Sarasota, FL)

UEP Invader II (Ultralight)
UEP Invader III-B (Ultralight)

UFAG (Ungarische Flugzeugwerke AG) (Austria-Hungary)

UFAG C.I
UFAG D.I
UFAG Parasol Monoplane
UFAG 60.01 (Modified C.I)
UFAG 60.03 (C.II Prototype)

UFM (Ultralight Flying Machines) (Cupertino, CA)

UFM Easy Rider
UFM Icarus II
UFM Icarus IIB
UFM Icarus V
UFM Solar Riser
UFM Sun Riser

UFM of Kentucky (Ultralight Flying Machines of Kentucky Inc) (Louisville, KY)

UFM of Kentucky Aeroplane XP

Ultimate HI (Poway, CA)

Ultimate Hi Trike (Ultralight/Hang Glider Power Unit)

***Ultra Efficient Products Inc** (See: UEP)*

UltraFlight Sales Ltd (Canada)

UltraFlight Lazair Series I
UltraFlight Lazair Series III
UltraFlight Lazair SS Surveillance Special, NASM

Ultralight Flight (Windsor, CT)

Ultralight Flight Mirage
Ultralight Flight Phantom

Ultralight Flying Machines (See: UFM)

Ultralight Flying Machines of Kentucky (See: UFM of Kentucky)

Ultralights, General

Ultralite Soaring Inc (Ft Lauderdale, FL)

Ultralite Soaring Wizard J-3
Ultralite Soaring Wizard T-38
Ultralite Soaring Wizard W-1

***Ultrasport** (See: American Sportscopter)*

Ultravia Aero Inc (Canada)

Ultravia Pelican PL
Ultravia Pelican Sport 450
Ultravia Pelican Sport 600
Ultravia Pelican 2-100 Tutor

***Umbaugh (Muncie, IN)** (See also: Air & Space Manufacturing Inc)*

Umbaugh U-18 Flymobil

Umbra (Aeronautica Umbra SA, Aeronautica Umbra Trojani, AUT) (Italy)

Umbra MB-902
Umbra T.18

Umeda, Yuzo (Japan)

Umeda 1910 Glider
Umeda 1914 Monoplane

Ung Aircraft Corp

Ung Model M32W

***Ungarische Allgemeine Maschinenfabrik AG** (See: MÁG)*

***Ungarische Lloyd Flugzeug und Motorenfabrik AG** (See: Lloyd)*

Union Aircraft Corp (New York, NY)

Union (Aircraft) Model 1 Rotorplane

Union Flugzeug-Werke GmbH (Flugzeug Union Sud) (Germany)

Union Flugzeug-Werke G.I
Union Flugzeug-Werke Pfeil

Unis Obchodni spol sro (Czech Republic)

Unis NA 40 Bongo

United Aircraft Corp (Wichita, KS)
United Aircraft Corp Cabin Biplane

United Helicopter Inc *(See: Hiller)*

United States Aircraft Corp (Van Nuys, CA)
United States Aircraft (CA) DC-3TR

United States Airplane Co (Oak Park, IL) *(See: US Airplane Co)*

United States Airplane Co of New Jersey (New Brunswick, NJ)
United States (NJ) SP7 Aristocraft
United States (NJ) SR4 Winged Bullet
United States (NJ) SR5 Winged Bullet
United States (NJ) US Flyer

United States Airplane & Motor Engineering Co (Upper Sandusky, OH)
United States Airplane (OH) Light Commercial (LC-7)

United States Aviation (South St Paul, MN) *(See: US Aviation)*

Univair Aircraft Corp *(See: Taylorcraft)*

Universal Aviation Co Ltd (UK)
Universal Aviation Birdling

University of Nebraska Glider Club (Lincoln, NE)
University of Nebraska Glider Club 1930 Training Type Glider

University of Utah (Salt Lake City, UT)
University of Utah Glider (U-1, 1929)

Upperçu-Burnelli *(See: Burnelli)*

Urban-Air sro (Czech Republic)
Urban UFM 10 Speed Lambáda
Urban UFM 11
Urban UFM 13 Lambáda

Ursinus, Oskar (Germany)
Ursinus Seaplane Fighter (1916)

US Airplane Co (Oak Park, IL)
US Airplane C1 U. S. Flyer
US Airplane 1928 3-place Open Land Biplane

US Aviation (South St Paul, MN)
US Aviation:
Cumulus Ultralight Motorglider
SuperFloater Ultralight Motorglider

Utah *(See: University of Utah)*

Utva (Fabrika Aviona Utva) (Yugoslavia)
Utva-56
Utva-60-AT
Utva-60-AG
Utva-60-AM
Utva-60H
Utva-65 Privrednik-Go
Utva-66

V

Vakhmistrov (Russia)
Vakhmistrov Zueno (Formation) Z-1
Vakhmistrov Zueno (Formation) Z-2
Vakhmistrov Zueno (Formation) Z-5
Vakhmistrov Zueno (Formation) Z-6
Vakhmistrov Zueno (Formation) Z-7
Vakhmistrov Zueno (Formation) SPB
Vakhmistrov Zueno (Formation) Ampvo (Aviamatka)

Valentin, Leo (France)
Valentin "Flying Suit" (1950)

Valentin (Germany) *(See: TWI)*

Valkyrie *(See: ASL)*

Valmet Aviation Industries (Valmet OY) (Finland) *(See also: VL)*
Valmet L-70 Miltrainer (Vinka)
Valmet L-90TP Redigo

Valsts Elektrotechniska Fabrika *(See: VEF)*

Valtion Lentokonetehdas *(See: VL)*

Van Anden, Frank (Islip, NY)
Van Anden 1909 Biplane

Van Berkel (N.V. Maatschappij Van Berkel's Patent) (Netherlands)
Van Berkel V.O.5
Van Berkel WA
Van Berkel WB Hydro Monoplane

Van Bezel Aircraft Corp, Ltd (San Diego, CA)
Van Bezel S-100 Sailplane

van den Born (France)
van den Born F.5 Trimotor

Van Gruneven, Richard *(See: Van's Aircraft Inc)*

Van Tuil, Norbert C. "Dutch" *(See: Williams, Art)*

Van Valkenburg, Eber H. (McKeesport, PA)
Van Valkenburg VM 11 Monoplane (See also: Phelps)

Van's Aircraft Inc (Richard Van Gruneven) (North Plains, OR)
Van's RV-3 (Homebuilt)
Van's RV-4 (Homebuilt)
Van's RV-5 Swinger (Homebuilt)
Van's RV-6 (Homebuilt)
Van's RV-6A (Homebuilt)
Van's RV-8 (Homebuilt)
Van's RV-8A (Homebuilt)
Van's RV-9 (Homebuilt)

Vance Aircraft Co (Clarence Vance) (Fresno, CA)
Vance Flying Wing (1933)
Vance Primary Training Glider

Vanek, Charles (Portland, OR)
Vanek Vancraft 3 (Gyroplane)

Vaniman, Melvin (France)
Vaniman 1907 Triplane No.1
Vaniman 1908 Triplane No.1*bis*
Vaniman 1908 Triplane (Voisin-Built)

Varga Aircraft Corp *(See: Morrisey)*

Varni, Alfredo (Italy)
Varni Helicopter (1931)

VAT (Voronezh Aerotechnical College) (Russia)
VAT 1965 "Flyying [sic] Motorcycle"

de la Vaulx-Tatin *(See: Clément-Bayard 1908 2-Boom Monoplane)*

VEB Dresden (Volkseigener Betriebe Flugzeugbau Dresden) (Germany)
VEB Dresden Baade Type 152
VEB Dresden Type 155

VEB Lommatzsch (Volkseigener Betriebe Apparatebau Lommatzsch) (Germany)
VEB Lommatzch:
Bönisch-Vogel Glider Towplane
Grunau Baby II (See: Grunau)
Kaplick-Kerl Jet Basic Trainer
Lehrmeister (Instructor) Training Glider
LOM 55/1 Libelle Sailplane
Meise Sailplane (See: DFS)

Vector Aircraft Corp (Turner Falls, MA)
In April 1981, **Vector Aircraft Corp** was established to acquire the rights to Klaus Hill's Humbug ultralight from **Sky Sports Inc**, renaming the aircraft the Vector 600. Vector produced and marketed this aircraft and derivatives until its assets were acquired by **Aerodyne Systems Inc** which, in 1984, sold them in turn to **Sky King International**. For products of Vector and later derivatives produced by Aerodyne Systems and Sky King, see **Vector**.
Vector 600 (Ultralight)
Vector 610 (Ultralight)

VEF (Valsts Elektrotechniska Fabrika) (Latvia)
VEF I-12
VEF I-15
VEF I-16
VEF I-17
VEF I-18

Vega Aircraft Corp (Burbank, CA)
In August 1937 the **Lockheed Aircraft Corp** established the **AiRover Co** as a wholly-owned subsidiary to bolster Lockheed's share of the commercial aviation market. In 1938 the company was renamed **Vega Airplane Co.** On 31 December 1941 Lockheed and Vega merged, with Vega's assets transferred to the **Vega Aircraft Corp.** On 30 November 1943 Vega was completely absorbed into Lockheed. For all products of Lockheed Aircraft, see **Lockheed.** For all AiRover and Vega products predating the 1941 merger, see **Vega.** For all Vega work postdating the merger, see **Lockheed.**
Vega Starliner
Vega 35 (North American NA-35)
Vega 40
Vega 80

Velocity Aircraft (Sebastian, FL)
Velocity Aircraft Velocity

Venables Harry J. (Fruitland, MD)
Venables HD-2 (Heath Parasol Variant)

Vendôme (Raoul Vendôme & Cie) (France)
Vendôme Ab2 Biplane
Vendôme A3 Biplane
Odier-Vendôme 1909 Biplane
Odier-Vendôme 1910 Biplane
Odier-Vendôme 1911 Monoplane
Vendôme II
Vendôme III
Vendôme V
Vendôme VI
Vendôme No.14
Vendôme 1914 Military Monoplane

***Vereinigte Flugtechnische Werke GmbH** (See: VFW)*

Vereiningung Stuttgart (Germany)
Vereiningung 1922 Sailplane

Verilite Aircraft Co Inc (Albuquerque, NM)
Verilite Sunbird

Vermilya-Huffman Flying Service (Wright "Ike" Vermilya, Jr & Stanley C. "Jiggs" Huffman) *(See: Metal Aircraft Corp)*

***Verrinst-Maneyrol** (See: Peyret 1922 Tandem Glider)*

***Versuchbau GmbH Gotha-Ost (VGO)** (See: Zeppelin-Staaken*

Vertak Corp (Troy OH)
Vertak S-220 Estol

***Vertidynamics Corp** (See: Nagler, Bruno)*

***Vertigyro Co** (See: Nagler, Bruno)*

***Vertol** (See: Piasecki, also: Boeing-Vertol)*

Verville Aircraft Co (Detroit, MI)
In 1928, Alfred Verville left the **Buhl-Verville Aircraft Co** to found his own business, the **Verville Aircraft Co.** The company failed in 1931. For aircraft built by Buhl-Verville, see **Buhl**; for products of Verville Aircraft, see **Verville Aircraft.** For aircraft designed by Verville during his tenure at the Air Service Engineering Division (1918-1925), see **Verville, Alfred.** For aircraft designed by Verville, but built for the Air Service by Lawrence Sperry Aircraft Co Inc, see **Verville-Sperry.** For aircraft designed by Verville while employed by other companies, see the appropriate company.
Verville Aircraft Aircoach Model 102 (104-W, Warner Engine)
Verville Aircraft Aircoach Model 104-C
Verville Aircraft Aircoach Model 104-P (Packard Diesel Coach)
Verville Aircraft AT Sportsman
Verville Aircraft AT Sportsman, NASM
Verville Aircraft Flying Boat (See: General Aeroplane Beta)
Verville Aircraft YPT-10
Verville Aircraft YPT-10A
Verville Aircraft YPT-10B
Verville Aircraft PT-10C
Verville Aircraft YPT-10D
Verville Aircraft Pusher Seaplane (See: General Aeroplane Gamma S)

Verville, Alfred V. (US Army Air Service Engineering Division, Dayton, OH)
Alfred Victor Verville worked as an aircraft designer for several companies, including **Curtiss Aeroplane and Motor Corp, Thomas Morse Airplane Co, General Airplane Co,** and **Fisher Body Corp** before joining the **Engineering Division** of the U.S Army Air Service as a civilian in 1918. In 1925, he left government service to co-found the **Buhl-Verville Aircraft Co.** In 1928, he left Buhl-Verville to establish **Verville Aircraft Co**, which failed in 1931. For aircraft designed by Verville during his tenure at the Air Service Engineering Division, see **Verville, Alfred.** For aircraft designed by Verville, but built for the Air Service by Lawrence Sperry Aircraft Co Inc, see **Verville-Sperry.** For aircraft designed by Verville and built by other companies by which he was employed, see the appropriate company.
Verville (Alfred) PW-1
Verville (Alfred) PW-1A
Verville (Alfred) R-3 (See: Verville-Sperry R-3)
Verville (Alfred) VCP-R (R-1)
Verville (Alfred) VCP-1
Verville (Alfred) VCP-1A
Verville (Alfred) VCP-2 (See: Verville (Alfred) PW-1)
Verville (Alfred) 1924 Pulitzer Racer Proposal

Verville-Sperry (Lawrence Sperry Aircraft Co Inc & Alfred V. Verville) (Farmingdale NY)
A number of the aircraft designed by Alfred Verville, but built for the US Army Air Service were produced by the **Lawrence Sperry Aircraft Co Inc**; these are listed under **Verville-Sperry.** Other aircraft designed by Verville during his tenure with the Air Service Engineering Division appear under **Verville, Alfred.** Aircraft designed by Verville but produced by other companies appear under those companies.
Verville-Sperry M-1 Messenger

(See: Sperry Messenger M-1)
Verville-Sperry R-3
Verville-Sperry Sport Plane (See: Sperry Sport Plane)

VFW-Fokker (Vereinigte Flugtechnische Werke GmbH (Germany)

At the end of 1963, **Focke-Wulf GmbH** and **Weser Flugzeugbau GmbH** merged to form **Vereinigte Flugtechnische Werke GmbH (VFW)**. They were joined in 1964 by **Ernst Heinkel Flugzeugbau GmbH.** In 1968, VFW acquired a 65% share in **Rhein-Flugzeugbau GmbH** and a 50% share in **Henschel Flugzeugwerke AG.** As of 1 January 1969, VFW merged with **NVKNV Fokker** of the Netherlands to form **Zentralgesellschaft VFW-Fokker GmbH**, with Fokker (renamed **Fokker-VFW NV**) and VFW (renamed **VFW-Fokker GmbH**) operating as independent subsidiaries. On 1 January 1980, VFW-Fokker was acquired by **Messerschmitt-Bölkow-Blohm GmbH (MBB)**. For all products of the various companies which merged to form VFW, see the respective companies. For all products of VFW and VFW-Fokker between 1963 and 1980, see **VFW-Fokker**. For all products of Fokker and Fokker-VFW, see **Fokker**.

VFW-Fokker H 2
VFW-Fokker H 3
VFW-Fokker SG 1262 (VTOL Testbed)
VFW-Fokker VAK 191 B
VFW-Fokker VC 400
VFW-Fokker VFW 614

***VGO** (See: Zeppelin-Staaken)*

Viale (France)

Viale T.E.P.2

Viberti (Dr Angelo Viberti) (Ali Viberti SpA) (Italy)

Viberti Musca 1
Viberti Musca 1*bis*
Viberti Musca 2
Viberti Musca 4

Vickers (UK)

In 1908, the armaments firm **Vickers, Sons and Maxim Ltd** established an aircraft office. In 1928, Vickers merged its steel, armaments, and shipbuilding interests with those of **Sir W. G. Armstrong, Whitworth & Co Ltd** to form **Vickers-Armstrong Ltd.** As neither company's aviation interests were involved, Vickers' aviation assets reformed as **Vickers (Aviation) Ltd.** Later in 1928, Vickers (Aviation) acquired **Supermarine Aviation Works Ltd**, although the two groups remained separate and independent. In 1938 Vickers (Aviation) and Supermarine were taken over by Vickers-Armstrong, becoming the **Vickers-Weybridge Works** and **Vickers-Supermarine Works** (respectively) of the **Aviation Division of Vickers-Armstrong.** (Staffs did not merge.) With the December 1954 reorganization of Vickers-Armstrong, the Aviation Division became **Vickers-Armstrong (Aircraft) Ltd**; in July 1955, Vickers-Weybridge and Vickers-Supermarine were absorbed into this company. In 1960 Vickers-Armstrong (Aircraft) merged with **Bristol Aircraft Ltd**, **English Electric Aviation Ltd**, and **Hunting Aircraft Ltd** to form **British Aircraft Corp (BAC)**.

For Vickers designs predating 1955, see **Vickers.** For Supermarine and Vickers-Supermarine designs, see **Supermarine.** For all products of Armstrong Whitworth Aircraft see **Armstrong Whitworth.** For products whose major production occurred following the BAC merger, see **BAC.**

Vickers (UK) E.F.B.1 Destroyer
Vickers (UK) E.F.B.2 (Type 18)
Vickers (UK) E.F.B.3 (Type 18B)
Vickers (UK) E.F.B.4
Vickers (UK) E.F.B.5 Gunbus
Vickers (UK) E.F.B.7
Vickers (UK) E.F.B.7A
Vickers (UK) E.F.B.8
Vickers (UK) E.S.1 Barnwell Bullet
Vickers (UK) E.S.2 (E.S.1 Mk.II) Bullet
Vickers (UK) F.B.5 Gunbus
Vickers (UK) F.B.6
Vickers (UK) F.B.9 Streamline Gunbus
Vickers (UK) F.B.11
Vickers (UK) F.B.12
Vickers (UK) F.B.12A
Vickers (UK) F.B.12B
Vickers (UK) F.B.12C
Vickers (UK) F.B.12D
Vickers (UK) F.B.14
Vickers (UK) F.B.14A
Vickers (UK) F.B.14D
Vickers (UK) F.B.14F
Vickers (UK) F.B.16 Hart Scout
Vickers (UK) F.B.16D
Vickers (UK) F.B.16E
Vickers (UK) F.B.19 Mk.I
Vickers (UK) F.B.19 Mk.II
Vickers (UK) F.B.23
Vickers (UK) F.B.24
Vickers (UK) F.B.24C
Vickers (UK) F.B.24E
Vickers (UK) F.B.24G
Vickers (UK) F.B.25
Vickers (UK) F.B.26 Vampire
Vickers (UK) F.B.26A Vampire II
Vickers (UK) G.F.B.1
Vickers (UK) G.F.B.2
Vickers (UK) G.F.B.3
Vickers (UK) Monoplane No.1, No.2
Vickers (UK) Monoplane No.3, No.4
Vickers (UK) Monoplane No.5
Vickers (UK) Monoplane No.6
Vickers (UK) Monoplane No.7
Vickers (UK) Monoplane No.8
Vickers (UK) Vagabond
Vickers (UK) Valentia (1918 Flying Boat)
Vickers (UK) Valentia (1934 Transport)
Vickers (UK) Valetta Prototype
Vickers (UK) Valetta C.Mk.1
Vickers (UK) Valetta C.Mk.2
Vickers (UK) Valetta T.Mk.3
Vickers (UK) Valetta T.Mk.4
Vickers (UK) Valetta Civil Conversions
Vickers (UK) Valiant (1927 Biplane)
Vickers (UK) Valiant Prototype No.1
Vickers (UK) Valiant Prototype No.2
Vickers (UK) Valiant B.Mk.1
Vickers (UK) Valiant B.Mk.1 (C)
Vickers (UK) Valiant B(K).Mk.1
Vickers (UK) Valiant B(PR).Mk.1
Vickers (UK) Valiant B.PR(K).Mk.1
Vickers (UK) Valiant B.Mk.2
Vickers (UK) Valparaiso Mk.I
Vickers (UK) Valparaiso Mk.II
Vickers (UK) Valparaiso Mk.III
Vickers (UK) Vampire (See: Vickers F.B.26 Vampire)
Vickers (UK) Vanellus
Vickers (UK) Vanguard (1921 Biplane)

Vickers (UK) Vanguard (1953 Turboprop)
Vickers (UK) (Vanox) Type 150 (B.19/27)
Vickers (UK) Vanox (Type 255; B.19/27)
Vickers (UK) Varsity T.Mk.1
Vickers (UK) Varsity T.Mk.1, Napier Eland Testbed
Vickers (UK) VC1 Viking Prototypes
Vickers (UK) VC1 Viking Mk.1
Vickers (UK) VC1 Viking Mk.1A
Vickers (UK) VC1 Viking Mk.1B
Vickers (UK) VC1 Viking C.Mk.2
Vickers (UK) VC1 Viking C.Mk.2, Royal Flight
Vickers (UK) VC1 Viking Nene-Viking
Vickers (UK) VC2 Viscount 600 series Prototypes
Vickers (UK) VC2 Viscount 630, Rolls Royce Tay Testbed
Vickers (UK) VC2 Viscount 700 series
Vickers (UK) VC2 Viscount 800 series
Vickers (UK) VC2 Viscount 900 series (See: Vanguard Turboprop)
Vickers (UK) VC7 (Type 1000)
Vickers (UK) VC10 (See: BAC VC-10)
Vickers (UK) Vellore Mk.I
Vickers (UK) Vellore Mk.II/IV
Vickers (UK) Vellore Mk.III
Vickers (UK) Vellox
Vickers (UK) Vendace Mk.I
Vickers (UK) Vendace Mk.II
Vickers (UK) Vendace Mk.III
Vickers (UK) Venom
Vickers (UK) Venture
Vickers (UK) Vernon Mk.I
Vickers (UK) Vernon Mk.II
Vickers (UK) Vernon Mk.III
Vickers (UK) Vespa Mk.I
Vickers (UK) Vespa Mk.II
Vickers (UK) Vespa Mk.III
Vickers (UK) Vespa Mk.IV
Vickers (UK) Vespa Mk.V
Vickers (UK) Vespa Mk.VI
Vickers (UK) Vespa Mk.VII
Vickers (UK) Viastra Mk.I
Vickers (UK) Viastra Mk.II
Vickers (UK) Viastra Mk.III
Vickers (UK) Viastra Mk.VI
Vickers (UK) Viastra Mk.VIII
Vickers (UK) Viastra Mk.X
Vickers (UK) Victoria Mk.I
Vickers (UK) Victoria Mk.II
Vickers (UK) Victoria Mk.III
Vickers (UK) Victoria Mk.IV
Vickers (UK) Victoria Mk.V
Vickers (UK) Victoria Mk.VI
Vickers (UK) Viget
Vickers (UK) Vigilant
Vickers (UK) Viking Mk.I
Vickers (UK) Viking Mk.II
Vickers (UK) Viking Mk.III
Vickers (UK) Viking Mk.IV
Vickers (UK) Viking Mk.V
Vickers (UK) Viking Mk.VI (See: Vickers Vulture)
Vickers (UK) Viking Mk.VII (See: Vickers Vanellus)
Vickers (UK) Viking (1944 Airliner) (See: Vickers VC1 Viking)
Vickers (UK) Vildebeest Prototype, Type 132
Vickers (UK) Vildebeest Prototype, Private Venture Aircraft
Vickers (UK) Vildebeest, Spain
Vickers (UK) Vildebeest Mk.I
Vickers (UK) Vildebeest Mk.II
Vickers (UK) Vildebeest Mk.III
Vickers (UK) Vildebeest Mk.IV
Vickers (UK) V.I.M. (Vickers Instructional Machine)
Vickers (UK) Vimy F.B.27 Prototypes
Vickers (UK) Vimy F.B.27A
Vickers (UK) Vimy Ambulance
Vickers (UK) Vimy Australian (1919 Smith Brothers Flight)
Vickers (UK) Vimy Commercial Prototype
Vickers (UK) Vimy Commercial
Vickers (UK) Vimy South African (1920 Van Ryneveld-Brand)
Vickers (UK) Vimy Transatlantic (1919 Alcock-Brown Flight)
Vickers (UK) Vimy Trainer
Vickers (UK) Vincent
Vickers (UK) Vireo
Vickers (UK) Virginia Mk.I
Vickers (UK) Virginia Mk.II
Vickers (UK) Virginia Mk.III
Vickers (UK) Virginia Mk.IV
Vickers (UK) Virginia Mk.V
Vickers (UK) Virginia Mk.VI
Vickers (UK) Virginia Mk.VII
Vickers (UK) Virginia Mk.VIII
Vickers (UK) Virginia Mk.IX
Vickers (UK) Virginia Mk.X
Vickers (UK) Vivid (Vixen Mk.VII)
Vickers (UK) Vixen Mk.I
Vickers (UK) Vixen Mk.II
Vickers (UK) Vixen Mk.III
Vickers (UK) Vixen Mk.IV
Vickers (UK) Vixen Mk.V
Vickers (UK) Vixen Mk.VI
Vickers (UK) Vixen Mk.VII (See: Vickers Vivid)
Vickers (UK) Vulcan (Type 61)
Vickers (UK) Vulcan (Type 74)
Vickers (UK) Vulture Mk.I
Vickers (UK) Vulture Mk.II
Vickers (UK) Warwick Prototypes (B.1/35)
Vickers (UK) Warwick ASR.Mk.I
Vickers (UK) Warwick B.Mk.I
Vickers (UK) Warwick C.Mk.I
Vickers (UK) Warwick B.Mk.II
Vickers (UK) Warwick GR.Mk.II
Vickers (UK) Warwick C.Mk.III
Vickers (UK) Warwick GR.Mk.V
Vickers (UK) Warwick ASR.Mk.VI
Vickers (UK) Wellesley Prototype (Type 246, G.4/31 Monoplane)
Vickers (UK) Wellesley
Vickers (UK) Wellesley, Long-Range Flight
Vickers (UK) Wellington Prototype (B.9/32)
Vickers (UK) Wellington Mk.I
Vickers (UK) Wellington Mk.IA
Vickers (UK) Wellington Mk.IC
Vickers (UK) Wellington Mk.II
Vickers (UK) Wellington Mk.II, 40mm Gun Testbed
Vickers (UK) Wellington Mk.II, Whittle Jet Testbed
Vickers (UK) Wellington Mk.III Prototype
Vickers (UK) Wellington Mk.III
Vickers (UK) Wellington Mk.IV
Vickers (UK) Wellington Mk.V
Vickers (UK) Wellington Mk.VI
Vickers (UK) Wellington Mk.VIII
Vickers (UK) Wellington Mk.X
Vickers (UK) Wellington Mk.X, Rolls-Royce Dart Testbed
Vickers (UK) Wellington T.Mk.X
Vickers (UK) Wellington Mk.XI
Vickers (UK) Wellington Mk.XII
Vickers (UK) Wellington Mk.XIII
Vickers (UK) Wellington Mk.XIV
Vickers (UK) Wellington Mk.XV
Vickers (UK) Wellington Mk.XVI
Vickers (UK) Wellington Mk.XVII
Vickers (UK) Wellington Mk.XVIII
Vickers (UK) Wellington Mk.XIX
Vickers (UK) Wellington DWI (Directional Wireless Installation)
Vickers (UK) Wibault Scout
Vickers (UK) Windsor
Vickers (UK) Type 123 Hispano Scout
Vickers (UK) Type 141 Scout

Vickers (UK) Type 143 Bolivian Scout
Vickers (UK) Type 151 Jockey
Vickers (UK) Type 161 (F.29/27; C.O.W. Gun Fighter)
Vickers (UK) Type 163 (C.16/28)
Vickers (UK) Type 177
Vickers (UK) Type 207 (M.1/30)
Vickers (UK) Type 253 (G.4/31 Biplane)
Vickers (UK) Type 432 (F.7/41)

Vickers (Canada)

In 1911, British firm **Vickers, Sons and Maxim Ltd** established **Canadian Vickers Ltd** as subsidiary in Montreal. An aircraft division was added in 1922. Canadian Vickers acquired **Canadian Associated Aircraft** in 1941, transferring its aircraft operations to a government-owned factory a year later. In 1944 Canadian Vickers' aircraft division was absorbed into **Canadair, Ltd.** For products of the parent (British) Vickers, see **Vickers**. For original designs of Canadian Vickers through 1944, see **Vickers (Canada)**. For all later development see **Canadair**.

Vickers (Canada) Vancouver Mk.I
Vickers (Canada) Vancouver Mk.II
Vickers (Canada) Vancouver Mk.II/SW
Vickers (Canada) Vanessa
Vickers (Canada) Varuna Mk.I
Vickers (Canada) Varuna Mk.II
Vickers (Canada) Vedette Mk.I
Vickers (Canada) Vedette Mk.II
Vickers (Canada) Vedette Mk.V
Vickers (Canada) Vedette Mk.V, Amphibian Prototype
Vickers (Canada) Vedette Mk.Va
Vickers (Canada) Vedette Mk.VI
Vickers (Canada) Velos
Vickers (Canada) Vigil
Vickers (Canada) Vista

Vickers Ltd (Engineering Group of Vickers-Armstrong Ltd) (UK)

In the 1960s the shipbuilding interests of **Vickers-Armstrong Ltd** (separate from the aerospace interests, **Vickers-Armstrong (Aircraft) Ltd**, which had already been absorbed into **British Aircraft Corp**) began working on air cushion vehicles (ACVs). In 1966 the British government's **National Research Development Corp (NRDC)** realigned British ACV work, merging Vickers' ACV interests with the **Sauders-Roe Division of Westland Aircraft Ltd** (much of Saunders-Roe had been acquired by Westland in 1959) to form **British Hovercraft Corp Ltd (BHC)** with NRDC holding 10% interest and Vickers and Westland dividing the remaining 90%. In 1970, Westland acquired Vickers' share of BHC, which then became a subsidiary of Westland. For all products of Vickers' ACV interests predating the 1966 merger, see **Vickers Ltd.** For all products of Westland and BHC having Saunders Roe (SR) designations see **Saunders-Roe**; for products with BH designations see **BHC**.

Vickers Ltd VA-2 (ACV)
Vickers Ltd VA-3 (ACV)

Vickers-Slingsby (See: Slingsby)

Victa Consolidated Industries (Victa Ltd) (Australia)

When Australian Henry Millicer won a lightplane design competition in 1958, his prototype was built by **Pacific & Western Aviation**. Full production fell to **Victa Consolidated Industries** (later **Victa Ltd**) until 1967, when the company's interests were purchased by New Zealand's **AESL**. AESL continued to develop and produce Millicer's basic designs until merging into **New Zealand Aerospace Industries Ltd** in 1973; Airtourer production ended that year. Millicer's original designs are found under **Victa**, with subsequent development under **AESL**.

Victa Aircruiser R-2
Victa Aircruiser 210
Victa Air Tourer Prototype
Victa Air Tourer AT.2C
Victa Air Tourer AT.2L
Victa Airtourer 100 (AT.100)
Victa Airtourer 115 (AT.115)
Victa Airtourer 150 (See: AESL)
Victa All-metal 4-Seater
Victa EP.9 Prospector (See: Edgar Percival)

Victor Aircraft Corp (Mt Holly, NJ)

Victor (Mt Holly, NJ) Biplane (c.1929)

Victor Aircraft Corp (Freeport, NY)

Victor (NY) Advanced Trainer
Victor (NY) Scout

Victor Metal Aircraft Corp (Camden, NJ)

Victor (Camden, NJ) G-1

Victory Aircraft Engineering Corp (North Hollywood, CA)

Victory SS-1

Viking Aircraft (Pensacola, FL)

Viking (Aircraft) Dragonfly

Viking Flying Boat Co (New Haven, CT)

Viking Flying Boat Co of New Haven, CT was established in 1928 for the licensed production of **FBA-Schreck** flying boats. In 1930, Viking bought **Bourdon Aircraft Corp** (est. 1928) and all manufacturing rights to the Bourdon Kittyhawk. All Viking production is listed under **Viking** except for Kittyhawks, which are filed under **Bourdon**.

Viking (Flying Boat) OO-1
Viking (Flying Boat) V-2

Villiers (Établissements François Villiers) (France)

Villiers 2 AMC.2
Villiers 4 C.2
Villiers 5 CAN.2
Villiers 9 C.1
Villiers 11
Villiers 24 CAN.2
Villiers 26

Vincent, Ernest Peter (New York, NY)

Vincent (E. P.) Triplane (1910 Patent)

Vincent, G. C. (Zanesville, OH)

Vincent (G. C.) 1927 Biplane

Vinet, Gaston (France)

Vinet 1910 Monoplane
Vinet 1911 Monoplane (Type D)

VisionAire Corp (Chesterfield, MO)

VisionAire Vantage

VL (Valtion Lentokontehdas; State Aircraft Factory) (Finland) (See also: Valmet)

VL Pyry (Snowfall)
VL Tuisku (Blizzard)
VL Vihuri (Flurry)
VL Viima (Cold Wind)

Vlaicu, Aurel (Rumania)
Vlaicu 1911 Monoplane

Vlasak-Husnik *(See: Aero, Prague)*

Vliegtuigbouw (Netherlands)
Vliegtuigbouw Sagitta

Vodochody *(See: Aero Vodochody)*

Voelker, Karl (Los Angeles, CA)
Voelker 1929 Monoplane

Vogt *(See: Burgfalke)*

Voiles, Clyde C. (Los Angeles, CA)
Voiles VR-1 Monoplane

Voisin, 1930s *(See: Ailes Enghiennoises)*

Voisin (France)

In 1905 Gabriel and Charles Voisin established **Appareils d'Aviation Les Frères Voisin**. Gabriel continued operations after Charles' death in 1912, eventually renaming the firm **Société Anonyme des Aéroplanes G. Voisin**. After the end of the First World War the company converted to automobile production and left the aviation field. For products of Frères Voisin and Aéroplanes Voisin, see **Voisin**.

Voisin Prewar Aircraft
Voisin 1897 Ornithopter
Voisin 1903 Glider
Voisin 1906 Filiasi Hydroplane
Voisin (1) 1907 Kapferer
Voisin (2) 1907 Delagrange No.1
Voisin (3) 1907 Delagrange No.2
Voisin (4) 1907 Farman No.1
Voisin (5) 1908 Farman No.1*bis*
Voisin (6) 1908 58-hp Renault
Voisin (7) 1908 Farman No.1*bis* Triplane
Voisin (8) 1908 Moore-Brabazon No.1
Voisin (9) 1908 50-hp Vivinus
Voisin (10) 1908 de Caters No.1 Triplane
Voisin (11) 1908 Zipfel No.1
Voisin (12) 1909 Moore-Brabazon No.2 Triplane
Voisin (13) 1909 60-hp ENV Type F Engine
Voisin (14) 1909 Moore-Brabazon No.3
Voisin (15) 1909 Moore-Brabazon No.4
Voisin (16) 1909 de Caters No.2 Biplane
Voisin (17) 1909 de Caters No.3
Voisin (18) 1909 Vivinus No.1
Voisin (19) 1909 52-hp Chenu Engine
Voisin (20) 1909 Delagrange No.3
Voisin (21) 1909 Farman No.2
Voisin (22) 1909 Koch No.1
Voisin (23) 1909 Fournier No.1
Voisin (24) 1909 Simms No.1
Voisin (25) 1909 Hein No.1
Voisin (26) 1909 Euler No.1
Voisin (27) 1909 Aero-Club d'Odessa
Voisin (28) 1909 Kaulbars No.1
Voisin (29) 1909 Fletcher
Voisin (30) 1909 Rougier
Voisin (31) 1909 58-hp Renault Engine
Voisin (32) 1909 Alsace
Voisin (33) 1909 Ile-de-France
Voisin (34) 1909 Daumont No.1
Voisin (35) 1909 Octavie No.3
Voisin (36) 1909 38-hp Gnome Engine
Voisin (37) 1909 Hansen No.1
Voisin (38) 1909 Bolotoff Triplane
Voisin (39) 1909 Gaudart
Voisin (40) 1909 Gobron No.1
Voisin (41) 1909 Buneau-Varilla No.1
Voisin (42) 1909 Sanchez-Besa No.1
Voisin (43) 1909 Sanchez-Besa No.2
Voisin (44) 1909 Da Zara
Voisin (45) 1909 Swendsen
Voisin (46) 1909 Cockburn No.1
Voisin (47) 1909 de Baeder
Voisin (48) 1909 Metrot-Marce No.1
Voisin (49) 1909 de Laroche
Voisin (50) 1909 Italian Aeronautical Society
Voisin (51) 1910 50-hp Gnome Omega Engine
Voisin (52) 1910 Champel
Voisin (53) 1910 Pauwels No.1
Voisin (54) 1910 Stoeckel No.1
Voisin (55) 1910 Mignot
Voisin (56) 1910 Ravetto
Voisin (57) 1910 Paul
Voisin (58) 1910 Chailley
Voisin (59) 1910 Antelme
Voisin (60) 1910 Allard
Voisin (61) 1910 Poillot
Voisin (62) 1910 Adorjan
Voisin (63) 1910 Croquet
Voisin (64) 1910 de Montigny
Voisin (65) 1910 Haeffely
Voisin (66) 1910 Cagno
Voisin (67) 1910 Economo
Voisin (68) 1910 Rigal
Voisin (69) 1910 Militaire (Military)
Voisin (70) 1910 Tourisme (Touring Plane)
Voisin (71) 1911 Militaire (Military)
Voisin (72) 1911 Militaire (Military)
Voisin (73) 1911 Militaire (Military)
Voisin (74) 1911 Astra Triplane
Voisin (75) 1912 Sanchez-Besa
Voisin (76) 1912 Bathiat-Sanchez

Vojenská Tovarna na Letadla *(See: Letov)*

Volaircraft Inc (Pittsburgh, PA)

Volaircraft Inc was established in 1961 to manufacture light aircraft. In 1965 it was purchased by **Rockwell Standard Corp**. Volaircraft was merged into the **Aero Commander Division** of Rockwell, which produced Volaires under the Aero Commander name. For production predating the 1965 merger, see **Volaircraft**. For all later production see **Aero Commander**.

Volaircraft Volaire 10

Volf, Christian A.
Volf Rotor Ship (c.1930)

Volkseigener Betriebe *(See: VEB Dresden, VEB Lommatzsch)*

Volland (Avions Volland) (France)
Volland V-10

Volmer Aircraft (Glendale, CA)

During the 1930s, Volmer S. Jensen designed and built a number of sailplanes under the **Volmer Sailplanes** logo. After World War II he and John Carssow established **Volmer-Carssow Aircraft Co.** By 1947 the company was known as **Volmer Aircraft.** As of 1995, Volmer was still offering kits, ultralights, and hand gliders. For Volmer Sailplanes, Volmer-Carssow and Volmer Aircraft, see **Volmer**.

Volmer VJ10 (Sailplane)
Volmer VJ14 (Sailplane)
Volmer VJ21 Jaybird
Volmer VJ22 Sportsman

Volmoller (Germany)
Volmoller III (1909)

Volmoller IV (1911)

Volpar Inc (Pacoima, CA)
Volpar Coin (T-28 Coln Modification)
Volpar Jetstream 3 (See: Century Aircraft Corp, Amarillo, TX)
Volpar Tri-Gear Mk.IV (Modified Beech 18)
Volpar Turbo 18 (Modified Beech 18)
Volpar Turboliner (Modified Beech 18)
Volpar Turboliner II (Modified Beech 18)

Von Baumhauer, Albert Gilles (Netherlands)
Von Baumhauer Helicopter (1930)

Von Hoffmann Aircraft Co (Anglum, MO)
Von Hoffmann X-1 Experimental Training Plane

von Schertel *(See: Schertel)*

Von Siemens, Werner *(See: Siemens)*

Von Wechmar, Ernst (Germany)
Von Wechmar "Flying Suit" (1888)

Voronezh Aerotechnical College *(See: VAT)*

Vought, Chauncey Milton "Chance" (USA)
Vought (Chauncey) PLV (Pontkowski-Lichorsik-Vought)

Vought (Hartford, CT; Dallas, TX)

In 1918 **Lewis & Vought Corp** began producing the military aircraft designs of Chauncey M. "Chance" Vought. In 1924 the company was renamed **Chance Vought Corp.** When Vought died in 1930, Chance Vought Corp became a division of **United Aircraft and Transport Corp.** In 1935 the division was renamed **Chance Vought Aircraft** and in 1939 it merged with the Sikorsky division to form **Vought-Sikorsky Division of United Aircraft Corp.** The divisions split again in 1943 with Sikorsky concentrating on helicopter development and **Chance Vought Aircraft Division** producing combat aircraft. In 1954 **Chance Vought Aircraft Inc** became a subsidiary (no longer a division) of United Aircraft. In 1960 the company became **Chance Vought Corp,** merging in 1961 with **Ling-Temco Electronics Inc** to form **Ling-Temco-Vought Inc.** LTV's aerospace division was called **Chance Vought Corp** initially, **LTV Aerospace Corp** after 1963, **Vought Corp** from 1976, **Vought Aero Products Division** from 1984, and **LTV Aircraft Products Group** from 1986. After the sale of several elements of the Group, it was redesignated **Vought Aircraft Division.** In 1992 LTV, Lockheed Corp, and Martin Marietta Corp reached an agreement which resulted in the sale of the Aircraft Division to **The Carlyle Group** and its minority (49%) partner, **Northrop Grumman Corp,** as **Vought Aircraft Co.** Northrop Grumman immediately exercised its option to acquire Carlyle's interest and absorbed Vought.

For all Chauncey Vought designs by predating the establishment of Lewis & Vought, see **Vought, Chance** or the appropriate manufacturer. For all Lewis & Vought, and subsequent Vought designs, see **Vought.** For aircraft designed by LTV divisions unrelated to Vought, see **LTV Electrosystems.** For Vought-Sikorsky aircraft, see Sikorsky or Vought, depending upon the origin of the design. For Temco aircraft designed before the formation of Ling-Temco Electronics, see **Temco.**

Vought AU-1 (F4U-6) Corsair
Vought YA-7A Corsair II
Vought A-7A Corsair II
Vought A-7B Corsair II
Vought A-7C Corsair II
Vought TA-7C Corsair II
Vought A-7D Corsair II
Vought A-7E Corsair II
Vought YA-7F Corsair II (A-7 Plus, A-7 Strikefighter)
Vought KA-7F Corsair II
Vought A-7G Corsair II (Switzerland)
Vought YA-7H Corsair II
Vought A-7H Corsair II (Greece)
Vought TA-7H Corsair II
Vought A-7K Corsair II
Vought A-7P Corsair II (Portugal)
Vought ADAM (Air Deflection and Modulation)
Vought ADAM II
Vought XC-142A
Vought (C-142) Downtowner (Civil Transport Proposal)
Vought FU-1 (UO-3)
Vought XF2U-1
Vought XF3U-1
Vought XF4U-1 Corsair
Vought F4U-1 Corsair
Vought F4U-1A Corsair
Vought F4U-1D Corsair
Vought F4U-1D Corsair, NASM
Vought F4U-2 Corsair
Vought XF4U-3 Corsair
Vought XF4U-4 Corsair
Vought F4U-4 Corsair
Vought F4U-4B Corsair
Vought F4U-4N Corsair
Vought F4U-4X Corsair
Vought XF4U-5 Corsair
Vought F4U-5 Corsair
Vought F4U-5N Corsair
Vought F4U-5NL Corsair
Vought F4U-6 Corsair (See: Vought AU-1 Corsair)
Vought F4U-7 Corsair
Vought (F4U) Corsair Mk.I
Vought (F4U) Corsair Mk.II
Vought (F4U) Corsair Mk.III (See: Brewster F3A)
Vought (F4U) Corsair Mk.IV (See: Goodyear FG)
Vought XF5U-1 Skimmer
Vought XF6U-1 Pirate
Vought XF6U-1 Pirate, Afterburner Prototype
Vought F6U-1 Pirate
Vought XF7U-1 Cutlass
Vought F7U-1 Cutlass
Vought F7U-3 Cutlass
Vought F7U-3 Cutlass, Delta Tip Ailevator Testbed
Vought F7U-3M Cutlass
Vought F7U-3P Cutlass
Vought (XF-8A) XF8U-1 Crusader
Vought (XF-8A) XF8U-1 Crusader, NASM
Vought (F-8A) F8U-1 Crusader
Vought (F-8A) F8U-1(M) Crusader
Vought (DF-8A) F8U-1D Crusader
Vought (RF-8A) F8U-1P Crusader
Vought (TF-8A) F8U-1T Crusader
Vought (F-8B) F8U-1E Crusader
Vought (YF-8C) YF8U-2 Crusader
Vought (F-8C) F8U-2 Crusader
Vought (F-8D) F8U-2N Crusader
Vought (F-8E) F8U-2NE Crusader
Vought [F-8E(FN)] F8U-2NE(FN) Crusader
Vought RF-8G Crusader
Vought F-8H Crusader

Vought F-8J Crusader
Vought F-8K Crusader
Vought F-8L Crusader
Vought F-8P Crusader (Philippines)
Vought (F-8) F8U-3 Crusader III
Vought F-8 DFBW (Digital Fly-by-Wire Testbed)
Vought F-8 Supercritical Wing Testbed
Vought Gluhareff Dart-Shaped Fighter Proposal
Vought UO (UF)
Vought UO-1 (UF-1)
Vought UO-1, Airship Hook Testbed
Vought UO-1C
Vought UO-2 (1922 Curtiss Marine Trophy)
Vought UO-3 (See: Vought FU-1)
Vought UO-4 (USCG)
Vought UO-5
Vought O2U-1 Corsair
Vought O2U-1A Corsair (Argentina)
Vought O2U-2 Corsair
Vought O2U-2 Corsair, USCG
Vought O2U-3 Corsair
Vought O2U-4 Corsair
Vought O2U-4 Corsair, Grumman Amphibian Float Testbed
Vought O2U-4A Corsair (Mexico)
Vought O3U-1 Corsair
Vought O3U-2 Corsair (See: Vought SU-1 Corsair)
Vought O3U-3 Corsair
Vought O3U-4 Corsair (See: Vought SU-2, SU-3 Corsair)
Vought XO3U-5 Corsair
Vought XO3U-6 Corsair
Vought O3U-6 Corsair
Vought XO4U-1
Vought XO4U-2
Vought XO5U-1
Vought O-28 Corsair
Vought XOSU-1
Vought XOS2U-1 Kingfisher
Vought OS2U-1 Kingfisher
Vought OS2U-2 Kingfisher
Vought OS2U-3 Kingfisher
Vought OS2U-3 Kingfisher, NASM
Vought (OS2U) Kingfisher Mk.I
Vought Pampa 2000 (JPATS Entry)
Vought QO-1 (Cuban UO-1)
Vought SU-1 (O3U-2) Corsair
Vought SU-1 Special (O3U-2) Corsair
Vought SU-2 (O3U-4) Corsair
Vought SU-2 Special (O3U-4) Corsair
Vought SU-3 (O3U-4) Corsair
Vought SU-3 Special (O3U-4) Corsair
Vought XSU-4 Corsair
Vought SU-4 Corsair
Vought XSBU-1
Vought SBU-1
Vought SBU-2
Vought XSB2U-1 Vindicator
Vought SB2U-1 Vindicator
Vought SB2U-2 Vindicator
Vought XSB2U-3 Vindicator
Vought SB2U-3 Vindicator
Vought XSB3U-1
Vought XSO2U-1
Vought XTBU-1 Seawolf
Vought V-50 Corsair
Vought V-65B Corsair
Vought V-65C Corsair (China)
Vought V-65F Corsair (Argentina)
Vought V-66E Corsair
Vought V-66S Corsair
Vought V-70 Corsair
Vought V-70A Corsair
Vought V-70B Corsair
Vought V-80 Corsair
Vought V-80F Corsair (Argentina)
Vought V-80P Corsair (Peru)
Vought V-85G Kurier
Vought V-90 Corsair
Vought V-93S Corsair (Thailand)
Vought V-97 Corsair (Mexico)
Vought V-99 Corsair
Vought V-99M Corsair (Mexico)
Vought V-100 Corsair Junior
Vought V-135 Corsair
Vought V-141
Vought V-142
Vought V-142A (Argentina)
Vought V-143
Vought V-148
Vought V-156 Vindicator
Vought V-156B-1 Chesapeake
Vought V-156F Vindicator (France)
Vought V-167
Vought V-169A
Vought V-173 ("Flying Pancake")
Vought V-173 ("Flying Pancake"), NASM
Vought V-530
Vought VE-7
Vought VE-7, Pulitzer Race
Vought VE-7, Retractable Landing Gear Testbed
Vought VE-7G
Vought VE-7GF
Vought VE-7H
Vought VE-7S
Vought VE-7SF
Vought VE-7SH
Vought VE-8
Vought VE-9
Vought VE-9H
Vought VE-10
Vought VE-19
Vought VS-44 (See: Sikorsky VS-44)
Vought VS-326
Vought VS-326A

***VPK Mapo** (See: MiG)*

VTOL, General

This topic includes material of a general nature on VTOL (Vertical Take-Off and Landing), V/STOL (Vertical/Short Take-Off and Landing) and STOL (Short Take-Off and Landing) aircraft. For information on specific VTOL, V/STOL, and STOL aircraft, see the specific type.

Vuia, Trajan (Hungary/France)

Vuia No.1 Monoplane (1906)
Vuia No.1*bis* Monoplane (1906)

Vuillemenot (France)

Vuillemenot Type A.E. 15 Glider (1933)

Vuitton, Louis (France)

Vuitton Helicopter No.1 (1909, Vuitton-Huber)
Vuitton Helicopter No.2 (1910)
Vuitton Monoplane No.2 (1911)

***Vulcan Aircraft Co (Portsmouth, OH)** (See also: Davis, Walter)*

Vulcan V-1 American Moth
Vulcan V-2 American Moth Parasol
Vulcan V-3 American Moth

Vultee (Downey, CA)

The first Vultee aircraft was produced in 1932 by **Airplane Development Corp.** In 1934 the company was reorganized as a division of **Aviation Manufacturing Corp**, an **AVCO** subsidiary. The division was retitled **Vultee Aircraft Division** in 1937; **Vultee Aircraft Inc** was established in 1939, acquiring the assets of Aviation Manufacturing Corp but remaining an AVCO subsidiary. Vultee gained control of **Consolidated Aircraft Corp** in 1941, forming **Consolidated Vultee Aircraft Corp (Convair)** in 1943. All company products which predate 1943 are listed as **Vultee**; all products from 1943 to 1961 are listed as **Convair**.

Vultee YA-19

Vultee XA-19A (Lycoming O-1230-1 Testbed)
Vultee XA-19B (P&W R-2800-1 Testbed)
Vultee XA-19C (P&W R-1830-51 Testbed)
Vultee XA-41 (Model 90)
Vultee BC-3 (Model 51)
Vultee (BT-13) Valiant (Model 54)
Vultee (BT-13) Valiant 54A
Vultee BT-13A Valiant
Vultee BT-13A Valiant, NASM
Vultee BT-13B Valiant
Vultee BT-13 Civil Conversions
Vultee BT-15 Valiant
Vultee BT-15 Civil Conversions
Vultee XP-54
Vultee SNV-1 Valiant
Vultee SNV-2 Valiant
Vultee SNV Civil Conversions
Vultee Vanguard Prototype (Model 61)
Vultee Vanguard Modified Prototype (Model 48-X)
Vultee Vanguard Production Prototype (Model 48-C)
Vultee Vanguard Model 48-A
Vultee Vanguard Model 48-C (P-66. J10, Vanguard Mk.I)
Vultee Vengeance Prototype (V-72)
Vultee Vengeance Mk.I (A-31, RA-31)
Vultee Vengeance Mk.IA (A-31, RA-31)
Vultee Vengeance Mk.II (A-31)
Vultee (Vengeance) XA-31A
Vultee (Vengeance) XA-31B
Vultee (Vengeance) XA-31C
Vultee (Vengeance) YA-31C
Vultee Vengeance Mk.III (A-31C)
Vultee Vengeance Mk.IV (A-35A)
Vultee Vengeance Mk.IV (A-35B)
Vultee (Vengeance) TBV-1 Georgia
Vultee V1
Vultee V1-A
Vultee V1-A "Lady Peace"
Vultee V11
Vultee V11-A
Vultee V11-G
Vultee V11-T
Vultee V12 (AB-2)
Vultee V12-C
Vultee V12-D
Vultee Model 56-A (2-Place Pursuit)

W

***W. B. Buethe Enterprises, Inc** (See: Buethe)*

***W. D. Waterman Aircraft Manufacturing Co** (See: Waterman)*

***W. G. Tarrant Ltd** (See: Tarrant)*

***W. H. Ekin Engineering Co Ltd** (See: Ekin)*

***W. Starling Burgess Co Ltd** (See: Burgess)*

Wabash Aircraft Co (Terre Haute)
Wabash WA-250X

Wackett (Australia)
Wackett Warbler (1923)
Wackett Warrigal I
Wackett Warrigal II
Wackett Widgeon (1928)

Waco (Troy, OH)
Between 1915 and 1919, H. C. Deuther, Clayton J. Bruckner, and Elwood J. Junkin experimented with various aircraft designs; the three formed the **DBJ Aeroplane Co** in September 1919. Two months later George E. Weaver formed the **Weaver Aircraft Co (Waco)**, eventually incorporating DBJ and its assets. In 1923 the company was renamed **Advance Aircraft Co**, though the aircraft were still known as "Wacos." In 1929 another reorganization brought the new name **Waco Aircraft Co.** The company ended production in 1959. All the aircraft produced by these individuals and companies are listed under **Waco**.

Waco Predecessors, 1915-1919
Waco 1915 Ice Sled (#1)
Waco 1915 Biplane (#2)
Waco 1917 Biplane Flying Boat (#3)
Waco (DBJ) 1919 Scout (#4)
Waco (DBJ) 1919 2-Seat Flying Boat (#5)

Waco 1919-1928 Projects (In production order)
Waco (DBJ) 1919 Modified Scout (#6)
Waco (DBJ) 1919 Modified Flying Boat (#7)
Waco Cootie No.1 Monoplane (#8)
Waco Cootie No.2 Biplane (#9)
Waco Cootie No.3 Biplane (#10)
Waco 4 (#11)
Waco 5 (#12)
Waco 4-1/2 (#13)
Waco 5 (#s 14-16)
Waco 6 (#s 17-20)
Waco 7 (#s 21-32)
Waco 8
Waco 9 (ATC #11)
Waco 9 (ATC #11), NASM
Waco 10 (Waco 90, GX, GXE) (ATC #13)
Waco 10-T Taperwing (Waco 90)

Waco 3-Letter Designations
A note on file organization: Numerical designations were applied to Waco series prior to 1930. A series of letter designations (corresponding to the airframe or fuselage style) replaced the numbers in 1930. Two prefix letters further delineated aircraft types - the first letter indicated the engine installed, the second stood for the wing design. Waco models are organized by the third letter (fuselage), then by the second letter (wing), and then the first (engine). Waco intermittently used a numeric suffix (e.g., PBF-2) to indicate the year that the model's production commenced (in this case, 1932). Also, a suffix letter could be appended to indicate some special function such as the "S" suffix used by some float-equipped aircraft (ZQC-S, etc).

Waco A Family
Waco BBA
Waco IBA (ATC #465)
Waco KBA (ATC #460)
Waco PBA (ATC #464)
Waco RBA (ATC #466)
Waco TBA (ATC #474)
Waco PCA
Waco UCA
Waco PLA (ATC #502)
Waco ULA (ATC #511)

Waco SFB
Waco NLB (See: Waco CG-13A)
Waco C Cabin Family (Standard Series) (See also KS Family)
Waco C Series
Waco C-6 Series
Waco C-7 Series
Waco BDC
Waco ODC
Waco PDC (ATC #2-388)
Waco QDC (ATC #412)
Waco UDC
Waco BEC (ATC #472)
Waco OEC (ATC #468)
Waco UEC (ATC #467)
Waco AGC-7 (ATC #664)
Waco AGC-8 (ATC #664)

Waco DGC-7 (ATC #639)
Waco EGC-7 (ATC #639)
Waco EGC-8 (ATC #665)
Waco UGC-7
Waco VGC-7
Waco YGC-7 (ATC #627)
Waco YGC-8 (ATC #664)
Waco ZGC-7 (ATC #627)
Waco ZGC-8 (ATC #664)
Waco UIC (ATC #499)
Waco UIC (ATC #499), NASM
Waco CJC (ATC #538)
Waco CJC-S (ATC #538)
Waco CJC-6 (ATC #538)
Waco DJC-6 (ATC #538)
Waco UKC (ATC #528)
Waco UKC-S (ATC #528)
Waco YKC (ATC # 533)
Waco YKC-S (ATC # 533)
Waco YKC-7
Waco ZKC
Waco ZKC-S
Waco QOC (ATC 412)
Waco UOC (ATC #568)
Waco YOC (ATC #569)
Waco YOC-1 (ATC #569)
Waco AQC-6 (ATC 598)
Waco CQC-6 (ATC #597)
Waco DQC-6 (ATC #597)
Waco EQC-6 (ATC #597)
Waco JQC-6
Waco SQC-6
Waco UQC-6 (ATC #598)
Waco VQC-6 (ATC #631)
Waco YQC-6 (ATC 598)
Waco ZQC-6 (ATC 598)
Waco CUC (ATC #575)
Waco CUC-1 (ATC #575)
Waco CUC-2 (ATC #575)

Waco HD Family
Waco CHD
Waco C2HD
Waco JHD (JHD-6) (ATC #512)
Waco SHD (SHD-6) (ATC #581)
Waco S2HD
Waco S2HD-A
Waco S3HD (S3HD-6) (ATC #543)
Waco S3HD-A (ATC #581)
Waco WHD (WHD-6) (ATC #512)
Waco WHD-A

Waco E Airistocrat Family
Waco ARE Custom (ATC #714)
Waco HRE Custom
Waco SRE Custom (ATC #714)
Waco WRE Custom

Waco GXE (ATC #13) (See: Waco 10)
Waco GXE Special

Waco F Family
Waco F Series
Waco F-2 Series
Waco F-3 Series
Waco F-5 Series
Waco F-6 Series
Waco F-7 Series
Waco OBF
Waco PBF-2 (ATC #491)
Waco TBF
Waco UBF-2 (ATC #473)
Waco PCF-2 (ATC #453)
Waco QCF-2 (ATC #416)
Waco UCF
Waco UMF-3 (ATC #546)
Waco UMF-5 (ATC #546)
Waco YMF-3 (ATC #542)
Waco YMF-5 (ATC #542)
Waco ENF
Waco INF (ATC #345)
Waco KNF (ATC #313)
Waco MNF (ATC #393)
Waco QNF
Waco RNF (ATC #311)
Waco CPF (ATC #583)
Waco CPF-1 (ATC #583)
Waco CPF-6 (ATC #583)
Waco DPF-6 (ATC #583)
Waco DPF-7
Waco EPF-6
Waco LPF-7
Waco UPF-6
Waco UPF-7 (ATC #642)
Waco VPF-6
Waco VPF-7 (ATC #642)
Waco YPF (ATC #586)
Waco YPF-6 (ATC #586)
Waco YPF-7 (ATC #586)
Waco ZPF-6 (ATC #586)
Waco ZPF-7 (ATC #586)
Waco CRG (ATC #362)
Waco FGH
Waco LAJ

Waco M Family
Waco JWM (Waco 300 Straightwing)
Waco JYM (Waco 300 Taperwing) (ATC #2-361)

Waco VN Family
Waco AVN-8 (ATC #677)
Waco YVN-8 (ATC #677)
Waco ZVN-8 (ATC #677)

Waco SO Straightwing Family
Waco ASO (Waco 220) (J-5-9-Powered Waco 10) (ATC #41)
Waco BSO (Waco 165) (J-6-5-Powered Waco 10) (ATC #168)
Waco CSO (Waco 225) (J-6-7-Powered Waco 10) (ATC #240)
Waco CSO, Armed for South American Military
Waco DSO (Waco 150) (Hispano-Suiza-Powered Waco 10) (ATC #32/42)
Waco HSO (Packard Diesel-Powered Waco 10) (ATC #333)
Waco PSO (Jacobs-Powered Waco 10) (ATC #339)
Waco QSO (Continental-Powered Waco 10) (ATC #337)

Waco TO Taperwing Family
Waco ATO (Waco 220) (J-5-9-Powered Waco 10) (ATC #123)
Waco CTO (Waco 225) (J-6-7-Powered Waco 10) (ATC #257)
Waco CTO Special
Waco HTO
Waco JYO
Waco NYQ (See: Waco XCG-3 Nightswift)
Waco NZR (See: CG-4 Hadrian)
Waco YZR & XZR (See: Waco Military PG Powered Gliders)

Waco KS Family
Waco DKS-6
Waco UKS-S (ATC #528)
Waco UKS-6 (ATC #528)
Waco UKS-7 (ATC #648)
Waco VKS-6 (ATC #648)
Waco VKS-7 (ATC #648)
Waco VKS-7F (ATC #648)
Waco YKS-S (ATC #533)
Waco YKS-6 (ATC #533)
Waco YKS-7 (ATC #626)
Waco ZKS-6 (ATC #533)
Waco ZKS-7 (ATC #626)

Waco RPT
Waco YST (See: Waco Military Powered Gliders)
Waco NEU (See: XCG-15)
Waco REU, WEU, & YEU (See: Waco Military Powered Gliders)
Waco NAZ Primary Training Glider

Other Waco designations:
Waco W Aristocraft
Waco 90 (See: Waco 10)
Waco 125 (Siemens-Halske-Powered Waco 10) (ATC #26)
Waco 150 (See: Waco DSO)
Waco 165 (See: Waco BSO)
Waco 220 (See: Waco ASO & ATO)
Waco 220 Special

Waco 225 (See also: Waco CSO & CTO)
Waco 300 (See: Waco JWM & JYM)

Waco Military designations
Waco UC-72 (SRE)
Waco UC-72A (ARE)
Waco UC-72B (EGC-8)
Waco UC-72C (HRE)
Waco UC-72D (VKS-7)
Waco UC-72E (ZQC-7)
Waco UC-72F (CUC-1)
Waco UC-72G (AQC-8)
Waco UC-72H (ZQC-6)
Waco UC-72J (AVN-8)
Waco UC-72K (YKS-7)
Waco UC-72L (ZVN-8)
Waco UC-72M (ZKS-7)
Waco UC-72N (YOC-1)
Waco UC-72P (AGC)
Waco UC-72Q (ZQC)
Waco XCG-3 (NYQ) Nightswift
Waco CG-3A (NYQ) Nightswift
Waco XCG-4 Hadrian (NZR) (Haig)
Waco CG-4A Hadrian (NZR) (Haig) (G-4)
Waco CG-4B Hadrian (NZR) (Haig)
Waco G-4C Hadrian (NZR)
Waco XCG-13 (NLB)
Waco YCG-13 (NLB)
Waco XCG-13A (NLB)
Waco YCG-13A (NLB)
Waco CG-13A (G-13A) (NLB)
Waco XCG-15 (NEU)
Waco XCG-15A (NEU)
Waco CG-15A (G-15A) (NEU)
Waco XPG-1
Waco XPG-2
Waco XPG-2A
Waco PG-2A
Waco XPG-2B (2-YZR)
Waco XPG-2C (2-XZR)
Waco XPG-2D (2-YST)
Waco XPG-2E (2-YST)
Waco XPG-3 (2-YEU)
Waco XPG-3A (2-XEU)
Waco XPG-3B (2-REU)
Waco XJW-1 (UBF)
Waco J2W-1 (EQC-6)
Waco XP-447
Waco YPT-14 (PT-14)
Waco PT-14A

Waco Aircraft Company (San Antonio, TX)

The second Waco company was formed in the late 1960s as a subsidiary of **Allied Aero Industries Inc**. Waco produced licensed-built versions of several **SIAI-Marchetti** and **SOCATA** aircraft before closing its doors in 1971.

Waco M-220-4 Minerva (Socata M.S.894 Rallye Commodore)
Waco Meteor (SIAI-Marchetti SF.260 Minerva)
Waco S-220 Vela II (SIAI-Marchetti S.205)
Waco T-2 Meteorite (SIAI-Marchetti SA.202 Bravo)
Waco TS-250-5 Taurus (SIAI-Marchetti S.205)

Wacyk & Tyrala (Stanislaw Wacyk & Tadeusz Tyrala) (Poland)
Wacyk & Tyrala WT-1

Wadsworth, John Washington (Leetsdale, PA)
Wadsworth 1911 Airship Patent

Wag-Aero (Richard H. Wagner) (Lyons, WI)
Wag-Aero Cuby Acro Trainer

Wagner (Helicopter Technik Wagner) (Germany)
Wagner Rotocar 1958
Wagner Sky-Trac 1
Wagner Sky-Trac 3

Wait, Wesley (Newburgh, NY)
Wait 1908 Aerial Vessel Patent

Waitomo Aircraft Ltd (New Zealand) *(See: Bennett Aviation Ltd)*

Walden Aircraft Corp (Dr Henry W. Walden) (Mineola, NY)
Walden I 1908 Tandem Biplane
Walden II 1909 Tandem Biplane
Walden III 1909 Monoplane
Walden IV 1910 Monoplane
Walden V 1910 Monoplane
Walden VI 1910 Monoplane
Walden VII 1911 Monoplane
Walden VIII 1911 Monoplane
Walden IX 1911 Monoplane
Walden X 1911 Monoplane
Walden XI 1912 Monoplane
Walden XII 1912 Monoplane
Walden 1930 Monoplane

Walden-Dyott (Mineola, NY) *(See: Walden Aircraft Corp)*

Walden-Hinners (Edgewater, NJ) *(See: Walden Aircraft Corp)*

Waldron Aircraft Manufacturing Corp *(See: Starling Aircraft)*

Walker, Thomas (USA)
Walker 1810 Aircraft Design

Walkair (Walkaircraft Corp, Walker Brothers) (Blencoe, IA)
Walkair Midwing (1931)

Wallace Aircraft Co (Chicago, IL) *(See also: American Eagle Aircraft Corp)*
Wallace A Touroplane (B Touroplane)
Wallace C-2 Touroplane (See also: American Eagle B-330)
Wallace C-3-1 Touroplane
Wallace Mohawk

Waller, Charles W. (Chicago, IL)
Waller 1911 Flying Machine Patent

Wallis (Poland)
Wallis (Poland) S-I Glider
Wallis (Poland) S-III Glider

Wallis Autogyros Ltd (W.Cdr. Ken Wallis) (UK)
Wallis (Autogyros) WA-116 Agile
Wallis (Autogyros) WA-116 "Little Nellie" (James Bond)
Wallis (Autogyros) WA-116-A
Wallis (Autogyros) WA-116-T/Mc Agile
Wallis (Autogyros) WA-116/F Agile
Wallis (Autogyros) WA-116/Mc Agile
Wallis (Autogyros) WA-116W (Vinten Wallis)
Wallis (Autogyros) WA-116X
Wallis (Autogyros) WA-117/R-R
Wallis (Autogyros) WA-118/M Meteorite
Wallis (Autogyros) WA-119 Imp
Wallis (Autogyros) WA-120/R-R (WA-117-S)
Wallis (Autogyros) WA-121/F
Wallis (Autogyros) WA-121/M Meteorite 2
Wallis (Autogyros) WA-121/Mc
Wallis (Autogyros) WA-122/R-R
Wallis (Autogyros) WA-201

Wally Timm Aerocraft (Glendale, CA)
Wally Timm Aerocraft 2SA

Walsh, Charles F. (Los Angeles, CA)
Walsh 1911 Silver Dart Biplane

Walters Brothers (Gilber M. & Alfred H. Walters) (San Leandro, CA)
Walters Brothers CGC (California Glider Club) Glider

Wanamaker, Rodney (USA)
Wanamaker 1914 Transatlantic Proposal (See also: Curtiss H)

Wancke (Germany)
Wancke 1914 Eindecker (Monoplane)

Wanzer, C. M. (Urbana, OH)
Wanzer Battleplane

War Aircraft Replicas (Santa Paula, CA)
War Aircraft Replicas SBD Dauntless

Warchalowski, Adolf (Poland, Austria)
Warchalowski Type I Landplane
Warchalowski Type II Landplane
Warchalowski Type III Landplane
Warchalowski Type IV Landplane
Warchalowski Type V Landplane
Warchalowski Type VI Landplane
Warchalowski Type VII Landplane
Warchalowski Type VIII Landplane
Warchalowski Type IX Landplane
Warchalowski Type X Landplane
Warchalowski Type XI Seaplane

Ward, John B. (San Francisco, CA)
Ward 1876 Aerial Machine

Warner (Richard Warner Aviation Inc) *(See: Anderson)*

Warner Aircraft Inc (Tampa, FL)
Warner Revolution I
Warner Revolution II

Warpath Aviation Corp (Piggott, AR; Ft Lauderdale, FL)
Warpath Aviation Mohawk
Warpath Aviation Mohawk X

Warren & Montijo (H. G. Warren & John G. Montijo) (San Luis Obispo, CA)
Warren & Montijo W.M.1 (Belmont, Glenmont Landau Sedan)

Warsztaty Szybowcowe (Poland)
Warsztaty Szybowcowe Bak (Horsefly) II
Warsztaty Szybowcowe Komar (See: Kocjan)

Warwick, Bill (Torrance, CA)
Warwick Bantam

Washburn, Beverly D. (Davenport, IA)
Washburn Bubplane

Washington Aeroplane Co (Washington, DC)
Washington Columbia Biplane
Washington Columbia Flying Boat
Washington Columbia Monoplane

Washington Navy Yard (Washington, DC)
Washington Navy Yard (Richardson) Twin Tractor Seaplane

W.A.S.P. Airplane Co (Oakland, CA)
W.A.S.P. 1928 High-wing Cabin Monoplane

Waspair Corp (UK; West Sacramento, CA)
Waspair H.M. 81 Tomcat Standard
Waspair H.M. 81 Tomcat Sport
Waspair H.M. 81 Tomcat XC Tourer
Waspair H.M. 81 Tomcat Super XC Tourer (2-Seat)

Wassmer Aviation (France)
Wassmer AV.36 (See: Fauvel)
Wassmer CE.43 Guépard (Cheetah)
Wassmer CE.44 Cougar
Wassmer CE.45 Léopard (Leopard)
Wassmer D.112 (See: Jodel)
Wassmer D.120 (See: Jodel)
Wassmer Javelot (Javelin) Glider
Wassmer Super IV Sancy
Wassmer Super IV Balladou
Wassmer Super IV-21
Wassmer WA.26 Squale Glider
Wassmer WA.28 Squale Glider
Wassmer WA.50
Wassmer WA.51 Pacific
Wassmer WA.52 Europa (Trans-Pacific)
Wassmer WA.54 Atlantic
Wassmer WA.70

Watanabe (KK Watanabe Tekkosho) (Japan)
In 1943 Watanabe was succeeded by **Kyushu Hikoki KK**; all subsequent Watanabe designs are filed under **Kyushu**.
Watanabe (Tekkosho):
- B3Y1 Carrier Attack Plane (See: Hiro)
- E9W1 (Experimental 9-Shi/Type 96 Small Recon Seaplane)
- E14W1 (Experimental 12-Shi Small Recon Seaplane)
- K4Y1 Trainer Seaplane (See: Kugisho)
- K6W1 (Experimental 11-Shi Advanced Trainer Seaplane)
- K8W1 (Experimental 12-Shi Primary Trainer Seaplane)
- K9W1 Momiji (See: Kyushu)
- MXY1 Experimental Research Plane
- MXY2 (XXY2) Experimental Research Plane
- Q1W1 Tokai (See: Kyushu)
- Thai Navy Recon Seaplane
- Type 3-2 Primary Trainer Biplane
- 14-Shi Primary Trainer Biplane

Watanabe, Takeo (Japan; Venice, CA)
Watanabe (Takeo) 1927 Sport

Waterhouse Aircraft Inc (Waterhouse & Royer) (Glendale, CA)
Waterhouse Cruzair Monoplane
Waterhouse Romair Biplane

Waterman (W. D. Waterman Aircraft Manufacturing Co) (Venice, CA)
Waterman (Aircraft) Gosling Racer
Waterman (Aircraft) (Curtiss) JN-4
Waterman (Aircraft) Rebuilt LePère Fighter
Waterman (Aircraft) 3-L-400

Waterman Arrowplane Corp (Santa Monica, CA)
Waterman (Arrowplane):
- W-4 Arrowplane (X13)
- Arrowbile Prototype (X262Y)
- W-5 Arrowbile (NR16332, NR18931, & NR18932)
- W-6F Arrowbile (NC262Y)

Waterman, Waldo Dean (San Diego, CA)
Waterman (W. D.):
- 1909 Chanute-Type Biplane Glider (Popular Mechanics)
- 1911 Curtiss Pusher "Lizzie"

1911/12 Tractor Biplane
1912/13 Flying Boat
1920s Boeing C Modification
1929 W-1 Flexwing (NC169W)
1932 B Whatsit (X12272)
1932 B Whatsit (X12272), NASM
1957 Aerobile (N54P), NASM
1965 Early Bird
1968 W-11 Chevy Bird
1969 Chevy Duck Flying Boat
Waterman-Kendall 1910 Pusher Biplane

Watkins Airplane Co (Wichita, KS)
Watkins Skylark

Watson, Gary (Newcastle, TX)
Watson (Gary) Windwagon GW-1

Watson, Jesse R. (Forest Grove, OR)
Watson (Jesse R.) Model L (Single-seat Land Monoplane)

Watters, Michael, MWT-1 *(See: TNCA)*

Weatherly-Campbell Aircraft Co (Dallas, TX)
Weatherly-Campbell Colt

Weaver Aircraft Co *(See: Waco)*

Weaver Air Service (Weaver, Goodwin K.) (Indianapolis, IN)
Weaver W.A.S. #5

Weber, Jan (Poland)
Weber 1910 Biplane

Weckler-Armstrong-Lillie Co (Chicago, IL)
Weckler-Armstrong-Lillie Air Boats

Wedell-Williams Air Service Corp (Patterson, LA)
Wedell-Williams:
Cirrus Derby Racer (NR 10337)
XP-34
Model 2 (NR 64Y) (Race #22)
Model 22 (NR 60Y) (Race #s: 22, 54)
Model 44 (44 III) (NR 61Y)
Model 44 I (NR 278V) (Race #s: 44, 91)
Model 44 II (NR 536V) (Race #92)
Model 44/92 (NR 9471) (Race #91)
Model 45 (NR 62Y) (Race #45)

Wedén, Paul (France)
Wedén Convertible Triplane

Wee Bee *(See: Bee Aviation Associates, Inc)*

Weedhopper of Utah Inc (Ogden, UT)
Weedhopper Two JC-24BL
Weedhopper-C JC-24C
Weedhopper-C JC-24C, NASM

Weick, Fred E. (Hampton Roads, VA) *(See also: Erco)*
Weick W-1
Weick W-1A

Weihmiller, H. E. *(See: Corman Aircraft Inc)*

Weil, Lehman (New York, NY)
Weil Aero-Plane

Weir (G. and J. Weir Ltd) (UK)
Weir W.2 Single-Seat Autogiro
Weir W.3 Jump-Start Autogiro
Weir W.5 Single-Seat Helicopter
Weir W.6 Two-Seat Helicopter
Weir W.7 (S.22/38) Helicopter
Weir W.8 Helicopter
Weir W.9 Helicopter

Weiss Manfred Repiülögép-és Motorgyár (WM) (Hungary)
Weiss Manfred WM.13 Biplane
Weiss Manfred WM.21
Weiss Manfred WM.23 Ezüst Nyíl (Silver Arrow)

Weiss, José (Weiss Aeroplane and Launcher Syndicate, Ltd) (UK)
Weiss (José):
1905 Model Glider
1909 Glider "Olive"
1909 Pusher Monoplane "Madge"
1909 Tractor Monoplane #1 "Elsie"
1910 Tractor Monoplane #2 "Sylvia"
1911 Glider "Joker"
1912 Aviette (See: Keith-Weiss)

Weiss-Keith *(See: Keith-Weiss)*

Welch Aircraft Industries (South Bend, IN)
Welch OW-6M Sport Monoplane
Welch OW-7M Sport Monoplane

Welford (Robert, George, & C. Welford) (UK)
Welford 1910 Tractor Monoplane

Wells, D. D. (Jacksonville, FL)
Wells 1910/11 Monoplane

Wels, Franz (Germany)
Wels-Etrich Batwing Glider

Weltensegler GmbH (Germany)
Weltensegler "Feldberg" (1921)

Wendt Aircraft Corp (Robert H. Wendt) (Niagara Falls, NY)
Wendt (NY) W-1 Falconer
Wendt (NY) W-2 Swift

Wendt Aircraft Engineering (Harold O. Wendt) (La Mesa, CA)
Wendt (CA) WH-1 Traveler

Wenham, Francis Herbert (UK)
Wenham 1866 Five-Winged Glider

Werkspoor NV (Netherlands)
Werkspoor Jumbo (Carley)

Weserflug (Weser Flugzeugbau GmbH) (Germany)
Weserflug We 271

West Virginia Airplane Co *(See: Kyle-Smith Aircraft Co)*

Westbrook Aeronautical Corp (New York, NY)
Westbrook Sportster

Western Airplane Corp (Chicago, IL)
Western (IL) Kingbird

Western Aircraft Corp (San Antonio, TX)
Western (TX) WestAir 204

Western Aircraft Supplies (Canada)
Western (Canada) PGK-1 Hirondelle (Swallow)

Western Airplane & Supply Co (Burbank, CA)
Western (CA) 1928 Sport Biplane

***Westfield Aircraft Co (Westfield, CT)** (See also: Miller, Howell)*
Westfield 1941 Trainer

Westlake, A. (UK)
Westlake 1913 Monoplane

Westland (UK)
Westland CL.20
Westland C.O.W. (Coventry Ordnance Works) Gun Fighter (F.29/27)
Westland C.29
Westland Dragonfly (Sikorsky S-51)
Westland Dreadnought
Westland EH101 (See: European Helicopter Industries)
Westland F.7/30
Westland Interceptor (F.20/27)
Westland Limousine Mk.I
Westland Limousine Mk.II
Westland Limousine Mk.III
Westland Lynx
Westland Lysander
Westland Lysander Mk.IIIA, NASM
Westland Lysander, Tandem Wing
Westland N.1B
Westland Pterodactyl Mk.IA
Westland Pterodactyl Mk.IB
Westland Pterodactyl Mk.II
Westland Pterodactyl Mk.III
Westland Pterodactyl Mk.IV
Westland Pterodactyl Mk.V
Westland (Houston-Westland) PV.3
Westland PV.6
Westland PV.7
Westland Scout
Westland Sioux
Westland SR.N2 Hovercraft
Westland SR.N3 Hovercraft
Westland SR.N5 Hovercraft
Westland Sea King (Sikorsky S-61)
Westland Wagtail
Westland Wallace
Westland Walrus
Westland Wapiti
Westland Wasp
Westland Weasel
Westland Welkin
Westland Wessex
Westland Wessex (Sikorsky S-58)
Westland Westbury
Westland Westminster
Westland WG.13 Helicopter
Westland Whirlwind
Westland Whirlwind (Sikorsky S-55)
Westland Widgeon (Helicopter)
Westland Widgeon (Parasol)
Westland Witch
Westland Wizard
Westland Woodpigeon
Westland (Sikorsky) WS-51 (Civil)
Westland (Sikorsky) WS-55 (Civil)
Westland Wyvern
Westland Yeovil
Westland IV
Westland 30 (WG.30)

Weston-Hurlin and Co (UK)
Weston-Hurlin 1911 Biplane

Weybridge (Weybridge Division, British Aircraft Corp Ltd) (UK)
Weybridge MPA (Man-Powered Aircraft)

Weymann (France)

Early in 1929 Charles Weymann and Captain G. Lepère formed **Avions Weymann-Lepère**. The company produced its own designs as well as license-built **Cierva** autogiros. Capt. Lepère left the company shortly after its establishment. Weymann continued in business until 1934 with the firm known variously as **Avions Weymann**, **Avions C. T. Weymann**, and **Sociéte Anonyme des Établissements Aéronautiques Weymann**. Aircraft produced prior to Lepère's departure are filed under **Weymann-Lepère**; aircraft after that period are filed under **Weymann**.

Weymann Type 52
Weymann Type 66 (W-66)
Weymann W.E.L. 80 R2
Weymann Type 100 R.B.L.
Weymann Type 130
Weymann Type 131
Weymann-Lepère W.E.L. 10
Weymann-Lepère C.18 (Autogiro)

Weymann, Charles (France)
Weymann (Charles) W.1 (1915 Monoplane; Cl)

***W-F-W Aircraft Corp** (See: Interstate S-1 Cadet)*

Wheelair Airplane Co (Seattle, WA)
Wheelair 111A

Wheeler, Ray B. (San Jose, CA)
Wheeler (Ray) 1910 Glider

Wheeler Aircraft Co (Ken Wheeler, Wheeler Technology Inc) (Tacoma, WA)
Wheeler (Aircraft Co) Express

Wheeling Aircraft Co, Inc (Pontiac, MI)
Wheeling PJ 1

Whigham, Eugene (USA)
Whigham GW-1
Whigham GW-2
Whigham GW-3
Whigham GW-4
Whigham GW-5
Whigham GW-7

White Aircraft Co, Inc (Donald G. White Aircraft Co) (Buffalo, NY)
White (Buffalo, NY) Aircraft A-R
White (Buffalo, NY) Aircraft Gull
White (Buffalo, NY) Aircraft Pirate
White (Buffalo, NY) Aircraft PT-2
White (Buffalo, NY) Aircraft PT-7
White (Buffalo, NY) Aircraft Tiger

White Aircraft Corp (Palmer, MA)
(See: White Aircraft Co, Inc)

White Aircraft Corp (Le Roy, NY)
White-Verville (Le Roy, NY) Advance Trainer WVAT-1

***White (Douglas) and Thompson (Norman) Co, Ltd (UK)** (See also: Norman Thompson)*
White and Thompson:
Bognor Bloater
No.1 Pusher Biplane
No.2 Pusher Biplane
No.3 Flying Boat
N.T.4 (See: Norman Thompson)
1914 Single-Engined Flying Boat
1914 Twin-Engined Flying Boat

White Eagle Aircraft and Motor Corp (Los Angeles, CA)
White Eagle White Eagle Monoplane

White Lightning Aircraft Corp (Walterboro, SC)
White Lightning Aircraft Corp White Lightning

***White, Burd S.** (See: Whites Aircraft Co)*

White, E. Marshall (Huntington Beach, CA)
White (E. Marshall) Der Jäger (Hunter) (1969)

White, George (St Augustine, FL)
White (George)1928 Ornithopter

White (George D. White Co) (Los Angeles, CA)
White (George D.):
1916 Baby White
1919 Sport Monoplane
1919 Trans-Pacific Seaplane

White, Grahame *(See: Grahame-White, Claude)*

White, Harold L. *(See: Whites Aircraft Co)*

White, John Samuel (UK) *(See: Wight)*

White, Van (Lubbock, TX)
White (Van) Whirlwind

White, William T. (Dallas TX)
White (William T.) Longhorn (1967)

White-Kremsreiter (Benjamin White & Hans Kremsreiter (Milwaukee WI)
White-Kremsreiter W-K Special

Whitehead Aircraft Co (UK)
Whitehead (UK) 1916 Scout

Whitehead, Gustave (Weisskopf) (Bridgeport, CT)
Whitehead (Gustave) Beach-Whitehead Biplane
Whitehead (Gustave) Glider
Whitehead (Gustave) No.21

Whites Aircraft Co (Burd S. & Harold L. White) (Des Moines, IA)
Whites Model A Special
Whites Burdette S-30
Whites Humming Bird
Whites Whitey Sport

Whitmar, W. C. (Grand Rapids, MI)
Whitmar B. W. Glider

Whittelsey Manufacturing Co, Inc (Bridgeport, CT)
Whittelsey (Avro) Avian

Whittemore-Hamm (Saugus MA)
Whittemore-Hamm L-2
Whittemore-Hamm L-3

Whittenbeck Special *(See: Folkerts SK-1)*

Whysall/Marion (George Whysall and Associates) (Marion, OH)
Whysall/Marion 3PL

WIB (Weltraum-Institut Berlin GmbH) (Germany)
WIB Stratolab

Wibault (France)
Wibault C1 Biplane
Wibault 1 C1
Wibault 2 BN.2
Wibault 3 C1
Wibault 5 C1
Wibault 6 C2
Wibault 7 C1 (70, 72, 73, 74)
Wibault 8 C2 (122)
Wibault 9 C1
Wibault 10 GR (123)
Wibault 12 C2 (121) Sirocco
Wibault 124 A2
Wibault 125
Wibault 13 C1 (130) Jockey
Wibault 170 C1 Trombe (Waterspout)
Wibault 210 C1
Wibault 220 RN.3
Wibault 260 R.2
Wibault 280 T.10
Wibault 281 T.10
Wibault 340 SE

Wibault-Penhoët (France)
Wibault-Penhoët 240 P.5
Wibault-Penhoët 282 T.10
Wibault-Penhoët 282 T.12
Wibault-Penhoët 283 T.12
Wibault-Penhoët 313 C1
Wibault-Penhoët 360 T.5
Wibault-Penhoët 365 T.7

Wickham, James M. (Seattle, WA)
Wickham A
Wickham B

Wicko (Foster, Wikner Aircraft Co, Ltd) (UK)
Wicko F.W.1
Wicko F.W.2
Wicko G.M.1

Wideröe Flyveselskap og Polar A/S (Norway)
Wideröe C-5 Polar

Wiederkehr, George (USA)
Wiederkehr GHW-1 CU-Climber

Wielkopolan Aeroplane Plant 'The Aeroplane" Co Ltd *(See: Samolot)*

Wielkopolska Wytwórnia Samolotów 'Samolot' Sp Akc *(See: Samolot)*

Wiencziers, Eugen (Germany)
Wiencziers 1911 Racing Monoplane

Wiener Karosserie und Flugzeugfabrik *(See: WKF)*

Wiener Neustädter Flugzeugwerke (Dobelhof) *(See: WNF)*

Wight (J. Samuel White & Co) (UK)
Aircraft of J. Samuel White & Co., including those designed for the firm by Howard Wright, are known by the name of Wight after the factory's location on the Isle of Wight. Serial numbers are listed in parentheses.
Wight Type A.I. Improved Navyplane Pusher Seaplane (171-177)
Wight Admiralty 840 Type Tractor Biplane Seaplane
Wight Admiralty A.D. 1000 Type Biplane Seaplane (1000, 1355-1361)
Wight Baby Single-Seater Tractor Biplane Seaplane (9097-9098, 9100)
Wight Biplane Flying Boat (N15)
Wight Biplane Seaplane (186, Canceled)
Wight "Converted" Tractor Biplane Seaplane (9841-9860, N1280-1289, N2180-2229)
Wight Elementary DC School Pusher Biplane Seaplane (8321-8322)
Wight Landplane Tractor Biplane (N501)
Wight Navyplane Pusher Biplane Float Seaplane (884)
Wight Pusher Biplane Seaplane (893-895)
Wight Quadruplane Fighter (N546)
Wight Twin Fuselage, 2-Engined Tractor Biplane Seaplane (187, 1450-1451)

Wight Type 4 Quadruplane (N14)
Wight 1913 Double-Camber No.1 Seaplane
Wight 1913 Double-Camber No.2 Navyplane
Wight 1914 Navyplane Pusher Biplane Seaplane (128-129, 155)

Wilber, George C. (Oakland, CA)
Wilber 1929 Homebuilt Glider

Wilcox, Philip W. (Columbia University, NY)
Wilcox 1910 White Ghost

Wild (Factory Wild Heerbrugg) (Switzerland)
Wild WT
Wild X

Wild-O'Meara (Horace B. Wild & J. K. "Jack" O'Meara) (USA)
Wild-O'Meara Chanute (ex-Darmstadt 1)
Wiley Post Aircraft Corp *(See: Straughan)*

Wilford Aircraft Corp (Narberth, PA) *(See also: Pennsylvania Aircraft Syndicate)*
Wilford Executive Transport

Wilkes (France)
Wilkes 1916 Tandem Wing

Willard, Charles F. (Melrose, MA)
Willard Banshee
Willard Express
Willard 1918 One Place Fighter

Willard Parker Aircraft Corp *(See: Parker, Willard)*

William Beardmore & Co *(See: Beardmore)*

Williams (USA)
Williams Suite 1 (Homebuilt)

Williams & Boose *(See: Williams, Israel)*

Williams Aircraft Design Co (Arthur L. "Art" Williams) (Alliance, OH; Northridge, CA)
Williams (Aircraft Design):
Midget Monocoupe
W-17 Stinger
W-18 Falcon
(Williams-Cangie) WC-1 Sundancer
1948 Special 2X Estrellita (Little Star)

Williams Airport Corp *(See: Aviation Engineering School)*

Williams Helicopter Corp
Williams (Helicopter):
Model UH-1H (See: Bell)
Model 205 (See: Bell)

Williams International Corp (Dr Sam Williams) (Walled Lake, MI)
Williams (International) Aerial System Platform (Wasp)
Williams (International) V-Jet II

Williams, Alexander (Nassau, NY)
Williams (Alexander) 1912 Biplane

Williams, Alford "Al" (Washington, DC) *(See: Mercury Flying Corp)*

Williams, Arthur L. "Art" *(See: Williams Aircraft Design Co)*

Williams, Beryl J. (Los Angeles, CA) *(See also: Tachibana)*
Williams (Beryl J.):
1911 Pusher Biplane
1911 Pusher Biplane, Reengined
1913 Tractor Biplane (Tsuboto)
1914 Tractor Biplane
1916/7 Training Biplane

Williams, Bob (Oceanside, CA)
Williams (Bob) W-1
Williams (Bob) W-2

Williams, C. W. (USA)
Williams (C. W.) 1908 Multi-Wing

Williams, Dale A. (Chugiok AK)
Williams (Dale A.) Willbird Model-1
Williams (Dale A.) Willbird 02
Williams (Dale A.) Willbird No.3

Williams, Donn L. (Springfield, IL)
Williams (Donn L.) 1A

Williams, Floyd (Eagle Grove, IA)
Williams, Floyd Biplane

Williams, Gary D. (Redding, CA)
Williams (Gary D.) Model B

Williams, George W. *(See: Texas Aero Manufacturing Co)*

Williams, Israel (Lima, OH)
Williams (Israel) 1925 Airship

Williams, James R. *(See: Niles Aircraft Corp)*

Williams, John M., Jr (Dayton, OH)
Williams (John M., Jr) 1910 Monoplane Glider

Williams, John Newton (Derby, CT)
Williams (John Newton) 1907 Helicopter

Williams, Myron Pell (Little Rock, AR)
Williams (Myron Pell) MP-1

Williams, (Miss) Olive Branch *(See: Aviation Engineering School)*

Williams, Osbert E. (O. E. Williams Aeroplane Co) (Fenton, MI; Scranton PA)
Williams (O. E.) 1911 Monoplane
Williams (O. E.) 1911/12 Curtiss-Type Pusher
Williams (O. E.) 1913 Tractor
Williams (O. E.) 1914 Hydroaeroplane
Williams (O. E.) Model 5 (1915)

Williams, Paul (Dayton, OH)
Williams (Paul) 750-PW

Williams, Robert F. (Houston, TX)
Williams (Robert F.) Skeeter Hawk

Williams, Robert L (Vancouver, WA)
Williams (Robert L) Model 1

Williams, Dr Sam *(See: Williams International Corp)*

Williams, Sylvester M. (San Francisco, CA)
Williams (Sylvester M.) 1909 High Speed Transport

Williams, Val D. (Apple Valley, MN)
Williams (Val D.) Mark IV

Williams, Vernon J. (Farmers Branch, TX)
Williams (Vernon J.) 601 HD

Williams, Walt (Perris, CA)
Williams (Walt) W (1937 Biplane)

Williamson (USA)
Williamson Parasol

Williamson, John Wesley (Montebello, CA)
Williamson (John Wesley) "Skeets"

Williamson, Roger L. (San Antonio TX)
Williamson (Roger L.) Wren

Willingham (Homebuilt) (USA)
Willingham V-W Sport

Willoughby, Hugh L. (Sewalls Point, FL; Newport, RI)
Willoughby (Hugh L.):
- Model A Warhawk
- Model B Pelican
- Cygnet
- Model F Pelican
- Model G Pelican III
- Model H Swan
- Ibis (Rebuilt Aeromarine 39-B)
- Sea Gull

Willoughby, Percival Nesbit (UK)
Willoughby (Percival Nesbit) Delta
Willoughby (Percival Nesbit) Delta F

Wilson and Co (W. E. Wilson) (Wichita, KS)
Wilson Cadet

Wilson-Smith (USA)
Wilson-Smith DSA-1 (Wilson Modification of Smith Kit-Plane)

Wimpenny, John C. (UK)
Wimpenny Puffin (Human-Powered)
Wimpenny Puffin II (Human-Powered)

Windecker Research, Inc (Midland, TX)
Windecker AC-7 Eagle 1
Windecker AC-7 Eagle 1, NASM
Windecker E-5 (YE-5) Eagle

Wing Aeronautical Corp (New York, NY)
Wing (Aeronautical) Dragonfly

Wing Aircraft Co (Torrance, CA) *(See also: Hi-Shear)*
Wing (Aircraft) Derringer

Wing Research Corp (Reno, NV)
Wing (Research) SA-82 Ultrawing
Wing (Research) Sadler Vampire (See: Sadler, Bill)

Winslow, S. V. (Riparia, WA)
Winslow 1904 Bicycle Aeroplane

Winstead, Guy (Wichita, KS)
Winstead Special (1932)

Winston, Charles (Topeka, KS)
Winston 1911 Flying Machine Patent

Winter, Edward
Winter (Edward) 1929 Ornithopter

Winter, H. (Germany)
Winter (H.) LF 1 Zaunkönig (Wren)

Wiseman, Frederick J. *(See: Wiseman-Peters)*

Wiseman-Cooke Biplane *(See: Wiseman-Peters #2 Biplane)*

Wiseman-Peters (Fred Wiseman & J. W. Peters) (Petaluma, CA)
Wiseman-Peters #2 Biplane
Wiseman-Peters #2 Biplane, NASM

Wisenant Longitudinal Aeroplane Co (Oscar H. Wisenant) (Colorado Springs, CO)
Wisenant Longitudinal Airplane

Wisniewski, Bronislaw (Poland)
Wisniewski 1911 Monoplane

Wissbaum, Leonard F. (Springfield, MO)
Wissbaum "Jenny"

Wissler Airplane Co (Bellefontaine, OH)
Wissler W.A.6
Wissler W.L.9

Wittemann (USA)

Twelve-year-old Charles R. Wittemann built his first kite in 1896, moving to man-carrying kites in 1898, monoplane gliders in 1900, and biplane gliders in 1901. In 1906 he built his first powered aircraft; later that year he teamed with his brother Adolph to form **C. & A. Wittemann**. Over the next few years they designed and manufactured a series of pushers and gliders, including the Red Devil biplanes flown by Thomas Baldwin. In 1914, as Adolph left the company, Samuel C. Lewis joined, and **Wittemann-Lewis Aircraft Co** was formed. Lewis left in 1917, and the company was renamed **Wittemann Aircraft Corp**. During the 1920s, control of the company passed to investors. Charles left in 1923, and the company was sold to **Atlantic** (**Fokker**) in 1925. Aircraft designs are organized below under the design firm or individual.

Wittemann Aircraft Corp (Teterboro, NJ)
Wittemann (Aircraft):
DH-4 Airmail Modifications
NBL-1 Barling Bomber (See: Engineering Division)

Wittemann Brothers (C. & A. Wittemann) (Staten Island, NY)
Wittemann Brothers:
- Red Devil Biplanes (See: Baldwin, Thomas)
- No.9 Biplane Hang Glider
- No.11 Biplane Hang Glider
- No.12 Biplane Hang Glider
- 1907 Pusher Biplane (Ailerons, 2 Prop)
- 1908 Biplane (C. W. Miller, 1 Prop)
- 1909 Helicopter
- 1909 Biplane

Wittemann, Charles R. (Staten Island, NY)
Wittemann (Charles R.):
- 1900 Monoplane Test Gliders
- 1901 Biplane Test Gliders
- 1902 Biplane (Man-Carrying) Glider
- 1905 Biplane (Man-Carrying) Glider
- 1906 Pusher Biplane (2 Prop)
- 1906 Triplane

Wittemann-Lewis Aircraft Co (Newark, NJ)
Wittemann-Lewis:
- FA-1
- FA-2
- FA-10
- George Lanzius' Aircraft

Sundstedt-Hannevig Transatlantic Seaplane
TT

Wittman, Sylvester "Steve" J. (Oshkosh, WI)
Wittman "A Witt Formula V"
Wittman Big X
Wittman "Bonzo"
Wittman "Bonzo II"
Wittman "Buster"
Wittman "Buster," NASM
Wittman "Buttercup"
Wittman "Chief Oshkosh"
Wittman W-8 Tailwind (Flying Carpet)
Wittman W-10 (See: Aircraft Spruce & Specialty)

Witzig-Lioré-Dutilleul (France)
Witzig-Lioré-Dutilleul Monoplane
Witzig-Lioré-Dutilleul 2
Witzig-Lioré-Dutilleul 1909 Biplane

WKF (Wiener Karosserie und Flugzeugfabrik) (Austria-Hungary)
WKF 80.01 thru 80.03
WKF 80.04
WKF 80.05 Triplane
WKF 80.06 and 80.06B
WKF 80.07 Two-Seater
WKF 80.08 (See: Aviatik D.I)
WKF 80.09
WKF 80.10
WKF 80.12

WNF (Wiener Neustädter Flugzeugwerke) (Austria)
WNF (Doblhof) Wn 16
WNF (Doblhof) Wn 342 V2
WNF (Doblhof) Wn 342 V4

Wolf Hirth *(See: Göppinger; See also: Schempp-Hirth)*

Wolf Pitts Samson *(See: Pitts)*

Wolfe Aviation Co (USA)
Wolfe Aviation Trike (WAT) Ultralight

Wolff, Paul (Ateliers Paul Wolff) (Luxembourg)
Wolff Flash-3

Wondra, Franz (Austria-Hungary)
Wondra 1908 Flying Machine Patent

Wood, Steve
Wood WG-1 (Ultralight)

Woodford Airplanes Inc (E. S. Woodford) (Portland, OR)
Woodford Special Arch Wing Monoplane (1927)

Woodhouse, Henry (Chicago, IL)
Woodhouse 1920 Electric Aircraft

Woodington, James W. (Folcroft, PA)
Woodington 1910 Flying Machine Patent

Woods, Bob (Niagara Falls, NY)
Woods Special Racer

Woods, Harris L. (Raleigh, NC) *(See: Aerosport Woody Pusher)*

Woodson Engineering Co (Bryan, OH)
Woodson Type 1A Foto
Woodson Type 2A Express
Woodson Type 3A Sport
Woodson Type 4B Transport

Woolsey-Frye-Whittier Aircraft Corp *(See: Interstate S-1 Cadet)*

Worldwide Ultralite Industries (Katy, TX)
Worldwide Ultralite Sky Raider
Worldwide Ultralite Spit Fire
Worldwide Ultralite Thunderbird
Worldwide Ultralite Water Moccasin Air Boat

Wren Aircraft Co Ltd (UK)
Wren (UK) Goldcrest

Wren Aircraft Corp (Fort Worth, TX) (Wren Aircraft Inc, AZ)
Wren (USA) 460 (Cessna 182 STOL Conversion)

Wright & Gingerich (Hervey C. Wright & S. W. Gingerich) (Iowa City, IA)
Wright & Gingerich HS

Wright Aeronautical Corp (Dayton, OH)
When Glenn L. Martin left the **Wright-Martin Aircraft Corp** in late 1917, Wright-Martin began concentrating on engine production rather than airframe design. Following the Armistice, the company was renamed **Wright Aeronautical Corp**, but continued its concentration on engine development, producing only a small number of original aircraft designs. In 1929, Wright merged with the **Curtiss Aeroplane and Motor Co** to form the **Curtiss-Wright Corp.** For aircraft designed by the predecessors of Wright Aeronautical, see **Wright Bros**, **Wright Co**, or **Wright-Martin**, as appropriate. For aircraft designed by Wright Aeronautical, see **Wright Aero.** For aircraft designed by Curtiss-Wright, see **Curtiss.**
Wright (Aero) F2W-1
Wright (Aero) F2W-2
Wright (Aero) F3W-1 Apache
Wright (Aero) F3W-1 Apache, Altitude Record Configuration
Wright (Aero) NW-1 (Racer)
Wright (Aero) NW-2 (Racer)
Wright (Aero) XO-3

Wright Brothers (Orville & Wilbur Wright) (Dayton, OH)
On 17 December 1903 Orville and Wilbur Wright became the first humans to fly a heavier-than-air craft which could take off under its own power, fly under the operator's control, and land at a point not lower than the take-off point. Up to this time, the brothers had designed and flown several gliders, and went on to design several other aircraft before formally establishing **The Wright Co** to manufacture aircraft in 1909. For information concerning the aircraft designed by the Wrights before the establishment of Wright & Co, see **Wright Brothers.** For aircraft designed by Wright & Co and its successors, see **Wright Co**, **Wright-Martin**, or **Wright Aeronautical.**
Wright (Brothers):
1900 Glider (No.1)
1901 Glider (No.2)
1902 Glider (No.3) Twin Fixed Rudders
1902 Glider (No.3) 1st Modification, Twin Steerable Rudders
1902 Glider (No.3) 2nd Modification, Single Steerable Rudder
1903 Flyer
1903 Flyer, NASM

1904 Flyer II
1905 Flyer III
1907 Hydroplane (Boat)
1911 Glider

Wright Co (Dayton, OH)

In 1908 Orville and Wilbur Wright began marketing their aircraft under the name **Wright & Co.** In early 1909 they formally organized **The Wright Co.** In 1915 Orville (Wilbur had died in 1912) sold the Wright's interests in the company to a syndicate, while remaining a consultant. The Wright Co acquired control of the **Simplex Automobile Co**, and in August 1916, merged all of various branches with the **Glenn L. Martin Co** to form the **Wright-Martin Aircraft Corp**. Orville Wright and the Dayton works of the Wright Co joined **Dayton Engineering Laboratory Co (Delco)** and **Dayton Metal Products Co** to establish the **Dayton Wright Airplane Co.** For aircraft designed by the Wrights before the establishment of the Wright Co, see **Wright Brothers**. For products of the Wright Co, see **Wright Co.** For products of Wright-Martin, see **Wright-Martin**. For Dayton Wright products, see **Dayton Wright**. For aircraft designed by Wright Aeronautical Corp, see **Wright Aeronautical**.

Wright (Co):
Type A
Type A, 1908-1909 European Tour
Type A, Fort Myer Trials (1908)
Type A, Hudson-Fulton Flight (1909)
Type A Military (Signal Corps No.1)
Type A Military (Signal Corps No.1), NASM
Type A Transitional
Model B
Model B, First Cargo Flight
Model B, US Navy Aircraft
Model B Hydroaeroplane
Model C
Model C, Wright Automatic Stabilizer
Model CH Prototype (Hydroaeroplane)
Model CH (Hydroaeroplane)
Model D
Model E
Model EX "Vin Fiz"
Model EX "Vin Fiz," NASM
Model F "The Tin Cow"
Model G Aeroboat
Model H
Model HS
Model HS, Modifications
Model J (Long Hull Hydroaeroplane)
Model K
Model L
Model R "Baby Wright" Racer, Short Span
Model R "Baby Wright" Racer, Long Span
Model R "Baby Grand" (Gordon Bennett Racer)
Model R "High Flyer" (Altitude Record)

Wright, A. F. (Dubuque, IA)

Wright (A. F.) Midwing (1932)

Wright, H. W. (H. W. Wright Co) (Wilmar, CA)

Wright (H. W.) 1928 Light Sport

Wright, Howard T. (Howard T. Wright Brothers Ltd) (UK) *(See also: Wight)*

Wright (Howard):
Avis Monoplane (1909)
Capone Helicopter (1908)
Capone Helicopter No.2 (1909)
Capone Ornithopter (1909)
Ornis Monoplane (1910)
1908 Pusher Biplane (Seaton-Karr)
1910 Biplane
1911 Modified Biplane

Wright-Bellanca *(See: Bellanca WB-1, WB-2)*

Wright-Martin Aircraft Co (Dayton, OH)

In 1916 the **Wright Aircraft Co**, **Glenn L. Martin Co**, **Simplex Automobile Co**, **Wright Flying Field, Inc**, and the **General Aeronautic Co of America** merged to form the **Wright-Martin Aircraft Corp**. Orville Wright and the Dayton works of the Wright Co soon joined **Dayton Engineering Laboratory Co (Delco)**, and **Dayton Metal Products Co** to form the **Dayton Wright Airplane Co.** Martin left the company in late 1917 and manufacturing emphasis began to favor engine production. With the armistice, the company reorganized as **Wright Aeronautical Corp**, which concentrated primarily on engine manufacture. For aircraft designed by the Wrights before the establishment of the Wright Co, see **Wright Bros**. For products of the Wright Co, see **Wright Co.** For products of Wright-Martin, see **Wright-Martin**. For aircraft designed by Wright Aeronautical Corp, see **Wright Aeronautical**.

Wright-Martin Model R (See: Martin Model R)
Wright-Martin Model V

W.R.K. Gyro *(See: Pennsylvania Aircraft Syndicate)*

Wróbel (Poland)

Wróbel 1910 Monoplane

Wróblewski Brothers (Piotr & Gabriel) (Poland)

Wróblewski 1914 Armored Aircraft

Wyandotte High School *(See: Porterfield)*

Wytwórnia Sprzetu Komunikacyjnego PZL Mielec SA *(See: PZL Mielec)*

X

X-Planes *(See: Research Aircraft)*

Xian Aircraft Co (China)
Xian F-7M Airguard
Xian H-6 (Hong-6, B-6) *Badger*
Xian J-7 (Jian-7, F-7) *Fishbed* (MiG-21F)
Xian JJ-7 (Jianjiao-7, FT-7) (MiG-21U)
Xian Y-7 (Yun-7) *Coke* (An-24)
Xian Y7-100
Xian Y7-200
Xian Y7-300

Y

Yackey Aircraft Co (Maywood, IL)
Yackey Aircruiser Biplane
Yackey Commercial Monoplane
Yackey Sport
Yackey Transport
Yackey Transport "The Ern" (1925 Stirling New Guinea Expedition)

Yakovlev (Russia)

Aleksandr Sergeyevich Yakovlev began designing gliders and aircraft while a worker at the Soviet **Akademiya Vozdushnogo Flota** (Academy of the Air Fleet, AVF). In 1927 he was admitted to the academy, which had been renamed **Voenno-Vozdushnaya Akademiya** (War-Air Academy, VVA) in 1925. Although disgraced and barred from aircraft design in 1932, a 1934 order from the Chairman of the Central Control Commission placed Yakovlev in control of an Experimental Design Bureau (OKB) associated with State Aircraft Factory GAZ 115. The bureau, generally called **OKB Yakovlev**, remained active in Soviet aviation through Yakovlev's death in 1989 until the collapse of the Soviet Union in 1991. At that time, the Yakovlev bureau in Moscow became **Moskovskii Mashinostroitelnyy Zavod "Skorost" Imieni A. S. Yakovleva** (Moscow Machine-building Factory "Speed" named after A. S. Yakovlev) while simultaneously organizing Skorost (Speed) Industrial Association, an association of ex-Soviet design offices and factories. In 1992 the original Yakovlev group (comprising the Moscow design office and factories in Saratov and Smolensk) reorganized as Joint Stock Company A. S. Yakovlev Design Bureau. Designs by A. S. Yakovlev, OKB Yakovlev, or Yakovlev Joint Stock Co, appear as **Yakovlev.**

Yakovlev Air-1 (VVA-3)
Yakovlev Air-2
Yakovlev Air-3
Yakovlev Air-4
Yakovlev Air-6 (Ya-6)
Yakovlev Air-7
Yakovlev Air-9
Yakovlev Air-11
Yakovlev Air-12
Yakovlev Air-15
Yakovlev Air-16
Yakovlev Air-17 (UT-3)
Yakovlev Air-19 (Ya-17)
Yakovlev Air-26 (Ya-26, I-26)
Yakovlev UT-1 (Air-14)
Yakovlev UT-2
Yakovlev Yak-1
Yakovlev Yak-1 MPVO
Yakovlev Yak-1M
Yakovlev Yak-3
Yakovlev Yak-3UA
Yakovlev Yak-4
Yakovlev Yak-5
Yakovlev Yak-6
Yakovlev Yak-7 Family
Yakovlev Yak-7 PVRD
Yakovlev Yak-7 PD
Yakovlev Yak-8 *Crib*
Yakovlev Yak-9 *Frank*
Yakovlev Yak-9B *Frank*
Yakovlev Yak-9D *Frank*
Yakovlev Yak-9M *Frank*
Yakovlev Yak-9P *Frank*
Yakovlev Yak-9U *Frank*
Yakovlev Yak-10
Yakovlev Yak-11 *Moose*
Yakovlev Yak-11U *Moose*
Yakovlev Yak-12 *Creek* Family
Yakovlev Yak-12 *Creek*
Yakovlev Yak-13 *Crow*
Yakovlev Yak-14 *Mare*
Yakovlev Yak-15 *Feather*
Yakovlev Yak-16 *Cork*
Yakovlev Yak-17 *Feather*
Yakovlev Yak-17UTI *Magnet*
Yakovlev Yak-18 *Max*
Yakovlev Yak-18 *Max*, NASM
Yakovlev Yak-18A *Max*
Yakovlev Yak-18P *Max*
Yakovlev Yak-18PM *Max*
Yakovlev Yak-18PS *Max*
Yakovlev Yak-18T *Max*
Yakovlev Yak-18U *Max*
Yakovlev Yak-19
Yakovlev Yak-20
Yakovlev Yak-21
Yakovlev Yak-23 *Flora*
Yakovlev Yak-23UTI
Yakovlev Yak-24 *Horse*
Yakovlev Yak-24K *Horse*
Yakovlev Yak-25K
Yakovlev Yak-25M *Flashlight-A*
Yakovlev Yak-25RV *Mandrake*
Yakovlev Yak-27 *Flashlight-C*
Yakovlev Yak-27K
Yakovlev Yak-27R *Mangrove*
Yakovlev Yak-28B *Brewer-A*
Yakovlev Yak-28L *Brewer-L*
Yakovlev Yak-28P *Firebar*
Yakovlev Yak-28PM
Yakovlev Yak-28PP *Brewer-E*
Yakovlev Yak-28R *Brewer-D*
Yakovlev Yak-28U *Maestro*
Yakovlev Yak-30 *Magnum*
Yakovlev Yak-32
Yakovlev Yak-36 *Freehand*
Yakovlev Yak-38 *Forger-A*
Yakovlev Yak-38U *Forger-B*
Yakovlev Yak-40 *Codling*
Yakovlev Yak-41 *Freestyle*
Yakovlev Yak-42 *Clobber*
Yakovlev Yak-50

Yakovlev Yak-52
Yakovlev Yak-100
Yakovlev Yak-130 (UTK-Yak)
Yakovlev Yak-200

Yates (P. B. Yates Machine Co) (Beliot, WI)
Yates 108

Yatsenko, V. P. (Russia)
Yatsenko I-28

Yeoman Aircraft Pty Ltd (Australia)
Yeoman YA1 (KS.3) Cropmaster 250
Yeoman YA5 Fieldmaster 285
Yeoman UA-1 (Commonwealth CA-6 Conversion)

Yermolaev (Russia)
Yermolaev Yer-2 (DB-240)
Yermolaev Yer-4

Yokell & Orton (Frank A. Yokell & Herbert Orton) (Gresham, OR)
Yokell & Orton Model A

***Yokosho** (See: Yokosuka)*

Yokosuka (Dai-ichi Kaigun Koku Gijitsusho) (Japan)
Yokosuka B3Y (See: Hiro)
Yokosuka B4Y (See: Kugisho)
Yokosuka D2Y (Navy Experimental Kusho 8-Shi Special Bomber)
Yokosuka D4Y (See: Kugisho)
Yokosuka E1Y (See: Nakajima)
Yokosuka E5Y (Navy Type 90-3 Recon Seaplane)
Yokosuka E6Y (Navy Type 91 Recon Seaplane)
Yokosuka E14Y (See: Kugisho)
Yokosuka H5Y *Cherry* (Navy Type 99 Flying Boat)
Yokosuka H7Y *Tillie* (Navy Experimental 12-Shi Flying Boat)
Yokosuka K1Y (See: Nakajima)
Yokosuka K2Y (Navy Type 3 Land-based Primary Trainer)
Yokosuka K4Y (See: Kugisho)
Yokosuka K5Y *Willow* (Navy Type 93 Intermediate Trainer)
Yokosuka L3Y *Nell* (Navy Type 96 Transport)
Yokosuka MXY1 (See: Watanabe)
Yokosuka MXY2 (See: Watanabe)
Yokosuka MXY7 *Ohka* (*Baka*) (See: Kugisho)
Yokosuka Navy Avro 504 Trainer
Yokosuka Navy F.5 Flying Boat
Yokosuka Navy Ha-go Small Seaplane (Sopwith Schneider Fighter Seaplane)
Yokosuka Navy I-go Recon Seaplane
Yokosuka Navy I-go Ko-gata Seaplane Trainer
Yokosuka Navy Short Type 184 Recon Seaplane
Yokosuka Navy Type Hansa Recon Seaplane
Yokosuka Navy Type Ka Seaplane (Curtiss 1912 Seaplane)
Yokosuka Navy Type Mo Large Seaplane (M. Farman 1914)
Yokosuka Navy Type Mo Small Seaplane
Yokosuka Navy Type 10 Recon Seaplane
Yokosuka Navy Type 91 Intermediate Trainer
Yokosuka Navy 1913 Experimental Seaplane
Yokosuka Navy 1914 Experimental Nakajima Tractor Seaplane
Yokosuka Navy 1916 Experimental Twin-Engined Seaplane
Yokosuka Navy 1916 Experimental Ho-go Otsu-Gata Seaplane
Yokosuka Navy 1917 Experimental Ho-go Small Seaplane
Yokosuka Navy 1925 Experimental Tatsu-go Recon Seaplane
Yokosuka Navy 1932 Experimental 6-Shi Special Bomber
Yokosuka P1Y (See: Kugisho)

Yomiuri (Japan)
Yomiuri Y-1

York Aircraft Corp (USA)
York (Aircraft Corp) XCG-12

York, C. H. (USA)
York (C. H.) XHC-22 Yorkopter

York, Jo (Santa Ana, CA)
York (Jo) Sunbeam

Young, Arthur M. (Buffalo, NY)
Young (Arthur M.) 1944 Pusher Aircraft Patent

Young, Edward W. (Los Banos, CA)
Young (Edward W.) Eddyo F-2

Young, Richard E. (Ypsilanti, MI)
Young (Richard E.) Model A

Youngman-Baynes (UK)
Youngman-Baynes High Lift Aircraft

Yunker Aircraft Corp (George C. Yunker) (Wichita, KS)
Yunker Y-2-165

Yur'ev-Cheramukhim (B. N. Yur'ev & A. M. Cheramukhim) (Russia)
Yur'ev-Cheramukhim 1-EA
Yur'ev-Cheramukhim 3-EA
Yur'ev-Cheramukhim 5-EA

Z

Zachartchenko, C. L. (USA)
Zachartchenko 1-Seat Monoplane Racer (1923)

Zajicek, Charles A. (Berwyn, IL)
Zajicek C-2 Midwing Sport
Zajicek C-3 Midwing Sport

Zaklady Mechaniczne E. Plage & T. Laskiewicz (E. Plage & T. Laskiewicz Engineering Establishments) *(See: Lublin)*

Zalewski, Wladyslaw (Poland)
Zalewski W.Z.I Glider
Zalewski W.Z.II Glider
Zalewski W.Z.III (Savieliev Zalewski Quadruplane No.1)
Zalewski W.Z.IV (Savieliev Zalewski Quadruplane No.2)
Zalewski W.Z.V (Savieliev Zalewski Quadruplane No.3)
Zalewski W.Z.VII
Zalewski W.Z.VIII DePeZe
Zalewski W.Z.X
Zalewski W.Z.XI Kogutek I (Cockerel)
Zalewski W.Z.XII Kogutek II (Cockerel)

Zander & Weyl (UK) *(See: Dart Aircraft Ltd)*

Zaparka, Eduard (Austria-Hungary)
Zaparka 1917/18 Biplane Fighter

Zaparka, Edward F. (Mamaroneck, NY)
Zaparka 1930 Magnus-Effect Rotorcraft

Zarkhi-Smirnov (Ya L. Zarkhi & M. V. Smirnov) (Russia)
Zarkhi-Smirnov LAKM-1

Zaschka (Germany)
Zaschka Helicopter
Zaschka Helicopter (1927)
Zaschka Human-Power Aircraft (1934)

Zaspl (Germany)
Zaspl SG-3

Zauner, Otto (USA)
Zauner OZ-4 (Schreder HP-14 Mod)
Zauner One Yankee (Schreder RS-15 Mod)

Zaunkönig *(See: Brunswick Zaunkönig II)*

Zavod Imenya Gorkova (Russia)
Zavod PS-84
Zavod PS-89 (See: Laville ZIG-1)

Zacco *(See: Zeebrugge Aeronautical Construction Co)*

ZASPL (Aviation Assoc of Technical University Students in Lwów) (Poland)
ZASPL Osa (Wasp)
ZASPL CW-8S

Zbaraz Secondary School (Poland)
Zbaraz Secondary School 1927 Glider

Zbieranski (Czeslaw) & Cywinski (Stanislaw) (Poland)
Zbieranski & Cywinski 1910/11 Biplane

Zeebrugge Aeronautical Construction Co (Zacco) (Belgium)
Zeebrugge A2 Flying Boat
Zeebrugge 2-Seat Cabin Monoplane
Zeebrugge 2-Seat Light Trainer
Zeebrugge 2-seat Recon Fighter

Zeise-Nesemann (Germany)
Zeise-Nesemann 1922 Glider "Senator"

Zelinka, Heinrich (Austria-Hungary)
Zelinka 1918 Fighter

Zenair Ltd (Canada); Zenith Aircraft Co (Mexico, MO)
Aeronautical engineer Chris Heintz worked for **Aérospatiale**, later becoming chief engineer at **Avions Robin**. Working in his spare time, Heintz designed and built his own aircraft, the Zenith (an anagram of his last name), which first flew in 1969. Heintz was soon selling blueprints and construction manuals to interested home-builders. In 1973, Heintz and his family moved to Canada, where Heintz worked for **de Havilland**. In 1974, Heintz formed **Zenair Ltd** to manufacture Zenith kits. In 1992, Heintz licensed the kit manufacturing and marketing rights to **Zenith Aircraft Co** in the USA. All of Heintz's independent designs are listed under **Zenair**.
Zenair Mini Z CH 50
Zenair Mono Z CH 100
Zenair Acro Z CH 150
Zenair Super Acro Z CH 180
Zenair Zenith CH 200

Zenair Zenith CH 250
Zenair Tri Z CH 300
Zenair Tri Z CH 400
Zenair Zodiac CH 600
Zenair Zodiac CH 601
Zenair Super Zodiac CH 601 HD
Zenair Super Zodiac CH 601 HDS
Zenair Zodiac CH 601 UL
Zenair Zodiac CH 601 XL
Zenair Gemini CH 620
Zenair Gemini CH 620 XL
Zenair Zodiac CH 640
Zenair STOL CH 701
Zenair STOL CH 701 AG
Zenair STOL CH 801
Zenair Zenith CH 2000 Alarus
Zenair Cricket MC-12
Zenair Zipper (Ultralight)
Zenair Zipper II (Ultralight)

Zenith (France)
Zenith (France) Monoplane (1910)

Zenith Aircraft Co (Mexico, MO)
(See: Zenair)

Zenith Aircraft Corp (Santa Ana, CA)
Zenith (CA) Albatross Trimotor
Zenith (CA) Z-6-A Cabin Biplane
Zenith (CA) Z-6-B Cabin Biplane

Zenith Aircraft Manufacturing Co (Uniontown, PA)
Zenith (PA) P-1 Sport

Zens Frères (Paul et Ernest) (France)
Zens Biplane (1908)
Zens Demoiselle-Type Monoplane
Zens Monoplane (1912)

Zentral Aviatik & Automobil GmbH (Austria)
Zentral Aviatik & Automobil B1
Zentral Aviatik & Automobil B2
Zentral Aviatik & Automobil Ehrlich V (1925 3-seat Touring Biplane)

Zeppelin Aircraft Companies (Germany)
With the beginning of World War I, **Luftschiffbau Zeppelin GmbH** began building airplanes, in addition to rigid airships. In 1914, Graf Ferdinand von Zeppelin established **Zeppelin-Lindau (Zeppelin Werke, Lindau GmbH)**, for designer Claude Dornier, and **Zeppelin-Friedrichshafen (Flugzeugbau Friedrichshafen GmbH)**. An entirely new corporation, **Versuchbau GmbH Gotha-Ost** (Experimental Works, Gotha-East, **VGO**), was financed by Zeppelin and the Robert Bosche Werke to build large bombers. [Aircraft built by VGO, though often described as "Gotha Bombers", should not be confused with aircraft produced by **Gothaer Waggonfabrik**, a separate concern which did not build giant bombers.] In 1916, Zeppelin established a subsidiary facility at Staaken, near Berlin. In August, the VGO works were transferred to Staaken, and the concern was renamed **Flugzeugwerft GmbH**, though the firm was commonly referred to as **Staaken** or **Zeppelin-Staaken**. In January 1918, Flugzeugwerke and Luftschiffbau Zeppelin merged to form **Zeppelin Werke GmbH**. When the Staaken works were closed in 1921 the company was reorganized as **Luftschiffbau Zeppelin GmbH**.

Files on aircraft built by the WWI Zeppelin aircraft subsidiaries will be found under **Dornier**, **Friedrichshafen**, and **Zeppelin-Staaken**. VGO aircraft are included under **Zeppelin-Staaken**. Staaken designs were also built under license by other companies; license-built aircraft are designated by an abbreviation: "Schül" for **Schütte-Lanz**, "Av" for **Aviatik**, and "Albs" for **Albatros**. Heavier than air aircraft designed by Luftschiffbau Zeppelin GmbH are included in this finding aid under **Zeppelin**.

Zeppelin (Luftschiffbau Zeppelin GmbH) (Germany)
Zeppelin "Rammer"
Zeppelin ZSO.523

***Zeppelin-Lindau** (See: Dornier)*

***Zeppelin-Friedrichshafen** (See: Friedrichshafen)*

Zeppelin-Staaken (Flugzeugwerke, Zeppelin Werke GmbH) (Germany)
Zeppelin-Staaken E.4/20
Zeppelin-Staaken L
Zeppelin-Staaken L Floatplane
Zeppelin-Staaken R-Plane Projects
Zeppelin-Staaken R.IV 12/15
Zeppelin-Staaken R.V 13/15
Zeppelin-Staaken R.VI 25/16
Zeppelin-Staaken R.VI (Schül) 28/16
Zeppelin-Staaken R.VI 30/16
Zeppelin-Staaken R.VI (Albs) 37/16
Zeppelin-Staaken R.VI (Av) 52/17
Zeppelin-Staaken R.VII
Zeppelin-Staaken R.VIII
Zeppelin-Staaken R.IX
Zeppelin-Staaken R.XIV 43/17
Zeppelin-Staaken R.XIVa
Zeppelin-Staaken R.XV
Zeppelin-Staaken R.XVI (Av) 50/17
Zeppelin-Staaken RML.1 (Rebuilt VGO I)
Zeppelin-Staaken VGO.I
Zeppelin-Staaken VGO II
Zeppelin-Staaken VGO.III 10/15
Zeppelin-Staaken 8301 Floatplane
Zeppelin-Staaken 8303 Floatplane

Zerbe, J. S. (Los Angeles, CA)
Zerbe Multiplane

Zeuzem, Ernst (Germany)
Zeuzem Rotor Wing Plane

Zheleznikov-Tomashevich (D. L. Tomashevich (Russia)
Zheleznikov-Tomashevich KPIR-5

Zherebtsov, B. Yu. (Russia)
Zherebtsov 1950 Pulsejet Helicopter

Zilina (Czechoslovakia)
Zilina Vazka (Dragonfly)

Zimer, Frederick W. (UK)
Zimer Improved Air-Ship (1890)

Zimmerman, Charles H. (Langley Field, VA)
Zimmerman 1935 Flying Wing
Zimmerman 1936 Flying Wing

Zingo (USA)
Zingo Intermediate Glider

Zinno, Joseph A. (North Providence, RI)
Zinno MPA Man-Powered Aircraft

Zintonec (USA)
Zintonec Monoplane

Zipfel, Armand (France)
Zipfel No.1 1908 Biplane (See: Voisin No.11)

Zipfel 1908 Powered Glider

***Zippy Sport** (See: Green Sky Adventures)*

Zlatoust Pioneers (Russia)

Zlatoust Malysh (Little One)

Zlin (Czechoslovakia)

During 1933, the **Masarykova Letecká Liga** (Masaryk Aviation League) apparently built aircraft in Zlìn, Czechoslovakia (now the Czech Republic). A second company, **Zlìnská Letecká Akciová Spolecnost** (Zlin Aviation Co Ltd), was organized there by **Bata, Ltd** in 1934. After WWII, the company was nationalized, renamed **Moravan Národni Podnik** (though its aircraft were referred to as Zlins), and operated as a subsidiary under various state-owned organizations. In 1998, the company became **Moravan Inc**, though the name **Moravan/Zlin** appears in company literature. **Zlin Aerospace** was also created as a Canada-based sales organization. All Masarykova, Zlìnská, and Moravan aircraft are listed under **Zlin**.

Zlin Z-I
Zlin Z-II
Zlin Z-III
Zlin Z-V
Zlin Z-VI
Zlin Z-VII
Zlin Z-IX Pošták
Zlin Z-X
Zlin Z-XI
Zlin Z-XII
Zlin Z-XIII
Zlin Z-XV
Zlin Z-XVa
Zlin Z-XVI
Zlin Z-XVIII
Zlin Z 20
Zlin Z 120
Zlin Z 22 Junák (Cadet)
Zlin Z 122
Zlin Z 23 Honza
Zlin Z 24 Krajánek
Zlin Z 124 Galánka
Zlin Z 25 Šohaj (LG25)
Zlin Z 125 Šohaj 2 (LG125)
Zlin Z 225 Medák
Zlin Z 325
Zlin Z 425 Šohaj 3
Zlin Z 26 Trenér
Zlin Z 126 Trenér 2
Zlin Z 226 A Akrobat
Zlin Z 226 B Bohatýr
Zlin Z 226 AS Akrobat Special
Zlin Z 226 MS
Zlin Z 226 T Trenér 6
Zlin Z 326 Trenér Master
Zlin Z 326 A Akrobat
Zlin Z 426
Zlin Z 526
Zlin Z 526 A
Zlin Z 526 AS Akrobat Special
Zlin Z 526 AFS Akrobat Special
Zlin Z 526 F Akrobat
Zlin Z 526 L Akrobat
Zlin Z 726 Universal
Zlin Z 30 Kmotr
Zlin Z 130 Kmotr
Zlin Z 32 Frajír
Zlin Z 35 Heli Trenér
Zlin Z 37 Cmelák (Bumble Bee)
Zlin Z 37A Cmelák (Bumble-Bee)
Zlin Z 37 T Agro Turbo
Zlin Z 137 T Agro Turbo
Zlin Z 41
Zlin Z 42
Zlin Z 42M
Zlin Z 142
Zlin Z 142 C
Zlin Z 142 CAF
Zlin Z 242 L
Zlin Z 242 LA
Zlin Z 43
Zlin Z 143
Zlin Z 143 L
Zlin Z 50 L
Zlin Z 50 LS
Zlin Z 50 M
Zlin Z 61 L
Zlin Z 90
Zlin Z 212 Mikron
Zlin Z 281 (Redesigned Bücker 181)
Zlin Z 381 (Redesigned Bücker 181)

Zlokazov, A. I. (Russia)

Zlokazov ARK-Z-1

Zmaj (Dragon) (Fabrica Aeroplani I Hydroplani Zmaj) (Yugoslavia)

Zmaj AF-2 Amphibian (See: Fizir)
Zmaj FN (See: Fizir)
Zmaj FP.2
Zmaj Observation Seaplane
Zmaj R.1
Zmaj Wright-Engined Observation Aircraft

Zodiac Aircraft Corp (Lodi, NJ)

Zodiac (NJ) ZAC-1 Libra-Det

Zodiac (Société Zodiac) (France)

Zodiac (France) First Monoplane (1910, Prince de Nissole)
Zodiac (France) l'Albatros (1910, Prince de Nissole)
Zodiac (France) Pusher Biplane (1911)
Zodiac (France) Tractor Biplane (2 S) (1911)

Zorn, R. R. (USA)

Zorn 1913 Curtiss Type

Zorn-Crume (Dayton, OH)

Zorn-Crume 1910 Aviette

***Zschach, R.** (See: Hannover)*

Zuck, Daniel R. (Los Angeles, CA)

Zuck Helicopter

***Zuck-Whitaker Plane-Mobile, Roadable** (See: Plane-Mobile)*

***Zunker, Norman W.** (See: Hegy & Zunker)*

A-Z OF AIRCRAFT NAMES

A

ABC Glider
Schultz

Abdul
Nakajima Ki-27

Abeille (Bee)
Robin
SAN D140R
SNCAC NC.2001

Abril
Levasseur

"Abyssinia"
Bellanca J-2

Academe
Grumman C-4

Acapella *(See: Air Acapella)*

Accord Jet
Avia (Russia)

Accountant
Aviation Traders

Ace *(See also: Baby Ace, Cabin Ace, Crossland Ace, Junior Ace, and Super Ace)*
Aircraft Engineering
Chrislea CH.3
Essig
Keane Aeroplanes
Keane Aircraft
Taylorcraft

Ace *(As: Asso)*
Fiat CR

Acey Deucy
Powell (J. C.)

"Achernar"
Latécoère LAT 611

Acro *(See also: Cuby Acro Trainer)*
PZL Bielsko SZD 59
Aircraft Technologies, Inc

Acro Z
Zenair

Acro-Dactyl
Pterodactyl

Acro-Pro
Davis (Robert N.)

Acrobat (As: l'Acrobat)
Robin

Acrobatique
Robin

Acrobin
Robin

Acroduster
Stolp SA-700
Aerovant

Acroduster Too
Stolp SA-750

Active
Arrow (UK)

Activian
Santa Ana VM-1

Adam
Nakajima

ADAM
Vought

Adastra
Johnson (Richard H.)

Adlershof (Eagle's Nest)
Albatros L74

Admiral
Consolidated PY

Adventure *(See also: Citabria Adventure)*
Adventure Air

Adventure *(As: Aventura)*
Arnet Pereyra

Adventurer
Adventure Air
Auster J.5

Aeolus *(As: l'Eole)*
Ader

Aéraptère
Domingo

Aerauto
Pellarini PL-5C

Aerial Coupé
Dayton Wright O-W

Aerial Steam Carriage
Henson

Aeriel
Durban
Genair
Robertson (South Africa)
Southern

Aérien (As: l'Aérien)
Delauries
Essort
Gonnel

Aero Cabin
Ohio

Aéro Ramo-Planeur
Baron

Aero Subaru
Fuji

Aero-Limousine
Grahame-White GWE.7

Aero-Sport
EAA

Aéro-Torpille (Torpedo)
Artois

Aerobat
Aerobat Aircraft
Bradley Aerospace
Aerial Service Corp

Aerobile
Waterman (W. D.)

Aeroboat
Loening (Corp)
Aeroboat
Wright (Co) Model G

Aerobot (Aeroboat)
Moller Merlin 200

Aerobus
Grahame-White Type 10

Aérobus (l'Aérobus)
Blériot XIII

Bourgoin et Kessels
D.N.F.
Sommer

"Aérobus"
Blériot 75

AeroCanard
Aerocad

Aerocar
Aerocar International
D.K.D.
Harriman (Aeromobile Co)
Portsmouth
Taylor (Molt)

Aerocoupe
Aero-Craft

Aerocycle
deLackner

Aerocycloid
Irvine

Aerodyne (VTOL)
Dornier

Aerofoilcraft
RFB X-114

Aérogyre
de Chappedelaine

Aerogyro
Lockheed H-51

Aerolite
Australian Aviation

AeroMaster AG
AeroLites

Aeromax
Australian Aviation

Aeromobil
D.K.D.
Maykemper

Aeromobile
Bertelsen
Granville, Miller, and De Lackner
Harriman (Aeromobile Co)

Aéromobile
Constantin

Aeromold
Timm S-160-K (PT-160-K)

Aeron
BRNO

Aeronaut
EÔA

Aeroneer
Phillips (Aviation Co)

Aéronef
Blériot XXIV

"Aeroplane Destroyer"
Martin 1914

Aeroscaphe
Ravaud

Aeroscooter
Partenavia P53

Aeroscout
Bell OH-58D

AeroSport
Thorp T-211

Aerostable
Grant-Morse

Aérostable
Moreau (Brothers)

Aerostar
Aerostar (ID) 700
Piper PA-60-290
Ted Smith

Aerostar 220 Executive
Mooney

AeroTiga
SME MD-3-160

Aerovan
Miles M.57, M.72

Aerovel
Miller (Howell) HM-4

Aérovoile
d'Andre

Aeroyacht
Borel

AeroYacht
Martin T Hydro

"Affordable Airplane"
de Vore

Ag Bearcat
AeroLites

Ag-Cat *(See also: Super Ag-Cat)*
Grumman G-164

Ag-plane
Adam (France) R.A.17

AgCarryall
Cessna 185

AgHusky
Cessna 188

Agile
Wallis (Autogyros) WA-116

Agile Eagle
McDonnell Douglas NF-15B

AgPickup
Cessna 188

Agricola
Auster B.8

Agro Turbo
Zlin Z 37 T

AgTruck
Cessna 188

Aguila
Enaer

AgWagon
Cessna 188

Aidea
Bergousi

Aigle
Druine

Aiglet
Auster J.1B

Aiglon (l'Aiglon) (Eaglet)
Chazal-Gourgas
Robin
Caudron C.600

Air Acapella
Option

Air Beetle
AIEP

Air Camper
Pietenpol

Air Car *(See: Amphibian Air Car)*

Air/Car
Airborn Utility Cars

Air Coach *(See also: Page Air Coach)*
Ford

International Aircraft F-18
Ketner
Kreutzer

Air Coupe
Coffman

Air Cruiser
AESL

Air Cycle
Air Cushion Vehicles, Inc

Air Express
Lockheed Model 3
Rogers (Construction)

Air Pullman
Ford
Stout 2-AT

Air Scooter (ACV)
Bell

Air Scout
McCarthy

Air Sedan
Bellanca C.F.
Stout AS-1

Air Shark
Freedom Master

Air Skimmer
Osprey

Air Sport
Johns

Air Tourer
Pacific & Western
Victa

Air Tractor
Lamson (Aircraft Co)

Air Transport
Bach
Ford

Air Yacht
Bach
Hall-Aluminum
Keystone
Loening (Corp)
Supermarine

Air-Bus
Shorts P.D.24

Air-Car
Benton
Convair Model 111

Air-King
National Airways System

Airabonita
Bell FL

Airacobra
Bell P-39, P-400

Airacomet
Bell P-59

Airacuda
Bell FM-1

Airaile
Rans

Airboat
Cooke (Weldon)
Kirkham
Sperry

Airboss *(See: Messer Airboss)*

Airbuggy
Airconcept
Ekin

Airbus *(See also: Air-Bus)*
Airbus Industries
Bellanca
Custer

"Airbus"
Garaux

Aircamper
Grega GN-1

Aircar *(See also: Air/Car, Air-Car)*

Aircar Hovercraft
Curtiss
Aircar Jet
Aircar System

Aircoach *(See also: Rota-Aircoach)*
Verville Aircraft

Aircoupe
Stinson (Jack)

AirCoupe
Alon A2
Elias
Fourney

Aircruiser
Bellanca
Victa
Yackey

Airedale
Beagle A.109, A.111
Blackburn

Airgeep *(See also: Jeep)*
Piasecki PA-59

Airguard
Chrislea LC.1
Xian F-7M

Airhorse
Cierva W.11

Airistocrat *(See also: Aristocrat)*
Waco E, W

Airliner *(See also: Midnight Airliner)*
Bristol
Lawson (Alfred)
Republic RC-2

Airmaster
Babcock
Cessna C-37, C-38, C-165

Airmaster Coupe
Ohio

Airmate
Schreder

Airmobile
Elias

Airone
Pasotti F.6

Airplume
Croses

AirRhoCar Hoverscooter
Rhodes

Airsedan
Buhl

Airspeed
Foster

Airsport
Bergholt
Elias

Airster
Babcock
Buhl
Kinner
Security National

Airtourer
AESL
New Zealand Aerospace
Victa

Airtrainer
AESL

Airtruck
Bennett (New Zealand)

Airvan
Gippsland GA-8

Ajax
Armstrong Whitworth

Ajeet
Hindustan

Akar
Karpinski S.L.1

Akigusa
Kugisho MXY8

"Akira-Go"
Itoh Emi 19

"Akita-Go"
Itoh Emi 24

Akro
Stephens (C. L.)

Akrobat
Zlin Z 226 A

Akrobat Special
Zlin Z 226 AS

Akromaster
Spinks Model 11

Akron Condor
Baker-McMillen

Al-Kahira
Helwan

"Al-Ma"
Aero (Prague)

Alarm Clock
Budil'nik

Alarus
Zenair Zenith CH 2000

"Alba-Iulia 1918"
Bellanca 28-92

Albacore
Fairey

Albatros (l'Albatros) (Albatross)
Aero (Vodochody) L 39
Aviasud
Zodiac (France)

"Albatros Artificiel" (Artificial Albatross)
le Bris

Albatross *(See also: Baby Albatross, Gossamer Albatross, Senior Albatross)*
Beriev Be-40
Bowlus
de Havilland D.H.91
Grumman G-111, G-262, SA-16, UF
Mantelli AM-11
Zenith (CA)

Albemarle
Armstrong Whitworth A.W.41

"Albert"
Fulda

Albis
Albatros L72c

Alcor
Lamson (Robert)

Alcotan
CASA

Alcyon (Halcyon)
Morane-Saulnier

Aldershot
Avro 549

Alerion
Damblanc

Alf
Kawanishi E7K1

Alibi
Trager-Bierens T-3

Alizé (Tradewind)
Aviasud
Breguet Bre.1050

Allege
Levasseur

Allegro
Mooney

Alliance
Fokker-Republic D.XXIV

Allievo Cantu Glider
Ambrosini

"Aloha"
Breese-Wilde

Alouette (Lark, Skylark)
Phillips (Lt Donald B.)
Republic (Sud)
Sud-Aviation SA 318
SNCASE SE 3120

Alouette Gouverneur (Governor)
SNCASE SE 3130

Alpha *(See also: Noralpha, Scion Alpha, Starship Alpha)*
Auster J.1N
General Aeroplane
Northrop
Northrop C-19
Partenavia P70
Pomilio
Robin
Robinson (Helicopter) R22

Alpha Jet
Alpha Jet
Lockheed

Alpha Sport
Robin

Alpine
Auster

Altair
Lockheed Model 8-D

Alti Cruiser
Aero Commander 720

Alula Fighter
Commercial Aeroplane Wing Syndicate

Aluminio-plane
Moisant (John)

Am Aerofoil Boat
RFB (Lippisch) X-113

Ambassadeur (Ambassador)
Robin

Ambassador
Airspeed
Collier CA-1
Douglas DA-1

America *(See: Large America, Small America)*

"America"
Curtiss Model H
Fokker C-2

American Falcon
ASC

American Flea Ship
Ace (NC)

"American Legion"
Keystone K-47 Pathfinder

American Moth
Moth
Vulcan V-1

"American Nurse"
Bellanca Skyrocket CH-400

American Spirit
ASC

Amiens
de Havilland D.H.10

Amor
Douglas DC-7

Amphibian
Towle

Amphibian Air Car
Spencer S-12-C

Amphibion
Sikorsky (USA) S-38, S-41A
Towle TA-2

Amphibion *(Baby Clipper) (See also: Sport Amphibion)*
Sikorsky (USA) S-43A

Amsoil
Rutan Model 68

Amur
Antonov

Anaconda
Aeroprogress T-710

Anade
Anatra

Anadis
Anatra

Anadva
Anatra

Analog
MiG MiG-21I

Anamon
Anatra

Anasal
Anakle
Anatra

Andover
Avro 561, 563
British Aerospace 780
Hawker Siddeley

Angel
Angel Aircraft
CH-7
Elisport CH-7
King's Engineering Fellowship

Angeles Monoplane
Summit

Anhanger
Kassel

Anjou
Boisavia B-260

Ann
Mitsubishi Ki-30

"Annelise"
Junkers J 13

Anser
Acme (Los Angeles)

Anson
Avro 652A

Ant (Mrówka)
PZL Warsawa-Okecie PZL-126P

Antarctic Baby
Avro 554

Antei
Antonov An-22

Antelope
Avro 604
Sopwith

Antheus
Antonov An-22

Antoinette
Albatros
Antoinette

"Anzac"
Lockheed Model 8-D

Apache *(See also: Longbow Apache)*
ACT
Hughes H-64
North American A-36
Piper PA-23
Wright (Aero) F3W

Ape
Armstrong Whitworth

Apollo
Armstrong Whitworth A.W.55

Aqua
Aquaflight

Aqua Gem Air Cushion Boat
National Research Associates

Aqua Glider
Explorer PG1
EAA

Aquarius
Baker (Marion)

Aquila
Angus
Schindler Brothers

Aquilon (North Wind)
SNCASE

Ara
Armstrong Whitworth

Aratinga
Instituto de Pesquisas Tecnologicas IPT-2

Arava
Israel Aircraft Industries

"Arc-en-Ciel" ("Rainbow")
Couzinet

Archer
Piper PA-28

Archer (As: Sagittario)
Aerfer
Reggiane RE 2005

Arctic Tern
Rocheville

Arcturus
Lockheed P-3

Ares
Scaled Composites Model 151

Argander Special
Nimmo Special Racer

Argentino, El Argentino (The Argentine)
Artigala
Borello
Ojea

Argo
Alliance Aircraft

Argonaut
Canadair CL-4
Thaden T-1

Argosy
Armstrong Whitworth
Hawker Siddeley

Argus
Canadair CL-28
Fairchild
Saab 100B

Ariel
Autogyro Design Bureau
SNCASO

Aries
Armstrong Whitworth A.W.17
Bellanca
Lockheed EP-3
Miles M.75

Ariete (Ram)
Reggiane RE 2002

Aristocraft
United States (NJ) SP7

Aristocrat (See also: Airistocrat)
GAC Model 102
Irish (Aircraft Corp)

Arkona
LFG V.20

Armadillo
Armstrong Whitworth F.M.4

Armagnac
SNCASE SE 2010

Army Mule
Piasecki H-25

Arrow (See also: Cherokee Arrow, Flying Arrow, Golden Arrow, Mercury Arrow, Phinn Arrow Biplane, Raw Arrow, Red Arrow, Silver Arrow, Sky Arrow, Speed Arrow, Super Arrow, Turbo Arrow)
Aeronca 9
Auster J.2
Avro (Canada) CF-105
Parks
Schneider (Edmund & Harry) ES-59
Spartan (UK)

Arrow (As: Pfeil)
Aviatik
Dornier Do 335
LFG Roland
Albatros Taube (Dove) Biplane

Arrow (As: Pulqui)
FMA

Arrow (As: Saeta)
Hispano HA-200

Arrow (As: Saetta)
Macchi M.C.200

Arrow Air Transport
Rogers (Aeronautical)

Arrow Flyer (As: Pfeilflieger)
Lohner

Arrowbile
Waterman

Arrowhead
Smith (B. L.)

Arrowplane
Waterman W-4

Arrowscout
Avro 511

"Artificial Albatross" (As: "Albatros Artificiel")
le Bris

Artillery Shell (As: l'Obus)
Borel

Asahi-Go
Shirato

"Asbestos"
Emsco

Ascender
Curtiss P-55
DFE Ultralights
Granville Q-1
Pterodactyl

Ashton
Avro 706

Asiago
Ambrosini

"Aspern"
Lohner 10.06

Ass
Albatros L75a

Asso (Ace)
Fiat CR

Astar
Aérospatiale AS 350

Astir Club
Grob

Astir CS (Club Standard)
Grob

Astor Seaplane
Burgess-Dunne

Astore (Goshawk)
Partenavia P48B

Astro
Robinson (Helicopter) R44

Astro-Rocket
Maule Air

Atalanta
Armstrong Whitworth A.W.XV

"Atalanta"
Fairey N.4

Atar
SNCASO SO 30

Athena
Avro 701

Atlantic
Auster C.6
Boulton Paul
Breguet Bre.1150
Sopwith
Wassmer WA.54

Atlas
Armstrong Whitworth
Cosmos

Atlas (As: Noratlas)
SNCAN 2503

Atom
Auster J.3
Strat

Attacker
Supermarine

Audax
Hawker

"Aula Volante" ("Flying Class-Room")
Fiat G.12 CA

Aurora (See also: Citabria Aurora)
Lockheed P-3

Auspicious Cloud (As: Zuiun)
Aichi E16A

"Austria" (See also: Standard Austria)
Kassel

Auto-plane
Smolinski

Autocar
Auster

Autocrat
Auster

Autodrome (l'Autodrome)
Leyat

Autoplane
Aimé et Salmson
Curtiss
McWorter

Autostable
Albessard

Ava
Avro 557

"Ava"
Avro 652

Avalon
Airmaster
"Avalon"
Avro 652

Avalon Twinstar
Airmaster

"Avatar"
Avro 652

Avenger
Avro 566, 567
GM (Eastern) TBM
Grumman TBF

Aventura (Adventure)
Arnet Pereyra

"Aviaarktika"
Tupolev ANT-6/AM-34R

Avian
Avro 581, 594, 605, 616, 625
Whittelsey

Aviastar
Avia (Italy) LM.5

Avid Flyer
Light Aero

"Avioneta"
CASA

Avis
Auster Model P
Avro 562
Wright (Howard)

Avitor
Marriott

Avitruc
Chase C-122
Fairchild C-123

Avocet
Aero Wood Specialties
Avro 584

Avrocar
Avro (Canada)

Avron
Simmonds

AWACS
Boeing 737-700 Wedgetail
Boeing E-3
Boeing E-767

Awana
Armstrong Whitworth A.W.I

Ayr
English Electric

Azor
CASA

Aztec
Piper PA-23-250

Azték
Letov Air ST-4

Aztek Nomad
Air Muskoka

Azure (As: Norazur)
SNCAN 2100

B

Baade
VEB Dresden

Babe
Bristol

Baboon
B.A.T.

Babs
Mitsubishi C5M, Ki-15

***Baby** (See also: Antarctic Baby, BeeGee Baby Sportster, Fly Baby, Grunau Baby, Hamble Baby, Heli-Baby, New Baby, Porte Baby, Sky Baby, Skybaby, Super Baby, Water Baby)*
Avro 534, 543, 544
Burgess E
EoN
Grahame-White
Grunau
Lippisch
McCabe (Ira)
Sopwith
Supermarine N.1B
Wight

Baby (As: Bébé)
Astra-Wright Type BB
Jodel

Baby (As: Bubi)
Messerschmitt S 16a

Baby Ace
Ace (NC)
Corben

Baby Albatross
Bowlus BA-100

Baby Amphibion
American Aeronautical

Baby Biplane
Laird ("Matty")

Baby Biplane Houperzine
Melton

Baby Bomber
Snyder

Baby Bullet
Heath

Baby Clipper
Fairchild 91
Sikorsky (USA) S-43A

"Baby Columbia"
Albert Avionnette

"Baby Grand"
Wright (Co) Model R

Baby Great Lakes
Oldfield

Baby Hawk
Smith-Cirigliano SC-1

"Baby Hawk"
Cirigliano

Baby Plane
Kimball (Asa) K8

Baby Pursuit
Prest

Baby Scout
Curtiss S-1

Baby Sesquiplane
Farman (Henri) H.F.24

Baby White
White (George D.)

"Baby Wright"
Wright (Co) Model R

Bachão
Instituto de Pesquisas Tecnologicas IPT-11

Bachstelze (Water Wagtail)
Focke Achgelis Fa 330

Backfin
Tupolev Tu-98

Backfire
Tupolev Tu-22M

Baddeck Biplane
Canadian Aerodrome

Badger
Beech
Bristol

Badger
Tupolev Tu-16
Xian H-6

Badminton
Bristol

Baffin
Blackburn B-5

Bagoas
Abrial

Bagshot
Bristol

Bak (Horsefly)
Warsztaty Szybowcowe

Baka (Crazy)
Kugisho MXY7

Balbuzard
AAT

Baldo
Alaparma AM-75

Balilla
Ansaldo A1
Fiat A1

Baljims
AAT

Ball
Austin

Balladou
Wassmer Super IV

Balliol
Boulton Paul

Balsan
Lioré

Baltimore
Martin 187

Balzac
Dassault

Bambi
Holleville RH-1

Bamby
Matra-Cantinieau

"Bamel"
Gloster Mars I

Banbi
Colomban
Dyn'Aero

Bandeirante (Pioneer)
EMBRAER

Bandit
Avid Aircraft

Banner (As: Étendard)
Dassault

Banshee
McDonnell F2H
Willard

Bantam
B.A.T.
Grahame-White GWE.6
Northrop X-4
Warwick

Banty
Butterfly

BAP
Curtiss

Bärbel
LFG V.8

"Barcenas, El" ("Dapple, The")
Dirección General de Reparaciones...

Bark
Ilyushin Il-2

Barling Bomber
Engineering Division NBL-1

Barnstormer
Fisher

Barnwell Bullet
Vickers (UK) E.S.1

Baron
Beech

Baroudeur (Fighter)
SNCASE SE 5000

Barracuda
Buethe
Fairey
Jeffair

Barrel (As: Tunnan)
Saab 29

BarrSix
Barr Aircraft

Basilisk
B.A.T.

Basset
Beagle B.206

Bat *(See also: Aerobat,* Foxbat, *Minibat, Mothbat, Sabre Bat, Sea Bat, Seabat, Ultrabat, Vampire Bat)*
Baynes
McDonnell P-67
Noran
Stout

Bat
Tupolev Tu-6, Tu-8, Tu-10

BAT
Curtiss

Bat (As: Pipistrello)
Savoia-Marchetti SM.81

Bat Apparatus (As: Fledermausfugelapparat)
Lilienthal (Otto)

Bat Boat
Sopwith

Bat Glider
Pilcher

Bathtub
Dormoy

Battle
Fairey

Battle Plane
Gallaudet Aircraft

Battleplane
Curtiss CB
Sturtevant
Wanzer

Battler
Lawson (Alfred)

Batwing
Stout

Bayern (Bavaria)
BFW (1926)
Caspar C.36
Maiss

Beacon (As: Jalor)
Fouga MC

Beagle
Blackburn

Beagle
Harbin H-5
Ilyushin Il-28

Bear
Tupolev Tu-95, Tu-142

Bearcat *(See also: Ag Bearcat)*
AeroLites
Grumman F8F

Beast
Ilyushin Il-10

Beaufighter
Bristol

Beaufort
Bristol

Beautiful Cloud (As: Keiun)
Kugisho R2Y-1

Beauty
Skandinaviska Aero BHT-1

Beauty
Tupolev Tu-22

Beaver
ASAP
Bristol
de Havilland (Canada) DHC-2

Bébé (Baby)
Astra-Wright Type BB
Jodel

Bébé Jodel
Jodel

Bee *(See also: Bowlus Bumblebee, Bumble Bee, Honey Bee, Little Bee, Queen Bee, Seabee, Wee Bee)*
Baynes
Sopwith

Bee (As: Abeille)
Jodel
SAN D140R
SNCAC NC.2001

Bee (As: Hachi)
Nippon

Bee (As: Pchela)
Aeroprogress T-203

Bee Dee
Maule Air

Bee Line Racer
Aerial Engineering Corp

Beechjet
Beech

BeeGee Baby Sportster
BG

Beetle *(See also: Air Beetle, "Silver Beetle")*
Pilcher

"Beetle"
Birchan

Beetle (As: Taupin)
SFCA

Beezle Bug
Braley (Aircraft)

Beija-Flor (Hummingbird)
Centro Técnico de Aeronáutica

Bekas
Dubna Z-7
Kasper

Belalang (Grasshopper)
Angkatan

Belfair *(See: Tipsy Belfair)*

Belfast
Shorts S.C.5/10

Bellatrix
Breguet Bre.731
SNCASO SO 30

Belmont
Warren & Montijo W.M.1

Belvedere
Bristol Type 192

Bengali
Nicollier HN.500

Bergfalke (Mountain Falcon)
Scheibe

Berkeley
Bristol

Berkut
Experimental Aviation

Berline (Coach)
Blériot XXIV

Bermuda
Brewster SB2A

Beryl
Piel

Bestiola
Beneš-Mráz

Bestmann
Bücker Bü 181

Beta
Filper
General Aeroplane
Robinson (Helicopter) R22

Beta Bird
Hovey

Beta-Junior
Beneš-Mráz

Beta-Major
Beneš-Mráz

Beta-Minor
Beneš-Mráz

Beta-Scolar
Beneš-Mráz

Betty
Messerschmitt S 16b

Betty
Mitsubishi G4M

"Betty Jo"
Johnson (Luther)

"Betty Joe"
North American P-82B

Beverley
Blackburn B-101

Bi-Craft
Greenwood-Yates

Bi-Mono Slip-Wing Aeroplane
Hillson

Bi-Rudder 'Bus
Grahame-White Type XV

Bi-Vittoria (Victory)
Magni

Bibi
Beneš-Mráz

Bichinho
Instituto de Pesquisas Tecnologicas IPT-0

Bicycle *(See: Flying Bicycle)*

Bidulm
Cosmos

Bies
TS-8

Big Bird
Clère

Big Dipper
Lockheed Model 34

Big Duster
Huff-Daland HD-31

"Big Flapper"
Project Ornithopter Model 1

Big X
Wittman

Bijou
Robin

Bimbus (Bimbo)
Jach

Bimonoplane
Downey

Binz
LFG V.14

Bird *(See also: Beta Bird, Big Bird, Blackbird, Blue Bird, Bluebird, Catbird, Chevy Bird, De Bruyne-Maas Ladybird, Early Bird, Firebird, Frigatebird, Hummingbird, Jaybird, King Bird, King-Bird, Kingbird, Ladybird, "Little Bird," Mechanical Bird, Paraplane Le Bird, Quiet Bird, Sea Bird, Seabird, Silverbird, Soaring Bird, "Song Bird," Storm Bird, Sunbird, T-Bird, Thunderbird, Water Bird, White Bird, Willbird, Young Bird)*
Lanier (Edward M.) Paraplane
Taylor (Aero Industries)

"Bird, The"
Kowalski

Bird Flight Machine
Farrar

"Bird of Paradise"
Fokker C-2

Birdling
Universal Aviation

Birdwing
Nicholas-Beazley

Bisley
Bristol

Bison
Avro 555

Bison
Myasishchev M-4

Bison (As: Zubr)
L.W.S.6

Bittern
Boulton Paul

Bitty Bipe
Bailey

Bizerte
Breguet Bre.52-1

Black Bullet
Northrop XP-56

Black Cat (As: Czarny Kot)
Drzewiecki (Jerzy) J.D.1

"Black Devil" (As: "Schwarze Teufel")
Aachen FVA-1

Black Diamond
Maupin-Lanteri

Black Hawk
Butler
Sikorsky (USA) S-70, H-60

Black Widow
Northrop P-61

Blackbird
Lockheed A-12, F-12, SR-71

Blackbird (As: Kos)
PZL Warsawa-Okecie PZL-102B

Blackburd
Blackburn

Blackburn
Blackburn

Blackhawk
Curtiss P-87
Sikorsky (USA) S-67

Blackjack
Tupolev Tu-160

Bleeker Helicopter
Curtiss

Blenheim
Bristol

Blinder
Tupolev Tu-22

Blitz (Lightning)
Arado Ar 234
Heinkel He 70

Blizzard (As: Tuisku)
VL

Bloodhound
Bristol

Bloudek
Aero Club du Royaume

Blowlamp
Ilyushin Il-54

Bluchón
Hispano HA-1109

Blue Bird
Ashmussen
Dorna
Hess
National Airplane & Motor

Blue Canoe
Cessna U-3

Blue Eagle
Beijing Keyuan

Blue Ibis (As: Ibis Bleu)
Delanne

Blue Mouse (As: Souris Bleue)
Aachen

Blue Sentinel
Lockheed P-3

"Blue Streak"
Bellanca Tandem

Blue Zephyr
Bartlett

Bluebird
Blackburn
Handley Page
Moisant (Co)
Ryan

Bluebird (As: Oiseau Bleu)
Farman F.180
Landes-Derouin

Boarhound
Bristol

Boat *(See: Airboat)*

Bob
Ilyushin Il-4
Kawasaki Ki-28

***Bob-O-Link** (See also: Boblink)*
Ashley
Keesling

Bobcat
Cessna AT-17, UC-78

***Boblink**(See also: Bob-O-Link)*
Boulton Paul

Bobsleigh
Reid and Sigrist

Bodmin
Boulton Paul

Bognor Bloater
White and Thompson

Bohatýr
Zlin Z 226 B

Bohem
Bohemia Air

Bohóc
Lampich L-4

Bolingbroke
Bristol

Bolo
Douglas B-18

Bolt of Light (As: Denko)
Aichi S1A

Bolton
Boulton Paul

"Bol'shoi Bal'tisky" (Great Baltic)
Sikorsky (Igor) S-21

Bombay
Bristol

Bomber
Sopwith

***Bonanza** (See also: Propjet Bonanza, Turbine Bonanza, Twin Bonanza)*
Beech

Bongo
Unis NA 40

Bönisch-Vogel
VEB Lommatzch

Bonzo
Mráz M-3

"Bonzo"
Wittman

Boo Ray
Baker (Marion)

Boomerang
Autogyro Design Bureau
CAC
Scaled Composites Model 202
Schneider (Edmund & Harry) ES-60

Boot
Tupolev Model 91

Booth Racer
Aerial Engineering Corp

Borea
Caproni Ca.308

Bosbok
Aerfer

Boston
Douglas A-20, DB-7

Bosun
Tupolev Tu-14

Botha
Blackburn B-26

Bottle Tuborg
Cameron

Bounder
Myasishchev M-50

Bourges
Boulton Paul

Bowlus Bumblebee
Nelson (Aircraft)

Boxcar
Miles M.68

Boxkite
Bristol
Challenger
Grahame-White Type XV

Boyero, El (The Cowboy)
FMA
Pertolini

Bozena
Selnaszka

Brabazon
Bristol

Braemar
Bristol

"Brandenburg"
Focke-Wulf Fw 200 S-1 Condor

Brandon
Bristol

Brasilia
EMBRAER

Brat (As: Smyk)
Moczarski, Idzkowski, and Ploszajski

Bravo
FFA AS202/18A
Mooney
Morrisey
SIAI-Marchetti SA.202

Brawny
Broughton Blayney

Breeze
Aeropract A-25
Aeroprakt A-25

Breeze (As: Zef)
Piel

"Bremen"
Junkers W 33b

"Brenner"
Stuttgart

Bretagne
SNCASO SO 30R

Brewer
Yakovlev Yak-28B

Bridget
Huff-Daland HD-4

Brigadier
Baumann

Brigadýr
Aero (Vodochody) XL 60

Brigand
Bristol

Brisfit
Bristol Fighter

Bristol Fighter
Bristol
Curtiss

"Britain First"
Bristol Type 142

Britannia
Bristol

Britannic
Shorts S.C.5

Bronco
North American OV-10

Brougham
Ryan B.1

Broussard
Max Holste

"Brown Special"
Brown (L. W.) B-1

Brownie
Bristol

Brownie (As: Elfe)
Pfenninger

Bubi (Baby)
Messerschmitt S 16a

Bubplane
Washburn

Buccaneer
Blackburn B-103
Brewster SB2A
British Aerospace
Lake (Aircraft)

Büchner
Gotha

Buck Private
American Helicopter XA-6

Buckaroo
Temco T-35, TE

Buckeye
North American T2J, T-2

Buckingham
Bristol

Buckmaster
Bristol

Buffalo
Avro 571, 572
Brewster F2A
de Havilland (Canada) DHC-5
Sopwith

Bugle
Boulton Paul

Bulgarian Tourer
Bristol

Bull
Tupolev Tu-4

Bull (As: Tauro)
Anahuac

"Bull Head" Biplane
Dayton Wright

Bull Pup
Buhl

Bulldog
Beagle B.125
Bristol
British Aerospace
Hall (Bob)
Itoh Emi 23
Nakajima
Scottish Aviation
Sopwith
Springfield

Bulldogg (Bulldog)
Siemens-Schuckert

Bulldogge (Bulldog)
Forssman
Sigismund

Bullet (See also: Baby Bullet, Barnwell Bullet, Black Bullet, "Drake Bullet," Taylor Bullet, Texas Bullet, Winged Bullet)
Aerocar, Inc
Alexander Eaglerock C-1, C-3
Bristol
Brokaw
Christmas
Gallaudet Aircraft A-1
Johnson (TX)
Vickers (UK) E.S.2

Bullet 2100
Aerocar, Inc

Bullfinch
Bristol

Bullpup
Bristol

Bullshot
Tupolev Tu-22

"Bull's-eye"
Avro Roe I Triplane

Bumble Bee
Lincoln (Aircraft)

"Bumble Bee"
Curtiss Model E Hydro Headless
Keith Rider R-2

Bumble Bee (As: Cmelák)
Zlin Z 37

Bumble Bee (As: Hummel)
Siebel Si 202

Bumblebee
Aircraft Designs (USA)
Bowlus
Johnson (OH)
Nelson (Aircraft) Bowlus

Burd
MIT

Burdette
Whites

Burga
Avro

Bus (See also: Aerobus, Aérobus, Air-Bus, Airbus, Bi-Rudder 'Bus, Gunbus, Helibus, Sigrist Bus, Skybus, Streamline Gunbus)
Longren

Bushmaster
Aircorp
Stout
Sylvaire

Business Speedster
Stearman (1927) C-3R

Businessliner
Cessna 402B

Bussard (As: Buzzard)
Focke-Wulf A 32

"Buster"
Wittman

"Buttercup"
Wittman

Butterfly
LWF L

Butterfly (As: Kolibrie)
Brügger

Butterfly (As: Motyl)
Bohatyrew

Butterfly (As: Papillon)
Caproni Bulgaria

Buzzard (See also: Rhine Buzzard)
Martinsyde F.4
Motor Gliders
Snyder MG-1

Buzzard (As: Bussard)
Focke-Wulf A 32

Buzzer
Haufe

Bydgoszczanka
Dzialowski

C

Cab
Lisunov Li-2

Caballero
Helio

Cabin Ace
Corben

Cabin Amphibian
Loening (Corp)

Cabinaire
Paramount

Caboré
Instituto de Pesquisas Tecnologicas IPT-12

"Cactus Kitten"
Curtiss-Cox Racer

Cadet (See also: Club Cadet, Kirby Cadet, Senior Cadet)
Aeronca C-1
Aerosystems
Avro 643
Babcock LC-11
Baker-McMillen
Culver
GAC Model 111-C
Interstate S-1
Kirby
Mooney
Piper PA-28-161
Roamair
Robin
Robin
Wilson

Cadet (As: Junak)
LWD

Cadet (As: Kadett)
Heinkel He 72

Calcutta
Short

CALF (Common Affordable Light-Weight Fighter)
Boeing

Calif
Caproni Vizzola

"California"
Belcher

Californian
Aircraft Manufacturing (CA)

"Californian"
Allenbaugh

Call Monoplane
Aerial Navigation

Calquin (Royal Eagle)
FMA

Camber
Ilyushin Il-86

Cambgul
SABCA

Camel
Slingsby T.57
Sopwith

Camel
Tupolev Tu-104

Camp
Antonov An-8

Campania
Fairey

Campeiro
Neiva

Camper *(See: Air Camper, Aircamper, Sky Camper)*

Canada
Curtiss

Canadian Duster
Keystone

Canadian Skyrider
Aero Tech

Canary
Johnson (OH)
Kinner

Canberra
English Electric
Martin B-57

Candid
Ilyushin Il-76

Canguro (Kangaroo)
Ambrosini
Savoia-Marchetti SM.82
SIAI-Marchetti SF.600TP

Cannon Ball Special
Heath

Canoe Plane
Curtiss

Canso
Boeing (Canada)
Canadair CL-1

Canuck
Avro (Canada) CF-100
Curtiss JN-4Can
Fleet (Canada)

"Cape Cod"
Bellanca J-300

Capone Helicopter
Wright (Howard)

Capronicino
Caproni Ca.100

Capstan
Slingsby T.49B

Captain (As: Captaine)
Robin

Car *(See: Aerocar, Air-Car, Air/Car, Aircar, AirRhoCar, Amphibian Air Car, Autocar, Avrocar, Boxcar, Convaircar, Flying Car, Rotocar, Sidecar, Sky Car, Sky-Car, Skycar)*

Carabao (ACV)
Bell SK-3

Carajá
EMBRAER
Neiva

Caravan
Cessna 208, 406, C-76

Caravelle (Caravel)
SNCASE SE 210

Carbon Dragon
Maupin

Cardinal *(See also: Super Cardinal)*
Cessna 177
Kentucky
St Louis

Careless
Tupolev Tu-154

Cargomaster
Douglas C-133

Caribou
Bell P-39
de Havilland (Canada) DHC-4

Carioca
CAP

Carley
Werkspoor

"Caroline"
Convair 240, 340 Convair-Liner

Carp (As: Karas)
PZL P.23, P.42, P.43

Carp (As: Karausche)
Fokker M 16E

Carrier Pigeon
Curtiss

Carrier Wing
Baynes

Cart
Tupolev Model 75, Tu-12

Carvair
Aviation Traders

Cash
Antonov An-28

Cassutt Special
National Aeronautics

Castor
Fizir

Cat *(See: Ag-Cat, Ag Bearcat, Bearcat, Black Cat, Bobcat, Hellcat, Hurricat, Jet-Cat, Super Ag-Cat, Super Tomcat, Supercat, "Texas Wildcat," Tigercat, Tomcat, "Wild Catfish," Wildcat)*

Cat
Antonov An-10

Catalina *(See also: Super-Catalina)*
Avid Aircraft
Boeing (Canada) PB2B-1
Consolidated OA-10, PBY

Catbird
Rutan Model 81

Catfish *(See: "Wild Catfish")*

"Cathedral"
Cody 1909 Biplane

Cavalier
Menzimer
Star

Cayuse
Hughes H-6

Cedar
Tachikawa Ki-17

Celerity
Mirage

Celstar
Celair GA-1

Centaur
Acme (Torrance)
Central (UK)
Central (Wichita, KS)
Longren

Centaur (As: Centuro)
Fiat G.55

Centaurus
Hawker F.18/37

Centennial
Found Model 100

Center-Drop Biplane
Dailey

Centurion
Century (MO)
Cessna 210

Centuro (Centaur)
Fiat G.55

"Century of Progress"
Lockheed Model 5

CenturyJet
Century Aerospace

Ceres
CAC

Chain Lightning
Lockheed XP-58

Chaka
Riga

Challenger *(See also: Solar Challenger)*
Canadair CL-600, CL-601, CL-604
Champion (Wisconsin)
Cook JC-1
Emsco
Fairchild KR-21, KR-31, KR-34
Mead
Quad City

Challenger Robin
Curtiss Robin

Chamois
Short S.3b

Champ
Bellanca

Champion
Aeronca 7
Avid Aircraft

Chancellor
Cessna 414A

Channel
Supermarine

Chanute
Wild-O'Meara

Chaos
Rans

Chaparral
Mooney

Chappedelaine
American Tool

Charabanc
Grahame-White Type 10

Charger
Convair
Marquart MA-5

Charger
Tupolev Tu-144

Charlie
Partenavia P66C

Charlotte
Breslau

Chaser
Aerial Arts

Cheekee Chipmunk
de Havilland (Canada)

Cheetah
Cheetah Light Aircraft

Chemet
Borel 1913 Hydravion

Cherokee
Hall (Stanley)
Piper PA-28

Cherokee Arrow
Piper PA-28R

Cherokee Challenger
Piper PA-28-180

Cherokee Cruiser
Piper PA-28-140

Cherokee Pathfinder
Piper PA-28-235

Cherokee Six
Piper PA-32-260

Cherokee Warrior
Piper PA-28-151

Cherry
BX-2

Cherry
Yokosuka H5Y

Cherry Blossom (As: Ohka)
Kugisho MXY7

"Cherry Blossom" (As: "Sakura")
Kawanishi K-12

Chesapeake
Vought V-156B-1

"Chester Special"
Falck

Chevalier (Knight)
Robin

Chevron
AMF

Chevy Bird
Waterman (W. D.)

Chevy Duck
Waterman (W. D.)

Cheyenne
Lockheed H-56
Piper PA-31T
Piper PA-42

Chiang Hung Seaplane
China Naval Air Establishment

Chic
Aerial Service Corp

"Chicago Javelin"
Locomotive

"Chicago"
Douglas World Cruiser

Chick
Aerial Service Corp

Chickasaw
Sikorsky (USA) H-19

Chief
Aeronca 11, 50, 65
Cessna DC-6A
Golden Eagle

"Chief Oshkosh"
Wittman

Chieh-Shou
AIDC PL-1B
Pazmany PL-1B

Ching-Kuo
AIDC

Chingolo
FMA

Chinook
ASAP
Birdman Enterprises
Boeing-Vertol H-47
Hermanspann

Chipmunk *(See also: Cheekee Chipmunk)*
de Havilland (Canada) DHC-1

Chiricahua
Applebay GA II
Pilatus UV-20

Chirri
Hispano HA-132-L

Chitral
Handley Page Clive Mk.I

Choctaw
Sikorsky (USA) H-34

Chorlito (Curlew)
IMPA

Chrysalis
MIT

Chrysanthemum *(See: White Chrysanthemum)*

Chum
Aeronca 12
Huntington (CT)

Chummy
Arrowing
Dayton Wright TA-3
Huntington (CT)
Taylor

Chummy Flyabout
Gallaudet Aircraft E-L-2

Chupirosa
Hegy

Churchill
Sopwith

Chuschin (Thrush)
Aer-Fabric AF-3

Cicada
Hall (Bob)

Cicogna (Stork)
Fiat BR.20

Cigale *(See: Super-Cigale)*

Cinema
Frankfort

Ciolkosz
PWS 46

Circuit
MacFie

Cirrus
Hüffer
Schempp-Hirth

Cirrus Moth
de Havilland D.H.60

Citabria
Bellanca
Champion (Texas)
Champion (Wisconsin)

Citabria Adventure
American Champion

Citabria Aurora
American Champion

Citabria Explorer
American Champion

Citabria Pro
Champion (Wisconsin)

Citation
Cessna, Cessna 500, 550, 551, 560, 650, 700, 750

Citationjet
Cessna 525

"City of Columbus"
Ford 5-AT-B

"City of New York"
Fairchild FC-2W2

"City of Peoria"
National Airways System Air-King

"City of Springfield"
Granville Z

"City of Tacoma"
Emsco

"City of Washington
Piper PA-12 Super Cruiser"

Clam
Ilyushin Il-18

Clank
Antonov An-30

Clara
Tachikawa Ki-70

"Clasina Madge"
Emsco

Classic *(See also: Kitfox Classic)*
Fisher FP-404
Taylorcraft

Classic
Ilyushin Il-62

Claude
Mitsubishi A5M4

Clear Sky Storm (As: Seiran)
Aichi M6A

Cleat
Tupolev Tu-114

Clen Antú (Sunray)
FMA

Cleveland
Curtiss SBC

Cline
Antonov An-32

Clinogyre
Odier-Bessiere

Clipper *(See also: Baby Clipper, Cub Clipper, Russia Clipper, Super Clipper, Tadpole Clipper)*
Boeing 314, C-40A
Cessna 303
Martin 130
Piper PA-16
Sikorsky (USA) S-40, S-42B
Spartan (UK)

Clive
Handley Page

Clobber
Yakovlev Yak-42

Clod
Antonov An-14

Cloud
Saunders-Roe A.19, A.29

Cloudboy
Stearman (1927) 6

Cloudbuster
Cloudbuster Ultralights

Cloudster
Aerocar Co of America
Douglas
Rearwin
Ryan
Ryson ST-100
deLackner

Club *(See also: Midgy-Club, Tampico Club)*
Jodel
Kucher
Robin
Robin

Club Cadet
Avro 638, 639, 640

Club Standard
Grob

Cmelák (Bumble Bee)
Zlin Z 37

Coach *(See also: Air Coach, Aircoach, Monocoach, Page Air Coach, Rota-Aircoach, Rotorcoach, Twin Monocoach)*
Butler
Lignel 46
Stearman (1927) CAB-1

Coach
Ilyushin Il-12

Coach (Berline)
Blériot XXIV

Coaler
Antonov An-72

Coastguarder
Hawker Siddeley HS.748

Cobham
Sopwith

Cobra *(See also: Airacobra)*
Advanced Aviation
Bell AH-1
Northrop P-530
Procaer F.400
PZL Bielsko SZD 36

Cobra (As: Petan)
Hughes H-64

Coccinelle (Ladybug)
SIPA 1000

Cochise
Beech T-42A

Cock
Colditz

Cock
Antonov An-22

Cockerel (As: Kogutek)
Zalewski W.Z.XI

Cockle
Short S.1

Codling
Yakovlev Yak-40

Coke
Antonov An-24
Xian Y-7

Cold Wind (As: Viima)
VL

Coléoptère
SNECMA

Colibrí (Hummingbird)
FMA

Collegian
Aeronca C-3

Collegiate
Porterfield Model 50
Timm K-90

Colmar
Breguet Bre.500

Colombe (Pigeon)
Ladougne

"Colorado, El" ("The Red One")
Borello

Colt
Piper PA-22-108
Weatherly-Campbell

Colt
Antonov An-2
Huabei Y-5

Columbia *(See also: "Baby Columbia")*
Lancair

"Columbine"
Lockheed VC-121A Constellation

Comanche *(See also: Twin Comanche)*
Boeing Sikorsky RAH-66
Eberhart
Piper PA-24

Combat Explorer
McDonnell Douglas Helicopter MD 900

Combat Scout
Bell 406CS

Combat Shadow
Lockheed HC-130N, HC-130P

Combat Talon
Lockheed MC-130E, MC-130H

Combination Wing
Alexander Eaglerock

Combowing Eaglerock
Alexander

Comet *(See also: Airacomet, "Continental Comet")*
B & F Technik
Cessna
de Havilland D.H.88, D.H.106
Hockaday CV-139
Ireland
Russell Convertible Monoplane

Comet (As: Komet)
Dornier
Kawasaki
Messerschmitt Me 163

Comet (As: Suisei)
Kugisho D4Y

Comet Robin
Curtiss Robin

Comète
Labaudieet et Puthet

Commander *(See also: Flight Commander)*
Air Command

Commando
Curtiss C-46, C-113, R5C

"Commando"
Consolidated (C-87) Liberator Mk.II

Commando Solo
Lockheed EC-130E

Commercial *(See also: Sport Commercial)*
Alco
Crawford (Venice, CA)
Laird ("Matty")
McCook
Meyers (C. W.)
Swallow (KS)
Thunderbird
Yackey

Commercialwing
Texas Aero

Commodore
Avro 641

Consolidated
Israel Aircraft Industries

Commuter *(See also: Super Commuter, Turbo-Commuter)*
Cessna 150L
Courtney CA-1
Douglas
Gulfstream Aerospace
Hiller UH-4
Keystone K-84
Lanier (Edward M.) Paraplane
Loening (Corp)
Moller XM-5

Comp Air
Aerocomp

Comp Monster
Aerocomp

Compass Call
Lockheed EC-130H

Compostela
Jodel

Concept 70
Berkshire

Concorde
Concorde

Concordia
Cunliffe-Owen

Condor *(See also: Akron Condor, Golden Condor, Gossamer Condor)*
Condor Aircraft

Condor
Akron
Curtiss, Curtiss B-2, C-30, CW-20T, R4C
Dittmar
Druine
Focke-Wulf Fw 200
Hannover
Rollason D.62
Schleicher
Schweizer SA 2-38A
TL Ultralight TL-132

Condor
Antonov An-124

Condor (As: Kondor)
Udet U 11

Condor Plus
TL Ultralight TL-232

Condor Wing
Burke (Carroll)

"Conejo"
Barrón

Conestoga
Budd RB-1
Skytrader ST1700

Conquest
Cessna 425, 441

"Conquest I"
Grumman F8F-2

"Conquistador del Cielo"
Granville, Miller, and De Lackner

Consort
Rock Island

"Constance"
Balmer

Constancia
FMA

Constellation
Lockheed C-69, C-121, Model 49

Constitution
Lockheed R6O

Consul
Airspeed

Continental
Bombardier

"Continental Comet"
Laird ("Matty") LC-D Speedwing

Convair-Jet
Convair 880

Convair-Liner
Convair 240

Convaircar
Convair 118
Hall (T. P.)

Convertiplane
Carlton
Leinweber-Curtiss
McDonnell XV-1
Reinke
Transcendental 2

Convertoplane
Herrick

Convoy Fighter
Seversky SEV-2PA

Cooker
Tupolev Tu-110

Cookpot
Tupolev Tu-124

Coot
Aerocar, Inc
Taylor (Molt)

Coot
Ilyushin Il-18

Cootie
Mummert
Waco

'Copter *(See: Little 'Copter)*

Cora
Fantasy Air

Cork
English Electric
Phoenix P.5

Cork
Yakovlev Yak-16

Cormoran (Cormorant)
SNCAC NC.211

Cornell
Aeronca PT-19
Fairchild PT-26

Coronado
Consolidated PB2Y
Convair 990

"Coronel Pringles"
Artigau

Corregidor
Consolidated P4Y

Corsair
Brewster F3A
Cessna 425
Goodyear FG
Historical Aircraft Corp
Vought A-7, AU, F4U, O-28, O2U, O3U, SU, V-50, V-65, V-66, V-70, V-80, V-90, V-93, V-97, V-99, V-135

Corsair Junior
Vought V-100

Corsaire (Corsair)
Breguet BR.11
Mauboussin

Corsaire (Corsair) Major
Mauboussin

Corse (Corsica)
SNCASO SO 95

Corsico
EMBRAER

Corvette
Aérospatiale SN 600

Cosmic Wind
Levier

Cosmopolitan
Canadair CL-66
Convair 540

Cossack
Antonov An-225

Cosy Classic
Co-Z

Cougar *(See also: Tri Cougar)*
CERVA CE.44
Grumman F9F
Wassmer CE.44

Counter Invader
Douglas A-26A B-26K
On Mark B-26K

Coupe *(See also: Aerocoupe, Air Coupe, Aircoupe, AirCoupe, Airmaster Coupe, Cub Coupe, Ercoupe, Junior Coupe, Midget Monocoupe, Mini Coupe, Monocoupe, Plymo-Coupe, Sioux Coupe Jr, Skycoupe, Sport Coupe, Sports Coupe)*
Curtiss CR-2, CW-19L, CW-19W
Dansaire
Driggs
Dunham
Eyerly
Kari-Keen
Kinner
Neilson
Swanson (Airplane) W-15

Coupé *(See also: Aerial Coupé)*
Bristol
Civilian Aircraft Co
Swallow (KS) Model C

Coupster
Alco

Courier *(See also: Hi-Vision Courier, Strato-Courier, Super Courier, Trigear Courier, Twin Courier)*
Airspeed
Alcor Aviation, Inc (TX)
Consolidated O-17
Crawford (Venice, CA)
Helio
Kinner
Rans

Courier (As: Kurier)
Focke-Wulf Fw 200

Courier (As: Kuryer)
Aeroprogress T-910

Courlis (Curlew)
SECAN S.U.C.10

Courtney Amphibian
Curtiss

Cowboy (As: El Boyero)
FMA

Cowboy (As: Gardian)
Dassault

Coyote *(See also: Hi-Nuski Coyote)*
Rans

Crane *(See also: Flying Crane)*
Cessna

Crane (As: Kranich)
DFS

Crane (As: Manazuru)
Kokusai Ku-7

Crane (As: Zaraw)
LWD

Crane Amphibian
Curtiss

Crate
Ilyushin Il-14

Crazy (Baka)
Kugisho MXY7

Credible Hawk
Sikorsky (USA) H-60

Creek
Yakovlev Yak-12

Crescent
Menefee

Crested Harpy
Sullivan (Corp) K-3

Cri-Cri (Cricket)
Colomban

Cri-Cri (Cricket) Major
CFA D.7

Crib
Yakovlev Yak-8

Cricket
Campbell
Zenair

Cricket (As: Cri-Cri)
Colomban

Cricket (As: Cri-Cri) Major
CFA D.7

Cricket (As: Cricri)
Salmson D.6

Cricket (As: Grille)
Krekel

Cricket (As: Sverchok)
Kuibyshev

Cricri (Cricket)
Salmson D.6

Criquet
Carlson

Criquet (Locust)
Croses
Morane-Saulnier

"Croix du Sud" ("Southern Cross")
Latécoère LAT 300

Cromarty
Short N.3

Cropmaster
Yeoman YA1

"Crosley Flea"
Crosley CF

Crossland Ace
Aviation Construction

Crow
B.A.T.

Crow
Yakovlev Yak-13

Crow (As: Wrona)
Kocjan

Croydon
G.A.L. Monospar ST-18

Cruisair
Bellanca

Cruisair Jr
Bellanca Junior

Cruisaire
Fernic T-9

Cruisemaster
Bellanca

Cruiser *(See also: Air Cruiser, Aircruiser, Alti Cruiser, Cherokee Cruiser, Cub Cruiser, Family Cruiser, Nine-Hour Cruiser, Sopwith Cruiser, Standard Cruiser, Stratocruiser, Super Cruiser, World Cruiser)*
Decatur T-W-1 Sopwith
Saunders-Roe/Percival A.24
Spartan (UK)

Cruizaire
Dunn

Crusader
Argonaut
Cessna T303
Gillis
Knepper
Short-Bristow
Vought F-8

Crusty
Tupolev Tu-134

Cruzair
Waterhouse

CU-Climber
Wiederkehr GHW-1

Cub *(See also: Tiger Cub)*
Piper J-3
Taylor

Cub
Antonov An-12
Shaanxi Y-8 (Yun-8)

Cub Clipper
Piper P-1

Cub Coupe
Piper J-4, J-5

Cub Cruiser
Piper J-5

Cub Flitfire
Piper J-3

Cub Special
Piper PA-11

Cub Sport
Piper J-3

Cub-Killer
Mooney

Cubaroo
Blackburn

CUBY
Aces High

Cuby Acro Trainer
Wag-Aero

Cuckoo
Sopwith

Cuff
Beriev Be-30, Be-32

Culex
Fisher

Cumulus
Reinhard
US Aviation

Cupid
British Aircraft Mfg

Curl
Antonov An-26

Curlew
Campbell

Curlew (As: Chorlito)
IMPA

Curlew (As: Courlis)
SECAN S.U.C.10

"Custom 800"
de Havilland D.H.104 Dove

Custombilt
Crown

Cutlass
Cessna 172
Vought F7U

Cutty Sark
Saunders-Roe A.17

D

Denhaut
Borel Aeroyacht

"Denison" Hydrofoil
Grumman

Denko (Bolt of Light)
Aichi S1A

DePeZe
Zalewski W.Z.VIII

"Der Dessauer"
Dessau

Der Jäger (Hunter)
White (E. Marshall)

Derby
de Havilland D.H.27

Derringer
Wing (Aircraft)

Desford
Reid and Sigrist

"Dessauer"
Dessau

Destroyer *(See also: "Aeroplane Destroyer")*
Douglas B-66, BTD
Vickers (UK) E.F.B.1

"Detroit Glider"
Gliders, Inc

"Detroit Gull"
Gliders, Inc

"Detroit News"
Lockheed Model 5

Detroiter
Stinson (Aircraft) SB-1, SM-1, SM-6

Deux-Ponts (Two Bridges)
Breguet Bre.76

Devastator
Douglas TBD

Devil *(See also: "Black Devil," Daredevil, Red Devil, Sea Devil)*
Aachen FVA-1

Devil Queller (Shoki)
Nakajima Ki-44

Devon *(See also: Sea Devon)*
de Havilland D.H.104

Diamant (Diamond)*(See also: Super Diamant)*
FFA
Piel

Diamond *(See also: Black Diamond, Diamant, Dimona)*
Mitsubishi MU-300

Diamond Star
Diamond

Dick
Seversky SEV-2PA-B3

Diesel Airsedan
Buhl Airsedan CA-3E

Dietz Special
General Aeronautical

Digby
Douglas B-18

Dimona (Diamond)
Hoffmann (Flugzeugbau) H 36

Dinah
Mitsubishi Ki-46

Dingo
Aero-Reek
Aeroric
de Havilland D.H.42

Dipper
Collins (Aero)

Discus *(See also: Duo Discus)*
Schempp-Hirth

Diskoplan
Sukhanov
Antonov

Disposalsyde
ADC Martinsyde A.D.C.1

Distributor Wing
Aerial Distributors

"Dixie Flyer"
Longren #1 Pusher

Dizzy Dog
Huff-Daland HD-7

Djinn (Genie)
SNCASO SO 1221

Dnepr
Abramov 48

Dog *(See: Dizzy Dog, Yellow Dog)*

Dog Ship
Huff-Daland AT-2

"Dogship"
Piasecki HRP

Dolphin
de Havilland D.H.92
Douglas
Douglas C-21, C-26, C-29, FP-1, FP-2, OA-3, OA-4
Sopwith

"Dolphin" Hydrofoil
Grumman

Dolphin (As: Dauphin)
Aérospatiale SA 360, SA 365

Dolphin (As: Delfin)
Aero (Vodochody) L 29
Czerwinski (Waclaw) W.W.S.3
Letov L.115

Dolphin (As: Delphin)
Dornier Do L

Dolphin (As: Haitun)
Harbin Z-9

Dominator
Consolidated B-32

Dominie
de Havilland D.H.89
Hawker Siddeley

Don
de Havilland D.H.93

Donald
Piel

Doncaster
de Havilland D.H.29

Donryu (Storm Dragon)
Nakajima Ki-49

"Doodle Bug"
Ryan X-1

Doodlebug
McDonnell

Doohickey
List

"Doolittle Hawk"
Curtiss Hawk 1

"Doolittle Special"
Seversky SEV-DS

Doppelraab Glider
Schempp-Hirth

Doppelsitzer (Two-Seater)
Kassel

Dormouse
de Havilland D.H.42

"Dorothy"
Bellanca 28-90 Flash

Dottie S
Bennett-Carter

Double Eagle
Amax
American Aerolights
British Aircraft Mfg

"Double Tractor"
Martin (J. V.)

"Double-Dirty"
Short S.46 Type Twin-Gnome Monoplane

Doubler
Myasishchev M-52D

Dove
de Havilland D.H.104
Sopwith

Dove (As: Ramier)
SNCAN 1101

Dove (As: Taube)*(See also: Steel Dove, Water Dove)*
Albatros
Aviatik
DFW
Etrich
Euler
Isobe
Jeannin
Kondor
Krieger
Lawrenz
Lohner
M-B
Roesner
Rumpler
Schwade
Siemens-Schuckert

Downtowner
Vought (C-142)

Drache (Kite)
Focke Achgelis Fa 223

Drachenflieger (Dragonfly)
Fahlbusch
Kress

Dracula
CPCA DK-10

Dragon *(See also: Carbon Dragon, Fire Dragon, Flying Dragon, Green Dragon, Puff the Magic Dragon, Sea Dragon, Storm Dragon)*
de Havilland D.H.84
Douglas B-23
North American B-21
Sopwith

Dragon (As: Draken)
Saab 35

Dragon Rapide
de Havilland D.H.89

Dragon Slayer (As: Toryu)
Kawasaki Ki-45

Dragonfly
Bowlus
Crouch-Bolas
de Havilland D.H.90
Glider Aircraft Corp
Martin-Handasyde 4.B
Nelson (Aircraft) BB-1
Rotor-Craft XR-11
Ryan O-51
Slip Stream
Stiles
Viking (Aircraft)
Westland
Wing (Aeronautical)

Dragonfly (As: Drachenflieger)
Fahlbusch
Kress

Dragonfly (As: Libel)
Hollandair HA-001

Dragonfly (As: Libelle)
Dornier Do 12, Do A

Dragonfly (As: Libellula)
Manzolini
Miles M.35, M.39, M.42, M.43 M.63

"Dragonfly" (As: "Libellule")
Blériot VI
Farcot
Hanriot

Dragonfly (As: Qingting)
Shijiazhuang

Dragonfly (As: Tombo)
Mitsubishi

Dragonfly (As: Vazka)
Zilina

"Dragonfly Snipe"
Sopwith

Dragster
Cosmos

Drake
Goodyear GA-22

"Drake Bullet"
Seversky SEV-S2

Draken (Dragon)
Saab 35

Dreadnaught *(See also: Flying Dreadnaught)*
Stástík
Westland

Dream
NAC

Dream (As: Mriya)
Antonov An-225

Dream Machine
Buckeye

Drifter
Austflight
Maxair

Dromader (Dromedary)
PZL Mielec M18

Drover
de Havilland (Australia) DHA.3

Drozd (Thrush)
ITS-7

Duce
Bakeng

Duchess
Beech
Saunders-Roe

Duck *(See also: Chevy Duck, "Dalton's Duck," Flying Duck, Monoduck)*
Goodyear
Grumman JF, J2F; OA-12
Loening (Corp)
Paulhan

Duck (As: Ente)
Focke-Wulf F 19a
Reissner

Duck (As: Utka)
MiG MiG-8
Tokaev

Duck Amphibian
Applegate

Duckling
Curtiss
International Aviation
Loening (Grover)

"Dud, The"
Short S.26 Pusher Biplane

Duet
Dubna Z-6

Duicker
Hawker

Duigan Biplane
Avro

Duke
Beech

Dumbo
GCA 2
Supermarine Type 322

Dunkirk Fighter
Curtiss HA

Dunstable Dart
Dart (UK)

Duo Discus
Schempp-Hirth

Duo-6
Alcor C.6.1

Durandal
SNCASE SE 212

Duster *(See also: Big Duster)*
BJ
Huff-Daland

DX'er
Champion (Wisconsin)

DYNAIRSHIP
Aereon

Dynamite
BJ

Dziaba
Malinowski

E

E-Z Flyer
Blue Yonder

Eagle *(See also: Agile Eagle, Blue Eagle, Double Eagle, Golden Eagle, Lone Eagle, Pave Eagle, Rhine Eagle, Royal Eagle, Sea Eagle, Silver Eagle, Streak Eagle, Strike Eagle, White Eagle Monoplane)*
Aeronca 10
Amax
Aviat
British Aircraft Mfg
Buckeye
Christen
Crane
Curtiss
GM (Fisher) P-75
McDonnell Douglas F-15
MIT
Mooney
Sikorsky (USA) S-76
Slingsby T.42
Windecker

Eagle Eyes (As: Washi)
Mitsubishi

Eagle Glider Rider
American Aerolights

Eaglerock
Alexander

Eaglet
American Eagle A-230
American Eaglecraft
Guttman
Itoh N-62

Eaglet (As: Aiglon)
Caudron C.600

Eaglet (As: Orlik)
Kocjan

Eagle's Nest (As: Adlershof)
Albatros L74

Early Bird
Huff-Daland HD-1
Waterman (W. D.)

"Early Bird"
Lockheed Model 9-D

Eastchurch Kitten
Port Victoria PV.8

Eastern Caravan (As: Vostochnyi Karavan)
Aeroprogress T-205

Eastern Sea (As: Tokai)
Kyushu Q1W1

Easy Rider
UFM

Echo
Costruzioni Aeronautiche Tecnam P92

Eclectic Transport Flying Boat
Phoenix P.7

Ecureuil (Squirrel)
Aérospatiale AS 350

Eddyo
Young (Edward W.)

Edelkadett (Senior Cadet)
Heinkel He 72B-3

Edelweiss
Duruble
Siren C-30S

Eggers Monoplane
Rumpler

Egrett
E-Systems-Grob
Grob

"Eight Ball"
Keith Rider R-6

Eightster
Granville C-8

Electra
Lockheed Model 10, Model 188

Electra Junior
Lockheed Model 12

Elephant
Martinsyde G.102

Éléphant Joyeux (Happy Elephant)
SNCASE S.55

Elf
Parnall

Elf (As: Norelfe)
SNCAN 1750

Elfe (Brownie)
Pfenninger

"Elida"
Kassel 28

Elizabethan
Airspeed

Elk (As: Los)
PZL P.37

Elli
Junkers Ju EF 126

Ello
Messerschmitt M 17

"Elmendorf Special"
Keith Rider R-5

"Elsie"
Weiss (José)

Elster-B (Magpie-B)
Pützer

Élytroplan (l'Élytroplan)
de Rouge

Emerald (As: Emeraude)
Piel
Scintex

Emerald (As: Smaragd)
Binder

Emeraude (Emerald) *(See also: Super Emeraude)*
Piel
Scintex

Emily
Kawanishi H8K

Empire Boat
Short S.23

Empire Type
Nieuport (UK)

Empress
MacFie

"Encanto, El"
Goddard/Imperial

Encore
Mooney

Enfin, le
Alkan

Enforcer
Piper PA-48

"Enna Jettick"
Bellanca K

"Enola Gay"
Boeing

Ensign
All American Aircraft
Armstrong Whitworth A.W.27

Ente (Duck)
Focke-Wulf F 19a
Reissner

Envoy
Airspeed
Fairchild Dornier
Interstate L-6
Kinner

Eole (l'Eole)(Aeolus)
Ader

Epervier (Sparrowhawk)
Renard

Epsilon
Socata TB30

Era
Iniziative Industriali Italiane Raw Arrow

Ercoupe
Erco

"Ericka"
Erickson

Erka
Raab-Katzenstein RK 25

"Ern, The"
Yackey Transport

"Errant"
Day (Chas)

Espadon (Swordfish)
SNCASO SO 6020

"Essor" (As: "l'Essor") ("Soaring")
Fabre

Estol
Vertak S-220

Estrellita (Little Star)
Williams (Aircraft Design)

Estudiantil (Scholar)
Marichal

Eta Beta
GCA 3

Étendard (Banner)
Dassault

Europa
Wassmer WA.52

"Europa"
Junkers W 33b

Eurostar
Evektor EV97

Evader
Skytrader ST1700 MD

Eve
Mitsubishi

Even-Keel Airplane
Lake (Aero)

"Evergreen"
Montgomery (John)

Examination (As: Prüfling)
Lippisch

"Example" (As: "Musterle")
Grunau

Excalibre
Lockheed Model 144

Excalibur
Swearingen

Excellence
Jodel
Robin

Exec
Rotorway

Executive *(See also: Turbo-Executive, Vega Executive)*
Aerostar 220
Bennett (TX)

Hughes 500E
Lanier (E. M.) Paraplane
Mooney
Shorts P.D.14
Spartan (OK)
Stits-Besler SA-4A

Executive Skyknight
Cessna 320

Executive Transport
Seversky
Wilford

Exocet
Iniziative Industriali Italiane Sky Arrow

Expediter *(See: Metro Expediter)*

Expeditor
Beech Beech C-45, JRB

Explorer *(See also: Citabria Explorer, Combat Explorer)*
Abrams
AEA (Australia)
de Chevigny
Lockheed Model 4, Model 7
McDonnell Douglas Helicopter MD 900
Patchen TSC-2
Thurston TSC-2

"Explorer I"
Boeing (Canada) Canso

Express *(See also: Air Express, Global Express, Honeymoon Express, Liberator Express, "Skyway Express")*
Avtek 419
Wheeler (Aircraft Co)
Willard
Woodson Type 2A

"Express"
Lohner 10.23

Express Airmail
Grahame-White GWE.6

Extender
McDonnell Douglas KC-10

Ezüst Nyíl (Silver Arrow)
Weiss Manfred WM.23

F

Faceplate
MiG Ye-2A

Fachiro
Partenavia P57

Fagot
MiG MiG-15
Shenyang

Faithless
MiG MiG-23PD STOL

Falcao
Avibras

Falco
Sequoia

Falco (Falcon, Hawk)
Aviamilano
Laverda F.8
Fiat CR.42
Reggiane RE 2000

Falcon *(See also: Falco, Falke, Falken, Fighting Falcon, Freedom Falcon, Mountain Falcon, Peregrine Falcon, Reed Falcon, Tandem Falcon)*
American Aerolights
ASC
Buckeye
Curtiss
Curtiss A-3, BT-4, CW-22, F8C, O-1, O-11, O-12, O-13, O-16, O-18, O-26, O-39, OC, SNC
Dassault Mystère
Friesley
Maranda
Miles M.3
Slingsby T.1, T.24
Williams (Aircraft Design) W-18

"Falcon"
Bowlus-duPont 1-S-2100

Falcon (As: Halcon)
Lavelli

Falcon (As: Sokol)
Aeroprogress T-401
Mráz M-1

Falcon (As: Sokól)
Kocjan
PWS 42
PZL Swidnik W-3

Falcon (As: Stösser)
Focke-Wulf Fw 56

Falcon Major
Miles M.3

Falcon Special
Lefevers

Falconer
Wendt (NY) W-1

Falke (Falcon)
Dornier Do H
Focke-Wulf Fw 43
Focke-Wulf Fw 187
Kahnt
Scheibe SF-25
Slingsby T.61A

Falken (Falcon)
Svenska

Family Air
LOK

Family Cruiser
Piper PA-14

"Family Plane"
Kaiser-Hammond Y

Fan Ranger
RFB
Rockwell-MBB

Fang
Northrop N-102

Fang
Lavochkin La-11

FanStar
American Aviation Industries

Fantail
Lavochkin La-15

Fantan
Nanchang Q-5

Fantôme (Ghost)
Fairey

Fantrainer
RFB

Farewell (As: Supervale)
Magni

Farewell (As: Vale)
Magni

Farfadet (Goblin)
SNCASO SO 1310

Fargo
MiG MiG-9

Farmer
MiG MiG-19
Shenyang

Fastback
Capella

"Fat Albert"
Lockheed KC-130F

Fatman
Gippsland GA-200

Favorit (Favorite)
Aviatehnologia

Favorite
Heath

Fawn
Fairey
Fleet (Canada)

Fazan (Pheasant)
Caproni Bulgaria

Feather
Heath

Feather
Yakovlev Yak-15, Yak-17

"Feldberg"
Weltensegler

Fencer
Sukhoi Su-19 Su-24

Féroce (Savage)
Fairey

Ferret
Fairey

Ferry
Airspeed

Fiddler
Tupolev Tu-128

"Field Kitchen"
Short S.47 Type Triple-Tractor Biplane

"Field Observation Post"
G.A.L. 47

Fieldmaster
Croplease
EPA
NDN
Yeoman YA5

Fighter
Bristol
Curtiss
Engineering Division
Fairey

Fighter (As: Baroudeur)
SNCASE SE 5000

Fighting Falcon
General Dynamics F-16

Fin
Lavochkin La-7

Final Effort (As: Tobi)
Mitsubishi

Finback
Shenyang

Finch
Fleet (Canada)

Fire Dragon (As: Karyu)
Nakajima Ki-201

Fireball
Ryan FR

Firebar
Yakovlev Yak-28P

Firebird
Kalinin K-12

Firebrand
Blackburn B-37

Firecracker *(See also: Turbo Firecracker)*
Hunting
NDN NDN-1
Norman
Parker (Francis)

"Firecracker"
Schoenfeldt Special

Firecrest
Blackburn B-48

Firefly
Capital Copter
Fairey
Slingsby T.67

Firefly (As: Luciole)
Caudron C.270

Firemaster
Croplease
EPA

Fishbed
MiG MiG-21
Xian J-7

Fishpot
Sukhoi Su-9 Su-11

Fitter
Sukhoi Su-7, Su-17, Su-20, Su-22

Five
Avro 619

Fizir Nastaun (Basic Trainer)
Fizir FN

FLAC
Striplin

Flaggship
Flagg

Flagon
Sukhoi Su-15

Flagship *(See: Turbo Flagship)*

"Flagstaff"
Grumman PG(H)-1

"Flak Bait"
Martin B-26B Marauder

Flamant (Flamingo)
Dassault

Flaming (Flamingo)
PZL Warsawa-Okecie PZL-105L

Flamingo *(See also: Flamant, Flaming)*
Aeroprogress T-433
BFW (1926)
de Havilland D.H.95
Metal
Simmering-Graz-Pauker M222
Udet U 12

Flandre
Borel-Boccaccio

Flanker
Sukhoi Su-27

Flarecraft
Ground Effect Craft Corp

Flash *(See also: "Green Flash," Thunderflash)*
Bellanca 28-90
Delgado
Tilbury-Fundy
Wolff

Flashlight
Yakovlev Yak-25, Yak-27

Flea *(See: American Flea Ship, "Crosley Flea," Flying Flea)*

Flea (As: Floh)
DFW

Fléchair
Payen Pa 112

Fledermausfugelapparat (Bat Apparatus)
Lilienthal (Otto)

Fledge
Manta

Fledgling
Curtiss, Curtiss AT-9, N2C
Fairchild VZ-5
Manta
Pterodactyl

Fledgling Guardsman
Curtiss

Fledgling Jr
Curtiss

Fleep
Ryan V-8

"Fleet Follower"
G.A.L. 38

Fleet Shadower
Airspeed

Fleetwing
Fairey
Nelson/Driscoll
Pitcairn

Fleuret (Foil/Sword)
Morane-Saulnier

Flex Wing
Ryan

Flexwing
Waterman (W. D.)

"Fliegende Bierdeckel" (Flying Beertray)
Sack AS 6

Fliegender Jeep (Flying Jeep)
Bölkow/MBB

Flight Commander
Swallow (KS)

Flightsail
Flight Dynamics

Flightstar
Flight Designs
Pioneer

Flipper
MiG Ye-152A

Flite Liner
Piper PA-28-140

Flittermouse
Dart (UK)

Flitzer
Bell Aeromarine

Fliver
Bolte

Fliverplane
Mattley

Flivver *(See also: Flying Flivver)*
Eastman
Ford
Mooney
Mooney (Albert)

Flivver Plane
Peterson (Albin)

Flogger
MiG MiG-23, MiG-27

Floh (Flea)
DFW

Flora
Yakovlev Yak-23

"Floyd Bennett"
Ford 4-AT-B

"Floyd Bennett Jr"
San Francisco Aero Club

Flugi
de Coster

Flugschoner
Oertz W 6

Flurry (As: Vihuri)
VL

Flut-R-Bug
Stits SA-5A

Fly *(See: Mono-Fly)*

Fly Baby
Bowers

Fly Cycle
Farnham

Flyabout *(See also: Chummy Flyabout)*
Alexander Eaglerock D-1, D-2
Howard (Benny)
Porterfield Model 35
Swanson (Carl)

Flycatcher
Fairey

"Flycycle"
Lane (Dick)

Flyer
Airbirde
Avid Aircraft
Kolb
Mathewson
Wright (Brothers)

Flying Arrow
Mix

Flying Automobile
Southern (TX)

"Flying Auto"
Michel

Flying Banana
Piasecki HRP

Flying Bathtub
Ramsey

Flying Bedstead
Rolls Royce

"Flying Beertray: (As: "Fliegende Bierdeckel")
Sack AS 6

Flying Bicycle
Juiseux

"Flying Bicycle"
Baudot

Flying Boxcar
Fairchild C-119

Flying Car
Convair 116

Flying Carpet
Wittman W-8

"Flying Class-Room" (As: "Aula Volante")
Fiat G.12 CA

Flyin' Dutchman
Ibbs

Flying Crane
Hughes H-17, H-28

"Flying Donut"
Myers Annular Multiplane

Flying Dragon
Herring-Burgess

Flying Dragon (As: Hiryu)
Mitsubishi Ki-67

Flying Dreadnaught
Sikorsky (USA) PBS

Flying Duck
AVCO Lycoming

"Flying Dutchman"
Bucher Special
Szekely SR-3

Flying Fish
Burgess A
Detroit Boat
Farman (Henri)
Herring-Burgess

"Flying Fish"
Curtiss Flying-Boat No.2

Flying Flea (As: Pou du Ciel)
Ojea
Mignet

Flying Flivver
Nelsch
Storms

Flying Fortress
Boeing B-17

Flying Jeep
Chrysler VZ-6

Flying Jeep (As: Fliegender Jeep)
Bölkow/MBB

Flying Life Boat
Fokker PJ

"Flying Mattress"
M. L. Utility Aircraft

"Flying Pancake"
Vought V-173

Flying Plank
Backstrom

Flying Platform
Hiller VZ-1

Flying Rail
Aerosport

Flying Ram
Northrop P-79

Flying Run-About
Clutton

"Flying Saucer"
Lee-Richards

Flying Tank (As: Kryl'yatanka)
Antonov

Flying Teatray
Grahame-White Type 14

Flying Yacht
Eastman
Loening (Corp)

Flymobil
Umbaugh U-18

Flyworm
Maiwurm

"Flyying Motorcycle"
VAT

Foil (As: Fleuret)
Morane-Saulnier

Foka
PZL Bielsko SZD 24C

Folder Seaplane
Sopwith Admiralty Type 807

Folgore (Thunderbolt)
Macchi M.C.202

"Foo, The"
Folkerts

Foo Fighter
Stewart (Donald)

Ford Pinto Auto-plane
Smolinski

Forger
Yakovlev Yak-38

Formation (Zueno)
Vakhmistrov

Forssman
Siemens-Schuckert

Fort
Fleet (Canada)

Fortress
Boeing B-17

Fortress (As: Pucara)
FMA

"Fortress Roland" (As: "Roland Festung")
Eisenlohr

Forwarder
Fairchild UC-61

Foto
Woodson Type 1A

Foursome
Ryan C 1

Fourster
Granville C-4

Fox *(See also: Jet Fox, Kitfox, Mono-Fox, Nightfox, Rofox, Seafox)*
Fairey
MDM-1

Fox Moth
de Havilland D.H.83

Foxbat
MiG MiG-25BM

Foxhound
MiG MiG-31

Foxtrot
DFL

Fragrance (As: Kaoru-Go)
Shirato

Frajír
Zlin Z 32

"France"
Bernard 191 T
Couzinet 1928 Hydravion

Frances
Kugisho P1Y

Frank
Nakajima Ki-84
Yakovlev Yak-9

"Frankfurt/M"
Schleicher

Fre-Wing
Cornelius

Freak Boat
Curtiss

FRED
Clutton

Free Spirit
Cabrinha
CEI

Freedom
Brutsche

Freedom Falcon
Champion (Ken)

Freedom Fighter
Northrop F-5

Freehand
Yakovlev Yak-36

Freelance
NDN

Freestyle
Yakovlev Yak-41

Fregat
Aeroprogress T-130, T-230

Frégate (Frigatebird)
Caudron C.480
de Lesseps (Robert)
Max Holste

Freighter
Bristol
Fleet (Canada)
Shorts P.D.15

Frelon (Hornet)
SNCASE SE 3200

Fremantle
Fairey

"French Quarter Special"
Levier

Fresco
MiG MiG-17
Shenyang

Fresh
TL Ultralight TL-532

Fresh I (ACV)
Boeing

Freshman
d'Apuzzo D-200

Friendship
Fairchild F-27 FH-227
Fokker C-31, F.27

Frigatebird (As: Frégate)
Caudron C.480
de Lesseps (Robert)
Max Holste

Fritz
Lavochkin La-9

Frog (As: Zaba)
Czerwinski (Waclaw)
W.W.S.2

Frogfoot
Sukhoi Su-25

Froggy (As: Zabus)
Jach
Naleszkiewicz J.N.1

Fugaku (Mt Fuji)
Nakajima G10N

"Fuji-Go"
Itoh Emi 16

Fulcrum
MiG MiG-29

Fulgur
Breguet Bre.47

Fulmar
Fairey

Fury (See also: Sea Fury)
Felixstowe
Hawker
North American FJ

Futura
Learjet

G

G/B Special
Beets

Gadfly
Glenny and Henderson

Gadfly (As: Taon)
Breguet Bre.1001

Gafanhoto
Instituto de Pesquisas Tecnologicas IPT-1

Gafhawk
Hawk

Galánka
Zlin Z 124

Galaxy
Lockheed C-5, L-500

Gale (As: Hayate)
Nakajima Ki-84

Gale (As: Wicher)
PZL.44

Galeb (Seagull)
Soko G2-A

"Galileo"
Convair 990

Gambit
Gloster

Gamecock
Gloster

Gamma
General Aeroplane
Northrop
Pomilio

Ganagoble
Lobet-de-Rouvray

Gander
Kokusai Ku-8

Gannet
Fairey
Gloster

Ganymede
Grahame-White

Gapa
DWLKK

Garaix/ACR Monoplane
Alexandre

Gardian (Cowboy)
Dassault

Gauntlet
Gloster

Gawron
PZL Warsawa-Okecie PZL-101

Gazelle
Aérospatiale SA 340, SA 341, SA 342
de Havilland D.H.15
Lanier (E. M.) Paraplane

Gebirgsflieger (Mountain Flyer)
Lohner

"Geheimrat"
Darmstadt

Gekko (Moonlight)
Nakajima J1N1-S

Gelber Hund (Yellow Dog)
Euler

"Gelitas, le" (Mechanical Bird)
Gellet

GEM
Princeton University

Gémeaux (Gemini)
Fouga CM 88-R

Gemini
Miles M.65
Zenair

Genesis
Group Genesis
Slip Stream

Genet Moth
de Havilland D.H.60

Genie (As: Djinn)
SNCASO SO 1221

"Gennariello"
SIAI S.16*ter*

Gentleman
Comte A.C.4

Geodetic
Thalman

George
Kawanishi N1K1-J, N1K2-J

Georgia
Vultee TBV

Gerfaut (Gyrfalcon)
SFECMAS 1402
SNCAN 1405

Ghibli
Caproni Ca.309

Ghost (As: Fantôme)(See also: Sky Ghost, White Ghost)
Fairey

Giant (See also: Gigant, Jolly Green Giant)
Kennedy
Supermarine Type 179

Giant Moth
de Havilland D.H.61

Gigant (Giant)
Messerschmitt Me 321

Giles
Akrotech

"Gilmore the Record Breaker"
Keith Rider R-3

Ginga (Milky Way)
Kugisho P1Y

Gipsy Moth
de Havilland D.H.60

Giravion
SNCASO

Giroptére
de Chappedelaine

Gitterschwanz (Lattice Tail)
DFW

Gizmo
Goodyear GA-400

Gladiator (See also: Sea Gladiator)
Gloster

"Glamorous Glennis"
Bell X-1

Glasair
Stoddard-Hamilton

Glass Goose
Quikkit

Glastar
Stoddard-Hamilton

Glen
Kugisho E14Y1

Glenmont Landau Sedan
Warren & Montijo W.M.1

Global Express
Bombardier

Global Trainer
Global GT-3

Globemaster
Douglas C-74, C-124
McDonnell Douglas C-17

"Glomb"
Pratt Read LBE-1

Gluhareff
Vought

Gnat
Folland
Gnat (As: Komar)
Kocjan

Gnatsnapper
Gloster

Gnatsum
Jurca MJ-7

"Gnome Sandwich"
Short S.27 Type Tandem Twin Biplane

Gnu
Sopwith

Goblin
Grumman G-23
McDonnell F-85

Goblin (As: Farfadet)
SNCASO SO 1310

Goblin (As: Kobold)
Albatros L79

"Goebel Special"
Cessna CPW-6

Goeland (Gull, Seagull)
Fabre

Goéland (Gull, Seagull)
Caudron C.440

Gold Bug
Curtiss No.1

Gold Duster
Goldwing

Goldbug
Clem (John)

Goldcrest
Wren (UK)

Golden Arrow
Convair 880

Golden Condor
Sorrell

Golden Eagle
Cessna 421

"Golden Eagle"
Lockheed Model 1

Golden Flier
Curtiss No.1

Golden Hawk
DSK

Goldfinch
Gloster

Goldfinch (As: Stieglitz)
Focke-Wulf Fw 44

Goldie
Goldhammer

Goldwing
Goldwing Ltd

Golf
Costruzioni Aeronautiche Tecnam P96

Goliath
Farman F.60 F.169

Golondrina (Swallow)
Mira

Gomhuria
Heliopolis

Goon
Chester (Art)

Goose (See also: Glass Goose, "Spruce Goose," Super Goose, Turbo Goose, Wild Goose)
Grumman G-21, J3F, JRF, OA-9, OA-13
McKinnon

Goral
Gloster

Gorcock
Gloster

Gordon
Fairey

Gordon England
Bristol

Goring
Gloster

Goshawk
Curtiss F11C-2, Hawk II Model 35
McDonnell Douglas T-45
Nieuport (UK)

Goshawk (As: Astore)
Partenavia P48B

Goshawk (As: Habicht)
DFS
Focke-Wulf A 20, A 28

Gosling Racer
Waterman (Aircraft)

Gosport
Avro 504R

Gossamer Albatross
MacCready

Gossamer Condor
MacCready

Gossamer Penguin
MacCready

Gouverneur (Governor)
SNCASE SE 3130 Alouette

Governor *(See also: Alouette Gouverneur)*
Huntington (CT)

Grace
Aichi B7A

Grain Griffin
Port Victoria

Grain Kitten
Port Victoria PV.7

Grand
Sikorsky (Igor) S-21

Grand Commander
Aero Commander 680FL

Grand SIM
Rogozarski

Grand Tourisme
Jodel

Grasmücke (Meadow Warbler)
Raab-Katzenstein RK 9

Grasshopper
Slingsby T.38
Sopwith

Grasshopper (As: Belalang)
Angkatan

Grasshopper (As: Sauterelle)
Potez 600 T2

Gratch (Rook)
Aeroprogress T-101

"Great Baltic" (As: "Bol'shoi Bal'tisky")
Sikorsky (Igor) S-21

Great Lakes Tourer
Martin T Hydro

Grebe
Gloster

"Greek" Seaplane
Sopwith Admiralty Type 880

Greek Tourer
Bristol

Green Dragon
Johns

"Green Flash"
Bellanca J

Green Post (As: Grüne Post)
Lippisch

"Green River Whiskey"
Chamberlin C-82

Greif (Griffin)
Dornier Do G
Heinkel He 177, He 277

Greyhound
Austin
Grumman C-2
Northrop Grumman

Greyhound (As: Veltro)
Macchi M.C.205

Grief (Condor)
Hannover

Griffin (See also: Grain Griffin)
Canada ARV

Griffin (As: Griffon)
SNCAN 1500

Griffon
Aeroprogress T-204
Agusta AB 412
Interplane

Griffon (Griffin)
SNCAN 1500

Grifo
Ambrosini

Grifone
SAI (Italy) SAI.10

Grille (Cricket)
Krekel

Gripen
Saab 39

Grizzly
Beech A-38
Rutan Model 72

Grognard (Grumbler)
SNCASE SE 2415

Ground Hog
Lawson (Alfred)

Grouse
Gloster

Growth Hawk
Sikorsky (USA) S-92IU

Grunau Baby
Schleicher
Schneider (Edmund)
Slingsby T.5

Grüne Post (Green Post)
Lippisch

Guan
Gloster

Guanabara
Centro Técnico de Aeronáutica

Guarani
FMA

Guardian
Airmaster A-1200
Dassault U-25
Grumman AF

"Guardian"
Sikorsky (USA) S-37-B

Guardsman
Seversky AT-12

"Guba"
Consolidated (PBY) 28 Catalina

Guépard (Cheetah)
CERVA CE.43
Wassmer CE.43

Guerrier (Warrior)
Socata Rallye 235G

Gufo
Meteor

Gugnunc
Handley Page H.P.39

Guión (Royal Standard)
Hispano HA-231

"Gulfhawk"
Curtiss Hawk 1A

"Gulfhawk II"
Grumman G-22

"Gulfhawk III"
Grumman G-32

"Gulfhawk 4"
Grumman G-58A

Gulfstream
Grumman G-159, G-1159
Gulfstream Aerospace

Gull *(See also: Detroit Gull, Kirby Gull, Mew Gull, Sea Gull, Seagull, Super-Gull, Vega Gull)*
Bonney
Gnosspelius
Kirby
Percival
Slingsby T.15, T.25, T.32
White (Buffalo) Aircraft

Gull (As: Goeland)
Fabre

Gull (As: Mewa)
Bilski
Grzeszczyk & Kocjan
L.W.S.3
PZL Mielec M20

Gull (As: Möwe)
Dittmar

Gull Amphibian
Royal (Milwaukee)

Gull Glider
Pilcher

Gunbus *(See also: Streamline Gunbus)*
Burgess O
Sopwith Admiralty Type 806
Vickers (UK) F.B.5, E.F.B.5
Vickers (UK)

Gunner
Fairchild AT-14, AT-21

Guppy
Aero Spacelines
Sorrell SNS-2

Gurnard
Short S.10

Gus
Nakajima

Gusty
Limbach

Gypsy Moth
Moth

Gyrfalcon (As: Gerfaut)
SFECMAS 1402
SNCAN 1405

Gyro-Copter
Bensen

Gyro-Glider
Bensen B-8, X-25B

Gyrodyne
Fairey
Gyrodyne

Gyronaut
Gyronautics

Gyropter
Lay (Charles E.)

Gyroscope
Lataste

Gzhelka (Little Gzhel)
Alfa-M A-211

H

"H & H Special"
Halpin & Huf

Ha-go
Yokosuka

Habicht (Goshawk)
DFS
Focke-Wulf A 20, A 28

Hachi (Bee)
Nippon

Hadrian
Waco CG-4

Haefelin Monoplane
Rumpler

Haifisch (Shark)
LFG Roland D.I

Haig
Waco CG-4

Haitun (Dolphin)
Harbin Z-9

Haiyan (Petrel)
Nanchang

Halcon (Falcon)
Lavelli

Halcón
CASA
Enaer

Halcyon
Hampshire Aero Club

Halcyon (As: Alcyon)
Morane-Saulnier

Halifax
Handley Page

Hallore
Halle Fh 104
Siebel Fh 104

Halo
Mil Mi-26

Halton
Handley Page

Hamble Baby
Fairey
Parnall

Hamilcar
G.A.L. 49, 58

Hamilton
Handley Page

Hamlet
Handley Page H.P.32

Hammerhead Shark (As: Maillet)
Fouga

Hamp
Mitsubishi A6M3

Hampden
Handley Page

Hampstead
Handley Page

Handcross
Handley Page

Handley Sport
Ricks

Hangwind
Lippisch

Hanley
Handley Page

Hansa
HFB
National Aerospace Labs
Taneja

Hanscat
Bougie

Hap
Mitsubishi A6M3

Happy Elephant (Éléphant Joyeux)
SNCASE S.55

Harbinger
Shenstone

Hardy
Hawker

Hare
Handley Page

Hare
Mil Mi-1

Harke
Mil Mi-10

Harp
Kamov Ka-25

Harpoon
Lockheed PV

Harpy *(See: Crested Harpy)*

Harrier *(See also: Sea Harrier)*
British Aerospace
Hawker
Hawker Siddeley
McDonnell Douglas AV-8

Harrow
Handley Page H.P.31, H.P.54

Hart
Hawker

Hart Scout
Vickers (UK) F.B.16

Hartbee
Hawker

Hartbees
Hawker

Hartebeeste
Hawker

"Harukaze" ("Spring Breeze")
Kawanishi K-6

Harvard
Beech T-6A
Noorduyn
North American

"Hasten slowly" (As: "Spiesz sie powoli")
Czechowski

Hastings
Handley Page

Have Blue
Lockheed

Havoc
Douglas A-20

Havoc
Mil Mi-28

Hawcon
Miles M.6

Hawfinch
Hawker

Hawk *(See also: Baby Hawk, Black Hawk, Blackhawk, Credible Hawk, Dale Hawk, "Doolittle Hawk,"*

Gafhawk, Golden Hawk, Goshawk, Growth Hawk, "Gulfhawk," Hunting Hawk, Jayhawk, JetHawk, Junior Hawk, Kitty Hawk, Kittyhawk, Minihawk, Mohawk, Nieuhawk, Night Hawk, Nighthawk, Ocean Hawk, Pacific Hawk, Pave Hawk, Redhawk, Rescue Hawk, Rhine Sparrow Hawk, Sea Hawk, Sea-Hawk, Seahawk, Skeeter Hawk, Sky Hawk, Skyhawk, Sparrow Hawk, "Spirit of Kitty Hawk," Super Skyhawk, Tomahawk, Treasure Hawk, Warhawk, White Hawk, Wichawk)
Aero Composite
British Aerospace
Cessna R172K
CGS Aviation
Curtiss, Curtiss F6C, F11C, P-1, P-2, P-3, P-5, P-6, P-11, P-17, P-20, P-21, P-22, P-23, P-36, PW-8
DSK
Haller-Hirth
Haufe
Maranda
Miles M.2
Pacific

Hawk (As: Falco)
Fiat CR.42
Reggiane RE 2000

Hawk (As: Jastreb)
Soko J1

Hawk (As: Jastrzab)
PZL P.50

Hawk (As: Taka)
Mitsubishi

Hawk Glider
Pilcher

Hawk Major
Miles M.2

Hawk Moth
de Havilland D.H.75

Hawk Trainer
Miles M.14

Hawk-Pshaw
Schroeder

Hawkeye
Grumman E-2

Hayabusa (Peregrine Falcon)
Manko MT-1
Mitsubishi
Nakajima Ki-43

Hayate (Gale)
Nakajima Ki-84

Hayhow
Slingsby T.40

Haze
Mil Mi-14

Headwind
Stewart (Donald)

Heart of Mountains (As: Shinzan)
Nakajima G5N

Heavenly Thunder (As: Tenrai)
Nakajima J5N

Heck
Hendy 3308
Parnall

Hector
Hawker

Hedgehog
Hawker

Heja
Mavag

Helen
Nakajima Ki-49

Heli Trenér
Zlin Z 35

Heli-Baby
Aero (Vodochody) HC-2

Heli-Jeep
Rotor-Craft

Heli-Porter
Fairchild

Heli-Star
Meger

Heli-Vector
deLackner DH-4

Helibus
Sikorsky (USA) S-92

Hélica
Leyat

Hélicat
Leyat

Helicogyre
Isacco

Helicon
Picken

Helicoplan
Bertin

Helicoplane
Caze

Hélicoptère
Pénaud (Alphonse)

Helicospeeder
Gazda

Helicostat
Œhmichen

Helidyne
Gyrodyne

Helioplane
Helio

HELIOT
Dragon Fly

Heliplane
Baynes

Helix
Deltour

Helix
Kamov Ka-27, Ka-32

Hellcat
Grumman F6F

Helldiver
CCF SBW-4E
Curtiss, Curtiss A-25, F8C, O2C, SBC, SB2C, SB3C
Fairchild SBF

Helvellyn
Hillson

Hen *(See also: Water Hen)*
Kamov Ka-15

Hendon
Fairey
Handley Page

Hengist
Slingsby T.18

Henley
Hawker

Herald
Handley Page H.P.R.3/5

Hercules *(See also: Herkules)*
de Havilland D.H.66
Hughes H-4
Lockheed C-130, Model 82

Hereford
Handley Page

Herkules (Hercules)
Junkers Ju 352

"Hermann Köhl"
Lippisch Delta II

Hermes
Aviakit
Boeing E-6
Handley Page
SPCA IV type

Hermit
Mil Mi-34

Heron *(See also: Sea Heron)*
de Havilland D.H.114
Grob G 115
Hawker
Heinkel He 57

Heron (As: Reiher)
DFS

Heron (As: Zwergreiher)
Burgfalke Lo 100

Hertfordshire
de Havilland D.H.95

Heyford
Handley Page

Hi Trike
Ultimate

Hi-Low
Long (Les)

Hi-Nuski
Advanced Aviation

Hi-Vision Courier
Helio

Hibari (Skylark)
Nihon (Hikoki KK)
Mitsubishi

Hickory
Tachikawa Ki-54

Hien (Swallow)
Kawasaki Ki-61

"High Flyer"
Wright (Co) Model R

"High Point"
Boeing Hydrofoil

Highclere
de Havilland D.H.54

Highly Maneuverable Aircraft Technology
HIMAT

Hiller-Copter
Hiller XH-44

Hilo
Hendy 321

HIMAT
HIMAT

Hinaidi
Handley Page

Hinazuru
Mitsubishi

Hind
Hawker

Hind
Mil Mi-24, Mi-35

Hip
Mil Mi-8, Mi-17

Hiperbipe
Sorrell SNS-6

Hiperlight
Sorrell SNS-8

Hippo
Sopwith

Hirondelle (Swallow)
Albert A.10
Dassault
Western (Canada) PGK-1

Hirundo
Agusta A 109

Hiryu (Flying Dragon)
Mitsubishi Ki-67

Hiryu-Go
Ueda

Hisso Jenny
Curtiss JN-4H

Ho-go
Yokosuka

Ho-go Otsu-Gata
Yokosuka

Hobby
Miles M.13

Hobo
Hendy 281

"Hochi-Hinomaru"
Bellanca J-3

Hocker-Denien Sparrow Hawk
Denien

Hog
Kamov Ka-18

Höhenflugzeug
Boerner

Hokum
Kamov Ka-50

"Hol's der Teufel"
Schleicher

Homer
Mil Mi-12

Honey Bee
Bee Aviation
Classic (Aero) H-2

Honeymoon Express
Dayton Wright DH-4K

Honza
Zlin Z 23

Hoodlum
Kamov Ka-26

Hook
Mil Mi-6, Mi-22

Hoop
Kamov Ka-22

"Hoople"
Gary

Hoopoe
Hawker

Hoplite
Mil Mi-2
PZL Swidnik V-2

Hoppi-Copter
Capital Copter

Horace
Farman F.40

Horizon
Hawker
Raytheon Hawker 4000

Hormone
Kamov Ka-25

Hornbill
Hawker

Hornet (See also: Hornisse, Super Hornet)
Brookland Mosquito Mk.1
Curtiss 18-B
de Havilland D.H.103
Fogle
Free Flight
Glasflügel
Hawker
Hiller H-32, HJ-1, HOE
McDonnell Douglas F/A-18
Northrop F-18
SR-1 Ultra Sport SR-1

"Hornet"
Hanes (Arnold L.)

Hornet (As: Frelon)
SNCASE SE 3200

Hornet Moth
de Havilland D.H.87

Hornisse (Hornet)
Focke Achgelis Fa 266
Messerschmitt Me 410

Horsa
Airspeed

Horse (See also: Airhorse)

Horse
Yakovlev Yak-24

Horsefly (Bak)
Warsztaty Szybowcowe

Horsley
Hawker

"Hot Shot"
Mullins (Modisette) H-S

Hotspur
G.A.L. 48
Hawker

Hound
de Havilland D.H.65

Hound
Mil Mi-4

Hound (As: Ogar)
PWS 35

Houperzine
Melton Baby Biplane

Hoverbug
Eglen Hovercraft

Hoverfly
Sikorsky (USA) R-4

Hu-Go Craft
Hugo

Huanquero
FMA

Hubbard Monoplane
Canadian Aerodrome

Hudson
Lockheed A-28, A-29, AT-18, B14S, PBO

Hudson Flyer
Curtiss

Huey
Bell H-1

Hummel (Bumble Bee)
Siebel Si 202

Hummer
Maxair

Humming Bird
Brea
de Havilland D.H.53
Whites

Hummingbird
Aerotek
Curtiss
Glider Aircraft Corp
Lockheed V-4
Nelson (Aircraft)
Tremaine

Hummingbird (As: Beija-Flor)
Centro Técnico de Aeronáutica

Hummingbird (As: Colibrí)
FMA

Hummingbird (As: Koliber)
PZL Warsawa-Okecie PZL-110

Hummingbird (As: Kolibri)
Aeroprogress T-311
Borgward
Flettner Fl 282
Udet U 7

Hummingbird (As: Kolibrie)
Netherlands Helicopter H-3

Hummingbird (As: Picaflor)
Farge

Hummingbird (As: Koliber) Junior
PZL Warsawa-Okecie PZL-112

Hummingbird (As: Koliber) Senior
PZL Warsawa-Okecie PZL-111

Hunter
Hawker

Hunter (As: Der Jäger)
White (E. Marshall)

Hunting Hawk (As: Jaktfalk)
Svenska

"Hurlburt Hurricane"
"Hurlburt Hurricane"

Huron
Beech C-12A

Hurricane *(See also: "Hurlburt Hurricane," Sea Hurricane)*
Hawker
R.A.E.
Sindlinger HH-1

Hurricane (As: Ouragan)
Dassault

Hurricat
Hawker

Huski - Hi-Nuski
Advanced Aviation

Huskie
Kaman H-43, K-1125

Husky *(See also: AgHusky)*
Aviat
Consolidated NY
Fairchild F 11

Husky Junior
Consolidated Fleet Model 1

Hustler
American Jet Model 500
Convair B-58
Gulfstream Aerospace

Hyderabad
Handley Page

Hydro-Air (ACV)
Crowley

Hydro-Copter
Bensen B-8MW

Hydro-Glider
Bensen B-8W

Hyena
de Havilland D.H.56

"Hyphen" (As: "Le Trait d'Union")
Dewoitine D.33

Hythe
Short S.25

I

I-go
Yokosuka

I-go Ko-gata
Yokosuka

Ibex
Hall (Stanley)

Ibis
Aero (Vodochody) Ae 270
Ross (Harland) R-2
Willoughby (Hugh L.)

Ibis Bleu (Blue Ibis)
Delanne

Icarus
UFM

Ida
Tachikawa Ki-36, Ki-55

Ikas (Icarus)
L.O.P.P.

"Ike"
Howard (Benny) DGA-5

Ikub
Kubicki

Ile-de-France
Voisin (33)

Ilse
Siegrist AS1

Il'ya Muromets
Sikorsky (Igor)

Imp *(See also: Micro-Imp, Mini-Imp, Ultra-Imp)*
Aerocar, Inc
Parnall
Wallis (Autogyros)

Imperial *(See also: Knight Twister Imperial)*
Starling Aircraft H-12
Bird Wing

"Independence"
Douglas C-118 Liftmaster

Infinité
Dumod

Inflatoplane
Goodyear

Inflexible
Beardmore
Rohrbach-Beardmore

Innovator
Bird (Corp)

Instructor
Fokker S.11 S.12

Instructor (As: Lehrmeister)
VEB Lommatzch

Integral
Breguet Bre.940, Bre.942

Interceptor
Hawker F.20/27
Westland

"Intestinal Fortitude"
Granville R-1/2

Intruder
Grumman A-6
Sorrell SNS-10

Invader
Douglas A-26, JD
Kinner
UEP

"Inverness"
Rohrbach Ro IV

Ione
Aichi AI-104

Ipanema
EMBRAER

Iridium (As: Iryda)
PZL Mielec M96

Iris
Blackburn

"Irish Swoop"
Bellanca 28-70

Iroquois
Bell
Eberhart

Irving
Nakajima J1N

Iryda (Iridium)
PZL Mielec M96

Iskierra (Little Spark)
PZL Mielec M26

Iskra (Spark)
PZL Mielec TS-11

Islander
Britten Norman BN-2A
PADC

Iwao-Go
Shirato

J

J-PATS
Cessna 526

J-Stars
Boeing E-8

Jabiru (Stork)
Farman (Henri) H.F.II
Farman F.3X F.4X F.170

Jack
Mitsubishi J2M

"Jackrabbit"
Keith Rider R-5

Jäger (Hunter)
White (E. Marshall)

Jaguar
Chrysostomides
Grumman F10F
Messerschmitt Bf 162
Sepecat

Jake
Aichi E13A

Jaktfalk (Hunting Hawk)
Svenska

Jalor (Landmark, Beacon)
Fouga MC

Jane
Mitsubishi Ki-21

Janin Patent Boat
Curtiss

Jantar
PZL Bielsko SZD 38

Janus
Schempp-Hirth

Jaragua
Instituto de Pesquisas Tecnologicas IPT-5

Jaskólka (Swallow)
ITS
Libanski

Jastreb (Hawk)
Soko J1

Jastrzab (Hawk)
PZL P.50

Javelin
Capella
Gloster
Peterson Max A. J-4

Javelot (Javelin)
Wassmer

Jaybird
Volmer VJ21

Jayhawk
Beech T-1A
Sikorsky (USA) H-60

Jean
Kugisho B4Y1

"Jean-Hubert"
Bernard 80 G.R

Jeep *(See also: Airgeep, Fliegender Jeep, Flying Jeep, Heli-Jeep, Jet Jeep)*
Chester (Art)
Curtiss AT-9
Northrop N-1M

Jenny *(See also: Hisso Jenny, Twin Jenny)*
Cloud Dancer
Curtiss JN-2, JN-3, JN-4, JN-5,
Early Bird

"Jenny"
Wissbaum

"Jesus del Gran Poder"
CASA-Breguet 19GR

Jet Commander
Aero Commander 1121

Jet Fox
Arnet Pereyra

Jet Gyrodyne
Fairey

Jet Jeep
American Helicopter XH-26

Jet Mentor
Beech

Jet Provost
Percival

Jet Ranger
Agusta AB 206

Jet Wasp
Texas Helicopter M79T

Jet Wing
Flight Designs

Jet-Cat
Continental Copters

Jetasen
Maule Air

Jetcruzer
Advanced Aerodynamics & Structures

Jetfoil
Boeing 929

JetHawk
Aero-K

Jetliner
Avro (Canada) C-102

Jetpacket
Steward-Davis

JetRanger
Bell 206

Jetstar
Lockheed, Lockheed C-140

Jetstream
British Aerospace
Century (TX)
Handley Page
Volpar

Jettoplane
Nishi

Jetwing
Ball-Bartoe JW-1

Jewel Box (Norécrin)
SNCAN 1203

"Jezebel"
Ibbs

Jill
Nakajima B6N

Jim
Nakajima Ki-43

Jindivik
Government Aircraft Factory (Australia)

"Jinny"
Foss

Jockey
SPAD S.91 C1
Vickers (UK) Type 151
Wibault 13 C1 (130)

Jockey (As: Zhokei)
Tupolev I-8

John
Nakajima Ki-44

John Doe STOL
American Homebuilts

"John Wanamaker"
Ford 2-AT

Johnny Reb
Pack

"Joker"
Weiss (José)

Jolly *(See also: Super Jolly)*
Partenavia P59

Jolly Green Giant
Sikorsky (USA) H-3

Jona
Magni

"Joseph le Brix"
Blériot 110

"Josephine Ford"
Fokker F.VIIA-3m

Jr *(See: Junior)*

Jubile (Jubilee)
Jodel

Judson Triplane
Curtiss

Judy
Kugisho D4Y

"Julia"
Heinkel P.1077

Jumbo
Werkspoor

"Jumping Jet"
Shorts P.D.46

Junák (Cadet)
LWD
Zlin Z 22

June Bug
Aerial Experiment Assoc

Jungmann
Bücker Bü 131
Jungmeister
Bücker Bü 133

Jungster *(See also: Papoose Jungster)*
Kaminskas

Junior *(See also:Beta-Junior, Corsair Junior, Cruisair Jr, Electra Junior, Fledgling Jr, "Floyd Bennett Jr," Husky Junior, Koliber (Hummingbird) Junior, Knight Twister Junior, Mac Jr, Mercury Jr, "Saginaw Junior," Sioux Coupe Jr, Speedwing Jr, Sperber (Sparrowhawk) Junior, Tipsy Junior)*
Bellanca
Bölkow/MBB Bo 208
Breese Aircraft
Curtiss
Instituto de Pesquisas Tecnologicas IPT-7, IPT-10
Nicholson KN-1
PZL Bielsko SZD 51
Rearwin
Stinson (Aircraft) Model R, Model S, SM-2, SM-7, SM-8
Stits (SA-1)

Junior Ace
Ace (NC)
Corben

Junior Airsedan
Buhl Airsedan CA-3B

Junior Coupe
Alco

Junior Hawk
Haller-Hirth

Junior Speedmail
Stearman (1927) 4

Junior Transport
Alcor C.6.1

Jupiter
Champion (Ken)
Jamieson
RAF College Halton
"Jupiter - Pride of Lemont"
Folkerts Speed King

Jutrzenka (Dawn)
Kozlowski WK 1

K

Kachina
Morrisey

Kaczka-Nadzieja
Bohatyrew

Kaibo Gikai
Rinji Gunyo Kikyu Kenkyu Kai

Kaje
Kjeller

Kamikaze-Go
Hino Number 4

Kangaroo
Blackburn

Kangaroo *(As: Canguro)*
Ambrosini
Savoia-Marchetti SM.82
SIAI-Marchetti SF.600TP

Kangourou (Kangaroo)
Caudron C.570
Fairey Fox

Kania (Kitty Hawk)
PZL Swidnik

Kansan
Beech AT-11, SNB

Kaoru-Go (Fragrance)
Shirato

Kaplick-Kerl
VEB Lommatzch

Karakorum
Nanchang K-8

Karas (Carp)
PZL P.23, P.42, P.43

Karatoo
Australian Aviation

Karausche (Carp)
Fokker M 16E
Karhu
Karhumäki

Karigane (Wild Goose)
Mitsubishi

Karyu (Fire Dragon)
Nakajima Ki-201

Kasatnik (Darling)
Aviacor

KASKR
Kamov-Skrzhinsky

Kasperwing
Cascade

Katana
Diamond

Katana Eclipse
Diamond

Katana Xtreme
Diamond
Hoffmann (Flugzeugbau) DV 20

Kate
Mitsubishi B5M
Nakajima B5N

Katy
Payen Pa 49/B

Kaunus
Antonov Pk-4

Kaydet
Stearman (1927) N2S, PT-13, PT-17, PT-18, PT-27

Kéa
B & L Hinz

Keiun (Beautiful Cloud)
Kugisho R2Y-1

Ken-Royce
Rearwin

Kent
Short S.17

Kestrel
Austin
Glasflügel
Hawker Siddeley
Miles M.9
Slingsby T.59C (Glasflügel)

Kestrel (Rooivalk)
Air Nova

Kfir (Young Lion)
Israel Aircraft Industries

Kiebitz (Lapwing)
Dornier Do 34
Focke-Wulf S 24

Kikka (Orange Blossom)
Nakajima

King *(See: Air-King, Sea King, Silver King, Sky King, Speed King, Super King Air)*

King Air
Beech

King Bird
Fokker F.50

King Kite
Slingsby T.9

King-Bird
King

Kingbird
Curtiss, Curtiss RC
Herring-Burgess
Western (IL)

Kingcobra
Bell 308, P-63

Kingfisher
Anderson EA-1
Taft
Vought OS2U

Kingsland
Auster J.1

Kingston
English Electric

Kiowa
Bell H-58

Kirby Cadet
Slingsby T.7, T.31B

Kirby Gull
Slingsby T.12

Kirby Kite
Slingsby T.6

Kirby Kitten
Slingsby T.10

Kirby Tutor
Slingsby T.8

Kirby Twin
Slingsby T.11

Kis
Tri-R

Kite *(See also: King Kite, Kirby Kite)*
Comper
Kirby
Simplex (OH)
Slingsby T.23, T.26

Kite (As: Drache)
Focke Achgelis Fa 223

Kite (As: Korshun)
Aeroprogress T-620

Kite (As: Milan)
Dassault

Kite (As: Weihe)
DFS
Focke-Wulf Fw 58
Jacobs Schweyer

Kitfox
Denney Aerocraft
Sky Star

Kitfox Classic
Sky Star

Kitfox Safari
Sky Star

Kitfox Speedster
Sky Star

Kitfox Vixen
Sky Star

Kitten *(See also: "Cactus Kitten," Delta Kitten, Eastchurch Kitten, Grain Kitten, Kirby Kitten, Mercury Kitten)*
Dart (UK)
Grumman G-63 G-72
Loening (Corp) M-2
Martin (J. V.) K.III

Kittiwake
Mitchell-Proctor
Saunders

Kitty Hawk (As: Kania)
PZL Swidnik

Kittyhawk
Bourdon
Curtiss P-40

Kmotr
Zlin Z 30

Knight *(See also: Phantom Knight, Sea Knight*
Deekay

Knight Twister
Payne

Knight Twister Imperial
Payne

Knight Twister Junior
Payne

Knuckleduster
Short S.18

Koala *(See also: Super Koala)*
Aerotechnik (Czech) P 220 S
Agusta A 119
Fisher

Kobold (Goblin)
Albatros L79

Kobra
Instytut Lotnictwea

Kocour
Letov MK-1

Kogutek (Cockerel)
Zalewski W.Z.XI

Koliber (Hummingbird)
PZL Warsawa-Okecie PZL-110

Koliber (Hummingbird) Junior
PZL Warsawa-Okecie PZL-112

Koliber (Hummingbird) Senior
PZL Warsawa-Okecie PZL-111

Kolibri (Hummingbird)
Aeroprogress T-311
Borgward
Brügger
Flettner Fl 282
Udet U 7

Kolibrie (Hummingbird)
Netherlands Helicopter H-3

Komar (Gnat)
Kocjan
Warsztaty Szybowcowe

Komet (Comet)
Dornier, Dornier Do V
Kawasaki
Messerschmitt Me 163

Kompress
CH-7

Komsomolets
DOSAAF/Komsomol

Kondor (Condor)
Klosterlein
Udet U 11

Konsul
Darmstadt D.9

Kopter-Kart
Bensen B-11M

Kornett
Bücker Bü 182

Korsa (Corsica)
Schmid

Korshun (Kite)
Aeroprogress T-620

Kos
Aero (Prague) A 34

Kos (Blackbird)
PZL Warsawa-Okecie PZL-102B

Kraft Super Fli
P. K. Plans

Krähe
Rock

Krajánek
Zlin Z 24

Kranich (Crane)
DFS

Krechet
Avgur

Kruk
PZL Warsawa-Okecie PZL-106

Kryl'yatanka (Flying Tank)
Antonov

Kudu
Atlas

Kuma-Go
Shirato 25

Kupe
SAI K.Z.II

Kurier (Courier)
AGO Ao 192
Focke-Wulf Fw 200
Vought V-85G

Kuryer (Courier)
Aeroprogress T-910

Kvant
MAI

Kvazimodo (Quasimodo)
Caproni Bulgaria

Kyofu (Mighty Wind)
Kawanishi N1K1

L

"Lady Peace"
Vultee V1-A

"Lady Southern Cross"
Lockheed Model 8-D

Ladybird *(See also: De Bruyne-Maas Ladybird)*
Curtiss

Ladybug (As: Coccinelle)
SIPA 1000

Lakes Water Bird
Avro

Laking
Clarke

Lama (Llama)
SNCASE SE 315B

Lambáda *(See also: Speed Lambáda)*
Urban UFM 13

Lampart (Leopard)
PZL P.48

LAMPS
Boeing-Vertol H-61

Lancair
PADC

Lancaster
Avro 683

Lancastrian
Avro 691

Lance
Piper PA-32-300

Lance (As: Lansen)
Saab 32

Lanceiro
Neiva

Lancer
Aerostar (Romania) MiG-21
Champion (Wisconsin)
Lancair
Marquart MA-4
Republic P-43
Rockwell B-1

Landau *(See: Glenmont Landau Sedan)*

Landmark (As: Jalor)
Fouga MC

"Langley"
Handley Page O/400
Standard Handley Page O/400

Languedoc
SNCASE SE 162

Lansen (Lance)
Saab 32

Lapwing (As: Czajka)
Kocjan

Lapwing (As: Kiebitz)
Dornier Do 34
Focke-Wulf S 24

Large America
Curtiss Model H-12

Lark *(See also: Metalark, Motor Lark, Skylark, Twin Lark)*
Curtiss
Helton
IAR Brasov IS-29D2
Keleher
Maranda
SAI K.Z.VII

Lark (As: Alouette)
Sud-Aviation SA 318

Lark (As: Lerche)
Heinkel

Lark (As: Tchutchuliga)
Caproni Bulgaria

"Lark of Duluth"
Benoist Type XIV

Lasconder
Lasco

Lascoter
Lasco

Lascowl
Lasco

LASH (Light Antisub Helicopter)
Bell 608

Lattice Tail (As: Gitterschwanz)
DFW

Lavadoux
Starck AS.80

Lazair
UltraFlight

Learstar
Lear
Pacific Airmotive

Lehrmeister (Instructor)
VEB Lommatzch

Leisure (Loisirs)
Adam (France) R.A.14

Lemon (Cytrynka)
Start Aviation Circle ZE-1

"Léonard de Vinci"
Blériot 165

"Leonardo da Vinci"
Bellanca J-300

Leone (Lion)
AUI

Leopard
Chichester-Miles

Léopard (Leopard)
CERVA CE.45
Wassmer CE.45

Leopard (As: Lampart)
PZL P.48

Leopard Moth
de Havilland D.H.85

Lerche (Lark)
Dätwyler
Heinkel

Lerwick
Saunders-Roe S.36

Leviathan
Breguet

Lexington
McCabe (Ira)

Libeccio
Caproni Ca.310

Libel (Dragonfly)
Hollandair HA-001

Libelle (Dragonfly)
Dornier Do A, Do 12
Glasflügel
VEB Lommatzch LOM 55/1

Libellula (Dragonfly)
Manzolini
Miles M.35, M.39, M.42, M.43 M.63

"Libellule" (Dragonfly)
Farcot
Hanriot

Liberator
Consolidated B-24, B-41, BQ-8, C-109, F-7, PB4Y, RY

Liberator Express
Consolidated C-87, RY

Liberty
Associate Air

Liberty Battler
Curtiss CB

Liberty Bell
North Star M6

Liberty Bell Biplane
Liberty

Liberty Tourist
Gallaudet Aircraft C-3

"Liberty"
Bellanca J-300

Libra-Det
Zodiac (NJ) ZAC-1

Lictor
Gabardini

Lietuva
Sportina LAK 12

"Lieutenant de Vaisseau Paris"
Latécoère LAT 521

Liftfan
Ryan VZ-11

Liftmaster
Douglas C-118, R6D

Light (Lumiére)
de Monge

Light Flyer
Pterodactyl

Light Intratheater Transport
Boeing LIT

Light Sport
Curtiss CW-16

Lightning *(See also: Chain Lightning, Magnificent*

Lumiére (Light)
de Monge

Lunar Rocket
Maule Air

Lutèce
SECM Type XX

Lwów
Grzeszcyk SG-21

Lynx
de Scórzewski
Westland

Lysander
Westland

M

M.G.M. Special
Ryan B.1

Maager (Seagull)
Orlogsvaerftet FB.III

Mabel
Mitsubishi B5M

Mac Jr
Pittsburgh

Mach Trainer
Fokker S.14

Madge
Beriev Be-6

"Madge"
Weiss (José)

"Madon-Carmier"
Arnoux

Maerch
Hirsch

Maestro
Yakovlev Yak-28U

Magister
Miles M.14

Magister (Schoolmaster)
Fouga CM 170

Magnet
Yakovlev Yak-17UTI

Magnificent Lightning (As: Shinden)
Kyushu J7W1

Magnum
Avid Aircraft

Magnum
Yakovlev Yak-30

Magpie (Sroka)
Kocjan

Magpie-B (Elster-B)
Pützer

Mahoney-Ryan Special
Ryan X-1

"Maia"
Short S.21

Maid
Delgado
Miami (Aircraft Corp)

"Maiden Dearborn"
Ford 2-AT

"Maiden Milwaukee"
Hamilton Metalplane H-18

Maiden Tulsa
Mid-Continent

Mail
Beriev Be-12

Mail Carrier
Boulton Paul

Maillet
SFCA

Mailplane
Avro 627, 654
Douglas DAM-1
GAC Model 107
Percival
Saunders-Roe/Percival A.24
Spartan (UK) A.24
Pitcairn PA-5

Mainstay
Beriev A-50

Maitsuru-Go
Iga

Major
Adam (France) R.A.15
Robin

Makalu
Fouga CM 171

Mako
Daimler-Benz AT-2000

"Maksim Gorkii"
Tupolev ANT-20

Malibu Mirage
Piper PA-46-350P

***Mallard** (See also: Turbomallard)*
Cornelius
Grumman G-73

Mallow
Beriev Be-10

"Malolo"
Edo

Malysh (Little One)
Zlatoust

Mamba
AIA MA-2

"Mammouth" ("Mammoth")
Blériot 74

"Mammy"
Long (Dave) Midget Mustang

Manazuru (Crane)
Kokusai Ku-7

Manchester
Avro 533, 679

Mandrake
Yakovlev Yak-25

Mangrove
Yakovlev Yak-27R

Mangusta (Mongoose)
Agusta A 129

Manx
Handley Page

Mañque
FMA

Maple (As: Momiji)
Kyushu K9W1

Maple Leaf
CCF

Marathon
Aviator Scientific
Handley Page
Miles M.60 M.69

***Marauder** (See also: Super Marauder)*
Martin B-26, JM

Marco
Aviation Farm

"Marcoux-Bromberg Special"
Keith Rider R-3

Mare
Yakovlev Yak-14

"Margon"
de Havilland D.H.52

Mariner
Martin PBM
Robinson (Helicopter) R22

Maritime Enforcer
Fokker F.50

Marketeer
On Mark

Marksman
On Mark

Marlborough
Miles M.55

Marlburian
Northern (Lowe) H.L.(M).9

Marlin
Martin P5M

Marquézy
Clerget

Marquise
Mitsubishi MU-2

Marreco
Instituto de Pesquisas Tecnologicas IPT-14

Mars
Breguet Bre.891 R
DFW

Meteorplane
Irwin

Methvin Safety Wing
Aircraft Methvin

Metro
Fairchild

Metro Expediter
Fairchild

Metropolitan
Convair 440

Mew Gull
Percival

Mewa (Gull)
Bilski
Grzeszczyk & Kocjan
Kocjan
L.W.S.3
PZL Mielec M20

Mezek
Avia (Czech) S 199

Michigander
Gliders, Inc

Mickl
Ikarus

Micro-Imp
Aerocar, Inc

Microjet
Microturbo

Microlight
Scaled Composites Model 97

"Middle River Stump Jumper"
Martin XB-26H Marauder

Midge
Folland

Midget
Kreider-Reisner
Meyers (C. W.)

Midget
MiG MiG-15

Midget Monocoupe
Williams (Aircraft Design)

Midget Mustang
Bushby MM-1
Long (Dave)

Midget Racer
Francis & Angell

Midgy-Club
MDG LD-261

Midnight Airliner
Lawson (Alfred) L-4

Midwest
Steinhauser

Mighty Wind (As: Kyofu)
Kawanishi N1K1

Mignet Pou-du-Ciel (Flying Flea)
Ojea

"Mike"
Howard (Benny) DGA-4

Mikron
Zlin Z 212

Milan
Akadflieg München Mü 10
Fournier SFS-31
Scheibe SFS-31
SETCA
Sportavia SFS-31

Milan (Kite)
Dassault

Miles-Atwood Special
Brown (L. W.)

Milky Way (As: Ginga)
Kugisho P1Y

Millennium
Buckeye

Millennium Swift
Aviat

Miltrainer
Valmet L-70

Minerva
Miles M.51
SIAI-Marchetti SF.260
Waco M-220-4

Mini Ace
Cvjetkovic CA-61

Mini Coupe
Johnson (Bill)
Sport Aircraft
Tena

Mini Guppy
Aero Spacelines

Mini Z
Zenair

Mini-Imp
Aerocar, Inc
Taylor (Molt)

Mini-Mac
McCarley

Mini-Nimbus
Schempp-Hirth

Minibat
Haig

Minicab
Béarn
Ord-Hume

Minicopter
Rotorcraft

Minihawk
Hawk

Minijet
SIPA 200

Minimoa
Göppingen Gö 3
Schempp-Hirth

Miniplane
Smith (Frank) DSA-1

"Minnow"
Levier

Minor
Miles M.30

Minuano
EMBRAER
Neiva

Minus
Halton HAC-2

Mir
Mráz XLD 40

Mirach 20 Pelican
Meteor

Mirage
CAC
Dassault
Leighnor W-4
Ultralight Flight

Mis (Teddy Bear)
Bohatyrew
PZL P.49

"Miss Columbia"
Bellanca WB-2

"Miss Dara"
Dayton Racer

"Miss Doran"
Buhl Airsedan CA-5

"Miss Kansas City"
Baker (Ray & Al)

"Miss Los Angeles"
Brown (L. W.) B-2

"Miss Silvertown"
Lockheed Model 5A

"Miss Teanek"
Lockheed Model 1

"Miss Veedol"
Bellanca Skyrocket CH-400

Missel Thrush
ANEC IV

Missileer
Douglas F6D
McDonnell Model 153A

"Mister Mulligan"
Howard (Benny) DGA-6

Mistral (Cold Northern Wind)
Armella-Senemaud
Aviasud
SNCASE SE 535

Mitchell
North American, North American B-25, F-10, PBJ

Mite
Mooney

"Mitey Mite"
Thompson (Carl) Special

Mixmaster
Douglas B-42

MMIRA
Kaman

Moazagotl
Hirth

Module
Aviastar

Modwing
Breuil

Mohawk
Curtiss Hawk H75A
Grumman OV-1

Miles M.12
Wallace
Warpath Aviation

"Moineau"
Clément (Louis)

Mojave
Piper PA-31P-350
Sikorsky (USA) H-37

Momiji (Maple)
Kyushu K9W1
Watanabe (Tekkosho) K9W1

Monarch
Marske
Miles M.17
MIT
Rock Island

Monerai
Monnett

Monex
Monnett

Mongol
MiG MiG-21

Mongoose (As: Mangusta)
Agusta A 129

Moni
Monnett

Monitor
Miles M.33
Neiva B-2

Mono Z
Zenair

Mono-Fly
Teman

Mono-Ford
Peters

Mono-Fox
Fairey

Monobiplane
Aircraft Improvement

Monobloc
Antoinette

Monocoach
Monocoupe

Monocoupe *(See also: Midget Monocoupe)*
Monocoupe

Monoduck
Loening (Grover)

Monogast
Quaissard GQ.01

Monomail
Boeing Model 200, 221

Monoped
Hall-Aluminum

Monoplane of the "Three"
Lwów

Monoprep
Monocoupe

Monosport
Monocoupe

Monowheel Amphibian
Loening (Corp)

Monster (See: Comp Monster)

"Monstro"
Boeing EB-29B

Monsun (Monsoon)
Bölkow/MBB Bo 209

Montagne
Marshall (Aircraft)

Monterosa
Fiat G.212

Montrose
Miles M.36

Moon Maid
Doyle

Moonbeam
Crosley

Moonlight (Gekko)
Nakajima J1N1-S

Moose
Yakovlev Yak-11

Moraa
Pützer

Morava
Let L 200

"Morelos"
Emsco

Morris Boat
Curtiss Model M

Moryson
Morisson and Nawrot

Moskito (Mosquito)
Comte A.C.12

Focke-Wulf Ta 154

"Moskva" ("Moscow")
Arzenkoff
Ilyushin

Mosquito
Aviolight
Brookland
de Havilland D.H.98
Glasflügel
Heinemann
Partenavia P86

Moss
Tupolev Tu-126

Moth *(See also: American Moth, Cirrus Moth, Fox Moth, Genet Moth, Giant Moth, Gipsy Moth, Hawk Moth, Hornet Moth, Leopard Moth, Puss Moth, Swallow Moth, Tiger Moth)*
Burgess F
Curtiss
de Havilland D.H.60
Loening (Corp)

Moth (As: Phalène)
Caudron C.280

Moth Major
de Havilland D.H.60

Moth Minor
de Havilland D.H.94

Moth Three
de Havilland D.H.80

Mothbat
Geske

Motor Lark
IAR Brasov IS-28M2

Motor Tutor
Slingsby T.29

Motorcycle *(See: "Flyying Motorcycle")*

Motorspatz (Motorized Sparrow)
Scheibe SF-24A

Motyl (Butterfly)
Bohatyrew

Mouette (Seagull)
Breguet Bre.901 S

Mouette, la (Seagull)
Bulot

Mountain Falcon (As: Bergfalke)
Scheibe

Mountain Flyer (As: Gebirgsflieger)
Lohner

Mountain Racer (As: Renn Gebirg)
Lohner-Etrich

Mountain Range (As: Renzan)
Nakajima G8N

Mouse *(See also: Blue Mouse)*
Comper

Mousquetaire (Musketeer)
Jodel

Moustique
Farman

Möwe (Gull, Seagull)
Dittmar
Focke-Wulf A 17, A 29, A 38
Geest
Nordflug

"Mr. Smoothie"
Pearson-Williams PW-1

Mriya (Dream)
Antonov An-225

Mrówka (Ant)
Agrolot PZL-126P
PZL Warsawa-Okecie PZL-126P

Mt Fuji (As: Fugaku)
Nakajima G10N

Mucha Standard
PZL Bielsko SZD 22C

Multi-Gun
Saunders-Roe A.10

Multiplane
Joachimczyk
John's

Musca
Viberti

Muscadet
Boisavia B-50

Mushshak
Pakistan Aeronautical Complex

"Music M-100"
Muessig P-2

Musketeer
Beech

Musketeer** (As: **Mousquetaire)
Jodel

Mussel
Short S.7

Mustang *(See also: Midget Mustang, Thunder Mustang, Twin Mustang)*
Bushby M-II
CAC
Cavalier
Historical Aircraft Corp
Jurca MJ-77
Mooney
North American, North American F-6, P-51
Sturgeon

"Musterle" ("Example")
Grunau

Myrt
Nakajima C6N

Mystère (Mystery)
Dassault

Mystère/Falcon
Dassault

Mystery Ship
Emsco

"Mystery Ship"
Travel Air Model R

N

NABA
Peters

Nacional
Sfreddo & Paolina SYP-I El

Nanok
Supermarine

Nanzan (Southern Mountain)
Aichi M6A1-K

Narwhal (As: Norval)
SNCASO SO 8000

Nasim
All AVA-101

Nate
Nakajima Ki-27

Natter (Viper)
Bachem Ba 349

Nautilus
Blackburn

Navaho
Indian Aircraft

Navajo
EMBRAER
Piper PA-31

Navajo Chieftain
Piper PA-31-350

Navigator
Beech AT-7, SNB

Navion *(See also: Twin Navion)*
North American L-17
Tusco

NEACP
Boeing E-4

Nebojsa
Fizir

Nell
Mitsubishi G3M
Yokosuka L3Y

Nemere
Rotter

Nene
SNCASO SO 30

Neova
Neoteric

Neptune
Lockheed P2V

Neringa
Antonov Bk-6

Nesmith Tri Cougar
EAA

New Baby
Grahame-White

New Swallow
Swallow (KS)

"New York Times"
Curtiss R-7

News
Robinson (Helicopter) R44

Nexus
Goldwing

Nibbio
Aviamilano F.14

Nick
Kawasaki Ki-45

Nieuhawk
Nieuport (UK)

Nifty
Morrisey

Night Hawk
Sikorsky (USA) H-60

Night Mail
Curtiss

Night Mail *(See also: Mercury Night Mail)*
Aeromarine AM-1, AM-2, AM-3

Nightfox
Hughes H-6

Nighthawk
General Aeronautical CX-12
Gloster Mars VI
Lockheed F-117
Miles M.7
Nieuport (UK) L.C.1
Pemberton Billing P.B.31E
Supermarine
Theis

"Nighthawk"
Cocke & Scott

Nightingale
Douglas C-9A
Howard GH

Nightjar
Gloster Mars X
Nieuport (UK)

Nightswift
Waco CG-3

Nikko
Fuji

Nile
Blackburn

Nimbus *(See also: Mini-Nimbus)*
Mitchell (Don)
Schempp-Hirth
Short

Nimbus Martinsyde
ADC

Nimrod
Hawker
Hawker Siddeley

Nine-Hour Cruiser
Dayton Wright DH-4R

Nipper
Dixon

Nippi
Nihon (Hikoki KK)

Nippon-Go
Tamai Number 2

"Nippon"
Mitsubishi G3M2

Niska
Fairchild 34-42

No-Tail Advanced Theater Transport
Boeing NOTAIL ATT

Nomad (See also: Aztek Nomad)
Government Aircraft Factory (Australia)
Naval Aircraft Factory PBN
Northrop Gamma 8A

Nomair
Hamilton (AZ)

Noralpha
SNCAN 1101

Noranda
Noury T-65

Noratlas (Atlas)
SNCAN 2503

Norazur (Azure)
SNCAN 2100

Noréclair (Lightning)
SNCAN 1500

Norécrin (Jewel Box)
SNCAN 1203

Norelfe (Elf)
SNCAN 750, 1750

Norelic
SNCAN 1700

Norm
Kawanishi E15K1

Normal Segelapparat (Sailing Apparatus)
Lilienthal (Otto)

Noroit
SNCAN 1402

Norseman
Noorduyn

North Star
Canadair CL-2

"North Star"
Bellanca J

North Wind (Aquilon) *(See also: Mistral)*
SNCASE

Norval (Narwhal)
SNCASO SO 8000

NOTAIL ATT
Boeing

Nucleon Glider
Schultz

Nugget
Laister LP-15

"Nungesser-Coli"
Breguet Bre.19 A2

Nymph
Britten Norman BN-3

Ñamcú
FMA

O

Oasis
Lockheed P-3C Orion

Observer
Hoffmann (Flugzeugbau) H 38
Partenavia P68
Patchen TSC-2
Taneja P68

Obus (l'Obus)(Artillery Shell)
Borel

Ocean Hawk
Sikorsky (USA) H-60

Ochoaplane
International Airship

"Octave Chanute"
Blériot 165

Ogar (Hound)
PWS 35
PZL Bielsko SZD 45

"Oguri-Go"
Itoh Emi 20

Ohka (Cherry Blossom)
Kugisho MXY7

Ohtori (Phoenix)
Kokusai Ki-105
Mitsubishi

Ohtori-Go
Narahara Number 4

Oiseau Blanc (l'Oiseau Blanc)(White Bird)
Levasseur 8

Oiseau Bleu (Bluebird)
Farman F.180
Landes-Derouin

"Oiseau-Tango"
Bernard 80 G.R

Okapi
de Havilland D.H.14

"Oklahoma"
Travel Air 5000

"Old Glory"
Fokker F.VIIA

"Olive"
Weiss (José)

Olympia
Champion (Wisconsin)
EoN

"Olympia"
Bellanca J-300
DFS Meise (Titmouse)

Olympian
Hughes 500E

"Olympic"
Loughead Brothers

Olympic Orlik
Kocjan Orlik III

Omar
Suzukaze 20 SSF

Omega
Bratukhin
ISAE
Sabreliner (Socata)
Socata TB31

Omnivator
Johnson (E. R.)

One Design
Aircraft Spruce & Specialty DR.107

One Yankee
Zauner

1½ (One-and-a-Half) Strutter
Sopwith

Onyx
Piel

Optica
Edgley (Aircraft)
FLS

Optica Scout
Brooklands
Lovaux

Optimist
Edgley (Sailplanes)

Opus
Hagfors

Orange Biplane
Poulain

Orange Blossom (As: Kikka)
Nakajima

Orbis
McDonnell Douglas DC-10

Orenburgskii-Osoaviakhim
Pavlov

Orenco
Curtiss
Fujinawa

Oriole
Curtiss
Doyle

Oriole (As: Wilga)
PZL Warsawa-Okecie PZL-104

Orion *(See Also: Orione)*
Aerodis America AA200
CAC
Lockheed P-3, 9, 9-F

Orion-Explorer
Lockheed Model 9-E

Orione (Orion)
Macchi M.C.205N

Orlicanu
Aero (Vodochody) L 60

Orlik (Spotted Eaglet)
Kocjan
PZL Warsawa-Okecie PZL-130, PZL-140

Ornicopter
Froebe

Orniplane
Kemph

Ornis Monoplane
Wright (Howard)

Orowing
Pitcairn PA-3

Oryx
CATA

Osa (Wasp)
ZASPL

"Osa's Ark"
Sikorsky (USA) S-38BS Amphibion

Oscar
Partenavia P64

Oscar
Nakajima Ki-43

Osprey *(See also: Pursuit Osprey, Super Osprey)*
Bell Boeing V-22
Curtiss CW-14
Hawker
Monaghan
Ogden

Osprey Triplane
Austin

Ostrovia
Morisson and Nawrot

Otter *(See also: Sea Otter, Turbo Otter, Twin Otter)*
de Havilland (Canada) DHC-3

Ouragan (Hurricane)
Dassault

Outlaw Hunter
Lockheed P-3C Orion

Ovation
Mooney

Overmount X
Dayton

Overstrand
Boulton Paul

Owl
Curtiss A-2, O-52
LWF H

Owl (As: Sova)
Kappa KP-2U

Owl (As: Uhu)
Focke-Wulf Fw 189
Heinkel He 219

Owlet
G.A.L. 45

Oxford
Airspeed
de Havilland D.H.11

P

P-Craft
Paup

"P-Shooter"
Long (Dave) LA-1

"Pabco Pacific Flyer"
Breese-Wilde

Pacemaker
Bellanca

Pacer
Pacer Aircraft Corp
Piper PA-20
Piper PA-22S-150

Pacific
Wassmer WA.51

Pacific Hawk
Timm

Packet *(See also: Jetpacket)*
Fairchild C-82, F 78, R4Q
Steward-Davis

Packplane
Fairchild C-120

"Pacusan Dreamboat"
Boeing YB-29J

Paddle-Wheel Flying Machine
Hérard

Page Air Coach
Lincoln (Aircraft)

Page Racer
Curtiss XF6C-6

Painted Cloud (As: Saiun)
Nakajima C6N

Pajtás
Lampich D-2

Pal
Parker (Willard)

Palmer
Southwestern

Pampa
FMA

Panda
Guangzhou

Pandora
Douglas DB-7

Panther
Aérospatiale SA 365M
Grumman F9F
Huff-Daland AT-2
Keystone LB-1, LB-6
Parnall

Panther Plus
Rotec

Papagal
Caproni Bulgaria

Papillon (Butterfly)
Caproni Bulgaria

Papoose Jungster
Kaminskas

Parabola
Horten

Parabole
Cheranovsky

Paraplane
Dietz (Howard J.)
Lanier (Edward M.)
Paraplane
Parisano

Paraplane Commuter
Lanier (Edward M.)

Paraplane Executive
Lanier (Edward M.)

Paraplane Gazelle
Lanier (Edward M.)

Paraplane Le Bird
Lanier (Edward M.)

Paraplane Sportster
Lanier (Edward M.)

Paraplane Super Commuter
Lanier (Edward M.)

Paras-Cargo
Croses

Parasol
Parnall

Parasol Sport
Storms

Paravan
Tupolev Tu-2

Paris
Morane-Saulnier

"Paris-Madrid"
Borel 1912 Monoplane

Park
Park Bros

Parker's Iron Balloon
Short S.4

Paroquet
Cleone 5M

Partner
Fokker P-1

Partridge
Boulton Paul

Passat Multoplane
RFB RW-3

"Passepartout" ("Master Key")
de Marçay

Pat
Tachikawa Ki-74

Pathfinder
Keystone K-47
NASA
Piasecki 16H-1

"Pathfinder"
Bellanca J

"Patoruzu"
Biro de Ditro

Patrician
Keystone K-78

Patsy
Tachikawa Ki-74

Paul
Aichi E16A

"Pauletta"
Nemeth (Helicopter Co)

Paulistinha
Centro Técnico Aerospacial
Neiva

Paulistinha Tourer
CAP

Pave Eagle
Beech U-22

Pave Hawk
Sikorsky (USA) H-60

Pave Low
Sikorsky (USA) H-53

Pawnee
Hiller VZ-1
Piper PA-25

Pawnee Brave
Piper PA-36

Pchela (Bee)
Aeroprogress T-203

Pchelka (Little Bee)
Antonov An-14

Peace Station
Boeing 707

Peacemaker
Fairchild AU-23

Pedco Special
Pfau

Pedro
GCA 1

Pegas (Pegasus)
Aeroprogress T-417
Tomashevich

Pegase (Pegasus)
Centrair

Pegass
Delta (Czech)

Pegasus (See also: Pegas, Pegase)
Aerodesign
Classic (Aero) H-3
Huff-Daland LB-1
Lippisch

"Pegasus"
Boeing Hydrofoil

Pegelow Monoplane
Rumpler

Peggy
Mitsubishi Ki-67

Pelegrim
Autogyro Design Bureau

Pelican
Commercial Aeroplane Wing Syndicate
Doman LZ-2A
Grumman UF
Huff-Daland
Keystone
Meteor Mirach 20
Sikorsky (USA) H-3
Ultravia
Willoughby (Hugh L.)

"Pelican"
Blackburn

Pélican (Pelican)
Caudron C.510

Pelican Sport
Ultravia

Pelican Tutor
Ultravia

Pelikan (Pelican)
Pilatus SB-2
PZL Warsawa-Okecie PZL-240
Raab-Katzenstein RK 2

Pellegrini
Libossart

Pellet
Blackburn

Pembroke
Percival

Penetrator
Perl PG-130

Penguin (See also: Gossamer Penguin)
Breese (Sydney S.)
Garber
Goodall
Penguin

Penguin Ground Trainer
Riverside

Penguin (Pingouin)
SNCAN 1002

Penteado Biplane
Chauvière

People's Fighter (Volksjäger)
Heinkel He 162

People's Plane (Volksplane)
Evans (Aircraft)

"Pepa, La"
Loring E.II

Pepsi Skywriter
Travel Air (4000) D-4-D

Peque
Iberavia I-11

Perch
Parnall

Peregrine
Gulfstream Aerospace
Miles M.8

Peregrine Falcon (As: Hayabusa)
Manko MT-1
Mitsubishi
Nakajima Ki-43

Pereira
Osprey

Performance 2000
Cagny

Perigee
Aerocar, Inc

Perry
Kawasaki Ki-10

Perth
Blackburn

Petan (Cobra)
Hughes H-64

Pete
Mitsubishi F1M

"Pete"
Howard (Benny) DGA-3

"Petit Brochet" ("The Little Brochet")
Brochet

Petit (Little) Prince
Robin

Petit SIM
Rogozarski

Peto
Parnall

Petrel
ATR
Billie
Huff-Daland
Percival
Slingsby T.13

Petrel (As: Haiyan)
Nanchang

Pétrel (Petrel)
Morane-Saulnier

"Petulant Porpoise"
Grumman J4F-2

Pfeil (Arrow)
Albatros Taube (Dove) Biplane
Aviatik
Dornier Do 335
LFG Roland
Union Flugzeug-Werke

Pfeilflieger (Arrow Flyer)
Lohner

Pfitzner Monoplane
Curtiss

Pfledge
Pterodactyl

"Pfttt"
Nimmo Special Racer

Phaeton
American Eagle A-251

Phalène (Moth)(See also: Super Phalène)
Caudron C.280, C.400

Phantom
Cessna Design #1
Flagg
Fokker F.26
Fowlie
Luscombe Model 1
McDonnell FD, F-4
Ultralight Flight

Phantom Knight
Phantom Knight

Pheasant (As: Fazan)
Caproni Bulgaria

Phinn Arrow Biplane
Phinn

Phoebus
Bölkow/MBB L-252

Phoenix
American Aero
Boulton Paul

Bryan
Heston

Phoenix (As: Ohtori)
Kokusai Ki-105
Mitsubishi

Phoenix Fanjet
Alberta

"Phoenix 6"
Lockheed R7V-1P

Phönix
Bölkow FS 24

Phrygane (Phrygian)
Salmson

Phryganet
CFA D.57

Picaflor (Hummingbird)
Farge

Picchiatelli
Breda Ba.201

Picchio
Procaer

Piernifero
Horten H Xb

Pigeon *(See also: Carrier Pigeon)*
Albree

Pigeon (As: Colombe)
Ladougne

Pijad
Avions de Colombia

Pika
Government Aircraft Factory (Australia)

Pike
Avro 523
Parnall

Pike (As: Shchuka)
Shcherbakov Shche-2

Pilgrim
Fairchild, Fairchild C-24, C-31
General Aviation GA-43

Pillán
Enaer T-35

Pillon
Piper PA-28R-300

Pine
Mitsubishi K3M

Pingouin (Penguin)
SNCAN 1002

Pinocchio
Piel

Pintail
Fairey

Pinto *(See also: Super Pinto)*
Mohawk
Smolinski
Temco

Pinwheel
Rotor-Craft RH-1

Pioneer *(See also: Prestwick Pioneer, Twin Pioneer)*
Budd BB-1
Marske
Northrop N-23
Scottish Aviation

Pioneer (As: Bandeirante)
EMBRAER

Pipistrello (Bat)
Savoia-Marchetti SM.81

Pipit
Parnall
Sisler SF-1

Piranha
Kahn

Pirat (Pirate)
Kucfir
PZL Bielsko SZD 30
Svenska

Pirate
Argonaut
Vought F6U
White (Buffalo, NY) Aircraft

"Pistol Ball"
Dawson

Pixie
Cranwell
Parnall
Pober

Plage Biplane
Rumpler

"Plainview" Hydrofoil
Boeing

Lockheed

Plaisant (Pleasant)
Lioré

Planalto
CAP
Instituto de Pesquisas Tecnologicas IPT-4

Planaphore
Pénaud (Alphonse)

Plank
Backstrom

Playboy
Kinner
Lincoln (Aircraft)
Stits SA-3A

Playmate
Stits SA-11A

Pleasant (As: Plaisant)
Lioré

Plebe
Temco Model 33

Plover
Huff-Daland HD-8
Parnall

Plymo-Coupe
Fahlin
Nicholas-Beazley

Pneumatic Glider
McDaniel (Taylor)

"Pobjoy Special"
Nicholas-Beazley

Pocono
Piper PA-35

Pogo
Convair FY

Pogo Stick
Lockheed FV

"Point d'Interrogation" ("Question Mark")
Breguet Bre.19 GR

Pointer (As: Wyzel)
PWS 33

Polar
Wideröe C-5

"Polar Star"
Northrop Gamma 2B

Police
Robinson (Helicopter) R22

Police Interceptor
Eipper

Poll Giant
Brüning

Polon
Blazynski

Pond Racer
Scaled Composites Model 76

Populair
Earl Aviation

Populaire (Popular)
Caudron

Popular
Grahame-White

Popuplane
Jodel

Porte Baby
Felixstowe

"Porteno"
Delest

Porter
Fairchild
Pilatus PC-6

Possum
Parnall

Pošták
Zlin Z-IX

Postjager
Pander

Postman
GAC Model 111-E

Pot-Bellied Brougham
Ryan B.1 Special

"Potlatch Bug"
Babcock

Pou du Ciel (Flying Flea)
Mignet

Pouplume
Croses

Pourquoi Pas? (Why Not?)
Berliaux et Salètes

Powered Capstan
Slingsby T.49C

Praga
Hillson

Preceptor
Percival

Predator
Advanced Technology Aircraft
Rutan Model 59
Scaled Composites Model 120

Prefect
Slingsby T.30

Pregnant Guppy
Aero Spacelines

Premier
Raytheon

Prentice
Percival

President
Percival
Robin

Pressurized Centurion
Cessna P210

Pressurized Grand Commander
Aero Commander 680FL(P)

Pressurized Skymaster
Cessna P337

Pressurized Super Skymaster
Cessna 337

Prestwick Pioneer
Scottish Aviation

Preti
Instituto di Aeronautica Politecnico di Milano

"Pride of Hollywood"
American Albatross

"Pride of Lemont"
Folkerts Speed King

"Pride of Los Angeles"
Catron and Fisk CF-10

Primer
Fairey

Prince *(See also: Petit Prince, Sea Prince, Survey Prince)*
Percival
Robin

Princess *(See also: Sky Princess)*
Saunders-Roe SR.45

Private Explorer
Private Explorer

Privateer
Consolidated PB4Y-2
Ireland

Privrednik-Go
Utva-65

Priwall
Caspar C.35

Procellaria
Reggiane CA 405

Proctor
Percival

Professor
Lippisch

"Professor"
Kassel

Promoter
Fokker F.25

Promyk (Small Radius)
Tarczynski & Stepniewski TS-1/34

Pronto
Keystone K-55

Prop-Copter
Bensen B-10

Propelloplane
Hiller X-18

Propjet Bonanza
Tradewind Turbines

Proposal
Miles M.36

Prospector
Bounsall (Curtis)
Cub (Piper)
Gemini
Percival (Edgar) EP.9
Victa EP.9

Protin Contal
Lioré

Provence
Breguet Bre.76

Provider
Fairchild C-123

Prowler
Grumman EA-6B

Prudden Transport
Curtiss

Prüfling (Examination)
Lippisch

Ptapta (Putt-Putt)
D.U.S.

Pterodactyl
Westland

Pterygoid
Lee (Cedric)

Ptiger
Pterodactyl

Ptraveler
Pterodactyl

Ptug
Pterodactyl

Pucara (Fortress)
FMA

Puchacz
PZL Bielsko SZD 50

Puchatek
PZL Krosno KR 03A

Puff the Magic Dragon
Douglas AC-47D

Puffin
Parnall
Wimpenny

Pulex
Phoenix P.6

Pullman *(See also: Air Pullman)*
Bristol

Pulqui (Arrow)
FMA

Pulsar
Aero Designs

Puma
Aérospatiale SA 330

Pup *(See also: Bull Pup, Bullpup, Roto-Pup, Sky Pup, Supapup)*
American Sunbeam LP-1
Beagle B.121
Culp Sopwith
Dart (UK)
Keystone NK-1
Neale
Sopwith

Pupil (As: Zogling)
Avia (France, Inter-War)

Pupil (As: Zögling)
Lippisch

Pupistrelle
Brochet M.B.50

Pursuit
Rans

Pursuit Osprey
Curtiss CW-17R

Pursuit Racer
Timm-Hughes

Pusher
Amphibious Aircraft

Pusher Safety Pin
Arpin A-1

Puss Moth
de Havilland D.H.80

Putbus
LFG Stralsund V.19

Putt-Putt (As: Ptapta)
D.U.S.

Pyry (Snowfall)
VL

Q

Q/STOL (QuietSTOL)
Grumman/Boeing

Q-Star
Lockheed

"QED"
Granville, Miller, and De Lackner

Qingting (Dragonfly)
Shijiazhuang

Quadrotor
Convertawings Model A

Quadruplane
Armstrong Whitworth F.K.10
Naglo

Quail
Aerosport

Quasar
Montel

Quasimodo (As: Kvazimodo)
Caproni Bulgaria

Queen Air
Beech
Swearingen

Queen Bee
Bee Aviation
de Havilland D.H.82

Queen Martinet
Miles M.50

Queen Wasp
Airspeed

"Question Mark"
Fokker C-2A

"Question Mark" (As: "Point d'Interrogation")
Breguet Bre.19 GR

Quetzalcoatl
TNCA 4-E-131

Quetzalcoatlus Northropi
MacCready

Quickie
Rutan Model 54

Quicksilver
Eipper

Quiet Bird
Quick Flight

QuietSTOL
Grumman/Boeing

R

Racak
Aerotrade

Raceabout *(See: Standard Raceabout)*

Racek
Ardea

Rafale (Squall)
Caudron C.530
Dassault
Robin

Raiden
Mitsubishi J2M

Raider *(See also: Sky Raider, Skyraider)*
Northrop C-125

Rainbow
Chadwick C-122S
Republic F-12
Sopwith

"Rainbow" (As: "Arc-en-Ciel")
Couzinet

Rally *(See also: Rallye)*
Rotec

Rally Sport
Rotec

Rally Spraymaster
Rotec

Rallye (Rally)(See also: Super Rallye)
Morane-Saulnier

Rallye (Rally) Club
Morane-Saulnier

Rallye Commodore
Morane-Saulnier

Rallye Minerva
Morane-Saulnier

Ram (Ariete)(See also: Flying Ram)
Reggiane RE 2002

Rambler
Reid (Aircraft Co)
Rinehart & Whalen

Rambler (As: Tulák)
Let-Mont

Ramier (Dove)
SNCAN 1101

"Rammer"
Zeppelin

Ramo-Planeur
Baron

Ranchwagon
Taylorcraft

Randy
Kawasaki Ki-102

Ranger *(See also: Lone Ranger, Rhon Ranger, Sea Ranger, Sherwood Ranger, Skyranger, Star Ranger)*
Babcock
Coffman
Consolidated P2Y
Mooney
Parker (W. L.)
Taylorcraft
Rearwin

Rangoon
Short

Ranquel
FMA

Rapace
Avia (France, Inter-War)

Rapide *(See also: Dragon Rapide)*
de Havilland D.H.84

Rapier
North American F-108

Raptor
Lockheed Martin F-22

Rason Warrior
Johns

Raven
Curtiss O-40
General Dynamics EF-111A
Hiller H-23

Raw Arrow
Iniziative Industriali Italiane

Raymor
Martinsyde F.2

Raz-Mut
St Germain

Ra'am
McDonnell Douglas F-15I

Rebel
Murphy (Aircraft Co)

Recruit
Ryan PT-22

Red Arrow
Simplex (OH)

Red Bare-Un
Miller (Merle B.)

Red Devil
Baldwin
Curtiss Baldwin

"Red One, The" ("El Colorado")
Borello

"Red Ship, The"
Grumman G-32A, G-58A

Red Wing
Aerial Experiment Assoc

Redhawk
Cessna 177

"Redhead"
Israel (Gordon)

Redigo
Valmet L-90TP

RediGO
Macchi M.290TP

Redkin (Shark)
PWS 102

Redskin
Mohawk

Redstone
Hull

Redwing
Robinson (UK)

Reed Falcon
Air Nova

"Regele Carolii"
Emsco

Regent
Robin
Neiva IPD 5901

"Regina"
Biro de Ditro

Regional Jet
Canadair
EMBRAER

Reiher (Heron)
DFS

"Reims Racer"
Curtiss-Herring No.1

Reisen
Mitsubishi A6M

Rekord
Opel

"Reliance"
Bellanca Pacemaker CH

Reliant
Stinson (Aircraft) AT-19, RQ, R3Q, SR

Remo
Robin

Remorqueur
Robin

Renegade *(See also: Turbo Renegade)*
Lake (Aircraft)
Lasher
Murphy (Aircraft Co)

Renegade Spirit
Murphy (Aircraft Co)

Renn Gebirg (Mountain Racer)
Lohner-Etrich

Renzan (Mountain Range)
Nakajima G8N

Reporter
Northrop F-15

Reppu
Mitsubishi A7M

Rescue Hawk
Sikorsky (USA) H-60

Rescuer
Piasecki HRP

Retriever
Piasecki HUP

Revelation
Slip Stream

Revolution
Warner Aircraft Inc

Rex
Kawanishi N1K1

Rheinland
Aachen FVA-10B

Rhino
Sopwith

Rhon Ranger
Mead

Rhönadler (Rhine Eagle)
Schleicher

Rhönbussard (Rhine Buzzard)
Schleicher

Rhönlerche (Rhine Lark)
Schleicher Ka-4

Rhönschwalbe (Rhine Sparrow)
Schleicher Ka-2

Rhönsperber (Rhine Sparrow Hawk)
Schleicher

Rigel
Aerodis America AA300

"Rigid Midget"
Culver

Rihn
Aircraft Spruce & Specialty DR.109

Riley
Temco

Ringwing
Piasecki PA-2

Ripon
Blackburn

Riser *(See: Solar Riser, Sun Riser)*

Rita
Nakajima G8N

Rival (As: Rywal)
Garstecki

"Rivets"
Falck

Riviera
Nardi FN.333

Ro-Go Ko-Gata
Kugisho

Road-Runner
Moore (Virgil)

Roadable
Bryan
Plane-Mobile

Roadable Autogiro
Pitcairn

Robbe
Rohrbach Ro VII

Robin
ABC Motors
Canadian Wooden Aircraft
Curtiss
de Havilland (Canada) CWA

Robur
Balaye

Roc
Blackburn B-25

Rocco
Rohrbach Ro V

Roche
Rohrbach Ro II

Rocket *(See also: Astro-Rocket, Cyclonic-Rocket, Lunar Rocket, Star Rocket, Steel Pier Rocket Plane, Strata-Rocket, Super Rocket, Turbo-Rocket)*
Gregg
Johnson (TX)
Maule Air
Regent
Republic P-44
Riley

Rocket Plane
Bugatti

Rodra
Rohrbach Ro IIIa

Rofix
Rohrbach Ro IX

Rofox
Rohrbach

Rogallo
Delta Wing Kites

Roland
Rohrbach Ro VIII

"Roland Festung" ("Fortress Roland")
Eisenlohr

Rolas
Rohrbach

Rollicopter
Kasmar

Roma
Lampich L-2

Romair
Waterhouse

Romar
Rohrbach Ro X

ROMAR
Kaman

"Roma"
Bellanca K

Rondone
Ambrosini

Ronix
Rohrbach Ro IV

Ron's I
Freiberger

Rooivalk (Kestrel)
Air Nova

Rook (As: Gratch)
Aeroprogress T-101

Rorex
Rohrbach Ro VI

Rossiya (Russia)
Tupolev Tu-114

Rostock
LFG V.16

Rostra
Rohrbach

Rota
Cierva C.30A
Turner (Barry)
Avro 671

Rota-Aircoach
RotaWings

Rotabuggy
Hafner

Rotachute
Hafner

Rotatank
Hafner

"Rote Presnia"
M.I.I.T.

Roteron
Thomas (W. H.)

Roto-Pup
Little Wing

Rotocar
Wagner

Rotodyne
Fairey

Rotor Ship
Volf

Rotor-Flyer
Chupp

Rotorchute
Prewitt

Rotorcoach
Saunders-Roe

Rotorcycle
Hiller ROE
Saunders-Roe 1033

Rotri
Rohrbach Ro III

Rovaire
Robertson (St Louis) RO-1504

Rover
States (TX)

Royal Eagle (As: Calquin)
FMA

Royal Standard (As: Guión)
Hispano HA-231

Royale
Robin

Rubber Bandit
Rubberworks

"Rubber" Glider
McDaniel (Taylor)

"Rubis"
Bassou

Rufe
Mitsubishi A6M2-N

Rukh
Pirie

Runabout
Crawford (Seal Beach)

Russia (As: Rossiya)
Tupolev Tu-114

Russia Clipper
Martin 156

"Russkiy Vityaz" (Russian Knight)
Sikorsky (Igor) S-21

"Rusty"
Gaffney-Haines Special

Rybka (Little Fish)
Mikhel'son-Korvin-Shishmarev

Ryusei (Shooting Star)
Aichi B7A

Rywal (Rival)
Garstecki

S

Sabre *(See also: Super Sabre)*
CAC
Canadair CL-13
North American F-86, F-95

Sabre (As: Tsurugi)
Nakajima Ki-115

Sabre Bat
North American

Sabre Commander
North American Sabreliner

Sabre Dog
North American F-86D

Sabreliner
North American, North American T-39

Sachem
Knoll Brayton

"Sacred Cow"
Douglas VC-54C Skymaster

Sadler Vampire
Wing (Research)

Saeta (Arrow) *(See also: Supersaeta)*
Hispano HA-200

Saeta Voyager
Hispano HA-230

Saetta (Arrow)
Macchi M.C.200

Safari *(See also: Kitfox Safari)*
Frye F-1
Robin
Saab MFI-15

Safety Pin
Arpin A-1

Safety Wing
Aircraft Methvin
Ford-Leigh

Safir (Sapphire)
Saab 91

"Sageburner"
McDonnell F-4A

"Saginaw Junior"
Carr

Sagitta
Vliegtuigbouw

"Sagittaire, le" ("Sagittarius")
Blériot 111

Sagittario (Archer)
Aerfer
Reggiane RE 2005

Sagittario (Sagittarius)
Ambrosini

Sahara
Breguet Bre.76

Saigai
Saito

Saïgon
Breguet Bre.53-0

Sailaire
Robinson (James Thomas)

Sailing Apparatus (As: Normal Segelapparat)
Lilienthal (Otto)

Saiun (Painted Cloud)
Nakajima C6N

Sakamoto
Tachibana

Sakota
Rans

"Sakura" ("Cherry Blossom")
Kawanishi K-12

Salamander
Sopwith

Salamandra (Salamander)
Czerwinski (Waclaw) W.W.S.1

Sally
Hiller H-32

Sally
Mitsubishi Ki-21, MC-21

Salto
Start + Flug H101

Sam
Mitsubishi A7M

Samaritan
Convair C-131 R4Y

"San Francisco"
Keith Rider R-1

Sancy
Wassmer Super IV

Sander
Opel

Sandringham
Short S.25

"Santa Clara"
Montgomery (John)

"Santa Lucia"
Bellanca J-3-500

"Santa Maria"
Keystone K-47 Pathfinder

"Santa Rosa Maria"
Bellanca J

Saphir (Sapphire)
Piel

Sapphire (As: Safir)
Saab 91

SARA
CSIR

Saracura
Instituto de Pesquisas Tecnologicas IPT-3

Sarafand
Short S.14

Saras
National Aerospace Labs

Saratoga
Piper PA-32R-301

Sarohale
Socata TBM60.000

Sassnitz
LFG V.18

Satellite
Planet
Short S.4
Superior

Saturn
Lockheed Model 75

Saturnian
Dumoulin

Satyr
Miles M.1

Sauro
Caproni-Trigona

Sausewind (Southwind)
Baümer B.IV

Sauterelle (Grasshopper)
Potez 600 T2

Savage
North American AJ

Savage (As: Féroce)
Fairey

SAVER
Kaman

Scamp
Aerosport
Agrocopteros
Mitchell-Prizeman

Scandia
Saab 90

Scapa
Supermarine

Scarab
R.A.E.
Supermarine

Scepter
Slip Stream
Roberts Sport Aircraft

Schatzie (Little Treasure)
Hoffelmann CH-1

Schelde Gull (As: Scheldemeeuw)
de Schelde

Schelde Sparrow (As: ScheldeMusch)
de Schelde

Schlacro
Akadflieg München Mü 30

"Schlesien in Not"
Breslauer

"Schloss Mainberg"
Kassel

Schneider
Sopwith

Scholar (As: Estudiantil)
Marichal

Schoolmaster (As: Magister)
Fouga CM 170

Schoolmaster (As: Schulmeister)
Burgfalke N.150

Schroeder-Meyer Special
Schroeder

Schudeisky Biplane
Rumpler

Schule
LFG V.17

Schulmeister (Schoolmaster)
Burgfalke N.150

Schwalbe (Swallow)
Etrich
Grade 1912 Monoplane
Messerschmitt Me 262
Raab-Katzenstein Kl I (K I)

"Schwarze Teufel" ("Black Devil")
Aachen FVA-1

Scimitar
Armstrong Whitworth A.W.35
Masak
Supermarine

Scion Alpha
Short S.16

Scion Senior
Short S.22

Scirocco
R.A.E.

Scooter *(See also: Aeroscooter, Air Scooter, Sky Scooter)*
Flaglor
Sopwith

Scorpion
Northrop
Peyret-Mauboussin
Rotorway

Scout *(See also: Aeroscout, Air Scout, Arrowscout, Baby Scout, Combat Scout, Hart Scout, Optica Scout, Sioux Scout, Speed Scout, Wibault Scout)*
Aeronca C-2, 11
Albree
Alcock A.1
American Champion
Bellanca
Boeing L-15
Bristol
Cessna DC-6B
Champion (Texas)
Curtiss CS, SC
Curtiss
Grahame-White Type 13
Pemberton Billing P.B.9, P.B.25
Pietenpol
Topa
Tunison
Victor (NY)
Westland

Scout-STOL
Skytrader

Scoutmaster
Lovaux

"Scrapheap, The"
Thompson (William)

"Screamin Wiener"
Culver

Scricciolo
Aviamilano P.19

SCUB
Aero-Kuhlmann

Scud
Baynes

Scylla
Supermarine

Sea Balliol
Boulton Paul

Sea Bat *(See also: Seabat)*
Piasecki PA-4

Sea Bird
Fleetwings
Lakes

Sea Devil
Century (IL)

Sea Devon
de Havilland D.H.104

Sea Dragon
Sikorsky (USA) H-53, S-80

Sea Eagle
Supermarine

Sea Falcon (As: Seefalke)
Dornier Do H, Do 26

Sea Fury
Hawker

Sea Gladiator
Gloster

Sea Gull
Willoughby (Hugh L.)

Sea Harrier
British Aerospace
Hawker Siddeley

Sea Hawk
Aero Composite
Hawker
Sikorsky (USA) H-60, S-70

Sea Hawker
Aero Composite
Leg-Air

Sea Heron
de Havilland D.H.114

Sea Hornet
de Havilland D.H.103

Sea Hurricane
Hawker

Sea King
Sikorsky (USA) H-3
Supermarine
Westland

Sea Knight
Boeing-Vertol H-46

Sea Lion
Supermarine

Sea Mosquito
de Havilland D.H.98

Sea Otter
Supermarine

Sea Pirate
Eastman

Sea Prince
Percival

Sea Ranger
Bell H-57

Sea Rover
Eastman

Sea Scan
Israel Aircraft Industries

Sea Stallion
Sikorsky (USA) H-53, S-65

Sea Tutor
Avro 646

Sea Urchin
Supermarine

Sea Vampire
de Havilland D.H.100, D.H.115

Sea Venom
de Havilland D.H.112

Sea Vixen
de Havilland D.H.110

Sea Wings Hydrofoil
Grumman CH-6

Sea-Hawk
Richmond

***Seabat** (See also: Sea Bat)*
Sikorsky (USA) HSS

Seabee
Republic RC-3

Seabird
Alliance P.2
Taylorcraft

Seacobra
Bell AH-1J, AH-1T

Seadart
Convair F2Y

Seafang
Supermarine

Seafire
Supermarine

Seaford
Short S.45

Seafox
Fairey

Seageep
Piasecki PA-59N

Seaguard
Sikorsky (USA) H-52

Seagull
Curtiss MF, O3C, SOC, SO2C, SO3C
Naval Aircraft Factory SON
Passat
Shenyang HU-1
Supermarine
Supermarine

Seagull (As: Galeb)
Soko G2-A

Seagull (As: Goéland)
Caudron C.440

Seagull (As: Maager)
Orlogsvaerftet FB.III

Seagull (As: Mouette)
Breguet Bre.901 S

Seagull (As: Möwe)
Focke-Wulf A 17, A 29, A 38
Geest
Nordflug

Seagull (As: Tchaika)
Beriev Be-12

Seahawk
Curtiss F7C, SC

Seahorse
Sikorsky (USA) HSS, HUS

Seal
Fairey
Supermarine

Sealand
Short S.44, S.B.7

"SeaLandAir"
Harriman (Aeromobile Co)

Seamaster
Martin P6M

Seamew
Curtiss SO3C-2C
Short S.B.6, S.C.2
Supermarine

Seamistress
Martin P6M

Seaquick
Eipper

Searchmaster
Ayres Corp
Government Aircraft Factory (Australia)

Seasprite
Kaman H-2

Seastar
Dornier CD2
Lockheed T2V

Seatarejo
EMBRAER

Seawind
Seawind/S.N.A.

Seawolf
Convair TBY
Lake (Aircraft)
Vought TBU

***Sedan** (See also: Air Sedan, Airsedan, Glenmont Landau Sedan)*
Aeronca 15AC
Alexander Eaglerock
Curtiss CW-15
Dayton Wright
Jovair
Kinner

Sedbergh
Slingsby T.21B

Seeadler (Sea Eagle)
Schleicher

Seefalke (Sea Falcon)
Dornier Do H, Do 26

Seely
Bristol

Segrave
Blackburn

Seicho-Go
Awazu

Seiran (Clear Sky Storm)
Aichi M6A

Seiran Kai
Aichi M6A1-K

Sekani
Fairchild 45-80

Selena
Dubna Z-2

Semi-Quaver
Commercial Aeroplane Wing Syndicate

Seminole
Beech L-23, U-8
Piper PA-44-180

Semiquaver
Martinsyde

Sempu-Go
Izaki

"Senator"
Zeise-Nesemann

Senden
Mitsubishi J4M

Seneca
Cessna H-1
EMBRAER
Neiva
Piper PA-34

Senior *(See also: Scion Senior, Koliber Senior, Sperber Senior)*

Senior Aero Sport
d'Apuzzo D-295

Senior Airsedan
Buhl Airsedan CA-8

Senior Albatross
Bowlus-duPont 1-S-2100

Senior Cadet (As: Edelkadett)
Heinkel He 72B-3

Senior Cruisair
Bellanca

Senior Pacemaker
Bellanca

Senior Skyrocket
Bellanca

Senior Sportster
Granville Y

Sentinel
Enstrom
Fokker F.50
Lockheed P-3 AEW
Stinson (Aircraft) OY-1, L-5
Stinson (Aircraft)

Sentry
Boeing E-3

Sep (Vulture)
PWS

Sequoya
Piper PA-60-290

Seraph
Bennett (IA)

Sergeant *(See: Top Sergeant)*

Serrania
Aero-Jean

Sesqui-Wing
Pitcairn PA-2

Seven Seas
Douglas DC-7C

Severn
Saunders-Roe A.7

Shackleton
Avro 696

Shadow *(See also: Streak Shadow)*
CFM
Fairchild AC-119G
Laron

"Shamrock"
Short N.1B Shirl

Shark *(See also: Air Shark, Dark Shark)*
Blackburn B-6
Enstrom

Shark (As: Haifisch)
LFG Roland D.I

Shark (As: Redkin)
PWS 102

Shavruske
Shavrov Sh-2

Shawnee
Piasecki H-21

Shchuka (Pike)
Shcherbakov Shche-2

Sheat-fish (As: Sum)
PZL P.46

Shekari
Rans

Sheldrake
Supermarine

Sheriff
Aircraft Designs (UK)

Sherpa
Sherpa Aircraft
Short S.B.4
Shorts 330-200

Sherwood Ranger
TCD Ltd

Shetland
Saunders-Roe S.35
Short
Short S.35

Shiden (Violet Lightning)
Kawanishi N1K1-J

Shiden Kai
Kawanishi N1K2-J

Shinden (Magnificent Lightning)
Kyushu J7W1

Shinzan (Heart of Mountains)
Nakajima G5N

Shiragiku (White Chrysanthemum)
Kyushu K11W1

Shirl
Short N.1B

Shiun
Kawanishi E15K1

Shoestring
Condor Aero
Mercury Air

Shoki (Devil Queller)
Nakajima Ki-44

Shooting Star
Lockheed F-80, TV, T-33

Shooting Star (As: Ryusei)
Aichi B7A

Shoreham
Radley-England

Shorthorn
Farman (Maurice) M.F.11

Shrike
Curtiss A-8, A-10, A-12, A-14, A-18, A-25, S2C

Shrike Commander
Aero Commander 500S

Shturmovik
Ilyushin Il-2, Il-10

Shusui (Sword Stroke)
Mitsubishi J8M, Ki-200

"Sibylla"
de Havilland D.H.52

Sicile
Robin

Sidecar
Blackburn

Sidestrand
Boulton Paul

Sierra
Advanced Aviation
Aircraft Industries (FL)
Beech

Sigma
Moskalev SAM-7

"Sign Carrier I"
Keystone K-47 Pathfinder

Sigrist Bus
Sopwith

Silbervogel (Silverbird)
Sänger

Silvaire
Luscombe Model 8, 10

Silvaire Deluxe
Luscombe Model 8

Silvaire Sedan
Luscombe Model 11A

Silvaire Trainer
Luscombe Model 8

Silver Arrow (As: Ezüst Nyíl)
Weiss Manfred WM.23

"Silver Beetle"
Jensen Helo

Silver Dart
Aerial Experiment Assoc
Canadian Aerodrome
Walsh

Silver Eagle
Hamilton Metalplane H-19
Mitchell (Aircraft) A-10

Silver King
Avro 528

Silver Sea-Dan
Hamilton Metalplane H-22

Silver Sixty
Rhodes Berry

Silver Star
Canadair CT-133
Lockheed T-33

"Silver Stirling"
Short S.37

Silver Streak
Brown Metalplane

Hamilton Metalplane H-21
Short

Silver Swan
Hamilton Metalplane H-20

Silverbird (As: Silbervogel)
Sänger

Silvercraft
SIAI-Marchetti SH-4

Silverstar
Canadair CL-30

Silverwing
Cessna

SIM
Rogozarski

Simoom (Simoun)
Caudron C.520, C.620, C.630

Sinaia
Armstrong Whitworth
Siddeley Deasy

Sinbad
Douglas

Singapore
Short

Sioux
Bell H-13, HTL
Westland

Sioux Coupe Jr
Kari-Keen

Sioux Scout
Bell 207

Sirius
Bénégent
Lockheed Model 8

Sirocco
Aviasud
Wibault 12 C2 (121)

Siskin
Armstrong Whitworth
Siddeley Deasy S.R.2

Sissit
Armstrong Whitworth F.K.1

Sisu
Arlington

Six
Avro 624

Skaut
Mráz M-2

Skeet
Pifer P-1

Skeeter
Cierva W.14
Curtiss CR-1
Fisher FP-505
Fokker
Saunders-Roe P.501

Skeeter Hawk
Williams (Robert F.)

Skeeto
Stits SA-8A

"Skeets"
Williamson (John Wesley)

Skimmer *(See also: Air Skimmer)*
Aeronautical & Automobile College
Colonial
Vought F5U

Skimmer Tach IV
Colonial

Skipper
Beech

Skua
Blackburn B-24

Skvorets (Starling)
Aeroprogress T-407

Sky
Slingsby T.34B

Sky Arrow
Iniziative Industriali Italiane
Pacific AeroSystem

Sky Baby
Clancy
Fisher FP-606
Stits SA-2

Sky Camper
Pietenpol

Sky Car
Kindree

"Sky Chief"
Northrop Gamma 2A

Sky Cycle
B W Rotor

Sky Ghost
Gross

Sky Hawk
Seibel YH-24

Sky King
Robin

Sky Princess
Rogers (Frank)

Sky Pup
Sport Flight

Sky Queen
Robin

Sky Raider *(See also: Skyraider)*
Worldwide Ultralite

Sky Scooter
Thorp T-11

Sky Sedan
Piper PA-6

Sky Trac
Champion (Wisconsin)

Sky-Car
Piasecki PA-59K

Sky-Mat
Bensen B-12

Sky-Scooter
Bensen B-4

Sky-Trac
Wagner

Skybaby
Kauffold

SkyBlaster
Slip Stream

Skybolt
Steen

Skyboy
Interplane

"Skyburner"
McDonnell F-4A

Skybus
Douglas DC-8 Proposal

Skycar
Convair Model 103
Stout

Skycoupe
Piper PWA-1
Stits

Skycraft
Skylark

Skycrane
Sikorsky (USA) S-64

Skycycle
Carlson
Piper PA-8

Skyfarer
General Aircraft Corp G1-80

Skyfly
Cvjetkovic CA-65

Skyhawk *(See also: Super Skyhawk)*
Cessna 172
Douglas A4D

SkyHawk
Buckeye

Skyhook
Australian Autogyro
Cessna H-1
Rotor-Craft

Skyhopper
Aviation Boosters

Skyjeep
Chrislea CH.4

Skyknight
Cessna 320
Douglas F3D

Skylancer
Douglas F5D

Skylane
Cessna 182

Skylark
Bethlehem
Cessna 175
Convair 880
Dietz-Schriber
Driggs
Hempstead
Ichimori
Pasped
Phillips (Aviation Co)
Robertson Development SRX-1
Shepherd

Slingsby T.37, T.41, T.43, T.50
Watkins

Skylark (As: Hibari)
Nihon (Hikoki KK)

Skymaster
Cessna 336
Douglas C-54, R5D

Skypirate
Douglas TB2D

SkyQuest
Slip Stream

Skyraider
Douglas AD

Skyranger
Aerodyn'
Rearwin
Striplin

Skyray
Douglas F4D

Skyride
Glenview GMP-1

Skyrider
Pasped

Skyrocket
Bellanca
Douglas D-558-2
Grumman F5F

Skyscooter
Saalfeld

Skyscraper Biplane
Gallaudet (E. F.)

Skyservant
Dornier Do 28

Skyshark
Douglas A2D
Skycraft (TX)

Skysport
Braley (Glider)

Skystreak
Douglas D-558-1

Skytrader
Dominion

Skytrain
Douglas C-9B, C-47, R4D, C-117

Skytrooper
Douglas, Douglas C-53

Skytruck
PZL Mielec M28
Steward-Davis Jetpacket 3400
Shorts

Skyvan
Fairchild YC-119H
Shorts, Shorts S.C.7

Skywagon
Cessna 180, 185, 207

Skywarrior
Douglas A3D

Skyway
Capen

"Skyway Express"
California Aero Glider

Slask (Silesia)
Gabriel

Slavutich
Antonov

"Slick"
Levier

Sluka
Letov Air LK-2M

Small America
Curtiss Model H

Small Radius (As: Promyk)
Tarczynski & Stepniewski TS-1/34

Smaragd (Emerald)
Binder

"Smith IV"
SPAD XIII

Smut
Scaled Composites Model 133

Smyk (Brat)
PZL Swidnik PW-5
Moczarski, Idzkowski, and Ploszajski

Snail
Sopwith

Snapper
Sopwith

Snargasher
Reid and Sigrist

Snark *(See also: De Bruyne Snark)*
Sopwith

Snipe *(See also: "Dragonfly Snipe")*
Philippine Aviation Corp
Pioneer (Aero Trades School)
Sopwith

Snoop
Eastern Ultralights

Snowfall (As: Pyry)
VL

Snyder Special
Flagg

Soaring Bird
Huffaker

"Soaring" (As: "l'Essor")
Fabre

Sociable
Sopwith

Sociable, The
Orenco Type P-19

Soga-Go
Ozaki

Šohaj
Zlin Z 25

Sohlmann Monoplane
Rumpler

Sokol (Falcon)
Aeroprogress T-401
Mráz M-1

Sokól (Falcon)
Kocjan
PWS 42, P.45
PZL Swidnik W-3

Sokolu
Beneš-Mráz

Solar Challenger
MacCready

Solar One
Solar Powered Aircraft Development TO

Solar Riser
UFM

Solent
Short S.45A
Supermarine

Solitaire
Mitsubishi MU-2
Rutan Model 77

"Solution"
Laird ("Matty") LC-D Speedwing

Sonerai
Monnett

"Song Bird"
Cessna 310B

Sonia
Mitsubishi Ki-51

Sooper-Coot
Aerocar, Inc

Sopocaba
Conal W-151

Sopwith Cruiser
Decatur T-W-1

Souris Bleue (Blue Mouse)
Aachen

Southampton
Supermarine

"Southern Cross"
Fokker F.VIIB-3m

"Southern Cross" ("Croix du Sud")
Latécoère LAT 300

Southern Martlet
Miles

Southern Mountain (As: Nanzan)
Aichi M6A1-K

"Southern Star"
Sikorsky (USA) S-37

Southernaire
Southern (TX)

Southwind (Sausewind)
Baümer B.IV

Sova (Owl)
Kappa KP-2U

Spacewalker
Australian Aviation

StarFire
Star Flight

Starflight
Monsted-Vincent MV-1

Starlet
Stolp SA-500

Starlifter
Lockheed C-141

***Starliner** (See also: Unitwin Model 2 Starliner)*
Lockheed Model 1649

Starling
Armstrong Whitworth A.W.XIV

Starling (As: Skvorets)
Aeroprogress T-407

"Stars and Stripes"
Fairchild FC-2W2

Starship
Beech

Starship Alpha
Davis Wing

Starwing
Massillon

Statesman
Mooney

Stationair
Cessna 206, 207

Stayer
Dubna Z-8

"Steam Air Liner"
de la Landelle

Steel Dove (As: Stahl Taube)
LFG Roland

Steel Dove (As: Stahltaube)
Fokker M 4
Jeannin

Steel Pier Rocket Plane
Swan (William)

Steel Truss Glider
Boeing (Canada)

Steffen
Siemens-Schuckert

Stein Monoplane
Rumpler

Steinmann (Stoneman)
Lippisch

Stella
Kokusai Ki-76

Stelling
Cox-Klemin

Stellite
Short S.1

Stemal
Malinowski

Step Glider
Setter

Sterkh
Aeroprogress T-201

Stieglitz (Goldfinch)
Focke-Wulf Fw 44

Stiletto
Douglas X-3

Stinger
Fairchild AC-119K
Williams (Aircraft Design) W-17

Stingray
Ace (Australia)

***Stirling** (See also: Super Stirling, "Silver Stirling")*
Short S.29

STOLmaster
Steward-Davis

Stomo
Möller

Stoneman (As: Steinmann)
Lippisch

Storch (Stork)
DFS
Fieseler Fi 156, Fi 256
Lippisch

Stork
Nihon University

Stork (As: Cicogna)
Fiat BR.20

Stork (As: Jabiru)
Farman F.3X F.4X F.170

Storm Bird (As: Sturmvogel)
Goedecker

Storm Dragon (As: Donryu)
Nakajima Ki-49

Storm Flyer (As: Sturmfügelmodell)
Lilienthal (Otto)

Storm-Bird (As: Sturmvogel)
Messerschmitt Me 262

Störtebeker
Junkers Ju 388 J

Stösser (Falcon)
Focke-Wulf Fw 56

Stove Bolt
Miller (M.) M-6

"Strana Sovyetov"
Tupolev TB-1

Stranraer
Supermarine

Strata-Rocket
Maule Air

Strato
Grob

Strato-Courier
Helio

Stratocruiser
Boeing 377

Stratocruzer
Advanced Aerodynamics & Structures

Stratoferic
Slingsby T.44

Stratofortress
Boeing B-52

Stratofreighter
Boeing C-97

Stratojet
Boeing B-47

Stratolab
WIB

Stratoliner
Boeing C-75, S-307

"Stratosphere"
Guerchais

Stratotanker
Boeing KC-135

***Streak** (See also: "Blue Streak," Silver Streak, Skystreak, Star Streak, Thunderstreak)*
Aero-Flight
CFM
Comper

Streak Eagle
McDonnell Douglas F-15

Streak Shadow
CFM
Laron

Streamline Gunbus
Vickers (UK) F.B.9

Strela
LFG V.13
Moskalev SAM-9

Strike Eagle
McDonnell Douglas F-15

Strikemaster
BAC 167

Strizh (Martin)
Aeroprogress T-420, T-501

Stromer
Möller

Student
Bücker Bü 180

Student Prince
American Aircraft Builders

Stuka
Junkers Ju 87

Sturgeon
Short S.6, S.38, S.B.9

Stürmer
Möller

Sturmflug
AGO

Sturmfügelmodell (Storm Flyer)
Lilienthal (Otto)

Sturmvogel (Storm Bird)
Goedecker

Super Viking
Bellanca

Super Vivat
Aerotechnik (Czech) L-13

Super Whale (Superwal)
Dornier Do R

Super Widgeon
McKinnon

Super Wing
Mitchell (Aircraft)

Super Ximango
Aeromot

Super Zodiac
Zenair

Super-Bibi
Beneš-Mráz

Super-Catalina
Steward-Davis

Super-Cigale
Aubert

Super-Goliath
Farman F.140

Super-Gull
Mill Basin

Super-Swallow
Swallow (KS)

Supercab
Béarn

Supercat
Bowdler

SuperCobra
Bell AH-1W

SuperFloater
US Aviation

Superfortress
Boeing B-29

Supersaeta
Hispano HA-220

Supervale (Farewell)
Magni

Superwal (Super Whale)
Dornier Do R

Supporter
Saab MFI-17

Survey
Gloster AS.31

Survey Prince
Percival

Surveyor
GAC Model 101

Susanne
LFG V.3a

Susie
Aichi D1A

"Suzie Jayne"
Robinson (Bill) Special

Suzuki
Tachibana

Svenska
Saab 22

Sverchok (Cricket)
Kuibyshev

***Swallow** (See also: New Swallow, Super-Swallow)*
British Aircraft Mfg
Laird ("Matty")
Slingsby T.45
Sopwith

Swallow (As: Golondrina)
Mira

Swallow (As: Hien)
Kawasaki Ki-61

Swallow (As: Hirondelle)
Albert A.10
Dassault

Swallow (As: Jaskólka)
ITS
Libanski

Swallow (As: Schwalbe)
Etrich
Messerschmitt Me 262

Swallow (As: Tsurubane)
Itoh Emi

Swallow Moth
de Havilland D.H.81

"Swallow" ("Tsubame-Go")
Itoh Emi 6

***Swan** (See also: Silver Swan)*
Supermarine
Willoughby (Hugh L.) Model H

Swastika
Babcock

Swati
Bharat

"Sweet P Field"
Porterfield Model 65

Swee'Pea
Chester (Art)

***Swift** (See also: Millennium Swift, Nightswift)*
Blackburn
Comper
Cranwell C.L.A.7
Curtiss P-31
Globe
Supermarine
Swift (KS)
Temco
Wendt (NY) W-2

Swift (As: Martinet)
SNCAC NC.701

Swiftfire
LoPresti (Piper)

Swiftfury
LoPresti (Piper)

Swing
S-Wing

SwissTrainer
Dätwyler MD-3

"Swoose"
Boeing B-17D Flying Fortress

Sword (As: Fleuret)
Morane-Saulnier

Sword Stroke (As: Shusui)
Mitsubishi Ki-200

Swordfish
Fairey
Lockheed P-38E Lightning

Swordfish (As: Espadon)
SNCASO SO 6020

Sycamore
Bristol

Sydney
Blackburn

Sylphe (Sylph)
Gassier
Fouga CM 8-R

"Sylvia"
Weiss (José)

Syrie (Syria)
SPCA Météore (Meteor) 63

Szpak
LWD

T

T-Bird
Parker (Raymon H.) RP-9

T-Boat
Patchen TSC-1
Thurston TSC-1

T-Raptor
Capella

Tabby
Nakajima L2D
Showa L2D

Tabloid
Sopwith

Tabor
Tarrant

TACAMO
Boeing E-6

Tacit Blue
Northrop

Tacuara
Baserga

Tadpole
Curtiss
Grumman G-65

Tadpole Clipper
Martin 162A

Tagak
Institute of Science and Technology XL-15

Pope (Leon)
Worldwide Ultralite

Thunderbolt
Fairchild A-10
Jurca MJ-11
Republic P-47

Thunderbolt (As: Folgore)
Macchi M.C.202

Thunderbolt (As: Viggen)
Saab 37

Thunderceptor
Republic F-91

Thunderchief
Republic F-105

Thunderflash
Republic RF-84F

Thunderjet
Republic F-84

Thunderstreak
Republic F-84, F-96

Thurston Monoplane
Aerial Engineering Corp

Tiara
Robin

Ticicolo
Maestranza Central de Aviación

Ticker (As: Cykacz)
Dabrowski

Tidal Wave (As: Mascaret)
Jodel

Tierra
Teratorn

Tiger *(See also: AeroTiga, Ptiger, Super Tiger)*
Eurocopter
Fieseler F 2
Grumman American AA-5B
Grumman F11F
Northrop F-5
Thorp T-18
White (Buffalo, NY) Aircraft

Tiger Cub
TCD Ltd

Tiger Moth
de Havilland D.H.71

Tiger Moth
de Havilland D.H.82

Tigercat
Grumman F7F

Tigereye
Northrop RF-5E

Tigerschwalbe (Tiger Swallow)
Raab-Katzenstein F 1

Tigershark
Northrop F-20

Tigrotto
Partenavia P52

Tillie
Yokosuka H7Y

"Time Flies"
Miller (Howell) HM-1

"Tin Cow, The"
Wright (Co) Model F

Tin Kettle
Short S.4

Tin Wind
Parker (C. Y.)

Tina
Mitsubishi Ki-33

"Tingmissartoq"
Lockheed Model 8

Tinker Toy
Taylor (Don)

Tiny Mite
Parker (Raymon H.)

Tipsy
Fairey

Tipsy Belfair
Fairey

Tipsy Junior
Fairey

Tipsy Nipper
Fairey

Titan
Aeroprogress T-274
Cessna 404

"Titania"
Fairey N.4

Titmouse (As: Meise)
DFS

TNT
Dornier

Tobago
Aérospatiale TB-10
Socata TB10

Tobi (Final Effort)
Mitsubishi

Tojo
Nakajima Ki-44

Tokai (Eastern Sea)
Kyushu Q1W1
Watanabe (Tekkosho) Q1W1

Tokubaku
Nakajima N-35

Tomahawk
Curtiss P-40
Historical Aircraft Corp
Kaman H-2
Piper PA-38-112

Tombo (Dragonfly)
Mitsubishi

Tomboy
Heath

Tomcat *(See also: Super Tomcat)*
Grumman F-14

Tomcat, El
Continental Copters

Tomcat Sport
Waspair H.M. 81

Tomcat Standard
Waspair H.M. 81

Tomcat Super XC Tourer
Waspair H.M. 81

Tomcat XC Tourer
Waspair H.M. 81

Tomtit
Hawker

"Tonatiuh II"
Dirección General de Reparaciones...

Tony
Kawasaki Ki-61

"Toots"
Folkerts

Top Sergeant
American Helicopter XA-5

"Topeka I"
Longren #1 Pusher

Topper
Taylorcraft

Topsy
Mitsubishi Ki-57
Mitsubishi L4M1, MC-20

Tora
Mitsubishi B1M

Tornado
Hawker
Johns
North American B-45
Panavia
Partenavia P55

Torpedoplane
Curtiss CT

Torpille, Aéro-Torpille (Torpedo)
Artois
Borel-Ruby
Paulhan-Tatin

Torpilleur
Farman

Tortoise (Tartuca)
CVV PM.280
Instituto di Aeronautica Politecnico di Milano

Toryu (Dragon Slayer)
Kawasaki Ki-45

Totem
Boeing (Canada)

Toucan
Canaero Dynamics

Toucan (As: Tucan)
Turbay T-1

Toucan (As: Tucano)
Alaparma AM-10

Tourabout *(See: Standard Tourabout)*

Tourbillon
Chasle YC-12

Tourer *(See also: Air Tourer, Airtourer, Tomcat XC Tourer)*
Bristol

Noury

Tourisme
Croses

Tourist
Avro 548
Taylorcraft

Tourister
Orenco Type F

Touroplane
Wallace

Tout-Terrain
Croses

Tracer
Grumman WF

Tracker
Grumman S-2

Tractor *(See also: Air Tractor, "Double Tractor")*
Benton
Taylorcraft

Trader
Grumman TF

Tradewind
Convair R3Y
Pacific Airmotive

"Tradewind"
Bellanca Pacemaker

Tradewind (As: Alizé)
Aviasud
Breguet Bre.1050

Trainer *(See also: Airtrainer, Mach Trainer, Universal Trainer*
Alexander Eaglerock
SAI K.Z.II

Trainor
Swift (KS) Model 19

"Trait d'Union, le" ("Hyphen")
Dewoitine D.33

Tramp
Bristol

Trans-Pacific
Wassmer WA.52

Transaereo
Caproni Ca.60

Transatlantic
Shorts P.D.20

Transport *(See also: Air Transport, Junior Transport)*
Alexander Eaglerock
Woodson Type 4B
Yackey

Transporter
Buxton
Piasecki H-16

Trasimenus
Ambrosini

Travel Air
Beech

Travel Aire
Cessna

Traveler *(See also: Ptraveler, Tri-Traveler)*
American Aviation AA-5
Champion (Wisconsin)
Wendt (CA) WH-1

Traveller
Beech, Beech C-43, GB, JB
Pheasant

Treasure *(See: Little Treasure)*

Treasure Hawk
Bounsall ("Eddie")

Trenér
Zlin Z 26

Trenér Master
Zlin Z 326

Tri Cougar
EAA

Tri Druga (Three Friends)
Semenov-Sutugin-Gorelov

Tri Turbo
Specialized Aircraft

Tri Z
Zenair

Tri-Con
Champion (Wisconsin)

Tri-Gear
Volpar

Tri-Pacer
Piper PA-22, PA-22-160

Tri-Traveler
Champion (Wisconsin)

Triad
C.A.L.
Curtiss
Paulhan-Curtiss
Saul 1000

Triana
Hispano HA-100

Tricap Sauro
Caproni-Trigona

Trident
British Aerospace
de Havilland D.H.121
Hawker Siddeley HS.121
SNCASO SO 9000

Trigear Courier
Helio

Trike
Delta Wing Kites
Wolfe Aviation

Trimmer
Allied Aviation Corp

Trinidad
Aérospatiale TB-20, TB-21
Socata TB20

Triphibian
Monte-Copter

Trislander
Britten Norman BN-2A-III

Tristar
Lockheed L-1011

TriStar
Star Flight

Triton
de Havilland (Canada) DHC-8
SNCASO SO 6000
Tupolev

Triumph
Scaled Composites Model 143

Trojan
Emigh
North American T-28

Trojan Horse
Laister-Kauffmann CG-10

Trojka
Cijan C-3

Trombe (Waterspout)
Wibault 170 C1

Trooper
Taylorcraft

Troopship
Fokker F.50

"Truck" (As: "Lore")
Grunau H2PL

"Truculent Turtle"
Lockheed P2V

Trusty
Consolidated PT-1

"Tsubame-Go" ("Swallow")
Itoh Emi 6

Tsurubane (Swallow)
Itoh Emi

Tsurugi (Sabre)
Nakajima Ki-115

Tsyklon
Tupolev Tu-16

Tu-Sa
IMPA

Tucan (Toucan)
Turbay T-1

Tucano (Toucan)
Alaparma AM-10
EMBRAER
Shorts S.312

"Tucumcari"
Boeing Hydrofoil

Tudor
Avro 688

Tug *(See: Ptug)*

Tugmaster
Auster 6A
Beagle A.61

Tuisku (Blizzard)
VL

Tulák (Rambler)
Let-Mont

Tundra
Back Forty

Tunnan (Barrel)
Saab 29

Tupi
EMBRAER

Turbi
Druine D-5

Turbine Bonanza
Allison

Turbine Islander
Britten Norman BN-2T

Turbine Legend
Performance Aircraft (KS)

Turbine Mentor
Allison

Turbinlite
Douglas DB-7

Turbo 18
Volpar

Turbo Arrow
Piper PA-28R-201T

Turbo Baron
Beech

Turbo Centurion
Cessna T210

Turbo Commander
Aero Commander 680T, 680V, 680W

Turbo Dakota
Piper PA-28-201T

Turbo Firecracker
NDN NDN-1T

Turbo Flagship
Allison

Turbo Otter
Cox

Turbo Renegade
Lake (Aircraft)

Turbo Skywagon
Cessna 206

Turbo Stationair
Cessna 206

Turbo Super Skywagon
Cessna 206

Turbo Viking
Bellanca

Turbo Ximango
Aeromot

Turbo-Commuter
Temco

Turbo-Executive
Riley

Turbo-Liner
Convair 240

Turbo-Porter
Fairchild, Fairchild OV-12
Pilatus

Turbo-Rocket
Riley

Turbo-Star
American Jet Model 400

Turbo-Thrush
Ayres Corp

Turbo Goose
McKinnon

Turbolet
Let L 410

Turboliner
Grumman G-159A
Volpar

Turbomallard
Grumman G-73

Turbulent
Druine D-31
Stark (Wilhelm)

Turcock
Blackburn

Turret Demon
Hawker

"Turtle, The"
Lockheed P2V

Tutor *(See also: Kirby Tutor, Pelican Tutor, Sea Tutor)*
Avro 621
Canadair CL-41
Timm N2T

Twin Astir
Grob

Twin Bonanza
Beech

Twin Comanche
Piper PA-30, PA-39

Twin Courier
Helio

Twin Delta
Bell 208

Twin Jenny
Curtiss JN-5

Twin Lark
IAR Brasov IS-28B2, IS-30

Twin Mistral
Aviasud

Twin Monocoach
Monocoupe

Twin Mustang
North American P-82

Twin Navion
Camair

Twin Otter
de Havilland (Canada) DHC-6

Twin Pioneer
Scottish Aviation

Twin Porter
Pilatus PC-8

Twin Quad
Beech

Twinecureuil (Twin Squirrel)
Aérospatiale AS 355

TwinRanger
Bell 406LT

Twinstar
Aérospatiale AS 355
Air & Space
Airmaster

Two Bridges (As: Deux-Ponts)
Breguet Bre.76

Two-Seater (As: Doppelsitzer)
Kassel

Twosome
Taylorcraft

Typhon (Typhoon)
Caudron C.640

Typhoon
Beardmore
Eurofighter
Hawker
TL Ultralight TL-32

Typhoon (As: Taifun)
Messerschmitt Bf 108

Tzu Chiang
AIDC AT-3

U

Ugur (Luck)
MKEK 4

Uhu (Owl)
Focke-Wulf Fw 189

Uirapuru
Aerotec

Ukraina
Antonov An-10

Ultra-Imp
Aerocar, Inc

Ultrasport
American Sportscopter

Ultrasport Vigilante
American Sportscopter

Ultrastar
Kolb

Ultrawing
Wing (Research) SA-82

Umbrellaplane
McCormick-Romme

Umino
Tachibana

"Uncle Sam"
C.A.L. Mailplane

V

Venture
Questair
Scheibe
Slingsby T.61E
Vickers (UK)

Ventus
Schempp-Hirth

Vernon
Vickers (UK)

Vertaplane
Herrick

Vertifan
Ryan V-5

Vertijet
Ryan X-13

Vertiplane
Ryan VZ-3

Vespa
Kirkham
Vickers (UK)

Vest-Pocket Pursuit
Granville, Miller, and De Lackner

Viastra
Vickers (UK)

Viceroy
Airspeed

Victor
Handley Page
Partenavia P68B
Taneja P68B

Victoria
Vickers (UK)

Victory (As: Bi-Vittoria)
Magni

Victory (As: Vittoria)
Magni

Victory Trainer
Morrow (Aircraft Corp)

"Vienna" (As: "Wien")
Kassel

Viget
Vickers (UK)

Viggen (Thunderbolt)
Saab 37

Vigil
Vickers (Canada)

Vigilant
Grob
Stinson (Aircraft) L-1
Vickers (UK)

Vigilante *(See also: Ultrasport Vigilante)*
Ayres Corp
North American A3J

Vihuri (Flurry)
VL

Viima (Cold Wind)
VL

Viking
ASL
Bellanca
Grob
Lockheed S-3
Vickers (UK)

Viking (As: Wiking)
Blohm und Voss Bv 222

"Viking" (ACV)
Bell (Canada) 7501

Vildebeest
Vickers (UK)

"Ville de Paris"
Sikorsky (USA) S-37

"Ville de Saint-Pierre"
Latécoère LAT 522

Vimy
Vickers (UK)

"Vin Fiz"
Wright (Co) Model EX

Vincent
Vickers (UK)

Vindicator
Vought SB2U
Vought V-156

Vinka
Valmet L-70

Vintokryl'
Kamov Ka-22

Violet Lightning (As: Shiden)
Kawanishi N1K1-J

Viper
Delta Wing Kites

Viper (As: Natter)
Bachem Ba 349

Vireo
Vickers (UK)

Virginia
Vickers (UK)

"Virginia"
Fokker Super Universal

Viscount
Vickers (UK)

Vista
Vickers (Canada)

Vittoria (Victory)
Magni

Vivat
Aerotechnik (Czech) L-13SL

Vivid
Vickers (UK)

Vixen *(See also: Kitfox Vixen)*
Vickers (UK)

Volaire
Volaircraft

Volant Solo
Lockheed EC-130E

Volksjäger (People's Fighter)
Heinkel He 162

Volksplane (People's Plane)
Evans (Aircraft)

Voodoo
McDonnell F-88, F-101

Vostochnyi Karavan (Eastern Caravan)
Aeroprogress T-205

Vought Special
General Aeronautical

Voyage
Aeroprogress T-610
Robin

Voyager *(See also: Saeta Voyager)*
Revolution Helicopter
Rutan Model 76
Stinson (Aircraft) L-9A, O-54, HW-90, Model 108

Voyageur
Boeing-Vertol CH-113
Cadillac

"Voyageur" (ACV)
Bell (Canada) 7380

Vulcan
Avro 698
Vickers (UK)

Vultur (As: Vulture)
Breguet Bre.46, Bre.960

Vulture
Vickers (UK)

Vulture (As: Sep)
PWS

Vulture (As: Urubú)
FMA

Vulture (As: Vautour)
Avia (France, Inter-War)
SNCASO SO 4050

Vulture (As: le Vautour)
Allard & Carbonnier

W

"Württemberg"
Stuttgart (WLV)

Wyvern
Westland

Wyzel (Pointer)
PWS 33

X

Xavante
EMBRAER

Ximango
Aeromot

Xingu
EMBRAER

Xtreme Motorglider
Diamond

Y

Yacht *(See: Aeroyacht, AeroYacht, Air Yacht)*

Yale
North American
North American BT-9, BT-14

"Yamagatakinen-Go"
Itoh Emi 22

Yamal Amphibian
AST

Yankee *(See also: One Yankee)*
American Aviation AA-1

Yankee Doodle
Fairchild AT-13

"Yankee Doodle"
Laister
Lockheed Model 5

"Yellow Peril"
Handley Page E

Yeovil
Westland

"Yippee"
Lockheed P-38J Lightning

Yokosho Ro-Go Ko-Gata
Nakajima

York
Avro 685

Yorkopter
York (C. H.) XHC-22

Young Bird (As: Wakadori-Go)
Shigeno

Young Lion (As: Kfir)
Israel Aircraft Industries

Youngstown Youngster
Ohio

Yukon
Canadair CL-44

Yvonette
le Prieur

Z

Zaba (Frog)
Czerwinski (Waclaw) W.W.S.2

Zabus (Froggy)
Jach
Naleszkiewicz J.N.1

Zak
LWD

Zanonia
Ross-Stevens RS-1

Zaraw (Crane)
LWD

Zaunkönig (Wren)
Brunswick
Winter (H.) LF 1

Zef (Breeze, Zephyr)
Piel

Zeke
Mitsubishi A6M

Zenith
Zenair

Zephyr
Atec Vos
Bartlett
Porterfield CP-40
R.A.E.
Taylorcraft

Zephyr (As: Zef)
Piel

Zéphyr (Zephyr)
Fouga CM 175

Zeppelin Destroyer
Pemberton Billing P.B.29E

Zeta
Miller (Howell)
Pomilio

Zhokei (Jockey)
Tupolev I-8

Zia
Applebay

Zipper
Buchanan (W. O.)
Christensen
Cleary CL-1
Zenair

Zippy Sport
Green Sky

Zodiac *(See also: Super Zodiac)*
Bristol
Mauboussin
Zenair

Zogling (Pupil)
Avia (France, Inter-War)
Lippisch

Zschach
Hannover

Zubr (Bison)
L.W.S.6

Zuch (Daredevil)
LWD

Zueno (Formation)
Vakhmistrov

Zugvogel
Scheibe

Zuiun (Auspicious Cloud)
Aichi E16A

Zùlù
Bulaero

Zuni
Applebay

Zvyeno
Tupolev I-4, TB-3

Zwergreiher (Heron)
Burgfalke Lo 100